International Scientific Conference Week of Science in SPbPU – Civil Engineering (SPbWOSCE-2015)

MATEC Web of Conferences Volume 53 (2016)

Saint Petersburg, Russia
3 - 4 December 2015

Editors:

Vera Murgul

ISBN: 978-1-5108-2272-6

Printed from e-media with permission by:

Curran Associates, Inc.
57 Morehouse Lane
Red Hook, NY 12571

Some format issues inherent in the e-media version may also appear in this print version.

Printed by Curran Associates, Inc. (2016)

For additional information, please contact EDP Sciences – Web of Conferences at the address below.

EDP Sciences – Web of Conferences
17, Avenue du Hoggar
Parc d'Activité de Courtabœuf
BP 112
F-91944 Les Ulis Cedex A
France

Phone: +33 (0) 1 69 18 75 75
Fax: +33 (0) 1 69 28 84 91

contact-edps@webofconferences.org

Additional copies of this publication are available from:

Curran Associates, Inc.
57 Morehouse Lane
Red Hook, NY 12571 USA
Phone: 845-758-0400
Fax: 845-758-2633
Email: curran@proceedings.com
Web: www.proceedings.com

TABLE OF CONTENTS

POLYTECH
Peter the Great
St. Petersburg Polytechnic
University

SPbWOSCE-2015
03-04 December 2015,
Saint-Petersburg, Russia

PREFACE

Institute of Civil Engineering of Peter the Great Saint-Petersburg Polytechnic University presents International scientific conference Week of Science in SPbPU - Civil Engineering (SPbWOSCE-2015)

The conference is supported by Saint Petersburg authorities: Housing Committee, Construction Committee, Committee on Energy and Engineering Support.

The Organizing Committee warmly welcomes all the distinguished authors and guests to SPbWOSCE-2015, held on December 03-04, 2015, in Ambassador hotel, Saint-Petersburg, Russia.

Scientific editor - Vera Murgul

The conference provides a platform to discuss contemporary engineering problems with participants from different countries. Its success is reflected in the papers received, with participants coming from several countries, allowing a real multinational multicultural exchange of experiences and ideas.

The present volumes collect accepted papers and represent an interesting output of this conference. This book covers these topics: History and Theory of Architecture, General Architecture, Urban Planning, Buildings and Structures, Foundations and Underground Space, Bridges, Roads and Tunnels, Building Materials and Products, Municipal Facilities and Building Services, Urban Environmental Engineering, Applied Mechanics in Construction and Material Science, Energy Efficiency, Renewable Energy, Green Buildings and Sustainable Development, Organization and Planning of Construction Works and Municipal Facilities.

On behalf of the organizing committee, I would like to especially thank Isaline Augusto and all the editors from EDP Sciences for their great support to SPbWOSCE-2015.

All the organizing members will be committed us to make SPbWOSCE more and more professional with fully and enjoyable academic research and discussion platform for authors and attendees.

Sincerely as always, we look forward to your attention to the next SPbWOSCE in 2016

Nikolai Vatin & Vera Murgul

edp sciences

MATEC Web of Conferences

Statement of Peer review

In submitting conference proceedings to MATEC Web of Conferences, I certify to the Publisher that I adhere to the **Policy on Publishing Integrity** of the journal in order to safeguard good scientific practice in publishing.

1. All articles have been subjected to peer review administered by the proceedings editors.
2. Reviews have been conducted by expert referees, who have been requested to provide unbiased and constructive comments aimed, whenever possible, at improving the work.
3. Proceedings editors have taken all reasonable steps to ensure the quality of the materials they publish and their decision to accept or reject a paper for publication has been based only on the merits of the work and the relevance to the journal.

Title, date and place of the conference

International scientific conference Week of Science in SPbPU - Civil Engineering
03-04 December 2015
Ambassador hotel, Saint-Petersburg, Russia

Proceedings editor(s):

Vera Murgul

Date and editor's signature

29.02.2016

MATEC Web of Conferences **53**, 01001 (2016)
DOI: 10.1051/matecconf/20165301001

Thermal Stressed State in Massive Concrete Structures in the Winter Building Period

Ivan Korotchenko[1,a], Ernest Ivanov[2], Kirill Semenov[3] and Yuri Barabanshchikov[4]

[1,2,3,4]*St. Petersburg State Polytechnical University, Institute of Civil Engineering, 29, Politechnicheskaya St., St. Petersburg, 195251, Russia*

Abstract. The article deals with issues of the thermal cracking resistance in massive concrete and reinforced concrete structures during the building period. The paper gives the calculation results of thermal stressed state in massive foundation slabs with different time organization of the thermal protection. It was found that the thermal crack resistance is fully guaranteed when the insulation is installed immediately after concrete has been placed or it has been placed with short delay. The delay can be allowed if the relative tensile strain of the concrete does not exceed the tensility limit. The acceptable delay ranges from 8 to 30% of time it takes to reach the maximum temperature in the concrete block. The surface insulation installation at the time of the temperature peak inevitably leads to the formation and development of cracks.

1 Introduction

In general practice the calculation of thermal fields is often based on the heat equation solution as well as thermal stresses definition [1-14].

During the building process, massive concrete and reinforced concrete structures are exposed to the temperature effect. A change in the thermal state of such structures occurs due to the heat liberation from cement hydration during the concrete hardening process, as well as outside temperature fluctuations, solar exposure, various technological factors, etc. Emerging thermal stresses may cause damage to the structural integrity [15-24, 26-30].

Due to a number of technological and manufacturing reasons, it is preferable to concrete massive foundation mats and other massive structures as a single block of equal height. However, it causes a considerable heat rise in the mass concrete as the result of an exothermic reaction during the concrete hardening. Consequently, the irregular temperature distribution arises along with the block height, which leads to the dangerous tensile strain first on the surface of the foundation slab and then in its central zones. In this regard, the winter period is especially unfavorable for construction [30].

Lowering the thermal field irregularity inside the block by raising its surface temperature prevents high tensile strain in the winter period. Both heat enclosure and an insulation layer on the concrete block are used in order to prevent high tensile strain [17, 22]. A peripheral concrete electric cable with or without minimum thermal protection is usually used as a more efficient and expensive measure. The regulation process of the hardening concrete thermal effect requires preliminary thermal

[a] Corresponding author : saenko84@yandex.ru

protection calculations or the electric heating management depending on the environmental conditions, thermal field inside the block and other factors. The papers [23, 24] deal with issues of safe replacement of passive thermal protection and lists the calculation results of electric-mode heating of massive concrete structures in the strict formulation (taking into account the influence of the hardening temperature on the thermal and deformation characteristics of concrete).

However, it is the time of the passive thermal protection installation that is also of great importance. There is an opinion that thermal protection installation with some delay is possible and even beneficial but not immediately after the placement of concrete. The time of the delay is assumed before the temperature reaches its peak in the core of the structure.

2 Purpose of article

The purpose of this article is to get calculation of the thermal protection installation time and research the influence of this factor on the thermal stressed state and provision of the cracking resistance of massive concrete structures. A distinctive feature of this article is to consider winter building period in the lifetime of the structure.

3 Statement of the problem

The justifying calculation of safe technological regimes in terms of thermal crack resistance is analyzed based on the example of 2 meters high foundation slabs mounted during the winter period.

This paper demonstrates calculation of the foundation slab thermal stressed state with the help of TERM software developed by the Institute of Civil Engineering at the St. Petersburg State Polytechnical University [18]. Considering horizontal mats sizes significantly exceed their height, we can study a one-dimensional structural model for the slab central part with the reasonable degree of accuracy. In this model, stress and temperature are functions of the vertical coordinate space.

In order to estimate the cracking resistance of the foundation slab, we would use the deformation criterion suggested by P.I. Vasiliev and provided by SP 41.13330.2012 [25]. According to this criterion, concrete elongation deformations, determined in view of the concrete creep factor and variable deformation modulus, should not exceed the ultimate concrete elongation.

4 Base line

As a first calculation example, we would consider the concrete hardening process B40 in the foundation slab of NPP's (Nuclear power plants) reactor hall with the cement consumption of 360 kg/m^3. The foundation slab is supported by the concrete bedding layer B12.5 with the grade foundation.

Thermal and physical characteristics of the concrete B40 are defined by the concrete thermal conductivity l=2.67 W/(m·^{0}C) and thermal capacity c=1.0 kJ/(kg·^{0}C).

The instantaneous elastic deformation modulus dependency on time is applied in accordance with N.A. Malinin's work [17]. The heat dissipation process follows the I.D. Zaporozhets equation [16].

The parameters of the equation of I.D. Zaporozhets were obtained based on the experimental data on heat dissipation of concrete.

The second example is pile foundation grillage of multifunctional building complex "Lahta Center". The concrete strength quality is B35 (modulus of elasticity is 34500 MPa), while the cement consumption is 340 kg/m^3. Piles, installed beneath the plate at regular intervals, were designed as a solid base with 10 GPa modulus of elasticity.

The following technological conditions of concreting were considered in our calculations: the concrete mixture is placed into the shelter as a 2 m high block, the temperature in the shelter is 7 °C, and the concrete mix temperature is 13 °C. After concreting the surface is covered with insulation, which thickness is determined by the cracking prevention condition. Installation time of thermal insulation varied from zero (the time of placing of concrete) to the moment of reaching the maximum temperature in the structure core. These examples vary for each temperature peak reached in 6-7 days and 2-3 days respectively.

5 Materials and methods

5.1 Foundation slab of NPP's reactor hall

We calculated that the temperature peak in the slab center is achieved 6 -7 days after laying the mixture. Figure 1 shows the calculation results of the concrete elongation for series with different installation time of surface insulation and with general reduced heat transfer coefficient β_{red} = 4.6 W/(m$^{2}\cdot{}^{0}$C). The curves in Fig. 2 are different in the delay duration, which varies from 0 to 144 hours with a 12 hour step. Red curve corresponds to the concrete limiting extensibility [25].

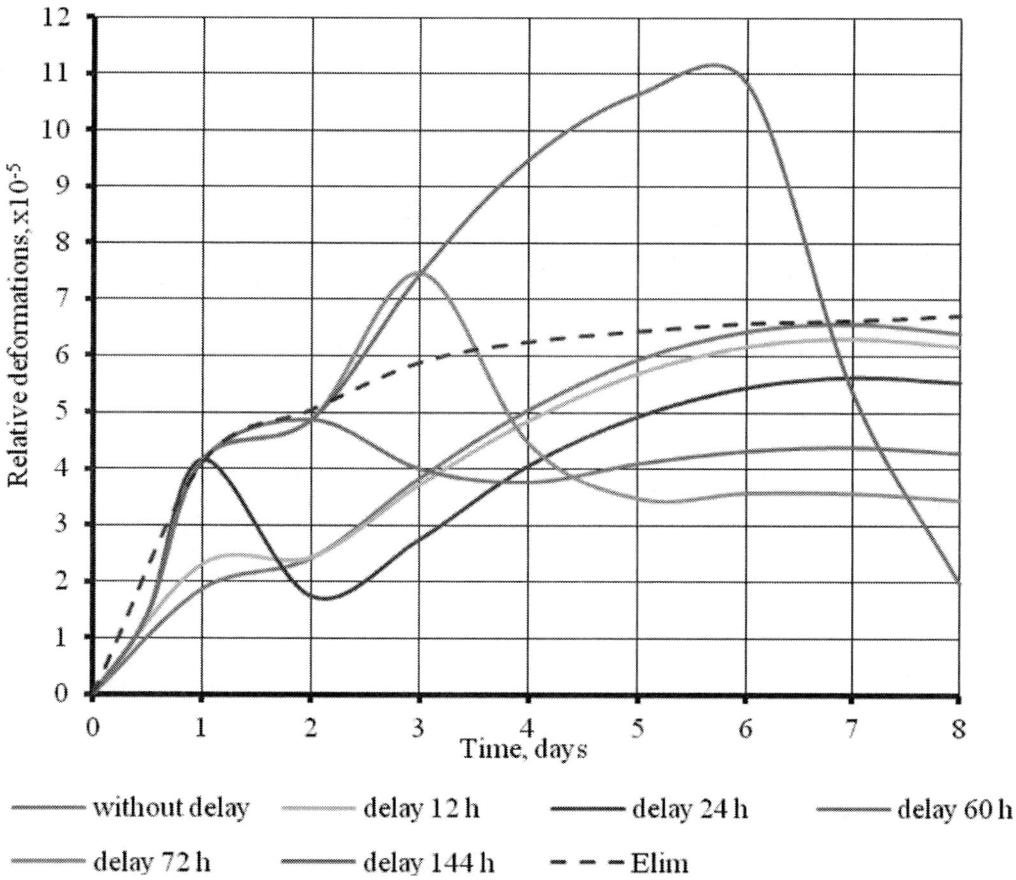

Figure. 1. Dependency the relative elongation strain on the upper surface of the slab on time for specific calculation cases.

Comparison of relative deformations to value limits shows that both the immediate insulation installation and a 12-hour delay satisfy the criterion of crack resistance. Installing insulation immediately after concrete has been placed provides greater crack resistance supply and a lesser supply after the 12-hour period. If the delay is more than 12 hours, cracks may appear on the first day.

The longer the delay, the greater and more prolonged are the relative deformations in comparison to the value limits.

5.2 Foundation slab of multifunctional building complex "Lahta -Center"

We calculated that the temperature peak in the plate center is achieved 2-3 days after laying the concrete. Figure 2 shows the calculation results of the concrete relative elongation strain for series of examples with different installation time of surface insulation and with general reduced heat-transfer coefficient $\beta_{red} = 2.0$ W/(m$^{2}\cdot{}^{0}$C). The delay duration varies from 0 to 48 hours with a 6 hour step. Red curve corresponds to the concrete limiting extensibility [25].

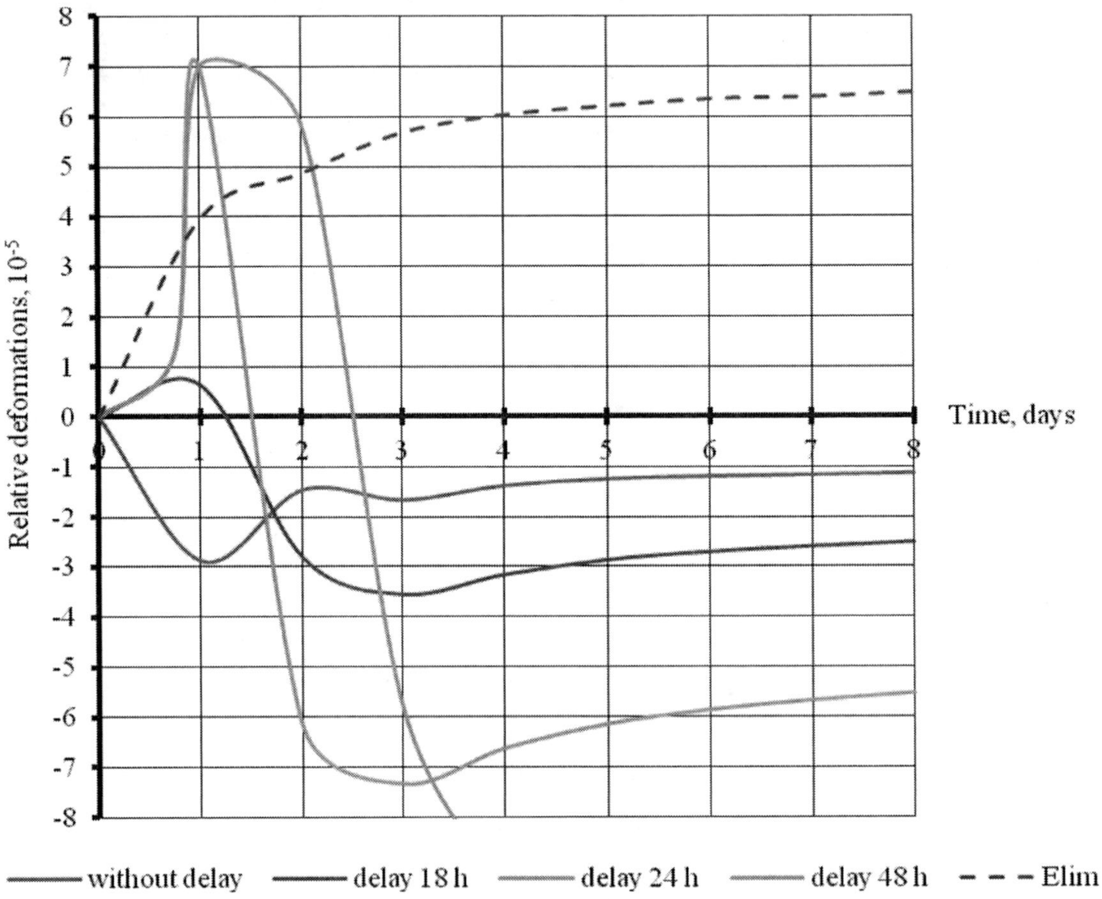

Figure 2. Dependency the relative elongation strain on the upper surface of the slab on time for specific calculation cases.

Comparison of tensile strain to value limits shows that the immediate insulation installation and a 18-hour delay satisfy the criterion of crack resistance. In cases of 24 and 48-hours delays cracks are formed on the second day. The longer the delay, the greater and more enduring are the relative deformations in comparison to the value limits.

As indicated by the NPP slab, the immediate insulation installation provides higher crack resistance during the first two days and lower resistance in case of a 18-hour delay.

6 Results and discussion

Investigation allows us to present the following results:

- Thermal crack resistance is fully guaranteed when insulation is installed immediately after concrete has been placed or it has been placed with small delay.
- Delay can be allowed, if the concrete relative tensile strain does not exceed its limit of tensility. In this case thermal stress state of the concrete block may be more beneficial because of lower tensile strains in the surface structure, when temperature reaches its peak in the core. This effect may take place primarily with temperature rising relatively slowly in the block.
- The acceptable delay ranges from 8 to 30% of time it takes to reach the maximum temperature in the concrete block.

7 Conclusion

Nowadays the construction using cast in-situ concrete is conducted almost all year round and, that especially important, in winter, when the air temperature is very low, and the temperature of the concrete core is high that leads to the formation of cracks. To avoid this it is necessary to insulate poured concrete, possibly with a certain delay, the value of which is discussed in this article. The results are particularly important in conditions of the far North, where delay in installation of insulation after concreting is especially dangerous. The results of research allow to determine the maximum acceptable delay of laying the thermal insulation on the surface of the laid concrete to avoid any cracking. As you can see, the acceptable delay should be between 8 and 30% of time it takes to reach the maximum temperature in the concrete block.

References

1. N.I. Vatin, D.V. Nemova, P.P. Rymkevich, A.S. Gorshkov, Mag. of Civ. Eng., **8**, 1 (2012)
2. D.V. Petrosova, N.M. Kuzmenko, D.V. Petrosov, Mag. of Civ. Eng., **8**, 31 (2013)
3. B. Kovačič, R. Kamnik, M. Premrov, Survey Review, **43 (320)**, 150-161 (2011)
4. D.V. Nemova, N.I. Vatin, A.S. Gorshkov, A.V. Kashabin, P.P. Rymkevich, D.N. Tseytin, Constr. of Un. Build. and Struct., **23**, 93 (2014)
5. N. Lipovac, A.M., Jandriček, Prostor, **19 (2)**, pp. 443-455 (2011)
6. A.S. Gorshkov, P.P. Rymkevich, N.I. Vatin, (2014) Mag. of Civ. Eng., **8** , 38 (2014)
7. M.A. Platonova, N.I. Vatin, D.V. Nemova, S.A. Matoshkina, D. Iotti, I. Togo, Constr. of Un. Build. and Struct., **19**, 83 (2014)
8. Z.S. Teplova, K.I. Solovyeva, D.V. Nemova, D.A. Trubina, D.V. Petrosova, Constr. of Un. Build. and Struct., **19**, 96 (2014)
9. A.S. Gorshkov, N.I. Vatin, D.V. Nemova, D.A. Tarasova, App. Mech. and Mater., **725 - 726**, 3 (2015)
10. N.I. Vatin, M.R. Petrichenko, D.V. Nemova, A.A. Staritcyna, D.A. Tarasova, App. Mech. and Mater., **633 -634**, 1023 (2014)
11. V.A. Korsun, A.V. Korsun, App. Mech. and Mater., **725-726**, 469 (2015)
12. V.A. Korsun, N.I. Vatin, A.V. Korsun, D.V. Nemova, (2014) App. Mech. and Mater., **633-634**, 1013 (2014)
13. M. S. Jaafar., Adv. in Eng. Soft., **38**, 886 (2007)
14. Y.G. Barabanshchikov, V.A. Sokolov, A.S. Vasiliev, M.V. Shevelev, (2012) ALITINF.: Tsem., Bet., Sukh. smes., **25**, 32 (2012)
15. C.B. Aleksandrovskiy, *Calculation of temperature change and humidity in terms of concerete creep in concrete and reinforced concrete structures* (Moscow, 1973)
16. I.D. Zaporozhets, S.D. Okorokov, A.A. Pariyskiy, *Heat Liberation by Concrete* (Stroyizdat, Leningrad, 1966)
17. N.A. Malinin, *Issledovaniye termonapryazhennogo sostoyaniya massivnykh betonnykh konstruktsiy s peremennymi deformativnymi kharakteristikami. Diss. na soisk. uchen. step. kan.*

teh. nauk: Spets 05.23.01 [*Research of thermal stressed state of mass concrete structures with changing deformations characteristics. Cand. tech. sci. diss.*] Leningrad (1977)

18. K.V. Semenov, *Temperaturnoye i termonapryazhennoye sostoyaniye blokov betonirovaniya korpusa vysokogo davleniya v stroitelnyy period: Dis. na soisk. uchen. step. kan. teh. nauk: Spets 05.23.01* [*Temperature and thermal stressed state of concreting blocks in a high pressure shell during the building period*] Leningrad (1990)
19. Y.G. Barabanshchikov, K.V. Semenov, (2007) Pow. Tech. and Eng., **41**, 197 (2007)
20. Y.G. Barabanshchikov, K.V. Semenov, M.V. Shevelev, Popul. beton., **1**, 70 (2009)
21. P.I. Vasilyev, D.A. Ivanov, Yu.I. Kononov, K.V. Semenov, O.P. Starikov, Probl. of atom. sc. and tech., **1**, 62. (1988)
22. L.P. Trapeznikov, *Temperaturnaya treshchinostoykost massivnykh betonnykh sooruzheniy* [*Thermal cracking resistance of mass concrete structures*] Moscow, (1986)
23. K.V. Semenov, Y.G. Barabanshchikov, Constr. of Un. Build. and Struct., **17**, 125 (2014)
24. K.V. Semenov, Y.G. Barabanshchikov, App. Mech. and Mater., **725 -726**, 431(2015)
25. SP 41.13330.2012. *Concrete and reinforced concrete structures of hydroengineering facilities* (2012).
26. M. Larson, Thermal *crack estimation in early age concrete -models and methods for practical application*, *Doctoral Thesis*, Lulea (2003)
27. S. Miyazawa, K. Koibuchi, A. Hiroshima, T. Ohtomo, T. Usui, CONCEC, **10**, 1487 (2010)
28. W. Shengxing, H. Donghui, Journ. of Therm. Anal. and Calorim., **105**, 171 (2011)
29. Z. Zhang, X. Zhang, X. Wang, T. Zhang, X. Zhang, App. Mech. and Mater., **94 -96**, 2107 (2011)
30. Y.G. Barabanshchikov, K.V. Semenov, Pow. Tech. and Eng., **41**, 197 (2007)

MATEC Web of Conferences **53**, 01002 (2016)
DOI: 10.1051/matecconf/20165301002

Steel and composite structural shells in the construction of high-rise building

Andrey Redkin [1,a], Mikhail Baranovskii [1] and Vladimir Tarasov [1]

[1] *St. Petersburg State Polytechnical University, Polytechnicheskaya st. 29, 195251, St. Petersburg, Russia*

Abstract. In modern construction thin-shell structures are very popular because they can be easily used to make different constructive and architectural shapes. The main value of such constructions is the ability to use them as load-bearing structures and as the element of architectural image of building at the same time. Thin-shell structures allow us to solve the problems of construction rigidness and stableness, therefore it is rational to use shells in high-rise building structures. At the same time construction of high-rise buildings is connected with a number of special requirements. Disregard of those requirements can cause destruction of the building. In this research there were considered some challenges of high-rise building design on the business complex example. The comparative analysis of two various bearing shells was made. The first case was a metal grid shell, and the second was the combination of metal elements and reinforced concrete ribs. The reinforced concrete ribs (stiffeners) were used to improve the rigidness of the whole structure and to increase the bearing capacity.

1 Introduction

High-rise buildings are different from ordinary buildings and have a number of features in design process and also during construction. The distinctive features are:

- A significant influence of horizontal dynamic loads (wind loads);
- Heavy load on bearing structures and foundations, in comparison with ordinary buildings;
- Increased requirements to account the anthropogenic and environmental factor influence. The most important factors are: earthquakes, floods, fires, terrorist acts and solar radiation;
- Difficult connections between steel and concrete bearing elements;

All these features must be considered in the process of high-rise building structural scheme selection.

[a] Corresponding author : redish132132@yandex.ru

2 Literature review

For the first time bearing shells were used in the construction by Russian engineer C.G. Shukhov in 1896. He developed and patented three types of load-bearing grid shells: hanging, convex and tower-shell (patents of the Russian Empire No. 1894, No. 1895, No. 1896; March 12, 1899).

Nowadays the theoretical basics of bearing shell mechanics are well studied. The research results are stated in monographs: V.Z. Vlasov - “Stroitelnaya mehanika obolochek”, A.P. Filin -“Elementyi teorii obolochek”, K.Z. Galimova - “Matematicheskaya teoriya obolochek”, H.M. Mushtari - “Nelineynaya teoriya obolochek”, V.V. Novozhilov - "Teoriya tonkih obolochek", A.L. Goldenveizer - “Teoriya tonkih uprugih obolochek” and others.

Researches of many scientists are devoted for the creation and development of methods: E.L. Axelrad - “Gibkie obolochki”, Yu.P. Artyuhin and P.G. Velikanov - “Raschyot anizotropnyih obolochek metodom granichnyih elementov”, D.V. Vaynberg - “Matrichnyie algoritmyi v teorii obolochek vrascheniya” and others.

The plasticity and creep theories are very important in bearing shell calculations. Researches in this area were made by scientists: R.A. Arutyunyan, G.I. Byikovtsev, R.A. Vasin, A.A. Vakulenko, M.A. Grekov, A.A. Gvozdev, G. Grinberg, O.Yu. Dinariev, A.S. Dehtyar and others.

In the twentieth century bearing shells were widely adopted in construction, and the number of projects with this scheme was increasing each decade. Therefore, new problems and difficulties in the construction appear in process of bearing shell technology development [1-29], but many of them are already well studied [29-37].

3 Issue definition

The aim of the research was to determine the rules of high-rise building structural scheme and materials selection. And also optimal structural scheme determination from two different variants.

4 Materials and Methods

The research object was 41-storey business complex with two various types of bearing shell. General view of the research object is shown on the Fig. 1.

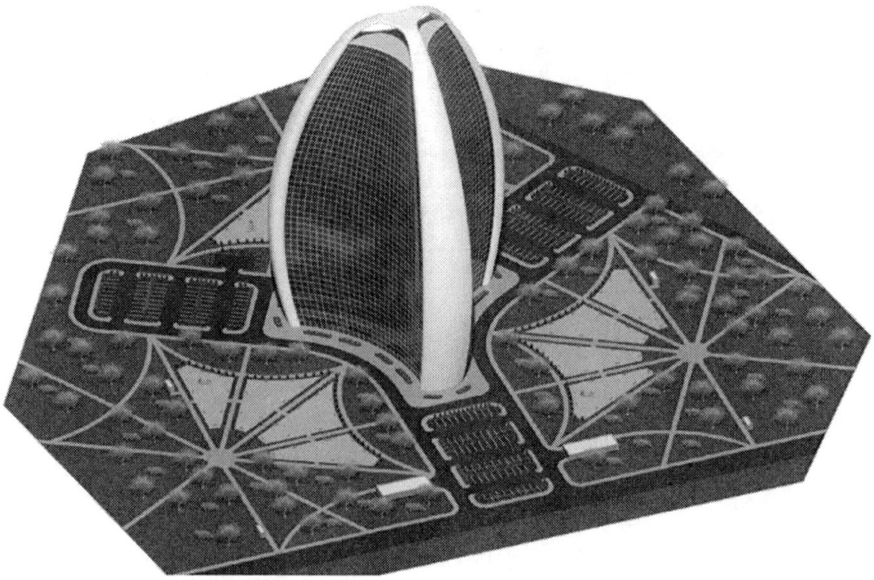

Figure 1. General view of the research object.

The building has a "shamrock" shape on the plan, also, the horizontal cross-section changes along the height of the building. Due to the architectural features and the height of the building it was decided to apply the "load-bearing outer shell in combination with a central core" [38]. Central load-bearing core (stiff trunk) are usually made of reinforced concrete, steel and its combinations. Stiff trunk perform the role of rigid vertical consoles fixed in the soils and perceives predominantly horizontal loads. The rigidity of the trunk depends on the shape and cross-sectional area of the stair-elevator site, which sizes are determined according to the number of elevators and stairs requirements.

The core is rigidly connected with the metal bearing shell by monolithic reinforced concrete slabs. Bearing structures can be prefabricated or monolithic. In the initial period of high-rise building development bearing structures were normally made from steel. Currently, bearing structures are normally made from reinforced concrete, because this material has better fire resistance and lower cost, and its strength characteristics under compression is quite the same. Nowadays B80 and B100 class concretes are created, although in construction practice are mainly used lower classes of high-strength concretes (B60 and B70), because with the concrete class growth increases its cost, fragility and decreased fire resistance.

The grid shell bearing system was selected because of insolation requirements for the building facades. The elements of the grid shell were tubes of square section with 200 mm side and 14 mm thickness. The steel was selected as a material for shell rods because of tensile stresses, which concrete cannot resist.

The difference between the two cases of structural schemes was: in the first case, each wing was locked by reinforced concrete elements to improve the stiffness and to make building heavier in these parts. In the second case, instead of the stiffeners on the building edges metal shell with the same configuration as in other parts of the building was used.

The analytical models of both cases were made in the SCAD Office software complex. Loads were applied according to the code “Nagruzki i vozdeystviya” [39], including equivalent wind load that was applied statically. Analytical models are shown on Fig. 2,3.

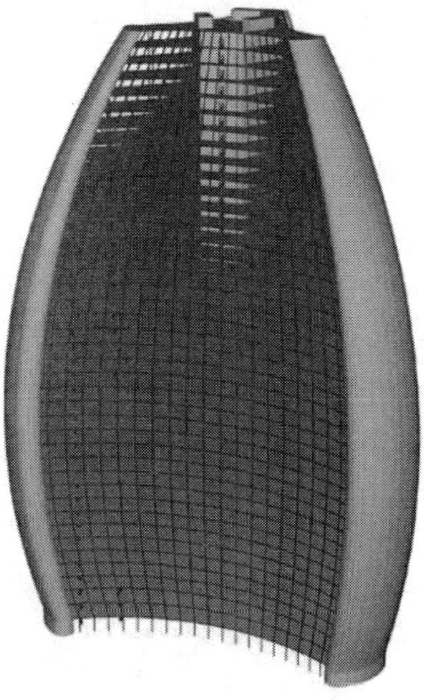

Figure 2. The analytical model in case with reinforced concrete stiffeners in SCAD

Figure 3. The analytical model in case with steel shell elements in SCAD

Reinforced concrete shell elements and the central core walls are made of B30class concrete, load-bearing columns – of B60 class concrete. The concrete ribs thickness changes along height from 400 to 200 mm. The columns diameter changes along height from 1000 mm in the lower floors to 300 mm at the top.

5 Results and Discussion

In the SCAD Office finite element environment two selected structural cases were calculated on the rigid foundation. 1-st and 41-st floors were selected for the calculation results comparison .They show the values that characterize the construction work: stresses in the elements, deformations. On the first floor compressive stresses in bearing structures from the vertical loads and bending from the wind are maximum. On the 41-st floor stresses in the elements from the force effects are very small, but there are large values of the total vertical deformations.

Deformations of the slab of the first floor on Z-axis and stresses in the shell elements are shown on Fig 4,5,6,7.

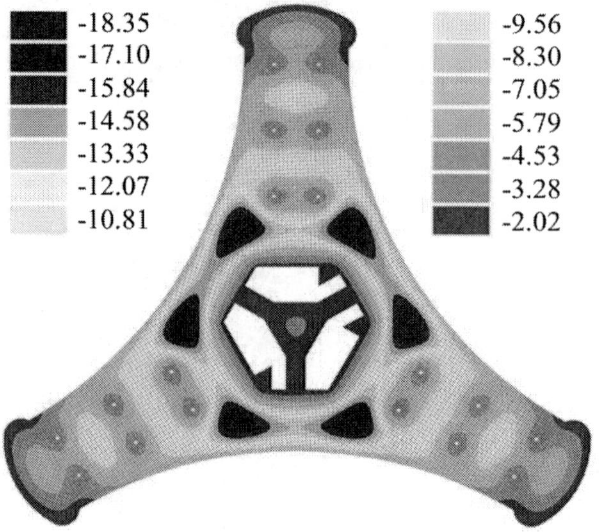

Figure 4. First floor slab deformations on axis Z (mm) in case with reinforced concrete stiffeners

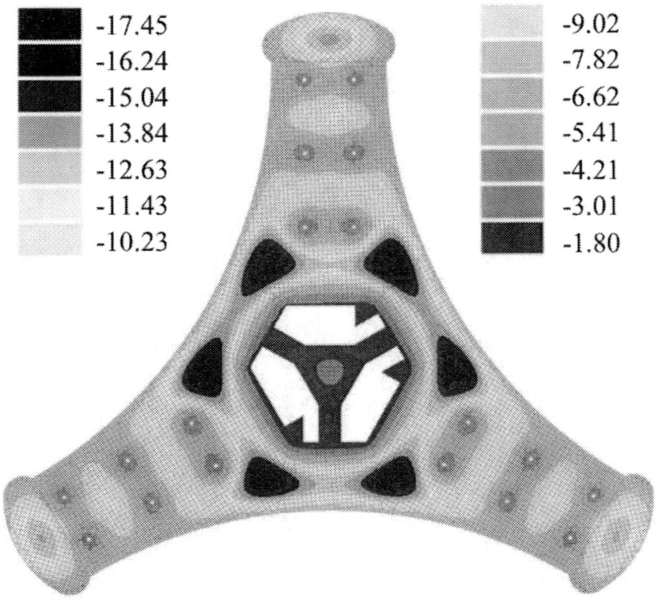

Figure 5. First floor slab deformations on axis Z (mm) in case with steel shell elements

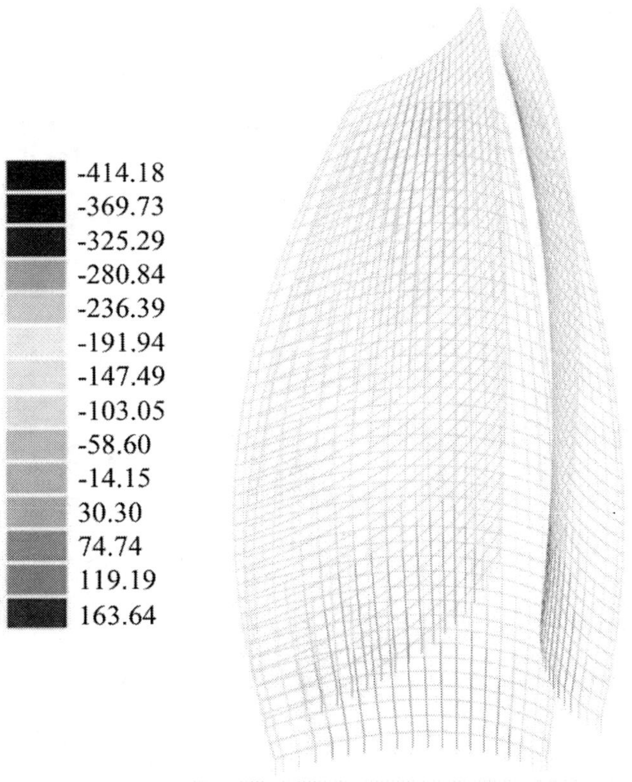

Figure 6. Stresses (N, [T]) in the shell elements in case with reinforced concrete stiffeners

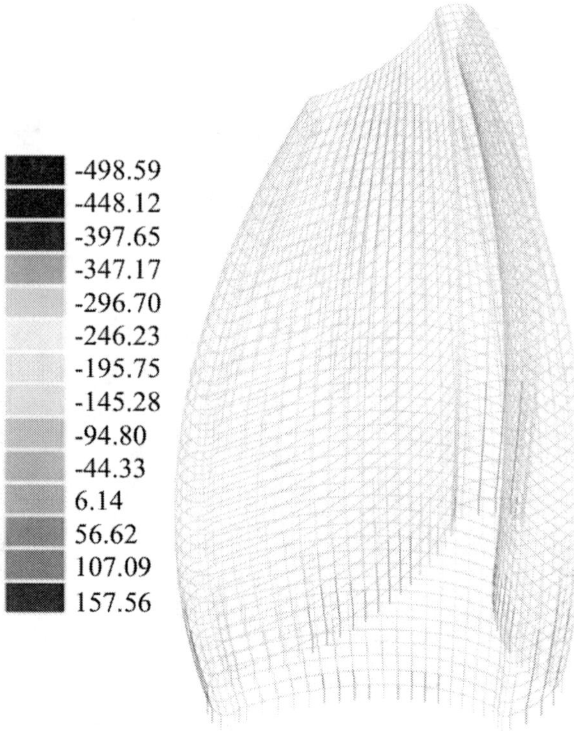

Figure 7. Stresses (N, [T]) in the shell elements in case with steel shell elements

First floor slab deformations in two different cases (Fig.4,5) are almost identical. In case with reinforced concrete stiffeners tension is less, the maximum compressive stress is 414.18 t, while in case with steel shell elements - 498.59 t. The difference is 20%. This shows that there is a stresses redistribution: in case with steel shell elements more weight falls on the shell rods along the glazed facades of the building.

41-st floor on Z-axis slab deformations in two different cases are presented on Fig. 8,9.

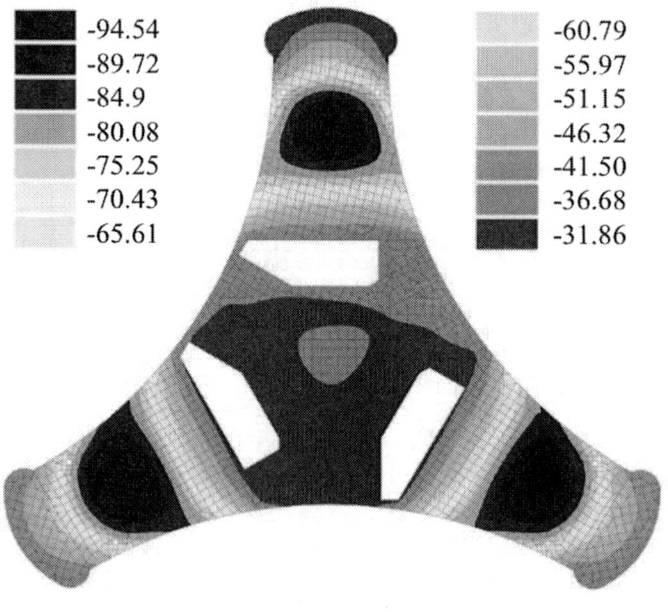

Figure 8. 41-st floor slab deformations on axis Z (mm) in case with reinforced concrete stiffeners

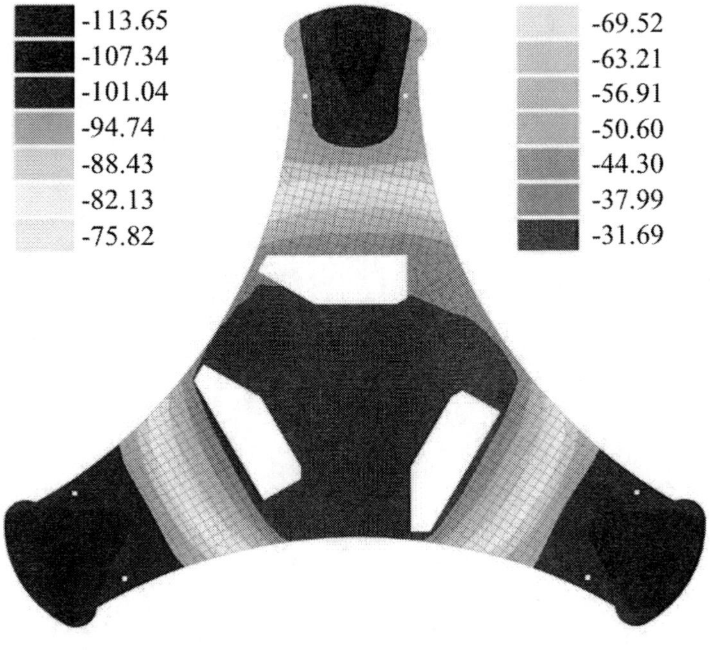

Figure 9. 41-st floor slab deformations on axis Z (mm) in case with steel shell elements

Deformations of the 41-st floor slabs in different cases are considerably different. Deformations in case with reinforced concrete stiffeners is less approximately on 60-75 mm (the difference is about 73%). Furthermore, it is obvious that in case with steel shell elements, all slab in the wing parts fall on 70-80 mm comparative the central core. With increasing height of the building slabs have increasing slope from the central core to the wing edges. As a consequence, loads on the intermediate supporting columns increases.

In case with reinforced concrete stiffeners stresses are redistributed due to changes in the stiffness of the whole building. Massive reinforced concrete ribs better than the metal grid shell perceive compressive loads. Because of their greater rigidity, despite the higher stresses, that they perceive, there are also smaller deformations. In this case slab doesn't fall from the central core to the edges of the building. Slab works as a plate on two relatively rigid supports and the maximum 55 mm deflection is in the span.

In case with steel shell elements the slab at the junction to the trunk, rests on a rigid support, and on the other side (on the edge of the wing) relies on the support with a certain deformability, that leads to the whole slab tilt.

In case with steel shell elements compressive stress in the most loaded central columns is 2773.79 t, in case with reinforced concrete stiffeners – 2761.40 t. The values are almost identical (difference less than 0.5%), the difference is negligible. First floor column stresses in both cases are shown on Fig. 10,11.

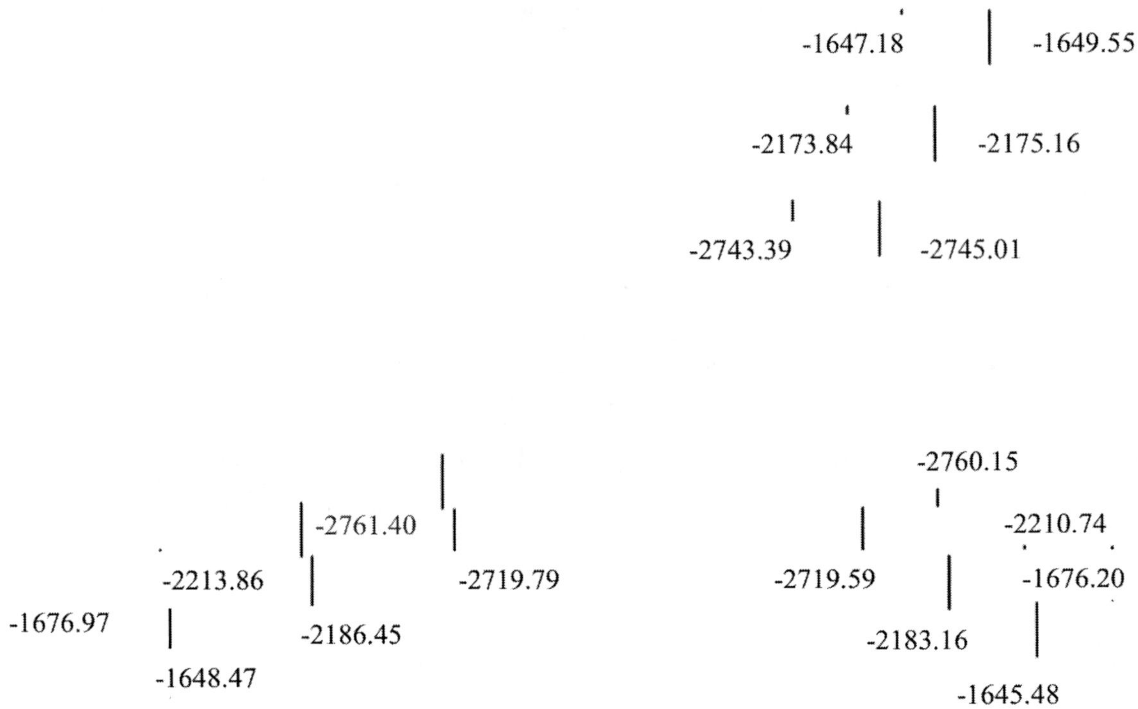

Figure 10. Stresses in the first floor columns in case with reinforced concrete stiffeners (T)

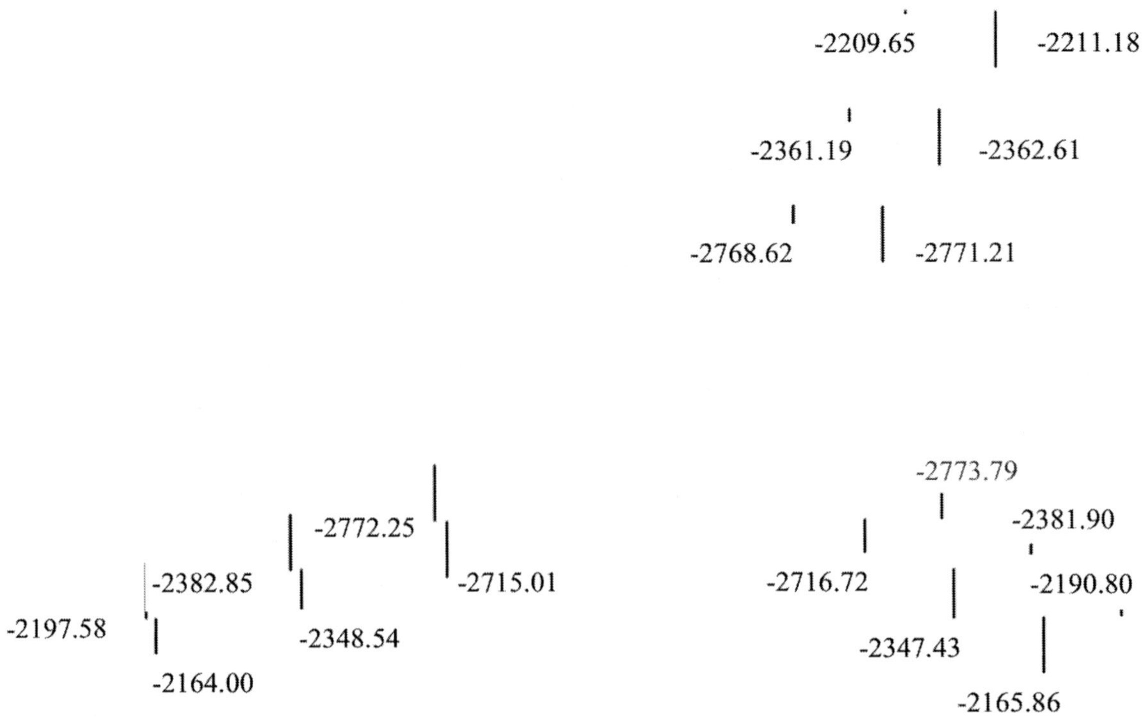

Figure 11. Stresses in the first floor columns in case with steel shell elements (T)

In columns along the edges of the wings difference is much more: in case with steel shell elements compressive stress in the column is 2211.18 t, In case with reinforced concrete stiffeners – 1676.97 t.

The difference is about 32 %. The first floor columns reinforcement calculation was made in program "Arbat" and it is presented in the Tab. 1.

Table 1. Layouts and diameters of the rods of reinforcement in both cases

Construction case	Longitudinal reinforcement	Cross reinforcement	Layout of a longitudinal reinforcement
In case with steel shell elements (compressive stress in the columns is 2773.79 [T])	Longitudinal reinforcement – 20 rods, diameter 32 [mm]	Cross reinforcement along the X-axis has 10 mm diameter, rod spacing - 200 [mm] Cross reinforcement along the Y-axis has 10 mm diameter, rod spacing - 200 [mm]	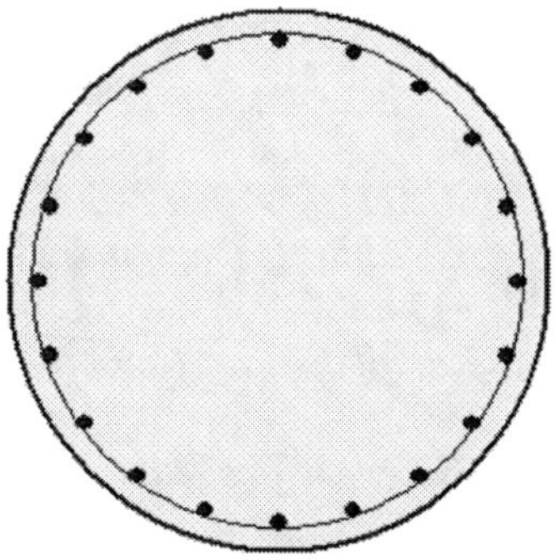
In case with reinforced concrete stiffeners (compressive stress in the columns is 2211.18 [T])	Longitudinal reinforcement – 15 rods, diameter 20 [mm]	Cross reinforcement along the X-axis has 10 mm diameter, rod spacing - 200 [mm] Cross reinforcement along the Y-axis has 10 mm diameter, rod spacing - 200 [mm]	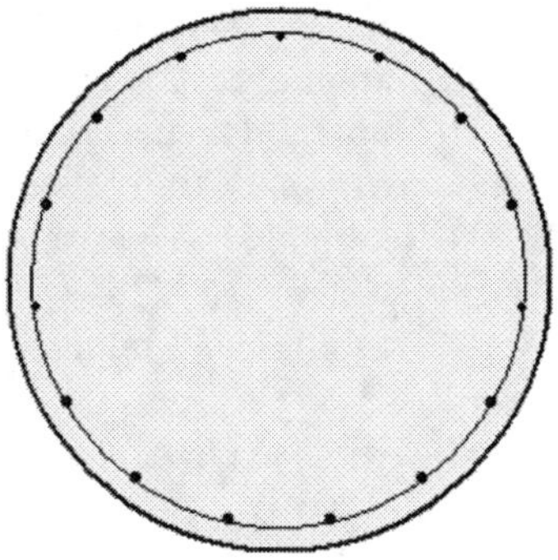

From the results it is obvious that the columns in the variation with the metal shell must be reinforced considerably stronger - required 20 rods, diameter - 32 mm, while in the construction with stiffeners – 15, diameter - 20 mm.

6 Conclusions

According to the calculation results and two variations of building structures comparison, it is possible to say that it is advisable to the use the combined shell of reinforced concrete and steel elements. Grid metal bearing shell, at first, reduces the weight of building bearing structures, and secondly, allows to satisfy the insolation requirements. Stiffeners play the role of additional rigid trunks. They unload metal shell and the intermediate columns inside the building. Connection of concrete ribs with the central core in a single spatial structure increases the building rigidity, and it is a favorable factor for the wind and other specific loads perception.

In conclusion, various combinations of massive concrete elements and steel grid shell solutions is rational for high-rise construction.

References

1. E. A. Golubeva, New Ideas of New Century, **2**, 66-70 (2014)
2. V. V. Lalin, V. A. Rybakov, A. Sergey, Appl. Mech. and Mater. **578-579**, 858-863 (2014)
3. N. I. Vatin, A. S. Sinelnikov, Des., fabric. and econ. of met. Str., 165-170 (2013)
4. N. I. Vatin, A. S. Sinelnikov, M. Garifullin, D. Trubina, Appl. Mech. and Mater. **633-634**, 1037-1041 (2014)
5. S. L. Chan, J. of Const. St. Res. **57**, 1217–1231 (2001)
6. M. Baranovskiy, V. Tarasov, S. Zimin, Appl. Mech. and Mater. **725-726**, 774-784 (2014)
7. Y. Hea, X. Zhoua, X. Zhanga, Thin-Wall. Str. **60**, 1–11 (2012)
8. V. S. Karpilovskiy, E. Z. Kriksunov, A. A. Malyarenko, *SCAD Office. Formirovaniye secheniy i raschet ikh geometricheskikh kharakteristik* (Izd. ASV, 2011)
9. V. S. Karpilovskiy, E. Z. Kriksunov, A. A. Malyarenko, *SCAD Office. Vychislitelnyy kompleks SCAD* (Izd. SKAD SOFT, 2011)
10. N. Vatin, J. Havula, L. Martikainen, A. Sinelnikov, A. Orlova, S. Salamakhin, Adv. Mater. Res. **945-949**, 1211-1215 (2014)
11. *Kompleks programm SCAD office - instrumentariy inzhenera-proyektirovshchika,* (PGS, **3**, 56, 2007) , 60,
12. *Kompleks programm SCAD Office - instrumentariy inzhenera – proyektirovshchika*, (PGS, **4** 2007)
13. Z. Shubber, *The development of the typical roof structure project based on trusses made of roll-welded rhs pipes: Bachelor's Thesis* (Civ. and Const. Eng., Lappeenranta: Saimaa University of Applied Sciences, 80, 2013)
14. R. Kamnik, B. Kovačič, B. Pribicević, A. Đapo, Geodetski List, **69 (3)**, 171-188 (2015)
15. N. Vatin, J. Havula, L. Martikainen, A. Sinelnikov, A. Orlova, S. Salamakhin, Proceedings of the International Conference "Innovative Materials, Structures and Technologies", 187 (2014)
16. D. Trubina, D. Abdulaev, E. Pichugin, V. Rybakov, M. Garifullin, O. Sokolova, Appl. Mechanics and Mater. **725–726**, 752–757 (2015)
17. D. Trubina, D. Abdulaev, E. Pichugin, V. Rybakov, Appl. Mech. and Mater. **633–634**, 982–990 (2014)
18. D. Trubina, D. Abdulaev, E. Pichugin, V. Rybakov, Appl. Mech. and Mater. **633–634**, 1133–1139 (2014)
19. M. V. Ananina, N. A. Beresneva, L. L. Shurovkina, Const. of Uniq. Build.s and Str. **7 (22)**, 54–70 (2014)
20. M. Garifullin, D. Trubina, N. Vatin, Appl. Mech. and Mater. **725–726**, 697–702, (2015)
21. V. Rybakov, A. Sergey, Appl. Mech. and Mater. **725–726**, 746–751 (2015)
22. N. Vatin, T. Nazmeeva, R. Guslinscky, Adv. Mater. Res. **941–944**, 1871–1875 (2014)
23. B. Kovačič, R. Kamnik, International Journal for Engineering Modelling, **20 (1-4)**, 77-84 (2007)
24. N. Vatin, A. Sinelnikov, M. Garifullin, D. Trubina, Appl. Mech. and Mater. **633–634**, 1037–1041 (2014)
25. V. Rybakov, A. Panteleev, G. Sharbabchev, E. Epshtein, Appl. Mech. and Mater. **584–586**, 1876–1880 (2014)
26. V. Lalin, V. Rybakov, A. Sergey, Appl. Mech. and Mater., **578–579**, 858–863 (2014)
27. D. Trubina, D. Abdulaev, E. Pichugin, M. Garifullin, Appl. Mech. and Mater., **633–634**, 1052–1057 (2014)
28. N. Vatin, J. Havula, L. Martikainen, A. S. Sinelnikov, A. Orlova, S. Salamakhin, Adv. Mater. Res. **945–949**, 1211–1215 (2014)
29. M. Al Ali, M. Tomko, I. Demjan, Appl. Mech. and Mater. (to be published)
30. M. Al Ali, Adv. Mater. Res. **969**, 93–96 (2014)
31. A. Anis, T. Björk, S.A. Heinilä, Int. J. of Mech. and Mechatr. Eng. **12 (4)**, pp. 94–98 (2012)
32. T. Björk, H. Saastamoinen, Tub. Str. **ISTS 2012**, 167–176 (2012)
33. T. Nykänen, T. Björk, H. Mettänen, A. V. Ilyin, M. Koskimäki, Fatig. & Fract. of Eng. Mater. & Str. **37 (3)**, 325–334 (2014)
34. S. Heinilä, T. Björk, G. Marquis, ASTM Spec. Tech. Publ. **1508 STP**, 200–215 (2009)

35. M. Pavlović, M. Spremić, Z. Marković, D. Budevac, M. Veljković, J. of Appl. Eng. Sc. **12**, 75–80 (2014)
36. M. Veljkovic, B. Johansson, J. of Const. St. Res. **64 (7–8)**, 816–821 (2008)
37. C. Rebelo, A. Moura, H. Gervásio, M. Veljkovic, L. Simões da Silva, Eng. Str. **74**, 283–291 (2014)
38. K. V. Rogalevich, Materialyi 66-y nauchno-tehnicheskoy konferentsii: Aktualnyie problemyi zhelezobetonnyih i kamennyih konstruktsiy, 48-52, (2010)
39. SP 20.13330.2011 (Izd. Minregion Rossii, Moscow, 2011)

MATEC Web of Conferences **53**, 01003 (2016)
DOI: 10.1051/matecconf/20165301003

Advanced Hygrothermal Performance of Building Component at Reconstruction of S. Radonezhskiy Temple in Volgograd

Sergey Korniyenko[1, a]

[1]*Volgograd State University of Architecture and Civil Engineering, Akademicheskaya st., 400074, Volgograd, Russia*
[a]*svkorn2009@yandex.ru*

Abstract. The paper presents new thermal design of external wall S. Radonezhskiy temple in Volgograd is developed according to author's concept. The three-layer brick wall, including a thermal insulation layer from concrete with polystyrene aggregates, is considered. Calculation of interstitial condensation in building component is carried out according to simplified calculation method developed by the author and harmonized to ISO 13788. Analysis of calculation results shows that condensation occurs at one interface during some months but there is no accumulation over the year as all the condensate is predicted to evaporate again. Thus, there is no systematic moisture accumulation at the building component within a year. The risk of run-off from non-absorbent materials will be very low. Analysis of the evaporation rates at the interface shows that duration of drying wetted layer in external wall during initial stage does not exceed admissible values.

1 Introduction

The priority direction of modern architecture and construction is energy efficiency and energy saving. Increase of hygrothermal performance of building components and optimum indoor thermal comfort are necessary during all life cycle of the building, i.e. at design, construction, operation and reconstruction of buildings. For objects of cultural heritage the issue as important as ever. Saving of the historical and cultural monuments which are national property is important both for present and future generation.

Now there are no evidence-based technical solutions on a hygrothermal performance of building envelope for new and reconstructed cultural buildings. It should be noted that the designer is faced by restrictions at the choice of thermal insulation of cultural buildings.

External insulation is an optimum thermal technology and therefore often used in mass construction.

As discussed by Murgul [1] in historical and cultural monuments case often is impossible to use as external walls have design and decorative features.

Indoor insulation is dangerous from the point of view of moisture conditions and often it is also impossible as liturgical rooms have specific internal finishing. Thermal insulation layer in external wall has numerous thermal bridges as connectors. Multidimensional temperature and moist fields are formed in heterogeneous building elements that considerably complicate carrying out thermotechnical calculation. It complicates design process, construction and reconstruction of cultural buildings and demands an individual approach to design with involvement of highly qualified specialists. Relevance of this scientific issue is defined by it.

In articles [2–11] the issue of thermal performance and indoor thermal comfort for cultural buildings is considered. Need of thermal reconstruction of some cultural buildings is shown in articles [4, 5, 9]. Heat and moisture conditions of some cultural buildings are in detail investigated in articles [3, 6]. In articles [3, 7, 8] modern issues of improvement of thermal comfort in temples by means of heating and ventilation systems are considered.

As discussed elsewhere [8, 9] issues of improvement of ventilation systems with heat recovery in the historical buildings are important for achieving comfortable microclimate. Increase of buildings thermal performance due to improvement of building components with thermal bridges is considered in articles [9–13, 21].

Most accurate assessment of moisture conditions for building components can be executed by final differences or final elements methods. In this case it is possible to receive very exact analysis of processes heat and moisture transfer in building envelope [14–19]. Simplified calculation methods, such as method ISO 13788 and others, have smaller accuracy in comparison with mathematical modeling [20, 22]. At the same time the engineering methods do not demand application of specialized computer programs and are available to designers.

The review of scientific literature showed lock of study of hygrothermal performance for building components at reconstruction of the cultural buildings. Based on the review of scientific literature the purpose and objectives of research are formulated.

2 Materials and Methods

The object of research is reconstruction of Sergey Radonezhskiy temple in Volgograd [9]. Temple was built in 1999. Reconstruction of the temple started in 2011. Reconstruction purpose is increase total structural volume of the temple. Reconstruction objective is construction of the heated addition to western facade of temple for liturgical rooms. Now reconstruction of the temple is almost completed (see Figure 1).

Figure 1. General view of temple (September, 2015)

Figure 2. Construction of the heated addition to temple (July, 2013)

The original project and construction of the heated addition to temple has essential defects as shown in an article [9]. The detail analysis of these defects is not the purpose of this article. New thermal design of external wall is developed according to author's concept. In this case the three-layer brick wall, including thermal insulation layer from concrete with polystyrene aggregates, is considered (see Figure 2).

As shown in article [9], comparing to initial design data, the offered thermal design of external wall is more effective. Framework metal columns are located from the heated rooms, as shown in Figure 2, practically excludes the risks of moisture condensation on internal surface. Application of brick connectors in external wall (see Figure 3) reduces influence of thermal bridges. Increase in external surface of corners demanded for architectural reasons increases additional heat losses. At the same time the increase in thermal resistance of a corner reduces additional heat losses. Offered external wall joints correspond to the principles of designing without thermal bridges. Application of monolithic concrete with polystyrene aggregates as a thermal insulation increases thermal inertia of external wall approximately for 33.4 %. Cross and internal air filtration in a thermal insulation layer is practically excluded. Offered thermal design of external wall increases thermal comfort and energy efficiency during the heating and cooling periods.

Calculation of interstitial condensation in building component is carried out according to simplified calculation method developed by the author [20]. Purpose of the method is one-dimensional moisture transfer on the mechanism of water vapor diffusion under stationary boundary conditions. Generally the procedure of research assumes that there is more than one condensation interface. We call such building components with *multizone moisture condensation* [20].

Calculations are carried out according to the following input data:

- Location and building type
- External environmental conditions such as monthly mean temperature, t_{ext}, °C, and relative humidity, φ_{ext}, of external air
- Internal environmental conditions such as monthly mean temperature, t_{int}, °C, and relative humidity, φ_{int}, of internal air
- Material and product properties used for analysis such as density, ρ_0, kg/m^3, thermal conductivity, λ, W/(m·K), and water vapour permeability, μ, mg/(m·h·Pa)
- Boundary conditions as surface resistances at heat and water vapour transfer in building components

Calculation of interstitial condensation in building component is carried out on the following steps:

a) Condensation interfaces in structure for the coldest month are defined. Condensation interface is the section of building component in a condensation zone where maximum difference between water vapour pressure and value at saturation. To define the condensation interfaces the analytical calculation of water vapour pressure profile and saturated water vapour pressure profile in building component is used only [20].

b) For every month there is an amount of the moisture accumulated in the interface of condensation on the basis of moisture balance for the section.

c) Moisture accumulation rate in building component is defined, starting any month.

d) Assessment of moisture conditions in building component on annual balance of moisture is carried out.

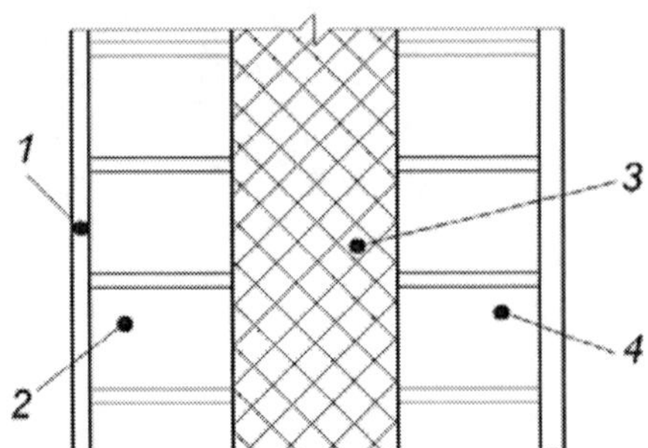

Figure 3. Scheme of external wall: *1* – plaster; *2* – inner brick layer; *3* – thermal insulation; *4* – outer brick layer; *5* – plaster

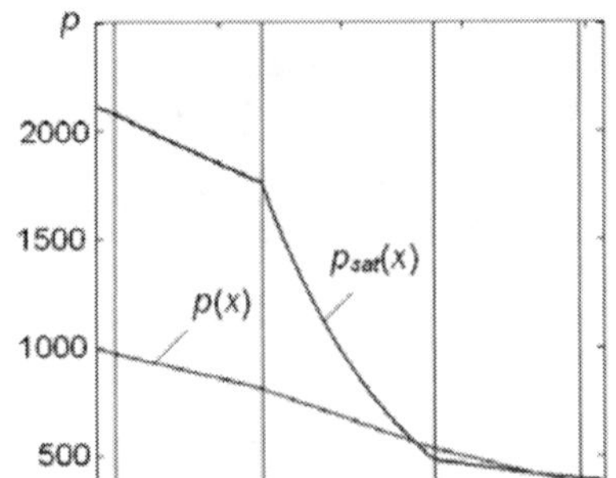

Figure 4. Schedules of $p(x)$ and $p_{sat}(x)$ in the cross section of external wall (key: x – coordinate; p – water vapour pressure; p_{sat} – saturated water vapour pressure)

The detailed calculation algorithm is given in article [20]. The method is harmonized with ISO 13788 and included in standards of the organizations in Russia. This method allows estimating moisture conditions of building components with multizone condensation. The calculation method is convenient in constructional development.

Calculation of drying of building component procedure assumes that there is an excess moisture content concentrated at the centre of a specified layer. And then moisture moves to the condensation interface and gradually evaporates from here. The monthly mean internal and external conditions are used to calculate the amount of evaporation in each of the 12 months of a year. This year is repeated until the excess moisture content of the specified layer reaches zero. During this time building component will dry out completely. Detailed calculation algorithm is given in ISO 13788.

Scheme of external wall of the heated addition to temple is shown in Figure 3. In this case it is predicted that the condensation interface located between thermal insulation and outer brick layer as shown in Figure 4.

Table 1 shows external and internal environmental conditions. Internal environmental conditions are derived from the external conditions using data ISO 13788, assuming high occupancy. Material properties for each material layer are shown in Table 2. Surface resistances for assessment of interstitial condensation are shown in Table 3.

Table 1. External and internal environmental conditions used for analysis

Month	t_{ext} [°C]	φ_{ext}	t_{int} [°C]	φ_{int}
January	–6.9	0.85	20	0.43
February	–6.5	0.85	20	0.44
March	–0.3	0.84	20	0.5
April	10.0	0.65	20	0.6
May	16.8	0.56	23.4	0.67
June	21.4	0.49	25	0.7
July	23.9	0.47	25	0.7
August	22.7	0.51	25	0.7
September	16.3	0.57	23.1	0.66
October	8.3	0.71	20	0.58
November	1.1	0.82	20	0.51
December	–4.4	0.86	20	0.46

Table 2. Material properties for external wall

Layer	Layer material	ρ_0 [kg/m^3]	λ [W/(m·K)]	μ [mg/(m·h·Pa)]
1	Lime mortar	1600	0.7	0.12
2	Hollow ceramic brick laying	1400	0.52	0.16
3	Concrete with polystyrene aggregates	250	0.085	0.11
4	Hollow ceramic brick laying	1400	0.52	0.16
5	Cement mortar	1800	0.76	0.09

Table 3. Surface resistances

Thermal resistance [m^2·K/W]		Water vapour resistance [m^2·h·Pa/mg]	
External	Internal	External	Internal
0.04	0.13	$3.7 \cdot 10^{-3}$	$11.1 \cdot 10^{-3}$

It is assumed that there is excess moisture content of 15 kg/m^2 as expected value at the moisture condensation interface.

If the building component is analyzed by using the environmental data for high occupancy internal climate without the wetted thermal insulation condensation is predicted at the interface between the thermal insulation and external laying in the coldest month of a year (see Figure 4).

3 Results and Discussion

Results of calculation of moisture conditions of external wall are shown in Table 4.

Table 4. Monthly condensation rate, g_c, and moisture accumulation, M_a, at condensation interface

Month	g_c [kg/m^2]	M_a [kg/m^2]
October	–0.276	0
November	–0.0433	0
December	0.043	0.043
January	0.0572	0.1
February	0.0551	0.155
March	–0.0119	0.143
April	–0.369	0
May	–0.739	0
June	–1.13	0
July	–1.46	0
August	–1.25	0
September	–0.684	0

Moisture accumulation in every month of a year is shown in Figure 5. Drying of wetted layer in external wall is shown in Figure 6.

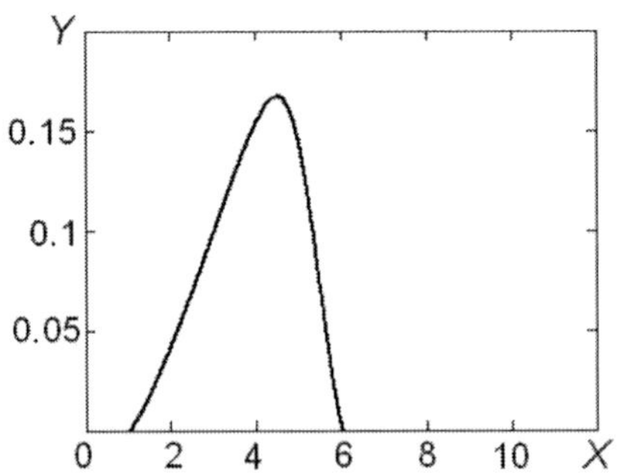

Figure 5. Moisture accumulation in external wall (key: x – month; y – condensate, expressed in kg/m^2; starting in October)

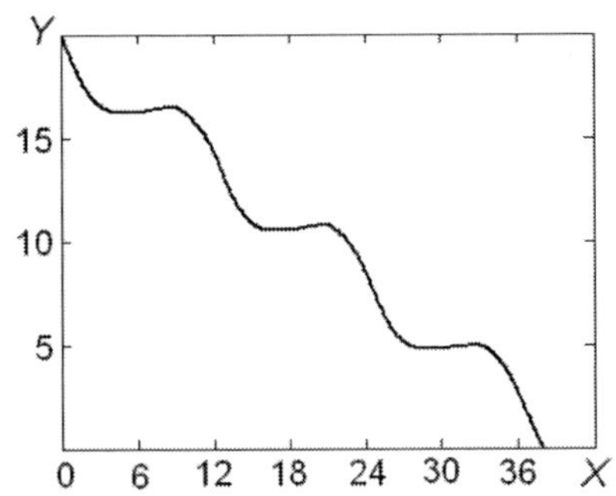

Figure 6. Drying of wetted layer in external wall (key: x – month; y – evaporated moisture, expressed in kg/m^2; starting in July)

As shown in Figure 5, the moisture accumulation at condensation interface rises to a highest point in February. From March onwards, the rate of condensation becomes negative (see Table 4), i.e. evaporation is occurring, and the accumulated moisture falls until it is close to zero in April.

Analysis of calculation results shows that condensation occurs at one interface during some months but there is no accumulation over the year as all the condensate is predicted to evaporate again. Thus, there is no systematic moisture accumulation at the building component within a year. As the maximum accumulation of condensate does not exceed 200 g/m^2 according to ISO 13788 the risk of run-off from non-absorbent materials will be very low.

Analysis of the evaporation rates at the interface, with the procedure specified in ISO 13788, starting in July, shows that mainly moisture evaporation from interface, until the amount of excess moisture reaches zero after 38 month (see Figure 6). Thus, duration of drying of wetted layer in external wall during initial stage does not exceed admissible values.

If necessary the received results can be specified by means of modeling of non-stationary moisture and heat transfer process in building components.

4 Conclusion

1. The object of research is reconstruction of Sergey Radonezhskiy temple in Volgograd. The reconstruction objective is construction of the heated addition to western facade of temple for liturgical rooms. Now reconstruction of the temple is almost complete.

2. New thermal design of external wall is developed according to author's concept. In this case the three-layer brick wall, including a thermal insulation layer from concrete with polystyrene aggregates, is considered. Offered thermal design of external wall is more effective, comparing to initial design data.

3. Calculation of interstitial condensation in building component is carried out according to simplified calculation method developed by the author and harmonized to ISO 13788. Analysis of calculation results shows that condensation occurs at one interface during some months but there is no accumulation over the year as all the condensate is predicted to evaporate again. Thus, there is no systematic moisture accumulation at the building component within a year. The risk of run-off from non-absorbent materials will be very low. Analysis of the evaporation rates at the interface shows that duration of drying of wetted layer in external wall during initial stage does not exceed admissible values.

5 Conflict of interests

The authors declare that there is no conflict of interest regarding the publication of this paper.

References

1. V. Murgul, Journal of Applied Engineering Science, **12 (1)**, 1–10 (2014)
2. G. Woroniak, J. Piotrowska-Woroniak, Energy and Buildings, **72**, 51–61 (2014)
3. M.J. Varas-Muriel, M.I. Martínez-Garrido, R. Fort, Energy and Buildings, In Press, Accepted Manuscript (2014)
4. A. Turk, M. Elif Celebi, Building and Environment, **41 (12)**, 1867–1871 (2006)
5. A.W.M. Van Schijndel, H.L. Schellen, M.H. De Wit, Building and Environment, **44 (1)**, 156–168 (2009)
6. D. D'Agostino, Building and Environment, **63**, 122–133 (2013)
7. S. Pitsch, S. Holmberg, J. Angster, Building and Environment, **45 (12)**, 2629–2643 (2010)
8. A.G. Kochev, O.V. Pasyakina, Privolzhsky Scientific Journal, **3**, 75–81 (2007)
9. S.V. Korniyenko, Construction of Unique Buildings and Structures, **5 (20)**, 39–53 (2014)
10. G.P. Vasilyev, V.A. Lichman, N.V. Peskov, M.M. Brodach, Y.A. Tabunshchikov, M.V. Kolesova, Energy and Buildings, **86**, 803–807 (2015)
11. B. Berggren, M. Wall, Energy and Buildings, **65**, 331–339 (2013)
12. K. Martin, A. Erkoreka, I. Flores, M. Odriozola, J.M. Sala, Energy and Buildings, **43 (2–3)**, 529–535 (2011)
13. G. Mao, G. Johannesson, Energy and Buildings, **26 (3)**, 233–240 (1997)
14. N.I. Vatin, D.V. Nemova, P.P. Rymkevich, A.S. Gorshkov, Magazine of Civil Engineering, **8 (34)**, 4–14 (2012)
15. S.V. Korniyenko, Applied Mechanics and Materials, **618**, 509–513 (2014)
16. S. Korniyenko, AER-Advances in Engineering Research, **22**, 529–532 (2015)
17. S. Korniyenko, Applied Mechanics and Materials, **725–726**, 1375–1380 (2015)
18. S.V. Korniyenko, Magazine of Civil Engineering, **8**, 25–37 (2014)
19. A.S. Gorshkov, P.P. Rymkevich, N.I. Vatin, Magazine of Civil Engineering, **8**, 38–48 (2014)
20. S.V. Kornienko, Energy-Safety and Energy-Economy, **4 (64)**, 12–17 (2015)
21. S.V. Korniyenko, Procedia Engineering, **117**, 191–196 (2015)
22. S.V. Korniyenko, N.I. Vatin, M.R. Petritchenko, A.S. Gorshkov, Construction of Unique Buildings and Structures, **6 (33)**, 19–33 (2015)

MATEC Web of Conferences **53**, 01004 (2016)
DOI: 10.1051/matecconf/20165301004

Chemical properties of reinforcing fiberglass in aggressive media

Arina Avdeeva[1,a], Inga Shlykova[1], Maikel Perez[1], Maria Antonova[1] and Svetlana Belyaeva[1]

[1]*St.Petersburg State Polytechnical University, 195251, St.Petersburg, Russian Federation*

Abstract. In this article describe reinforced fiberglass. Fiberglass – fiber or complex yarn, formed from glass. The main application area of fiberglass and fiberglass textile materials is their use as reinforcing elements of fiberglass plastics and composites. For its consideration, conducted laboratory tests were conducted. As a result of the research was chosen the best of the characteristics of fiberglass.

1 Introduction

The modern level of technological development has led to the creation of new materials, with special properties not inherent in the natural materials [1-3]. Such new synthetic materials includes glass fibre (fiberglass) [4]. Fiberglass has valuable properties: incombustibility, resistance to corrosion and bio effects, sufficiently high strength and relatively low density in combination with excellent optical, electrical, heat and sound insulation properties [5, 6]. It is increasingly used in various industries [7-9]. Industrial production of continuous fiberglass in the Soviet Union, was first organized in 1942.

Fiberglass is used in the production of insulating materials, which ensures reliable protection for people and equipment during prolonged exposure to high temperatures (long-term up to 700 °C, short-term up to 900°C) and protection from vibration for its entire lifetime [10, 11]. The strength of the fiberglass reinforcement is not inferior to steel reinforcement, however due to its higher cost, it is mainly used in critical structures which demand special requirements [12]. The basic design principles of reinforced concrete structures are applicable to concrete elements with fiberglass [13, 14]

2 Materials and Methods

Fiberglass is a fiber or yarn, formed from glass. In this form, the glass demonstrates unusual properties: it will not crack and not break, and is instead easy to bend without breaking [15, 16].

Fiberglass is made in three different types:

- Reinforced fiberglass - consists of polymer compounds, reinforced with fiberglass, which can be made in different types, depending on the needed composition and on the qualities of tasks performed;
- Fiberglass reinforced plastic - consists of a plastic materials, reinforced by glass;

[a] Corresponding author : arina.avdeeva@rambler.ru

- Carbon plastic or carbon fiber-reinforced polymer. This is the most expensive form of all the available types of fiberglass, and yet it is also the most durable and the lightweight.

This article will examine reinforced fiberglass and will carry out experiments to identify the best among several samples.

Selecting the right type of fiberglass (composite) is important to ensure a long service life and to reduce the risk of fracture of parts made from fiberglass plastics [17-20].

If the wrong type of fiberglass is selected, corrosive substances may come into contact with the fiberglass. This can lead to deterioration of the characteristics of the fiberglass, can cause a disruption in the bond with the resin, and lead to a reduction in structural properties [21].

A part or construction made of fiberglass-plastics can consist of materials in the form of fabric, roving, mat and veil. Roving is a bundle of glass fibers which is spliced from several glass strands. Roving can be straight or assembled. Straight roving provides the durability of parts and structures made of fiberglass-plastics. Assembled roving can be chopped. It replaces mats of chopped fiberglass and is used in the processes of impregnation by spraying for forming laminates, which are put to work in corrosive environments. The mat is made from chopped fibers or continuous fibers. The former provides strength if overlapping layers manually. The latter provides lateral resistance, high mechanical properties, and resistance to corrosion. The veil is used for forming the layer that follows the mat. It serves for the protection of fiberglass-plastic against an aggressive chemical environment.

Veils can be non-woven veils or surface veils. Non-woven veils enhance the strength of the corrosion barrier and also provide a bond with the layered material. Surface veils isolate the internal fibers from abrasion and corrosion. To ensure the highest level of corrosion resistance, it is important to specify the type of veil, independent of the specification of the mat. In Figure 1, it shows an example of structures of (FRP), including the corrosion barrier and the structural element.

When using reinforcement from fiberglass materials, you must specify the desired type of fiberglass for the structural element and the corrosion barrier.

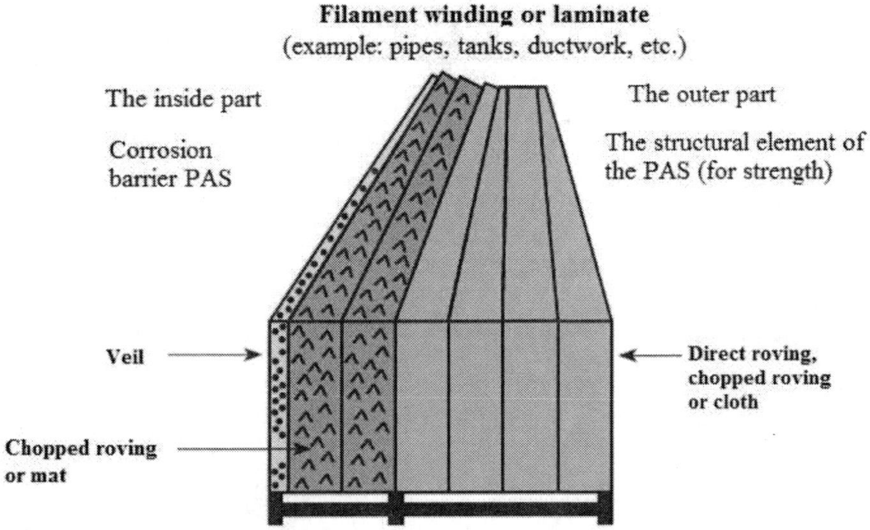

Figure 1. The winding of fibers or layered material.

Most of the products from fiberglass-plastic for applications in corrosive environments (pipelines, tanks, ducts, chimneys) are manufactured using the processes of winding the fibers and pultrusion and contains from 60% to 70% reinforcing fiberglass by weight.

Pultrusion is the process for producing profiles by continuous extrusion through a heated shaping nozzle of reinforcing glass materials impregnated with a thermosetting resin. [22, 23]

For fibre production, the raw materials used are those which are used in glass technology. [24, 25] The fiberglass is determined by composition and properties of glass from which it is made. Depending on its composition there are several brands of such glass (table 1).

Table 1. Typical composition of fiber-glass,% by weight.

Glass component	Glass mark				
	A	C	E	S	Quartz
	Strongly alkaline	Chemikal-resistant	Elektroinsulated	Extra-strong	
SiO_2	70.5	64	53	64.2	99.95
Al_2O_3	3.1	5.5	15	24.8	-
Fe_2O_3	0.2	1	0.1	0.21	-
CaO	8.7	12	17	0.01	-
MgO	3.1	2	4	10.27	-
Na_2O	12	9.5	0.3	0.27	-
B_2O_3	-	2	10	0.01	-
BaO	-	2	-	0.2	-
Other	2.4	2	0.6	0.03	0.05

A-glass is also called calcosodic, C-glass is called sodium borosilicate, E-glass is called aluminosilicate, and S-glass is called magnesia-alumina-silica. A number of application areas, such as the oil and gas industry, power plants, mining industry, heavy industry, water, sewage, marine work, tidal energy, wave energy and other areas require during construction the use of the best performance materials. The purpose of the research was the selection of the best characteristics of fiberglass along with samples.

3 Results and Discussion

Together with the Owens Corning Company a laboratory study of four samples were carried out. The tested glass types were the following: E-glass, C-glass, glass Advantex®, and AR-glass.

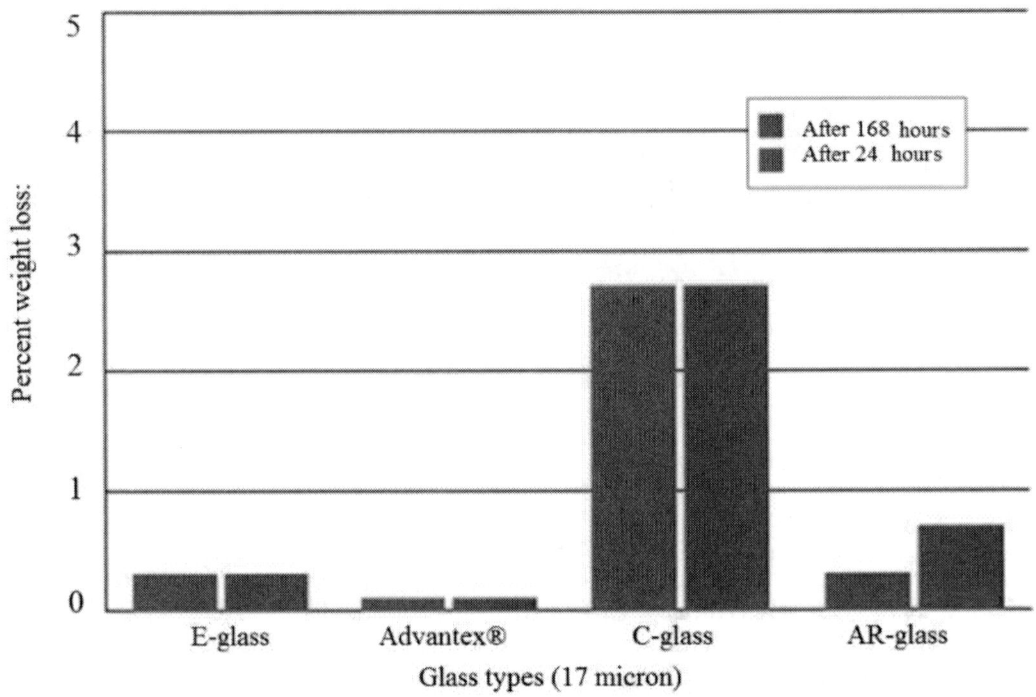

Figure 2. Tests results of the 80% acetic acid at 50°C.

Table 2 presents the main properties of the selected samples.

Table 2. Properties of glass-reinforced plastics

Glass	Main properties	Durability
E-glass	electrotechnical	middle
C-glass	corrosion resistance	low
glass Advantex®	corrosion resistance	middle
AR-glass	alkali resistance	low

Results for ammonium hydroxide (high pH) show little effect on the glass Advantex®. C-glass received significantly more damage than other types of fiberglass. Veil of C-glass is not the best choice as a corrosion barrier element.

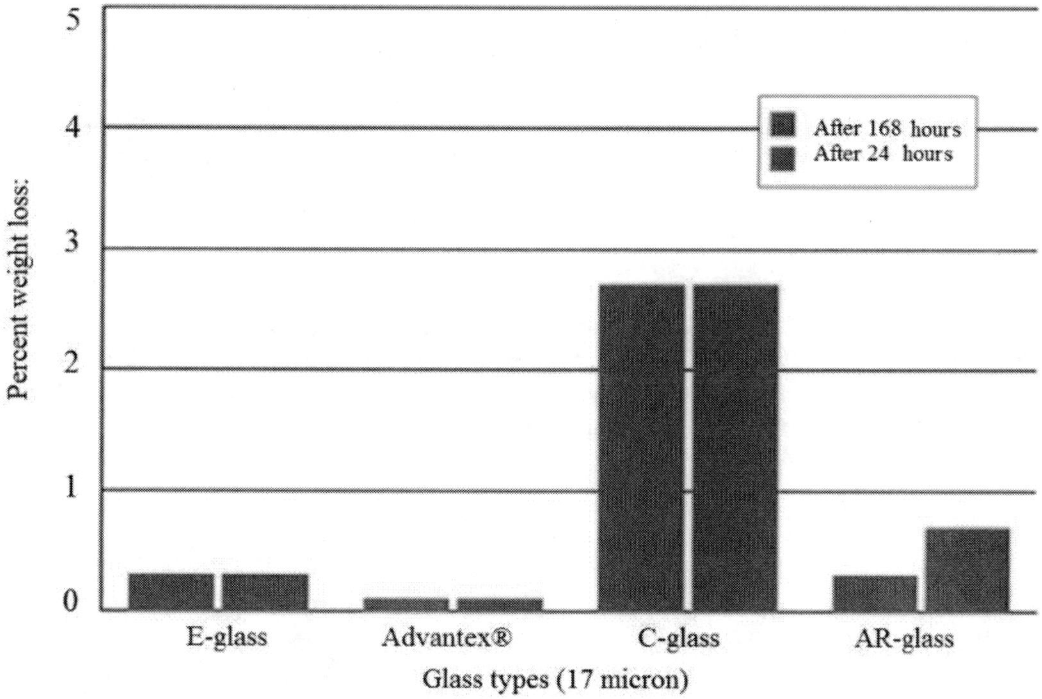

Figure 3. Tests results of the 28% ammonium hydroxide at 30°C.

Aqua regia is a mix of hydrochloric and nitric acids. It is a strong acid. The use of E-glass is not recommended due to the high percentage of weight loss. Veil of C-glass is also not the best choice.

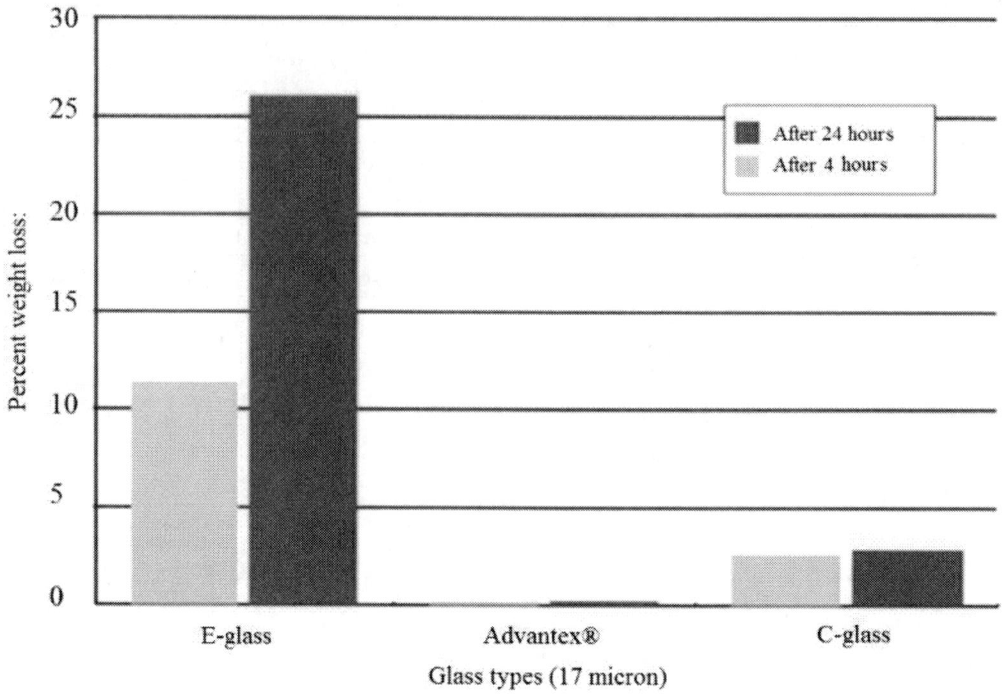

Figure 4. Tests results of the 5% aqua regia at 30°C.

Citric acid is unique. It forms a chelate (complex) with defined elements, etched from fiberglass. Fiberglass Advantex® does not contain so many elements preferred by citric acid, as do E-glass or C-glass. Other studies (R. L. Jones, University of Curtin, Australia) showed that several other organic acids, cause corrosion of the glass in a similar way.

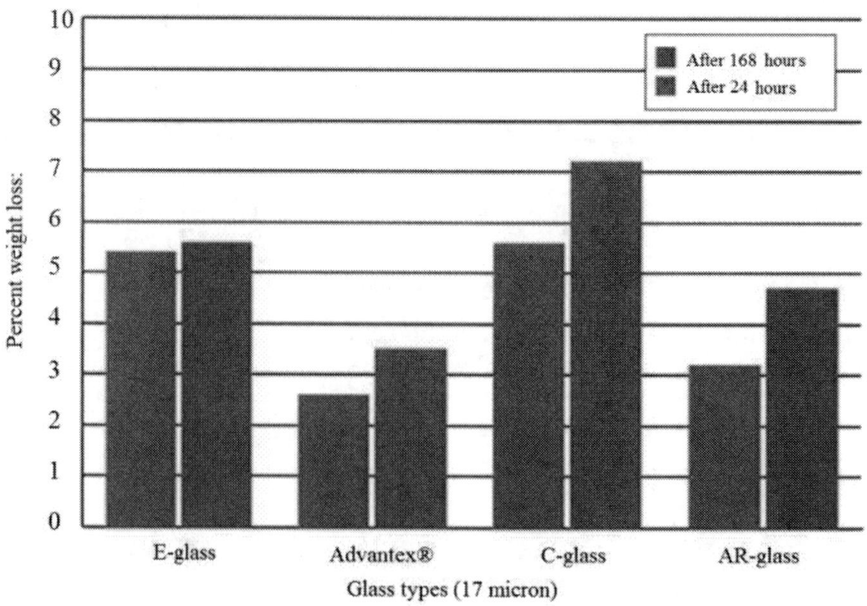

Figure 5. Tests results of the 50% citric acid at 96°C.

Long-term testing of composite cores for durability in terms of corrosion show that the glass Advantex® in deionized water at 25°C has a higher durability than E-glass. Glass Advantex® after 50 years retains 41.0% of its original strength. E-glass is able to withstand the same load for only 100 hours.

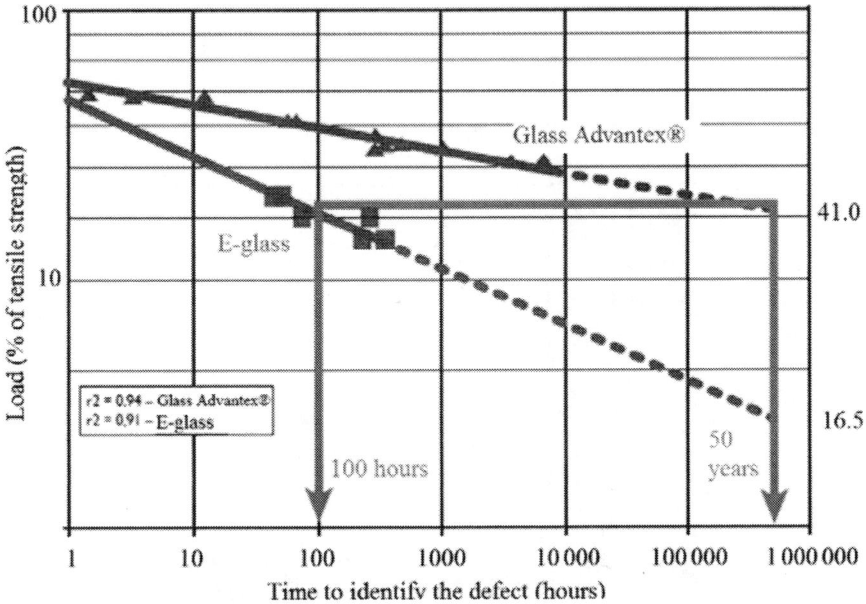

Figure 6. Tests results of the fracture properties under load, the composite rods in deionized water.

The difference between the glass Advantex® and E-glass is not reflected in the results of the measurement of weight loss of pure glass. Measurements of weight loss in deionized water do not identify differences to the same extent as the more expensive tests for durability in terms of corrosion. However, measurements of weight loss indicate that the veil of C-glass is a bad choice for a corrosion barrier.

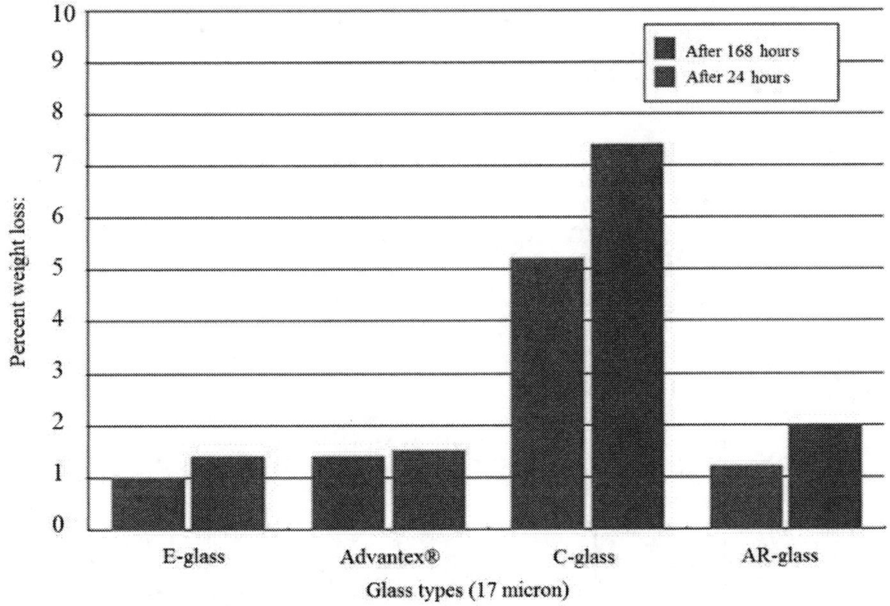

Figure 7. Tests results of the deionized water at 96°C.

E-glass is not recommended for any structural member that comes in contact with ferric chloride. AR-glass does not possess sufficient mechanical properties for a structural element in a fiberglass-plastic construction.

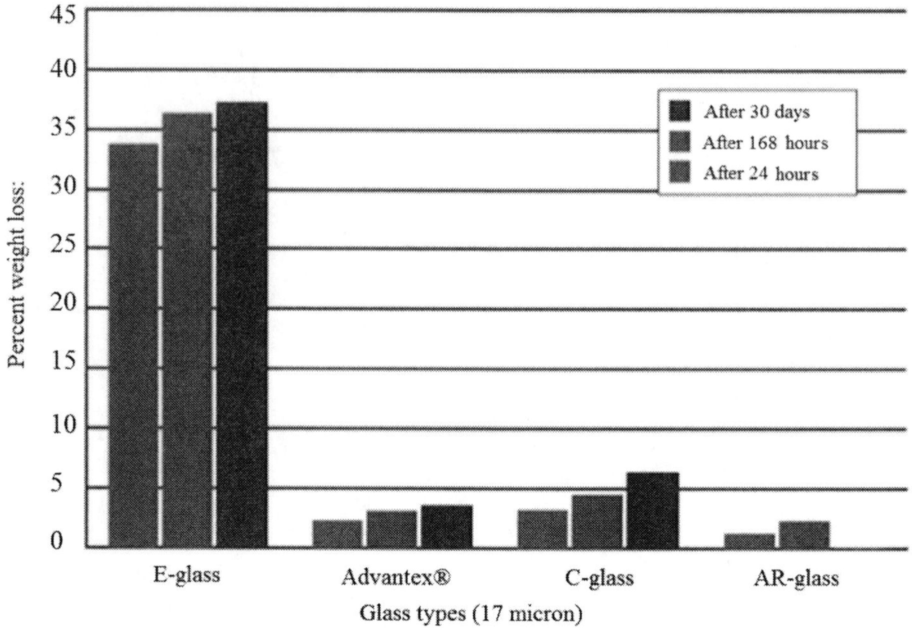

Figure 8. Tests results of the 40% ferric chloride at 96°C.

Analysis of the results of tests for durability under conditions of corrosion shows that the composite rods with E-glass loses strength with considerably faster speed than composite rods with the glass Advantex®. After 50 years, composite rods with glass Advantex® are still able to withstand 12.1% of the original load, while the composite rods with E-glass can withstand this load only for 73 hours. Etching of the glass fiber of E-glass leads to the formation of a porous outer surface and to subsequent spiraled cracking. The glass Advantex® remains intact.

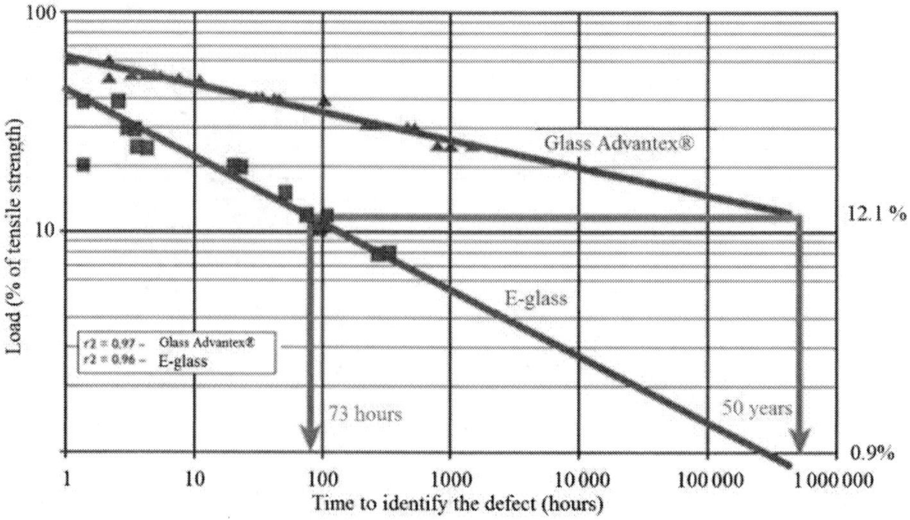

Figure 9. Tests results of the composite layered material under the influence of solution of acids ($HCl - H_2SO_4$).

Long-term strength tests in terms of corrosion show that the glass Advantex® has the advantage over E-glass in the sea. The test results show that after 50 years of immersion in seawater, a composite rod with fiber glass Advantex® can withstand 43.1% of the ultimate load, while the composite rods with E-glass can withstand this load only for 75 days.

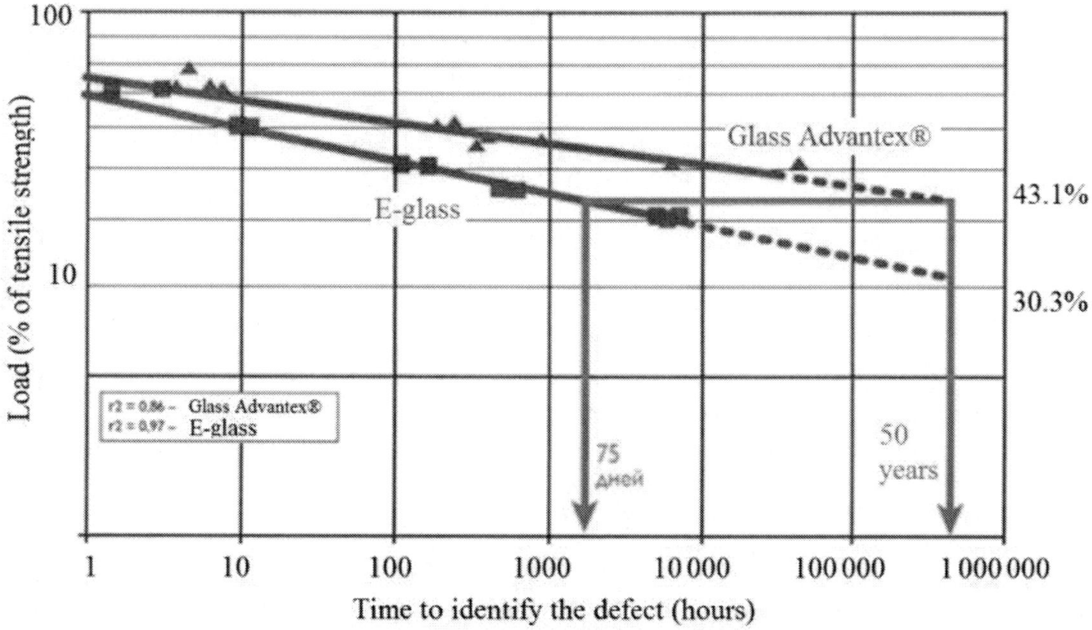

Figure 10. Tests results of the tear strength of the composite rods under loading in 5% hydrochloric solution.

Short-term weight loss measurements do not reflect an important difference in the characteristics between E-glass and Advantex glass®. Nevertheless, they clearly show that the veil of C-glass is a poor choice for use in seawater.

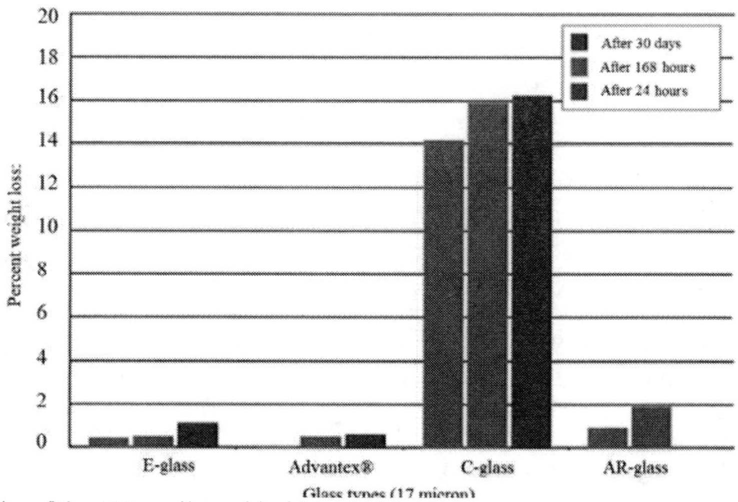

Figure 11. Tests results of the 10% sodium chloride at 96°C.

4 Conclusion

Acetic acid to some extent affects fiberglass, and to a greater extent C-glass. Ammonium hydroxide severely damaged C-glass, and had a negligible impact on the glass Advantex®. In aqua regia, E – glass was shown to prove the worst. In citric acid, all the samples behaved about the same with medium damage. Under load in deionized water, E-glass is less durable than Advantex®. In ferric chloride, AR-glass had a lower weight loss, but this glass is only available for specific fields of use. Under load in hydrochloric acid, E-glass loses strength much earlier Advantex® glass. In sea water, C-glass is the worst result in comparison with E-glass and Advantex®.

The results obtained with glass having the highest rates in all environments, is Advantex®.

Glass Advantex® can be recommended for use as a structural element and for the mat in corrosion barrier composition. Also C-glass can be considered for use as a veil for the inner surface of a corrosion barrier.

References

1. R.V. Lesovik, S.V. Klyuyev, Magazine of Civil Engineering, **3**, 41-98 (2012)
2. Dzh. Lyubin, *Handbook of composites* (Metallurgy, Moscow, 1988)
3. R. Kamnik, B. Kovačič, B. Pribicević, A. Đapo, Geodetski List, **69 (3)**, 171-188 (2015)
4. N.K. Tsirlin, *Continuous inorganic fibers for composite materials* (Metallurgy, Moscow,1992)
5. A. Abbasi, P.J. Hogg, *Temperature and environmental effects on glass fibre rebar: modulus, strength and interfacial bond strength with concrete* (Engineering, Moscow, 2005)
6. A.R. Gizdatullin, V.G. Khozin, A.N. Kuklin, A.M. Khusnutdinov, Magazine of Civil Engineering, **3**, 40-50 (2014)
7. S.V Klyuyev, Magazine of Civil Engineering, **8**, 61-103 (2012)
8. Dzh. Piatti, *Advances in composite materials*, (Metallurgy, Moscow,1982)
9. O.N. Stolyarov, A.S. Gorshkov, Magazine of Civil Engineering, **4**, 21-25 (2009)
10. N.Yu. Makusheva, N.B. Kolosova, Construction of Unique Buildings and Structures, **10 (25)**, 60-72 (2014)
11. T.S. Morozov, V.D. Kuznetsov, Magazine of Civil Engineering, **3**, 35-38 (2010)
12. B.A. Muralidhar, Materials and Design, **52**, 835-840 (2013)
13. Z.S. Teplova, S.S. Kiski, D.V. Nemova, A.V. Sokolov, Construction of Unique Buildings and Structures, **4 (19)**, 62-74 (2014)
14. Interstate standard GOST 31938-2011 "Reinforcing composite polymer for the reinforcement of concrete structures. General specifications"
15. Dzh Uiktn, E. Skal, *Fiber composite materials*, (Metallurgy, Moscow, 1978)
16. B. Kovačič, R. Kamnik, A. Štrukelj, N. Vatin, Procedia Engineering, **117** (1), 800-806 (2015)
17. Glass Fiber Reinforcement Chemical Resistance Guide (Metallurgy, Moscow, 2011)
18. S.A. Elders, A.A. Sundukova, *Strengthening masonry composite materials and screw rods* (State Technical University, Moscow, 2011)
19. J. Frank, *Reinforced plastics composites* (Univ of Sheffield, London, 1992)
20. M. Grosse, R. Hartnack, S. Lehmann, K. Rautenstrauch, *Modelling of timber-concrete composites connected only at a few discrete points* (Bauhaus-Universität Weimar, Berlin, 1992)
21. B. Kovačič, R. Kamnik, M. Premrov, Survey Review, **43** (320), 150-161 (2011)
22. Y. Barabanshchikov, S. Belyaeva, A. Avdeeva, M. Perez, Applied Mechanchanics and Materials, 475-481 (2015)
23. P.I. Silin, *Development of technology for concrete samples, using a modified polymer reinforcement* (State Technical University, Moscow, 2010)
24. Z.-Q. Ying, C.-B. Du, Y.-Y. Wang, *Numerical simulation of particle reinforced composite using extended finite element method Fourth Harbor Engineering Institute Co* (Bauhaus-Universität Weimar, Berlin, 1992)

MATEC Web of Conferences **53**, 01005 (2016)
DOI: 10.1051/matecconf/20165301005

The Concept of Economic Growth of the Construction Industry in St. Petersburg

Anna Teslya[1,a], Alexey Cherepovitsyn[2], Elena Vyboldina[1], Sergey Fedoseev[1] and Sergey Kozmenko [3]

[1] *St. Petersburg State Polytechnical University. Politechnicheskaya, 29, St. Petersburg, 195251, Russia*
[2] *St. Petersburg National Mineral Resources University (University of Mines). 2, 21 linya V.O., St. Petersburg, 199106, Russia*
[3] *Institute of Economic Problems. G.P. Luzina KSC RAS, st. Fersmana 24a, Apatity, Murmansk region, 184209, Russia*

Abstract. This article defines the concept of economic growth focusing on construction industry's needs in St. Petersburg. Authors analyse a broad range of conditions that impact competitiveness of this industrial sector. In addition to that, the article defines criteria for economic growth and sustainable development in relation to construction sector of St. Petersburg. Furthermore, authors provide a significant number of practical recommendations to ensure the rapid industry's economic growth, including suggestions regarding scientific and technological basis, impact of institutional reforms on national economy, corporate measures and measures related to environment.

1 Introduction

It is difficult to overestimate the importance of economic growth of the construction industry's sector for Russian economy and in particular, for St. Petersburg region. Being an important industry's sector, the competitive construction industry can significantly improve the industrial basis of national economy, enabling the country to solve its socio-economic problems and better meet its new national economic needs. The main strategic goal of the construction industry's structural transformation in St. Petersburg is the creation of prerequisites and conditions to neutralize and eliminate unfavourable trends constraining the sector's economic growth, while pushing for dynamic and sustainable development.

2. Materials and Methods

The modern economic science has put forward a new theory of the quality of economic growth, according to which economic growth can be indicated not only by rapid production increase, but also by other criteria like, for example, reduction of production costs, improvement of products' competitiveness due to resource-saving technologies, as well as by expansion to new markets due to implementation of successful marking strategies [1, 2]. Indeed, the modern economic growth's definition is broader than the classical one. In fact, nowadays theory and methodology of economics do not separate production growth from production quality and other production-related conditions.

---

[a] Corresponding author : anntes@list.ru

More precisely, economic growth can be better defined not solely as an increase in production, but rather through a new quality of production [3]. Thus, while defining the essence of the construction industry's transition in St. Petersburg to a new quality of economic growth, it would be rather simplistic to reduce it to a purely quantitative increase related to production and services [4].

As it is well observed by economists, structural changes in economy may lead to a situation where decline in output of some products can be accompanied by rapid growth of other types of products, which happens due to structural changes in demand. In economically developed countries, the general pattern of structural economic change can be characterized by significant decrease in raw materials sector, simultaneous increase in services and general industrial modernisation. Major changes are taking place in less material- and energy intensive sectors.

Strategy of economic growth includes not only scientific rationale of a new quality, but also identifies ways and methods how to achieve it. Economic growth can be measured in physical and monetary terms. The first method can be regarded as more reliable, since it allows eliminating effects of inflation. However, in this case by calculating economic growth's rate it is difficult to derive to a general indicator. The second method is more used by researchers, but it does not exclude the inflationary changes.

Economic growth can be defined by a number of factors. By defining the industrial economic growth it is important to establish by which factors it has been determined. The latter can be distinguished by extensive or intensive characteristics, by one-time factor or permanent impact, different nature of production sources: renewable or non-renewable. The dynamics of development is only a relative indicator of economic growth. However, by assessing the quality of economic growth is necessary to consider combination of all factors as well as added value of each of them, including product quality, technical level of production, product's consumer-related properties, demand and supply situation in a particular market and many others. Thus, the most suitable economic growth's definition, which can be applied to analysis of the construction industry in St. Petersburg, is result of interaction of a significant number of factors and conditions that determine quality of economic growth [5].

Transition to a qualitatively new type of economic growth of any industrial sector is mainly determined by the main production factors- supply, demand and distribution factors. These factors significantly contribute to economic growth. Another important pre-condition of economic growth of many industrial sectors is developed scientific and technological basis which also plays a very important role for construction industry [6].

Supply factors can be defined as quantitative and qualitative characteristics of natural and human resources, level of investments or capital stock, as well as scientific basis and level of technological development. These supply factors enable the physical ability of industry to economic growth. In this context it is important to distinguish between capacity for economic growth and real growth, which depends on factor of demand. In case of production factor's increase one should ensure adequate increase of total production expenses. This will enable to use all additional production factors available.

Economic growth is also impacted by factors of distribution and allocation [7]. Indeed, rational distribution of resources can boost production ensuring economic growth.

In terms of the construction industry the last two factors - sufficient level of total production expenses and efficiency of production resources' distribution - are crucial conditions to a new quality of industry's economic growth.

Analysis of factors affecting economic growth of the city allows to indicate general characteristics that enable economic growth to the industrial sector:

- maximum use of available natural resources, material and production potentials;
- recurring and renewable sources of economic growth.

Based on data provided for US economy, it would be useful to perform structural analysis of factors that affect economic growth (Table 1).

Table 1. The share of factors affecting economic growth (Growth in real national income for US 1929-2010) [8]

Factors of economic growth	In percentage [%]
Increase in labour costs	32
Increasing productivity:	68
Including as follows:	
Due scientific and technological progress	28
Due increasing cost of capital (investment)	19
Due improvement in education and training	14
Due legal improvements and other factors	9
In total	100

The table shows that only 32% of growth in real national income was due to increase in labour costs. 68% of increase was due to increased productivity. As it been demonstrated by the example of US economy, the main factors that positively affected productivity include [9]:

- acceleration of scientific and technological progress (introduction of new technology by updating the fixed assets)
- investment growth;
- improving education and training.

Therefore, the main permanent and renewable sources of economic growth of any economy, and surely this statement can be also applied to the construction sector- are scientific and technical progress, investment growth, increased educational and vocational training of employees [10, 11].

Scientific and technological progress is the main engine of economic growth. It includes not only transformation of industrial manufacturing basis but also implementation of new forms of management and organization. Scientific and technological progress includes discovery of new knowledge, provides a new combination of limited production resources in order to increase production of high quality competitive products. In practice, scientific and technological progress and investments are closely linked: technological progress requires investment growth, which in turn is the basis for development of scientific and technical progress.

In modern society goals and directions of scientific and technical progress are changing. If earlier scientific and technological progress has been seen as a form for introduction of new equipment, technology and logistics, its modern concept is associated with need to reduce production costs, production of competitive products, implementation of energy-saving technologies as well as with increase in production efficiency. [12] Thus, in modern society scientific and technological progress is considered as a major, permanent and renewable source of economic growth. This is very true for construction industry for which investments in scientific and technological sector are indispensible. [5] Thus, one of the main conditions for transition of construction industry to a new level of economic growth is formation and development of national economy with a significant proportion of high-tech industries.

One of the conditions to ensure rapid economic growth is to take advantage of corporate integration in development of industrial sectoral structure that corresponds to national priorities.

The main activities of corporate structures that define trends, rates, proportions, and, ultimately, the prospects for their further development of the construction industry are as follows [13]:

- ensuring stable growth of output production and technical and consumer demand, satisfaction of domestic market, as well as access to foreign markets;
- technical and technological modernisation of production base, development of innovative capacity, based on production of high-tech products, including the high-tech production that is unique worldwide;
- reconstruction of deteriorated industrial, technological and financial relations with CIS and other foreign countries, development of integration and co-operation processes;
- financial recovery of economic agents and restoration of regular commodity-money circulation;

- improving reliability and increasing investment flows to real economy through integration of banking and industrial capital;
- development of corporate management mechanisms based on assets and capitals consolidation.

3 Results and Discussion

Heading towards those objectives can create real conditions for transferring of construction industry to a new level of economic growth.

However, qualitative improvement in construction industry cannot be limited only to production factors and conditions. Transition of construction industry to a new level of economic growth requires formation and development of institutional structure of national economy. Transition of production to a new level of economic development requires restructuring of organizational and economic mechanism of production management, introducing of new legal organisation of economic activity. [14, 15] The process of implementing the strategy to accelerate industrial growth generates a new set of macro- and micro-actors in each industrial sector that interacts with traditional institutions and creates new connections between actors. Consequently, development of construction industry is closely linked to its re-organisation of its economic actors. [16]

Task to improve competitiveness of regions and cities can be solved with help of instruments for sustainable development [17]. The concept of sustainable development means achieving sustainable economic growth that enhances quality of life in a city or territory, without putting environment and non-renewable natural resources at risk. The concept of a "sustainable development" encompasses the following characteristics [14, 18, 19]:

1. Ensuring basic needs of population, including access to safe drinking water, presence of residential areas, access to quality health services and education.
2. Resource efficiency in consumption of water, electricity and recycling of industrial waste.
3. Maintaining clean environment, limiting impact of dangerous and polluting substances.
4. Creating urban infrastructure that provides access to green zones and services of public transport.
5. Heading towards sustainable development in the future, confirming volume of resources devoted to area of sustainable development.

According to the article "New models of urban development" [8], there is no clear relationship between economic growth and sustainable development. More important role plays the safeguarding of basic needs of population in infrastructure development, environmental security, access to quality health services and education. Following this consideration, we can follow the practical example of St. Petersburg Region and make practical recommendations (Table 2).

Table 2. The gross regional product of St. Petersburg by economic activity (at current prices, bln. Rubles)

Period	2004	2005	2006	2007	2008	2009	2010	2011	2012
GRP	542.3	666.3	825.1	1119.6	1431.8	1475.8	1699.5	2092.9	2292
including types of economic activity									
production and distribution of electricity, gas and water	21.9	23.2	29.5	40.1	51.9	61.9	63.4	81.6	82.1
building	27	37.8	54.6	81.7	110.6	111.2	133.1	134.2	124.7
education	17.3	28.5	33.7	42.8	52.7	63.3	67.5	76.4	86.5
health and social services	21.4	34.7	44.4	52.6	66.1	76.7	85	102.9	119.6
other community and personal services	14.4	21.5	27.8	38.6	43.9	38.8	38.2	43.6	48.3

This table demonstrates that gross regional product of St. Petersburg increased in the period concerned in 4.23 times, while at the same time, the volume of production in construction industry increased in 4.63 times. In particular, the share of construction industry increased from 5.0% in 2004 to 7.8% by 2010 and further decreased by 5.4% in 2012. The share of health and education amounted to 5.2% and 3.8% in 2012 and 4.0 % and 3.2 % in 2004.

It should be noted that the growth rate data is not stable since 2009 (Table. 3).

Table 3. The growth rate of gross regional product of St. Petersburg by economic activity (in % to the previous period) [19]

Period	**2005**	**2006**	**2007**	**2008**	**2009**	**2010**	**2011**	**2012**
GRP	1.23	1.24	1.36	1.28	1.03	1.15	1.23	1.10
production and distribution of electricity, gas and water	1.06	1.27	1.36	1.29	1.19	1.02	1.29	1.01
building	1.40	1.44	1.50	1.35	1.01	1.20	1.01	0.93
education	1.65	1.18	1.27	1.23	1.20	1.07	1.13	1.13
health and social services	1.62	1.28	1.18	1.26	1.16	1.11	1.21	1.16
other community and personal services	1.49	1.29	1.39	1.14	0.88	0.98	1.14	1.11

Given the dynamics of price increase it is useful to evaluate dynamics of physical volume of gross regional product of St. Petersburg by economic activity. This indicator demonstrates sharp decline in output of construction industry. Negative growth trends in construction are due to slow pace of GDP growth as a whole. Difficult situation persists, so in 2013 turnover of construction companies in St. Petersburg was 94% in comparison to the same indicator in 2012, and 100.8% in comparison to the same indicator of 2014 (Jan-Nov 2014). In the same time, in the period from January till November 2014 617.0 thousand square meters of residential properties became operational. This indicator is 100, 4% compared to the same period in 2013.

4 Conclusions

Hence, it is important to note that construction industry in St. Petersburg is essential to achieve quality of life improvement, sustainable socio-economic development of territories through maintaining and creating of housing, infrastructure and jobs. On the other hand, development of construction industry in St. Petersburg can be a significant source of environmental pollution, environmental degradation through pollution of dangerous chemicals substances harmful to human health and environment, as well as depletion of natural resources.

Therefore, in order to ensure sustainable development of St. Petersburg, it is important to use legal tools that will enable development of construction industry, while avoiding the negative effects on human health and environment in the city. This can be achieved by means of close cooperation between St. Petersburg government and private sector.

Thus, the necessary condition for the St. Petersburg construction industry's transition to a qualitatively new level of economic growth is to increase effectiveness of institutional reforms and management of state property.

References

1. A. Pekuri, H.Haapasalo, M. Herrala, International Journal of Performance Measurement, **1**, 39-58 (2011)
2. E. Kasimovskaya, M. Didenko, Journal of Applied Business Research, **2**, 37-51 (2013)
3. S. Rutešić, J. Ćetković, M. Knežević, M. Žarcković, N. Vatin, Procedia Engineering, **117** **(1)**, pp. 642 - 650 (2015)

4. S. Dlamini, International Congress on Construction Management, **34**, 213-225 (2012)
5. O. Gamayunova, N.Vatin, Applied Mechanics and Materials, **725-726**, 1640-1645 (2015)
6. J. Ćetković, S. Rutešić, M. Zarković, M. Knežević, N. Vatin, Procedia Engineering, **117** (**1**), pp. 780 - 790 (2015)
7. R. Macek, Review of Economic Perspectives, **4**, 309-328 (2014)
8. S. Bouton, J. Woetzel, M. Lindsay, McKinsey on Sustainability and Resource Productivity, 54-63 (2012)
9. B. Ark, M. O'Mahony, M. Timmer, Journal of Economic Perspectives, **22**, 25–44 (2008)
10. P. Chen, D. Partington, International Journal of Project Management **22**, 397–406 (2004)
11. T. Becker, E. Jaselskis, C. McDermott, Proceedings of the Associated Schools of Construction 2011 International Conference, **3**, 345-355, (2011)
12. S. Rutešić, J. Ćetković, M. Žarcković, M. Knežević, N. Vatin, Procedia Engineering, **117** (**1**), pp. 905 - 915 (2015)
13. S.Galperin, M. Dorodneva, Y. Mishin, E.Pukhov, *Mechanism of analysis and forecasting activities of corporate structures.*, (Institute of microeconomics. Moscow, 2001)
14. O. Gamayunova, N.Vatin, Advanced Materials Research, **1065-1069**, 2459-2462 (2015)
15. G. Peshkova, Corporate Governance and innovative economic development of the North: Bulletin of Research Center of Corporate Law, Management and Venture Investment of Syktyvkar State University, **4**, 53-64 (2014)
16. N.Vatin, O.Gamayunova, D. Nemova, Applied Mechanics and Materials, **638-640**, 2460-2464 (2014)
17. K. Negacz, Economic and Environmental Studies, **12**, 165-182 (2012)
18. N.Vatin, O.Gamayunova, D.Petrosova, Applied Mechanics and Materials, **635-637**, 2085-2089 (2014)
19. J. Ćetković, S. Rutešić, M. Zarković, M. Knežević, N. Vatin, Procedia Engineering, **117** (**1**), pp. 780 - 790 (2015)

MATEC Web of Conferences **53**, 01006 (2016)
DOI: 10.1051/matecconf/20165301006

Reinforcement of concrete structures by fiberglass rods

Arina Avdeeva[1,a], Inga Shlykova[1], Maria Antonova[1], Yuri Barabanschikov[1] and Svetlana Belyaeva[1]

[1]*St.Petersburg State Polytechnical University, 195251, St.Petersburg, Russia*

Abstract. In this article we introduced conducted experiments to determine the basic characteristics of composite reinforcement fiberglass reinforcement on the example of the same diameter, but with a different number of rovings. We have established strength along the reinforcing fiber and the corresponding class of the steel reinforcement.

1 Introduction

Corrosion is the main drawback of steel. Fiberglass solves this problem. Corrosion resistance is one of the main advantages of composite reinforcement. The others are: increased resistance to tearing, light weight (fiberglass is up to 9 times lighter than steel), chemical resistance, good elastic properties, frost (up to -70 C°) and heat (up to 100 C°) resistance, low thermal and zero electrical conductivity, radio transparency [1-4]. These features make possible to use this type of reinforcement in the aggressive environment [5].

Building armature is used to manufacture all the types of concrete structures. Fittings in cooperation with concrete resist tensile load [6, 7].

The usages of composite reinforcement can significally reduce the weight of structures, increase to aggressive environments, expand architectural features, and reduce labor costs [8 ,9].

The application of the framework structures improves physical and mechanical properties, reduces strains on building elements [10, 11].

The advantages of fiberglass reinforcement in comparison with conventional metal alloys the higher values of tensile strength and fatigue limits (10 - 50%), better elasticity, stiffness and reduced crack tendency [12, 13]. The usage of composite reinforcement increases rigidity of structure and saves metal [14-16].

Non-metallic reinforcing consists of the fiber bundle situated along the rod and interconnected by the polymer resin (cured thermoset) [17]. Bundle of fiber of smaller diameter is wrapped around the main stem in order to increase adhesion [18-20]. There are two types of non-metallic fittings depending on the type of fiber material that was used in manufacturing: glass and basaltic ones. Every type has its own distinctive features. Fittings are available in the form of different shaped rods. They are subdivided by the outer diameter and the rod length [21].

The need to ensure magnetic and dielectric properties of some products and facilities as well as an expanding use of reinforced concrete operated in aggressive environments stimulated interest to fiberglass reinforcement in the 60s [22].

[a] Corresponding author : arina.avdeeva@rambler.ru

Composite rebar is currently becoming an irreplaceable building material thanks to its distinctive properties. It is used to reinforce concrete and ferroconcrete, thin-wall structures, to repair structures damaged by exposure to corrosive environments [23]. Composite rebar is also used along with slag cement; pozzolanic cement; mixed binder (with the high content of active mineral additives); antifreeze additives; porous and coarse-pored concrete; light and cellular concrete [24]. Its field of usage also comprises monolithic construction, reinforced brickwork, construction of seaports and tunnels [25, 26].

Buildings can be reinforced both internal (fiberglass rebar (FRP) is embedded in the concrete) and external (fiberglass lattice is situated in the surface of construction and covered by sprayed plaster). The two types of reinforcement can be implemented simultaneously. As a result, the service life of construction lasts up to 80 years [21].

Fiberglass in the forms of non-woven geotextiles, geogrid and lattice is used in road construction. These components make roadbed more durable and reliable [27, 28]. They also reduce the thickness of paving [29].

The work is based on experimental investigation of physical, chemical and mechanical properties of two specimens of concrete (reinforced with polymer armature and concrete without reinforcement). Investigations show that toughness of specimens reinforced with polymeric beams is two times higher, flexural strength increases by 40% [30].

2 Materials and Methods

Fiberglass reinforcement produced by OOO "PROF". Test material was reinforced in one direction. Four compositions of reinforcement were tested. They had 17, 18, 19 and 20 rovings respectively. All the samples had a diameter of 8 mm. The results obtained from the tests (Tab. 1) show that increasing number of rovings increases weight content of fiber glass the sample. The volume content of filler in composition with 18 rovings is the same as in composition with 19 due to porosity of material. Ignition of sample leads to weight reduction (burned binder) and volume reduction (binder and pores). Porosity also causes discrepancy between volume and mass content of fiberglass in these samples. It is shown by the results of the experiments (Tab. 1).

Early fracture of rod ends hindered the investigation. The following methods helped us to overcome this problem. The diameter of 10 mm length middle part of the sample was reduced to 5 mm. Ends of the sample were wound with several layers of glass fiber with polyester resin.

Table 1. Indicators properties of fiberglass reinforcement a diameter of 8 mm

Name indicator	Number roving			
	17	**18**	**19**	**20**
Weight content of glass fiber, [%]	78.3	81.5	82.6	85.6
Volume content of glass fiber Z, [%]	58.5	64.1	64.1	70.0
FRP density γ, [g / cm3]	1.971	2.076	2.048	2.160
Water Absorption by weight Ww, [%]	1.55	0.54	1.76	0.94
Water Absorption by volume Wv, [%]	3.05	1.12	3.61	2.03
Common porosity Pc, [%]	7.86	5.67	7.86	5.47
Semi-closed porosity Pcl, [%]	4.82	4.55	4.26	3.44
Breaking strength Rp, [MPa]	881	922	870	967
Relative breaking elongation, [%]	1.97	1.80	1.73	1.71
Tensile modulus, [GPa]	46.0	52.8	51.8	58.5
Compression strength Rcom, [MPa]	221	311	260	273
Flextural strength Rben, [MPa]	267	310	287	340
Tensile strength (fracture), [MPa]	15.0	15.0	14.6	15.6
Shear strength, [MPa]	30.5	35.4	34.2	35.2
LTEC, 10-6 /[°C]	4.51	4.74	4.83	4.57

3 Results and Discussion

From Table 1 it is seen, that tensile strength by tension increases with the percentage of reinforcement. However, the tensile strength by break of samples 19 was less than that 18, which appears to be due to a lower density 19. Point 19 for breaks quite clear linear dependence of the strength of the density fiberglass reinforcement. With the exclusion of the point correlation coefficient, which is defined by the formula:

$$K^2 = 1 - \frac{\Sigma(R_i - R_i{}^e)^2}{[\frac{(\Sigma R_i{}^2) - (\Sigma R_i)^2}{n}]} \tag{1}$$

It increases from 0.751 to 0.992. Similarly, the dependence of tensile strength by break of the volume content fiber becomes sufficiently strict linearity ($K^2 = 0.9999$) with the exception of it to the point 19. When accounting for 19 of the correlation coefficient decreases to a value of 0.6453.

Character discontinuity bars fiberglass reinforcement has its own characteristics. In areas of stress concentration consistently produces local discontinuities of individual fibers or bundles, accompanied by a brief drop and subsequent increase of the load. This character is due to unequal tension of the fibers when the stress is perceived only the most intense of them, and as they discontinuity force is transmitted to other fibers.

Tensile strength of rods with reduced diameter (5 mm) is slightly less than the same characteristic of full diameter rods (8 mm). It corresponds to the well known idea of influence of scale factor, which equeals to the ratio between strength of samples with the diameter of 8 mm ad samples with the diameter of 5 mm.

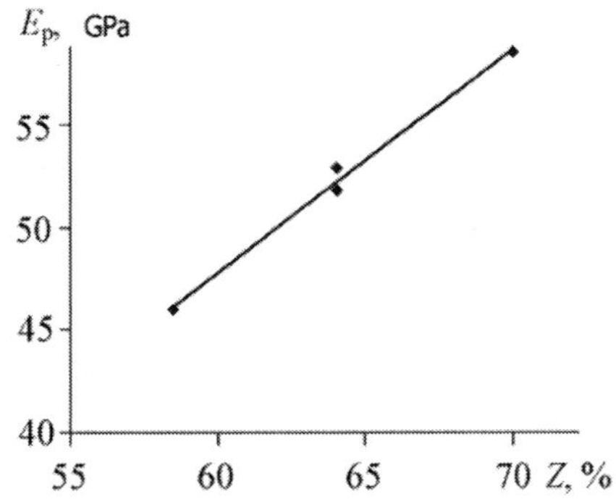

Figure 1. The dependence modulus of elasticity by tension FRP on the degree of reinforcement

In determining the elastic modulus of the samples was carried out loading steps of 250-300 kg. The deformation was measured with a device on the indicator of hour type with an accuracy of 1 micron. At each stage of loading or excerpts for 25 minutes. Exposure due to the need to stress relaxation characteristic of fiberglass reinforcement. Modulus of elasticity of volume content of the reinforcing filler is shown in Fig. 1. When exposed to a particular stage of the loading tension decay over time as shown in Fig. 2 wherein the ordinate the proportion relative to the initial tension value σ_0, i.e. relation σ/σ_0. As can be seen from Fig. 2, after 25 min. tension in samples average about 70 % of σ_0. In the presented experiments σ_0 for different samples varied from 344 to 466 MPa. Despite this, the relative values of σ/σ_0, as for the individual samples and the average for the series of largely coincided. At the same time, a decrease in the tension drop across the length of the sample, so that the elastic modulus in the relaxation process is kept constant.

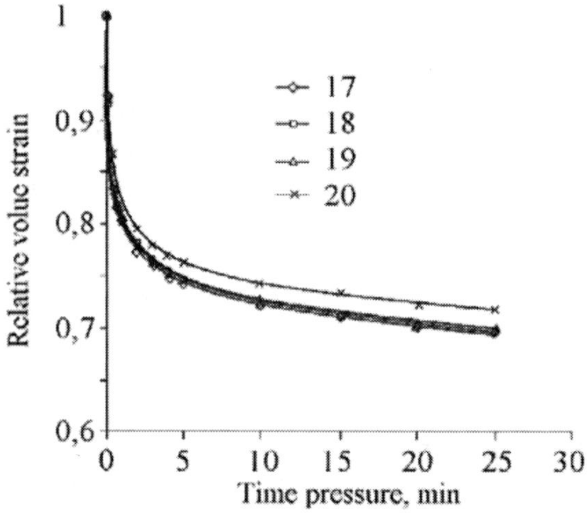

Figure 2. Relaxation in FRP

When the samples were compressed in parallel to grain, the destruction was accompanied by formation of longitudinal cracks, passing through the bond between fibers. Lack of reinforcement explains constant character of splitting strength. Similar results were obtained in case of chipping (in parallel to grain). In both cases only the resin resists the destruction of sample.

4 Conclusion

LTEC was determined by a quartz dilatometer. The samples were 100 mm long. The temperature was varied from 20 ° C to 80 ° C. The linear dependence is observed for elongation of the samples. Also, the elongation does not depend on the nature and density of the reinforcing filler. The content of the reinforcing filler in percent and the porosity affect the strength as a result of incomplete infiltration of the filler. By increasing the reinforcing fibers in the reinforcement reduces deformation properties. The break the binder between fibers caused by tensile stresses. Therefore, the strength of the resin acting as a binder characterize experiments on shear fracture along the fibers and tensile by shear fracture. The strength along the of the reinforcing fibers above tensile strength across the fibers. The degree of reinforcement does not depend on water absorption.

FRP strength corresponds to the class steel reinforcement Am-IV, at the same time characterized by low modulus values (3.5 times), LTEC (2.5 times) and elongation (4-5 times).

References

1. A.S. Boldyrev, P.A. Zolotov, A.N. Lusov, *Building materials: directory* (Stroyizdat, Moscow, 1989)
2. R. Kamnik, B. Kovačič, B. Pribicević, A. Đapo, Geodetski List, **69** (**3**), 171-188 (2015)
3. S.A. Elders, A.A. Sundukova, *Strengthening masonry composite materials and screw rods* (State Technical University, Moscow, 2011)
4. J. Frank, *Reinforced plastics composites* (Univ of Sheffield, London, 2013)
5. N.P. Frolov, *Fiberglass reinforcement and fiberglass concrete construction* (Stroyizdat, Moscow, 1980)
6. N. Gubeljak, J. Predan, D. Kozak, J. Tuma, B. Kovačič, P. Konjatić, J. Sertić, Strojarstvo, **51** (4), 263-271 (2009)
7. S.V. Klyuyev, Magazine of Civil Engineering, **8**, 61-103 (2012)
8. O.N. Stolyarov, A.S. Gorshkov, Magazine of Civil Engineering, **4**, 21-25 (2009)
9. Dzh. Uiktn, E. Skal, *Fiber composite materials* (Metallurgy, Moscow,1978)
10. T.S. Morozov, V.D. Kuznetsov, Magazine of Civil Engineering, **3**, 35-38 (2010)

11. A.L. Shagin, No. 924317 Bull. Inventions, **16,1**, 50 (1982)
12. R.V. Lesovik, S.V. Klyuyev, Magazine of Civil Engineering, **3**, 41-98 (2012)
13. N.Yu. Makusheva, N.B. Kolosova, Construction of Unique Buildings and Structures, **10** (25), 60-72 (2014)
14. Interstate standard GOST 31938-2011 "Reinforcing composite polymer for the reinforcement of concrete structures. General specifications"
15. Dzh. Lyubin, *Handbook of composites* (Metallurgy, Moscow, 1988)
16. *Standart. Rulebook Construction of concrete with composite non-metallic reinforcing. Design Rules*
17. M. Grosse, R. Hartnack, S. Lehmann, K. Rautenstrauch, *Modelling of timber-concrete composites connected only at a few discrete points*(Bauhaus-Universität Weimar, Berlin, 2011)
18. B.A. Muralidhar, Materials and Design, **52**, 835-840 (2013)
19. B. Kovačič, R. Kamnik, International Journal for Engineering Modelling, **20** (1-4), 77-84 (2007).
20. N.K. Tsirlin, *Continuous inorganic fibers for composite materials* (Metallurgy, 1992)
21. Z.S. Teplova, S.S. Kiski, D.V. Nemova, A.V. Sokolov, Construction of Unique Buildings and Structures, **4** (19), 62-74 (2014)
22. B. Kovačič, R. Kamnik, Z. Kapović, Geodetski List, **63** (4), 315-327 (2009)
23. B. Kovačič, R. Kamnik, A. Štrukelj, Promet - Traffic - Traffico, **22** (1), 15-22 (2010)
24. Y. Barabanshchikov, S. Belyaeva, A. Avdeeva, M. Perez, Applied Mechanics and Materials, **726**, 475-481 (2015)
25. B. Kovačič, R. Kamnik, M. Premrov, N. Gubeljak, J. Predan, Z. Tišma, Strojniski Vestnik/Journal of Mechanical Engineering, **54** (5), 364-371 (2008)
26. B. Kovačič, R. Kamnik, M. Premrov, Survey Review, **43** (320), 150-161 (2011)
27. Abbasi, P.J. Hogg, Composites Part B:Engineering, **36**, **5**, 394-404 (2005)
28. A.R. Gizdatullin, V.G. Khozin, A.N. Kuklin, A.M. Khusnutdinov, Magazine of Civil Engineering, **3**, 40-50 (2014)
29. B. Kovačič, R. Kamnik, A. Štrukelj, N. Vatin, Procedia Engineering, **117** (1), 800-806 (2015)
30. B . Kovačič, B. Geodetski Vestnik, **48** (1), 32-39 (2004)

MATEC Web of Conferences **53**, 01007 (2016)
DOI: 10.1051/matecconf/20165301007

The using feasibility Russian and European software products at thermal calculations

Daria Tarasova [1,a], Anastasiia Staritcyna [1], Darya Nemova [1] and Konstantin Andreev [1]

[1]*St. Petersburg State Polytechnical University, Politekhnicheskaya, 29, Saint-Petersburg, 195251, Russia*

Abstract. Modern software products for thermal calculations make possible to speed up the design process and to improve the energy efficiency of new and renovated buildings. This article provides a complex evaluation, comparison and analysis of the feasibility of using Russian and foreign programs in thermal calculations.

1 Introduction

One of the main buildings elements that are in contact with the environment is a facade. The primary task of the facade is to protect the building from weather, temperature extremes, drafts, heat loss [1-8]. The economy of energy resources is considered by the developed countries as the major national environmental and economic issue because of energy saving measures higher profitability and environmental safety than increasing energy resources [9-15].

The most effective use of energy resources of the country for the revival and subsequent recovery and provide a decent life for the population declared main goal of the energy strategy of Russia since 1995.

Thermal energy losses depend on the level of thermal protection of external envelope. In all countries there are regulatory requirements for the insulation level of enclosing structures. It depends on the climatic conditions of the country and its state energy saving policy. The norms of energy buildings consumption is constantly are decreasing and the requirements to the insulation level of external envelope are increasing because of constant growth of energy resource prices, the growth of thermal energy tariffs and reduction of stocks of non-renewable hydrocarbons (oil, gas). Thus, one of the most important ways of economy of energy resources is the reduction of heat losses through the building envelope of buildings [16-23]. One of the problems is low awareness of designers about the global experience of design of energy-efficient buildings. Using comprehensive software you can develop optimal conditions for the design of energy efficient buildings, because of engineering solutions are calculated to determine their contribution to the achievement of the overall reduction of energy consumption.

As discussed elsewhere [24-27] optimization of the energy performance of a building is similar to the solution of the equation with several unknowns. To help in its decision we can use a modern program of thermotechnical calculation. This allows you to visualize the wall structure of the building, to calculate various arrangements and to bring them into accordance with the norms and standards – and all this before construction starts.

[a] Corresponding author : tarasovads@gmail.com

Thus, designers can choose the least expensive option for energy upgrading, optimize the energy consumption and other indicators of the building and use technologies that surpass in the speed and cost of traditional approaches for energy modeling and auditing [28, 30].

2 Input data

Design conditions and heat energy parameters of the buildings for the climatic conditions of St. Petersburg adopted according to SNiP 23-01-99*are presented at the table 1.

Table 1. The design conditions.

The characteristics	Symbol parameter	Unit of measure	Value
Design temperature of inside air	t_{int}	°C	20
Design temperature of outdoor air	t_{ext}	°C	-26
The duration of the heating period	z	day/years	220
Heating degree day	D_d	°C· day/years	4796
Mean temperature of outdoor air	t	°C	-1.8

We will consider three types of various exterior structures. Structures of walls and its characteristic are given at the table 2.

Table 2. Structures of walls and its characteristic.

Type of structure	Structures of walls	λ, W/mK	ρ, kg/m3	δ, mm
Type I	Concrete	2.0	2400	500
	Foam polystyrene URSA XPS	0.033	35	120
	Air (ventilated layer)	0.024	0,001	30
	Granite tile	1.5	2600	12
	Brick	0.960	2000	250
Type II	Rock wool facade insulation	0.035	100	150
	Air (ventilated layer)	0.024	0,001	30
	Granite tile	1.5	2600	12
	Light weight concrete	1.3	1800	300
Type III	Polyurethane foam MUSTOVO	0.022	50	200
	Air (ventilated layer)	0.024	0,001	30
	Granite tile	1.5	2600	12

3 Materials and Methods

Two programs, which make thermomechanical calculation of the multilayered enclosure structure, were chosen for the complex analysis.

The program TeReMOK 0.8.5 "Thermomechanical calculation of the multilayered enclosure structure " is based on the requirements and calculation procedures stated in Construction Norms and Regulations 23-02-2003 "Thermal protection of buildings", design and construction specifications 23-101-2004 "Design of thermal protection of buildings" and Construction Norms and Regulations 23-01-99 "Construction climatology".

U-value.net is the english translation of u-wert.net which is a very successful website in Germany. U-value.net is a private website that aims to support you on the planning of your insulation. After entering all layers of a building component into the U-value calculator you will receive information about the expected heat loss, eventual moisture problems and many other things.

3.1 Thermomechanical calculation of the multilayered enclosure structure TeReMOK 0.8.5

After creation of three types of an enclosure structure design, calculation in this program was executed on check on the normalized value of heat transmission resistance.

Let's consider the received results for all types of the construction (table 3):

Table 3. Value of heat transmission resistance for different type of the enclosure structure.

Type of structure	R, (m²·°C)/W
Type I	5.213
Type II	6.012
Type III	10.787

By results of calculation, it is visible that the enclosure structure with insulation from polyurethane foam has the highest heat protection.

3.2 Thermomechanical calculation of the multilayered enclosure structure U-value.net

According to initial data, let's execute thermal calculation of the three types of multilayered enclosure structure. The following is a report on thermal calculations of the first type of the wall (figures 1-6).

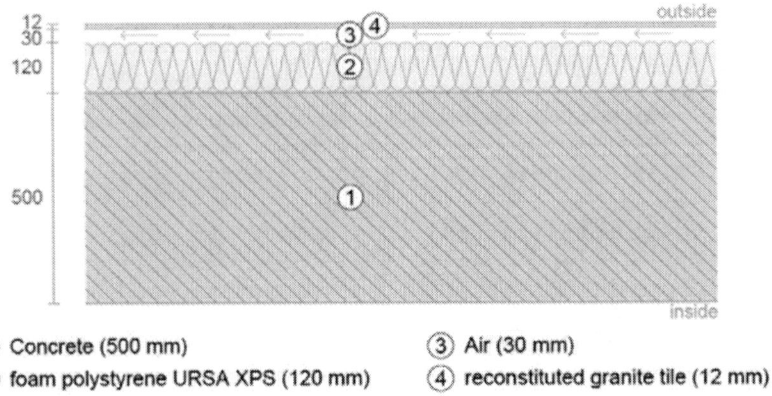

Figure 1. Cross-section of the enclosure structure.

By means of this program, we could receive contribution of different layers to the overall insulation, heat protection schedule, humidity schedule and temperature gradient schedule.

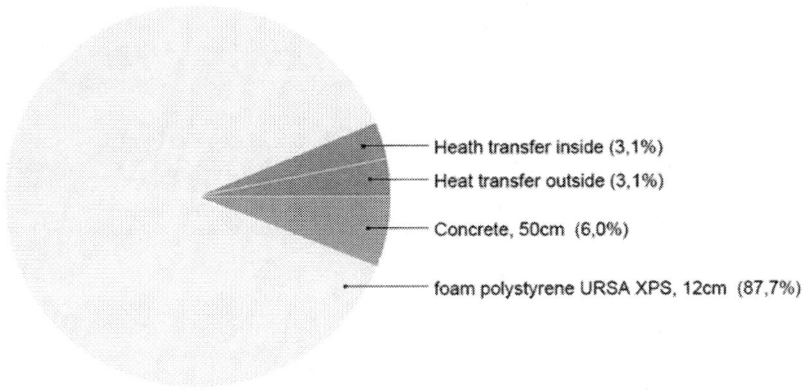

Figure 2. Contribution of different layers to the overall insulation.

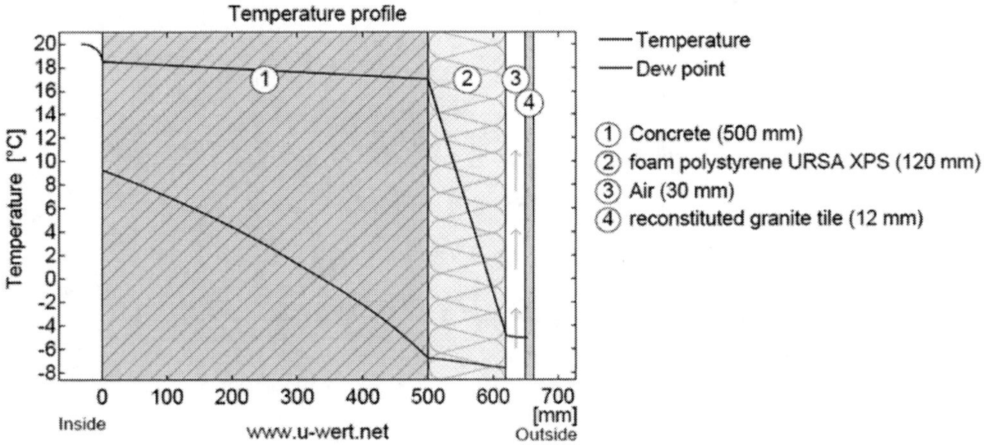

Figure 3. Temperature gradient. Condensing zone.

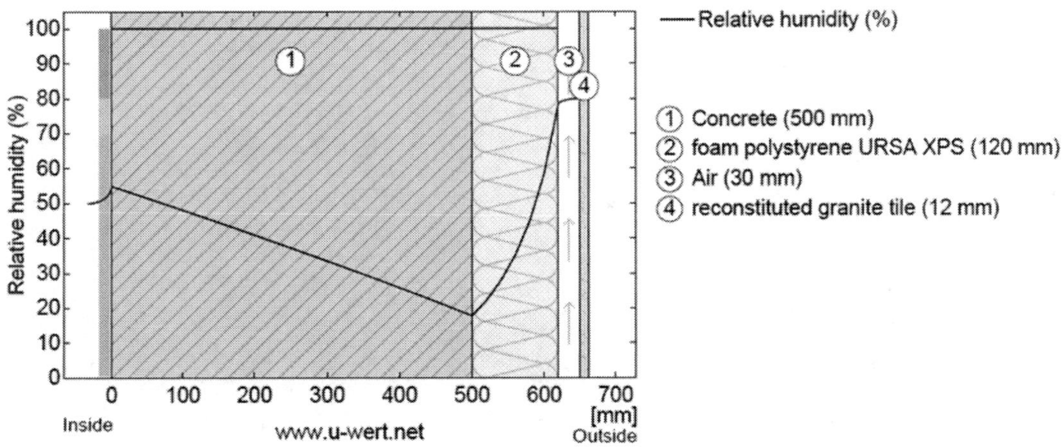

Figure 4. Humidity schedule.

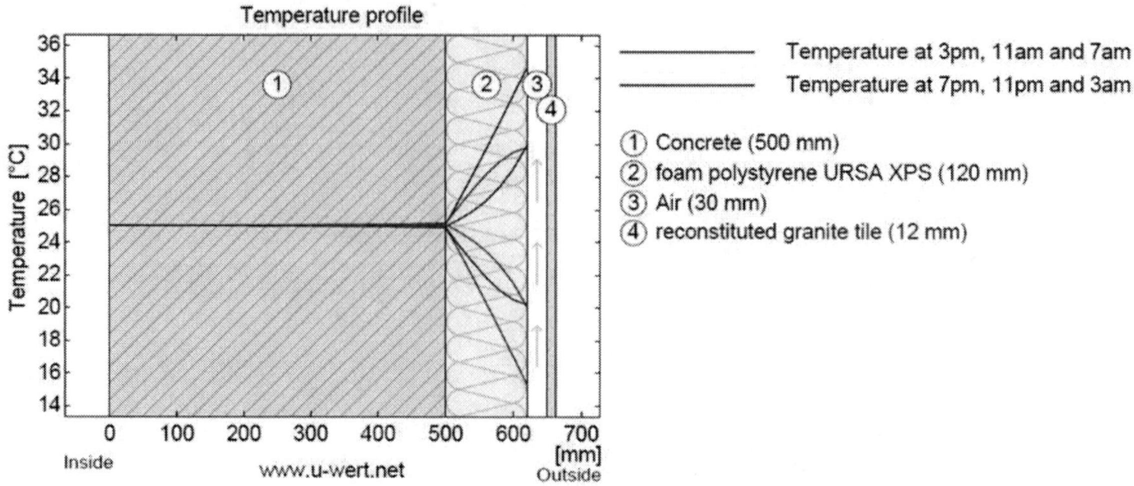

Figure 5. Temperature schedule.

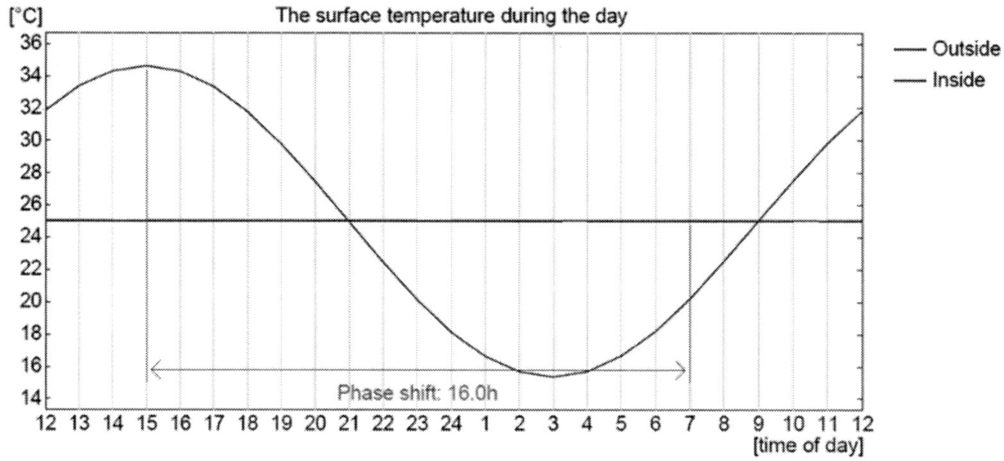

Figure 6. The surface temperature during the day.

The report on thermal calculations of the second type of the wall (figures 7-12).

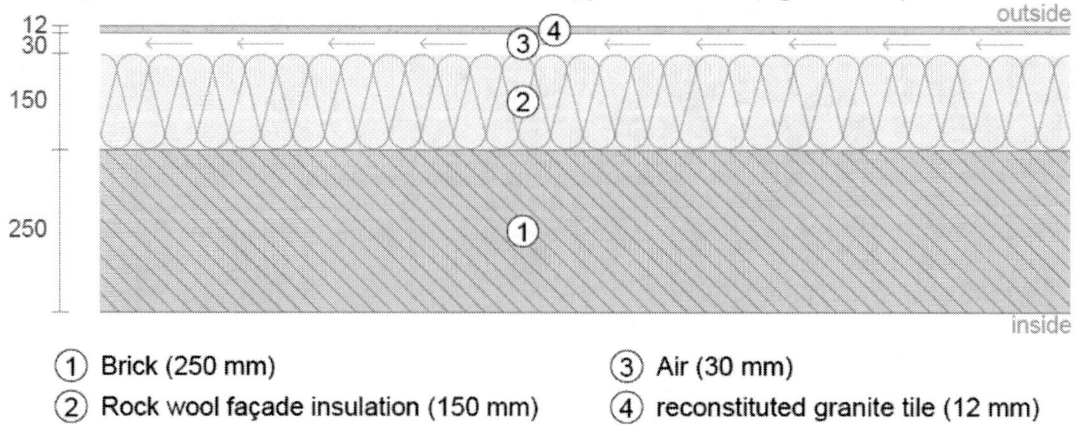

Figure 7. Cross-section of the enclosure structure.

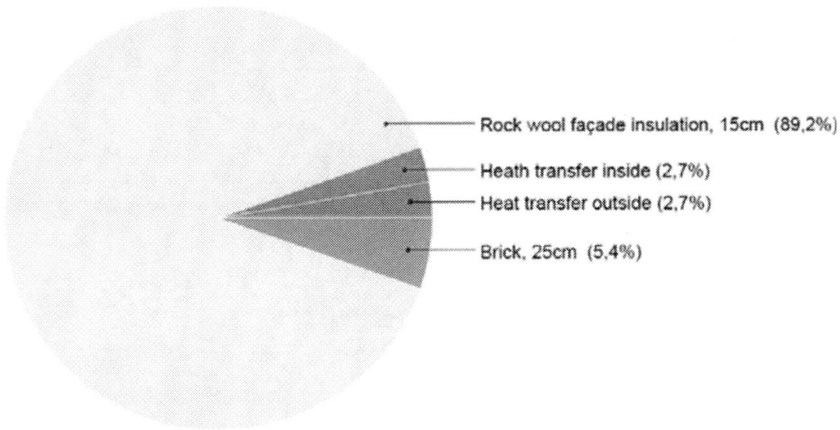

Figure 8. Contribution of different layers to the overall insulation.

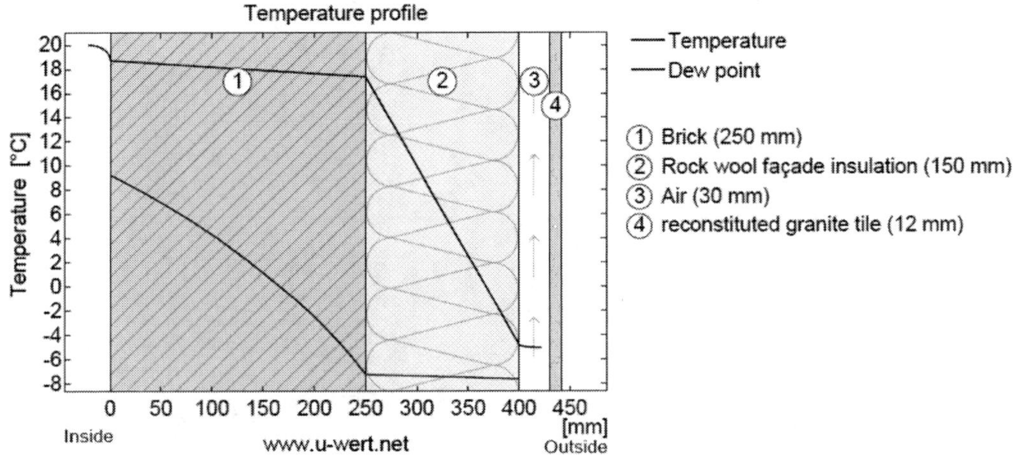

Figure 9. Temperature gradient. Condensing zone.

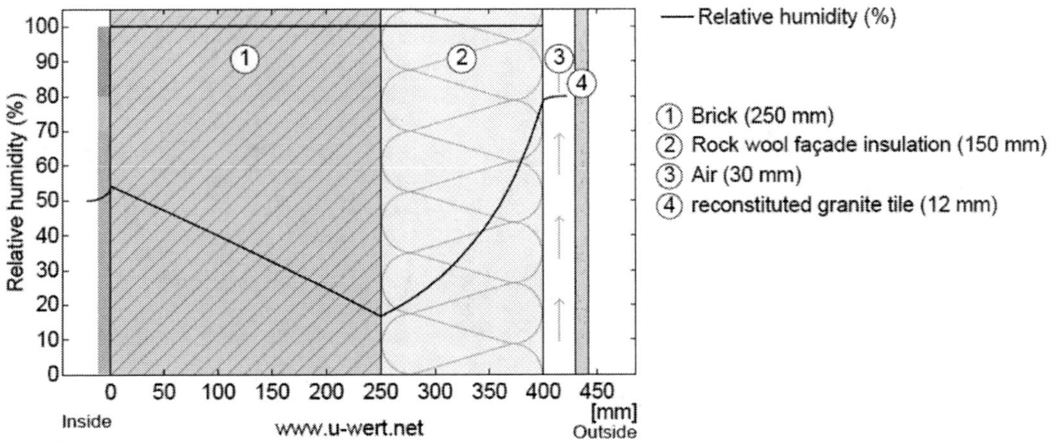

Figure 10. Humidity schedule.

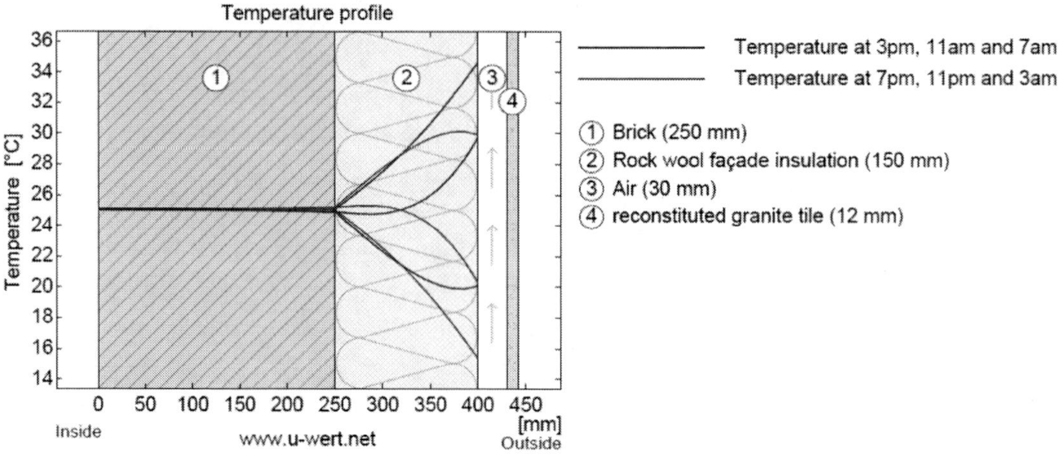

Figure 11. Temperature schedule.

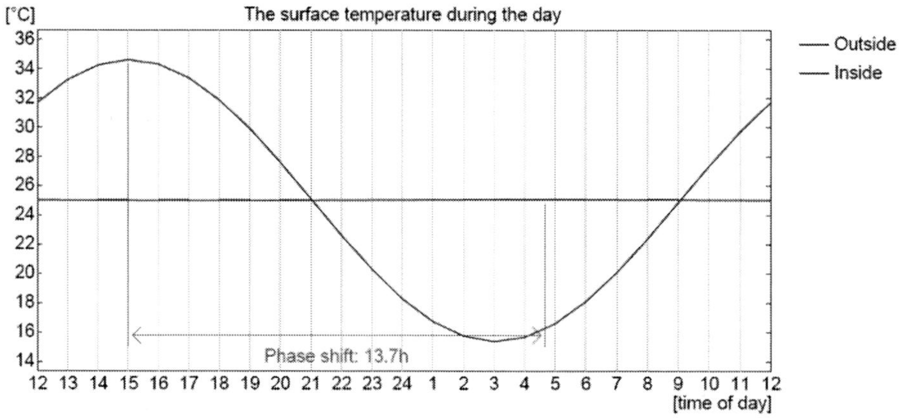

Figure 12. The surface temperature during the day.

The report on thermal calculations of the second type of the wall (figures 13-18).

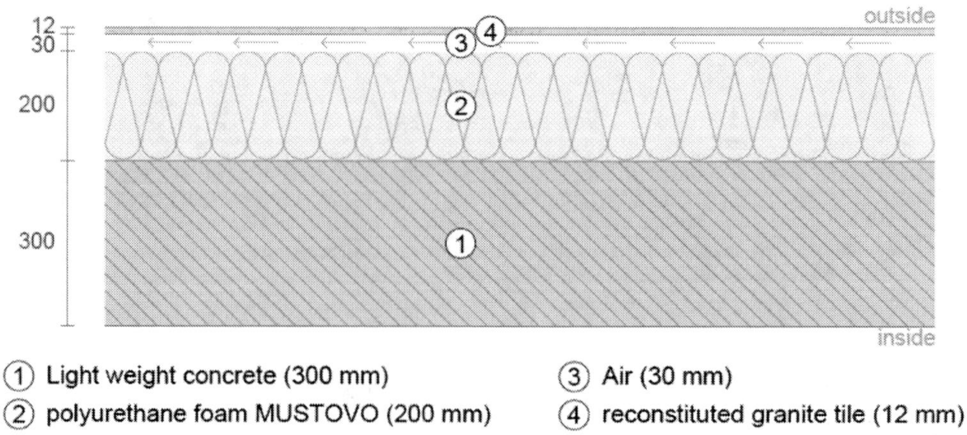

Figure 13. Cross-section of the enclosure structure.

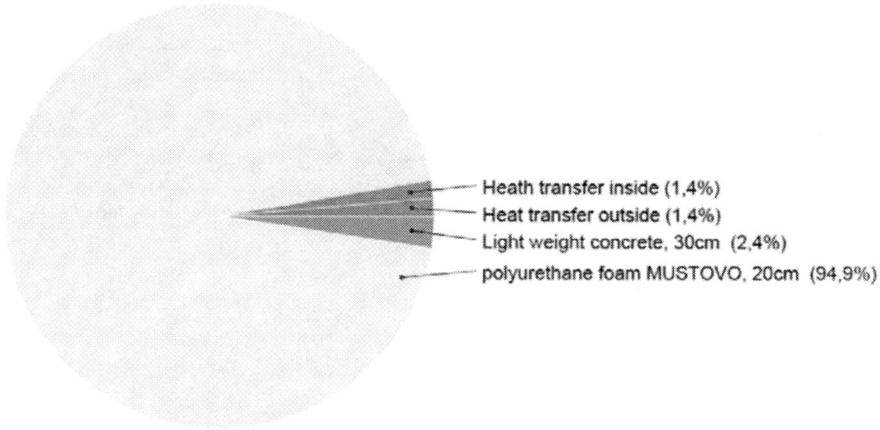

Figure 14. Contribution of different layers to the overall insulation.

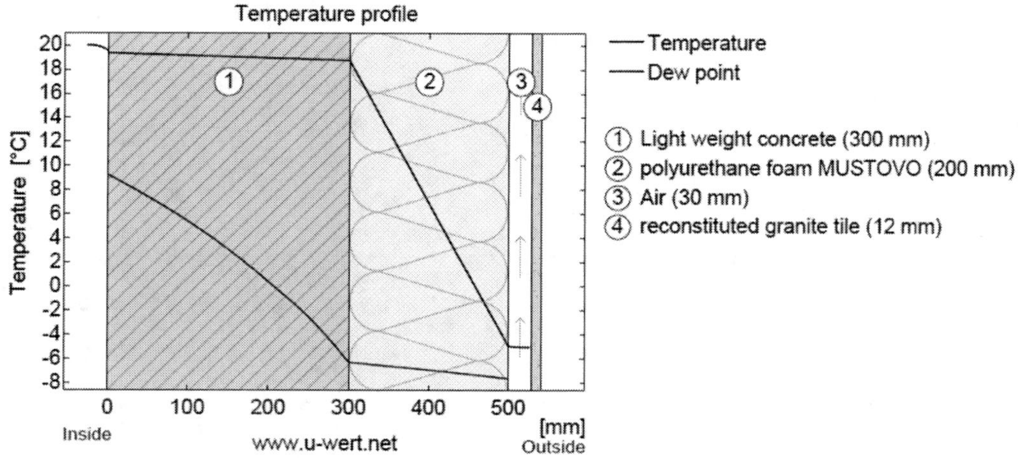

Figure 15. Temperature gradient. Condensing zone.

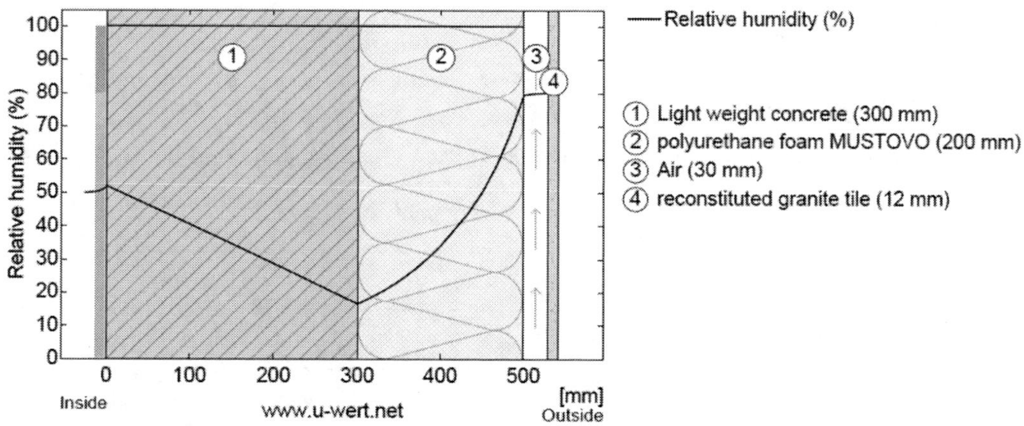

Figure 16. Humidity schedule.

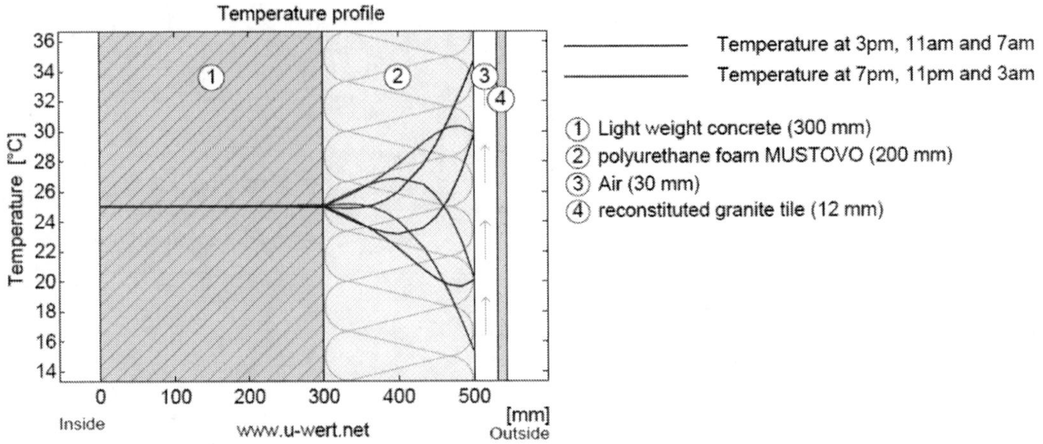

Figure 17. Temperature schedule.

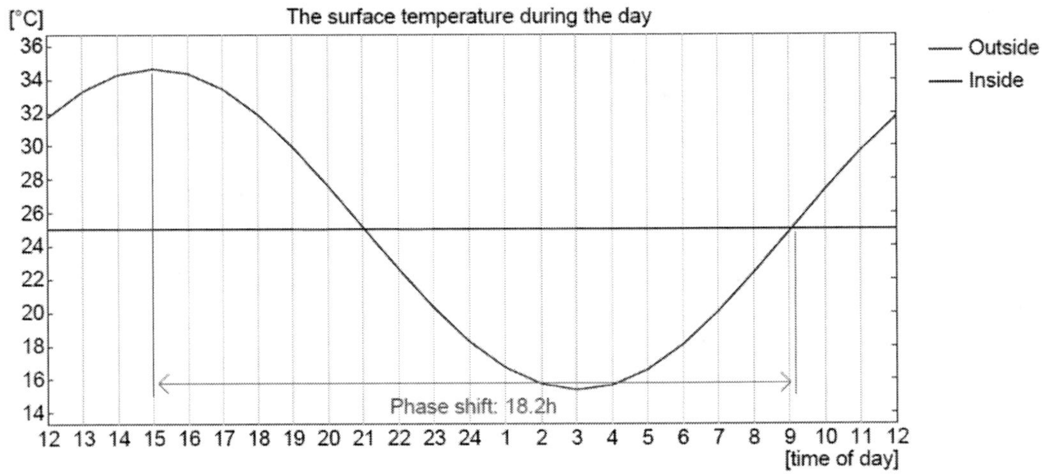

Figure 18. The surface temperature during the day.

4 Results and Discussion

The analyses and estimation of quality attributes for software is presented at table 4-7.

Table 4. Estimation of functionality of software.

TeReMOK	U-wert
Calculation of the demanded thickness of a required layer (a layer with not set thickness)	U-value calculator (graphical
Check on rated value of reduced resistance heat transmission	U-value calculator (tabular)
	Heat demand calculator
	Cross-section of exterior structure
	Diagram of contribution of different layers to the overall insulation
	Temperature gradient / Condensing zone Schedule

	Humidity schedule
	Heat protection schedule
	Possibility of creation of multy-layer cross-section of exterior structure
	Heat losses calculation

Table 5. Estimation of character of the user interface of software.

TeReMOK	**U-wert**
Simple interface	Difficult various interface
Possibility of a choice of the normative document	Without possibility of a choice of the normative document
The minimum quantity of the carried-out functions	A large number of the carried-out functions
The minimum percent of information entered in the on-line mode	A large number of information entered in the on-line mode
Ease of data processing	Ease of data processing
Without possibility of a choice of language (only one)	Possibilities of a choice of language (five languages)
	The minimum efforts on modification for elimination of mistakes
	Possibility of obtaining the report in the PDF-format
	Fast data processing

Table 6. Estimation of complexity of software online use.

TeReMOK	**U-wert**
Registration isn't obligatory	Registration is necessary for obtaining the individual report
Free in use	Free in use

Table 7. Estimation of the correctness of software.

TeReMOK	**U-wert**
There are no failures and mistakes in the program	There are no failures and mistakes in the program
Correctness of results, compliance to the Russian norms of construction	Correctness of results, compliance to the European norms of construction

5 Conclusion

Merits and demerits of each program were revealed according to the carried-out analysis. These modern software products for thermal calculations undoubtedly accelerate process of design and are additional check for the designers. By means of such programs it is possible to choose and calculate very quickly and reliably a energy-efficient construction of the enclosure structure. According to calculations in both programs the third type of a construction of the enclosure structure is the most energy-efficien. Having executed calculations by means of these programs, it is possible to draw a conclusion that on the Russian standards of the requirement to a building thermal protecting very low. Thus the Russian standard documentation requires modernization of regulatory base, development of new normative documents on a thermal protecting of buildings.

References

1. V. Savin, N. Savina, Gradostroitel'stvo, **1 (23)**, 82-84 (2013)
2. A. Avsjukevich, Construction of Unique Buildings and Structures, **2 (7)**, 40-54 (2013)
3. N. Vatin, A. Gorshkov, D. Nemova, Construction of Unique Buildings and Structures, **3 (8)**, 1-11 (2013)
4. S. Kaz'min, A. Fintisov, Construction of Unique Buildings and Structures, *Solar battery,* **2 (7)**, 40-54 (2013)
5. N. Vatin, O. Gamayunova, Applied Mechanics and Materials, **633-634**, 972-976 (2014)
6. E. Aronova, G. Radovic, V. Murgul, N. Vatin, Applied Mechanics and Materials, **587-589**, 348-354 (2014)
7. V. Murgul, V. Pukhkal, Procedia Engineering, **117**, 891-899 (2015)
8. Z.Gaevskaya, X. Rakova, Advanced Materials Research, **941-944**, 825-830 (2014)
9. V. Zegarac Leskovar, M. Premrov, WIT Transactions on the Built Environment, **118**, 541-552 (2011)
10. V. Zegarac Leskovar, M. Premrov, Journal of Asian architecture and building engineering, **11**, 71-78 (2012)
11. V. Pukhkal, V. Murgul, N. Vatin, Applied Mechanics and Materials, **633-634**, 1077-1081 (2014)
12. N. Vatin, D. Nemova, D. Nabiullina, A. Ovdienko, Applied Mechanics and Materials, **725-726**, 1434-1443 (2015)
13. D. Nemova, E. Reich, S. Subbotina, F. Khayrutdinova, Applied Mechanics and Materials, **725-726**, 26-33 (2015)
14. N. Vatin, D. Nemova, L. Khazieva, D. Chernik, Applied Mechanics and Materials, **635-637**, 2057-2062 (2014)
15. N. Mingottia, T. Chenvidyakarn, A.W. Woods, Energy and Buildings, **58**, 237–249 (2013)
16. E Bazhenova, J. Bykova, D. Bryus, D.Tseytin, Applied Mechanics and Materials, **725-726**, 1445-1456 (2015)
17. T. Zadvinskaya, A. Gorshkov, Advanced Materials Research, **953-954**, 1570-1577 (2014)
18. V. Murgul, Procedia Engineering, **117**, 808-818 (2015)
19. N. Bolshakov, S. Krivoy, X. Rakova, Advanced Materials Research, **941-944**, 895-900 (2014)
20. N. Vatin, D. Nemova, A. Kazimirova, K. Gureev, Advanced Materials Research, **953-954**, 1537-1544 (2014)
21. N. Vatin, D. Nemova, Y. Ibraeva, P. Tarasevskii, Applied Mechanics and Materials, **725-726**, 1395-1401 (2015)
22. N. Vatin, D. Nemova, A. Kazimirova, K. Gureev, Applied Mechanics and Materials, **725-726**, 1402-1407 (2015)
23. A. Kaklauskas, J. Rute, E. Zavadskas, A. Daniunas, V. Pruskus, J. Bivainis, R. Gudauskas, V. Plakys, Energy and Buildings, **50**, 7-18 (2012)
24. A. Fidrikova, O. Grishina, A. Marichev, X. Rakova, Applied Mechanics and Materials, **587-589**, 287-293 (2014)
25. V. Murgul, D. Vuksanovic, N. Vatin, V. Pukhkal, Applied Mechanics and Materials, **635-637**, 370-376 (2014)
26. V. Murgul, D. Vuksanovic, V. Pukhkal, N. Vatin, Applied Mechanics and Materials, **633-634**, 977-981 (2014)
27. V. Murgul, D. Vuksanovic, N. Vatin, V. Pukhkal, Applied Mechanics and Materials, **680**, 524-528 (2014)
28. Wilmer Pasut, Michele De Carli, Applied, **37**, 267-274 (2012)

29. A. Milajić, D. Beljaković, N. Davidović, N. Vatin, V. Murgul, Procedia Engineering, **117**, 916-923 (2015)
30. E. Aronova., N. Vatin, V. Murgul, Procedia Engineering, **117**, 771-779 (2015)

MATEC Web of Conferences **53**, 01008 (2016)
DOI: 10.1051/matecconf/20165301008

Structural Identification Problem

Aleksei Suvorov[1, a], Giulio Palamá[1] and Ekaterina Lysiakova[2]

[1]*Politecnico di Milano, Piazza Leonardo da Vinci 32, Milan, 20133, Italy*
[2]*Peter the Great St.Petersburg Polytechnic University, Polytechnicheskaya 29, St.Petersburg, 195251, Russia*

Abstract. The identification problem of the existing structures though the Quasi-Newton and its modification, Trust region algorithms is discussed. For the structural problems, which could be represented by means of the mathematical modelling of the finite element code discussed method is extremely useful. The nonlinear minimization problem of the L^2 norm for the structures with linear elastic behaviour is solved by using of the Optimization Toolbox of Matlab. The direct and inverse procedures for the composition of the desired function to minimize are illustrated for the spatial 3D truss structure as well as for the problem of plane finite elements. The truss identification problem is solved with 2 and 3 unknown parameters in order to compare the computational efforts and for the graphical purposes. The particular commands of the Matlab codes are present in this paper.

1 Introduction

The problem of definition of the current condition of the existing structure is always one of the most important problems, which arises during the life cycle of the building. An estimation of the material properties is an everyday work of the surveying firms and companies, which deal with the renovation and repair of the buildings.

There exist two approaches for the evaluation of the material properties. The first is a destructive way in which the consistency is interrupted. For instance, taking samples and consecutive testing in the lab or the well-known in-situ pulling test belong to the first category of the surveys. In the other hand the non- destructive techniques are also widely used in such services and moreover for the buildings of cultural heritage.

The method, which is described here can be applied in both cases as for the entire structure as well for the part of the interest. The field of application has a wide range: from the simple truss structures to the planar surfaces, and here we would describe the cases for the 3D truss and CST finite element code.

Depending on the scope of the survey it can be utilized for the evaluation of the condition of the damaged structures or the current load-carrying state.

Nomenclature

{}	vector
[]	matrix
3D	three-dimensional
CST	constant strain triangle

[a] Corresponding author: catherinewoolf@mail.ru

2 The identification problem

In the general framework the identification problem can be extremely useful in the cases when the points of interest are actual stress state or the material properties.

The basic concept of the identification problem for the 3D truss structure consists in the procedure of application of the static load with the known value and the following estimation of the nodal displacements {u}. Due to the absence of the possibility to test the real structure, we simulate the measurement by the direct procedure of the Finite Element code. In this case we are able to obtain the displacements which are provoked by the certain load condition.

After the construction of the stiffness matrix and definition of the solution which is now and later called as known. We need to repeat the entire procedure again with inserting instead the known quantities the symbolic variables.

As the consequence, we can construct the vector of unknowns {uUnknown}, which in each string contains the expression by means of symbolic variables, as it is written in the Equation (1).

$$\{u_{diff}\} = \begin{Bmatrix} f_1(E,..,E_n) \\ \cdot \\ \cdot \\ f_k(E,..,E_n) \end{Bmatrix} - \begin{Bmatrix} u_1 \\ \cdot \\ \cdot \\ u_k \end{Bmatrix} = \begin{Bmatrix} 0 \\ \cdot \\ \cdot \\ 0 \end{Bmatrix} \tag{1}$$

Finally, we can obtain the equality $\{uUnknown\} - \{uKnown\} = \{0\}$, which describes the compatibility of the nodal displacements in the certain directions. In order to obtain the values of unknown parameters we need to apply the minimization procedure [10, 16].

3 Minimization problem

3.1 Least squares method

The basic formulation of the minimization problem can be defined through the least squares method, which finds the parameters that minimize the sum of the squares of the errors between experimental data {uUnknown} and model predictions {uKnown}. In the other words in our problem we are identifying material properties by means of formula (2):

$$MIN\,E = \frac{1}{2}\sum_{i=1}^{k}\left|\{u_{Unknown}\} - \{u_{Known}\}\right|^2 \tag{2}$$

Where, k is the number of load cycles,

E is an objective function to minimize.

It is should be noted that the previous expression is nonlinear and depends on the number of desired parameters to identify.

3.2 Advanced formulation of least squares method

In addition to the basic least squares method there exists more advanced formulation, which takes in account the variance-covariance matrix which takes in account the statistical information and denoting by x the desired parameters, so then the objective function $E(x)$ becomes:

$$E(x) = \frac{1}{2}\{u_{Unknown}(x) - u_{Known}\}^T \cdot [C(x)]^{-1} \cdot \{u_{Unknown}(x) - u_{Known}\} \tag{3}$$

The challenge of construction $[C(x)]$ matrix arises the problems of definition of statistical data based on the observations or calculated using a model, which for the sake of simplicity are omitted in the further calculations [14].

3.3 Unconstrained nonlinear optimization

If to denote by x our desired values then the problem can be formulated as MIN $E(x)$ which means that we are trying to identify a vector x that is a local minimum to a scalar function $E(x)$, without any restrictions to the place of x.

[A] In our case we can consider the unconstrained minimization problem to minimize $E(x)$ by means of trust region. Suppose you are at a point x in k-space and you want to improve, i.e., move to a point with a lower function value.

The basic idea is to approximate $E(x)$ with a simpler function q, which reasonably reflects the behaviour of function f in a trust region (neighbourhood) N around the point x. A trial step s is computed by minimizing (or approximately minimizing) over N. This is the trust-region sub-problem, MIN $\{q(s), s \in N\}$.

The current point is updated to be $x + s$ *if* $E(x + s) < E(x)$; otherwise, the current point remains unchanged and N, the region of trust, is shrunk and the trial step computation is repeated.

The sequence of steps in the unconstrained minimization using trust-region can be shown as:

- Formulate the trust-region sub-problem
- Solve MIN $\left\{\frac{1}{2} s^T Hs + s^T g \text{ ; such that } \|Ds\| \leq \Delta\right\}$

where g is the gradient of E at the current point x,
H is the Hessian matrix (the symmetric matrix of second derivatives),
D is a diagonal scaling matrix,
Δ is a positive scalar, and $\|.\|$ is the 2-norm.

This is done in order to determine the trial step s

- If $E(x + s) < E(x)$, then $x = x + s$
- Adjust Δ

These four steps are repeated until convergence. The trust-region dimension Δ is adjusted according to standard rules. In particular, it is decreased if the trial step is not accepted, which can be stated in the Equation (4).

$$E(x + s) \geq E(x) \tag{4}$$

The basic requirement of this optimisation procedure is the presence of the gradient of the function, which we wish to minimize [2].

[B] Another way to define the minimum of the function is the possibility to use Newton or Quasi-Newton algorithm. They require higher computation effort, but the derivative of function is not needed.

These methods are described by the formulation of Equation (5) and build up curvature information at each iteration to formulate a quadratic model problem of this form.

$$MIN\left\{\frac{1}{2} x^T Hx + c^T x + b\right\} \tag{5}$$

where, H is the Hessian matrix (positive definite symmetric matrix),
c is a constant vector,
b is a constant,

The optimal solution for this problem occurs when the partial derivatives of the Equation (6) with respect to x go to zero

$$\nabla f(x^*) = Hx^* + c = 0 \tag{6}$$

And so the optimal solution x^* becomes of the form of the Equation (7) [15].

$$x^* = -H^{-1}c \tag{7}$$

3.4 Bayesian approach

This method does not provide a single value, for the identified parameter, but probability distribution is of material properties is to be estimated. In Bayesian approach we are trying to identify the distribution of material properties X having an instance x. This vector is characterized by probability density function $\pi_x(x)$ [1].

The vector of measurements u_{Known} is assumed to be contaminated by uncertainty errors e^{meas}, stemming from the random variable Σ^{meas}. Then the true but unknown response is represented by the Equation (8).

$$u_{Unknown} = u_{Known} - e^{meas} \tag{8}$$

and we can denote by U_{true}^{meas} the random variable of true response, so that the Equation (8) can be translated by the Equation (9).

$$U_{true}^{meas} = u_{Known} - \Sigma^{meas} \tag{9}$$

Recalling the definition of x – our seeking random value, we can say state that the Equation (9) can be converted to the Equation (10), using the definition [7].

$$U_{true}^{model}(\mathrm{x}) = u_{po}(\mathrm{x}) - \Sigma^{model} \tag{10}$$

where, Σ^{model} is a random variable due to the approximations involved in the input parameters p at the modelling stage, $u_{po}(x)$ is the function of material properties and also other p input parameters [3]. Bayesian identification involves calculating, for given material properties -x, the likelihood that the true response is the same whether derived from the measurement or from the model.

$$E = U_{true}^{model} - U_{true}^{meas} = 0 \tag{11}$$

Such that the Equality (11) should be valid [8, 9].

4 Truss 3D finite element procedure

4.1 Code for the direct problem

In the problem solution we have used the code, which solves every kind of truss structures under the hypothesis of linear elastic and first order approach [13].

In order to simulate the measurements of the nodal displacements we need to run the direct procedure, which estimates the values by means of the equilibrium.

The models are drawn by means of a pre-processor software GMesh in which the user inserts the nodes and the rods of the truss structure. The drawn data are contained in a file with a .geo extension that can be read by Matlab like a text file.

Through the script read_input_analysis.m Matlab acquires from the file .geo the number of nodes, number of elements, Coordinates Vector and connectivity vector. Every node and every beam have an identification number.

Young Modulus and area of every beam, the number of restrained nodes and loaded nodes, the degree of restrained of every restrained node, the load on every loaded node are also required data.

Following the directions in Matlab Command Window, these data can be inserted manually and can be saved in a file ".mat" that can be loaded in order to run the analysis.

Once input analysis phase is concluded, the code starts to initialise vectors and matrixes that will be used in the computation.

The main part of the code is the assembly of stiffness matrix, obtained performing a loop over the elements.

It starts with the identification of the number of nodes and their coordinates then computes the length of the element and the transformation matrix.

The transformation matrix is a rotation matrix to transform a vector from local reference system to global reference system.

Let us consider a beam in the 3D description of the rotation matrix, represented by the Equation (12).

$$\begin{Bmatrix} U_1^L \\ U_2^L \end{Bmatrix} = \begin{bmatrix} S(\gamma)C(\theta) & S(\theta)S(\gamma) & C(\theta) & 0 & 0 & 0 \\ 0 & 0 & 0 & S(\gamma)C(\theta) & S(\theta)S(\gamma) & C(\theta) \end{bmatrix} \begin{Bmatrix} U_x^1 \\ U_y^1 \\ U_x^2 \\ U_y^2 \\ U_x^3 \\ U_y^3 \end{Bmatrix} \quad (12)$$

By *C* and *S* are denoted corresponding trigonometric functions *Sine* and *Cosine*.

Knowing the rotation matrix, we can write the stiffness matrix of the element in the global reference system. After that identifying the degrees of freedom of the element, the code assembles the global stiffness matrix *K*.

In order to consider the restrained nodes, the code computes the load vector and modifies the stiffness matrix after the assembling.

Defining an α parameter such as a big number and summing it in *K* matrix terms, it is possible to take into account the infinite rigidity given by restrains.

Finally, it is possible to obtain the solution of the linear system in terms of the vector *{u}* representing the displacement field (Table 1).

Table 1. Code. Part 1.

```
%Solution of the linear system
u=K\F;
```

Knowing *{u}* vector it is possible to calculate the reaction forces and the strain in every beam [4].

4.2 Code for the inverse identification problem

The identification procedure starts with the initialization of a vector of unknowns in which there are the Youngs modulus written in symbolic form (Table 2).

Table 2. Code. Part 2.

```
YoungUn=sym('E',[1 length(Young)]);%Initialization of vector of unknowns
```

After that, the same procedure of direct problem is performed until the solution of linear system, by doing this it is possible to obtain a vector of displacement that depends by the modulus of elasticity of the elements.

At this point, a vector of displacement as a function of Young modulus and the numerical results given by the direct procedure are available.

In order to find the value of Young modulus it is possible to state the Equation (13) should be admitted.

$$\{uUnknown\} - \{uKnown\} = \{0\} \quad (13)$$

that in explicit form can be converted to Equation (14).

$$\{u_{diff}\} = \begin{Bmatrix} f_1(E,..,E_n) \\ \cdot \\ \cdot \\ f_k(E,..,E_n) \end{Bmatrix} - \begin{Bmatrix} u_1 \\ \cdot \\ \cdot \\ u_k \end{Bmatrix} = \begin{Bmatrix} 0 \\ \cdot \\ \cdot \\ 0 \end{Bmatrix} \tag{14}$$

By estimating the norm in L^2 of the vector $\{u_{diff}\}$ it is possible to obtain the function to minimize [4, 17, 18].

5 CST finite element procedure

5.1 Code for the direct analysis

The code solves linear elastic analyses for plane problems in plane strain or plane stress formulations. The models are drawn by means of a pre-processor software GMesh in which the user inserts the nodes and the contour of the section. The drawn data are contained in a file with a .geo extension that can be read by Matlab like a text file. The finite element mesh is obtained by using GMesh and is stored in a .msh file [20].

From a text file prepared by the user, and through the script read_input_analysis.m, Matlab acquires information about the material, such as Young modulus and Poisson coefficient, and obtains boundary conditions [21].

Once vectors and matrix are initialized, the main goal of the code is the assembly of stiffness matrix [19], obtained performing a loop over the elements.

Then, the element stiffness matrix is computed by means of a script function stiff_t3.m which takes in account the choice of plane strain or plane stress [4].

The last step before the solving system is to consider the boundary conditions and the nodal forces.

Finally, it is possible to obtain the solution of the linear system in terms of the vector u representing the displacement field. Stresses are computed as post-processing [4].

5.2 Code for the inverse identification problem

The identification procedure consists in a repetition of the direct procedure with an initialization of an unknown elastic modulus written in symbolic form and inserted into the element stiffness matrix script function, which becomes stiff_t3un.m.

After that, the same procedure of direct problem is performed until the solution of linear system, by doing this it is possible to obtain a vector of displacement that depends by the modulus of elasticity of the elements (Table 3).

Table 3. Code. Part 3.

```
%% solving linear system
disp 'solving linear system';
uUn=KKun\FF;
```

At this point, a vector of displacement as a function of Young's modulus and the numerical results given by the direct procedure are available.

In order to find the value of Young modulus it is possible to state the Equation (13) should be admitted, that in explicit form leads to the Equation (14).

By doing the norm in L^2 of the vector $\{u_{diff}\}$ it is possible to obtain the function to minimize (Table 4) [16,11,12].

Table 4. Code. Part 4.

```
LML=uUn-u;
NorLML=norm(LML);
```

5.3 Minimization procedure using Matlab Toolbox

In order to estimate the desired values the unconstrained minimization procedure was used. The chosen algorithm for the solution is Quasi-Newton (Table 5) [6].

Table 5. Code. Part 5.

```
options = optimoptions(@fminunc,'Algorithm','quasi-newton');
x0 = [150000,150000,150000]; % Starting initial guess
options.TolFun =1e-100;
options.TolX =1e-100;
%fun2min_...=function to miminize
[ELA,fval,exitflag,output] = fminunc(@fun2min_trepiediZ,x0,options)
```

From the previous steps by the solution of direct and inverse problems we were able to obtain the function which we want to minimize. It is required to insert the function into “fun2min_...m" and run the minprocQN file [5, 6, 12].

6 Numerical results

6.1 Truss 3D

Considering a structure composed by three beams and four nodes at the Figure 1 (a), it is possible to illustrate the way of computations.

The coordinates of nodes are

P_1=(333,333,333) P_2=(0,0,0) P_3=(1000,0,0) P_4=(0,1000,0).

The value of nodal load applied to the top is V=10000000 N

The elastic parameters are E1=2.0e+05 E2=3.0e+05 E3=2.5e+05

All quantities are expressed in the SI units (N, mm).

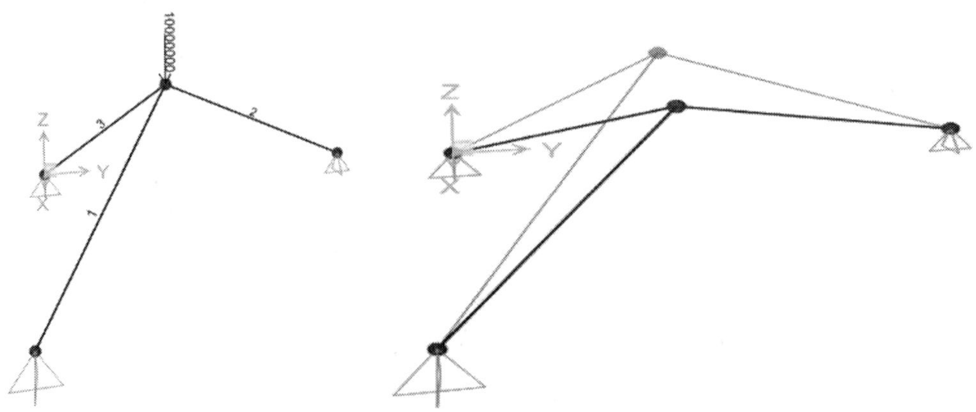

Figure 1. a) 3D Truss model in SAP; b) Obtained displacement in SAP.

Performing a structural linear elastic analysis of this model with the commercial finite elements software SAP2000, the following result in terms of displacements are obtained (Figure 2).

Joint Text	OutputCase	CaseType Text	U1 mm	U2 mm	U3 mm	R1 Radians	R2 Radians	R3 Radians
1	DEAD	LinStatic	2,175075	1,157613	-6,231133	0	0	0
2	DEAD	LinStatic	0	0	0	0	0	0
3	DEAD	LinStatic	0	0	0	0	0	0
4	DEAD	LinStatic	0	0	0	0	0	0

Figure 2. Results of calculation in SAP.

At this point the displacement field is related the following deformed shape is displayed at Figure 1 (a). The results obtained through the Matlab direct code are collected to the Table 6 and the deformed shape is displayed at Figure 3.

Table 6. Results of Matlab direct procedure.

Node	U1 [mm]	U2 [mm]	U3 [mm]
1	2.4852	1.3301	-6.7586
2	0	0	0
3	0	0	0
4	0	0	0

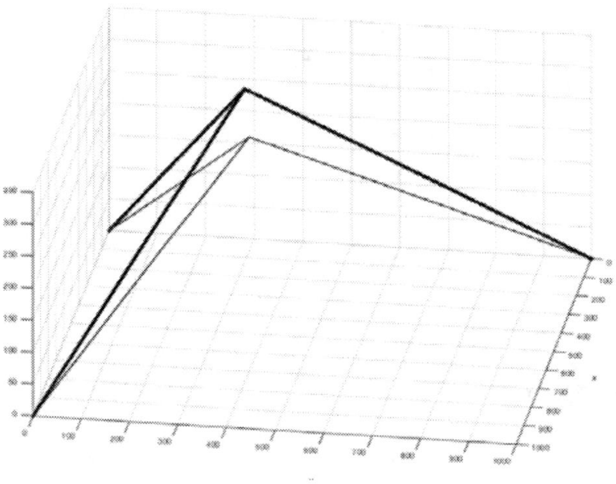

Figure 3. Deformed configuration in Matlab;

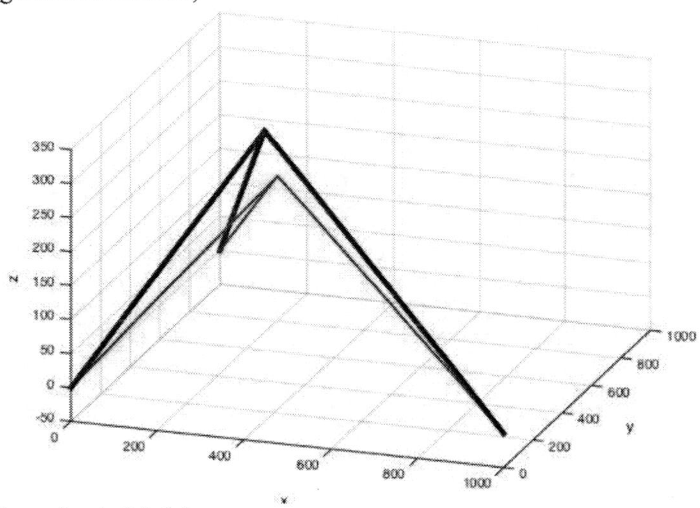

Figure 4. Deformed configuration in Matlab.

The vector *{UKnown}* = *{2.4852, 1.3301, -6.7586}* mm is obtained.

Then, using the inverse procedure it is possible to get a function, which depends on three elastic modules *E1 E2 E3*. This is exactly the function, which is needed for the further computations and estimation of the unknowns in the next step.

In order to use the Quasi-Newton algorithm of the Matlab Optimization Toolbox it is needed to set the initial guess. For the physical problems the range of the output values can be predicted, that is why starting guess [1.5e+05 1.5e+05 1.5e+05] was chosen. The results of the solution of our nonlinear minimization problem by means of Matlab code are listed at the vector ELA = [2e+05 3e+05 2e+05] are obtained.

Finally, it can be easily noticed that vector ELA of the desired parameters perfectly matches to the vector of the inserted elastic moduli during the direct simulation procedure.

6.2 Truss 3d with 2 unknown parameters

Using the geometric parameters of the problem discussed before we can solve the problem with two unknowns.

Setting the initial vector of elastic moduli for the direct procedure as [2, $0*10^5$, $3,0*10^5$, $2,5*10^5$] we obtain the deformed configuration shown at Figure 4.

The vector of simulated measurements is presented at the Table 7.

Table 7. Results of Matlab direct procedure.

Node	U1 [mm]	U2 [mm]	U3 [mm]
1	2.4852	1.3301	-6.7586
2	0	0	0
3	0	0	0
4	0	0	0

For the inverse procedure we set the symbolic vector of unknowns as [E1 , E2 , $2,5*10^5$] and solve the problem with two unknowns.

Obtaining the function which we need to minimize it is possible to plot the 2D plot depicting the point of the minimum and 3D surface (Figure 5) from which is clearly seen the coordinates of extreme point E1=2, $0*10^5$ and E2=3, $0*10^5$.

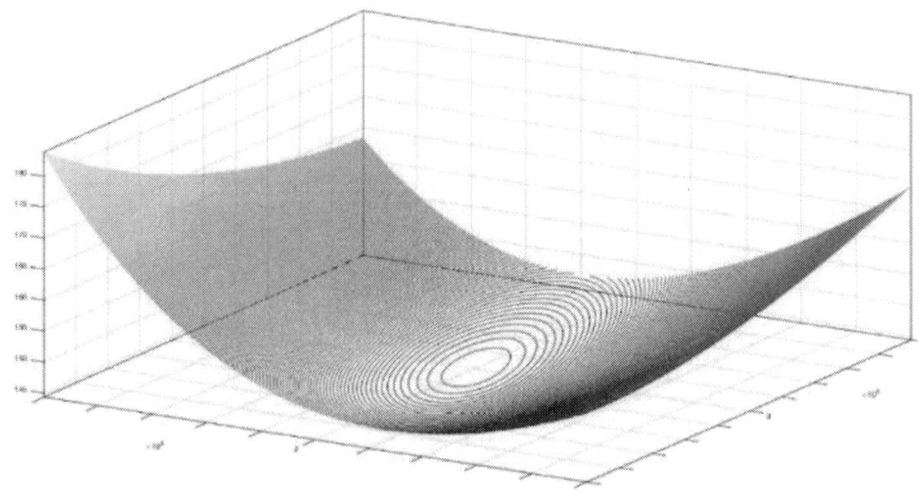

Figure 5. 3D surface in E1 and E2 axis.

By means of the inverse procedure of Matlab code we can observe in the Figure 9 the value of the function that goes to the minimum point and the number of iteration.

Matlab performs 32 iteration to obtain the values of the modules of elasticity which are slightly different from the exact ones E1=2.0005*10^5 and E2=3.0001*10^5.

The variation between exact value of the input and the result of minimization can be provoked by the threshold for the iteration procedure.

6.3 CST Finite element

Given the high requirement of computational resources needed for inverting the system in symbolic manner, the CST finite element identification problem has been performed on simple structure and with a very reduced number of finite elements. Nevertheless, after several tests performed on different samples, a rather good result has been observed.

Let us consider a concrete block with unknown elastic modulus subjected to a vertical distributed load q=100 N/mm, shown at the Figure 6.

In the real case procedure, it should be supposed to measure directly the nodal displacements under the load condition. Although, assuming a Poisson ratio 0.3 and a square cross section 400mm x 400mm, it is possible to obtain the displacements field from the direct CST procedure by imposing a value to the Young modulus (in this example YOUNG=24500 N/mmq).

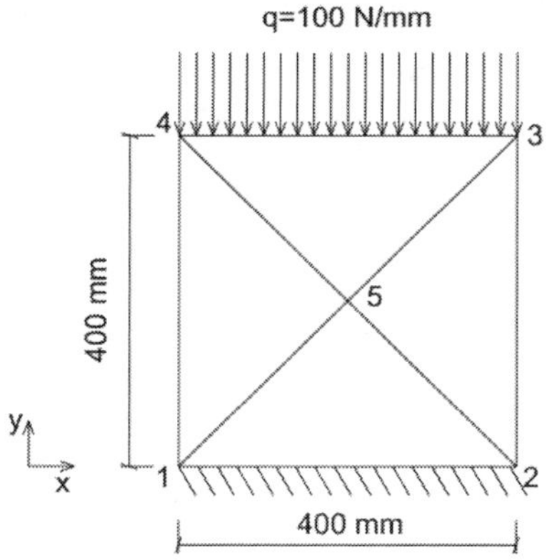

Figure 6. CST example.

Therefore, by means of a plane strain analysis the following displacement field is obtained in Table 9.

Table 8. Displacement field of CST model.

Direction	Value [mm]
U1x	0.0000
U1y	0.0000
U2x	0.0000
U2y	0.0000
U3x	0.4122
U3y	-1.3895
U4x	-0.4122
U4y	-1.3895
U5x	0.0000
U5y	-0.5802

Figure 7. a) Deformed configuration CST. b) Displacement diagram in the direction u1 and u2

At this point, by means of the inverse code it is possible to obtain a displacement vector as function of the unknown parameter E.

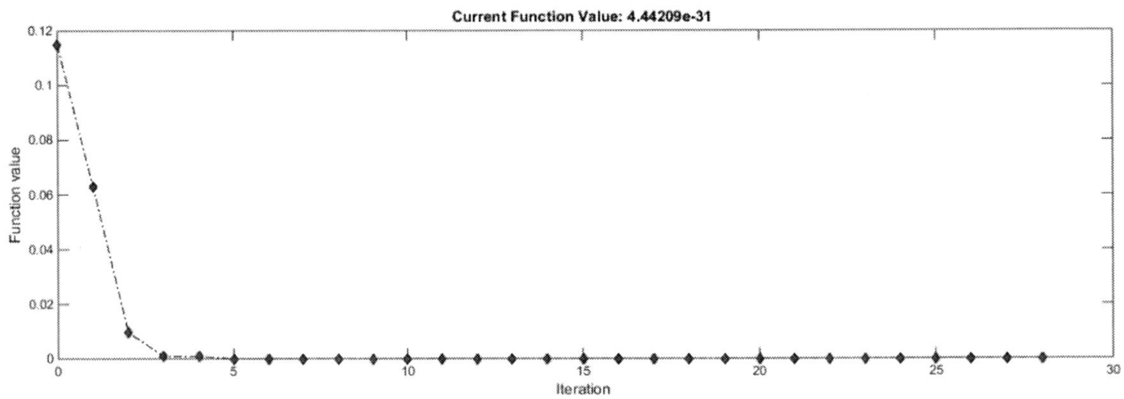

Figure 8. Iterations in CST model.

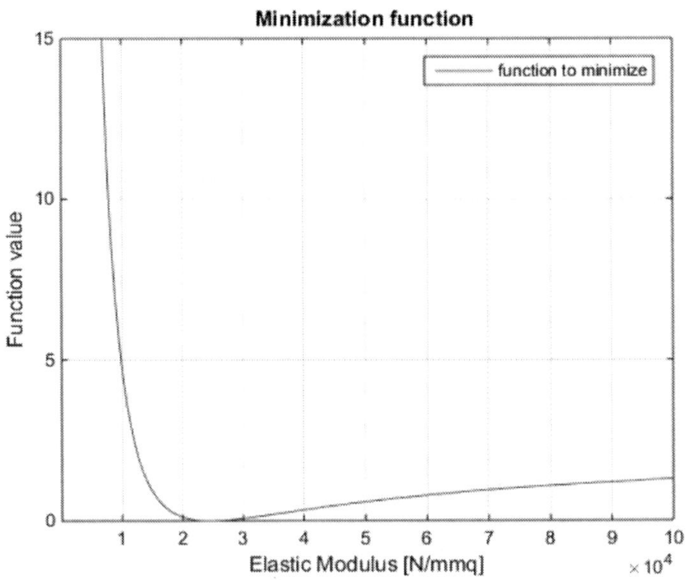

Figure 9. The function to minimize.

Using the Quasi-Newton algorithm of the Matlab Optimization Toolbox with a starting guess = [20000] and after 28 iterations (Figure 8), the correct value of the desired parameter ELA = [2.4500e+04] is obtained through the minimisation of the function drown at Figure 8.

7 Conclusions

In the real cases of the truss structures, the parameters of the most components are defined from the very beginning. The solution is the subject for identification is the matter several parameters. For instance having a complex structure as the object for survey we can definitely name the values related to the substructures. At least the groups of elements, which have the same properties can be defined. What is important is to mention that the structures that have uniform or classified distribution of sectional and material parameters require lower computational effort. This can significantly decrease the number of unknowns and as the consequence the complexity of the problem.

During the testing procedures of the Matlab code, we tried to run the structure with great number of rods, where each parameter is different and unknown and we mentioned that it requires the high computational capacity in order to construct the matrix of unknowns.

The surveying procedures can vary depending on the possibility to achieve the important parts of the structure and the used equipment. Sometimes the measurements are easier to perform in the terms of strains, in this case, the slight modification in the procedure are required, but the philosophy of the solution remains the same.

References

1. C. Gogu, R. Hatìkat, Introduction to the Bayesian Approach Applied to Elastic Constants Identification. *AIAA JOURNAL* **48** 5 (2010)
2. J.J. Moré, and D.C. Sorensen, Computing a Trust Region Step. *SIAM Journal on Scientific and Statistical Computing* **3** 553–572 (1983)
3. J. Idier, *Bayesian Approach to Inverse Problems* 392 (Wiley-ISTE, Paris, 2008)
4. Alberto Corigliano, Alberto Taliercio, *Meccanica Computazionale: Soluzione del Problematico Elastico Lineare* (Esculapio Editore, Italy, 2006)
5. The MathWorks. *Optimization Toolbox: User Guide* 634 (Natick, 2015)
6. J. E. Dennis, J. M. Jorge, Quasi Newton Methods, Motivation and Theory. *SIAM Review* **19** 46-89 (1977)
7. T. Marwala, S. Sibisi, Finite Element Model Updating Using Bayesian Framework and Modal Properties. *Journal of Aircraft* **42** 275-278 (2005)
8. K. F Alvin, Finite element model update via Bayesian Estimation and minimization of dynamic residuals. *AIAA Journal* **35** 879-886 (1997)
9. C Gogu, Facilitating Bayesian Identification of Elastic Constants through Dimensionality reduction and response surface methodology. Ph.D. Dissertation. University of Florida. *Gainesville, FL, and Ecole des Mines de Saint Etienne* (France, 2009)
10. M. S. Agbabian, S. E Masri, R. K. Miller, T. K. Caughey, System identification approach to detection of structural changes. *J. Engrg. Mech., ASCE* **117(2)** 370-390 (1991)
11. M. R. Banan, K. D. Hjelmstad, Parameter estimation of structure from static response I. Computational aspects. *J. Struct. Engrg., ASCE* **120(11)** 3242-3258 (1994)
12. M. R. Banan, K. D. Hjelmstad, "Parameter estimation of structures from static response. II. Numerical simulation studies. *J. Struct. Engrg., ASCE* **120(11)** 3259-3283 (1994)
13. M. Baruch, I. Y. Bar-Itzhack, Optimal weighted orthogonalization of measured modes. *AIM J* **16(4)** 346-351 (1978)
14. A. Bjorck, Numerical Methods for Least Squares Problems. *Society for Industrial and Applied Mathematics* (Philadelphia, 1996)
15. C.L. Lawson, R. J Hanson, *Solving Leàst Squares Problems* (NJPrentice Hall, Englewood Cliffs, 1974)

16. A. Tarantola, Inverse. Problem Theory and Model ParameterEstimation. *Society for Industrial and Applied Mathematics* (Philadelphia, 2005)
17. J. Isenberg, Progressing from Least-Squares to Bayesian Estimation. *Proceedings of the ASME Design Engineering Technical Conferences* 71–82 (American Society of Mechanical Engineers. New York, 1979)
18. O.C. Zienkiewicz, R.L. Taylor, *The Finite Element Metod for Solid and Structural Mechanics* (Butterworth-Heinemann, 2013)
19. A. Munjiza, E. Rougier, E. E. Knight, *Large Strain Finite Element Method: A Practical Course* (Wiley, 2015)
20. S. S. Rao, *The Finite Element Method in Engineering* (Butterworth-Heinemann, 2010)
21. A.J.M. Ferreira, *MATLAB Codes for Finite Element analysis. Solids and Structures* (Portugal, 2009)

MATEC Web of Conferences **53**, 01009 (2016)
DOI: 10.1051/matecconf/20165301009

Stress-Strain State of Brick Masonry Vault with an Aperture

Vladimir Bespalov[1,a], Romuald Orlovich[2] and Sergey Zimin[1]

[1]*Peter the Great Saint-Petersburg Polytechnic University, 29 Polytechnicheskaya st., St.Petersburg, 195251, Russian Federation*
[2]*West Pomeranian University of Technology Szczecin, al. Piastów 17, Szczecin, 70-310, Poland*

Abstract. A study of the mechanical behavior of brick masonry vaults with cut aperture is presented. The vault in New Holland complex in St. Petersburg was assumed as an illustrative example. This vault was tested and simulated in order to evaluate its stress-strain state. Comparison of the results of numerical analysis and experimental test data has been performed. The optimal way to perform an aperture was defined: aperture should be cutting parallel to the lines of action of the principal stresses minimizing direction transversal to these lines. As a further development of the research, the investigation of the various types of aperture in vaults has been recommended to perform using different simulating models and methods.

1 Introduction

It frequently needs to adapt historical buildings to modern requirements during the restoration process. Such adaptation usually includes applying modern materials, structural changes in certain building elements and the installation of new systems – air ducts, water-supply facilities, elevators etc. Existent apertures are often inadequate for these systems and it requires to cutting-out additional units. As a consequence, significant changes in stress state and analytical model of structures are taking place. Since the historical buildings are built mainly of stone and brick masonry, its analysis is the most difficult, especially analysis of brick masonry vaults.

Simulation and analysis of brick masonry perform either by masonry continuum model [1-4] or by discrete model [5, 6, 7]. The continuum model is mainly used for the analysis of the stress state of massive masonry structures. Otherwise, the discrete model is applicable for studying the crack propagation phenomena, interaction between brick and mortar under load and simulating tests of masonry samples. In addition, there are intermediate models, combining the first two [8, 9, 10].

Masonry vaults simulation is a more complex task. The review of manuscripts recently performed [11] demonstrate the lack of Russian works devoted to this subject. Foreign investigations can be divided into two groups. The first one is devoted to searching for numerical simulation techniques allowing the behavior of vaults in different configuration to be the most correct [12-15]. The second one is experimental study of brick arches and vaults failure phenomena [16-19]. Studies describe in detail the sources of the vaults failure, loss of stability and theirs development. However, these studies consider ordinary and unmodified arches and vaults. While behavior of vaults, which contain variety of additional apertures, and ways of their stresses redistribution is of practical concern.

---

[a] Corresponding author : chanchullero@yandex.ru

Authors had tested the vault of restored Forge building in New Holland complex located in St. Petersburg. During the restoration, a need arose for a large aperture in one of vaults to bring out the air duct. After cutting the aperture, strains were measured on adjacent areas during the removing of arch falsehood. Besides, the vault was simulated in ABAQUS software to evaluate its stress state. The results of experimental tests and numerical analysis are presented in the paper.

2 Initial data and task description

It was necessary to complete the following tasks to define the aperture influence on the stress-strain state of the vault:

- to measure the strain on the adjacent areas to the aperture of the existent vault;
- to simulate and analyze this vault in ABAQUS software;
- to compare the measurement results and the data of numerical simulation.

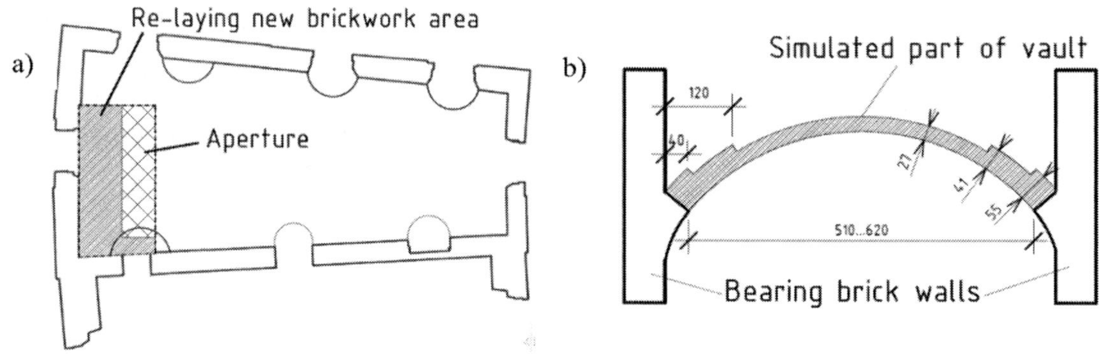

Figure 1. (a) Vaulted ceiling plan including spandrels, aperture and re-laying area; (b) The structural scheme of vault, dimensions in cm.

An aperture was cut out from the vault of Forge building. An aperture has dimensions of about 88 × 422 cm. For that scope, old brickwork has been removed from wide area of vault and the new has been laid out (Fig. 1, a). Vault is cylindrical with variable radius and additional cross-section thickness on abutments (Fig. 1, b). The brickwork has following main features: volume weight γ=1800 kg/m3, modulus of elasticity E=1680 MPa, Poisson's ratio μ=0.2.

3 Experimental tests

External surface of the vault was the most applicable to measurements. The strains were measured by dial gauges installed on one arc of vault in proximity to the aperture at three points (Fig. 2). Dial gauge supports were anchored in the bricks in such a way as to the reference indicator base was in each case about 20 cm.

Measurements were taken during wedges easing. At that moment, vault begins to take up the dead load. The first measurement check was made before wedge easing process had started. The second measurement check was taken when all the wedges were stroke out and the vault was settled in the operational position. Further measurements were not carried out, therefore, only the elastic strains have been taken into account, without creep strains.

Figure 2. Location of the dial gauges on the target vault.

The accuracy of indicators was about 0.001 mm. The strains were defined under results of measurements: the first point (-0.007) mm, second point (-0.004) mm, third point (-0.013) mm. Indications of the third indicator allows to deduce that the new-laying brickwork (areas 1 and 2) is more rigid. The corresponding relative strains to the 20 cm reference base: the first area 3.5·10-5; second 2·10-5; the third 6.5·10-5. For adopted elastic modulus 1680 MPa principal compressive stresses in these areas will be: 58.8 kPa, 33.6 kPa and 109.2 kPa, respectively.

4 FE simulation

By the structural scheme (Fig. 1, b) estimated on the base of [20] that the vertical support reaction and horizontal thrust of its dead load is transmitted to the foundation by its abutments directly on the bearing wall axis. In this case, most accurate results of the analysis would have been if simulated vault is slit transverse to its cross section at the point adjoining to the supporting wall.

Vault static analysis is performed using ABAQUS. The three-dimensional model was generated by volume finite elements. Connection between spandrel and vault was assumed as rigid fixing. The transmits of vertical and horizontal reaction forces from spandrels canopy to the adjacent wall were excluded. It was also assumed that the vault does not bear against the adjacent transversal brick walls. The analysis is performed from its dead load only with a partial safety factor of 1.1. The analysis does not take into account the filling which is covering abutments because of its insignificant influence on the target area.

The vault was simulated at two phases - the initial state and the current state with cut aperture (Fig. 3 a, b).

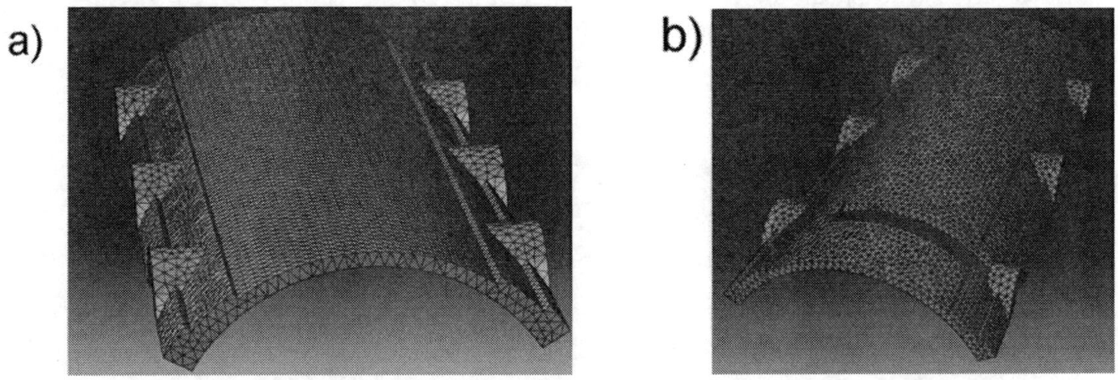

Figure 3. Finite-element models of vault. (a) the initial state, (b) a vault with a cut aperture.

Considered vault in the normal state has a critical area in the typical [21, 22] regions: minor tensile stresses are located in crowns of spandrels and concentration of compressive stresses at their abutments, and the stress concentration in the area of cross section differentials (Fig. 4, 5).

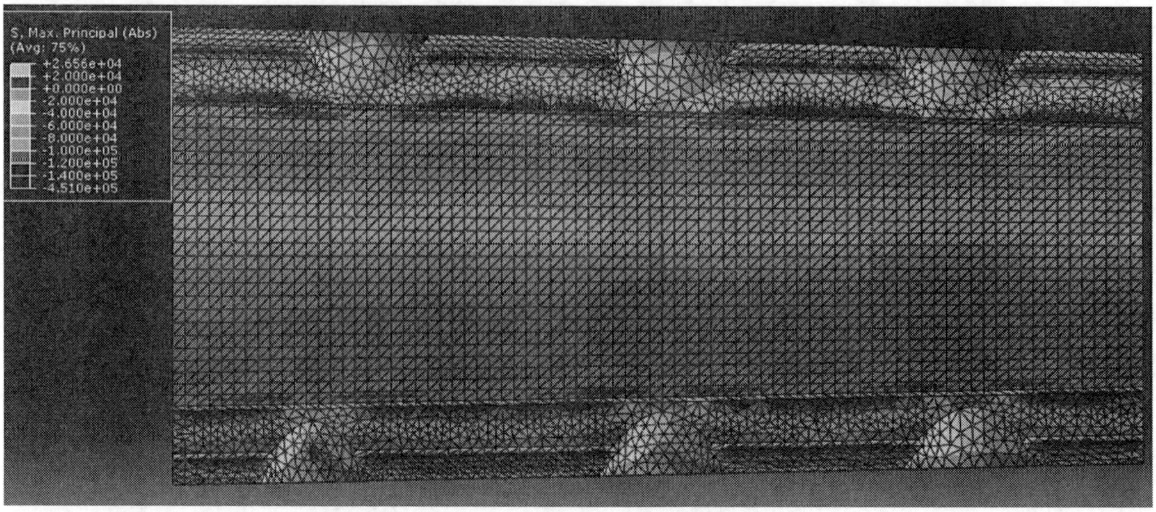

Figure 4. Maximum principal stresses on an external surface of the vault, Pa.

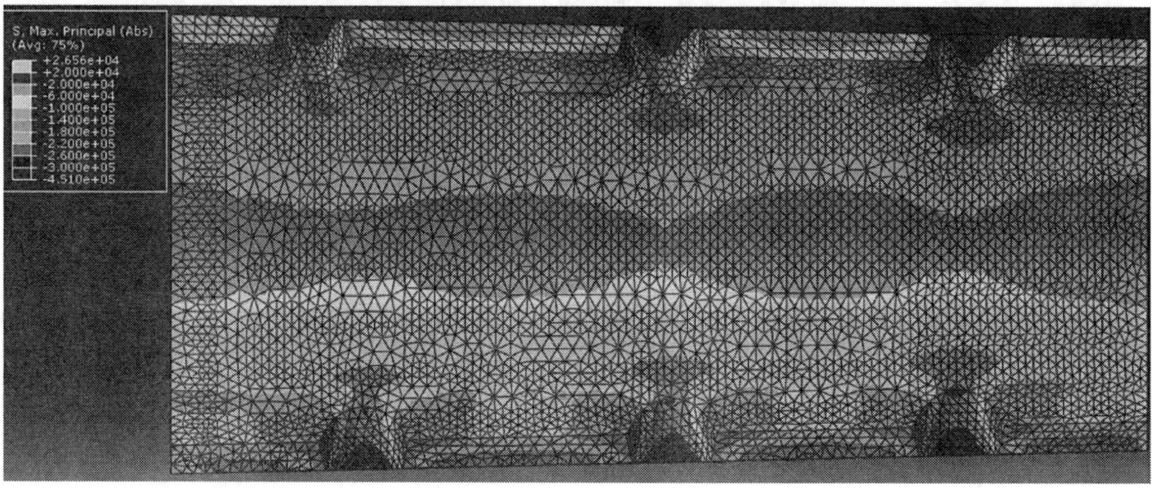

Figure 5. Maximum principal stresses on an internal surface of the vault, Pa.

Critical areas rose on the border of the aperture after cutting it out (Fig. 6). A danger tensile stresses acting perpendicular to the main stream of the compressive stresses rose at the end of the opening (Fig. 7). On the contrary, the spandrel adjacent to the opening has become partially unloaded; tensile stresses on its crown has disappeared and the magnitude of compressive stresses in its abutments has decreased (Fig. 8).

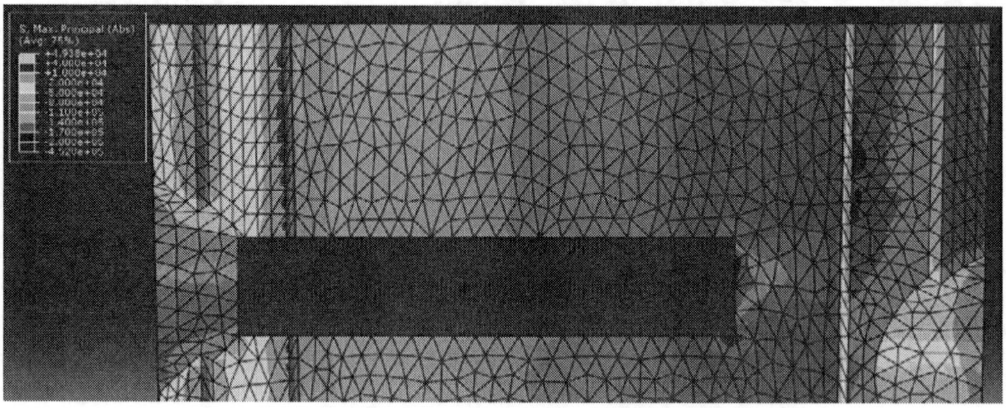

Figure 6. The maximum principal stresses in the apertures area at the external side, Pa.

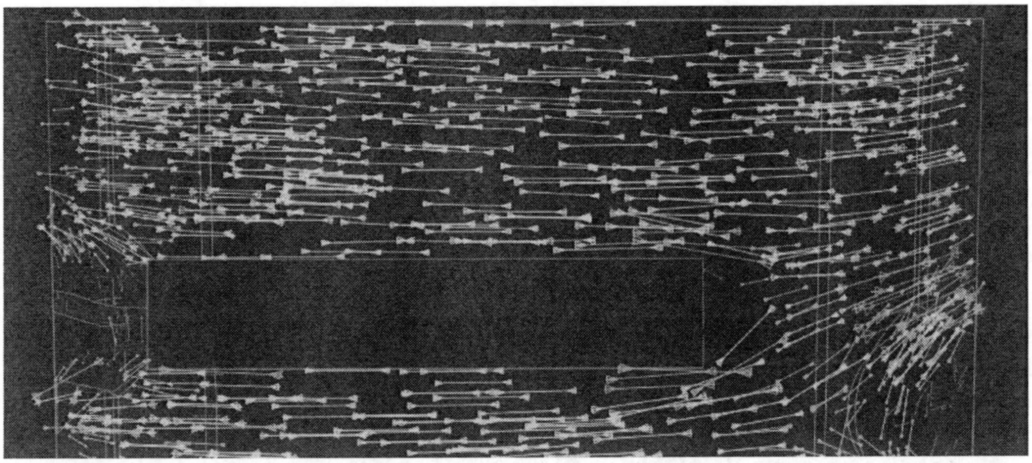

Figure 7. The paths of the maximum principal stress in the apertures area.

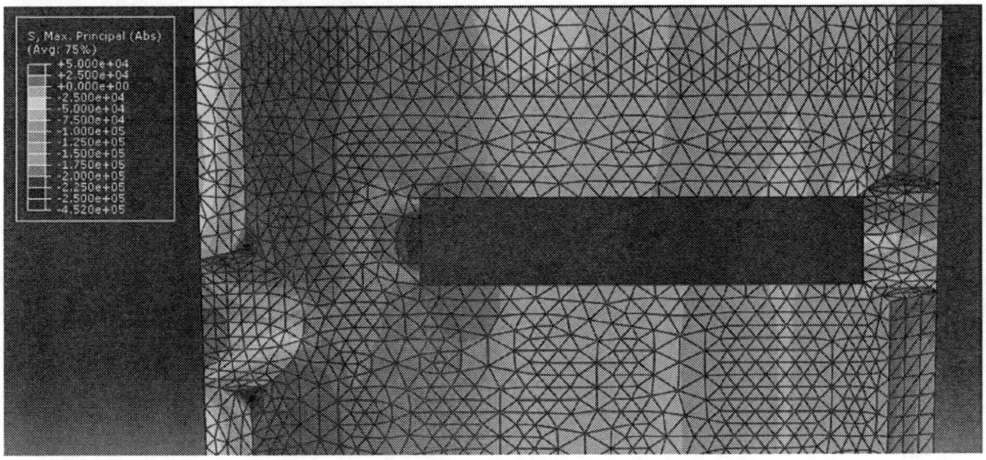

Figure 8. The maximum principal stresses in the apertures area at the internal side, Pa.

Compressive stresses are as follows: 1 - (0.07 ... 0.08 MPa), 2 - (0.08 ... 0.09 MPa), 3 - (0.11 ... 0.12 MPa) at points corresponding to points of the vault considered in natural measurements. Thus, stresses in old brick masonry coincide numerically with the stresses in the analytical model. Inconsistency stresses in FE model to the stresses in re-laying brickwork were presumably caused by the greater stiffness of the masonry in that area.

5 Conclusions

Evaluation of the paths of the maximum principal stresses allows to conclude that the aperture preferably cut parallel to the lines of action of the principal stresses. On the other hand, the aperture performing in a direction transversal to these lines causes dangerous tensile stress.

Comparison of the results of numerical analysis and experimental test data showed the permissibility of using masonry continuum model in a numerical analysis of the vault. The discrepancy of results at the new masonry was caused by its greater stiffness, as characteristics were given from the old brick masonry in the simulation for the entire vault.

The aperture performing in the vaults should make strength analysis based on numerical simulation as a prediction in the context of a minor knowledge about the behavior of masonry vaults, as well as a variety of their structural version.

As a further development of the research, the investigation of the various types of aperture in vaults is highly recommended to perform using different simulating models and methods.

References

1. M. Como, Meccanica, **27**, 185-194 (1992)
2. O.A. Kamal, G.A. Hamdy, T.S. El-Salakawy, HBRC J., **10**, 235-246 (2014)
3. S. Galassi, M. Paradiso, G. Tempesta, Open J. of Civ. Eng., **3**, 18-26 (2013)
4. G. Creazza, R. Matteazzi, A. Saetta, R. Vitaliani, J. of Str. Eng., **128**, 646–654 (2002)
5. S. Galassi, M. Paradiso, IERI Proc., **7**, 62-70 (2014)
6. S. Di Pasquale, Meccanica, **27**, 173-184 (1992)
7. I.E. Demchuk, A.V. Halaliuk, Proc. of the BSTU, **1**, 122-128 (2013)
8. S.S. Zimin, V.V. Bespalov, A.S. Kazimirova, Proc. of the DonNACEA, **3**, 33-37 (2015)
9. M. Gilbert, C. Melbourne, The Str. Eng., **72**, 356-361 (1994)
10. A. Penna, S. Lagomarsino, A. Galasco, Earthquake Eng. Str. Dyn., **43**, 159-179 (2014)
11. S.S. Zimin, O.D. Kokotkova, V.V. Bespalov, CUBS, **2**, 57-72 (2015)
12. E. Rizzi, F. Rusconia, G. Cocchettia, Eng. Str., **60**, 241-257 (2014)
13. G. Milani, A. Tralli, IJSS, **49**, 808-834 (2012)
14. G. Milani, Adv. in Eng. Software, **79**, 91-110 (2014)
15. P. Block, L. Lachauer, MRC, **56**, 53-60 (2014)
16. C. Melbourne, M. Gilbert, M. Wagstaff, The Str. Eng., **75**, 297-305 (1997)
17. L. Anania, G. D'Agata, C. Giaquinta, A. Badala, APCBEE Proc., **9**, 401-406 (2014)
18. J. Witzany, T. Cejka, R. Zigler, Proc. Eng., **14**, 2086-2093 (2011)
19. V. Sarhosis, S.W. Garrity, Y. Sheng, Eng. Str., **88**, 1-11 (2015)
20. R. Ahnert, *Typische Baukonstruktionen von 1860 bis 1960 zur Beurteilung der vorhandenen Bausubstanz* (Verlag für Bauwesen, Berlin, 1991)
21. P. Block, L. Lachauer, IJAH, **8**, 312-335 (2014)
22. A. Tralli, C. Alessandri, G. Milani, The Open Civ. Eng. J., **8**, 272-287 (2014)

MATEC Web of Conferences **53**, 01010 (2016)
DOI: 10.1051/matecconf/20165301010

Moisture Transfer in Ventilated Facade Structures

Vyacheslav Olshevskyi[1], Elena Statsenko [1,a], Tatiana Musorina [1], Darya Nemova[1] and Anastasia Ostrovaia [1]

[1]*St. Petersburg State Polytechnical University, Politekhnicheskaya, 29, Saint-Petersburg, 195251, Russia*

Abstract: This article discusses the phenomenon of moisture transfer in the designs of ventilated facades (VF). The main ways of moisture transfer are defined. The negative factors connected with moisture accumulation and excessive moistening of insulation are given. The physical processes occurring in the gap of the building envelope due to saturation of air with water vapor are described. The dependence of the intensity of the mass transfer on the air velocity in the layer is considered. Much attention is paid to the selection of the optimum design of the facade, namely a system with or without grooved lines. The dependence of velocity and temperature on the width of the ventilated gap is established empirically for the constructions with open and closed grooves. Expediency of a design without grooves to effectively remove moisture is determined.

1 Introduction

High heat-shielding properties are considered to be one of the main reasons for the use of ventilated facades (VF) in modern construction. Proper identification of sources of moisture and moisture transfer mechanisms in protecting structures building envelope can significantly improve the designs of buildings and air conditioning system. After all, one of the conditions for increasing the durability of buildings and improving the indoor climate is to control the level of humidity [1, 2, 3].

The problem of moisture transfer is extremely important. Moisture transfer represents the process of free and physically bound water movement due to gravity force and sorption, both molecular and capillary [4].

Thus, a quantitative calculation of moisture transfer is of great importance in the design of a multilayer building envelope, this exactly proves the relevance of this article. This article examines the impact of the design of ventilated facades on the air velocity in the gap, which the moisture transfer is directly dependent on [5, 6].

2 Literature survey

Moisture transfer in ventilated facade structures is covered extensively in papers.

In the article [1] conditions under which cold air can be filtered into an air gap are considered. In the publication [4] the optimum width of an air gap is found. In [7] the author considers the main problems of ventilated facades. In [8] analysis and assessment of the phenomenon of mass transfer in

[a] Corresponding author : staclena@mail.ru

porous and disperse materials is carried out. In [11] the authors analyze the sufficient conditions of existence of free convective flow in a vertical parallel-plate duct. The authors of the publication [22] analyze the questions of natural ventilation, heat transfer and protection from overheating of ventilated facades. In the paper [27] the application of these designs in regions with a frigid climate is considered. In [28] the author investigates the methods for assessing thermal insulation of a wall with a ventilated facade, taking into account longitudinal air filtration.

3 Setting goals and objectives

As a result of the analysis the following research objectives were formulated:

- Identification of the optimum design parameters
- Numerical simulation of processes occurring in ventilated facades
- Clarification of the grooved lines influence on air exchange and moisture transfer in VF gap

4 Materials and Methods

The prerequisite for determining expediency of application of the ventilated facade is favorable moisture conditions. This circumstance defines design features of a façade [9, 10]. By intensity increase the main processes of moisture transfer are divided into:

- Water vapor diffusion (and surface diffusion in some porous materials)
- Convective transfer of water vapor with an air flow
- Gravitational and capillary water transfer through porous materials, leakages under hydrostatic pressure of water through openings such as cracks and macropores.

Convective transfer of water vapor is one of the main ways how it gets inside a construction gap together with air. In the same way rain water can get inside the facade structure through grooved lines. To prevent or reduce these effects the building envelope should have low air permeability.

Despite this, one of the key features of calculating the humidity conditions of facade systems with air gap is their increased air permeability. One of the main barriers to air filtration is facing, which should be denser than other layers of structure as it withstands climatic and anthropogenic impact [12-15].

Dense facing is one of the main parts of the ventilated facade structure, but since it is separated from the wall by the air gap, no obstacles for air movement will arise. Significant impact on the humidity conditions has transverse (extending perpendicular to the plane of the wall) air filtration. As external facing does not present an obstacle for the air movement and the internal design is often carried out without taking into account possible filtration, in some buildings the filtration has considerable impact on moisture conditions [16].

It is known that the ever-existing difference in temperatures inside and outside the building causes difference in partial pressure and, as a result, water vapor diffusion through the building envelope [17]. In the course of diffusion there is movement of water vapor from a region of high concentration to a region of low concentration. Diffusive transfer is characteristic of air, including air in pores of materials. Therefore, water vapor doesn't diffuse through nonporous materials such as steel, glass, some plastics, etc [18, 19]

The calculation results and studies of heat and moisture transfer in the building envelopes prove that the zone of the greatest humidification is the layer of mineral wool insulation adjacent to the gap. Hence, it follows that the key factor is saturation rate of water vapor in the humid zone, not insulation thickness average humidity [20, 21].

If the temperature drops to the dew point (the saturation temperature of water vapor), the condensation of water vapor occurs.

Lack of an air gap or its insufficient width under some conditions can cause moisture accumulation and insulation overwetting. Humidification of the insulant leads to:

- swelling and formation of microcracks in internal layers,

- degradation of heat-insulating properties
- destruction of a porous body of insulation due to frost heaving
- adverse effect on its coefficient of heat conductivity
- corrosion of metal fastenings
- during the winter period due to a high temperature difference the humidity of insulant reaches its maximum. Because of this formation of an ice layer on mineral wool, a windproof membrane and on facing elements of a facade is possible.

It is therefore necessary to determine the conditions under which the insulation gap in the facade will remain unsaturated with moisture.

Removal of moisture from a humid zone may occur by diffusion of water vapor through the cover surface of the mineral wool insulation, the windproof membrane, the air gap and the facade facing layer. In addition, water vapor can be passed by an air flow through the air gap to the environment. Removal of moisture in the first case is small in comparison with the second one. Therefore it is necessary to provide possibility of removing all vapor from the zone of the greatest humidification by an air flow. Intensity of this technique depends on characteristics of an air gap (its width, height, etc.), which determine the air velocity in it.

Thus, for investigation of moisture transfer in the air space it is necessary to know the velocity and temperature of the air in the gap. It directly depends on the chosen design of the ventilated facade.

In designing ventilated facades much attention is paid to existence of grooved lines, which are a necessary operating condition of facades for diffusion removal of water vapor. The rustication grooves of the facing layer provide sufficient exchange with external air; therefore in most cases a vapor-permeable air compartment is optional.

However, due to the existence of these grooves it is impossible to prevent rain water and meltwater from getting on the back surface of the facing.

In addition, suction of cold outside air reduces the air velocity in the gap. This slows down the process of water removal. Drying rate of the moisture is influenced first and foremost by sufficient air motion and thermal radiation in the air gap.

Let us consider a stylized scheme of the ventilated gap between the "hot" plane y = 0 (with Th temperature) and the cold plane y = h (with Tc temperature). The height of the gap is equal to L, and L ≫h. To determine the influence of air velocity on moisture transfer it is expedient to run the following experiment: humidified to saturation insulation is placed into the ventilated gap of the structure; velocity is measured at various width of the air gap, with grooved lines and without them, difference in mass of the saturated and dried insulant is determined at the highest velocity of FCF (free convective flow).

Producing a model of a facade, we consider the temperature difference between external and internal layers substantiated by work of heating systems. Both heating of a wall and heat transfer to the air gap are taken into account. It is supposed that external air comes into the air space through the bottom air holes. It rises in the air space and leaves through the top air holes. While the air rises, its temperature increases and its relative humidity changes. The humidity will change monotonously along the full height of the layer.

5 Results and Discussion

Intensity of a mass transfer is estimated by criterial dependence:

$$N_D = const\sqrt[4]{\sigma}\sqrt[4]{Gr} = const\sqrt[4]{\sigma}\sqrt{Re} \tag{1}$$

where N_D- the diffusion Nusselt number, σ – the Prandtle number, Re – the Reynolds number, const=O(1) – a constant; for FCF condition in a confined gap we accept const≈0.5.

Let us carry out the experiment and measure air velocity and temperature in the air gap depending on its width. The most widespread device for measurement of air velocity in technical systems is the hot-wire anemometer.

The results of measurements with open (bottom lime) and closed (top line) grooved lines respectively:

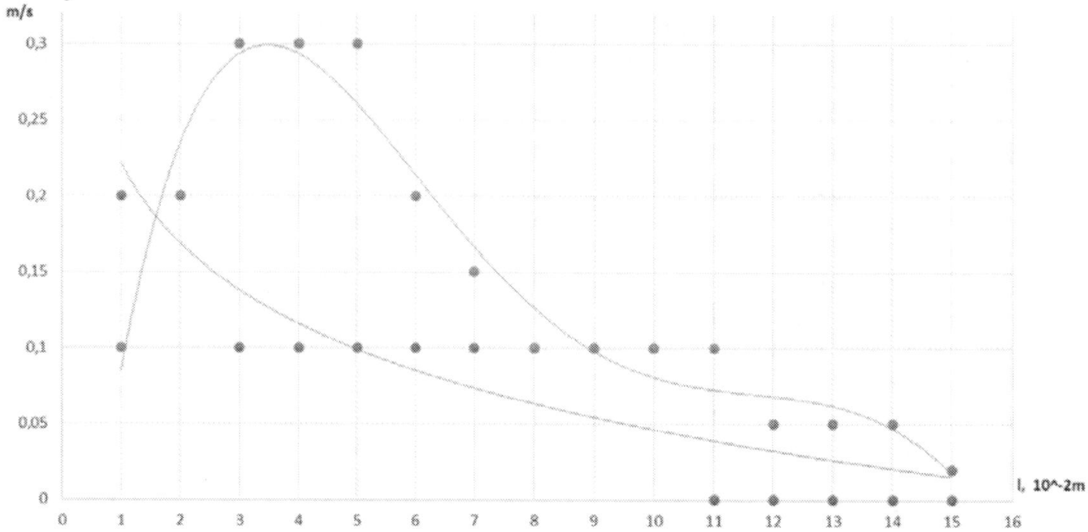

Figure 1. Air velocity as function from the geometrical sizes of the VF channel

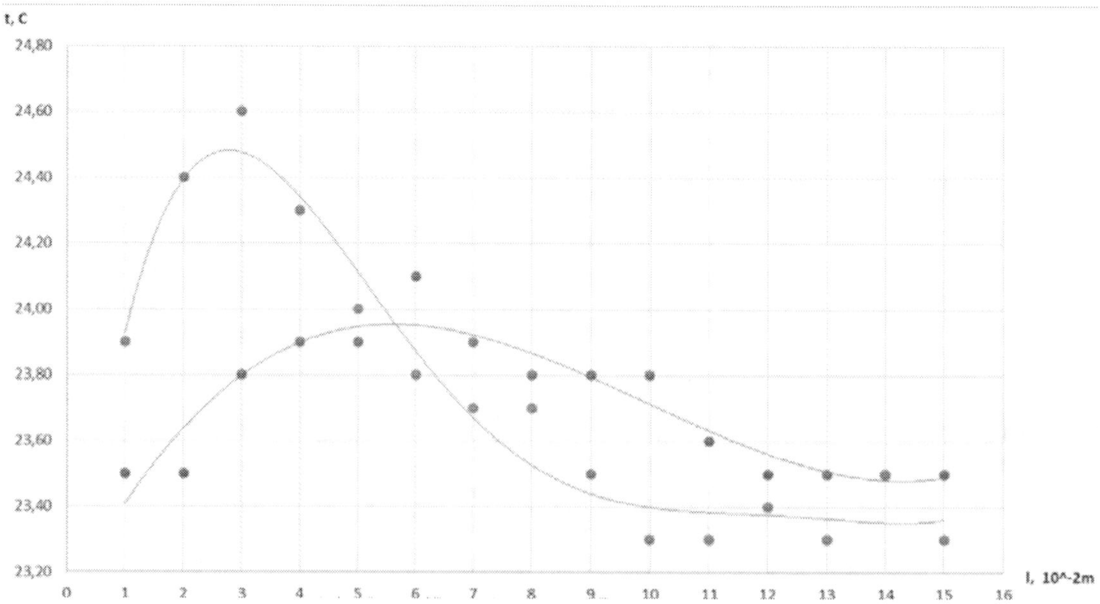

Figure 2. Air temperature in the gap as function from the geometrical sizes of the VF channel

On the basis of the data obtained experimentally it is possible to draw the following conclusions:

- maximum draught is observed in the "ideal" air gap
- grooved lines "overturn" FCF in a vertical gap
- the "ideal" layer is necessary for the normal work of insulation
- in a design with the closed grooved lines the insulation won't be saturated since the higher velocity and temperature values promote moisture removal.

6 Conclusions

Broad use of ventilated facade systems resulted in need of developing methods of forecasting moisture conditions of a building envelope.

Quantitative calculation of moisture transfer is one of the most important problems in designing the building envelope.

Rather high velocity values in the gaps of VF designs without grooved lines are one of the prerequisites for determining expediency of their application for effective moisture removal.

References

1. D.V. Nemova, V.Ja. Olshevskyi, D.N. Cejtin, Nauchno-tehnicheskie vedomosti SPBGPU, Jekonomicheskie nauki, 295–301 (2013)
2. S. Kosanović, A. Fikfak, Energy and Buildings, (2015)
3. I. Čerpes, N. Pandol, A. Fikfak, European Countryside, **6** **(3)**, 225-242 (2014)
4. V.A. Emel'janova, D.V. Nemova, D.R. Miftahova, Inzhenerno-stroitel'nyj zhurnal, **6**, 67-74 (2014)
5. A.F. Ostrovaja, M.R. Petrichenko, E.A. Stacenko, Applied Mechanics and Materials, 725-726 (2015)
6. M.R. Petrichenko, Second Edition St. Petersburg State Polytechnical University Journal. Physics and Mathematics, 143–149 (2012)
7. N. Vatin, , M. Petrichenko, D. Nemova, Applied Mechanics and Materials, **633-634**, 1007-1012 (2014)
8. A. Fikfak, Journal of Architecture and Urbanism, **37** **(4)**, 257-267 (2013)
9. N.I. Gamajunov, S.N. Gamajunov, Vestnik TvGTU, 128, **22**, 46-54 (2010)
10. S.V. Kornienko, Inzhenerno-stroitel'nyj zhurnal, **2**, 18-23, (2012)
11. S.V. Kornienko, Inzhenerno-stroitel'nyj zhurnal, **7**, 43-49, (2012)
12. M.R. Petrichenko, M.V. Petrochenko, Nauchno-tehnicheskie vedomosti SPBGPU, **2**, (2012)
13. D.V. Petrosova, Sankt-Peterburg, 118 (2012)
14. E.B. Yevtushenko, M.V. Petrochenko, Magazine of Civil Engineering, **8**,38-45
15. V.G. Gagarin, *Moskva*, 396 (2000)
16. N. Lipovac, Prostor, **19** **(1)**, 2-17 (2011)
17. A.S. Gorshkov, D.Ju. Popov, Magazine of Civil Engineering, **8**, 5-9 (2010)
18. A.A. Emel'janov, Promyshlennoe i grazhdanskoe stroitel'stvo, 62-71 (2013)
19. D.A. Zhukov, Stroitel'stvo: nauka i obrazovanie, 2-15 (2012)
20. N. V. Shilkin, AVOK, **1**, 12-20 (2005)
21. M.R. Petrichenko, D.V. Petrosova, M.V. Petrochenko, NTV SPbGPU. Fiziko-matematicheskie nauki, 68-72 (2012)
22. V.G. Gagarin, V.V.Kozlov, *Stroitel'naja fizika v XXI veke*, 73-80 (2006)
23. V. G. Gagarin, V. V. Kozlov, I. A. Mehnecov, AVOK, **8**, 71-80 (2005)
24. M. R. Petrichenko, X. Rakova, M. Vyatkin, , T.A. Musorina, D. Kuznetsova, Applied Mechanics and Materials, 1101-1106 (2015)
25. J.Li, Y.Chow, D.H.C. , *Centre for Sustainable Energy Technologies (CSET)*, 198-203 (2014)
26. Jun Hana, Lin Lua,Jinqing Penga, Hongxing Yanga, *Energy and Buildings*, 204–209 (2013)
27. V.N. Bukhartsev, M.R. Petrichenko, Power Technology and Engineering, **5**, 374-377 (2011)
28. Alin V. Rosca, Ioan Pop, International Journal of Heat and Mass Transfer, 355-364 (2013)

MATEC Web of Conferences **53**, 01011 (2016)
DOI: 10.1051/matecconf/20165301011

The feasibility of energy efficiency class improving of a building from economic point

Daria Tarasova[1,a], Kostantin Andreev[1] and Slobodan Lakić[2]

[1]*St. Petersburg State Polytechnical University, 195251 Politekhnicheskaya, 29, Saint-Petersburg, Russia*
[2]*Faculty of Economics, University of Montenegro, 81000 Podgorica, Montenegro*

Abstract. In this article 4 complexes energy efficiency measures for different energy efficiency classes are chosen, it is defined the thermal heating costs of the building during the heating period, it is defined the cost and the payback period of energy efficiency measures. The conclusion about the feasibility of capital repair of buildings by increasing energy efficiency class from an economic point of view is made.

1 Introduction

In Russia at the present time the potential of energy savings is 35 % to 45 % of total modern energy consumption, which 350-460 million tons is conditional fuel. Among the reasons indicated huge potential energy saving the main are technological imperfection of fixed assets, the traditional energy squandering and mismanagement, poorly managed and do not have modern technical means consumption of energy resources, lack of standardization and certification of energy consumption, lack of application of methods of energy management [1-3].

The most effective use of energy resources of the country for the revival and subsequent recovery and provide a decent life for the population declared main goal of the energy strategy of Russia since 1995. If we consider that the Russia total area of operated buildings is about 4.8 billion m2 and heating annually is 400 million tons of fuel equivalent, or 25% of the annual energy countries, it is quite obvious priority role of energy efficiency measures [4, 5].

The efficiency of the economy is the key to sustainable economic development, preservation of the natural environment. Improving energy efficiency leads to energy savings, the cost to them. One of the possible ways of improving energy efficiency is a capital repair and modernization of existing buildings by improving energy efficiency measures [6, 7].

The purpose of this article is to analyze the feasibility of capital repair of buildings by improving energy efficiency class from an economic point of view.

2 Input data

For calculation and the analysis the building of preschool institution was taken.

Design conditions and heat energy parameters of the buildings for the climatic conditions of St. Petersburg adopted according to SNiP 23-01-99*are presented at Table 1.

[a] Corresponding author : tarasovads@gmail.com

Table 1. The design conditions.

The characteristics	Symbol parameter	Unit of measure	Value
Design temperature of inside air	t_{int}	°C	-26
Design temperature of outdoor air	t_{ext}	°C	-0.9
The duration of the heating period	z	day/years	239
Heating degree day	D_d	°C·day/years	5473
Mean temperature of outdoor air	t	°C	22

Areas of external structure of building heated volume are presented at the table 2.

Table 2. Areas of external structure.

Type of structure	Areas of structure Ai, m2
External structure As	807.2
Windows Aw	457.8
Entrance external doors Ad	50.7
The combined covering Ac	1218
The overlapping Ao	1186
The total area Atot	3719.7

Design structure of the building is presented at table 3.

Table 3. Areas of external structure.

Type of structure	Characteristic	Materials			
		gypsum cement	slag brick	mortar	ceramic wall tile
External structure	δ, mm	0.005	0.51	0.02	0.01
	λ, (W/m2·°C)	0.35	0.58	0.3	1.7
Covering		rubberoid	mortar	foamed concrete	ferro-concrete
	δ, mm	0.02	0.02	0.125	0.18
	λ, (W/m2·°C)	0.17	1.2	0.3	1.7
Flooring		linoleum	floor batts	ferro-concrete	
	δ, mm	0.01	0.025	0.2	
	λ, (W/m2·°C)	0.33	0.041	1.7	
Door	λ, (W/m2·°C)	wood door			
		0.26			
Window	λ, (W/m2·°C)	triple-pane glass			
		0.3			

3 Thermotechnical calculation

By calculation we received:

$Q_h = 3441704.09\,(MJ)$

$Q_{int} = 417390\,(MJ)$

$Q_s = 170739\,(MJ)$

Then the amount used of energy on heating of buildings during heating season is calculated by formula 1.

$$Q_h^y = \left[Q_h - (Q_{int} + Q_s)\cdot \nu \cdot \xi\right]\cdot \beta_h \tag{1}$$

$Q_h^y = 3324145.916\,(MJ)$

Then the specific drain of energy on heating of building is calculated by formula 2.

$$q_h^{des} = \frac{10^3 \cdot Q_h^y}{V_h \cdot D_d} \tag{2}$$

$q_h^{des} = 76{,}76\left(kJ/(m^3 \cdot {}^oC \cdot day)\right)$

For the two-storeyed building of preschool institution required specific drain of energy on heating of building equal to= 38 кJ/ (m3·°C·day).

Then value of defection calculated specific drain of energy on heating of building from standard requirements is equal 102 %. It means this building corresponds to a class E («very low»).

4 Improvement of building structures for the purpose of energy efficiency increase

We modernize the building characteristics; we will improve its class of energy efficiency by the way of selection necessary energy effective measures.

Advanced building structures are chosen to improve the building characteristics to class D:

- Add insulation for wall - Venti Batts, δ =0.1m, λ=0.04
- Add insulation for roofing - Roof Batts, δ =0.1m, λ=0.041
- Increase insulation for flooring - Floor Batts, δ =0.05m, λ=0.041
- Door is the same wood door
- Change window construction to the glass cellular block 244x244x98mm, λ=0.33
- Advanced building structures are chosen to improve the building characteristics to class C:
- Increase insulation for wall - Venti Batts, δ =0.2m, λ=0.04
- Increase insulation for roofing - Roof Batts, δ =0.2m, λ=0.041
- Increase insulation for flooring - Floor Batts, δ =0.1m, λ=0.041
- Change door construction to the plastic door with one chamber glass, λ=0.35
- Change window construction to the triple-pane glass in single transom of ordinary glass with 8mm distance between glass panes, λ=0.50

Advanced building structures are chosen to improve the building characteristics to class B:

- Change the wall configuration is presented at table 4.
- Change insulation s for roofing - Polyurethane foam, δ =0.15m, λ=0.019
- Change insulation for flooring - Polyurethane foam, δ =0.15m, λ=0.019
- Change door construction to the plastic door with the triple-pane glass, λ=0.49
- Change window construction to the two one-chamber glasses in twin sashes, λ=0.80

Table 4. Construction of the new wall.

Material	λ, (W/m2·°C)	δ, mm
Plaster	0.9	0.01
Gas concrete	0.12	0.3
Polyurethane foam	0.019	0.15
Air gaps	0.022	0.05
Ceramic granite tile	0.2	0.07

5 The technical-economic analysis

5.1 The calculation of thermal energy cost

To similarly previous calculation thermal losses for each type of designs of a certain energy efficiency class were calculated and costs of the thermal energy for the heating during the heating period are defined. The costs of the construction are defined. Results are presented at Table 5.

Table 5. The annual thermal energy cost on building heating at the accounting of losses only through external enclosure structures.

Class of energy efficiency	Costs of the construction, millions	Cost of heating, millions/year
E	1.32	1.00
D	2.60	0.67
C	4.88	0.49
B	6.05	0.37

The change of construction characteristic and heating cost by energy efficiency class upgrade is presented at Figure 1. As a characteristic we choose transmission factor of heat transmission (Ktr). It depends on the construction layers and determinate by Russian standards. There is obvious fact than cost of the construction directly correlated with that factor. In case of cheaper structure it will be increasing of transmission factor and, accordingly, increasing heating cost [8-11].

Total costs of heating of the building and of a design taking into account growth of tariffs for each class on energy efficiency are presented at Figure 2. The minimum values for each year show the relevance of using the energy efficiency class for each period of time. Thus, during the first three years changing of class does not entail a sum cost benefit. In the span of 3-7 years, becoming the most profitable using of Class D [12, 13]. In considering the service life of more than 7 years it should be used a class B.

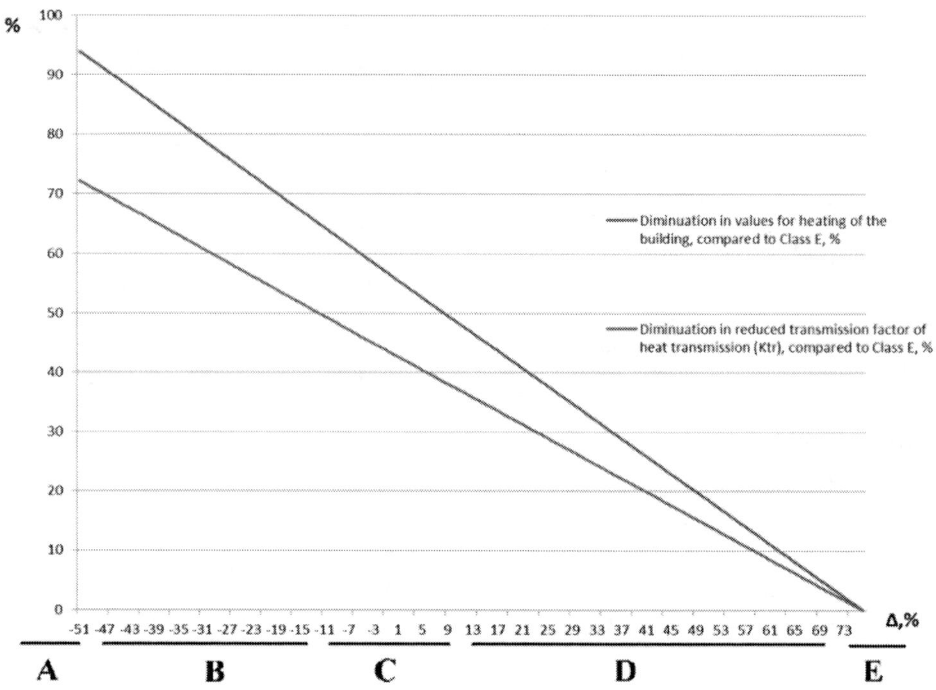

Figure 1. The schedule of change of construction characteristic and heating cost by energy efficiency class upgrade.

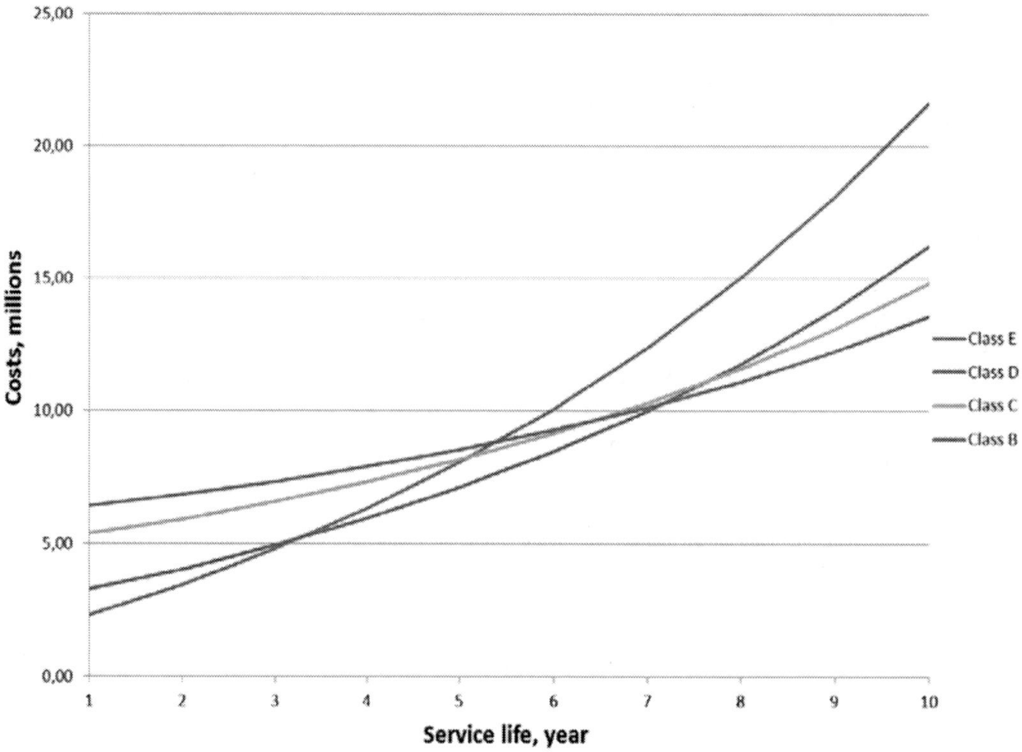

Figure 2. The dependence of total costs from service life.

5.2 The calculation of payback period

We will calculate a payback period of additional costs for a structure due to economy of thermal energy [14]. Additional costs are calculated as a difference between costs of construction (class E) and construction (other class). The economy of thermal energy is a difference between costs of the thermal energy used at application of construction (class E) and construction (other class). Costs of thermal energy are expressed in millions rubles. The payback period is calculated taking into account the average annual growth of tariffs 15% a year [15, 16]. Data for construction of a graph are presented in the Table 6.

Table 6. The annual thermal energy cost on building heating at the accounting of losses only through external enclosure structures.

Class of energy efficiency	Service life of the building , years/ Balance with accumulation, millions rub.							
	0	1	2	3	4	5	6	7
D	3.27	3.94	4.61	5.28	5.95	6.62	7.29	7.96
C	5.37	5.86	6.35	6.84	7.33	7.82	8.31	8.8
B	6.42	6.79	7.16	7.53	7.9	8.27	8.64	9.01

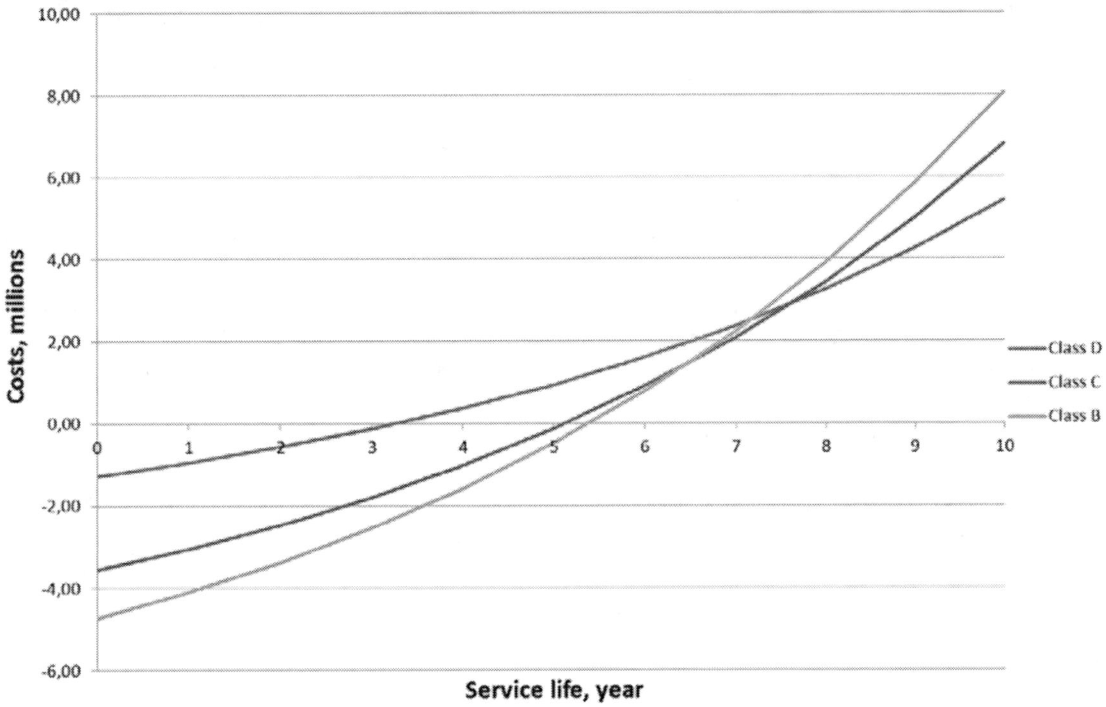

Figure 3. The schedule of payback period.

6 The conclusion

The thermal energy losses through the external structures of the building with the requirements of energy efficiency class C will be 33% less in comparison about the building constructed according to requirements of energy efficiency class E. And thermal energy costs will increase almost twice. The thermal energy losses through the external structures of the building with the requirements of energy efficiency class D will be 51% less in comparison about the building constructed according to requirements of energy efficiency class E. And thermal energy costs will increase to 3.5 times. The thermal energy losses through the external structures of the building with the requirements of energy efficiency class B will be 63% less in comparison about the building constructed according to requirements of energy efficiency class E. And thermal energy costs will increase to 4.5 times.

Modernization of the building by improve of a class of energy efficiency is expedient in case of long-term investment of capital. The payback period of the modernized designs due to economy of the thermal energy spent for heating of the building during the heating period at capital repairs of the building by improvement of a class with E on a class D makes 3 years, on a class C - 5 years, on B class - 6 years.

Using of a class C isn't relevance because the general costs of a class C aren't minimum in comparison with other classes for each year.

References

1. V. Savin, N. Savina, Gradostroitel'stvo, **1 (23)**, 82-84 (2013)
2. A. Avsjukevich, Construction of Unique Buildings and Structures, **2 (7)**, 40-54 (2013).
3. Z. Gaevskaya, X. Rakova, Advanced Materials Research, **941-944**, 825-830 (2014)
4. A. Gorshkov, D.Nemova, N. Vatin, Construction of Unique Buildings and Structures, **7 (12)**, 49-63 (2013)
5. N. Mingottia, T. Chenvidyakarn, A.W. Woods, Energy and Buildings, **58**, 237–249 (2013)
6. E Bazhenova, J. Bykova, D. Bryus, D.Tseytin, Applied Mechanics and Materials, **725-726**, 1445-1456 (2015)
7. N. Lipovac, Prostor, **5 (1)**, 1-33 (199
8. V.V Okrepilov, M.V. Leonidovich, Asian Social Science, **11 (7)**, 312-325 (2015)
9. V.V Okrepilov, V.L. Makarov, A.R. Bakhtizin, S.N Kuzmina, Economy of Region, **2**, 301-313 (2015)
10. V.L. Kvint, V.V Okrepilov, Herald of the Russian Academy of Sciences, **84 (3)**, 188-200 (2014)
11. V.V Okrepilov, V.N. Krutikov, G.I. El'kin, Measurement Techniques, **57 (2)**,109-116 (2014)
12. T. Zadvinskaya, A. Gorshkov, Advanced Materials Research, **953-954**, 1570-1577 (2014)
13. N. Bolshakov, S. Krivoy, X. Rakova, Advanced Materials Research, **941-944**, 895-900 (2014)
14. A. Kaklauskas, J. Rute, E. Zavadskas, A. Daniunas, V. Pruskus, J. Bivainis, R. Gudauskas, V. Plakys, Energy and Buildings, **50**, 7-18 (2012)
15. A. Fidrikova, O. Grishina, A. Marichev, X. Rakova, Applied Mechanics and Materials, **587-589**, 287-293 (2014)
16. Wilmer Pasut, Michele De Carli, Applied Thermal Engineering, **37**, 267-274 (2012)

MATEC Web of Conferences **53**, 01012 (2016)
DOI: 10.1051/matecconf/20165301012

Ventilated Facades: Insulation Materials of Different Manufacturers

Alexandra Shatornaya[1,a], Maria Chislova [1], Anna Chislova[1], Marina Drozdetskaya[1] and Daria Trubina[1]

[1] *Peter the Great St.Petersburg Polytechnic University., 195251, St. Petersburg, Russian Federation*

Abstract. The article analyzes the various manufacturers of thermal insulation materials that are used in hinged ventilated facades and there is a comparison them by the same criteria for identifying the advantages and disadvantages of their using in civil engineering. In the article there are the results of the carrying out calculations of wall construction with ventilated facade to resistance of heat transfer by using materials from different manufacturers for identical wall construction. As a result, has chosen material, with the best price, energy efficiency and safety of use.

1 Introduction

The Law N 261-FZ [1] was passed in 2009 to increase in the requirements to the energy efficiency of buildings and stimulation manufacturers in the market of thermal insulation materials. Researches show that 45% of the heat is lost through the enceinte walls [2-5]. The system of hinged ventilated facades (HVF) is used to improve the energy efficiency of buildings, and high-quality thermal insulation materials take on enormous importance in such type of systems. The choice of material affects the heat losses of the building

2 Materials and Methods

2.1 Studding of literature about the subject

In modern engineering, more attention is paid to the study of thermalphysic properties of enclosure constructions, their influence on the energy efficiency of buildings, severization requirements for them. These kinds of researches were conducted by Gagarin V.G. [6], A.S. Gorshkov [7,8].

The system of ventilated facades had been studied comprehensively in articles [9-12], and also the fire safety of such systems [13], the models of their work during the exploitation[14-16], and the construction of energy efficient houses [17,18].

Comparison of different modern insulating materials and study of their properties were carried out by Gorelik P.I., Zolotova Y.S. [19], Vatin N.I. [20], in works [21-22] there were compared the individual characteristics of materials. In Berdyugin's I.A. research [23] was made comparative analysis of glass wool and stone wool insulation as thermal covering.

---

[a]Corresponding author, e-mail: alexandrashatornaya@gmail.com

In previously published articles describe only the variety of existing thermal insulation materials, their properties and the field of application. However, a comparative analysis of the materials, that are used for the hinged ventilated facades (HVF), from the specific manufacturers for their actual properties and prices have never been made earlier.

2.2 Statement of Purpose of research

To carry out a comparative analysis of market of thermal insulation materials with the purpose the most favorable ratio between energy efficiency and economic feasibility for using them in hinged ventilated facade systems.

Owing from the objectives it is necessary to solve a number of problems:

1) Elucidate the market of modern insulating materials in Russia, to establish the most used types of thermal insulation materials for the HVF, their advantages and disadvantages.

2) Calculate total the resistance of heat loss for the enceinte wall with ventilated facade system using the selected thermal insulation materials.

3) Compile a comparison table of heaters for organizing data that was made during the research. Important criteria are: safety of use, thermal conductivity, and the cost of production.

4) Perform a final analysis of the results; identify the most energy-efficient, economical and safe to use insulation material.

2.3 Market Research of modern insulating materials and manufacturers

It should pay attention in choosing the material to the safety from the standpoint of fire safety (ignitability, toxicity) and ecologically cleanses. Investigation of these criteria is necessary for the construction of high-quality energy-efficient structures. There are several varieties of mineral wools: rock wool, glass wool and slag wool.

Slag wool is produced from blast-furnace slag, but can't find extensive use in Russia because of the low maintenance properties.

Glass wool is cheaper, less volume, occupied less space during transportation. A disadvantage of using it in the HVF is the flexibility, elasticity and low imporosity, it can lead to the shrinkage loss and the need to use a windproof film. However, glass wool creates less stress on the structure, because of low imporosity.

Rock wools advantages are: nonsusceptibility, deformations almost complete absence of changes, application temperature (1000 C) higher than that of glass wool (400 C). The stone wool is not used toxic substances. The main drawback - the price is above the glass wool.

We have considered thermal insulation materials for the HVF system from well-known manufacturers such as ISOVER, ROCKWOOL, URSA, IZOROCK, Technonikol, Paroc, Ecover and

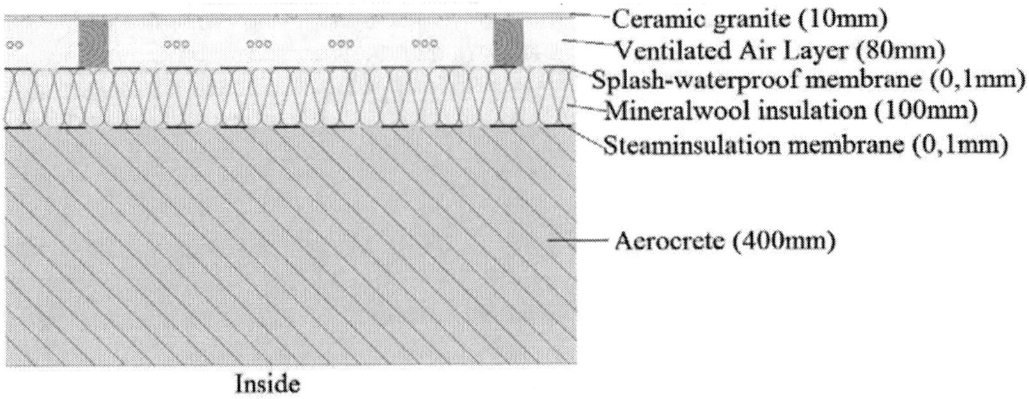

Figure 1. Construction of ventilated facade

others. As a result was chosen one material for each type of manufacturer. All insulation materials are non-combustible. To analysis of the data classification was compiled in comparative table (Table 1).

For calculations there was used the identical structure of the hanging ventilated facades: Aerated Concrete (400mm), vapor barrier sheet, insulation (100 mm), house wrap sheet, ventilated air layer (80mm), porcelain stoneware slabs (10mm) (Fig. 1).

The calculation is performed in accordance with the requirements of SNIP 23-02-2003 [24], SNIP 1.23.09, by the method of SP-23-1001-2004 for the North-West region, which is been in application condition B. It was taken into account the thermal conductivity index λB because thermal conductivity in operating conditions B has the greatest value.

For the analysis of the data was made several comparative tables.(tabl 1-3)

Table 1. Comparison of fibered glass from different manufacturers

Manufactor and product	fibered glass		
	URSA Pure One (eco-friendly material)	URSA (GEO P-30)	ISOVER (VentFasad Mono)
Main characteristics:			
possible thickness, [mm]	50, 100	50-100	50, 100
density, [kg/m3]	18-22	30	70
flammability group	NF	NF	NF
thermal conductivity λ at 10 C [W/mK]	0.034	0.032	0.034
thermal conductivity λ at 25 C [W/mK]	0.037	0.034	0.036
λ upon conditions A [W/mK]	0.039	0.036	0.038
λ upon conditions B [W/mK]	0.041	0.039	0.04
Thermal resistance of ventilated facade construction, [m2*C/W]	3.82	3.94	3.88
Heat loss per heating season [kWh]	28.53	27.63	28.08
m3 in package	0.45	0.75	0.493
cost for 1 m3	1900	2736	4746
Additional characteristics:			
Water absorption ability at short-term and not full immersion for 24 hours, no more, [kg / m2]	1	1	1
vapor permeability, no less [mg/(m*h*Pa)]	0.51	0.59	-
operating temperature [C]	from -60 to +220	from -60 to +290	-

Table 2. Comparison of mineral wool from different manufacturers

Manufactor and product	rock-wool				
	Rock-wool (Venti Butts)	Techno-Nikol (techno-vent Optima)	Paroc WAS 35	Isoroc Isovent	ISOVER (Venti)
Main characteristics:					
possible thickness, [mm]	50, 100	100 мм	30-150	40-160	50 , 100
density, [kg/m3]	90	81-99	85	90	82-103
flammability group	NF	NF	NF	NF	NF
thermal conductivity λ at 10 C [W/mK]	0.035	-	0.033	0.034	0.035
thermal conductivity λ at 25 C [W/mK]	0.037	0.036	0.034	0.036	-
λ upon conditions A [W/mK]	0.038	0.038	0.04	0.039	0.038
λ upon conditions B [W/mK]	0.04	0.04	0.043	0.041	0.039
Thermal resistance of ventilated facade construction, [m2*C/W]	3.88	3.88	3.7	3.82	3.88
Heat loss per heating season [kWh]	28.08	28.08	29.41	28.53	28.08
m3 in package	0.18	0.216	0.216	0.2	0.216
cost for 1 m3	3790	2764	3754	3920	3983.6
Additional characteristics:					
Water absorption ability at short-term and not full immersion for 24 hours, no more, [kg / m2]	1	1,5 % in volume	1	1	1
vapor permeability, no less [mg/(m*h*Pa)]	0.3	0.3	1	-	0.3
Compressive strength at 10% deformation, no less [kPa]	20	12	-	20	20
operating temperature [C]	-	-	-	-	-
organic matter content, no more than [%]	-	3	-	4	-

Table.3. Comparison of bazalt wool from different manufacturers

Manufactor and product	basalt fibre plates				
	Tekhno-Nikol Rocklife	Termo-bazalt PR-80 (VENTI)	Tizol EURO-Vent	Ecover Vent-Facade	Termo-life EcoVe nt Facade
Main characteristics:					
possible thickness, [mm]	50, 100	50, 100	40-200	30-250	50-200
density, [kg/m3]	30-40	80	80	80	60
flammability group	NF	NF	NF	NF	NF
thermal conductivity λ at 10 C [W/mK]	-	0.033	0.034	0.035	0.036
thermal conductivity λ at 25 C [W/mK]	0.039	0.035	0.036	0.036	-
λ upon conditions A [W/mK]	0.04	0.041	0.038	0.038	0.04
λ upon conditions B [W/mK]	0.041	0.044	0.04	0.039	0.044
Thermal resistance of ventilated facade construction, [m2*C/W]	3.82	3.65	3.88	3.94	3.65
Heat loss per heating season [kWh]	28.53	29.83	28.08	27.63	29.83
m3 in package	0.432	0.25	0.2	0.36	0.3
cost for 1 m3	1211	2792	2320	2681	2350
Additional characteristics:					
Water absorption ability at short-term and not full immersion for 24 hours, no more, [kg / m2]	2% in volume	1.5	1	0.75	2% in volume
vapor permeability, no less [mg /(m*h*Pa)]	0.3	0.37	0.3	-	0.3
Compressive strength at 10% deformation, no less [kPa]	-	20	18	16	12
operating temperature [C]	from -60 to +200	To +650	To +400	More 1000	-
organic matter content, no more than [%]	2.5	1.5	-	3.5	3

3 Results and discussion

After analyzing the data of Table 1, glass wool URSA (GEO P-30) has a higher rate on the main characteristics than glass wool URSA Pure One and ISOVER. It is necessary to take into account that the material URSA Pure One is environmentally friendly and has a lower cost, but it has the highest heat loss during the heating season (28.53 kW * h). In this regard, we have chosen material URSA (GEO P-30), which is higher in value, but is more energy efficient in view of the high performance in all other criteria.

Next, we examined the materials of mineral wool: stone wool (Table 2) and a basalt slab (Table 3).

From Table 2, there are two differing material: Rock wool Technonikol and ISOVER. Rock wool by ISOVER manufacturer has a variouse thickness, high density, low thermal conduction of the application conditions A and B. Wool Technonikol have only one thickness, thermal conduction of the application conditions B is slightly worse. Wool of this manufacturer has the lowest cost.

Index heat leakage and heat transmission resistance is the same for bouth manufacturers. As the second most important criterion is profitability, we chose rock wool of company Technonikol.

After reviewing Table 3, it can be concluded that the basalt slab Ecover is the best in terms of heat transmission resistance, heat loss and thermal conduction of the application condition B; the material is environmentally clean and has the lowest water absorption. The cost of the Technonikol slab is the cheaper, but it loses Ecover slab in all other indicators.

For the final choice of material was made the final consisting table of three materials in one of each type of material.(table 4)

Prices of the selected materials are in the same range and depend on suppliers and the region of building, but we should note the need of using a fuel windproof membrane for glass wool that could increase the total cost of construction. HVF with using materials from URSA GEO and Ecover have the smallest heat losses during the heating season, which will save significant sum of money in high-rise constructions. Ecover Material of HVF has an ecological certificate and is safe to use, so that the final choice was stopped on it.

Fig. 2 presents a distribution graph of the temperature in the construction of HVF with selected thermal insulation material.

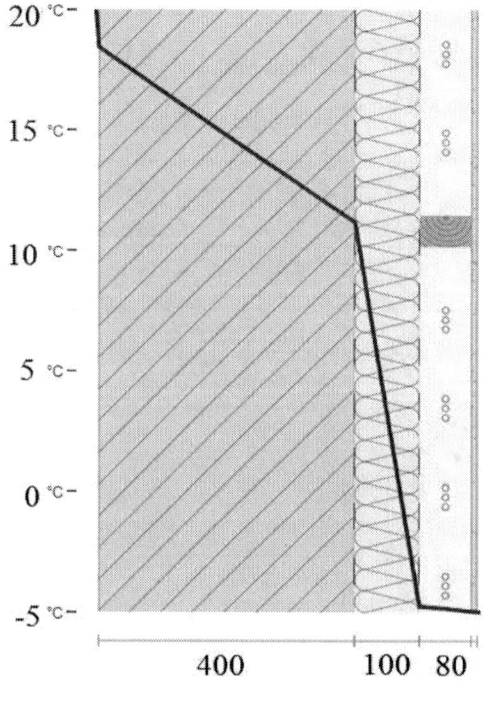

Figure 2. Distribution graph of the temperature (in °C).

Table 4. Comparison of bazalt wool from different manufacturers

	fibered glass	rock-wool	continuous basalt fibre plates
Manufactor and product	URSA (GEO П-30)	TechnoNikol (TECHNOVENT Optima)	Ecover VentFasad
Main characteristics:			
possible thickness, [mm]	50-100	100	30-250
density, [kg/m3]	30	81-99	80
flammability group	NF	NF	NF
thermal conductivity λ at 10 C[W/mK]	0.032	-	0.035
thermal conductivity λ at 25 C[W/mK]	0.034	0.036	0.036
λ upon conditions A [W/mK]	0.036	0.038	0.038
λ upon conditions B [W/mK]	0.039	0.04	0.039
Thermal resistance of ventilated facade construction, [m2*C/W]	3.94	3.88	3.94
Heat loss per heating season [kWh]	27.63	28.08	27.63
m3 in package	0.75	0.216	0.36
cost for 1 m3	2736	2764	2681
Additional characteristics:			
Water absorption ability at short-term and not full immersion for 24 hours, no more, [kg / m2]	1	1,5 % per volume	0.75
vapor permeability, no less [mg/(m*h*Pa)]	0.59	0.3	-
Compressive strength at 10% deformation, no less [kPa]	-	12	16
operating temperature [C]	from -60 to +290	-	More than 1000

4 Conclusion

As a result of a comparative analysis there was made a conclusions:

In the article was explored the market of modern insulating materials for NVF systems in Russia and installed the most common insulation materials: glass wool, rock wool and basalt slabs.

According to SNIP 23-02-2003 was calculated the HVF for the heat transfer resistance by using selected heaters.

Was compiled and analyzed the comparative table of using heaters.

Basalt slabs Ecover is the most appropriate ratio of energy efficiency and price per m3 as compared with products from other manufacturers. It is an environmentally clean insulation. For choosing of insulation material for the HVF system recommended to use products with similar characteristics.

In the conditions of the cold climate, limited resources and rising electricity prices it becomes a priority to reduce heat loss through the enclosing wall in modern construction. Choosing of high quality insulation material is extent the energy efficiency of buildings. In today's market there is a wide range of heaters, and therefore it is important to understand exactly what criteria to consider when designing.

Savings on heating by reducing the amount of heat loss through the building structure inclosing walls should be available in terms of pay for the costs of high-quality insulation, which may be higher than with other materials. In addition, you must take into account the fire and environmental safety, which can be a determining factor in the final selection.

5 Conflict of interests

The authors declare that there is no conflict of interest regarding the publication of this paper.

References

1. The Law of the Russian Federation "On energy saving and on energy efficiency and on amendments to certain legislative acts of the Russian Federation" of 23.11.2009
2. N. I. Vatin, A. S. Gorshkov, D. V. Nemova, Construction of Unique Buildings and Structures Journal, **3,** 8 (2013)
3. V. G. Gagarin, Journal Academia. Architecture and construction, **5,** 297-305 (2009)
4. A. S. Gorshkov, D. V. Nemova, N. I. Vatin, Construction of Unique Buildings and Structures Journal, **7** , 12 (2013)
5. O. I. Lobov, A. I. Ananyev, V. P. Abarykov, A. E. Sinyutin, II all-Russian scientific and technical conference: Building Thermophysics and energy effective design of walling of buildings, 80-92 (2009)
6. V. G. Gagarin, II all-Russian scientific and technical conference: Building Thermophysics and energy effective design of walling of buildings, 33-44 (2009)
7. A.S. Gorshkov, V.I. Livchak, Construction of Unique Buildings and Structures Journal, **3**, 30 (2015)
8. A. S. Gorshkov, I. A. Voilokov, II all-Russian scientific and technical conference: Building Thermophysics and energy effective design of walling of buildings, 45-48 (2009)
9. M.V. Petrochenko, K.I. Strelets, M.R. Petrichenko, E.B. Yavtushenko, Applied mechanics and materials, **672-674**, 567-570 (2014)
10. M.I. Nizovtsev, A.N. Sterlygov, V.T. Belyi, Construction of Unique Buildings and Structures, **6(** 33), 19-33 (2014)
11. G. M. Vasilchenko, Donbas National Academy of Civil Engineering and Architecture, **4**, 203-211(2013)
12. V. Bogdanovic, S. Samardzie, Architecture and Civil Engineering, **1**, 57 – 68 (2004)

13. A. P. Samar, E. J. Onokhov, O. V. Chalupova, Fire safety study of buildings shed-governmental facades, **1,** 357-362 (2013)
14. Balocco C., Energy and Buildings, **34**, 469-475 (2002)
15. L. Mei, D.G. Infield, U. Eicker, and V. Fux., Energy and Buildings, **32**, 569-575 (2011)
16. C. Marinosci, G. Semprini, G.L. Morini, P.A. Strachan, Energy And Buildings, **43**, 853-863 (2011)
17. A. Fikfak, Journal of Architecture and Urbanism, **37 (4)**, 257-267 (2013)
18. A. Fikfak, M.Z. Senegačnik, Glasnik SED, **48 (1-2)**, 2-67 (2008
19. P. I. Gorelik, Yu. S. Zolotova, Construction of Unique Buildings and Structures, **3**, 18 (2014)
20. N. Vatin, A. S. Gorshkov, A. Kazimirova, K. Gureev, D. V. Nemova, JAES, **3**, 207-216 (2014)
21. G. Chaykovskiy, Bachelor thesis, Mikkeli University of Applied Sciences (2010)
22. M. J. Duijve, MS Thesis, Utrecht University (2012)
23. I. A. Berdugin, Civil Engineering Journal, **1,** 345-366, (2010)
24. Construction rules and regulations "Thermal protection of buildings" of 23.02.2003

MATEC Web of Conferences **53**, 01013 (2016)
DOI: 10.1051/matecconf/20165301013

Mathematical Modeling of Hydrophysical Properties of Soils in Engineering and Reclamation Surveys

Vitaly Terleev[1], Ekaterina Petrovskaia[1,a], Natalia Sokolova[1], Anastasiia Dashkina[1], Inna Guseva[1], Vladimir Badenko[1], Yulia Volkova[1], Olga Skvortsova[1], Olga Nikonova[1], Sergey Pavlov[1], Aleksandr Nikonorov[1], Vitaly Garmanov[2] and Wilfried Mirschel[3]

[1]Peter the Great St.Petersburg Polytechnic University, Polytechnicheskaya, 29, St. Petersburg, 195251, Russia
[2]St.Petersburg State Agrarian University, Peterburgskoye shosse, 2, St. Petersburg-Pushkin, 196601, Russia
[3]Leibniz-Centre for Agricultural Landscape Research, Eberswalder Straße 84, Müncheberg, 15374 Germany

Abstract. All structures are built of, or through, or supported by earth materials including soil. Thus soil surveys in civil-engineering field are an important step in the construction. A lot of things depend on their results, especially the cost of construction, durability. The hydraulic conductivity needs one of the most laborious, and expensive surveys, thereby it is necessary to reduce these disadvantages. Were considered different methods (a laboratory, field and a mathematical model), and conclusions were drawn about economic feasibility of using the least laborious method. This method can be used under the various constructions, when engineering or reclamation surveys take place.

1 Introduction

Soil surveys are an important part of construction, because the cost of construction, durability and strength of structures depends on the results of these surveys. This is especially urgent in the design of complex structures erected in dense urban areas, or to previously unexplored territories. The lack of reliable methods for predicting the effects of human impact has led to negative results (soil consolidation, degradation, and so forth). Also, a significant problem is the high labor intensity of the survey work to determine the hydraulic conductivity of the soil [1-4]. One such property is the hydraulic conductivity. There are various methods for determining the hydraulic conductivity of the soil: a laboratory, field and a mathematical model. All existing laboratory methods for measuring coefficient of hydraulic conductivity of soil can be grouped into two major groups: the first includes methods for the analysis of stationary flows of moisture in the soil, wherein the moisture and capacity distribution throughout the system, as well as the flux density in any profile of the system perpendicular to the direction of flow, are not time dependent; the second group-methods of analysis of unsteady flows, i.e., flows, in which the distribution of humidity, density and capacity changes over time and space.

Field methods of the determination of the coefficient of hydraulic conductivity involve the formation of a stationary flow of moisture at constant soil moisture. For these purposes designed soil infiltrometers, which allow providing water to the soil surface at a negative pressure so that the flow of moisture in the soil is determined not hydrohead, but the suction force of the soil. A significant

[a] Corresponding author : kate-petro@yandex.ru

disadvantage is that in both methods the character of current and flux density are significantly different, which may affect the value of the coefficient of hydraulic conductivity.

On the basis of existing theoretical concepts, using data obtained in a few, but in a certain way planned and ordered full-scale experiments, formulates equations that describe the patterns of interaction of the elements forming the system under study, with each other and with the environment. These equations are summarized in a single system, form a mathematical (simulation) model.

Mathematical models allow solving a wide range of tasks, for example to predict the dynamics of soil moisture and groundwater level changes, as well as deformation of soils during construction and operation. Mathematical modeling of hydrophysical properties of soils enables the use of computer for the experiments, which have several advantages [5-7]: for example, to perform these experiments do not require much labour and time, while laboratory studies often involves a labour-intensive field experiments with a large number of replicates. Adequate mathematical models allow investigating the subject, even in the hypothetical circumstances [8, 9].

With regard to the problems of engineering and land reclamation surveys of particular practical importance is the opportunity to evaluate such important hydrophysical indicators such as water retention and hydraulic conductivity of soils, taking into account the hysteresis phenomenon [10-12]. In addition, the use of soil and hydrophysical model proves to be very productive in the agricultural research for precision farming technology [13], as well as a comprehensive assessment of agro-reclamation systems in the planning of investments in the construction of such systems and their reconstruction [14-17].

2 Materials and methods

Mualem-Van Genuchten method has been used successfully for quite a long time [18, 19]. The advantages of the method are obvious. It provides an estimating the parameters of the function to interpolate the measured data of soil water retention capacity (WRC) in the form of the dependence of volumetric soil water content θ on capillary pressure of moisture ψ. Then these parameters are used for estimating the ratio of the hydraulic conductivity of the soil k to the coefficient filtration of soil moisture k_S (relative hydraulic conductivity of the soil). The WRC model proposed by Van Genuchten is following [19]:

$$\bar{\theta} = 1/\left[1 + (-\alpha\psi)^n\right]^m, \tag{1}$$

where $\bar{\theta} = (\theta - \theta_r)/(\theta_S - \theta_r)$ - the normalized volumetric soil moisture content; θ_S - volumetric saturated soil moisture content; θ_r - minimum specific volume of liquid water in the soil; α, n and m - empirical parameters.

Van Genuchten used the formula (1) to calculate the ratio $\bar{k} = k/k_S$:

$$\bar{k} = \sqrt{\bar{\theta}}\left(\int_0^{\bar{\theta}} d\bar{\theta}/\psi \Big/ \int_0^1 d\bar{\theta}/\psi\right)^2. \tag{2}$$

In the case $m = 1 - 1/n$ Eq. 2 allows a simple analytical calculation:

$$\bar{k} = \sqrt{\bar{\theta}}\left[1 - \left(1 - \bar{\theta}^{1/m}\right)^m\right]^2. \tag{3}$$

The successful application of WRC (1) as an approximation of the dependence $\bar{\theta}(\psi)$ for a wide range of varieties of soil hampered by lack of physical interpretation of used soil and hydrophysical parameters. This deficiency manifests itself: firstly, the impossibility to estimate these parameters according to the indirect measurement of the physical properties of the soil; secondly, that the use of the parameters of the function approximating the relationship $\bar{\theta}(\psi)$, which obtained by interpolation of the experimental data in a certain range of capillary pressure, can be used to calculate k/k_s only the same range of capillary pressure.

Among the most successful attempts to overcome the above drawback relates approach proposed Kosugi [20, 21]. The basic idea of this approach is that by integrating the Mualem's formula it is not used the dependence $\bar{\theta}(\psi)$, but normalized differential moisture capacity of the soil $\bar{\mu} = d\bar{\theta}/d\psi$, with respect to which the function $\bar{\theta}(\psi)$ describing the dependence is a primitive. Similarly with Kosugi, in author's method is based on the hypothesis of lognormal distribution of the effective radius of the soil pores is a random variable $\bar{r} = (r - r_{\min})/(r_{\max} - r)$, where r - the pore radius; $r_{\min}$ - the radius of the smallest pore and $r_{\max}$ - the radius of the largest pore. This random variable is described by lognormal distribution density functions: $f(\bar{r}) = \exp[-1/(2\sigma^2)\ln^2(\bar{r}/\bar{r}_0)]/(\bar{r}\sigma\sqrt{2\pi})$, where $\ln\bar{r}_0$ - the most probable value and σ -the standard deviation of the logarithms of the effective pore radius [20-22].

From Laplas' law $\psi = P - P_a = -\beta/r$, where $\beta = 2\gamma\cos\varphi/(g\rho_W)$ - the coefficient of proportionality; γ - measure the surface tension of soil moisture; φ - angle of the contact of the surface of soil particles; g - gravity acceleration; ρ_w - density of water.

During the physical modeling of desiccation of the soil, using a pneumatic press, displacement of its water is achieved from excessive pressure of atmospheric gases in the initially saturated with moisture of (without entrapped air in the pores), a soil sample, which is placed on a porous membrane. With the increase in gas pressure above the water surface in the pores of the soil from normal external atmospheric pressure P_a to P_{ae} air dissolves in water, but the water content of the pore space practically does not change and the volume of the soil moisture is maximum. The excess gas pressure is transferred to the water; the absolute pressure of water is also increased. When the excess of gas pressure reaches value $\Delta P_{ae} = P_{ae} - P_a$ the interaction force between water molecules and the surface of soil particles is weakened so that the water goes into the category of free gravitational water from the soil and flows through the membrane into the pan, and an empty volume of pores takes air. The excess of gas pressure in the pneumatic press relative to the external atmospheric pressure corresponds to the value of the capillary pressure of moisture in the soil of the largest pore $\psi_{ae} = -\Delta P_{ae} = -\beta/r_{\max} < 0$. This value is called the bubbling pressure and is considered here as the "initial basis" of the capillary pressure of moisture. In the case of $r_{\min} << r$ accept $r_{\min} = 0$; in this case following relations are valid: $\psi - \psi_{ae} = -\beta/(r_{\max}\bar{r})$ and $\psi_0 - \psi_{ae} = -\beta/(r_{\max}\bar{r}_0)$, where $\bar{r}_0 = r_0/(r_{\max} - r_0)$, r_0 - the radius of the pores of the soil that corresponds to the most probable value $\ln\bar{r}_0$ of random variable $\ln\bar{r}$.

In [22] it has been described the $\bar{\theta}(\psi)$ approximation in the class of elementary functions:

$$\begin{cases} \bar{\theta} = \begin{cases} (1/2)\operatorname{erfc}\{(n\sqrt{\pi}/4)\ln[-\alpha(\psi - \psi_{ae})]\}, \psi < \psi_{ae}; \\ 1, \psi \ge \psi_{ae}; \end{cases} & (4a) \\ \bar{\theta} \approx \begin{cases} 1/\{1 + [-\alpha(\psi - \psi_{ae})]^n\}, \psi < \psi_{ae}; \\ 1, \psi \ge \psi_{ae}, \end{cases} & (4b) \end{cases}$$

where $\alpha = -1/(\psi_0 - \psi_{ae}) = r_{\max}\bar{r}_0/\beta$ and $n = 4/(\sigma\sqrt{2\pi})$ - parameters. It has been described the function of the normalized differential moisture capacity of the soil:

$$\bar{\mu} = d\bar{\theta}/d\psi = \begin{cases} -(n/4)/(\psi - \psi_{ae})\exp\{-\pi(n/4)^2\ln^2[-\alpha(\psi - \psi_{ae})]\}, \psi < \psi_{ae}; \\ 0, \psi \ge \psi_{ae}. \end{cases} \quad (5)$$

Thus was obtained the physically based model of differential conductivity of soil moisture. As a result the formulas that describe the ratio of the hydraulic conductivity of the soil to the coefficient of filtration of soil moisture, as well as an approximation of this ratio, are following [22]:

$$\begin{cases} \bar{k} = \begin{cases} \left[1/\left(4\sqrt{2}\right)\right]\sqrt{\operatorname{erfc}\left\{\left(n\sqrt{\pi}/4\right)\ln\left[-\alpha\left(\psi-\psi_{ae}\right)\right]\right\}}\left(\operatorname{erfc}\left\{\left(n\sqrt{\pi}/4\right)\ln\left[-\alpha\left(\psi-\psi_{ae}\right)\right]+2/\left(n\sqrt{\pi}\right)\right\}\right)^2, \psi<\psi_{ae}; \\ 1, \psi \geq \psi_{ae}. \end{cases} & (6a) \\ \bar{k} \approx \begin{cases} 1/\sqrt{1+\left[-\alpha\left(\psi-\psi_{ae}\right)\right]^n} \Big/ \left\{1+\exp\left[8/(n\pi)\right]\left[-\alpha\left(\psi-\psi_{ae}\right)\right]^n\right\}^2, \psi<\psi_{ae}; \\ 1, \psi \geq \psi_{ae}. \end{cases} & (6b) \end{cases}$$

In the particular case (under $\psi_{ae} = 0$ condition) formula (6*a*) is transformed into formula offered Kosugi [21]. Formulas (4*a*), (5) and (6*a*) form a closed system of the soil hydrophysical functions with common parameters, which are physically interpreted [23].

3 Results and discussion

Here is just one result of the verification of the parameters of a closed system of soil and hydro-physical functions. The data is used from the Van Genuchten's paper [19] for the clay soil *Beit Netofa clay*, for which Mualem-Van Genuchten method proved the least acceptable. The blue curves (1) and (2) in the fig.1 and fig.2 presents a model of Van Genuchten compared with experimental data (circles), and the red curves (3) and (4) in the fig.1 and fig.2 shows the results of calculation using the method described above in this paper. Visual analysis reveals much more exact coincidence of the red curves with the experimental data.

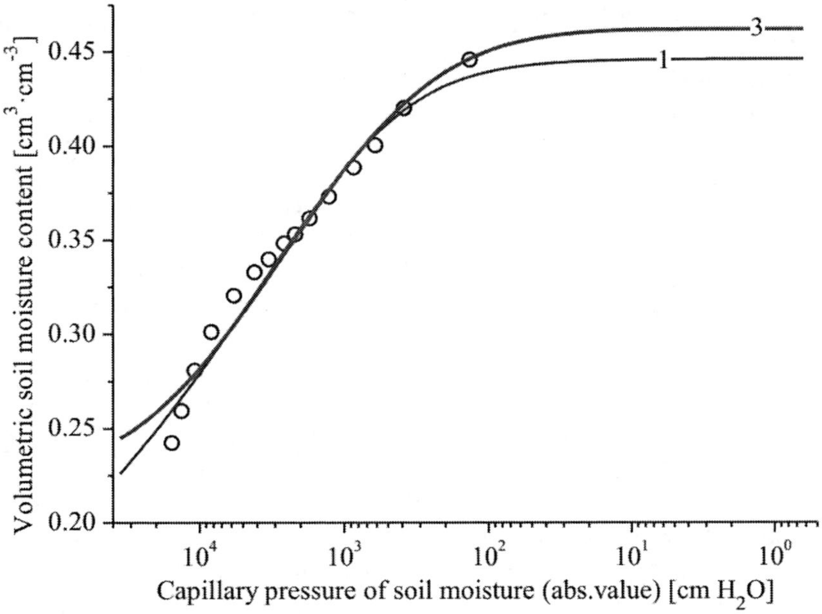

Figure 1. Water retention curve of the soil.

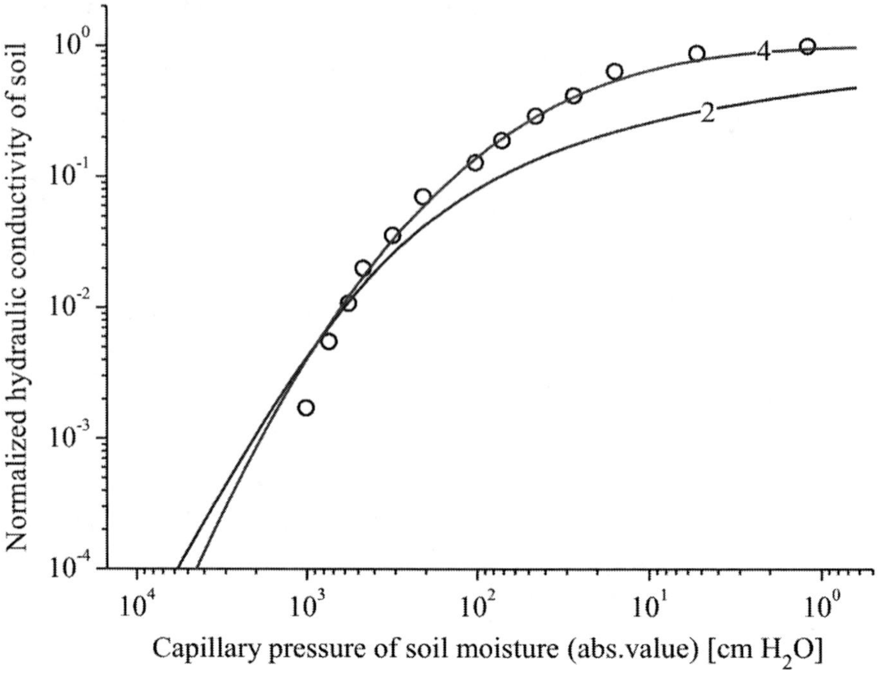

Figure 2. Normalized hydraulic conductivity of the soil.

Note that Van Genuchten predicted the hydraulic conductivity relatively accurately only in the range of capillary pressure, which is common to the soil moisture characteristic, and the hydraulic conductivity. The suggested model predicts well the hydraulic conductivity even in the range of capillary pressure, where there is no measurement of the soil moisture characteristics. It is no wonder, since the Van Genuchten's model has no physical interpretation, so it can only be an interpolation. Proposed model has a physical basis, so it has the property of extrapolation: soil moisture characteristics define parameters in a range of capillary pressure and hydraulic conductivity predicts a wider range.

Before it is not considered the economic feasibility of the use of mathematical modeling methods that are available for predicting the hydraulic conductivity of the soil according to its water-retention capacity. Direct measurement of the hydraulic conductivity of soils under laboratory conditions is very time-consuming and requires highly skilled staff. For these reasons, a small number of organizations involved in direct measurements of hydrophysical index.

The average cost of equipment, which used for measuring hydraulic conductivity of soil and fine soil is about 2000$ [24]. Moreover, it should take into account the cost of this work (the cost of factors of production: drilling, sampling, transportation, etc.). Using the method described in [12], can significantly reduce the cost of engineering and survey works, using available data on the relative density of the addition and the density of the solid phase of the soil as well as soil texture and soil-hydrological (soil hydrological) constants [10].

4 Conclusions

It follows that this method can find wide application because of its effectiveness and operational evaluation of a number of possibilities is very important to the reclamation construction of hydrophysical index - hydraulic conductivity of soils. This method is widely applicable under the planning preconstruction surveys, when the geological research of the landscape is done. In addition, the soil surveys are produced in landscape construction or as the part of reclamation arrangement. Soil surveys are very important, because they are the basis of all the construction [25-26].

Acknowledgements

The research was supported by DAAD, DFG, and Russian Foundation for Basic Research (#16-04-01473-*a*).

References

1. R.A. Poluektov, V.V. Terleev, Russ meteorol hydro+ **12**, 73-77 (2005)
2. R. Duan, C. Fedler, J. Borrelli, J. Irrig. Drain Eng. **138**, 322-327 (2012)
3. D.G. Fredlund, H. Rahardjo, M.D. Fredlund, *Unsaturated Soil Mechanics in Engineering Practice* (John Wiley & Sons, 2012)
4. K.S. Perkins, *Measurement and Modeling of Unsaturated Hydraulic Conductivity* (INTECH Open Access Publisher, 2011)
5. R.A. Poluektov, V.V. Terleev, *In: Modelling water and nutrient dynamics in soil-crop systems*, 75-89 (Springer, 2007)
6. R.A. Poluektov, I.V. Oparina, V.V. Terleev, Russ meteorol hydro+ **11**, 61-67 (2003)
7. R.A. Poluektov, V.V. Terleev, Russ meteorol hydro+ **11**, 70-75 (2002)
8. S. Medvedev, A. Topaj, V. Badenko, V. Terleev, IFIP AICT, **448**, 252-261 (2015)
9. V.V. Terleev, A.G. Topaj, W. Mirschel, Russ meteorol hydro+ **40,** 278-285 (2015)
10. V.V. Terleev, W. Mirschel, U. Schindler, K.-O. Wenkel, INT AGROPHYS. **24**, 381-387 (2010)
11. R.A. Poluektov, S.M. Fintushal, I.V. Oparina, D.V. Shatskikh, V.V. Terleev, E.T. Zakharova, Arch. Acker- Pfl. Boden. **48**, 609-635 (2002)
12. V. Badenko, V. Terleev, A. Topaj, AMM, **635-637,** 1688-1691 (2014)
13. V. Badenko, V. Terleev, N. Arefiev, J. Volkova, O. Nikonova, *Proceedings of AASRI International Conference on Industrial Electronics and Applications* (IEA 2015). Book Series: *AER-Advances in Engineering Research*, **2,** 452-455 (2015)
14. N. Arefiev, V. Terleev, V. Badenko, Procedia Eng. **117,** 39-44 (2015)
15. N. Arefiev. M. Mikhalev, D. Zotov, K. Zotov, N. Vatin, O. Nikonova, O. Skvortsova, S. Pavlov, T. Chashina, T. Kuchurina, V. Terleev, V. Badenko, Y. Volkova, V. Salikov, K. Strelets, M. Petrochenko, A. Rechinsky, Procedia Eng. **117**, 32-38 (2015)
16. N. Arefiev, V. Badenko, A. Nikonorov, V. Terleev, Y. Volkova, Procedia Eng. **117**, 20-25 (2015)
17. A. Nikonorov, S. Pavlov, V. Terleev, N. Arefiev, V. Badenko, Y. Volkova, Procedia Eng. **117**, 258-263 (2015)
18. Y. Mualem, Water Resour. Res. **12**, 513-522 (1976)
19. M. Th. Van Genuchten, Soil Sci. Soc. Amer. J. **44**, 892-989 (1980)
20. K. Kosugi, Water Resour. Res. **30**, 891-901 (1994)
21. K. Kosugi, Water Resour. Res. **32**, 2697-2703 (1996)
22. V. Terleev, V. Badenko, I. Guseva, W. Mirschel, AMM, **725-726**, 355-360 (2015)
23. V.V. Terleev, M.A. Narbut, A.G. Topaj, W. Mirschel, Agrofizika (Rus) **2,** 35-44 (2014)
24. Information from http://www.cv.nctu.edu.tw/chinese/teacher/Ppt-pdf/AGTwk6HydraulicCon.pdf
25. R.A. Usmanov, N.I. Vatin, V.A. Murgul, AMM, **680**, 474-480 (2014)
26. V. Chechevichkin, N. Vatin, AMM, **641-642**, 409-415 (2014)

MATEC Web of Conferences **53**, 01014 (2016)
DOI: 10.1051/matecconf/20165301014

The Classification of Accidental Situations' Scenarios on Hydropower Plants

Olga Skvortsova[1], Anastasiia Dashkina[1,a], Ekaterina Petrovskaia[1], Vitaly Terleev[1], Aleksandr Nikonorov[1], Vladimir Badenko[1], Yulia Volkova[1] and Sergey Pavlov[1]

[1]*Peter the Great St.Petersburg Polytechnic University, Polytechnicheskaya, 29, St. Petersburg 195251, Russia*

Abstract. The collapse of hydropower plant could lead to significant financial, social and environmental damages. Accordingly the issue of safety and reliability is one of major problems of designing, maintenance and operation. The proposed solution consists in evaluation of implementation frequency of emergencies scenario, which can be defined by suggested classification of accidental situations' scenarios based on systems analysis to implement incidents on hydropower plants. Emergency scenarios on hydropower plants were proposed to classify by three criteria: the cause, place and character

1 Introduction

The issue of reliability and safety of hydraulic structures is paid much attention both in Russia and abroad. Particularly this problem became acute because of the biggest hydro accident in the history of modern Russia, which took place 17 August 2009 on Sayano-Shushenskaya Hydroelectric Power Station.

At present, the main task for the energy is to ensure the safe and reliable operation of hydraulic structures), hydropower plants (complex and unique objects), the collapse of which could lead to significant financial, social and environmental damages [1-9].

This task includes the development by power plants of "Declaration of safety of hydraulic structures". In drawing up the declarations of safety, as stated above, the law requires a quantitative assessment of the level of safety of hydraulic structures. We know that measure of security is a level of risk:

$$R = P \cdot Y, \tag{1}$$

where P - the probability of an emergency at the facility, and Y - the value of the damage caused as a result of an emergency. Therefore, for assessing the safety of the facility, the actual task is determining the scenarios of possible emergency situations on the specific hydraulic structures and determination of the probability of their realization.

A review of literature on the issues of the safety assessment hydropower plant, and determining the number and probability of potential accidents at these sites, can be used to allocate the next methods to resolve it:

- The deterministic method based on the calculation method of limiting states [10-12].

[a]Corresponding author: dashkinanastia@mail.ru

- A probabilistic approach. It is based on a deterministic method of limiting states, but taking into account the random variation, as the parameters of materials and possible impacts and loads [13-14].
- The method of statistical analysis. The basis of this method is the determination of the number of accidents to implement in relation to the number of objects and considered their life [15-17, 7].
- The expert method. It is based on the knowledge and experience of experts to assess the risk of emergency on hydropower plants [18-22].

2 Materials and methods

From our point of view, the closest to the problem of determining the number and the probability of accident scenarios on a particular object a method, based on a complex three-dimensional association, to carry out statistical analysis of actual accidents and expert assessment with use of the general laws of a particular object. To adapt the method of expert assessment to the definition of the probability of the causes of accidents at a particular site needs to develop scenarios of possible emergencies [23-26].

All possible scenarios of hydroelectric facilities on emergency situations in the future are treated as a single set (system). Representing the whole set of particular groups, i.e. classification, is necessary to select the features that would be independent and common to $\rho\left(X^1_{j^1} = const; X^2_{j^2} = const; X^3_{j^3}\right) = \dfrac{N\left(X^1_{j^1} = const; X^2_{j^2} = const; X^3_{j^3}\right)}{N_0}$ all scenarios analyzable set. Then, the principle is the principle of allocation of classification groups of scenarios with the same values of these attributes:

$$X^1, X^2, \dots, X^i, \dots, X^n. \tag{2}$$

Signs of the scenarios under consideration may be quantitative or qualitative (i.e. their value can be expressed either a number or a set of quality values). Furthermore, they may be discrete or continuous values. An important property is the sign and the range of their possible values.

Then, if the sign of X^i is characterized like the discrete value, it receives a set of discrete values:

$$X^i_1,\ X^i_2, \dots, X^i_{j^i}, \dots, X^i_{j^i_{\max}}, \tag{3}$$

where $J^i_{\max}$ specifies the maximum value $X^i_{j^i}$ attribute and therefore the range of change of X^i attribute. And $X^i_{j^i}$ may be a number, if the attribute belongs to quantitative estimation, or it could be qualitative parameter on an attribute scale if it is quality.

Based on the above, the whole set of scenarios can be divided into separate groups, all emergencies within these groups are characterized by the same value of the parameters of classification, i.e.

$$X^i_j = const (i = 1,., n, where\ j^i = 1, \dots, j^i_{\max}). \tag{4}$$

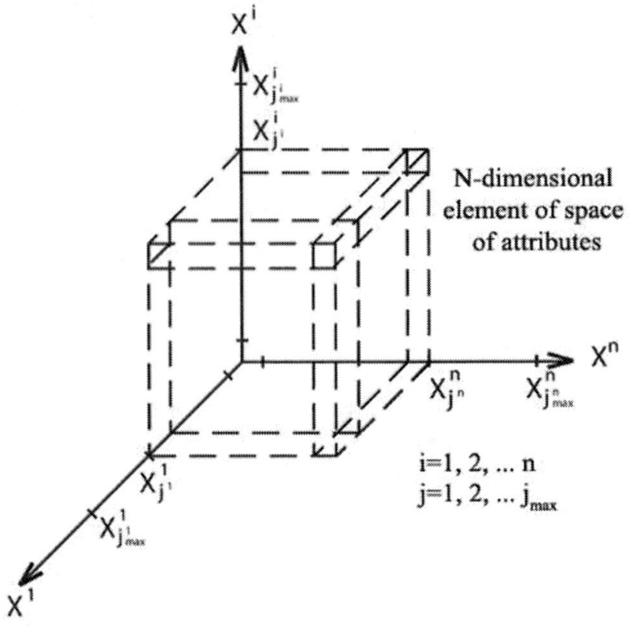

Figure 1. Discrete n-dimensional space of emergencies attributes.

Setting the range of change parameters (J_i = 1,..., J_{imax}, for i= 1,...,n) allows you to calculate the volume V_n (number of elements) all n- dimensional space of attributes:

$$V_n = J^1_{\max}, J^2_{\max}, \ldots, J^i_{\max}, \ldots, J^n_{\max} = \prod_{i=1}^{n} J^i_{\max}. \tag{5}$$

The introduction of the concept of space of attributes allows determining the coordinates of each element of this space: $(X^1_{j^1}, \ldots, X^i_{j^i}, \ldots, X^n_{j^n})$ and thereby determine the position of any class of accident scenarios. The latter allows implementing the mathematical formalization of operations with such phenomena as the development of processes of accidents. Introduce the function for the discrete space of attributes:

$$P_0\,(X^1_{j1}, \ldots X^{i''}_{j''}, \ldots X^n_{jn}). \tag{6}$$

This determines the probability of implementation of groups of scenarios of accidental situations, belongs to an element of the space of coordinates $X^1_{j^1}, \ldots, X^i_{j^i}, \ldots, X^n_{j^n}$. Then, knowing the value of n and $J^{i''}_{\max}$, we can estimate the probability of the realization of all accident scenarios on hydropower plants. Obviously, the implementation of accident scenario depends on so many factors, that any attempt to obtain an analytic expression for the function P_0 is not feasible at the moment.

Therefore, the only possible solution of the problem is the way of collecting data on the practice of accidental situations, their classification by attributes X^i and calculating the number of implementations belongs to the elements of PP (the path of experienced tabulation of the function P_0).

Let $N(X^1_{j1}, \ldots X^i_{ji}, \ldots X^n_{jn})$ be the number of accident scenarios with attributes $X^1_{j1}, \ldots X^i_{ji}, \ldots X^n_{jn}$ belong to the PP element with same coordinates. Then the relation:

$$\rho = \frac{N(X^1_{j1},...X^i_{ji},...X^n_{jn})}{N_0}, \tag{7}$$

shows the relative frequency implementations of process with coordinates ($X^1_{j1},..., X^i_{ji},..., X^n_{jn}$) to the total number N_o. of considered accidental scenarios. Obviously, the value of the function ρ will strive to values of function $P_0(X^1_{j1},...,X^i_{ji},...,X^n_{jn})$ by increasing the number N_o:

$$P_0(X^1_{j1},..., X^i_{ji},...,X^n_{jn}) \Rightarrow \lim_{N \Rightarrow \infty} \frac{N(X^1_{j1},..., X^i_{ji},...,X^n_{jn})}{N_0}. \tag{8}$$

3 Results and Discussion

Thus, to solve the problem of estimating the frequency of implementation of accident scenarios it is necessary to:

Identify the main attributes, which will define the classification of the scenarios of processes of accidental situations; determine the number of the scenario of accident situation that belongs to one or another element of space of attributes (practically, it is an empirical tabulation of function P_0).

Since the considered information from data bank is the expert (mixed) mode (requires non-algorithmic knowledge), then for analysis it is extremely important to have an emergency circuit, as well as photographs of the failure or accidents and their consequences. The analysis of the distribution of emergency situations, depending on their scene, mode, etc. is important for designers and operating personnel sector. Distribution of accidents according to the main reasons for their occurrence is shown in the corresponding histogram.

From the histogram it can be seen that the main reasons are faults in design (17%), river floods (12%), filtering (9%), defects in construction and equipment (6%). Therefore, it is extremely important to raise the quality of the design and provide additional measures of precaution against floods and filtering [27, 28]. These studies are, of course, give a far from complete picture, however, it is possible to objectively identify any critical points in the design, construction and exploitation of hydroelectric power plants.

To carry out the systematic analysis of real accidents to identify common, independent marks of classification three following attributes can be selected:

1) $X^{i=1}: X^1$ — cause of accident (i=1);
2) $X^{i=2}: X^2$ — accident scene (i=2);
3) $X^{i=3}: X^3$ — accident mode (i=3).

Thus, the system of accidental scenarios on the hydroelectric power plants is proposed to be classified according to three criteria: the cause, scene and mode. The numbers J^1_{max}, J^2_{max} and J^3_{max} at the figure define the number of values taken by the attributes X^1, X^2, X^3. In accordance with the classification selected features, scenarios of emergency situations of the entire system will be described by the three-dimensional discrete elements, orthogonal space (see. figure 1). The total amount of attribute's space, containing all the possible scenarios of real emergencies, according to the formula (6) is equal to $V_3 = J^1_{max} \cdot J^2_{max} \cdot J^3_{max} = 22 \cdot 17 \cdot 19 = 7106$.

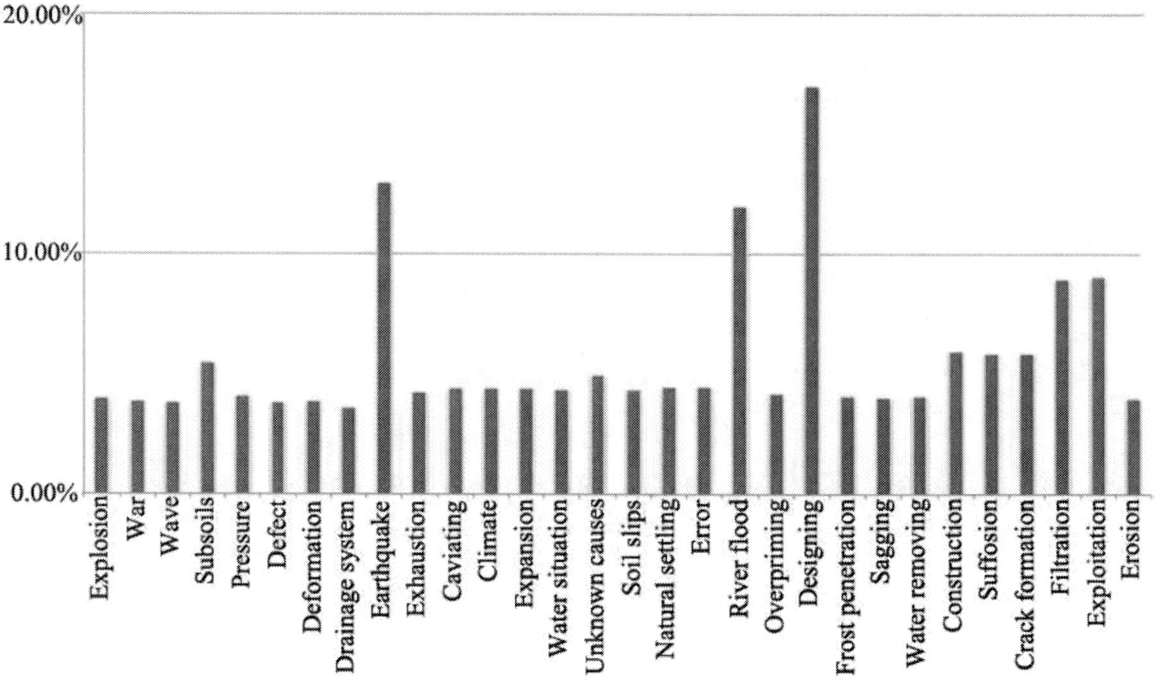

Figure 2. Histogram distribution of accidental situations on hydropower plants depending on the cause of the accident.

The distribution function of the probability of accidents on the three-dimensional feature space elements will look like $P_0\left(X^1_{j1}, X^2_{j2}, X^3_{j3}\right)$, where $X^1_{j-1}, X^2_{j-2}, X^3_{j-3}$ — the coordinates, including all the scenarios of accidents with these values of classification.

$$\rho(X^1_{j1}, X^2_{j2}, X^3_{j3}) = \frac{N(X^1_{j1}, X^2_{j2}, X^3_{j3})}{N_0}. \tag{9}$$

Here $N(X^1_{j1}, X^2_{j2}, X^3_{j3})$ — the number of actual ongoing accident scenarios with the attributes: $X^1_{j1}, X^2_{j2}, X^3_{j3}$; N_0 — the total number of accidental scenarios; $\rho(X^1_{j1}, X^2_{j2}, X^3_{j3})$ — the relative number (frequency) of unrealized accidental scenarios with the attributes: $X^1_{j1}, X^2_{j2}, X^3_{j3}$.

Numerical values of N and N_0 are determined by analyzing a database of real emergencies, "GTA". The algorithm of this calculation is represented by the following operations.

- All cases with one of the characteristic values X^1 (cause): X^1_{j1} are selected.
- From the resulting array at paragraph 1 accident scenarios all the processes with the value X^2_{j2} attribute X^2 (the accident site) are selected.
- From the array of implementation of accident scenarios, received from operations carried out in paragraph 2, with the help of an automated data system distribution of processes emergencies histogram can be built on the basis of X^3 (mode of the accident). The result is a number of emergency scenarios implemented in each element of the column $X^1_{j1} = const; X^2_{j2} = const$ at the space of attributes distributed by characteristic values of attribute X^3: $X^3_{j3}(J^3 = 1, 2, \ldots, J^3_{max})$, which means that numbers $N(X^1_{j1} = const; X^2_{j2} = const; X^3_{j3})$ are obtained.

- By dividing the numbers N, obtained in step 3 by the total number of pending emergency situations N_0 can be founded the desired frequency of implementations of processes emergencies:

$$\rho(X^1_{j1} = const; X^2_{j2} = const; X^3_{j3}) = \frac{N(X^1_{j1} = const; X^2_{j2} = const; X^3_{j3})}{N_0}, \tag{10}$$

i.e. distribution of processes' mode is taking place at scene $X^2_{j^2} = const.$ due to the cause $X^1_{j^1} = const.$

- To get the frequency ρ distribution of the character and place of accidents that have occurred due to $X^1_{j^1} = const$ it is necessary to perform operation of paragraphs 3 and 4 of this algorithm cyclically for all characteristic values X^2: $X^2_{j^2}(j^2 = 1, 2, \ldots, j^2_{\max})$.
- For the implementation of the processes of frequencies emergencies around the three-dimensional space of attributes it is necessary to cyclically repeat steps 1-5 of the algorithm for all the characteristic values of $X^1 : X^1_{j^1}(j^1 = 1,2,\ldots,j^1_{\max})$.

The result of this algorithm gives a list of really existent classes of accident scenarios and the frequency of their implementation. The algorithm described above allows receiving function $\rho\ (X^1_{j^1}, X^2_{j^2}, X^3_{j^3})$ in three-dimensional space of attributes, which defines as the class of scenarios of emergency situations as the frequency of their implementation [29-34].

The function $\rho\ (X^1_{j^1}, X^2_{j^2}, X^3_{j^3})$ determines general regularities of processes of accidental situation's scenario development on hydroelectric power stations. But, obviously, it can't take in account the individual distinctive features of each particular object.

4 Conclusions

So here is the question of how to complement existing information about typical patterns with data that take into account the specific features of the object. It is impossible to reveal the statistical regularities of emergencies at a particular object, as if it has a large number of accidents it has to be closed. Therefore, the only way to obtain the necessary baseline information is to use the experience and knowledge of adepts, treated by the expert evaluation method.

References

1. G.T. Aronica, A. Candela, P. Fabio, M. Santoro, Phys. Chem. Earth, **42-44**, 119-129 (2012)
2. O.S. Skvortsova, *Proceedings of the Chetvertaya Sankt-Peterburgskaya assambleya molodykh uchenykh i specialistov*, (1999) (rus)
3. V. Chechevichkin, N. Vatin, AMM, **641-642**, 409-415 (2014)
4. A Bayburin, Procedia Eng. **117**, 73-79 (2015)
5. A. Nikonorov, S. Pavlov, V. Terleev, N. Arefiev, V. Badenko, Y. Volkova, Procedia Eng. **117**, 258-263 (2015)
6. V. Badenko, V. Terleev, A. Topaj, AMM, **635-637**, 1688-1691 (2014)
7. S. Medvedev, A. Topaj, V. Badenko, V. Terleev, IFIP AICT, **448**, 252-261 (2015)
8. A. Makarov, A. Mihailova, N. Arefiev, S. Pavlov, T. Chashchina, V. Terleev, V. Badenko, Procedia Eng. **117**, 225-231 (2015)
9. N. Arefiev, V. Garmanov, V. Bogdanov, Y. Ryabov, V. Terleev, V. Badenko, Procedia Eng. **117**, 26-31 (2015)
10. L. You, C. Li, X. Min, T. Xiaolei, Procedia Eng. **28**, 382-38 (2012)
11. H. Apel, A.H. Thieken, B. Merz, G. Blöschl, Nat. Hazards Earth Syst. Sci. **4**, 295-308 (2004)

12. M. Balzannikov, Procedia Eng. **111**, 65-71 (2015)]
13. Robin Fell, *Simposium in Stokholm, June 5-7*, 567-576 (1996)
14. H. Apel, A.H. Thieken, B. Merz, G. Blöschl, Nat. Hazards Earth Syst. Sci. **38**, 79-100 (2006)
15. D.N.D. Hartford, International Journal on Hydropower and Dams **18**, 66-70 (2011)
16. N.V. Arefiev, V.L. Badenko, G.K. Osipov, Power Technology and Engineering **32**, 660-663 (1998)
17. S. Ahmad, S. Simonovic, J COMPUT CIVIL ENG. 190-198 (2000)
18. O.S. Skvortsova, N.V. Arefiev, *Proceedings of the XXIX nedelya nauki SPbGTU* (2001) (rus)
19. O.S. Skvortsova, *Aquaterra Sbornik materialov konferencii* (2000) (rus)
20. N. Arefiev, V Terleev, V. Badenko, Procedia Eng. **117**, 39-44 (2015)
21. N. Arefiev, V. Badenko, A. Nikonorov, V. Terleev, Y. Volkova, Procedia Eng. **117**, 20-25 (2015)
22. Y.G. Motovilov, V.I. Danilov-Danilyan, E.V. Dod, A.S. Kalugin, Water Resources 42, 580-593 (2015)
23. L. Muravyeva, N. Vatin, AMM, **635-637**, 468-471 (2014)
24. L. Muravyeva, N. Vatin, AMM, **633-634**, 1042-1046 (2014)
25. X. Zhou, J. Zhou, X. Du, Y. Wang, Journal of Hydroelectric Engineering **34**, 63-72 (2014)
26. N. Arefiev. M. Mikhalev, D. Zotov, K. Zotov, N. Vatin, O. Nikonova, O. Skvortsova, S. Pavlov, T. Chashina, T. Kuchurina, V. Terleev, V. Badenko, Y. Volkova, V. Salikov, K. Strelets, M. Petrochenko, A. Rechinsky, Procedia Eng. **117**, 32-38 (2015)
27. V. Badenko, V. Terleev, N. Arefiev, J. Volkova, O. Nikonova. Proceedings of the AASRI International Conference on Industrial Electronics and Applications (IEA 2015). Book Series: *AER-Advances in Engineering Research*. **2**, 452-455 (2015)
28. V.V. Terleev, A.G. Topaj, W. Mirschel, Russ meteorol hydro+ **40**, 278-285 (2015)
29. R.A. Poluektov, S.M. Fintushal, I.V. Oparina, D.V. Shatskikh, V.V. Terleev, E.T. Zakharova, Arch. Acker- Pfl. Boden. **48**, 609-635 (2002)
30. R.A. Poluektov, V.V. Terleev, Russ meteorol hydro+ **11**, 70-75 (2002)
31. R.A. Poluektov, I.V. Oparina, V.V. Terleev, Russ meteorol hydro+ **11**, 61-67 (2003)
32. R.A. Poluektov, V.V. Terleev, Russ meteorol hydro+ **12**, 73-77 (2005)
33. V.V. Terleev, W. Mirschel, U. Schindler, K.-O. Wenkel, INT AGROPHYS. **24**, 381-387 (2010)
34. V. Terleev, V. Badenko, I. Guseva, W. Mirschel, AMM, **725-726**, 355-360 (2015)

MATEC Web of Conferences **53**, 01015 (2016)
DOI: 10.1051/matecconf/20165301015

Relationship Between Structure and Viscoelastic Properties of Geosynthetics

Irina Loginova[1,a], Daria Artamonova[1] and Oleg Stolyarov[1]

[1]*Peter the Great St.Petersburg Polytechnic University, Polytechnicheskaya 29, 195251 St.Petersburg, Russia*

Abstract. In this work, a study on viscoelastic properties of geosynthetic materials used in civil engineering is presented. Six samples of geofabrics and geogrids with different structures including woven geotextile fabric, nonwoven geotextile fabrics, warp-knitted geogrids and extruded geogrid were investigated. The tensile properties of geosynthetics including tensile strength, strain at maximum load and tensile load at specified strain have been determined. The creep and relaxation tests were carried out. The structure type was found to significantly affect the viscoelastic properties of the geosynthetics materials. In the article some results of numerous conducted tests are presented, analyzed and may be used to preselection of geosynthetics materials.

1 Introduction

Geosynthetic materials are widely used in civil engineering applications presenting a special class of modern building structural materials [1-2]. They have found a number of applications and have a great potential in various fields of civil engineering, including the construction of engineering structures, road construction, railway construction, hydraulic engineering, building of sports facilities, etc. Historically this type of materials used in the largest quantities in road construction. Most geosynthetics are man-made materials made from various types of polymers both natural and synthetics used in environmental, transportation and civil engineering projects. Geosynthetics are used in one or more of the following functions: separation; reinforcement; protection; filtration; drainage; erosion control; barrier. Along with the traditional structural materials such as steel, concrete and other building materials, they have high strength properties. However, apart from this positive structural characteristic, geosynthetics have exceptional viscoelastic properties. These ones become a problem when assessing the long-term strength of the structure due to variability force and deformation of material over time. The viscoelastic properties are developed in the area of non-destructive mechanical effects and affect the mechanical performance of geosynthetic materials.

Geosynthetics are a class of materials that have exceptional time-dependent mechanical properties. In many practical cases, geosynthetics and articles made of them are subjected to loading or elongation and such a stressed-strained state may remain for a long time. As experience has shown, the deformation stiffness and consequently their response to subsequent mechanical effects are essentially a function of time and therefore these properties must be taken into account in various tasks. The time-dependent mechanical properties of geosynthetics can be determined in various modes

[a] Corresponding author: iraloginova8@gmail.com

of loading and elongation: creep and recovery tests, stress relaxation and inverse stress relaxation tests as shown in Fig. 1 [4].

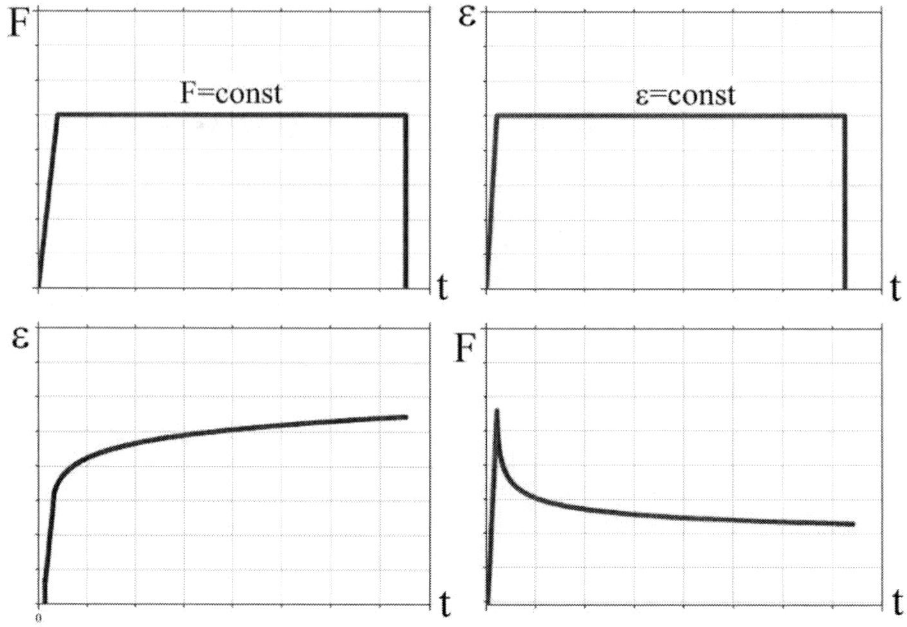

Figure 1. Test mode: (a) creep; (b) stress relaxation.

The viscoelastic properties of geosynthetics have been investigated in detail by many authors. Geosynthetics are characterized by the elastic and viscoelastic properties that bring some difficulties to the design of structures using geosynthetic materials [11]. The study of the properties of geosynthetic materials, which appear over time, is not less important than short-term tests. Viscoelastic properties are also one of the most important assessment criteria of evaluating long-term strength of geosynthetic materials used as reinforcement [12, 13]. The greatest interest in estimation long-term structural durability of geosynthetic materials represents the process of creep, because in real operating conditions a geosynthetic material is subjected to a constant load [14, 15]. Although in some applications the functions of geosynthetics can experience the processes of reducing the internal forces due to the fixation of their size, for example, by anchoring. To determine the stress relaxation throughout the experiment, a constant deformation must be set [16]. In articles [17-20] an investigation of creep based on the measurement of the sample elongation as a function of external tensile force was done. Tests were carried out at different levels of an applied load. Generally, the load steps are chosen as a percent value by maximum tensile strength of geosynthetic material in a particular range, for example, in 5, 10, 20, 30, 40, 50 and 60 % of maximum tensile strength [21].

There are a wide variety of methods in practice for the determination of creep of geosynthetic materials. The simplest method to determine creep of geosynthetic materials is the direct measurement of specimen deformation over a long period of time. The duration of measurements, as a rule, is thousands and tens of thousands of hours [1, 2]. In [14], the creep experiment was conducted during 6 years. Data obtained from such experiments can be used to assess the long-term deformation of geosynthetic material that haven't exceed the maximum allowable deformation [13,14]. The main disadvantage of such measurements is the period of time that requires special equipment, allowing conducting such tests with a big number of samples. The use of such systems significantly increases the cost of testing and doesn't always guarantee the achievement of the reliable result.

Along with direct measurements of the creep accelerated test methods become popular. They allow to activate deformation processes in time by load or elevated temperature [22-25]. Creep measurements carried out with the use of such techniques can significantly reduce the test time by several orders of magnitude and obtain satisfactorily predicted curves. The use of such methods provides a significant advantage in experimental studies, however, requires using of expensive specialized equipment – thermochamber. Besides, these methods depend on the testing conditions and the type of investigated geosynthetic material. So, for example, varying temperature in the step of 1-2 °C leads to substantial differences in the shape of the resulting creep curves.

Methods of creep determination are rather contradictory. On the one hand the direct measurement of creep has an advantage in determining the long-term strength, because it eliminates the influence of various factors, e.g. temperature. On the other hand such tests are time-consumption. In addition, sometimes a large part of materials, which is intended for use in the project, may be rejected on the basis of experimental data. At the same time, application of accelerated tests doesn't always guarantee an adequate outcome. In our opinion, one of the major disadvantages in determining the viscoelastic properties of geosynthetic materials is the lack of express method that would allow identify the viscoelastic properties of geosynthetics for a short period of time. Such information may be used for preliminary selection of geosynthetic material and for more optimal planning of experimental work on the determination of viscoelastic properties. This also may allow reducing cost of tests significantly.

2 Experimental

2.1 Materials

The objective of this work was studying the viscoelastic properties of geosynthetic materials of various structures at short-term measurements in creep and relaxation stress. The main factors that affect the viscoelastic properties of geosynthetics depend on the structure of the material, type of weave in the fabric, raw materials used, etc. For the experiment, six different types of geosynthetics including geotextile fabrics and geogrids have been investigated. Table 1 summarizes the designation and characteristic of samples investigated.

Table 1. Experimental data.

№	Sample Designation	Structure	Raw materials	Surface density, g/m^2
1	1-GTX-PET	Woven geotextile fabric	Polyester	400
2	2-GTX-PP	Nonwoven needle-punched geotextile fabric	Polypropylene	160
3	3-GTX-PP	Nonwoven thermobonded geotextile fabric	Polypropylene	160
4	1-GGR-PET	Warp-knitted geogrid with layer of nonwoven fabric, mesh size: 35*35	Polyester	285
5	2-GGR-GL	Warp-knitted geogrid, mesh size: 40*40	Glass, Treatment: Bitumen polymer	300
6	3-GGR-PP	Extruded geogrid, mesh size: 40*40	Polypropylene	530

They include three geotextile samples and three geogrid samples. Geotextile samples consist of woven fabric (denoted as 1-GTX-PET), which have a mass per area of 480 g/m^2 functioning as separation and reinforcement fabric and made of polyester and two types of nonwovens including needle-punched fabric (denoted as 2-GTX-PP) and thermobonded fabrics (denoted as 3-GTX-PP) which have a mass per area of 160 g/m^2 and made of polypropylene functioning as filter and protection layer. Geogrid samples consist of warp-knitted geogrid (denoted as 1-GGR-PET) with layer of nonwoven fabric, the components are sewn together, which have a mass per area of 285 g/m^2 functioning as reinforcement and made of polyester; warp-knitted geogrid (denoted as 2-GGR-GL) made of glass fibers treated with bitumen polymer and extruded plastic geogrid (denoted as 2-GGR-PP) made of polypropylene functioning as reinforcement.

2.2 Tensile test

The tensile test was performed on an Instron 5965 tensile machine. The determination of tensile behavior of geotextiles under uniaxial loading is performed by standard method named strip test. The stress-strain diagrams were obtained with a 100-mm sample base and clamp movement of 100 mm/min. Testing of geogrids was conducted on single ribs. No fever than five specimens in the machine direction (MD) were tested. It should be noted that tensile strains were calculated from crosshead displacement. The tensile diagrams of samples investigated in MD and CMD are presented in figures 1 and 2 respectively. Using the test results, the relative values of the tensile strength normalized to the sample width; strain at maximum load and tensile load at specified strain (2%) were determined from stress-strain curves.

The tensile strength of a geosynthetic is expressed in kilonewtons per meter (kN/m) directly from the data obtained from the tensile testing machine as follows:

$$T_{max} = F_{max} \cdot c \tag{1}$$

Where F_{max} is the recorded maximum load, in kilonewtons; c is the specimen width.
For woven and nonwoven c is determined as follow:

$$c = \frac{1}{B} \tag{2}$$

where B is the specimen nominal width, in meters.
For geogrids c is determined as follow:

$$c = \frac{N_m}{N_s} \tag{3}$$

where N_m is the minimum number of tensile ribs within a 1 m width of the geogrid; N_s is the number of tensile elements within the test specimen.

2.3 Creep and stress relaxation tests

Creep tests and stress relaxation tests are usually carried out at several levels of a given load in percentage of the maximum tensile strength of the sample [26]. In this work the samples were tested for creep and stress relaxation at the load level of 30% of T_{max}, that provides a comparable test for materials of different structures. The test time was 3600 s.

3 Results and discussion

3.1 Tensile properties

Test results of tensile strength, strain at maximum load and tensile load at specified strain (2%) are shown in Figure 2a, Figures 2b, and 2c respectively. The obtained data show that almost all of the investigated structure geosynthetics are isotropic in strength, except for sample 1-GTX-PET. This sample has the greatest strength in the machine direction. Samples of nonwoven fabrics exhibit highest elongation. This sample is thermally bonded fabric (3-GTX-PP) has a similar tensile strength to the needle-punched fabric (3-GTX-PP) at a lower surface density. It should also be noted that glass-fiber geogrid (2-GGR-GL) has the lowest elongation. As it can be seen from the results geosynthetics have a wide range of mechanical properties.

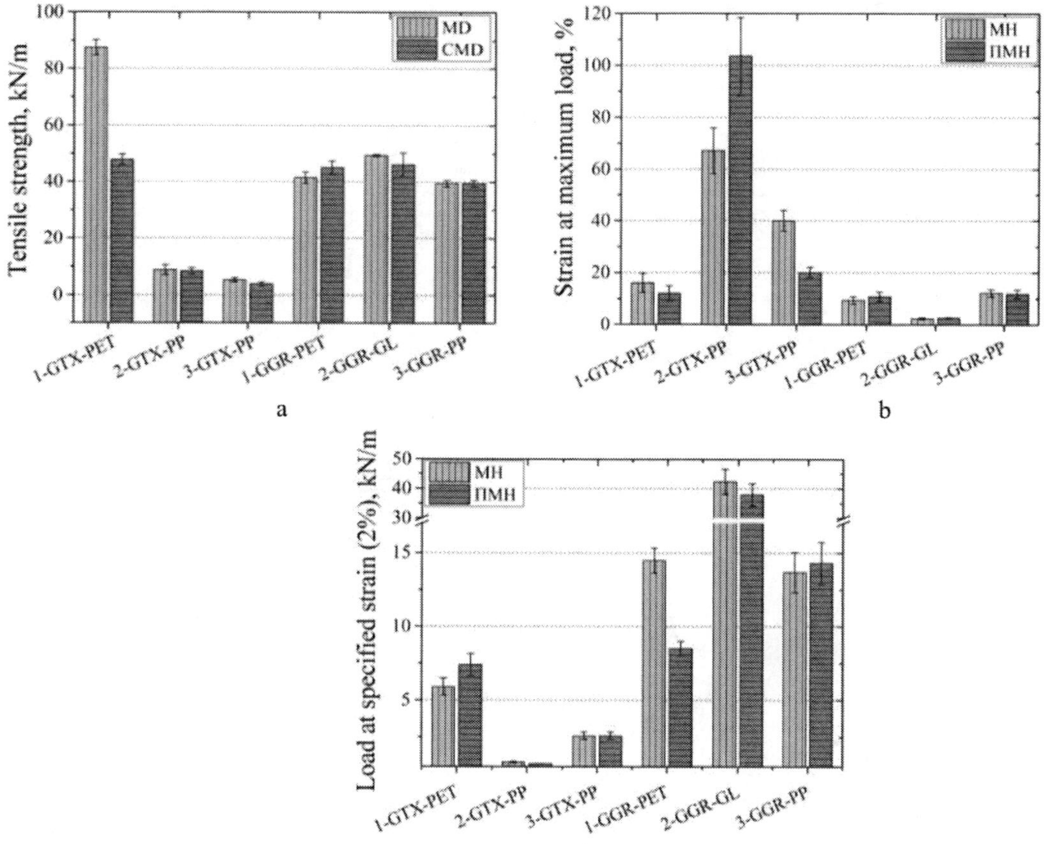

Figure 2. Results of tensile test: tensile strength (a), strain at maximum load (b) and tensile load at specified strain of 2% (b).

3.2 Viscoelastic properties

Creep and stress relaxation curves under load level of 30% of the T_{max} for all of the samples are shown in Figure 3a and 3b, respectively. From these curves it is clear that all of the materials exhibit viscoelastic properties. Samples of nonwoven fabrics show the greatest creep due to their relatively high compliance. At the same time the stiffest sample of glass-fiber geogrid (2-GGR-GL) showed no change in strain with time. Its slight increase in strain may be caused by creep of a bituminous coating adhesion to glass fiber. The behavior of the samples during stress relaxation test was found similar.

In order to evaluate the effect of the structure on the viscoelastic properties of geosynthetic materials. Two characteristics of creep and stress relaxation were analyzed. The creep behaviour was characterized by creep rate (s^{-1}), increment of deformation over time and determined by the following equation:

$$\dot{\varepsilon} = \frac{\varepsilon(t_2) - \varepsilon(t_1)}{t_2 - t_1} \tag{4}$$

where $\varepsilon(t_i)$ is the creep strain at i - measured point, %; t_i is the time, h.

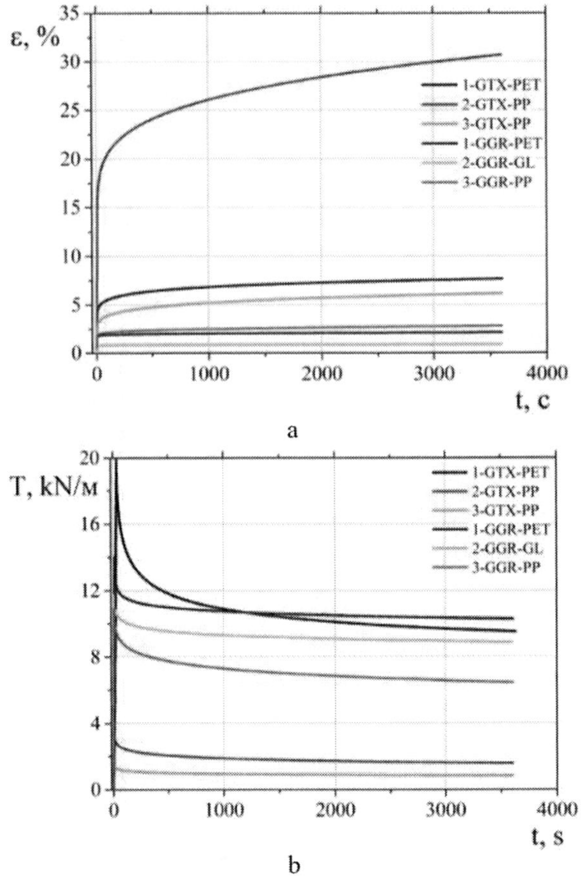

Figure 3. (a) test creep; (b) stress relaxation.

To characterize the behavior of the samples in the relaxation, the residual stress was determined by the following equation:

$$T_r = \frac{T_\kappa}{T_0} \tag{5}$$

whereT_κ –force in the sample at t = 3600 s (kN / m), T_0 - initial sample load (kN/m).

The obtained results have been plotted compare to the load at an elongation of 2%. Figure 4a shows the variation in creep rate as a function of $T_{2\%}$. The obtained data show that with increasing the stiffness of the material, creep rate decreases dramatically. The more compliance material, for example, samples of the nonwoven fabrics (3-GTX-PP and 3-GTX-PP), have less resistance to creep. The minimum creep rate, and as a result, the maximum stiffness has glass-fiber geogrid (2-GGR-GL). Also samples of other geogrids have a good creep resistance, which determines their reinforcement function. An exponential curve can be fit to resulting plot. The plateau was found at an infinitely large stiffness of the material. Figure 4b shows the results of calculations of residual relaxation stresses of the experiment, depending on the $T_{2\%}$. The decreasing loss of stress in time was found with the increase of the stiffness of material. Samples of geotextile show greatest loss of stress. Further, with an increase in stiffness of the sample, a decrease in the relaxation time was observed. The test results on the relaxation are well-correlated with the creep test. Samples with high compliance have a high creep rate and a significant reduction in stress at a given constant strain, and vice versa, the most severe examples of geogrids well withstand mechanical stress. It should also be noted that the samples of nonwoven fabrics used as a function of separation and/or filtration, and not reinforcement. However, an excessively high strain can lead to fracture in their structure.

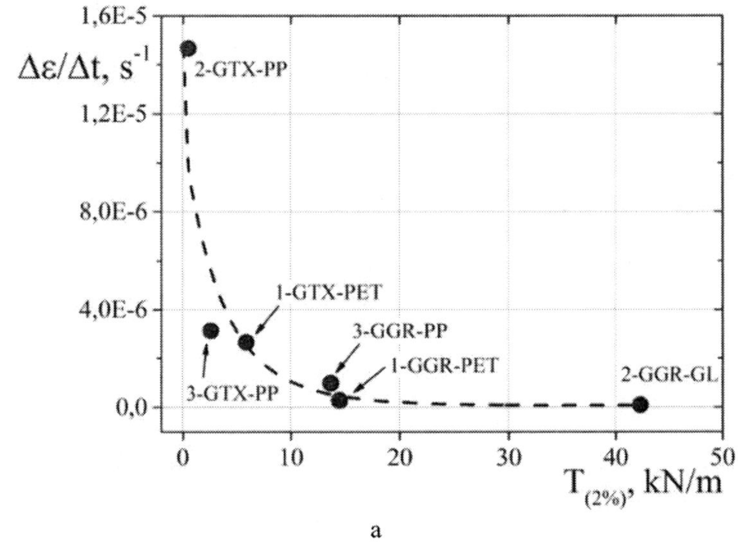

a

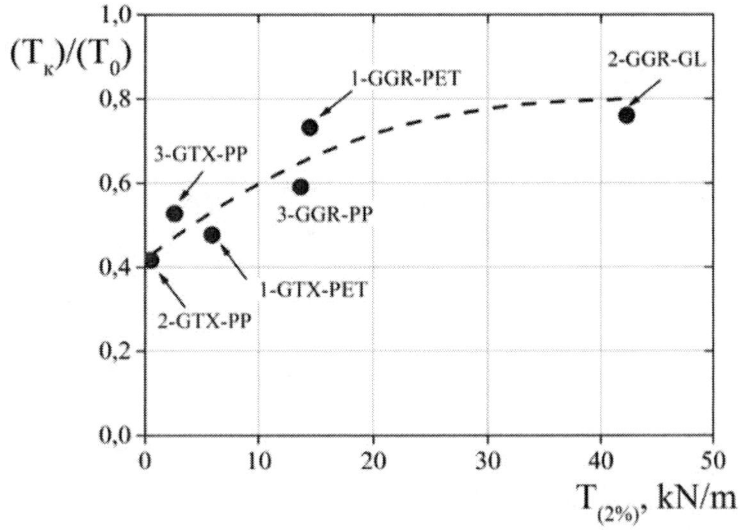

b

Figure 4. Dependences of: (a) the creep rate; (b) residential stresses relaxation of the experiment depending on the $T_{2\%}$

4 Conclusions

The tensile and viscoelastic properties of various geosynthetic materials including woven geotextile fabrics, nonwoven geotextile fabrics, warp-knitted geogrids and extruded geogrids were determined.

The viscoelastic properties were investigated with laboratory measurements in creep and stress relaxation processes at the load level of 30% of maximal tensile strength. The tensile properties of geosynthetics including tensile strength, strain at maximum load and tensile load at specified strain have been determined. It was found that the viscoelastic properties of various geosynthetic materials significantly vary.

Dependences of the creep rate and residential stresses relaxation of the experiment depending on the $T_{2\%}$ was determined. Results shows that the structure and polymer type significantly affect the viscoelastic properties of the geosynthetics materials. It can be used for goal-oriented solution of problems of the manufacturing technology of geosynthetic materials and the study of their properties in use.

References

1. O. Stolyarov, A. Gorshkov, Magazine of Civil Engineering, **4**, 21-25 (2009)
2. N. Ustyan, Magazine of Civil Engineering, **4**, 22-25 (2011)
3. V.V Okrepilov, Studies on Russian Economic Development, **24(1)**, 35-42 (2013)
4. V.V Okrepilov, Standarty i Kachestvo, **10**, 52-55 (2005)
5. F.A.N. França, B.S. Bueno, Proceedings of the 9th International Conference on Geosynthetics – Geosynthetics, 789-792 (2010)
6. V.V Okrepilov, Standarty i Kachestvo, **3**, 94-96. (2003)
7. B.S. Bueno, M.A. Costanzi, J.G. Zornberg, Geosynthetics International, **12(6)**, 276-287 (2005)
8. R.M. Koerner, G.R. Koerner, Geosynthetics Research and Development in Progress. American Society of Civil Engineers, **34**, 4259-4264 (2005)
9. J.T.H. Wu, S.M.B. Helwany, Geosynthetics International, **3(1)**, 107-124 (1996)
10. S. Xu, Zhang, Y.Z. Wang, 2011 International Conference on Structures and Building Materials. Trans Tech Publications, 1572-1576 (2011)
11. A.M.V. Paula, M. Pinho-Lopes, M.L. Lopes, 9th ICG, **45**, 781-784 (2010)
12. C.T Gnanendran, G. Manivannan, S.R. Lo, Canadian Geotechnical Journal, **43(2)**, 134-154 (2006)
13. R.M. Koerner, Y.G. Hsuan, G.R. Koerner, D. Gryger Geotextiles and Geomembranes, **28(6)**, 503-513
14. S.W. Perkins, Geotextiles and Geomembranes, **18(5)**, 273-292 (2000)
15. A.M. Stalevich, *Straining of oriented polymers St.Petersburg* (SPBGUTD, St.Petersburg (2002)
16. J.G. Zornberg, B.R. Byler, J.W. Knudsen, Journal of Geotechnical and Geoenvironmental Engineering, **130(11)**, 1158-1168 (2004)
17. J.T.H. Wu, S.M.B. Helwany, Geosynthetics International, **3(1)**, 107-124 (1996)
18. H.-Y. Jeon, Proceedings of the Institution of Civil Engineers: Ground Improvement, **163(4)**, 189-195 (2010)
19. S.-S. Yeo, Y.G. Hsuan, Geotextiles and Geomembranes, **28(5)**, 409-421(2010)
20. H. Yoo, H.-Y. Jeon, Y.-C. Chang, Textile Research Journal, **80(2)**, 184-192(2010)
21. S.E. Scarborough, T. Fredrickson, D.P. Cadogan, G. Baird, International SAMPE Symposium and Exhibition, **52**, 18(2008)
22. R.M. Koerner, G.N. Richardson, N.E. Wrigley, D.I. Bush, G. den Hoedt, Geotextiles and Geomembranes, **5(4)** , 304-306(1987)
23. T.L. Baker, J.S. Thornton, Geosynthetics Conference , pp. 729-740(2001)
24. Chiwan Wayne Hsiehl , Kwangyeol Lee , Han Kyu Yoo , Hanyong Jeon, Fibers and Polymers. The Korean Fiber Society, **9(4)**, 476-480 (2008)
25. A.M. Stalevich, E.V. Kikets, O.N. Stolyarov, E.D. Saidov, Fibre Chemistry, **35(2)**, 164 (2003).
26. GOST R 56339-2015 Automobile roads of general use. Geosynthetic for road construction. Determination of the Tensile Creep and Creep Rupture Behavior (2015)

MATEC Web of Conferences **53**, 01016 (2016)
DOI: 10.1051/matecconf/20165301016

Flexural Behavior of Textile-Reinforced Concrete

Anna Volkova [1, a], Alexey Paykov[1], Sergey Semenov[1], Oleg Stolyarov[1] and Boris Melnikov[1]

Peter the Great St.Petersburg Polytechnic University, Polytechnicheskaya 29, 195251 St.Petersburg, Russia

Abstract. This paper deals with the flexural behaviour of textile-reinforced concrete (TRC). Two samples of TRC made of high strength reinforcing fabrics made of glass and carbon rovings were produced. Three-point bending test was carried out to examine the flexural performance of the developed samples. The maximum flexural strength and reinforcement efficiency were calculated. Experimental results showed that that all types of applied fabric reinforcement contributed to increases strength as compared to nonreinforced concrete. Furthermore, the deformation behavior of reinforced concrete was analyzed. The advantage is in higher residual load-bearing capacity, which allows maintaining the integrity of the structure.

1 Introduction

Different types of high-strength textile materials are finding new applications for the reinforcement of structural parts [1-4]. For their manufacture, high strength and high modulus fibers and filaments, such as basalt, glass, carbon and others are generally used [2]. In combination with the cement matrix, they present a new class of structural building materials - Textile-reinforced concrete. Appeared relatively recently, textile-reinforced concrete is becoming one of the most promising materials used in construction. Compared to traditional building materials textile-reinforced concrete has a number of advantages including high corrosion resistance, minimal concrete covering, light weight, etc. However, its use is often limited by the high cost of initial reinforcement materials, whereby the use of such concrete structures in massive constructions is not economically feasible. Therefore, such reinforcement usually finds effective use in various structural elements, which combine lightweight, small size and strength, for example, sandwich panels, and also can be used in the reconstruction of various concrete structures.

The use of high-strength textile materials for the reinforcement of concrete structures has many advantages, one of which is the production of concrete elements sufficiently thin compared to conventional concrete. In addition to this entire textile reinforcement has greater flexibility and drape.

[a] Corresponding author: volkova-anna-andreevna@mail.ru

The reinforcing fibers for reinforced concrete include alkali-resistant glass fibers (AR-glass), also basalt and carbon fibers. These fibrous fillers may be used in the form of chopped (staple) fibers or in the form of fabric with open cells structure. Unlike discrete fibrous fillers, the yarn and rovings made of continuous filaments in the form of textile fabric can fully implement the properties of high strength and high-modulus filaments. The mechanism of the reinforcing fibers in the cement matrix is greatly different from that which occurs in the polymer matrix because of the lower ultimate elongation of the cement matrix, which is significantly less than the elongation of the fibers. Therefore, in the cement composites, the matrix breaks up before the strength properties of the fibrous reinforcement will be fully realized. In polymer composites, the matrix is generally more flexible than the fiber. A significant advantage is also the fact that the fiber will be able to sustain the load after the formation of cracks and fracture of the matrix.

Fabric reinforced concrete is a planar or spacer structure consisting of a particular geometrical mesh shape to facilitate penetration of the cement into the structure. The form of mesh cells of reinforcing fabric may have different shapes: rectangle, diamond, triangle, and others. The size and shape of cells may be different. The most common rectangular cells have sizes of 8, 10, 16 and 20 mm. Bonding to the concrete matrix, and as a consequence, the mechanical properties of reinforcement depend on the mesh size of the fabric. The larger cell size leads to greater probability that there will not be any cavities inside of the concrete reducing its strength characteristics. Over the past two decades, many works have been devoted to this topic. In work [3], the systematic approach to the study of concrete reinforced fiber composites is described. In general, it comprises:

1) The test material;
2) Analysis of the properties;
3) The calculation of the best combination of strength and fracture toughness; Statistical analyses and calculations of strength.

In [4-6], the method of measurement and application of structures of reinforced concrete composites and method of production of reinforcement fibers and fabrics are presented. The dependence of the strength characteristics of the composite on the structural parameters of the reinforcement rovings and fabrics were considered in [7-9]. In [10, 11], the use of polymers in a composite concrete is described. The possibility of using carbon fiber materials and composites based on them in concrete structures was considered in [12]. Various methods for the reinforcement of composites, such as the casting of concrete into the mold of the fabric and using a mixture of steel and non-metallic fibers are discussed in [13, 14]. In studies [15-18], various models, such as the computer model of the composite structure of concrete, models describing the behavior of the structure under the influence of prolonged stress and corrosion were analysed. Existing methods to strengthen reinforced concrete superstructure, including the strengthening of composite materials based on carbon fibers in the polymer matrix are investigated in [19]. Increased durability of concrete structures can also be accomplished using prestressed textile fabrics described in [20, 21].

A review of the literature has shown that the main advantages of using high-strength reinforcing textiles consist in the increasing strength characteristics, particularly in cases of large loads on thin elements); improving durability and reducing the costs of structures and other materials used in the construction to obtain the same strength characteristics as in the application of composites; an increase in the carrying capacity of the structural elements; increased fracture toughness, and smaller cracks.

2 Experimental

2.1 Materials

The main aim of this research is to study the possibility of using high-strength reinforcing fabrics made of glass and carbon rovings for textile reinforced concrete. In this work, two types of warp-knitted reinforcing fabrics made of glass and carbon rovings were used. The characteristics of the specimens are presented in Tab. 1.

Table 1. Characteristics of reinforcing materials.

№	Reinforcing roving				Fabric	
	Machine direction	Cross-machine (transverse) direction	Tensile strength, MPa	Young's Modulus, MPa	Cell size	Surface density, g /m2
1	AR-glass, 4K, 2400 tex	AR-glass	1024,0±60,3	66,5±2,0	8×8 mm	620
2	Carbon (PAN-based fiber), 24K, 1600 tex		2093,0±73,4	173,0±4,3	8×8 mm	510

The structure of the reinforcing fabric presents an open-grid structure with rectangular cells created by the roving laid in two directions connected with thin knitting polyester yarn with a linear density of 15.6 tex. The fabric included a glass (specimen 1) or carbon (specimen 2) rovings in longitudinal direction, and only glass rovings in the transverse direction. The mesh size was 8x8 mm, with 50 mm accounted for 6 rovings. The structures used in the reinforcing fabrics of glass and carbon rovings are shown in Fig. 1a and 1b, respectively.

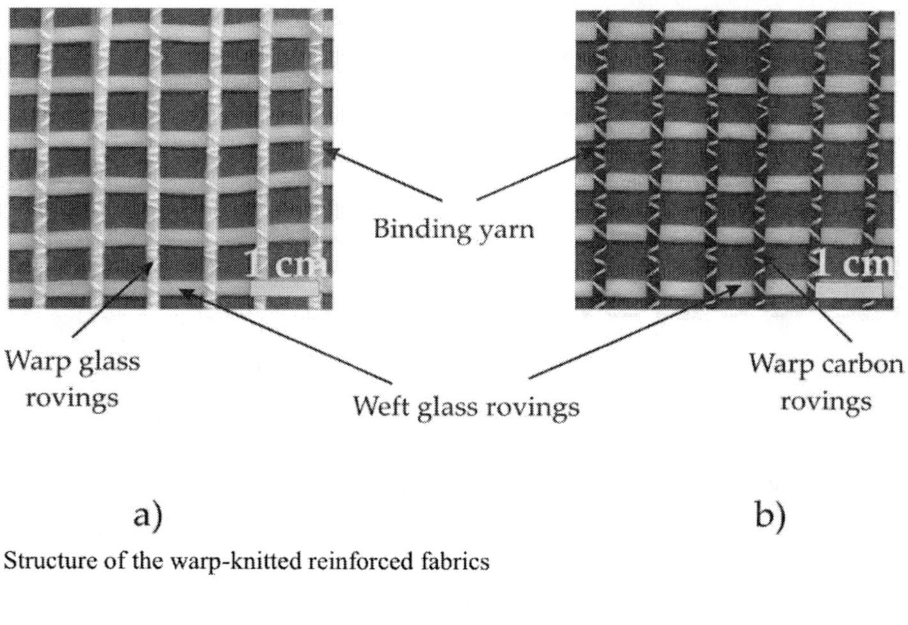

Figure 1. Structure of the warp-knitted reinforced fabrics

2.2 Fabrication of textile-reinforced concrete

Fine-grained concrete with a maximal sand fraction of 0.63 mm was used to manufacture the concrete composite samples. The specimens were casted in the mold (56x200x20 mm). Two samples of the reinforced fabrics were placed in the tension and compression areas at a distance of 2 mm from each edge. The test samples were stored for 28 days at 23°C and 95% RH before testing.

2.3 Three-point bending test of concrete

Samples of textile-reinforced concrete were tested in a three-point bending test (Fig. 2). The flexural tests were carried out in a Instron 5965 testing machine at a clear span of 150 mm. The specimens were constantly loaded at a rate of 1mm/min.

Figure 2. Three-point bending test of concrete

According to the test results the bending strength was determined as the ratio of the maximum bending moment of the specimen to the axial section modulus in bending by the equation

$$\sigma = \frac{M_b}{W_z}, \tag{1}$$

where M_b - maximum bending moment, kN * m; W_z – moment of inertia, m^3.

3 Results and discussions

3.1 Flexural behaviour

Fig. 3 shows the stress-deflection curves of three-point bending test of reinforced cement composites. From the diagrams it is clear that the behavior of reinforced and non-reinforced specimens is different. Moreover, the flexural behaviour greatly depends on the type of the reinforcing roving. The samples of textile-reinforced concrete show similar flexural behavior. The initial part of the deformation is characterized by a linearity of properties with the subsequent transition to a section with a slight non-linearity. After this there is a first transverse crack. In the case of AR-glass-reinforced specimens, magnitude of this load is the maximum on the stress-deflection curve. A specimen reinforced with carbon rovings shows flexural load corresponding to the first crack of approximately equal to the strength of nonreinforced concrete. Then, an increase in bending stress due to the load-bearing capacity of carbon fibers is observed. This process occurs in steps to produce a few more of the maximum peaks.

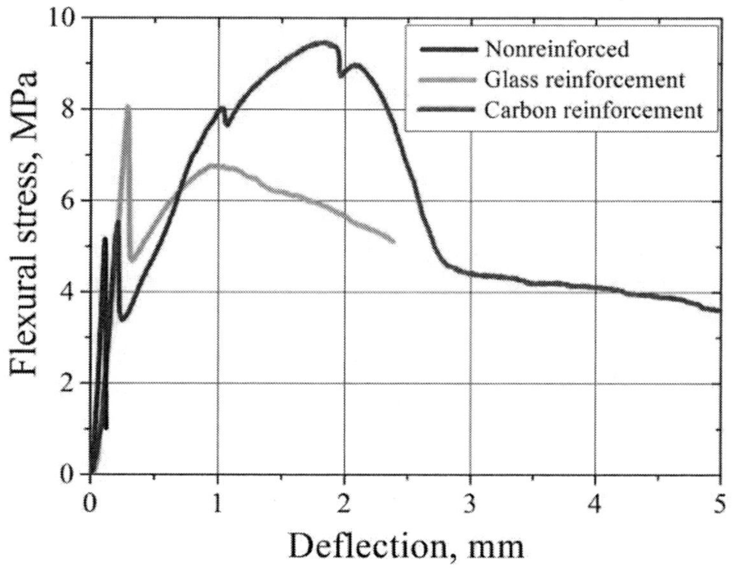

Figure 3. The stress-deflection curve of textile reinforced concrete with different roving type

Results of flexural strength are shown in Fig. 4. During the analysis of the results, it can be noted that the bending strength of reinforced concrete is 5.2 MPa. Application of reinforcing fabric gives a certain increase in flexural strength. The strength of the specimens reinforced with AR-glass is 8.1 MPa, and for specimens reinforced with carbon rovings is 9.5 MPa. However, as can be seen from the curves presented in Fig. 3, the first peak in the stress-deflection curve of the AR-glass specimen is the same as the flexural strength of the non-reinforced concrete specimen. While in the specimen reinforced with carbon roving, the first peak value was about 0.5 of the flexural strength. Further, with increasing deflection, an increase in load due to resist rovings is observed. This ability to maintain the shape and performance without breaking at very large strains can be used in critical areas, such as those associated with the seismic resistance of buildings and structures.

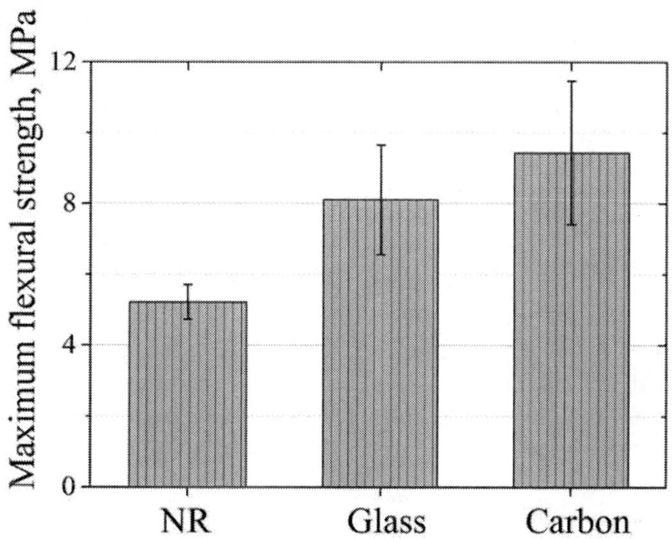

Figure 4. Maximum flexural strength

3.2 Reinforcement efficiency

The reinforcement efficiency of the developed samples may be evaluated by a factor indicating the ratio of reinforced concrete's flexural strength to the flexural strength of the nonreinforced reference sample. This ratio is calculated as

$$k_{RE} = \frac{\sigma_{rd}}{\sigma_{nr}}, \quad (2)$$

where σ_{rd} is the flexural strength of a reinforced concrete (MPa) and σ_{nr} is the flexural strength of a reinforced concrete (MPa).
The results obtained from the Eq. 2 are shown in Fig. 5. As can be seen from the data, the reinforcement efficiency depends on the type of roving. For the specimen manufactured of AR-glass rovings, the flexural strength increased by 1.55 times, and for the specimen manufactured of carbon rovings, an increase in flexural strength is 1.81 times. It is evident that there is a certain effect of reinforcement with applying the high-strength textile fabrics. However, by the results of tests it is difficult to evaluate the validity of the results, because the scatter of values for flexural strength for reinforced specimens is very significant. This is well illustrated by the standard deviation at the histograms for specimens of reinforced concrete.

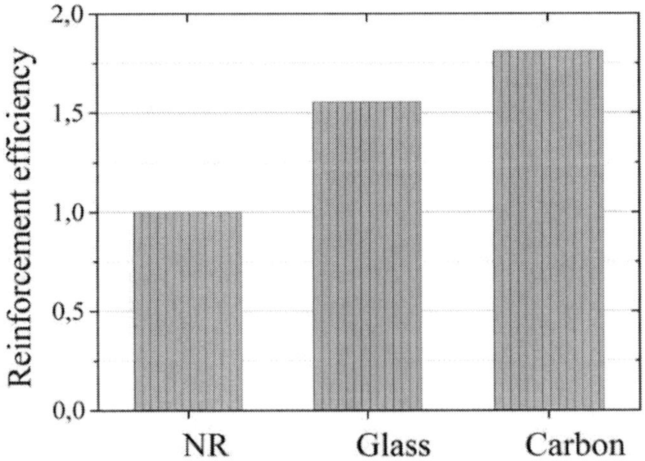

Figure 5. Maximum flexural strength (a) and the reinforcement efficiency

3.3 Statistical analysis

In order to assess the accuracy of the results a statistical analysis for the significant difference based on the Student t-test was carried out. Results are presented in Tab. 2. Analyzing the statistics of the average values of the maximum flexural strength, a significant difference is observed in some pairs (bold font) at the 0.05 level. The results show that between series of nonreinforced concrete and reinforced concrete on the basis of glass or carbon rovings the significant difference is observed. However, a comparison between the results of a series of specimens with glass and carbon rovings differences are not significant.

Table 2. Statistical analysis.

Parameter	The reinforcing component	Unreinforced	AR-glass	Carbon
Flexural strength	Unreinforced		**0,0119**	**0,0051**
	AR-glass			0,3168
	Carbon			

4 Conclusions

It is proved that the specimens with textile reinforcement have higher strength properties than the original specimens. Reinforcement efficiency was calculated using the proposed coefficient k_{re} and is given depending on the type of reinforcing fabric. The flexural behavior of reinforced concrete was analysed. It is shown that the specimens are reinforced with high-strength textile fabrics have greater residual load-bearing capacity so as to maintain the structural integrity is not destroyed after the maximum applied loads. A significant difference between the results of a series of reference specimens and reinforced specimens was found. However, the comparison of flexural strength between the reinforced specimens made of glass and carbon rovings showed no significant differences. However, in general it may be noted the obvious advantages of using textile reinforcement consisting mainly of high residual load-bearing capacity of the material.

References

1. O. Stolyarov, A. Gorshkov, Magazine of Civil Engineering, **4**, 21-25 (2009)
2. O. Stolyarov, T. Quadflieg, T. Gries, Textile Research Journal, **85 (18)**, 1934–1945 (2015)
3. I. Löfgren, *Fibre-reinforced Concrete for Industrial Construction. Department of Civil and Environmental Engineering, Structural Engineering*, (Chalmers University Of Technology, Göteborg, (2005)
4. V.V Okrepilov, A.Yu. Smirnov, Measurement Techniques, **56 (1)**, 54-60 (2013)
5. S. Voss, J. Hegger, ICTRC, 1-10 (2006)
6. S. Voss, 6th International PhD Symposium in Civil Engineering, 1-8 (2006)
7. A. Peled, Z. Cohen, Y. Pasder, A. Roye, Cement & Concrete Composites, **30**, 174-183 (2008)
8. M. Horstmann, A. Shams, J. Hegger, CTRS6, 329-340 (2011)
9. C. Kulas, J. Hegger, M. Raupach, U. Antons, CTRS6, 329-340 (2011)
10. D. Kurlapov, A. Kuvaev, A. Rodionov, R. Valeev, Magazine of Civil Engineering, **3**, (2009)
11. A. Keil, M. Raupach, 12th International Congress on Polymers in Concrete, 873-881 (2007)
12. N. Paranicheva, T. Nazmeeva, Magazine of Civil Engineering, **2**, (2010)
13. M. Al A. Ghaib, J. Go´rski, Cement and Concrete Research, **31**, 1459-1465 (2001)
14. A. Sivakumar, M. Santhanam, Cement & Concrete Composites, **29**, 603-608 (2007)
15. S. Holler, C. Butenweg, S.-Y. Noh, Computers and Structures, **82**, 1971–1979 (2004)
16. J. Orlowsky, *Modeling the Long-term Behavior of Textile Reinforced Concrete.* (IBAC, (2010)
17. J. Orlowsky, M. Raupach, *Corrosion Modeling of AR-glass Reinforcement in Textile Reinforced Concrete* (IBAC, (2010)
18. J. Orlowsky, M. Raupach, Materials and Structures, **39**, 635-643 (2006)
19. I. Ovchinnikov, S.Valiev, I. Ovchinnikov, V. Zinov'ev, A. Umirov, Internet-magazine «Science studies», **4**, (2012)
20. H. W. Reinhardt, M. Kruger, C. U. Crobe, Journal of Advanced Concrete Technolog, **3**, 231-239 (2003)
21. A. Peled, Cement and Concrete Research, **37**, 805–813 (2007)

MATEC Web of Conferences **53**, 01017 (2016)
DOI: 10.1051/matecconf/20165301017

The Effect of Conjunction Flexibility on the Local Stability of Steel Thin-walled Slab Beams

Vladimir Rybakov [1,a], Natalia Molchanova [1], Vladimir Laptev [1], Anna Suslova [1] and Aleksandr Sivokhin [1]

[1] *St. Petersburg State Polytechnical University, 29, Polytecknycheskaya st., St. Petersburg, 195251, Russia*

Abstract. Influence compliance of node connections are not taken into account fully in the existing calculation methods. Basically, this is because of insufficient three-dimensional interference knowledge of elements of thin-walled steel frames. In practice, the structural analysis of the LSTC is usually made with the help of design diagrams with articulated or rigid joints connection elements, therefore, real behavior of structure is not taken into account, as the compliance units may lead to substantial redistribution efforts. Researches in clarifying the design diagrams allow to calculate the steel thin-walled structures with more accuracy and to improve the economic efficiency of design solutions. Thus, the purpose of the study is to determine the qualitative and quantitative characteristics on the basis of experimental and numerical studies that affect the loss of local stability in joints, taking into account the compliance of node connections.

1 Introduction

Nowadays there is a significant increase of thin-walled molded sections from galvanized steel in the construction industry, which shows that there is an evolution of new technologies in Russia. These new technologies provide many things such as research, development, engineering, production and building of light steel thin-walled constructions (LSTC). Construction technology based on the use of thin-walled molded sections can be used independently or in cooperation with other types of building structures. This technology allows to build high-quality and cost-effective buildings in a short of time, which allows production in remote areas. Moreover this is particularly relevant in emergency response and recovery. LSTC-technology allows you to build additional structures and mansard roofs without foundations strengthening due to the small weight of the structures, as well as residential, public and industrial buildings [1, 2]. The most common theory of thin-walled calculation was developed by V.S. Vlasov [3]. Theoretical developments listed in his works [3, 4], confirmed by numerous experimental studies that uses this theory to build an engineering method for calculating thin-walled open section. Equations of sustainability have been clarified by Broude B.M. [5] and Vorobiev L.N. [6]. Calculation of thinwalled bars of the deformed scheme were studied by Vyazmenskiy S.P. [7], Beilin E.A. [8], Meshcheryakov V.B. [9]. Influence of initial imperfections on the behavior of thin-walled structure is investigated by Broude B.M. [5]. Sustainability issues of thin-walled bars were considered by Repmanom Y.V. [10]. Stability of plane bending of thin-walled, studied by Reut V.I. [11], Moulin S.M. [12], Obrazcov I.F. [13]. Development of calculation methods are devoted to Bychkov D.V. [14-

[a] Corresponding author : author@e-mail.org

16], Gorbunov B.N. [17], Strelbitskaiya A.I. [17], Beliy G.I. [18, 19], Tusnin A.R. [20], Lalin V.V. [21, 22], Rybakov V.A. [21-24]and other scholars. Application of thermoprofiles is also studied widely [25-28]. There are a lot of impressive studies related to thin-walled bars made by hand of Russian [29-34] and foreign authors [35-42]. A large number of studies are also related to the study of node connections of thin-walled structures. Research in bolted connections and in the development of methods for their calculations are devoted to Streleckiy N.N. [43, 44], Babushkin V.M. [44, 45], Velikhov V.P. [45]. Influence on zinc covering on the work of the node connections is considered by Sotskov N.I. and Yakubov G.P. [46]. Issues of the work of node connections using rivets and screws are devoted to Orlov I.V. [47, 48], Ayrumyan E.L. [49] Ganichev S.V. [50]. Classification of node connections is shown in the article [51]. The significant gap in node connections of thin-walled structures requires research into the deformation of node connections to the thread-cutting screws [52, 53]. Due to the deformability in nodes of construction there are additional displacement and a redistribution of internal forces in the structural elements caused by deformations in the local connection area.

2 Materials and Methods

Behavior of used test pieces was equal to work of a thin walled steel slab beams. The experimental model consists of a thin walled steel beam (Type PS-150, length 800 mm), 2 profiles (type PN-154, each length 400 mm) and timber inserts. Beams of type PS-150 simulate a work of the floor slab. Profiles of type PN-154 act as a guides. Timber bars were used as an inserts to provide more stiffness for the construction and imitate angles in the beam connections. Some of the test pieces were carried out using angles for strengthening a wall of the profile. Angles fixation was situated on the distance of 200 mm from beam verges (type PS-150). All elements are connected by thread forming screw.
11 test pieces of 4 types were examined. The difference between test pieces is in profile thickness and angle stiffener (its existence or absence).
-Test pieces with the beam of a type PS-150-1.5 without angle stiffeners
- Test pieces with the beam of a type PS-150-2.0 without angle stiffeners
- Test pieces with the beam of a type PS-150-1.5 with angle stiffeners
- Test pieces with the beam of a type PS-150-2.0 with angle stiffeners

The research model is a single-span beam with a hinge support. The following hypothesis was taken for an experiment simplification: distributed load is substituted by two single forces, applied at a distance 200 mm from the supports. As a result, we have a beam situated in the conditions of pure bending, because the bending moment between two points is constant. Load increase is being carried out by steps of 100 KGS. Load scheme is shown on the Figure 1.

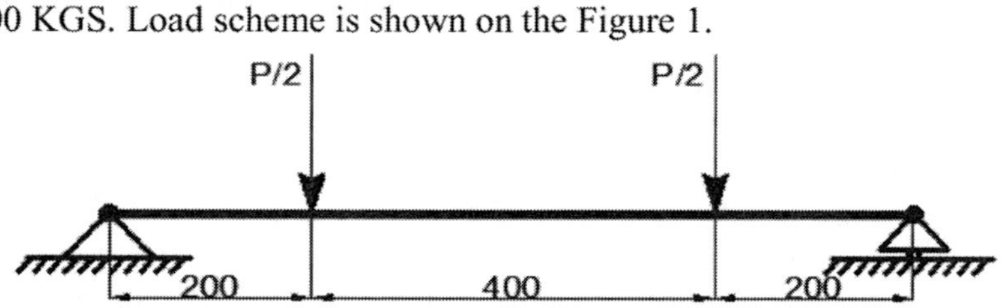

Figure 1. Loading scheme

The aim of the experiment is to determine the critical load. Appearance of a load curve plateau (its decrease or permanence in the time of the deflection increase) is a characteristic feature of the bearing depletion and stability loss. Outward signs of the stability loss are:

The experimental unit and test pieces view is shown on the Figure 2.

Figure 2. Experiment set-up

Obtained values of the critical forces are shown in the table 1.

Table 1. Critical loadings

Test piece's type	$F_{кр}$, daN	$\overline{F}_{кр}$, daN
Test pieces with the beam type PS-150-1,5 without angle stiffeners	1800	1776
	1700	
	1830	
Test pieces with the beam type PS-150-2,0 without angle stiffeners	2500	2360
	2100	
	2480	
Test pieces with the beam type PS-150-1,5 with angle stiffeners	1830	1920
	2150	
	1780	
Test pieces with the beam type PS-150-2,0 with angle stiffeners	2400	2500
	2600	

Critical force analysis shows that the metal thickness increase of the profile leads to bearing capacity increase (approximately 25%). Deformation type is the same in the time of the steel thickness increase. The angle stiffener installation influences a geometry change and stability loss type of the test piece, but it increases bearing capacity quite unessentially (not much that on 10%).

Let's consider deformations which appear in the each of the different test pieces.

Test pieces with the beam type PS-150-1, 5 without angle stiffeners (Figures 3, 4): right after the load is applied, the upper board profile has a deformation and deflection, then, there is deformation under the bars, which transmit loading and profile crumple occurs. Otherwise, the lower board incurves up into the profile. At the same time, the wall under the bars is being buckled. Local stability loss starts to develop. Maximum loading is situated under the bars, transmitting the loading. Then, there are two options of the deformation development: in the direction of a central part of bended profile or in the direction of obstacles.

As deformation develops from the walls under the bars to the central part of a beam, all central parts of the beam are being buckled. In case of the deformation development towards the obstacles, in the profile connections there is a crumple of a lower wall part. Destruction is possible because of a pulling out of the thread forming screw (Figure 4). Deformation of obstacles is obtained and it is expressed by the board deformation of the profile guides.

Asymmetric deformation development is possible as well. On the one hand deformations are developing in the direction of the beam's central cross-section; on the other hand deformations are developing in the direction of the joint.

Figure 3. View of the deformed test piece

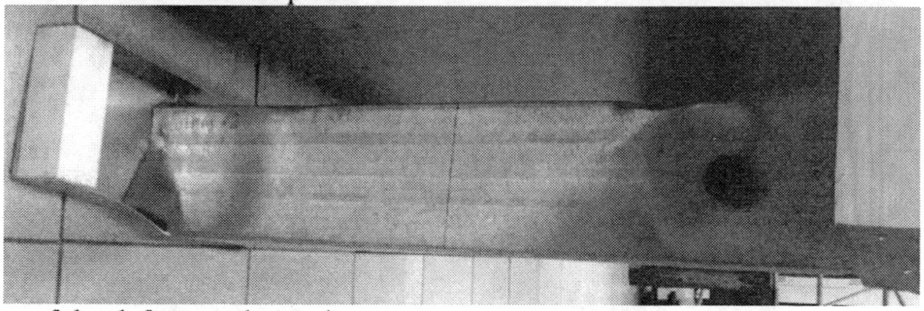

Figure 4. View of the deformated test piece

Test pieces with the beam type PS-150-2,0 without angle stiffeners: deformations are similar to previous test pieces deformations but they are less expressed. Essential obstacle obliquity is observed as well.

Test pieces with the beam type PS-150-1, 5 with angle stiffeners: (Figures 5, 6, 7): angle stiffeners essentially change the deformations behavior.

After applying load, buckling occurs under the bars, which transmit the load. By the way, deformations caused by profiles buckling essentially decrease in the beam's central part. There is unessential bulging of the upper part of the wall. Lower part of the wall is almost not deformed.

Deformations in the direction of the obstacles have a similar behavior to the test pieces without stiffer angles: profile's wall stays straight in the area which is close to angle stiffeners. But in the area which is far from angle stiffeners there is a fast deformation development: buckling of the lower wall part, deformation of the guide profiles, and sometimes the destruction because of a pulling out of the thread forming screws.

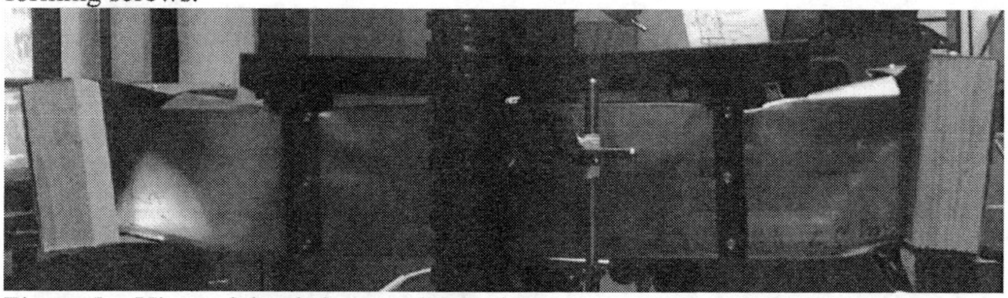

Figure 5 – View of the deformated test piece

Figure 6. View of the deformed test piece

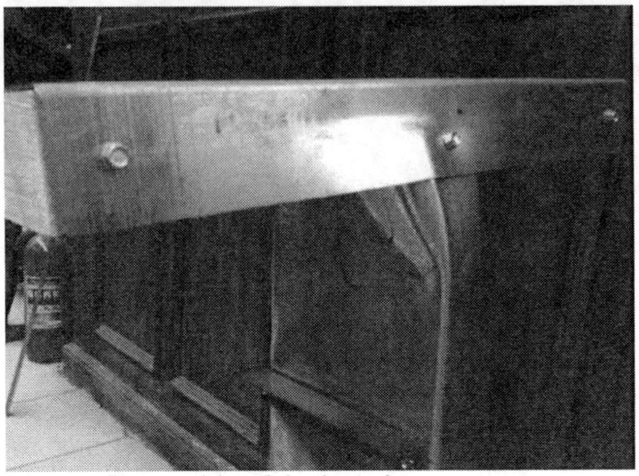

Figure 7. Deformation of the profile's guide of the board

Test pieces with the beam type PS-150-2 with angle stiffeners: deformations are similar to those described previously, but less expressed. Essential obliquity of the obstacles is obtained.

3 Results and discussion

Intensive numerical method development of calculations allows simulation of the interaction processes of the steel thin walled constructions and make three-dimensional calculations with the required accuracy. Finite elements method is the most widely spread method [54].
Calculation was carried out in the ANSYS software. The geometrical model was created using Design Modeler, which use modern methods of the modeling plane and 3D geometry. Geometrical models are shown on the Figures 8 and 9.

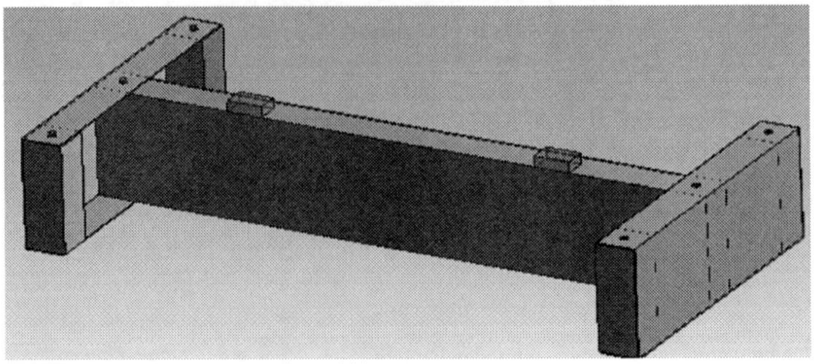

Figure 8. Geometrical model without angle stiffeners

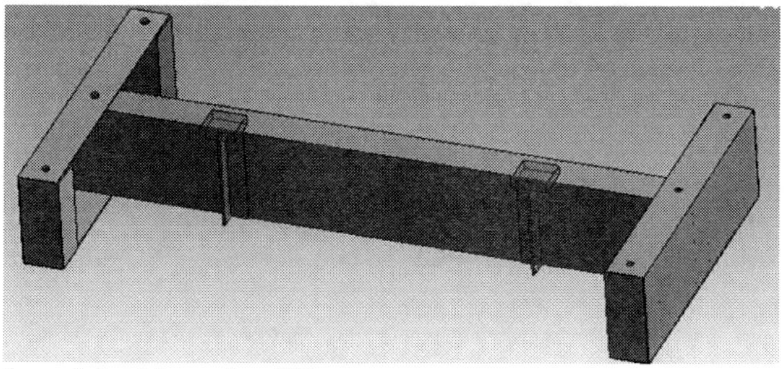

Figure 9. Geometrical model with angle stiffeners

Deformation behavior in the geometrical models is detailed in the table 2. Deformations of the test pieces with the different profile's thicknesses are equal.

Table 2. Appearance of the deformed test pieces

№.	Type	Calculated deformations
1	Test pieces with the beam type PS-150-1,5 without angle stiffeners; Test pieces with the beam type PS-150-2,0 without angle stiffeners	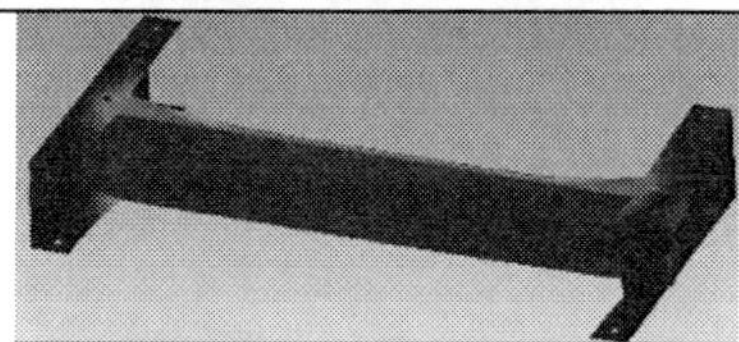 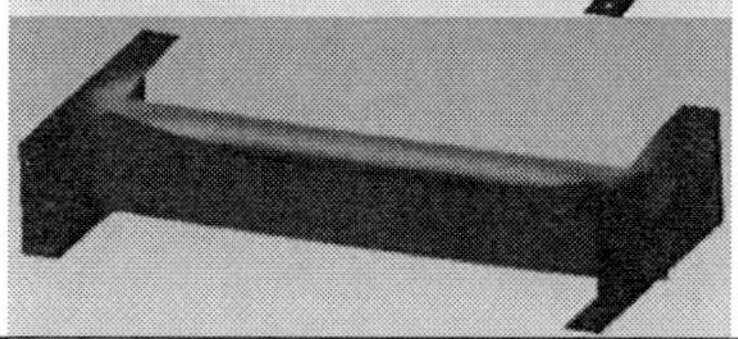
2	Test pieces with the beam type PS-150-1,5 with angle stiffeners; Test pieces with the beam type PS-150-2,0 with angle stiffeners	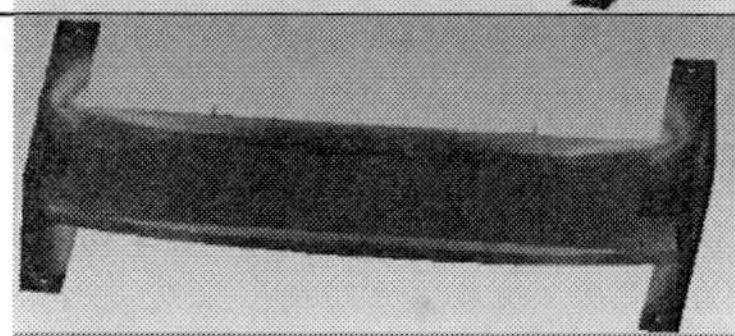 

Analysis of calculations showed that collected calculated deflections and deformations are the same as measured ones.

Deflections in the central sections of test pieces are similar to measured ones. Values of the same magnitude are received. Differences in values of deflections are the idealized model. Moreover it doesn't reflect initials imperfections of test pieces. Thus there is a differences in connections: in the calculated model they are ideal but in the measured model, some test pieces a displacement of supports with an increase of stress was obtained. To increase guided profile's roughness special bodies were set. They imitate wooden insertions but imperfections of wooden bars, used in the installation of test pieces (some of wooden bars were destroyed during the experiment), are not considered.

Deformation behavior in calculated models is similar to deformations from the experiment. Deformation of sections appeared in the time of load application – in central section and near the obstacles. The largest deformation of wall appears in the place of load transmission. Deformation behavior of profile's boards and deformation of test profile is similar. Deformation of obstacles' profiles are observed as well.

Thus, collected calculated forms of test pieces' deformation allow to make a conclusion that the calculated form with high accuracy describes a deformation which appeared in the time of bearing capacity of thin-walled profiles' loss.
Since calculated and measurable materials demonstrate obstacles' deformability, due to guided profile boards' deformation, that obstacles are resilient. Moreover force distribution is not the same for rough or hinge scheme of supports. The scheme of thin-walled beam's supported in the time of node connections' deformability is shown in the Figure 10.

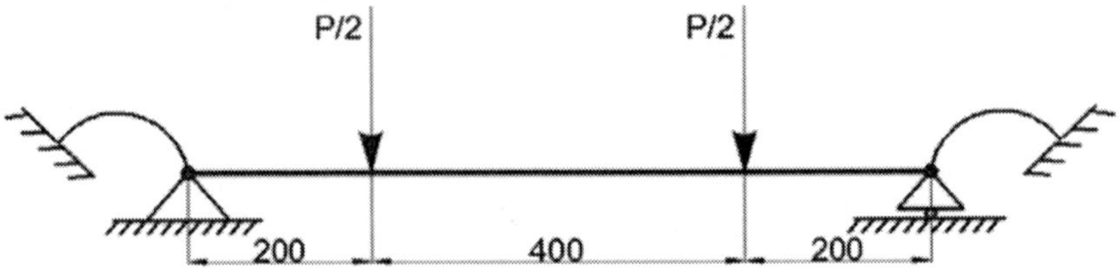

Figure 10. Scheme of thin-walled beam's supporting with obstacle's deformability

Let's express the relation of stresses in the end of steel thin-walled beam's section through the coefficient k_n considering deformability of node connection:

$$\frac{\sigma_{fl}}{\sigma_{st}} = k_n \qquad (1)$$

Where σ_{fl} и σ_{st} – direct stressed in a way of resilient and rough supporting accordingly.

Values of direct stresses in both ways could be made from the result in calculations using finite elements method. The whole construction is calculated to determine direct stresses in a way of resilient support. In a way of rough obstacles calculation for which the following is provided: besides current fixation rough support of low profile's boards defines; its deformation is about deformability of node connection.
Typical points of thin-walled profile's cross-section are in the Figure 11. For these points values of node connections in sections were determined. Then value of coefficient k_n is calculated. In the Figure 12 there is a graph of coefficient k_n for all points of section.

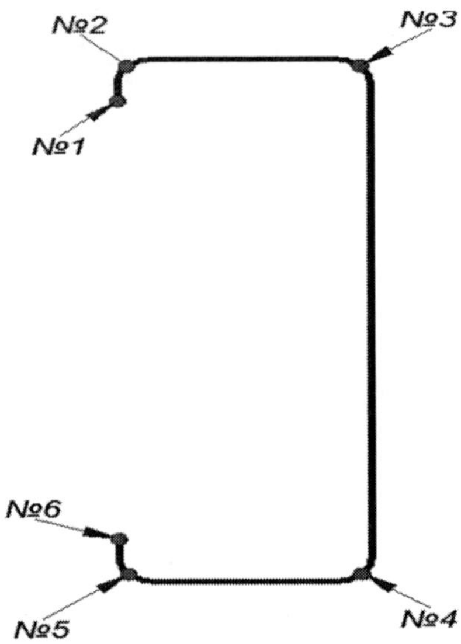

Figure 11. Typical profile's points PS

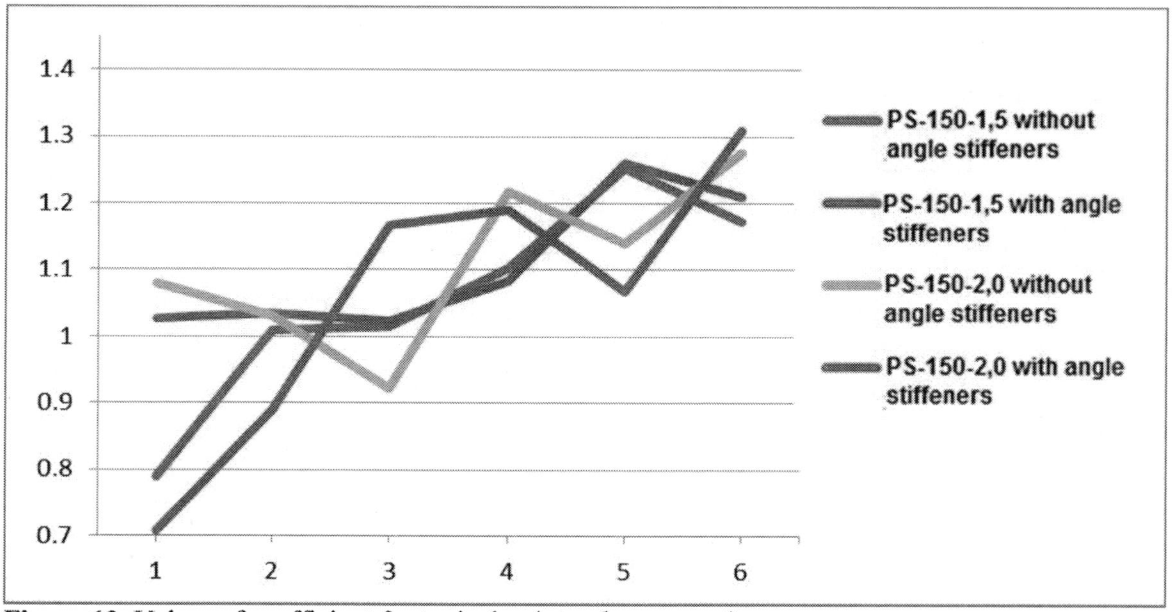

Figure 12. Values of coefficient for typical points of cross-section

According to calculations, the values of direct stresses in the end of thin-walled beam's section is increasing allowing for the deformability of supporting profile.

Total direct stress for thin-walled beams with bend is determined from the formula which takes into account bar's rotation and warping [55]:

$$\delta_{sum} = \delta_{My} + \delta_{Mz} + \delta_W \qquad (2)$$

Where:

δ_{My} – direct stress from bending moment M_y;

δ_{Mz} – direct stress from bending moment M_z;

δ_W – direct stress from warping.

Only bending moments arise in a case of rough supporting of considering thin-walled beams in the end of profile's section. The amount of stress from warping and rotation is not essential. Deformability of obstacle due to guided profile's board deformation determines a big compliance of bending beam's cross-section. In other words stress of rotation and warping has higher role.

Let's say that coefficient k_n for each considering model is an arithmetical average of profile ambit's coefficients (Table 3).

Table 3. Coefficient k_n for considering test pieces

Type of profile	$1/k_n$	k_n
PS-150-1,5 without angle stiffeners	0.852	1.172
PS -150-2,0 without angle stiffeners	0.823	1.215
PS -150-1,5 with angle stiffeners	0.828	1.208
PS -150-2,0 with angle stiffeners	0.838	1.192

Relation between direct stresses in a way of rough and resilient supporting is almost the same. This allows to suppose that node connection's deformability doesn't depend on the type of beam and existence or absence of angle stiffener.

Calculations allow us to make a conclusion that deformability's accounting of bearing construction provides to the increase of collected direct stresses of 15-20%. Therefore, when calculating we should use a coefficient which takes into account an effect of node connections' deformability on the state of stress of thin-walled beam. It equals to 1,2. Coefficient of node connections' deformability doesn't depend on the type of bending beam and existence of angle stiffener because a deformability is about a deformation of guided profiles' board.

4 Conclusions

Analysis of test results showed that the girders' sustaining capacity strongly depends on the thickness of steel - increase in profile's thickness of 0.5 mm gave an increase in the girders' sustaining capacity of about 25%.
The use of strengthening ribs has a significant impact on the local loss of stability and allows to save the section of the central profile's part almost unchanged, but doesn't produce a significant effect when used in floor beams - sustaining capacity increased by only 10%.
It was considered a deformability of node connections expressed by deformations of shelves guide's profiles. According to the calculation, the normal tensioning is in the end sections of thin-walled beams, taking into account that the deformability increases by 1.2 times.
Compliance coefficient of node connection does not depend on the type of beam's bend and the presence or absence of ribs.

References

1. N.I. Vatin, Ye.N. Zhmarin, V.G. Kurazhova, K.Yu. *Konstruirovaniye zdaniy i sooruzheniy. Legkiye stalnyye tonkostennyye konstruktsii: ucheb. rukovodstvo* (Saint-Petersburg, 2012)
2. K.V. Petrov, Ye.A. Zolotareva, V.V. Volodin, Vatin N.I., Zhmarin Ye.N. Magazine of Civil Engineering, **2**, 59-64 (2010)
3. V.Z. Vlasov, *Tonkostennyye uprugiye sterzhni* (Moscow, Fizmatgiz, 1959)
4. V.Z. Vlasov, *Izbrannyye trudy. Part 3.* (Moscow, Nauka, 1964)
5. B.M. Broude, Raschet prostranstvennykh konstruktsiy, **4**, 5-35 (1958)
6. L.N. Vorobyev, Novocherkasskiy politekhnicheskiy institut.-Novocherkassk: NPI, **69(83)**, 3-48 (1958)
7. S.P. Vyazmenskiy Leningr. Inzh.-stroit. in-t.- L.: LISI, **60**, 20-29 (1969)
8. Ye.A. Beylin, Stroitelnaya mekhanika i raschet sooruzheniy, **4**, 1-3 (1967)
9. V.B. Meshcheryakov, Sb.nauchn.tr., Moscow, Transzheldorizdat, **236**, 77-85 (1967)
10. Yu.V. Repman, Trudy laboratorii stroitelnoy mekhaniki, 1, 154-211 (1941)
11. V.I. Reut, Problemy ustoychivosti v stroitelnoy mekhanike: Tr. Vsesoyuz. konf. po problemam ustoychivosti, **1**, 466-483 (1965)
12. S.M. Mulin, *Issledovaniye ustoychivosti ploskoy formy izgiba* (Tomsk, 1950)
13. T.V. Nazmeyeva, Magazine of Civil Engineering, **5(40)**, 44-51 (2013)
14. Ye.A. Beylin, Stroitelnaya mekhanika i raschet sooruzheniy **4**, 1-3 (1967)
15. D.V. Bychkov *Raschet balochnykh i ramnykh sterzhnevykh sistem iz tonkostennykh elementov,* (Moscow, 1948)
16. D.V. Bychkov, *Structural Mechanics of stalky thin-walled structures* (Moscow, 1962)
17. B.N. Gorbunov, A.I. Strelbitskaya, *Teoriya ram iz tonkostennykh sterzhney* (Moscow, 1948)
18. G.I. Belyy, Metallicheskiye konstruktsii i ispytaniya sooruzheniy LISI, 1, 26-30 (1984)
19. G.I. Belyy Sbornik trudov LISI, 11, 18-25 (1974)
20. A.R. Tusnin, *Chislennyy raschet konstruktsiy iz tonkostennykh sterzhney otkrytogo profilya* (Moscow, MGSU, ASV, 2009)
21. V. V. Lalin, V. A. Rybakov, Magazine of Civil Engineering, **8(26)**, 69-80 (2011)
22. Lalin V. V., Rybakov V. A., Morozov S. A., Magazine of Civil Engineering, **1(27)**, 53-73 (2011)
23. V. A. Rybakov *Osnovy stroitelnoy mekhaniki legkikh stalnykh tonkostennykh konstruktsiy* (Saint-Petersburg, SPbGPU, 2011)
24. V. A. Rybakov, O. S. Gamayunova, Construction of Unique Buildings and Structures, **7(13)**, 79 (2013)
25. N. I. Vatin, Ye. N. Popova, *Termoprofil v legkikh stalnykh stroitelnykh konstruktsiyakh* (Saint-Petersburg, SPbGPU, 2006)
26. A. O. Gordeyeva, N. I. Vatin, Magazine of Civil Engineering, **3**, 36-46 (2011)
27. Obraztsov I.F., Sb. nauch. tr. Mosk. aviats. in-t, **26**, 85 (1953)

28. A. V. Orlova, Ye. N. Zhmarin, K. O. Paramonov, Construction of Unique Buildings and Structures. **6(11)**, 1-13 (2013)
29. M.R. Garifullin, N.I. Vatin, Construction of Unique Buildings and Structures, **6(21)**, 32-57 (2014)
30. D.N. Smaznov, Politematicheskiy Setevoy Elektronnyy Nauchnyy Zhurnal Kubanskogo Gosudarstvennogo Agrarnogo Universiteta, **67**, 101-113 (2011)
31. V.V. Lalin, V.A. Rybakov, A. Sergey, Applied mechanics and materials, **578-579**, 858-863 (2014)
32. D. Trubina, D. Abdulaev, E. Pichugin, V. Rybakov, Applied mechanics and materials, **633-643**, 982-990 (2014)
33. D. Trubina, D. Abdulaev, E. Pichugin, V. Rybakov, Applied mechanics and materials, **633-643**, 1133-1139 (2014)
34. D. Trubina, D. Abdulaev, E. Pichugin, M. Garifullin, Applied mechanics and materials, **633-643**, 1052-1057 (2014)
35. K. M. Bajoria, K. K. Sangle, R. S. Talicotti, International Journal of Advanced Structural Engineering, **1(2)**, 153-181 (2014)
36. J. Becque, M. Lecce, J. Rasmussen, R. Kim, Journal of Constructional Steel Research, 64(11), 1231-1238 (2008)
37. J. H. Bickford *An introduction to the design and behavior of bolted joints, 3rd edition* (2009)
38. Ju Chen, B. Young, Engineering Structures, **10(29)**, 1161-1174 (2007)
39. C. N. Louise, A. M. Md Othuman, M. Ramli, Applied Science Research, **3(5)**, 2847-2859 (2012)
40. A. G. Mamalis, D. E. Manolakos, M.B. Ioannidis, P.K. Kostazos, C. Dimitriou, Thin-Walled Structures, **41(10)**, 891-900 (2003)
41. J. R. Rasmussen, J. Kim, Thin-Walled Structures, **44(11)**, 1175-1184 (2006)
42. Silvestre N., Young B., Camotima D, Engineering Structures, **10(30)**, 2613-2630 (2008)
43. N.N. Streletskiy, V.V. Kalenov A.B. Pavlov, Mezhdunarodnyy kollokvium. Boltovyye i spetsialnyye montazhnyye soyedineniya v stalnykh stroitelnykh konstruktsiyakh, **1**, 14-22 (1989).
44. V.M. Babushkin, Mezhdunarodnyy kollokvium. Boltovyye i spetsialnyye montazhnyye soyedineniya v stalnykh stroitelnykh konstruktsiyakh, 1, 105 (1989)
45. V.P.Velikhov, V.M. Babushkin, *Provedeniye issledovaniy boltovykh montazhnykh soyedineniy, rabotayushchikh na srez, rastyazheniye i smyatiye, i razrabotka 135 predlozheniy po izmeneniyu i dopolneniyu SNiP P-23-81. Otchet OTIZM-872* (Moscow, 1985)
46. N.I. Sotskov, Montazhnyye i spetsialnyye raboty v stroitelstve, **11**, 34-37 (2005)
47. I.V. Orlov, Tekhnologii stroitelstva, **7(41)**, 5 (2005)
48. I.V. Orlov, Tekhnologii stroitelstva, **6(47)**, 14 (2008)
49. E.L. Ayrumyan, Montazhnyye i spetsialnyye raboty v stroitelstve, **3**, 19-20 (2009)
50. S.V. Ganichev, Tekhnologii stroitelstva, **3(65)**, 22 (2009)
51. V. G. Kurazhova, T. V. Nazmeyeva, Magazine of Civil Engineering, **3**, 47-52 (2011)
52. Yu.S. Kunin, A.I. Kolesov, I.A. Yambayev, D.A. Morozov, Herald MGSU, **11**, 74-81 (2012)
53. S.V. Salamakhin, A.S. Sinelnikov, Construction of Unique Buildings and Structures, **4(9)**, 53-63 (2013)
54. V.A. Bruyaka, V.G. Fokin, Ye.A. Soldusova, H.A. Glazunova, I.Ye. Adeyanov, Samar. gos. tekhn. un-t., **1**, 271 (2010)
55. *Eurocode 3: Design of steel structures Part 1-3. General rules. Supplementary rules for cold-formed members and sheeting* (2006)

MATEC Web of Conferences **53**, 01018 (2016)
DOI: 10.1051/matecconf/20165301018

Improvement of Energy Efficiency of Schools and Kindergartens

Danica Stankovic[1,a]
[1]*University of Nis, Faculty of Civil Engineering and Architecture, Aleksandra Medvedeva st. 14, Nis, Serbia*

Abstract. The paper exposes the reasons and methodology of energy efficiency improvement of schools and kindergartens as public facilities, which will give children the understanding of importance of such measures and it's positive impact on our planet. Modernization of educational institutions is targeted at ensuring positive dynamics for fuel and energy consumption reduction.

1 introduction

The significance of school and kindergartens for overall energy efficiency may be questioned. They are relatively small and rare as compared to other types of buildings. As such, they consume small part of overall energy which is spent in buildings. However, there are several good reasons for the improvement of energy efficiency in schools and kindergartens. Schools and kindergartens, as public facilities, should serve as an example to other buildings in the area of design and materialization, but also in promoting energy efficiency. Promoting the use of energy efficiency systems in public facilities is one of the best ways to implement such systems in private households and multifamily buildings [1-3]. The main occupants of the schools and kindergartens are children at the beginning of their journey through knowledge about the world. If we can explain meaning of renewable energy sources to children followed with real life examples that are put into their close environment we can be sure that earth is in good hands. Installation of energy efficiency systems is a process that can imply a lot of work [4-8]. Therefore, it would be a good idea to remodel schools and kindergartens along with the setting up of these systems, if their current ambient, functional or comfort properties are not satisfactory. So far, reconstruction of this type of buildings has been limited to necessary technical interventions, like repainting the walls and repairing the roofs. Functional organization and psychological factors have been mostly neglected, as well as improving energy efficiency.

2 Reasons and methods of modernization

Modernization of educational institutions is targeted at ensuring positive dynamics for fuel and energy consumption reduction. Reasons for the reconstruction [9-11]:

- the building's conditions are the technical reasons for the reconstruction. There are overheat energy losses through the walls, the windows, doors and roof of the school or kindergarten building;

[a]Corresponding author: danica0611@gmail.com

- the socio-economics reasons for the reconstruction to be carried out are increased energy consumption in the building and municipality deficit budget stipulated inefficient comfort conditions in the school or kindergarten.
- The opportunities for energy efficiency in existing building are vary widely, depending on:
- the climate-zone where the building is located,
- the age of the building and the type of construction and
- the requirements and incentives involved [12, 13].

Educational institutions under control of Education Committee are located in 2 278 buildings [14]. According to an analysis of technical and engineering status 80% of buildings have structural members, which need repairing. Since the requirements for energy consumption have been tightened practically all the existing schools have low energy efficiency class and need to be streamlined to reach significant reduction of thermal energy consumption.

The most significant reasons for the high energy consumption in school and kindergarten buildings are follows:

- the large losses in transportation through the systems for heating - through the dilapidated and inadequate isolated tube;
- awareness of energy as a valuable resource that should spend rational;
- low price does not contribute to the motivation for less consumption;
- factor of the heat loses through the windows;
- a lot of heat failure - inadequate external thermal insulation;
- lack radiator valves to regulate consumption in accordance with the temperature;
- inadequate electrical appliances that consume more electricity [12, 13].

The major measures that can be taken to achieve this aim are as follows:

- insulation of building envelopes;
- heat recovery ventilation systems [15, 16]
- optimization of engineering equipment operation;
- installation of any types of meters and control of resources consumption.

The plan of the reconstruction includes:

- to change the windows,
- to change the outside doors,
- to install heat reflecting barriers and
- to install radiator's temperature regulators [13].

Let's take the school № 258 in Saint-Petersburg, which was repaired and where envelopes were insulated, as an example (Fig. 1) [14].

Figure 1. Secondary school № 258, Kolpinskiy district, Saint-Petersburg, before and after reconstruction, Kolpino, Pavlovskaya 80 A

Expected results:

- increase thermal resistance of the windows,
- decrease heat loses factor of the windows,
- increase thermal resistance of the outside doors,
- decrease heat loses factor of the outside doors,
- decrease energy consumption for heating and hot water and
- improve comfort conditions of children in the school or kindergarten.

The basic purpose of the reconstruction:

- to improve energy efficiency,
- to reduce energy consumption by reducing the heat looses of the building and
- decrease expenses for energy in the budget [13].

2.1. Exterior walls insulation in case of buildings under reconstruction

The most popular insulation method is given in Fig. 2 [14].

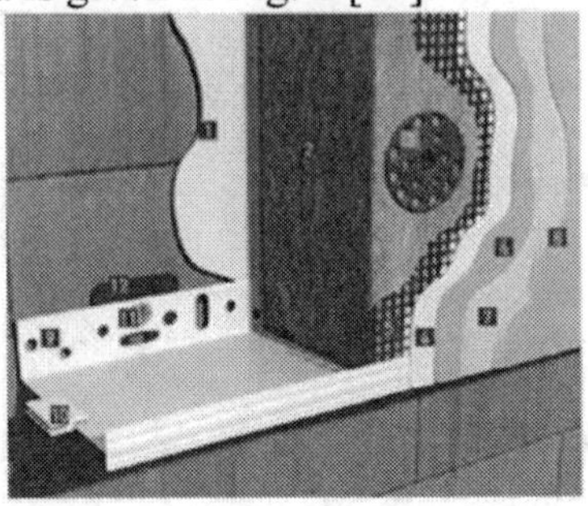

Figure 2. Insulation method for exterior walls; 1 - special adhesive mixture; 2 - rock wool insulated panels; 3 - frontal expansion anchor stud; 4 - reinforcing filler; 5 – exterior glass cloth grid; 6 - water-dispersion primer; 7 - decorative mineral plaster; 8 - frontal silicone paint; 9 - ground profile; 10 - mating element; 11 - ground anchor; 12 – roughness compensator

This method implies that a number of façade layers should be fixed to an exterior wall with the slabs made of rock wool serving as an insulator, and a thin finishing mortar with a reinforced pre-layer appears to be a façade layer. This method implies that a number of façade layers should be fixed to an exterior wall with the slabs made of rock wool serving as an insulator, and a thin finishing mortar with a reinforced pre-layer appears to be a façade layer [14].

2.2. Improvement of energy efficiency through implementation of solar energy systems

The use of solar energy systems is one of the primary strategies in modern sustainable architectural design. There are many possibilities for using active solar energy systems in both newly designed and reconstructed buildings. Architectural structures with integrated systems for renewable energy production have greater autonomy in energy usage and a higher degree of independence from public energy grid. Building-integrated solar system is ranked as one of the most prospective active green energy systems. The required pre-requisites for zero energy building construction can be created by the using of active building-integrated solar systems for independent energy supply of schools and preschool facilities. Active building-integrated solar systems can be divided into three groups, according to the type of energy produced: photovoltaic (PV) systems (BIPV) (Figure 3); thermal systems (BIST) (Figure 4); PV and thermal systems (BIPVT) [6, 7, 17, 18].

The use of solar energy systems is becoming predominant and essential part, not only of new building design, but also the existing building reconstruction design. These principles are especially important in the process of revitalization of the existing school and kindergartens which do not meet contemporary standards. Flexibility and wide range of possible use make these systems very convenient for preschool facilities. The use of solar collectors is fully justified and valid because of relatively high demands for sanitary hot water and low cost of solar collectors, which makes the return on investment in these systems very quick [19].

Figure 3. Example of photovoltaic system

Figure 4. Example of thermal system

The result of significant spatial development in architectural plan is a large roof area which could provide great capacity in terms of space that could be used for installation of solar systems. Using roof for solar energy system installation could be a good strategy for transforming building disadvantage (unfavorable volume to building envelope area) into an advantage in terms of energy efficiency. Therefore, multifunctional roof treatment is desirable [12].

2.3. Guidelines in the process of reconstruction of the internal environment – comfort improvement

By implementing new technologies, existing materials of improved quality and the introduction of new environmentally friendly materials and equipment in the area of internal environment of schools and preschool facilities, comfort of the children is rising at a significantly higher development level.

Air comfort. The indoor quality of air that children breathe is of vital importance for their general health and development. Ventilation system is crucial for correct air comfort of the space occupied by children. Thus, comfort can be enhanced by improving ventilation system. Internal air comfort depends on indoor absence of pollutants i.e. ventilation systems which pump fresh air, and the power of filters to remove polluting substances, activities of kindergarten users and properties of materials applied to structural elements and accessories [18-20].

Special caution is needed when selecting proper surface and equipment finishing materials for interior surfaces such as floors, walls, ceilings and every other surface. It is strongly recommended avoiding using materials that release off-gas volatile organic compounds into the air, particularly formaldehyde-based finishes, adhesives, carpeting and particleboard.

When the given assignment refers to designing or revitalizing old schools and kindergarten, in order to ensure higher air quality, the key thing is predicting ventilation system with enough power and equipment, as well as devices for mechanical ventilation of certain areas (in the lobby area, toilets, etc.). It is also important to test all materials in order to determine the degree of their impact on air quality in rooms where children reside, prior to applying them as a final finish. Furthermore, materials which are used for making kindergarten furniture and other preschool accessories must undergo these tests.

Light comfort. Light comfort is defined by the following parameters: daylight, window position, window area, glazing types, etc. [19- 22]. Apart from the above mentioned parameters, light comfort is certainly affected by additional installed artificial lighting that comes from built-in light fittings. As for parameters of colour and intensity, artificial light should certainly be as close as possible to daylight. Additionally, it is advisable to install lighting system that will help improve visual comfort without any negative effects on children's vision. Avoiding the use of fluorescent lamps is strongly advisable as well. Apart from providing soft shadows and uniform light, fluorescent lamps produce negative flickering that has negative effect on children's health and their work abilities because it creates sense of embarrassment and discomfort.

Achieving visual comfort and more pleasant stay is reflected in correct amount of light that has been introduced to a room. Regarding the amount of luminance required in various indoor spaces in preschool facilities, current legislation and regulations in Serbia stipulate that parent units, i.e. rooms for group care, room lighting with artificial light sources should be 150lux for incandescent bulb and 250lux for fluorescent lamps, and 60lux for toilets with artificial lighting. In order to achieve more environmentally friendly building, replacement of old light bulbs with new and energy-efficient blubs that have significantly lower consumption of electricity is necessary [12].

Aesthetic comfort. Preschool children have the need for facilities that will aid them in physical and mental development. Space designed for children must meet certain requirements. In terms of aesthetic comfort, interior should be imaginative and attractive, abundant with content, with purposely designed areas (calm or dynamic interior depending on the purpose of the area purpose, e.g. sleeping or playing zones). Aesthetic comfort may be affected by a combination of: how the whole space appears to the eye, how materials feel to the touch, impression of whole kindergarten in combination with the surroundings, how it interacts with everyday users, how it is remembered and how children can manipulate it in order to create different space of their own [23, 24].

3 Conclusion

Achieving cheerful and pleasing experience of school and kindergarten interior areas is imperative for creating good educational environment and must be carried out through proper application of colour and texture, pleasant to the eye and touch. Therefore, the place where they spend a significant amount of time learning and having fun needs to meet their requirements for beautiful, imaginative and attractive space. Beside the need for aesthetically pleasing experience in space, interior environment should respond to children's needs for safe and secure environment. Complete experience of design has to be organized in a way which helps children experience and accept school and kindergarten as a second home. Viewed from the perspective of a child, interior structure of preschool building should be in proportion with children's abilities to percept space. Identification and recognition of interior space elements, their shapes, colours, and size play key role in appropriate perception of space [12].

The accompanying positive aspects of reconstruction could be:

- the close co-operation with municipality representatives involved in energy conservation and efficient use sector;
- recommendation of the reconstruction could be identified as it may be of service to implement energy efficiency reconstructions in public building to ensure the energy efficiency, energy conservation and acceptable payback time of the reconstructions [13].

References

1. J. Ćetković, S. Rutešić, M. Zarković, M. Knežević, N. Vatin, Procedia Engineering, **117 (1)**, 780-790 (2015)
2. S.G. Popović, S. Vukanić, D.F. Komatina, E. Alihodzic-Jasarevic, N. Vatin, Procedia Engineering, **117 (1)**, 614-620 (2015)
3. M. Tanic, D. Stankovic, V. Nikolic, M. Nikolic, D. Kostic, A. Milojkovic, S. Spasic, N. Vatin, Procedia Engineering, **117 (1)**, 924-937 (2015)
4. V. Murgul, N. Vatin, E. Aronova, Applied Mechanics and Materials, **635-637**, 2029-2035 (2014)
5. G. Radovic, V. Murgul, N.I. Vatin, Applied Mechanics and Materials, **584-586**, 564-569 (2014)
6. V. Murgul, Journal of Applied Engineering Science, **12 (2)**, 121-128 (2014)
7. R. Alihodzic, V. Murgul, N. Vatin, E. Aronova, V. Nikolić, M. Tanić, D. Stanković, Applied Mechanics and Materials, **624**, 604-612 (2014)
8. G. Radovic, V. Murgul, N. Vatin, E. Aronova, Applied Mechanics and Materials, **627**, 357-364 (2014)
9. V. Murgul, Procedia Engineering, **117**, 808-818 (2015)
10. V. Pukhkal, V. Murgul, M. Garifullin, Procedia Engineering, 117, 624-627 (2015)
11. M. Penić, N. Vatin, V. Murgul, Applied Mechanics and Materials, **680**, 534-538 (2014)
12. D. Stankovic, Proceedings of 10th International Multidisciplinary Scientific Geo-Conference SGEM 2010**,** Sofia: STEF92 Technology Ltd., **2**, 725- 731, (2010)
13. D. Stankovic, M. Tanic, et al., *Revitalization of preschool facilities in Serbia*, (Faculty of Civil Engineering and Architecture, University of Nis, (2014)
14. S. Golovina, Procedia Engineering, 117, 476-486 (2015)
15. V. Murgul, D. Vuksanovic, N. Vatin, V. Pukhkal, Applied Mechanics and Materials, **635-637**, 370-376 (2014)
16. V. Pukhkal, N. Vatin, V. Murgul, Applied Mechanics and Materials, **680**, 529-533 (2014)
17. E. Aronova, N. Vatin, V. Murgul, Procedia Engineering, **117**, 771-779 (2015)
18. V. Murgul, D. Vuksanovic, N. Vatin, V. Pukhkal, Applied Mechanics and Materials, **680**, 524-528 (2014)
19. V. Murgul, D. Vuksanovic, V. Pukhkal, N. Vatin, Applied Mechanics and Materials, **633-634**, 977-981 (2014)
20. A. Milajić, D. Beljaković, D. Davidović, N. Vatin, V. Murgul, Procedia Engineering, **117**, 916-923 (2015)
21. V. Murgul, N. Vatin, I. Zayats, Procedia Engineering, **117**, 824-829 (2015)
22. S. Kosanović, A. Fikfak, Energy and Buildings, (2015)
23. I. Čerpes, N. Pandol, A. Fikfak, European Countryside, **6 (3)**, 225-242 (2014)
24. Leskovar, V.Ž., Premrov, M. Energy and Buildings, 43 (12), 3410-3418 (2011)

MATEC Web of Conferences **53**, 01019 (2016)
DOI: 10.1051/matecconf/20165301019

Method of Hydrogenous Fuel Usage to Increase the Efficiency in Tandem Diverse Temperature Oxidation System

Marina Zubkova[1,a], Alexander Stroganov[2], Alexander Chusov[1] and Dmitry Molodtsov[1]

[1]*Peter the Great St.Petersburg Polytechnic University, Polytechnicheskaya str. 29, St. Petersburg 195251, Russia*
[2]*Lumex Instruments Canada, Unit 207, 31510 Gill Avenue, Mission, B.C., V4S0A1 Canada*

Abstract. This paper presents the results of estimation energy efficiency, the collation data of thermodynamic calculations and data on material balance for an assessment of electric and thermal components in considered ways to use convention products, performance enhancement in the tandem system containing the high-temperature fuel cell and the low-temperature fuel cell with full heat regeneration for hydrogenous fuel (CH_4). The overall effective efficiency ($\eta_{\Sigma ef}$. making full use of the recovered heat) considered tandem system depends on the efficiency of its constituent fuel cells. The overall effective efficiency of the tandem installation including the fuel converter, separating system, high-temperature oxidation system, and hydrogen disposal system in case of fuel use in the low-temperature fuel cell, is higher than for each of the fuel cell elements separately.

1 Introduction

An integrated approach to the task of creation the economic systems of power supply involves obtaining electricity and heat in systems with heat recovery, which will reduce losses and improve the reliability of the process of ensuring energy consumer [1].

Natural gas is currently the main energy resource in a global sense, and the cheapest source for hydrogen production [1,2], but the task of the active involvement of renewable energy, it becomes actual at the present moment [2-5]. Energy resource biomethane is of interest for autonomous energy supply and basic [3-8] in problems of urban civil engineering and municipal facilities [9-12]. Use for these purposes fuel cells, offering direct energy conversion supplied fuel into electrical energy, is a promising direction for the solution of such tasks [1,3-6,13].

Special equipment and arrangements of the fuel reforming and heat recovery process proposed in [14] solves the problem of the heat exchange, but it does not fully ensure the heat recovery due to its partial dissipation. However, even with the structural chain simplification, i.e. in the flow and reforming conditions and electrochemical reactions in a one system (in the fuel cell) ensuring the heat exchange conditions for the heat recovery [15] has some essential faults:

- non-equal anode thermal conditions due to the endoergic reaction input stream zone cool down;
- incomplete use of the fuel cell performance thermal component (emergence of the excess heat in the fuel cell exothermic reaction compared with the endothermic reforming requirements);

[a] Corresponding author: marina19zubkova@list.ru

- CO oxidation on the anode is carried out under conditions of competing dynamic hindrance relating to the H_2 degrading on the same electrode.

The hydrogen produced by the reforming can be used directly in the low-temperature fuel cell (LTFC) to produce the electrical power or transferred to storage [16]. This requires CO and H_2 separation, which in the high-temperature conditions is achieved by pressure swing absorption [17,18] or by a hydrogen elicitation using the palladium membranes [19].

Construction solution concept of the heat exchange arrangement with several interaction types of reformer and fuel cell assemblage, considered by the US 7.008.711 [20], offers the possibility to use the thermal resource of the high-temperature fuel cell exothermic process in the better way.

2 Experimental studies

It is known that it is possible to produce the electric power by carrying-out the electrochemical change in the fuel cell using the hydrogenous fuel only if there is a fuel internal conversion that can take place in the high-temperature fuel cell, operating on the temperature range of 800 – 1200K (in this paper the process is considering at T = 1000K). Hydrogenous fuel (CH_4) reforming into a syngas (CO and H_2 mixture) within the said temperature range is an endothermic process (Eq. 1), and the further anode reaction of the hydrogen and carbon monoxide oxidation (Eq. 2,3) give an exothermic effect [21].

$$CH_4 + H_2O = CO + 3H_2 \qquad \Delta H_{conv.}(CH_4) = 225.82\ kJ \tag{1}$$
$$H_2 + 0.5\cdot O_2 = H_2O \qquad \Delta H_{H2} = -247.92\ kJ \tag{2}$$
$$CO + 0.5\cdot O_2 = CO_2 \qquad \Delta H_{CO} = -282.96\ kJ \tag{3}$$

In the high-temperature conditions thermal component augments, therefore system performance depends on the heat efficiency.

Heating of the conversion agents to the required temperature and levelling of the endothermic conversion reaction input is a step-ahead solution for the high-temperature fuel cell performance thermal component use. In some cases, the thermal component implementation problem is solved by of hybrid systems installation paired by the heat engines [17,22,23], which certainly balances to a varying degree the use of heat, but only concerning the exergy, and thus results in the losses characteristic of every heat engine.

The object of the present invention is hydrogenous fuel estimation energy efficiency, the collation data of thermodynamic calculations and data on material balance for an assessment of electric and thermal components in considered ways to use convention products, performance enhancement in the tandem system containing the high-temperature fuel cell (HTFC) and the low-temperature fuel cell (LTFC) with full heat regeneration.

The fuel processing method according to the scheme [24] shown on the Figure 1 is carried out in the following way: hydrogenous fuel and the conversion agent flow to the converter (C) where at the T=800-1000K as the result of endothermic reactions (1) the syngas (gaseous mixture ($CO + H_2$)) is formed. The syngas is transferred from the converter to the separating system. In the separating system a part of H_2 (released H_2) is extracted from the incoming syngas by separating the incoming syngas stream into at least two outgoing streams: depleted gas mixture (CO and H_2), with the H_2 content depleted to the set level, and the H_2 stream, released from the initial syngas stream and transferred to the hydrogen disposal system.

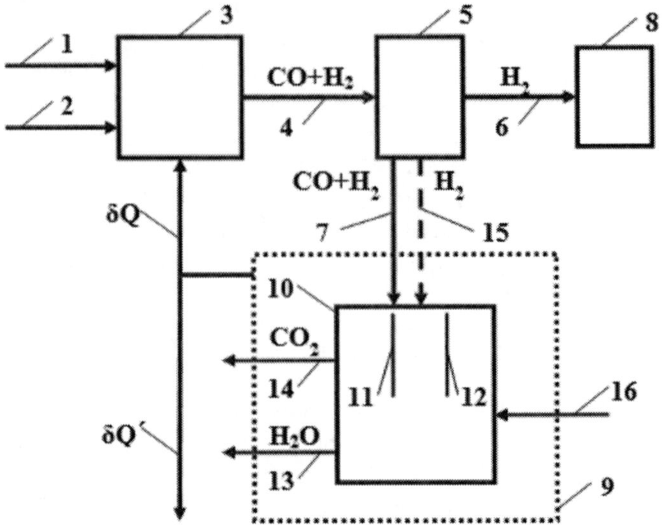

Figure 1. The fuel processing method: 1 – fuel; 2 – conversion agent; 3 – converter; 4 – syngas (CO and H_2); 5 – separating system; 6 – released H_2; 7 – depleted gas mixture (CO and H_2); 8 – hydrogen disposal system; 9 – high-temperature oxidation system (HTOS); 10 – high-temperature fuel cell (HTFC); 11 – fuel cell anode; 12 – fuel cell cathode; 13 – output streams H_2O 14 – output streams CO_2; 15 – separate stream of H_2; 16 – oxidizing gas mixture (e.g. oxygen or air oxygen).

The depleted gas mixture is transferred to the high-temperature oxidation system (HTOS), comprising high-temperature fuel cell (HTFC). In the high-temperature oxidation system, including the high-temperature fuel cell, the CO and H_2 are oxidized at the $T>T_{conv.}$. The hydrogen depleted gas stream is directed to the high-temperature fuel cell anode. To the high-temperature fuel cell cathode an oxidizing gas mixture (e.g. oxygen or air oxygen) is conducted. On the high-temperature fuel cell anode CO electrochemical oxidation takes place according to reaction:

$$CO + 0.5 \cdot O_2 - 2e \rightarrow CO_2 + \delta Q + \delta Q' \tag{4}$$

H_2 electrochemical oxidation according to reaction:

$$H_2 + 0.5\ O_2 - 2e \rightarrow H_2O + \delta Q + \delta Q' \tag{5}$$

with electric energy production (δQ) and emission of heat (δQ′).

The edges of the calculation are determined the values of the electric component η_{HTFC} and η_{LTFC} from 50 to 60%. Unattainable ideal limits the effectiveness for LTFC and HTFC equal to $\eta = 83\%$ (ideal) and $\eta = 73.2\%$ (ideal) respectively [21].

Table 1 presents data on the electrochemical oxidation heats ($\Delta H^{el.}$) required amount of H_2 in stages HTFC processes in accordance with the efficiency (η) of the fuel cell system under consideration tandem.

Heat transfer from the high-temperature fuel cell to the converter is carried out by way of heat and mass transfer, heat exchange, emission or their combinations. Part of the heat carried over by the output streams (H_2O, CO_2) is regenerated in the counter flow heat exchanger to heat the input streams [24] (Fig. 2).

At the high-temperature fuel cell output H_2O and CO_2 are formed as a result of the electrode reaction, which can be used as a conversion agent of the vapor and carbon dioxide conversion. It is suggested to use the solid oxide fuel cell or a molten-carbonate fuel cell or its combination as the high-temperature fuel cell with the anion conductance. The heat emitted in the process (δQ) is transferred to the converter. The amount of heat emitted in the high-temperature oxidation system during the CO and the specified H_2 amount oxidation reaction must be at least enough to maintain the prescribed temperature in the converter.

Table 1. Electrochemical oxidation heats of H_2 on stages of the process in HTFC and LTFC.

LTFC	η=50%	η=55%	η=60%	η=83%(ideal)
	ΔH^{el} [kJ]	ΔH^{el} [kJ]	ΔH^{el} [kJ]	ΔH^{el} [kJ]
HTFC, η=50%, α=0.29				
CO (HTFC)	-141.5			
H_2 (HTFC) 0.68 moles	-84.3			
H_2(LTFC) 2.32 moles	-331.5	-364.7	-397.8	-550.3
B=C-H	11.22			
HTFC, η=55%, α=0.42				
CO (HTFC)	-155.6			
H_2 (HTFC) 0.88 moles	-120.0			
H_2(LTFC) 2.12 moles	-302.6	-332.8	-363.1	-502.3
B=C-H	12.73			
HTFC, η=60%, α=0.61				
CO (HTFC)	-169.8			
H_2 (HTFC) 1.14 moles	-169.0			
H_2(LTFC) 1.86 moles	-266.4	-293.1	-319.7	-442.3
B=C-H	14.6			
HTFC, η=73.2 % (ideal), α=3.04				
CO (HTFC)	-207.1			
H_2 (HTFC) 2.26 moles	- 409.7			
H_2(LTFC) 0.74 moles	-106.1	-116.7	-127.4	-176.2
B=C-H	23.0			

B=C-H – heat balance cooling minus heating streams.

α= H_2(HTFC)/H_2(LTFC), depends only on the efficiency HTFC.

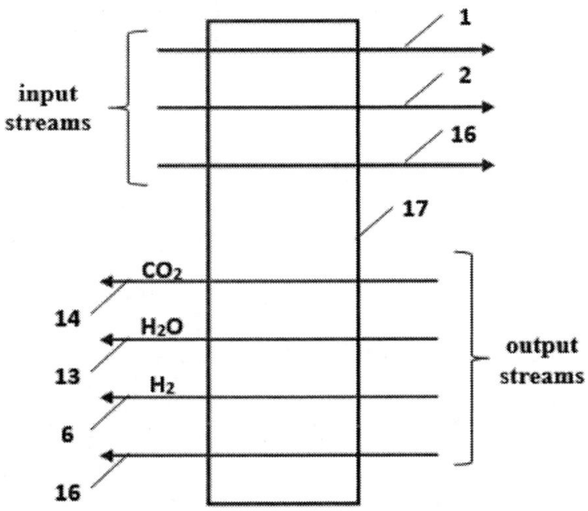

Figure 2. Counter flow heat exchanger scheme: 1 – fuel; 2 – conversion agent; released H_2; 13 – output streams H_2O; 14 – output streams CO_2; 16 – oxidizing gas mixture (e.g. oxygen or air oxygen); 17 – counter flow heat exchanger.

The heat from the hot gases released from the high-temperature oxidation system is transferred to the high-temperature oxidation system input oxidizer stream as well as to the carbon-containing fuel and the conversion agent at the converter input.

Adjusting of the amount of H_2 transferred to the high-temperature oxidation system can be done by way of H_2 extraction rate adjustment from syngas in the separating system. It is preferable to maintain the steady H_2 extraction rate from syngas in the separating system. Adjusting of the amount of H_2 transferred to the high-temperature oxidation system is performed by means of additional H_2 amount from the extracted part, which is transferred the high-temperature fuel cell either in a separate stream of or joining the depleted gas mixture stream (CO and H_2).

The amount of hydrogen directed to the high-temperature fuel cell must be at least enough to maintain the prescribed temperature in the converter by means of the heat emitted in the high-temperature oxidation system during the CO and the specified H_2 amount oxidation reaction. To raise the converter temperature, the amount of heat (δQ) transferred from the high-temperature oxidation system to the converter is increased by way of increasing the amount of the hydrogen fed to the high-temperature fuel cell. To lower the converter temperature accordingly the amount of heat (δQ) transferred from the high-temperature oxidation system to the converter is decreased by way of increasing the amount of the hydrogen fed to the high-temperature fuel cell. Thus in case of the small heat losses the heat (δQ) emitted in the high-temperature oxidation system is nearly fully absorbed by the endothermic reaction (1) in the converter.

Table 2 presents data per 1 mole of hydrogenous fuel (CH_4) on thermodynamic balance on thermal streams of heating initial substances and cooling reaction products

The H_2 share transferred to the high-temperature fuel cell after the separating system, which is the H_2 proportion in the syngas, is further characterized by the γ parameter – quantity of hydrogen moles (from the 1 mole of the hydrogenous fuel formed in the syngas according to reaction (1)) transferred to the high-temperature fuel cell. The γ value is defined by the high-temperature fuel cell coefficient of efficiency value. The higher is the high-temperature fuel cell coefficient of efficiency, the bigger is the prescribed γ for the converter temperature regulation. Increase of the heat leakage ($\delta Q'$) (through the heat insulation) as well as the heating of the input streams ($\delta Q'$) of the hydrogenous fuel, conversion agent and the oxidizer-rich gas results in the need for γ increase.

Table 2. Thermodynamic balance on thermal streams of heating initial substances and cooling reaction products.

Heating [kJ]		Cooling [kJ]		
HTFC	**HTFC**	**HTFC**	**HTFC**	**C→LTFC**
O_2/H_2	**O_2/H_2**	**CO_2**	**H_2O**	**H_2**
HTFC η= 50%				
(Δ = 0.3)				
7.7	11.4	33.1	50.2	48.2
$\sum$heat. =131.2		$\sum$cool. =131.5		
HTFC η = 55%				
(Δ =8.7)				
10.0	11.4	33.1	65.2	44.0
$\sum$heat. =133.5		$\sum$cool. =142.2		
HTFC η = 60%				
(Δ = 19.3)				
12.9	11.4	33.1	83.9	38.7
$\sum$heat. =136.4		$\sum$cool. =155.7		

When considering the data presented in Table 2, it should be borne in mind that:

heating 1 mole of fuel (CH_4) for the conversion (38.3 kJ) is it added to the calculation sum of expenses for heating (Σheat.);

heating 1 mole of converting agent (H_2O) (73.8 kJ) is it added to the calculation sum of expenses for heating (Σheat.);

$\Delta = \sum$cool. - $\sum$heat.;

C→LTFC – heat flow from the converter to LTFC.

Table 3 based on the results of the calculations presented in Tables 1 and 2 shown the distribution of hydrogen produced by the conversion of (Eq. 1) is supplied to HTFC for expenses endothermic converter reaction and for input streams heating losses compensation, as well as the amount of hydrogen released (for LTFC), ignoring losses during the heat recovery for the compensation of losses of the amount of heat necessary for the input streams heating. For all considered values of efficiency HTFC there is a positive thermal balance and flow of H_2 in HTFC not required for input streams heating losses compensation [24].

Table 3. Distribution of H_2 on the electrochemical oxidation systems.

Efficiency HTFC, [%]	**(γ) Amount of hydrogen, fed to HTFC, including:**		**Released hydrogen (LTFC)**	
	For the converter reaction, [mol]	**For input streams heating losses compensation, [mol]**	**[mol]**	**contained in the syngas, [%]**
50	0.68	not required	2.32	77.3
55	0.88	not required	2.12	70.6
60	1.14	not required	1.86	62.1

The data analysis presented in Tables 1-3 shows that for all values considered HTFC efficiency observed positive heat balance, and feed H_2 HTFC not required to fill the heating expenses of input flows, as high temperature heat of the electrochemical oxidation of CO in HTFC enough to ensure expenses converter process. Oxidation of H_2 remained quantity in LTFC increases a contribution of useful work on this stage with growth of LTFC efficiency.

When the temperature of the conversion is reduced (lower than the minimal prescribed – 800K) the rate of H_2 extraction from the mixture is lowered, and when the temperature of the conversion raises (higher than the maximal prescribed – 1100K) it is increased. The amount of H_2 transferred to the high-temperature oxidation system is changed in the separating system using the selective membrane [19] (permeable for H_2 and impermeable for CO) or pressure swing absorption (PSA) [17] by way of adjusting the rate of H_2 extraction from the mixture. The rate of H_2 extraction is adjusted by changing the pressure swing on the membrane or by changing the frequency of cycles or the pressure swing in the PSA cycle. The amount of H_2 extracted from the mixture in the separating system is adjusted by adjusting its quantity and transferred to the high-temperature oxidation system.

To ensure the maximum methane fuel efficiency the hydrogen released from the separating system is transferred to the low-temperature fuel cell included in the hydrogen disposal system.

3 Results and discussion

Estimation of the effective efficiency tandem diverse temperature oxidation system was carried out in two ways: based on the Gibbs energy change (ΔG) stages helpful electrical work is done in a low-temperature fuel cells and high-temperature fuel cells [21] and heat balance burnt fuel (ΔQ) making full use of the recovered heat.

$$\eta(\Delta G) = \frac{\Delta G_{CO}(HTFC) + \gamma \Delta G_{H2}(HTFC) + (3-\gamma)\Delta G_{H2}(LTFC)}{\Delta H_{comb.CH4}}, \tag{6}$$

where:

$\Delta G_{CO}(HTFC)$ – the Gibbs energy change oxidation reaction CO (Eq.3) in HTFC;

$\Delta G_{CO}(HTFC)$ = -196.6 kJ;

$\Delta G_{H2}(HTFC)$ – the Gibbs energy change oxidation reaction γ moles H_2 (Eq.2) in HTFC to repay the expenses of the endothermic reforming reaction;

$\Delta G_{H2}(HTFC)$ = -193.1 kJ;

$\Delta G_{H2}(LTFC)$ – the Gibbs energy change oxidation reaction (3- γ) moles H_2 (Eq.2) in LTFC;

$\Delta G_{H2}(LTFC)$ = -237.3 kJ;

$\Delta H_{comb.CH4}$ – heat combustion CH_4 in oxygen (Eq.7).

$\Delta H_{comb.CH4}$ = -800.9 kJ.

$$CH_4 + 2O_2 = CO_2 + 2H_2O \qquad (7)$$

In calculating the effective efficiency of the heat balance burnt fuel (ΔQ.) (Eq. 8) it is believed that helpful work is all the heat except the water heat formation at T = 300 K in LTFC.

$$\eta(\Delta Q) = \frac{\Delta H_{comb.CH4} - \Delta H_{form.H2O(l)} \cdot (3-\gamma)\cdot(1-\eta^{el}(LTFC))}{\Delta H_{comb.CH4}}, \qquad (8)$$

where:

$\Delta H_{form.H2O(l)}$ – liquid water heat formation (Eq. 2);

$\Delta H_{form.H2O(l)}$ = -241.81 kJ;

$\eta^{el}(LTFC)$ – electrical efficiency of LTFC;

$\frac{\Delta H_{form.H2O(l)} \cdot (3-\gamma)\cdot(1-\eta^{el}(LTFC))}{\Delta H_{comb.CH4}}$ – the amount of non-convertible heat.

Table 4 presents data on the effective values of the efficiency of the tandem and multi-temperature oxidation system. The edges of the calculation are determined the values of the electric component η_{HTFC} and η_{LTFC} from 50% to 60%.

Table 4. The overall effective efficiency ($\eta_{\Sigma ef}$) tandem diverse temperature oxidation system.

HTFC	**LTFC**			
	η=50%, α=0.293		**η=60%, α=0.604**	
	0.68 moles H_2 (HTFC) 2.32 moles H_2 (LTFC)		1.13 moles H_2 (HTFC) 1.87 moles H_2 (LTFC)	
	based on ΔG	based on ΔQ	based on ΔG	based on ΔQ
	$\eta_{\Sigma ef}$ [%]	$\eta_{\Sigma ef}$ [%]	$\eta_{\Sigma ef}$ [%]	$\eta_{\Sigma ef}$ [%]
η =50%	61.7	63.3	68.5	69.7
η =60%	69.1	70.6	74.6	75.8

Considering the data presented in Table 4, it should be noted that the overall effective efficiency ($\eta_{\Sigma ef}$, making full use of the recovered heat) considered tandem system depends on the efficiency of its constituent fuel cells.

When η_{HTFC} = 50% the overall effective efficiency ($\eta_{\Sigma ef}$) increases from 61% to 69% (making full use of the recovered heat). Step CO oxidation in HTFC contributes ~ 15% into helpful work, H_2 oxidation step in HTFC ~ 9%, H_2 oxidation step in LTFC ~ 37% to 45% (depending on the efficiency of LTFC). When η_{HTFC} = 60% the overall effective efficiency ($\eta_{\Sigma ef}$) increases from 69% to 75% (making full use of the recovered heat). Step CO oxidation in HTFC contributes ~ 19% into helpful work, H_2 oxidation step in HTFC ~ 19%, H2 oxidation step in LTFC ~ 31% to 37% (depending on the efficiency of LTFC).

Integrated approach addressing this issue involves getting electricity and heat in the area of consumption, which will reduce losses and improve the reliability of power supply. Thus, the appropriateness of hydrogen production from organic waste, followed by its usage for production of electrical energy and heat may be implemented in the considered approach.

4 Conclusions

1. With the η_{HTFC} growth, the amount of hydrogen transferred to the high-temperature fuel cell to compensate the losses of the endothermic reaction in the converter increases and accordingly the amount of hydrogen transferred to the high-temperature fuel cell to compensate the losses connected with the input streams heating decreases. The additional hydrogen for the compensation of the losses connected with the input streams heating in the high-temperature fuel cell is not necessary for all the considered η_{HTFC} values.
2. With the η_{LTFC} growth when burning the H_2 moles left the useful yield (ΔG) contribution of this stage increases, and $\eta_{\Sigma ef}$ rises.
3. To compensate the losses connected with the input streams heating in the high-temperature fuel cell it is required to transfer additional hydrogen, and the heat emitted during its burning quenches the cold input streams heating creating positive heat balance.
4. The overall electric efficiency of the tandem installation including the fuel converter, separating system, high-temperature oxidation system, and hydrogen disposal system in case of fuel use in the low-temperature fuel cell, is higher than for each of the fuel cell elements separately.
5. The suggested solution based on the preliminary conversion out of the high-temperature fuel cell eliminates the high-temperature fuel cell input section cooling by the endothermic reaction, which is one of the most important reasons for the direct fuel-fired high-temperature fuel cell accelerated degradation.

References

1. D.O. Dunikov, V.I. Borzenko, S.P. Malyshenko, D.V. Blinov, A.N. Kazakov, Teploenergetika [Thermal Engineering], **60(3)**, 202-211 (2013)
2. S.P. Malyshenko, O.V. Nazarova, Yu.A. Sarymov, Atomic hydrogen energy and technology, **7**, 105-126 (1986)
3. G. Gahleitner, International Journal of Hydrogen Energy, **38(5)**, 2039-2061 (2013)
4. A.N. Chusov, M.Yu. Zubkova, V.V. Korablev, V.I. Maslikov, D.V. Molodtsov, St. Petersburg State Polytechnical University Journal, **4-1(183)**, 78-85 (2013)
5. M.Yu. Zubkova, V.I. Maslikov, D.V. Molodtsov, A.N. Chusov, Advanced Materials Research, **941-944**, 2107-2111 (2014)
6. M.Yu. Zubkova, V.I. Maslikov, D.V. Molodtsov, A.N. Chusov, Applied Mechanics and Materials, **587-589**, 330-337 (2014)
7. V.V. Zhazhkov, M.Yu. Zubkova, V.I. Maslikov, D.V. Molodtsov, A.N. Chusov, D.V. Semenenko, Applied Mechanics and Materials, **725-726**, 1602-1607 (2015)
8. R. Chaubey, S. Sahu, O.O. James, S. Maity, Renewable and Sustainable Energy Reviews, **23**, 443-462 (2013)
9. V.V Okrepilov, M.V. Leonidovich, Asian Social Science, **11 (7)**, 312-325 (2015)
10. N.I. Vatin, A.S. Gorshkov, D.V. Nemova, A.A. Staritcyna, D.S. Tarasova, Advanced Materials Research, **941-944**, 905-920 (2014)
11. V.V Okrepilov, V.L. Makarov, A.R. Bakhtizin, S.N Kuzmina, Economy of Region, **2**, 301-313 (2015)
12. R. Kamnik, B. Kovačič, B. Pribicević, A. Đapo, Geodetski List, **69 (3)**, 171-188 (2015)
13. S. Giddey, F.T. Ciacchi, S.P.S. Badwal, Ionics, **11**, 1-10 (2005)

14. U. Bossel, inventor; Asea Brown Boveri Ltd., assignee. Device for conversion of chemical energy from hydrocarbons into electric energy by an electrochemical high-temperature process. United States patent US 5,079,105 (January 7, 1992)
15. B.G. Keefer, M.L. Babicki, M.H. Kirby, inventors; QuestAir Technologies Inc., assignee. Enhanced solid oxide fuel cell systems. United States patent US 7,285,350 (October 23, 2007)
16. R.A. Geisbrecht, M.C. Williams, inventors; The United States of America as represented by the Department of Energy, assignee. Fuel cell-fuel cell hybrid system. United States patent US 6,623,880 (September 23, 2003)
17. B.G. Keefer, inventor; QuestAir Technologies Inc., assignee. High temperature fuel cell power plant. United States patent US 7,097,925 (August 29, 2006)
18. B.G. Keefer, D.J. Connor, C.F. Hunter, inventors; QuestAir Technologies Inc., assignee. Energy efficient gas separation for fuel cells. United States patent US 7,087,331 (August 8, 2006)
19. A.I. Livshits, *Proc. Reports. 4-th Russian Conference "Physical Problems of Hydrogen Energy"*, 1-3 (2007)
20. J.M. Pondo, D.J. Goodwin, C.P. Sishtla, J. Randy, R.J. Remick, inventors; Gas Technology Institute, assignee. Thermally integrated fuel cell power system. United States patent US 7,008,711 (March 7, 2006)
21. A.A. Stroganov, M.U. Zubkova, *Ivanovo: XVI International conference "Chemical Thermodynamic in Russia (RCCT 2007)" and X International conference "Problems Solvatic Complex Formation in Solution"*, **II**, 382-384 (2007)
22. P.L. Micheli, M.C. Williams, F.A. Sudhoff, inventors; The United States of America as represented by the United States Depratment of Energy, assignee. Indirect-fired gas turbine dual fuel cell power cycle. United States patent US 5,541,014 (July 30, 1996)
23. M.J. Skowronski, inventor; Southern California Edison Company, assignee. Power generation system utilizing turbine and fuel cell. United States patent US 5,811,201 (September 22, 1998)
24. A.A. Stroganov, M.Yu. Zubkova, inventors and assignee. Carbon-bearing fuel use in the system containing the high-temperature fuel cell. Russian Federation patent 2475899 RU (February 20, 2013)

MATEC Web of Conferences **53**, 01020 (2016)
DOI: 10.1051/matecconf/20165301020

Measuring the thermal resistance of typical multi-layer building walls using the accelerated measurement method

Daniil Zubkov[1], Alexander Chusov[2] and Dmitry Molodtsov[2,a]

[1]*Saint Petersburg National Research University of Information Technologies, Mechanics and Optics, Kronverksky pr. 49, St. Petersburg 197101, Russia*
[2]*Peter the Great St.Petersburg Polytechnic University, Polytechnicheskaya str. 29, St. Petersburg 195251, Russia*

Abstract. This paper considers the application of accelerated thermal resistance measurement to typical constructions of multi-layer walls of buildings with thermal insulation. It is shown that it is possible to reduce the measurement time considering the results of preliminary modelling. The required calculations for some typical constructions of building walls were made. The examples of simulation results shown in this paper allow to reduce the thermal resistance measurement time by 15-20%. There can be made calculations for other structures of building walls, following the presented method.

1 Introduction

The lack of thermal resistance of the main building structures is one of the main reasons for low energy efficiency of buildings built in the USSR in 1960-1990 years [1-4]. According to the research, in a traditional apartment house up to 40% of heat is lost through walls, 18% through windows, 10% through basement, 18% through roof and 14% through ventilation [5-8]. Heat loss is especially noticeable in the northern regions [9-11]. Usually buildings built before 1995 in central Russia have values of thermal resistance 2-3.5 times smaller than it should be according to the modern standards [12]. In 1995 in Russia the nation-wide regulatory requirements for thermal protection of buildings were established, increasing the resistance requirements for enclosing structures. Now it is regulated by SNiP 23-02-2003 "Thermal protection of buildings". Additional insulation on facades of buildings pays off in about 3.5 years [13-15], but it requires estimating of thermal resistance of the existing walls.

Methods for determination of thermal resistance, approved in Russia, require long-term measurements. For building walls, it takes more than 15 days, according to GOST [16]. Reduction of the measurement time will significantly simplify such work, reduce its cost and will make it easier to increase the energy efficiency of the buildings [17-19].

[a] Corresponding author: kovu@bigcats.ru

2 Materials and methods

One method of measuring the thermal resistance of the wall is the thermal stabilization of its sides at different temperatures and measuring the heat flux through the wall. A noticeable drawback of it is that you must wait for the system to steady state, which can take up to 15 days.

A method of accelerated measurement of thermal resistance of homogeneous walls is considered in the patent [20]. It is shown that when considering homogeneous walls, the use of the half-sum of incoming and outgoing heat flows becomes constant much earlier than the whole system. That allows to take measurements significantly earlier than the system achieves a steady state. The essence of this method is clearly demonstrated in Figure 1.

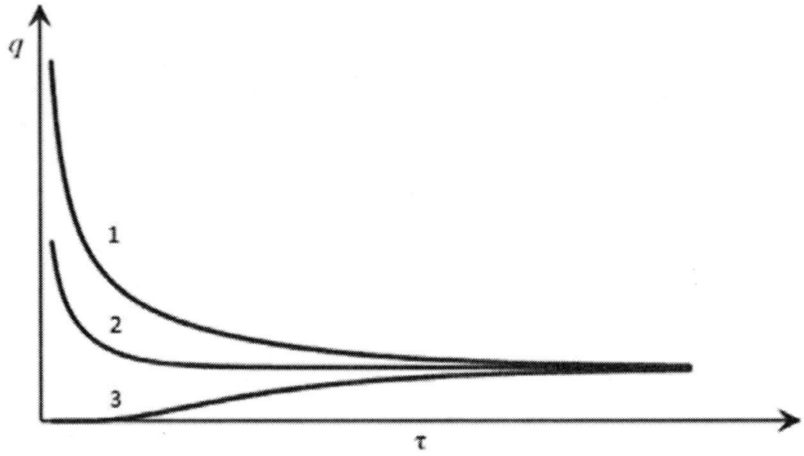

Figure 1. The essence of the original method. Time dependence of the incoming (1), outgoing (3) heat flows and their half-sum (2).

To determine the minimum measurement time with an error of less than 5% is used the following relationship:

$$a\tau/h^2 > 0.1 \quad (1)$$

where
a - thermal diffusivity of the wall;
h – its thickness;
τ – the measurement time.

As shown in [21], the use of this method for measuring the thermal resistance of multi-layer walls made of materials with significantly different thermal properties allows to reduce the measurement time by about 15%, but it is hard to estimate the minimum measurement time in advance with sufficient precision.

Since the structural elements the walls of buildings are built of (bricks, concrete blocks, sheets of insulation, etc.) have a number of standard dimensions, it is possible to produce a computer simulation of the measurement of thermal resistance for a number of typical constructions of walls [22]. Taking into account the results of the simulation, it is possible to make accelerated measurement of deviations of the expected thermal resistance values, taking the measurement time with some excess to take into account the possible influence of changed material properties.

3 Results and discussion

As was shown in [21], for the walls of brick and concrete without thermal insulation layers the relationship (1) works completely. In this paper are considered some models of building walls consisting of brick, concrete, polystyrene and mineral wool in the form of blocks in various combinations.

Let us consider in detail the case of a brick wall with a layer of polystyrene foam insulation and gypsum layers on both sides. The simulation was performed in the program Heat 2. A two-dimensional model represents a part of a vertical wall with the temperature of t = 20 °C at the inner surface and t = 0 °C on an outer surface. Adiabatic conditions are applied to the horizontal surfaces [21]. The material properties are taken from the built-in library of the program. Figure 2 shows the structure of the model, Figure 3 - the temperature field in a steady state, and Figure 4 – the specific heat flux field.

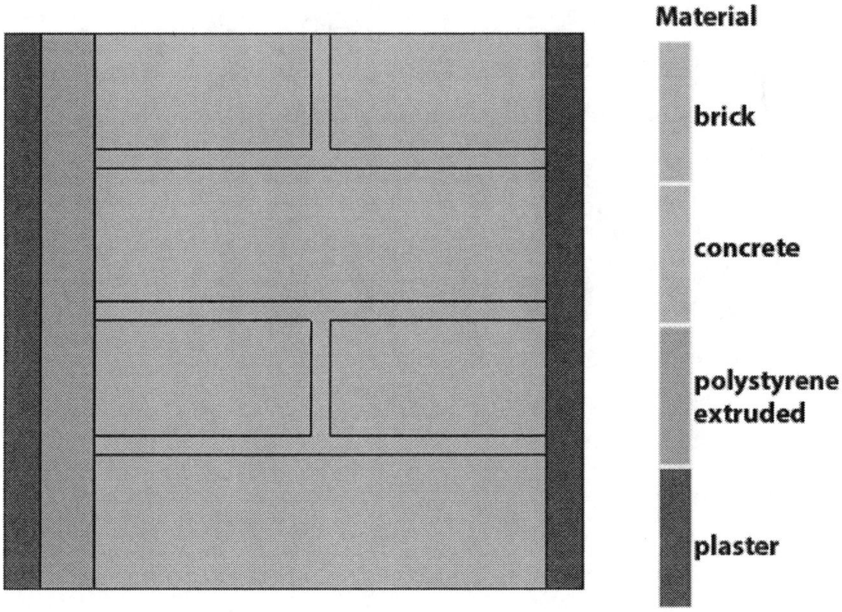

Figure 2. Structure of the model, brick wall with a layer of thermal insulation and two layers of plaster.

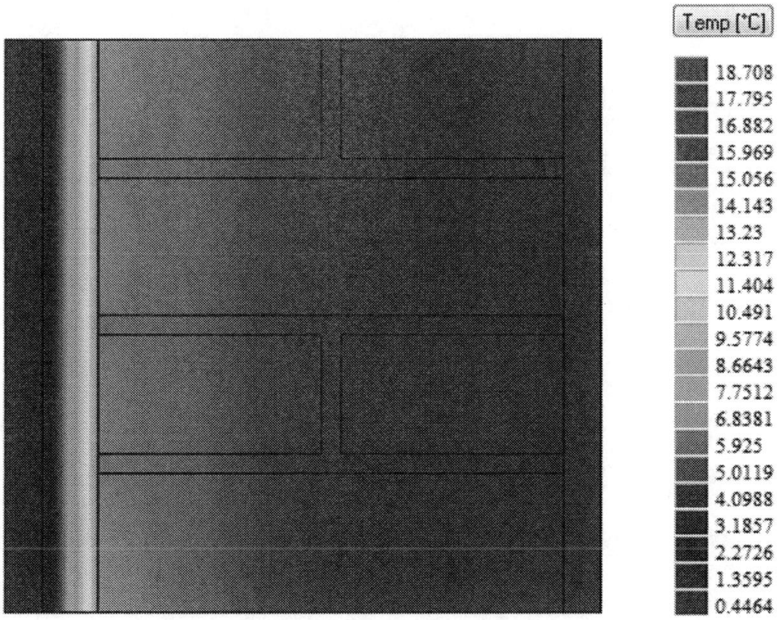

Figure 3. Results of the simulation. Temperature field in a steady state.

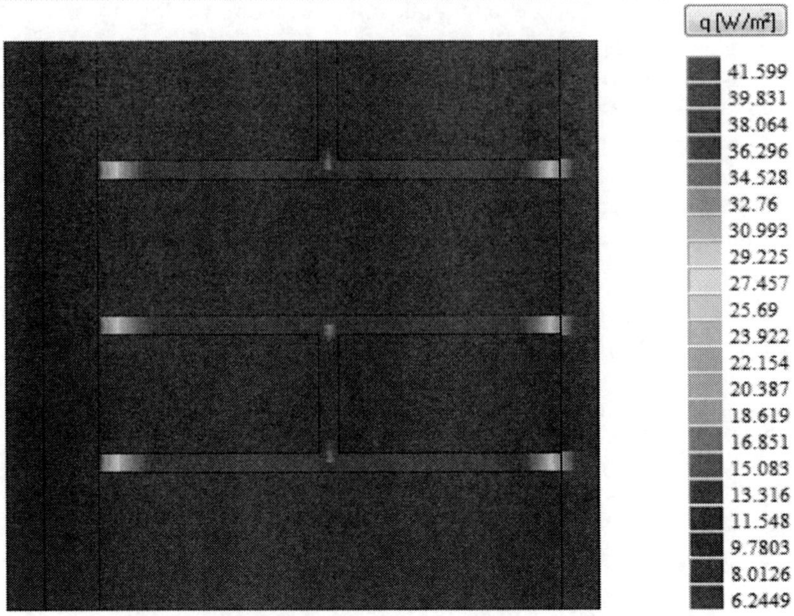

Figure 4. Results of the simulation. Specific heat flux field in a steady state.

During the simulation the program tracked heat flux on both of the vertical surfaces within 15 days of virtual time in increments of 15 minutes. The results are presented in Figure 5.

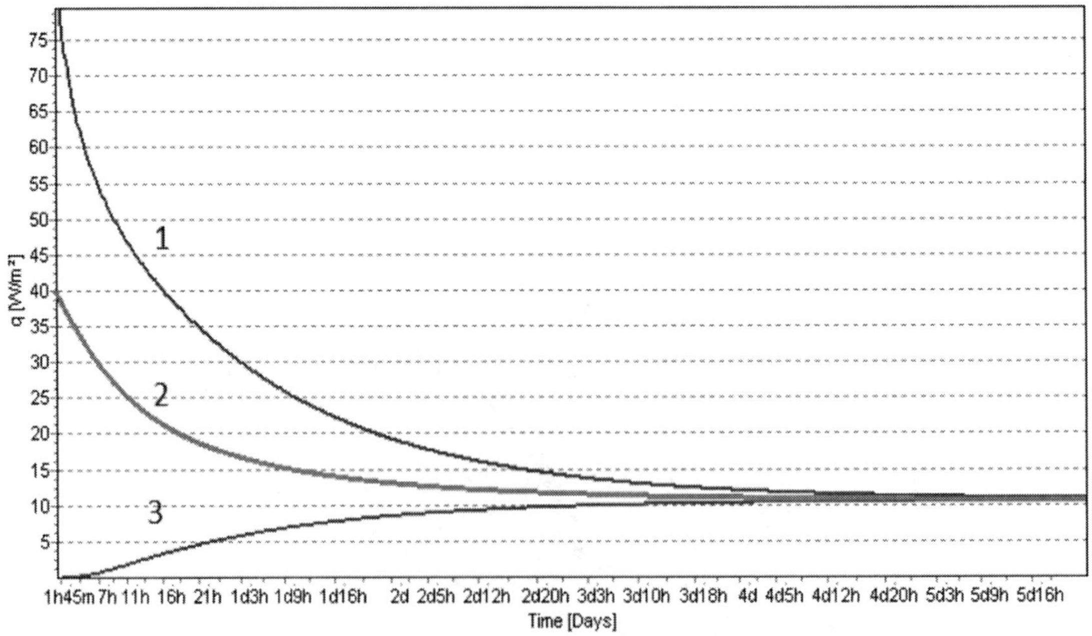

Figure 5. Results of the simulation. Time dependence of the incoming (1), outgoing (3) heat flows and their half-sum (2).

It was found that with a deviation of 5% the values of incoming and outgoing heat flows become equal to the flows obtained after 15 days in 5 days and 2 hours (τ_1), and half the sum of these flows becomes equal to the same flows in 3 days and 20 hours (τ_2). Thus, the reducing of measurement time is Δ=24,6%. Similarly, were considered some other types of building wall structures. The results are shown in Table 1.

Table 1. Simulation results for various wall designs.

Wall structure	τ_1, h	τ_2, h	Δ, %
One brick (250 mm), polystyrene foam (20 mm), plaster (2 x 20 mm)	77	62	19.5
One brick (250 mm), polystyrene foam (30 mm), plaster (2 x 20 mm)	122	92	24.6
One brick (250 mm), polystyrene foam (50 mm), plaster (2 x 20 mm)	142,5	117	17.9
One brick (250 mm), polystyrene foam (20 mm), plaster (2 x 20 mm)	172,5	140	18.8
One brick (250 mm), mineral wool (50 mm), plaster (2 x 20 mm)	93,5	80	14.4
One brick (250 mm), mineral wool (100 mm), plaster (2 x 20 mm)	128	112	12.5

Thus, the use of the presented simulation results allows to reduce the thermal resistance measurement time by 15-20% (18% on the average for the provided examples). The results for other combinations obtained by a similar method also can be used to simplify the renovation of old buildings to increase their energy efficiency.

4 Conclusions

1. The results obtained by the presented method can be used to simplify the process of renovation of old buildings.
2. For acceleration the measuring of thermal resistance of typical design multilayer insulated wall can be used the results of calculations presented in this paper, or obtained by the same method.
3. The use of the calculations presented in this paper allows to reduce measurement time by 15-20%.

References

1. D. Nemova, V. Murgul, V. Pukhkal, A. Golik, E. Chizhov, N. Vatin, Journal of Applied Engineering Science, **12(1)**, 37-44 (2014)
2. A. Borodinecs, B. Gaujena, *10th International Conference on Healthy Buildings 2012*, **2**, 1715-1722 (2012)
3. O. Smorygo, V. Mikutski, A. Marukovich, V. Sadykov, V. Usoltsev, N. Mezentseva, A. Borodinecs, O. Bobrenok, *IOP Conference Series: Materials Science and Engineering*, **23** (2011)
4. V. Murgul, Journal of Applied Engineering Science, **12 (1)**, 1-10 (2014)
5. A. Borodinecs, J. Zemitis, A. Kreslins, B. Gaujena, *10th International Conference on Healthy Buildings 2012*, **2**, 1114-1119 (2012)
6. A. Borodinecs, B. Gaujena, J. Zemitis, A. Kreslins, *12th International Conference on Indoor Air Quality and Climate 2011*, **2**, 973-978 (2011)
7. A. Borodinecs, Z. Budjko, *9th International Conference and Exhibition - Healthy Buildings 2009* (2009)
8. A. Borodiņecs, A. Kresliņš, E. Dzelzitis, A. Krumiņš, *IAQVEC 2007 Proceedings - 6th International Conference on Indoor Air Quality, Ventilation and Energy Conservation in Buildings: Sustainable Built Environment*, **1**, 361-368 (2007)
9. A. Borodinecs, J. Zemitis, A. Prozuments, *World Renewable Energy Forum, WREF 2012, Including World Renewable Energy Congress XII and Colorado Renewable Energy Society (CRES) Annual Conferen*, **6**, 4181-4186 (2012)

10. A. Borodinecs, A. Kreslins, *Proceedings - 28th International PLEA Conference on Sustainable Architecture + Urban Design: Opportunities, Limits and Needs - Towards an Environmentally Responsible Architecture, PLEA 2012* (2012)
11. V.V Okrepilov, V.N. Krutikov, G.I. El'kin, Measurement Techniques, **57 (2)**,109-116 (2014)
12. A. Borodinecs, A. Kreslins, *Proceedings of Conference: Air Conditioning and the Low Carbon Cooling Challenge - Windsor 2008 Conference* (2008)
13. D.V. Nemova, N.I. Vatin, A.S. Gorshkov, A.V. Kashabin, P.P. Rymkevich, D.N. Tseitin, Construction of Unique Buildings and Structures, **8(23)**, 93-115 (2014)
14. E. Aronova, Ž. Radovanović, V. Murgul, N. Vatin, M. Shvarts, Applied Mechanics and Materials, **725-726**, 1505-1511 (2015)
15. V.L. Kvint, V.V Okrepilov, Herald of the Russian Academy of Sciences, **84 (3)**, 188-200 (2014)
16. GOST [Russian government technical requirements] 26254-84. Buildings and structures. Methods for determination of thermal resistance of enclosing structures.
17. N. Vatin, M. Petrichenko, D. Nemova, A. Staritcyna, D. Tarasova, Applied Mechanics and Materials, **633-634**, 1023-1028 (2014)
18. N.I. Vatin, D.V. Nemova, D.S. Tarasova, A.A. Staritcyna, Advanced materials research, **953-954**, 854-870 (2014)
19. V.V Okrepilov, A.Yu. Smirnov, Measurement Techniques, **56(1)**, 54-60 (2013)
20. V.A. Korablev, A.S. Nekrasov, D.S. Nechaev, inventors and assignee. Measurement of heat conductivity and heat resistance of construction Structure. Russian Federation patent 2527128 RU (August 08, 2014)
21. D.V. Zubkov, K.S. Tikhomirov, V.A Korablev, *Proseedings of best papers of XLIII Week of Science in St.-Petersburg Polytechnic University*, 306-312 (2014)
22. D. Zubkov, A. Chusov, D. Molodtsov, *Proceedings SPbUCEMF 2015, Urban Civil Engineering and Municipal Facilities, Saint-Petersburg, Russia; 18-20 March 2015. CD-ROM* (2015)

MATEC Web of Conferences **53**, 01021 (2016)
DOI: 10.1051/matecconf/20165301021

Master degree in Construction area (main results of the admission company)

Eliza Gumerova[1,a], Olga Gamayunova[1], Sofiia Pospelova[2]

[1] *St. Petersburg State Polytechnical University, Polytechnicheskaya, 29, Saint-Petersburg, 195251, Russia*
[2] *Plekhanov Russian University of Economics (Sevastopol branch), Vakulenchuka st., 29, Crimea, Sevastopol, 299053, Russia*

Abstract. Russia enacted a move to two-tier education in line with Bologna Process model more than 10 years ago. Nowadays after the bachelor's degree student can earn a master's degree (usually two years). Master degree programs give advanced knowledge of a specialized body of theoretical and applied topics and prepare graduates for a career in their chosen field. Master degree programs in the field of Construction Engineering are popular among enrollees. The article is about results of the admission to master's degree programs of the Institute of Civil Engineering of Peter the Great Saint-Petersburg Polytechnic University in 2015.

Introduction

Peter the Great St.Petersburg Polytechnic University is the national research facility which was founded in 1899 [1]. SPbPU is one of 15 Russian leading universities who entered the Ministry of Education and Science Program "5-100-2020" [17]. Its aim is to ensure at least 5 Russian universities in the TOP-100 of the world's leading universities according to the QS World University Rankings by 2020 [3].

The Institute of Civil Engineering is the legal successor of the Faculty of Civil Engineering founded in 1907 and considered one of the most respected construction faculties of Russia. The institute aims to prepare managers and leaders of construction companies and designing bureaus, heads of departments, chief engineers of projects, project managers and other highly qualified staff. [4]

Each student has opportunity to go abroad for internship and practice in foreign universities [20]. The institute carries out the "Double Degree" program; after a year training at a foreign partner institution a student receives both Russian and European diplomas at once [21].

Our Institute is proud of its graduates who have made the career of famous builders, scientists, heads of ministries and major politicians.

The Institute of Civil Engineering carries out the admission to the bachelor's degree, master's degree and engineer degree.

The Institute of Civil Engineering has been on the first place in Russia by the average result of the exam of students enrolled the programs in the field of construction engineering for the last 3 years. The Institute has the international accreditation of educational programs.

---

[a] Corresponding author : eliza_gumerova@mail.ru

The teachers of the Institute are engaged in training and research activities. Some of them are active managers and employees of the leading construction companies of the city [19].

Cooperation with construction companies is one of the foreground directions of the activities in the Institute. The majority of students from the 3rd course have already been working in partner-organizations of the Institute [13].

1 Master's degree programs of the Institute of Civil Engineering

The Institute of Civil Engineering (ICE) carries out the admission to the master's degree programs in two fields of study such as «Construction Engineering» and «Environmental and Water Use». Today the Institute of Civil Engineering offers a wide range of master's degree programs in the field of Construction Engineering and one in the field of Environmental and Water Use (table 1).

Table 1. Master's degree programs of the ICE [5].

Program name	Code of the program	Program profile
08.04.01 Construction Engineering	02	Marine hydraulic engineering constructions and constructions of waterways
	06	Organization and management of construction and investment projects
	08	Construction of landscape architecture facilities
	10	Computer-aided design of buildings and structures
	11	Engineering systems of buildings and structures
	12	Civil Engineering (master's degree program in English)
	13	Design, construction and management of renewable energy facilities
	14	Energy Efficient and Sustainable Buildings (master's degree program in English)
	15	Innovative methods of design and construction of hydraulic structures
	16	Design and construction of facilities of the housing and public utility sector
	17	Urban Construction and Economy
	18	Environmental engineering protection in construction

	19	Survey and technical audit of buildings and structures
	20	Technology of calculation, design and construction of industrial and civil buildings and structures with the use of modern software
20.04.02 Environmental and Water Use		Landscape arrangement of the territory

2 International master's degree programs (in English)

Nowadays there are two international master's degree programs in the Institute of Civil Engineering: 08.04.01.12 «Civil Engineering» and 08.04.01.14 «Energy Efficient and Sustainable Buildings». Training in international groups allows to improve knowledges of the English language, to get acquainted and learn more about other cultures of foreign students [4].

Graduates will get experience and practical skills in the field of Civil Engineering in the key subjects of structural and hydraulic engineering, project management and design. Major areas of expertise [10]:

- Structural design and analysis;
- Building services and facilities;
- Capital Investment Planning;
- Building Information Modeling skills;
- Energy efficient solutions;
- Construction Project Management;
- Publishing and scientific Research work.

The following teaching methods are applied to deliver the pertinent knowledge in civil engineering and to develop practical skills in the most efficient way [4]:

- Lectures and discussions. A lecturer explores a problem on a relevant topic to be considered and studied in a detailed way. Students are welcome to ask any pertinent questions at the end of a lecture.
- Seminars. A lecturer provides general guidance for a group working on an advanced study or research project. This approach is aimed at exchanging information and developing new solutions to problems under study by the group.
- Study assignments. This method is implied to orient students to a topic prior to classroom or laboratory work. Students are normally asked to read periodicals or any relevant teaching materials, to do some research work or to exercise for more practice.
- Demonstration. This method implies actual performing an operation or doing a job which shows students what to do and how to do it. Students might get valuable empirical knowledge in the field of expertise and develop vocational skills.
- Scientific research activities and publishing. These activities imply working in research labs as well as writing scientific articles under a professor's supervision. Students can get their research and scientific articles published on global citation websites.
- Brainstorming and role plays. These approaches make it possible for a lecturer to engage students in teamwork and to stimulate more active participation in order to fix the knowledge acquired during studies.

Major areas of program «Energy Efficient and Sustainable Buildings»:

- Up-to-date energy efficient technologies and solutions
- Civil engineering insights;
- Structural design and analysis;
- Construction Projects Management in the context of energy efficiency and sustainability;
- Investment and Budgeting of energy efficient buildings;
- Publishing and scientific research work.

Key Points of program «Energy Efficient and Sustainable Buildings»:

- individual approach and on-going assistance;
- online distance learning using MOODLE (www.dl.spbstu.ru) [9];
- communication with experts from leading construction and engineering companies;
- scholarships to study abroad in English in one of the European partner universities (a semester or a summer school)
- the course is fully taught in English;
- scientific research activities, regular participation in seminars, conferences and forums;
- joint projects and research activities with teachers' engagement;
- using advanced computer technology in construction;
- online communication between students and teachers via intranet web-site based on MOODLE system.

3 Results and Discussion

Each student should pass Multiple Choice Test in a relevant field of studies to be enrolled for Master's Degree Programs. The Institute of Civil Engineering ranked first on admission to training engineers, bachelor and master degrees in the field of Construction Engineering on budgetary basis in the Northwestern Federal District (table 2).

Table 2. Plan of the admission to baccalaureate, specialty, magistracy in the field of Construction Engineering in the universities in the Northwestern Federal District (2015).

	ICE (SPbPU)	SPbGASU [6]	The National Mineral Resources University [7]	St. Petersburg State University of Communication [8]
Academic baccalaureate («Construction Engineering»)	198	150	-	-
Practical baccalaureate («Construction Engineering»)	-	150	-	-
Specialty («Construction of Unique Buildings and Constructions»)	57	40	-	-
Magistracy «Construction Engineering»	200	80	-	17

In table 3 there are results of the admission to master's degree programs in full-time or correspondence form of education in 2015.

Table 3. Results of the admission to master's degree programs of the Institute of Civil Engineering in 2015.

	Construction Engineering		Environmental and Water Use	
	Full-time form	Correspondence form	Full-time form	Correspondence form

Plan of the admission	200	60	20	-
Quantity of submitted applications	274	38	14	-

The quantity of submitted applications are graduates from more than 10 universities of Russia and foreign countries. Nevertheless, the main part of enrolees completed the bachelor degree at SPbPU (42%) and SPbGASU (24%) (fig.1).

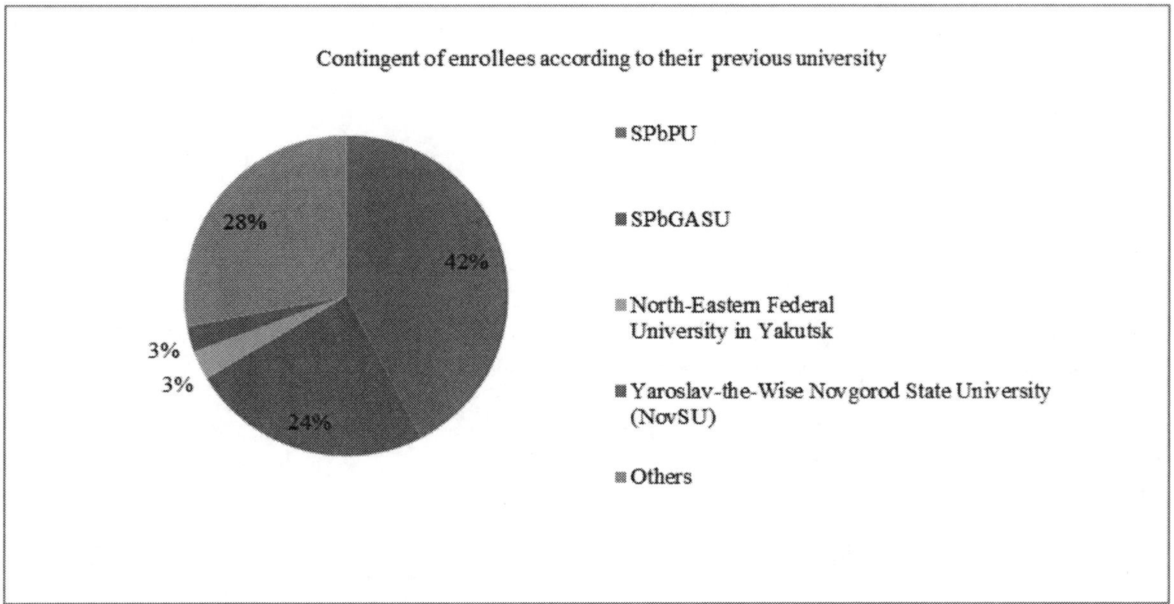

Figure 1. Contingent of enrollees according to their previous university.

Information about the master's profile program and the plan of admission is on the figure 2. The competition is based on the profile that was listed as prior to others in the application. There are five profile programs that are popular among enrollees:

- 08.04.01.06 «Organization and management of construction and investment» (competition 1.92)
- 08.04.01.10 «Computer-aided design of buildings and structures» (competition - 2.67)
- 08.04.01.12 «Civil Engineering» (competition – 2.1)
- 08.04.01.17 «Urban Construction and Economy» (competition - 3.92)
- 08.04.01.20 «Technology of calculation, design and construction of industrial and civil buildings and structures with the use of modern software» (competition - 1.75)

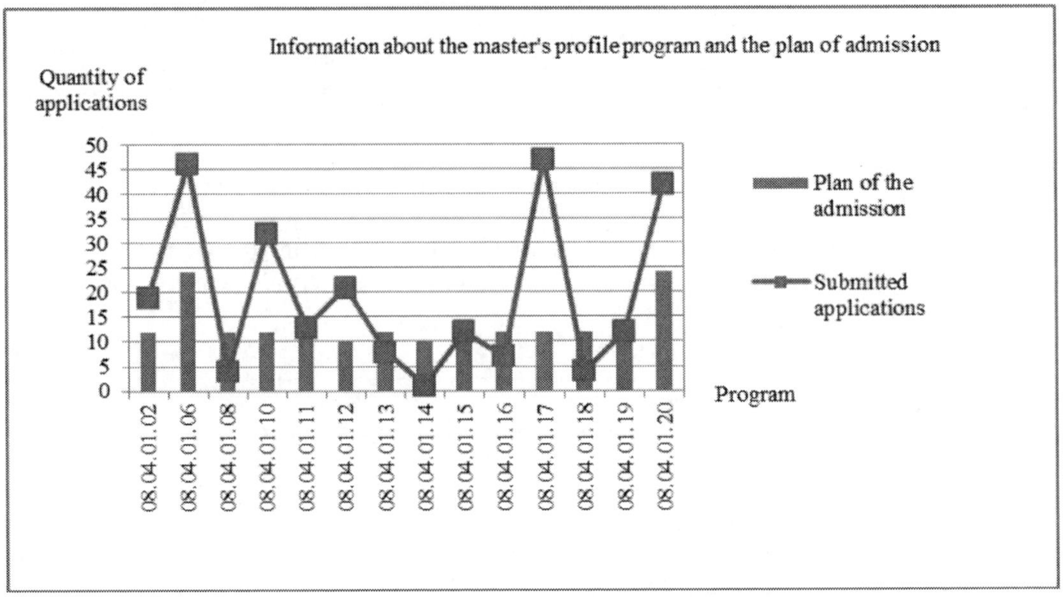

Figure 2. Information about the master's profile program and the plan of admission.

The vast majority of bachelors with the diploma with honors successfully passed the exam and entered the magistracy of the Institute of Civil Engineering. 77% and 63% of enrollees were enlisted to master's degree programs in full-time and in correspondence form of education in 2015. (Fig.3)

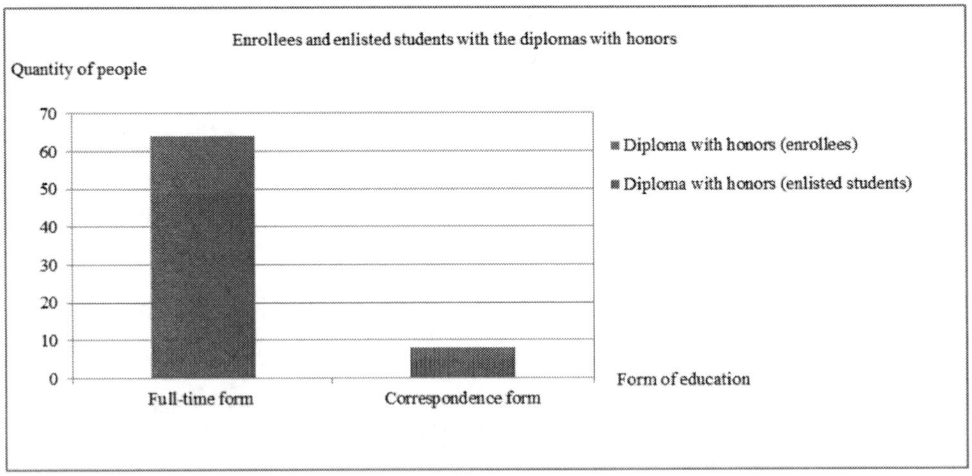

Figure 3. Enrollees and enlisted students with the diplomas with honors.

The percentage ratio of enrollees with the diplomas with honors depends on their previous place of study. Mostly, there are graduates from SPbPU (38%) and SPbGASU (16%). (Fig.4)

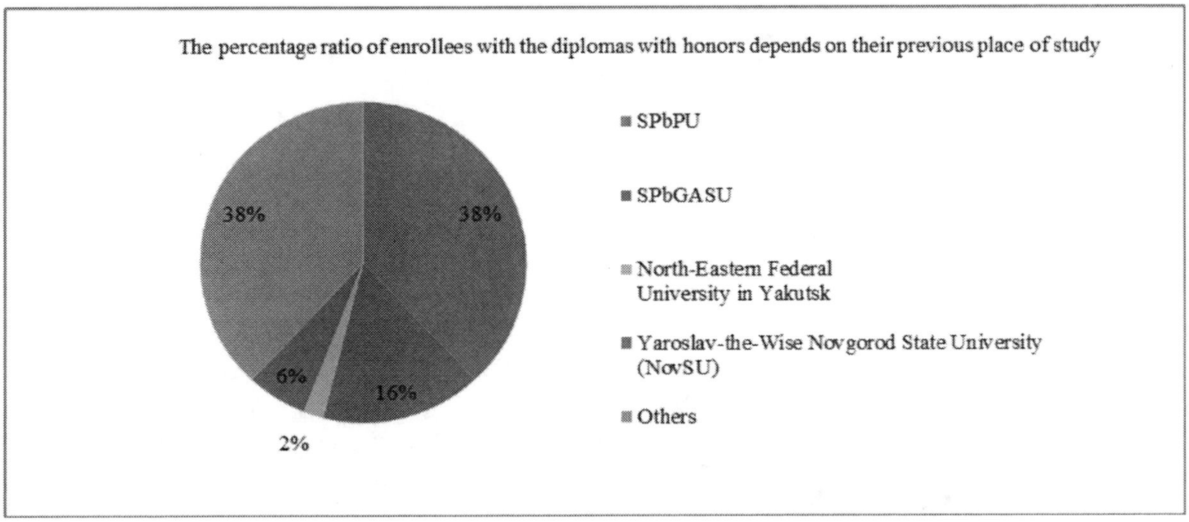

Figure 4. The percentage ratio of enrollees with the diplomas with honors depends on their previous place of study.

Each enrollee was to write multiple choice test in field of Construction Engineering. This test took into consideration not only students' theoretic knowledges, but also their scientific works and publications. The results of the exam showed that graduates from SPbPU showed better results than students from other universities (Fig.5). However, according to the chart it is impossible to make an unambiguous conclusion about the quality of students. Graduates of SPbPU showed the best results due to their publications, while the majority of graduates of other universities did not have them at all.

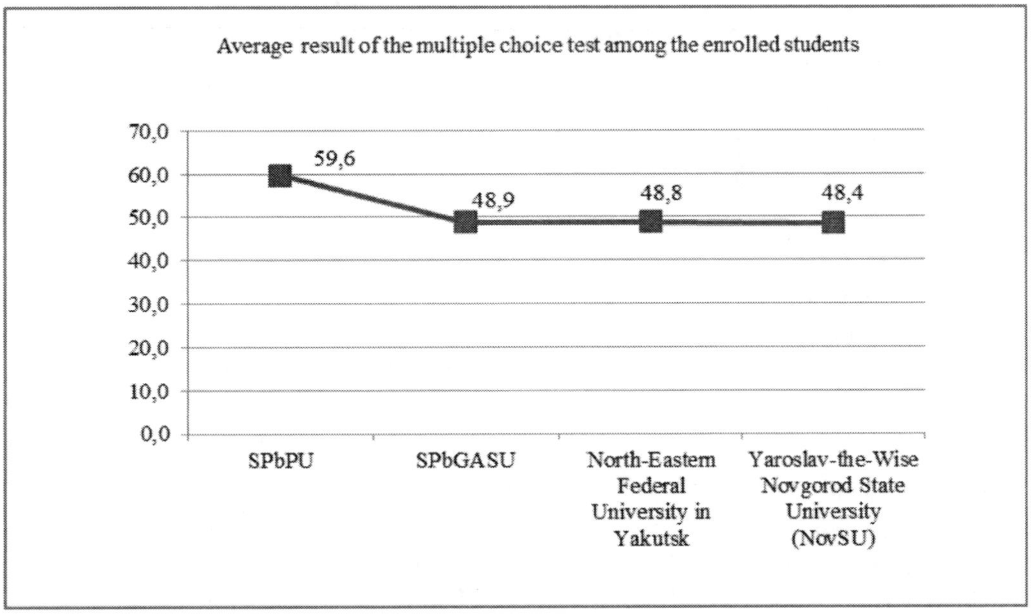

Figure 5. Average result of the multiple choice test among the enrolled students.

The maximum amount of points for the test is 100 points. According to the figure 6 only the graduates from SPbPU wrote the exam and scored above 60 points. Graduates from other universities only have crossed the threshold value of success (40 points) and a some of them got 50-60 points.

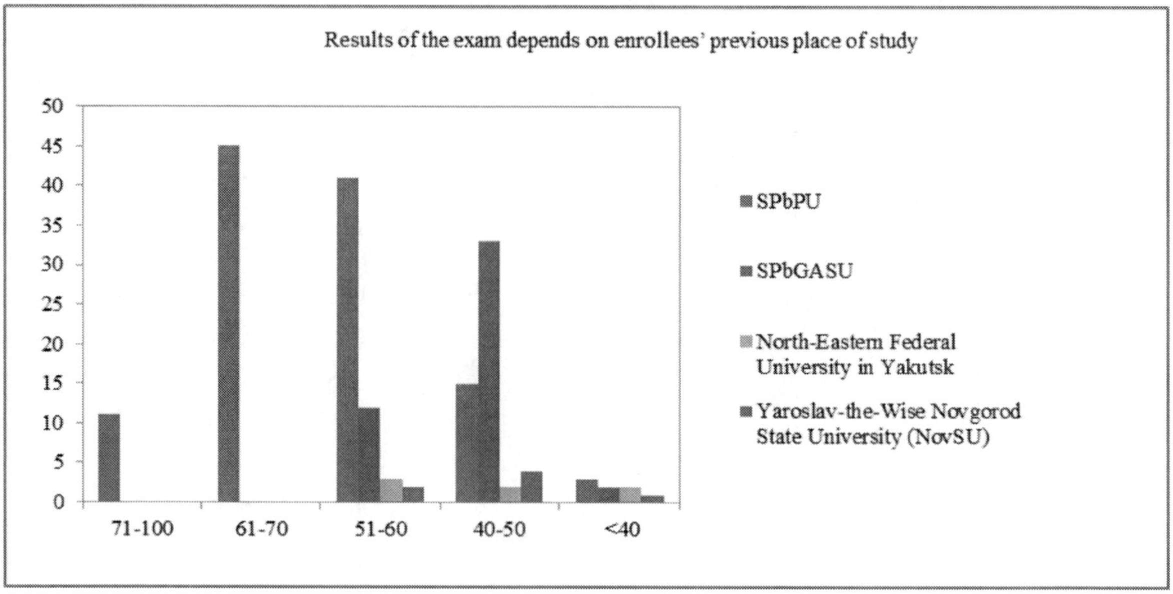

Figure 6. Results of the exam depends on enrollees' previous place of study.

Foreign students are interested in the master's degree program in the field of Constriction Engineering [14]. Primarily, there are citizens of Uzbekistan, Georgia, Iran, Syria, Egypt, China, India and Africa. The division of foreign students into profile programs is showed on the figure 7.

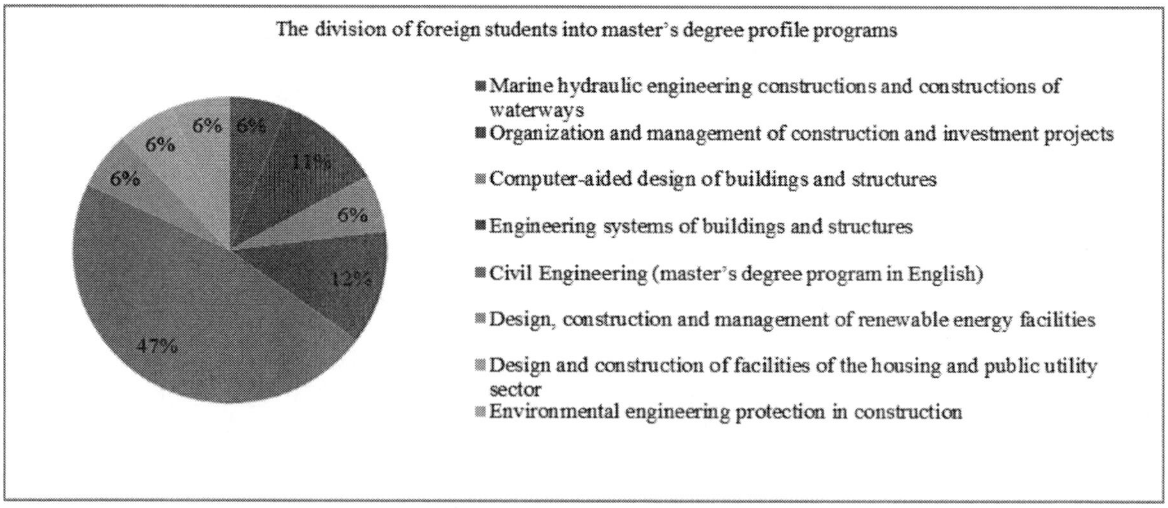

Figure 7. The division of foreign students into master's degree profile programs.

Almost the half of the foreign students (47%) prefers international master's degree profile program "Civil Engineering". This program is in the English language and it is easier for international students to master the educational program. Nevertheless, in SPbPU for many years the system of training, where foreigners learn Russian language [16]. That is why 53% of foreign students according to figure 7 are enrolled in master's degree programs in Russian language (fig.7).

Conclusion

The magistracy provides preparation of scientific and pedagogical staff, helps to identify the ability of students to the research work and thus, to determine the future direction of its activities. A master's degree gives the opportunity to feel confident in life and to find a high-paying job. The Institute of Civil Engineering is the leader in the admission to the master's degree programs in the field of

Construction Engineering among universities in the Northwestern Federal District of Russia. Graduates from SPbPU enrolling the magistracy of the ICE show the better results than graduates from other universities. International master's degree programs attract foreign students, which improves the position of SPbPU in the world rankings of universities.

References

1. Wikipedia. *Peter the Great St. Petersburg Polytechnic University.* Retrieved from http://www.wikipedia.org/
2. *Peter the Great St. Petersburg Polytechnic University.* Retrieved from http://www.alluniversitycourses.com/
3. Peter the Great St. Petersburg Polytechnic University. *Program «5-100».* Retrieved from http://5-100.spbstu.ru/en/about-program/info/
4. Peter the Great St. Petersburg Polytechnic University. *Programs in English.* Retrieved from http://eng.spbstu.ru/
5. Civil Engineering Faculty. Retrieved from http://www.cef.spbstu.ru/
6. SPbGASU. *Plan of the admission in 2015.* Retrieved from http://www.spbgasu.ru/Abiturientam/Priem/
7. The National Mineral Resources University. *Plan of the admission in 2015.* Retrieved from http://priem-univer.ru/plan-priema
8. St. Petersburg State University of Communication. *Plan of the admission in 2015.* Retrieved from http://pgups.ru/abitur/pk/plan_priyema/
9. A.V. Rechinskiy, N.I. Vatin, O.S. Gamayunova, K.Y. Usanova, Construction of Unique Buildings and Structures, **2**, 6-17 (2012)
10. O.S. Gamayunova, V.V. Ershov, A.A. Iljin, S.I. Li, B.V. Sokolov, Construction of Unique Buildings and Structures, **5**, 26-30 (2012)
11. D.G. Arseniev, A.V. Rechinskiy, N.I. Vatin, O.S. Gamayunova, Construction of Unique Buildings and Structures, **2 (17)**, 5-24 (2014)
12. N.I. Vatin, O.S. Gamayunova, D. Petrosova, Applied Mechanics and Materials, **635-637**, 2085-2089 (2014)
13. N.I. Vatin, O.S. Gamayunova, D. Nemova, Applied Mechanics and Materials, **638-640**, 2460-2464 (2014)
14. D.G. Arseniev, A.V. Rechinskiy, K.V. Shvetsov, N.I. Vatin, O.S. Gamayunova, Applied Mechanics and Materials, **635-637,** 2076-2080 (2014)
15. O.S. Gamayunova, N.I. Vatin, Advanced Materials Research, **1065-1069**, 2459-2462 (2015)
16. O.S. Gamayunova, K.Y. Usanova, Construction of Unique Buildings and Structures, **7 (12)**, 10-19 (2013)
17. D.G. Arseniev, A.V. Rechinskiy, K.V. Shvetsov, N.I. Vatin, O.S. Gamayunova, Construction of Unique Buildings and Structures, **8 (23),** 21-35 (2014)
18. N.I. Vatin, O.S. Gamayunova, Advanced Materials Research, **1065-1069,** 2622-2625 (2015)
19. V. Kolokoltsev, E. Razinkina, The Vysshee Obrazovanie v Rossii (Higher Education in Russia), **5**, 3-10 (2011)
20. D.G. Arseniev, N.I. Vatin, Construction of Unique Buildings and Structures, **2**, 1-5 (2012)
21. D.G. Arseniev, N.I. Vatin, A.E. Vysotskiy, Construction of Unique Buildings and Structures, **5,** 1-5 (2012)

MATEC Web of Conferences **53**, 01022 (2016)
DOI: 10.1051/matecconf/20165301022

Composite Concrete Modifier CM 02-10 and Its Impact on the Strength Characteristics of Concrete

Luka Akimov[1,a], Nikita Ilenko[1], Roman Mizharev[1], Artemiy Cherkashin[1], Nikolay Vatin[1] and Ludmila Chumadova[1]

[1]*Peter the Great Saint-Petersburg Polytechnic University, 29 Polytechnicheskaya st., St.Petersburg, 195251, Russian Federation*

Abstract. The main aim of this research is the investigation of effect of the composite additive (a mixture of hyperplasticizer and microsilica) CM 02-10 (10 % of MS and 0.2 % of HP) on strength characteristics of concrete. Interaction of concrete with plasticizing and siliceous additives individually depending on their percentage content by the weight of cement is also investigated. Results of experiments with such additives on the fluidity of concrete mix and strength of concrete are presented. Additives used in the experiment: superplasticizer S-3, fused microsilica FMS-85, hyperplasticizer PENTAFLOW AC 2, and composite additive CM 02-10. In this paper results of tests of samples aged 3, 7 and 28 days are presented.

1 Introduction

The feature of life nowadays is an erection of different high-rise and unique buildings and constructions [1]. According to this, materials, concrete in particular, have to solve many problems which arise in the process of building. Filler materials-modifiers were aimed to help in solving these problems.

An advisability of using filler materials-modifiers comes from technical requirements and economic gain, which includes qualitative characteristics of materials adding to concrete or mortar, in condition of saving expensive concrete.

A decrease of using cement is the government support of Russian Federation "Energy conservation and energy efficiency for the period up to 2020".

The main goal of this research is to compare known plasticizing and siliceous additives with the new combined additive CM 02-10 in their impact on the concrete properties. It is extremely important to determine which of these additives could provide the highest mobility and strength to a concrete mix.

It is expected that the new additive will be applied at all stages of construction, where high workability and strength of concrete are necessary.

To achieve these goals series of experiments with modified concrete samples were held.

---

[a] Corresponding author: lukas-ak@mail.ru

2 A brief characteristics of additives-components of the composite additive and an overview of previous researches

1.1 Hyper and superplasticizer

Plasticizing additives are used to increase a mobility of concrete. These additives were called dewatering (water reducing) additives because of their ability to reduce mixing rate. By using these additives, it becomes possible to reduce water-cement ratio to 40-45 % while maintaining the necessary mobility [2-6]. These materials are classified as surfactant-active substances (surfactants) [7]. One of the most popular plasticizer additive is superplasticizer S-3.

Superplasticizer S-3 is a reliable water reducing additive, which has been used for a long time. S-3 is used in all types of concreting: monolithic and precast. S-3 is a mixture of sodium salts poly-metil-methylene-naphthalenes-sulfonic of various molecular weights [8, 9].

According to the data given in [4], applying S-3 allows to get the following indicators:

- increase of mobility of the concrete mix from P1 to P2;
- reduce of water demand during mixing with the binder by 20-25 %;
- increase of final strength characteristics by 25 % and more.

With the introduction of this additive the heat generation also reduces in the early period up to 20 % [10].

Hyperplasticizer HP, which is made on the basis of polycarboxylate ethers, is another example of plasticizers. HP has larger water reducing properties than superplasticizer, therefore it is could be used in a smaller amount [11, 12]. This additive is used to provide the maximum possible strength of the concrete by reducing water-cement ratio.

In [5] is described how the effect of plasticizers influences on the physical and mechanical properties of normal weight concrete hardening at the age of 1, 3, 7 and 28 days. The main conclusion of the research:

- the introduction of the investigated additives in a concrete mixture enables a significant increase in density, strength, water resistance and frost resistance of concrete

1.2 Siliceous mineral additive

Amorphous microsilica may be treated as the most effective mineral modifier and, therefore, it is used as the most popular additive in a concrete. It makes cement stone more durable and long-lasting [13-15]. Microsilica is a product of oxidation of gaseous silicon generated during high temperature manufacturing.

Microsilica is an ultrafine material which is produced from spherical particles which are byproducts in the process of gas purification furnaces in the production of silicon-containing alloys, that is confirmed by the specification 5743-048-02495332-96. The permissible content of an amorphous modification of silicon dioxide which is the main component of microsilica should be not higher than 30 % by the weight of cement.

Microsilica differs from other additives by the small size of particles (100 to 500 nm) and the high specific surface [15-17].

Pozzolanic MS activity and high dispersion of particles cause positive impact of MS on structure, physical and mechanical properties of concrete.

The silica reacts with the calcium hydroxide formed during hydration of silicate phases of cement to form low-basic calcium hydrosilicates. High specific surface area of microsilica (13000-25000 m^2 / kg) intensifies the formation of the products of hydration, therefore, the sharing of microsilica in an amount of 5 % by the weight of cement with plasticizer should increase the strength of fine concrete [18, 19].

According to the experimental data [20], consumption of cement in the control part of sand concrete was 500 kg per 1 m^3 of mix. Cement mark is M500 with normal consistency 25 %. Fillers, that were introduced into the sand concrete, have replaced fixed amount of cement in the mixture; the dosage of superplasticizer was 0.5 % by the weight of cement. Fluidity of cement paste was a constant in all formulations. The combined use of crushed quartz sand in an amount of 5% by the weight of cement with a plasticizer S-3 cause a significant increase in the strength of fine concrete [21].

1.3 Composite concrete modifier CM 10-01

Concrete modifier, which is described in [22], is able to increase the strength of concrete in 2-2.5 times. This modifier is called CM-01, it was engaged in the development of modifier Kaprielov S. S. Concrete modifier CM-01 is an organo-mineral-based powdered product, containing a fused microsilica and superplasticizer S-3. Series 10-01 CM of such modifiers are the most widely used in modification of concrete mixture. It consists of 90 % of amorphous microsilica and 10 % of superplasticizer-3. Under normal conditions Portland cement hydration results to forming up to 40 % of portlandite in the structure of the cement paste-crystalline calcium hydroxide.

Portlandite includes large crystals, which are comparably easy-soluble in water (1.65 g / l at 20 ° C) and have low strength and hardness (2 in accordance with Mohs' scale). Portlandite forms mostly when a concrete streams. There is a variety of amorphous silicon dioxide, which can react with calcium hydroxide in pozzolanic reaction at ordinary temperatures:

$$Ca(OH)_2 + SiO_2 + (n-1)H_2O = CaOSiO_2 \cdot nH_2O = CaOSiO_2 \cdot nH_2O \qquad (1)$$

This kind of amorphous silica is called amorphous (or active) silica or pozzolan. In accordance with [20], it is possible to improve the strength of cement stone by reducing the amount of portlandite in the total mass of growth, it happens because this mineral is the weakest in comparison with others and also degrades the contact zone between the aggregates and the cement. One of the ways to reduce the portlandite in the system is to use active mineral components, namely microsilica brand FMS-85.

Basing on the experience of previous researches, using the standard documentation and own experimental data, we can develop an additive that will be used in the composition of concrete or grout, as well as pick up raw materials from which additives will be produced. By analyzing the effect of additives the efficiency of improving the strength and plasticizing properties becomes clear. Data requirements are met by the addition of CM-01, described in [22], which has no difficulty in obtaining, in view of the prevalence of plasticizer S-3 and the wide availability of microsilica.

3 Materials and Methods

Basing on the [20-22], a modifying composition similar to a CM 10-01 [20] is proposed in this research. It is composed of 98 % of microsilica MS-85 and 2 % of hyperplasticizer Pentaflow AC2. It is used in conjunction with a cement at a weight fraction of 0.2 % of hyperplasticizer and of 10 % of microsilica by the weight of a cement. Saving a plasticizer for almost similar values of strength with increasing content of the HP in relation to the cement is the rationale for minimum dosage of HP, which is used in the range from 0.2 to 0.6 % by the weight of cement. The amount of MS-85 in this composition is the same as in the CM 10-01 for the subsequent comparison of the strength results of researches [20, 23], in which 10 % of silica is also used.

In the course of the experiments materials described in Table. 1 were used.

Table 1. Materials used in the course of experiments.

Material	Specification
Cement	Portland cement of Sukholozhskoye's cement plant with strength class 42.5 normally- hardening CEM I 42.5H GOST 31108-2003
Sand	I class sand with middle size of particles (depending on GOST 8736-93)
Water	depending on GOST 23732-2011
Rubble	depending on GOST 8736-93
Plasticizing additives	Superplasticizer S-3 (specification 5870-34-00369171-02) and Hyperplasticizer PENTAFLOW AC 2
Siliceous additive	Fused microsilica FMS-85 (specification 5743-048-02495332-96)

On the first step of determination of the efficiency on the strength characteristics of the additive in concrete 5 kinds of cement mortars with different additives were prepared, in accordance with GOST 30459-2008. One of the mortars was controlling which was prepared without additives for subsequent tests at the age of 3, 7, 28 days. We were using forms depending on GOST 22685-89 7 to prepare samples of a prism form (bar, which is the equivalent of cube) with dimensions 50x50x150mm. The controlling composition – composition No.1 without additives. Composition and consumption of materials with witch the bars were prepared are shown in Table. 2.

Table 2. Consumption of materials for the prism bars of cement mortar.

No.	Sand, (kg/m^3)	Cement, (kg/m^3)	Additive and its percentage by the weight of cement	W/C	Flow of cone, (mm)
1	1500	500	–	0.50	110
2	1500	500	HP(0.2 % C)	0.40	107
3	1500	500	S-3(0.7 % C)	0.45	108
4	1500	500	MS(10 % C)	0.51	107
5	1500	500	CM 02-10(10.2 % C)	0.43	110

With the same flow of cone (±3 mm) different strength is expected due to the difference in water-cement ratio. The bars of cement and sand were stored and tested under normal conditions in water in accordance with GOST 5802-86.

Calorimetric studies of the kinetics of hydration of cement [24-27] allow us to assess the activity of additives. On the basis of previous calorimetric studies of the kinetics of hydration of cement with additives and on the basis of studies of the strength of cement bars, in accordance with GOST 27006-86 two compositions of heavy concrete were developed. The first was a controlling, CM 02-10 was added in the second. Composition of concrete mixtures provided the same mobility (FoC). Amount of additive CM 02-10 is the same as in the case of cement-sand samples – 10.2 % by the weight of cement.

For the experiment some samples were prepared (concrete cubes with the size of the ribs 100 mm) in accordance with GOST 10180-2012 for determining the strength at 3, 7, 28 days of 2 different compositions. Composition №1 is controlling (Table. 3).

Table 3. The composition of the concrete cubes.

The composition of the concrete mix, [kg/m^3]						W/C	Flow of Cone, (cm)	Mark of the workability	Density of the mixture, (kg/m^3)
No	cement	CM 02-10	sand	rubble	water				
1	550	–	595	870	242	0.44	19	W4	2257
2	550	56.1	650	950	215	0.39	21	W5	2421

Concrete cubes were stored in normal conditions of temperature (20 ± 2) ° C and relative humidity of (95 ± 5) %.

Next, the test of concrete samples was held by destructive method in a hydraulic press. The test results of concrete samples under compression are shown in Table. 5 and in Fig. 2.

4 Results and Discussion

Compression-force test was held on 50-ton press. According to test results on the compressive strength of bars (Table 4), the graphs of strength changes over time were constructed (Fig. 1).

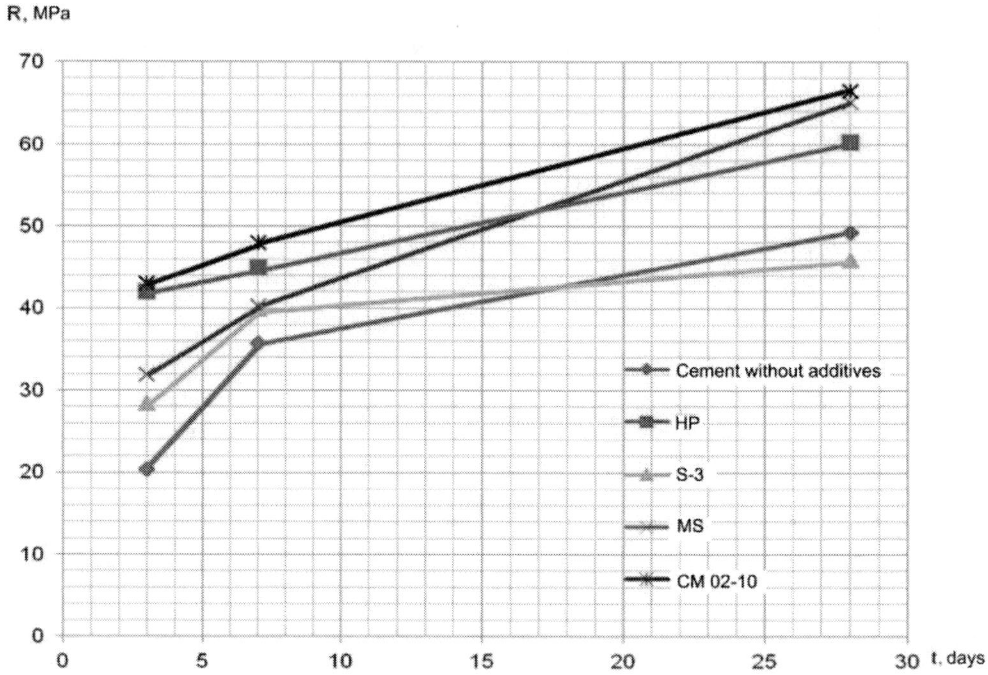

Figure 1. Strength variation (R) of samples over time at 3, 7, 28 days.

Table 4. The results of the strength test of the bars.

No.	Additive	W/C	FoC, (mm)	Strength of the bars, (MPa)					
				3 days		7 days		28 days	
				$R_{bend.}$	$R_{comp.}$	R_{bend}	$R_{comp.}$	$R_{bend.}$	$R_{comp.}$
1	–	0.50	110	5.4	20.4* 100 %	8.5	35.7* 100 %	7.8	49.2* 100 %
2	HP	0.40	107	8.8	42* 206 %	6.7	45* 126 %	7.5	60.2* 122 %
3	S-3	0.45	108	8.2	28.4* 139 %	5.7	39.9* 112 %	6.2	45.9* 93 %
4	MS	0.51	107	5.5	32* 157 %	6.5	40.3* 113 %	8.3	65.1* 132 %
5	CM02-10	0.43	110	6.7	43* 211 %	7.8	48* 134 %	8	66.5* 135 %

*Note: above the line – the average value of the index; below the line – the relative value of the index in % of controlling composition.

According to the received information, there is a significant increase of strength of cement with CM 02-10 to the third day unlike cement without additives.

On the 7th day of hardening the cement strength of bars with the MS and S-3 is almost the same (1 % difference). The value of strength of the bars with HP is 13 % more than the bars with MS, which polar changes to the 28th day, when strength value of the bars with MS is 10 % more than the bars with HP. Studying strength of samples with S-3 leads to the conclusion that as the amount of unhydrated cement decreases the SAS membrane becomes thicker, thus, the hardening speed decreases.

Strength curve for the bars with S-3 shows the worst results, so in the future it will not be discussed.

Thus, the combination of 10 % of MS and 0.2 % of HP (CM 02-10) in the cement solution gives a strength by 35 % higher in comparison with a cement without additives.

With the addition of CM 02-10 the strength of cement bars, compared to other prototypes of the same cement content with the same mobility, increases, it serves as a basis for the development of heavy concrete with the use of this additive.

Table 5. The results of compression test of the concrete cubes.

No.	CM 02-10, (kg/m^3)	Amount of cement in concrete, (kg/m^3)	W/C	Density of mix, (kg/m^3)	Compressive strength R, [MPa]			Class of concrete	Specific consumption of binder per unit of designed strength, (kg/MPa)
					3 days	7 days	28 days		
1	–	550	0.44	2257	28.2	37.6	44	C35	12.5
2	56.1	550	0.39	2421	35.3	38.9	56.4	C45	9.75

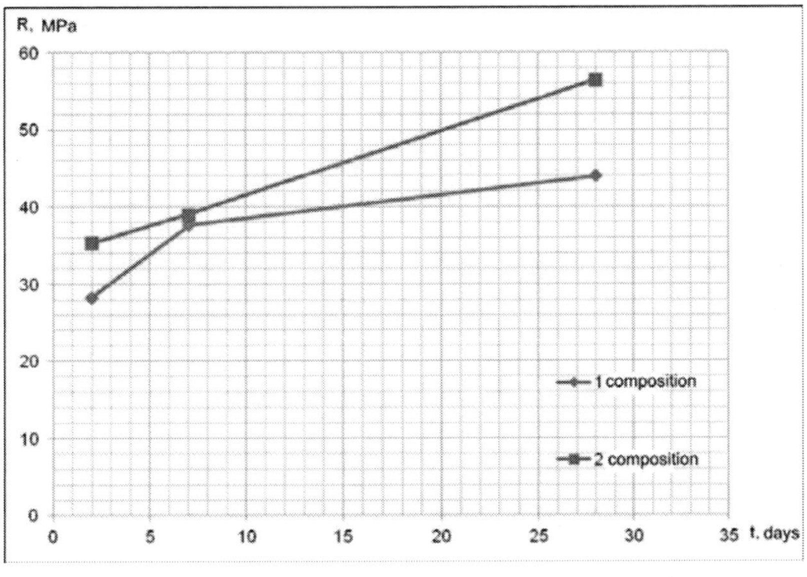

Figure 2. Change in strength values of the concrete cubes over time.

As we can see in the results of experiment (Fig. 2), the samples of the concrete, modified by CM 02-10, have a stable increasing in the strength. This fact allows us to predict more gains of it over time. Compressive strength of concrete sample No.2 is 128 % higher than the strength of the control sample in the age of 28 days. It illustrates the effectiveness of using an additive CM 02-10 when its content is 10.2 % by the weight of cement with the same cement consumption.

5 Conclusions

After analyzing the effect of various additives on the concrete strength characteristics, it is possible to find out the following conclusions:

1. By adding CM 02-10 to a concrete mix we could conclude, that this additive has the most positive influence on the strength characteristics of the cement beams. At the age of 28 days we could observe the increasing of strength compare to a controlling sample:

- 122 % using HP;
- 132 % using MS;
- 135 % using CM 02-10.

2. Cement mortar have similar rates of mobility comparing to cement mortar without any additives by reducing W/C:

- 20 % using HP;
- 10 % using S-3;
- 14 % using CM 02-10.

It is possible to conclude considering the results of experiments (Fig. 1 and Fig. 2) that the concrete prepared with the mixture of hyperplasticizer and microsilica shows the best strength values, thereby, using of an additive CM 02-10 as a modified additive to concrete is reasonable.

References

1. EN 1992 Eurocode 2. Design of concrete structures.
2. I.N. Akhverdov, *Osnovy fiziki betona* (M., Stroyizdat, il. UDK 666.972.017:53, 1981)
3. A.V. Usherov-Marshak, M. Tsiak, *Khimicheskiye i mineralnyye dobavki v tekhnologii tsementa i betona* (Zaporozhye: Khortitsa, 2002)
4. Company "Polylayer", Pop. conc., **2**, 70-74 (2004)
5. V.S. Izotov, R.A. Ibragimov, Izvest. KGASU, Kazan, **1 (13)**, 287-291 (2010)
6. Electronic directory of a builder [web source] (20.12.2013). URL: http://www.bibliotekar.ru/spravochnik-70/26 (date of reference: 18.09.2015)
7. P.G. Vasilik, I.V. Golubev, *Publication of electronic resource of company ZAO "EuroChem- 1". Features of the application of polycarboxylate hyperplasticizer Melfluxâ* [web source] (16.02.2014). URL: http://www.chem.eurohim.ru/catalog/DryMortarAdditives/articles/art52 (date of reference: 15.09.2015)
8. Yu.G. Barabanshchikov, M.V. Komarinskiy, Adv. Mat. Res., **941 (944)**, 780-785 (2014)
9. S.S. Kiski, I.V. Ageyev, A.N. Ponomarev, A.A. Kozeyev, M.E. Yudovich, Civil Eng. J., **8**, 42-46 (2012)
10. Ta Van Fan, *Samouplotnyayushchiyesya vysokoprochnyye betony s zoloy risovoy shelukhi i metakaolinom* (Abstract Ph.D., Rostov-na-Donu, 2013)
11. N.P. Bleshchik, N.N. Kalinovskaya, *Modern concrete* (IX international scientific conference, Zaporozhye, 2007)
12. N.P. Bleshchik, N.N. Kalinovskaya, Constr. Sc. and Tech., **1**, 30-40 (2006)
13. V.I. Kalashnikov, O.V. Tarakanov, Yu.S. Kuznetsov, V.M. Volodin, Ye.A. Belyakova, Civil Eng. J., **8**, 47-52 (2012)
14. X. Zhang, S. Yang, B. Diao, Q. Zhang, Y. Li, J. of Build. Str., **30 (2)**, 324-327 (2009)
15. A. Sprince , G. Fischer, L. Pakrastinsh, A. Korjakins, Adv. Mat. Res., **842**, 470-476 (2012)
16. U.G. Distanov, *Mineralnoye syrye. Opal-kristobalitovyye porody: spravochnik* (M. ZAO "Geoinformmark", 1998)
17. G.V. Nesvetayev, G.S. Kardumyan, Ta Van Fan, L.A. Khomich, A.M. Blyagoz, Nov. tekh., **4**, 125-128 (2012)
18. G.V. Nesvetayev, G.S. Kardumyan, Ta Van Fan, L.A. Khomich, A.M. Blyagoz, Nov. tekh., **4**, 122-125 (2012)

19. G.V. Nesvetayev, G.S. Kardumyan, Ta Van Fan, L.A. Khomich, A.M. Blyagoz, Nov. tekh., **4**, 118-121 (2012)
20. N.M. Morozov, Kh.G. Muginov, V.G. Khozin, A.B. Antakov, Izvest. KGASU, Kazan, **2 (20)**, 183-188 (2012)
21. L.A. Veshnyakova, A.M. Ayzenshtadt, Proc. IV Int. Sem.-cont. of young sc. and grad. stud. work. in the f. of bind., conc. and dry m., **5**, 164 (2013)
22. *Publication of the electronic version of the journal "StroyPROFIl"* [web source] (05.06.2007). URL: http://stroyprofile.com/archive/2652 (date of reference: 20.09.2015)
23. Ye.G. Matveyeva, *Povysheniye effektivnosti betona dobavkoy nanodispersnogo kremnezema,* (Abstract Ph.D., Belgorod, 2011)
24. Yu.G. Barabanshchikov, A.V. Churillo, *Heat dissipation binders. Methodical instructions,* (A.: LPI, St.Petersburg, 1988)
25. L.I. Chumadova, et al., *Posobiye po teplovoy obrabotke sbornykh zhelezobetonnykh konstruktsiy i izdeliy.(podgotovka k SNiP 3.09.01-89),* (M. Stroyizdat., St. Petersburg, 1989)
26. I.D. Zaporozhets, A.A. Pariyskiy, S.D. Okorokov, L.I. Chumadova, Concrete and Reinforced Concrete, **6**, 52-62 (1977)
27. A.S. Vasilyev, Yu.G. Barabanshchikov, Civil Eng. J., **8**, 67-77 (2012)

MATEC Web of Conferences **53**, 01023 (2016)
DOI: 10.1051/matecconf/20165301023

Quality Improvement of Concrete Articles

Larisa Svatovskaya[1], Anastasiya Sychova[1], Maxim Sychov[2,a], Vladimir Okrepilov[3]

[1]*Petersburg State Transport University, St.Petersburg, Russia*
[2]*Saint Petersburg State Technological Institute, St.Petersburg, Russia*
[3]*State Regional Centre for Standardization, Metrology and Testing in St. Petersburg and Leningrad Region (State Centre "Test-St. Petersburg"), Kurlyandskaya, 1, St. Petersburg, 190103, Russia*

Abstract. In the paper it is shown that quality of concrete articles and structures may be significantly improved by silica sol solution absorption. Improvements include increase of compressive strength, resistance to low temperatures, coefficient of constructive quality, decrease of water sorption, contraction. The reason of improvement is discussed.

1 Introduction

It is well known that surface properties of construction materials, including acid-basic ones are very important [1-5]. The main idea of the paper is the possibility of improvement of quality of concrete articles by means of saturation and modification of their surface with silica sol solution. Due to the presence of capillary pores concrete articles and structures possess ability to soak silica sol solution. After soaking process silica sol appears in cement based materials' pores and may participate in the reaction:

$$5Ca(OH)_2+6(SiO_2{\cdot}H_2O)=5CaO{\cdot}6SiO_2{\cdot}5{,}5H_2O+5{,}5H_2O,\ \Delta G^0{}_{298} = -584.8\ kJ \qquad (1)$$

Reaction (1) is characterized by negative value of Gibbs energy, therefore new calcium silicates are formed, which is good for improvement of the construction article quality [6-12]. Due to presence of certain amount of capillary pores, their sizes and peculiarities of reaction (1), each cement based article can intake only certain limited quantity of silica sol. Thus we introduce new parameter to characterize concrete articles' ability to intake silica sol – absorption capacity C (kg/m3), mass of silica sol absorbed by unit volume of article.

2 Experiment, results and discussion

According to the preliminary experiment [8] it was chosen to perform experiments on quality improvement of construction articles using samplers of foam concrete with the following average density values D: 0.4; 0.5 and 0.6 Ton/m3. For the modification of samples we used 1.5% silica sol solution. Characteristics of solution: pH ~ 9, density of solution – 1.02 g/cm3, particle diameter of silica sol dispersions – 12 nm.

[a] Corresponding author : msychov@yahoo.com

Samples were saturated by the silica sol solution on the third day of hardening process and then were examined. The process of saturating with silica was described using suggested above new parameter – absorption capacity, C, kg/m3, calculated according to equation (2).

$$C = \frac{m_1 - m_2}{V}, kg/m^3, \tag{2}$$

where between m1 and m2 are masses of sample before and after the modification process, kg; V – volume of sample, m3.

After that modified sample of concrete article was tested according to common protocols such as water absorption, compressive strength, resistance to low temperatures, contraction and coefficient of constructive quality. Overall quality of cement articles was characterized by the value of specific compressive strength [10, 13, 14].

The main results are shown in the Table 1. One can see that quality of modified articles was improved and their properties are superior compering to ones of article without modification. Some two-fold increase of main characteristics of concrete articles was achieved by modification. Explanation of such kind of quality improvement is the healing of the pore structure and decrease of total volume of pores by means of silica sol saturation and occurrence of reaction (1). Parameter C reflects absorption of silica sol and formation of new phases in the pore space of articles according to the reaction (1). From the data of table 1 it is clear that value of parameter C increases with the rise of density of concrete article [15-18].

Table 1. Improvement of characteristics of foam concrete samples after modification.

D, Ton /m3	C, kg/m3	Improvement of sample quality									
		σ, MPa		Water absorption, %		Resistance to low temperatures, cycles		Contraction, mm/m		Specific strength, σ/D	
		Before modification	After modification	Before modification	After modification	Before modification	After modification	Before modification	After modification	Before modification	After modification
0.4	0.24	0.5	1.04	24.2	15.0	15	25	4.2	1.3	1	2
0.5	0.30	1.2	2.08	22.8	11.2	15	35	3.5	1.1	2	4
0.6	0.36	1.56	2.98	21.3	9.3	25	50	2.9	0.9	2.6	5

In the Table 2 one can see results of the experiment on modification of concrete samples using different concentrations of silica sol. The level of the improvement of characteristics of modified heavy concrete samples is lower comparing to that of foam concrete, nevertheless 40…60% increase of strength was achieved [19, 20].

The main reason for the improvement of properties is the reaction (1) providing formation of additional amount of calcium silicates. That amount may be calculated taking into account reaction (1) and values of absorption capacity C. Examples of such kind of calculations are presented in the Table 3.

Table 2. Increase of strength of concrete samples

D, Ton/m3	Days of hardening	Strength of sample before modification, MPa		Concentration of silica sol, %	Strength of sample after after modification, MPa		Increase of strength after modification, %	
		Compressive	Bending		Compressive	Bending	Compressive	Bending
2.0	10, heat treatment	14.97	3.92	10	24.38	5.53	63	41
2.0	28, normal hardening (n.h.)	14.08	4.23	5	20.65	5.95	47	41
0.6	28 n.h.	1.07	0.86	15	1.89	1.23	74	43
0.5	28 n.h.	0.93	0.73	15	1.65	1.14	78	56

Table 3. Hydrate calcium silicates in concrete and absorption capacity

Average density of concrete samples D, Ton/m3	Absorption capacity C, kg/m3	Calculated quantity of calcium silicates, formed according to reaction (1), kg/m3
0.4	0.38	0.60
0.5	0.47	0.74
0.6	0.57	0.90
2.0	0.76	1.20

3 Conclusions

It was shown that such characteristics of foam concrete and heavy concrete articles as compressive strength, water absorption, resistance to low temperatures and contraction may the significantly improved by modification with silica sol solutions. Due to the presence of capillary pores intake of solution takes place resulting in formation of new silicate phases and healing of pores.

References

1. M.M. Sychov, Applied Surface Science, **244**, 461-464 (2005)
2. M.M. Sychov, Journal of the SID, **11**, 33-38 (2003)

3. O.A. Cheremisina, M.M. Sychev, S.V. Myakin, V.G. Korsakov, V.V. Popov, N.Y. Artsutanov, Russian Journal of Physical Chemistry, **76 (9)**, 1472-1475 (2002)
4. S.V. Myakin, V.G. Korsakov, T.I. Panova, E.A. Sosnov, Yu.C. Fomchenkova, M. M. Sychov, O.A. Shilova, Glass Physics and Chemistry, **37(6)**, 624-628 (2011)
5. V. G. Korsakov, S. A. Alekseev, M. M. Sychev, M. N. Tsvetkova, E. V. Komarov, B. Lee, S. V. Myakin, and I. V. Vasil'eva, Russian Journal of Applied Chemistry, **80(11)**, 1931–1935 (2007)
6. L.B. Svatovskaya, M.M. Sychov, *Aktivirovannoye tverdeniye tsementa [Activated hardening of cement]*, (Leningrad, Building-published house, 1983)
7. L.B. Svatovskaya, L.L. Maslennikova and others *Thermodynamical and electronic aspects of the building materiel properties*, (St.Peterburg, Building published house, 2004)
8. M. Hammady, The method of the modification of the concrete by by means of nanosolution. www.spgasu.ru
9. P. Perrot P, *A to Z of Thermodynamics* (Oxford University Press. 2004)
10. W. Greiner, L. Neisse, H. Stocker, *Thermodynamics and statistical mechanics* (Springer-Verlag, 1995)
11. J. Ćetković, S. Rutešić, M. Zarković, M. Knežević, N. Vatin, Procedia Engineering, **117** (1), 780 - 790 (2015).
12. S. Rutešić, J. Ćetković, M. Žarcković, M. Knežević, N. Vatin, Procedia Engineering, **117** (1), 905 - 915 (2015).
13. S. Rutešić, J. Ćetković, M. Knežević, M. Žarcković, N. Vatin, Procedia Engineering, **117** (1), 642 - 650 (2015).
14. N. Gubeljak, J. Predan, D. Kozak, J. Tuma, B. Kovačič, P. Konjatić, J. Sertić, Strojarstvo, **51** (4), 263-271 (2009)
15. B. Kovačič, R. Kamnik, M. Premrov, N. Gubeljak, J. Predan, Z. Tišma, Strojniski Vestnik/Journal of Mechanical Engineering, **54** (5), 364-371 (2008)
16. B. Kovačič, R. Kamnik, International Journal for Engineering Modelling, **20** (1-4), 77-84 (2007).
17. M.M. Sychov, N.V. Zakharova, S.V. Mjakin, Ceramics International, **39,** 6821–6826 (2013)
18. V.V. Bakhmet'ev, M.M. Sychev, V.G. Korsakov, Russian Journal of Applied Chemistry, **83(11)**, 1903-1910 (2010)
19. B. Kovačič, B. Supej, (2004) Geodetski Vestnik, **48** (1), 40-49 (2004)
20. B . Kovačič, B. Geodetski Vestnik, **48** (1), 32-39 (2004).

MATEC Web of Conferences **53**, 01024 (2016)
DOI: 10.1051/matecconf/20165301024

New Parameter of Geoecological Protective Ability of Construction Articles

Larisa Svatovskaya[1], Anastasiya Sychova[1], Maxim Sychov[2,a] and Vladimir Okrepilov[3]

[1] *Petersburg State Transport University, St.Petersburg, Russia*
[2] *Saint Petersburg State Technological Institute, St.Petersburg, Russia*
[3]*State Regional Centre for Standardization, Metrology and Testing in St. Petersburg and Leningrad Region (State Centre "Test-St. Petersburg"), Kurlyandskaya, 1, St. Petersburg, 190103, Russia*

Abstract. New parameter of geoecological protective ability of construction materials and articles is suggested. Effect is based on the fact that any solid surface possess centers of adsorption. It is shown that Broensted basic surface centers are active in adsorption of heavy metal ions from water solutions and materials having such centers on their surface are able to remove heavy metal ions from water and thus to protect soil from contamination with such ions. For example, foam concrete with ability to adsorb heavy metal ions may be used to protect soil near railway tracks.

1 Introduction

Construction articles should possess such properties as high strength and resistance to low temperatures, low water sorption etc.

In addition to those well-known characteristics we suggest to consider the new one - geoecological protective ability characterizing their ability to adsorb and immobilize heavy metal ions and/or oil products from storm flows, rein waters, waste waters etc. [1-5]. This ability provides detoxication of lithosphere from pollutions which is very important because of modern ecological problems.

In this paper geoecological protective ability of construction articles against heavy metal ions is studied. Important information on such ability is provided by the method of study of surface properties of solids which is called "distribution of active centers" (DAC, [6-12]).

This method provides information about amounts and types of active centers existing on the surface of construction materials including cement based ones.

There are four types of surface active centers: acidic and basic Broensted centers and acidic and basic Lews ones. Particularly Broensted basic centers represent -OH functional groups on the surface.

This groups are capable to interact chemically with heavy metal ions contained in water to form metal hydroxides with very low solubility immobilized on the surface of material [13-16]l.

As a result heavy metal ions are removed from water which was in contact with construction article and thus detoxication effect is achieved.

2 Experiment, results and discussion

For the experiments foam concrete articles were used with average density, D, 0.3; 0.5 and 0.7 ton/m3. Samples were ground up to from particles with diameters d = 0.63...1 mm. [17] Then ground samples were saturated with solutions of heavy metal ions Fe(III) and Mn(II). Such kind of ions occur in water for example as a result of railway transport exploitation therefore there is a need to preserve soil near to railway tracks from contamination with heavy metal ions [18, 19].

After saturation amounts of adsorbed metal ions were measured and adsorption capacities were calculated. Capacity was designated as C, kg/T. Then we suggest to introduce parameter K, reflecting ability of studied construction material to adsorb heavy metal ions and thus its geoecological protective ability, equation (1).

$$K = D \cdot C, \tag{1}$$

Table 1 presents data on several studied cement based materials. According to obtained results, the highest value of geoecological protective ability has material with density value of 0.5 T/m3 and such kind of result is connected with structure of materials and their surface properties [20-22].

Figures 1 and 2 shows distribution of adsorption centers for the studied samples of foam concrete. Distributions show amounts of centers (Y axis) in micromole per kilogram of material depending on pKa value of each center (X axis), pKa being value of negative logarithm of dissociation constant. Centers with pKa values in the range 7...14 represent Broensted basic centers capable to interact with heavy metal ions with formation of metal hydroxides like Fe(OH)3 and Mn(OH)3. From the data of figures 1 and 2 one can see that after the contact with water solutions containing heavy metal ions amounts of Broensted basic centers on the surfaces of studied samples significantly decreased since those centers interacted with ions and thus were blocked by new phases of hydroxides formed on the surface of materials [23].

Thus foam concrete materials may be used in railway construction to protect environment form the pollution with heavy metal ions originated from exploitation of railway transport.

Table 1. Geoecological protective ability of foam concrete materials

Heavy metal ions	Average density of material, T/m^3					
	0.3		0.5		0.7	
	C, kg/T $\cdot 10^3$	K, kg/m^3	C, kg/T $\cdot 10^3$	K, kg/m^3	C, kg/T $\cdot 10^3$	K, kg/m^3
Fe(III)	0.70	0.21	0.5	0.25	0.22	0.15
Mn(II)	2.71	0.81	2.4	1.20	1.50	1.05

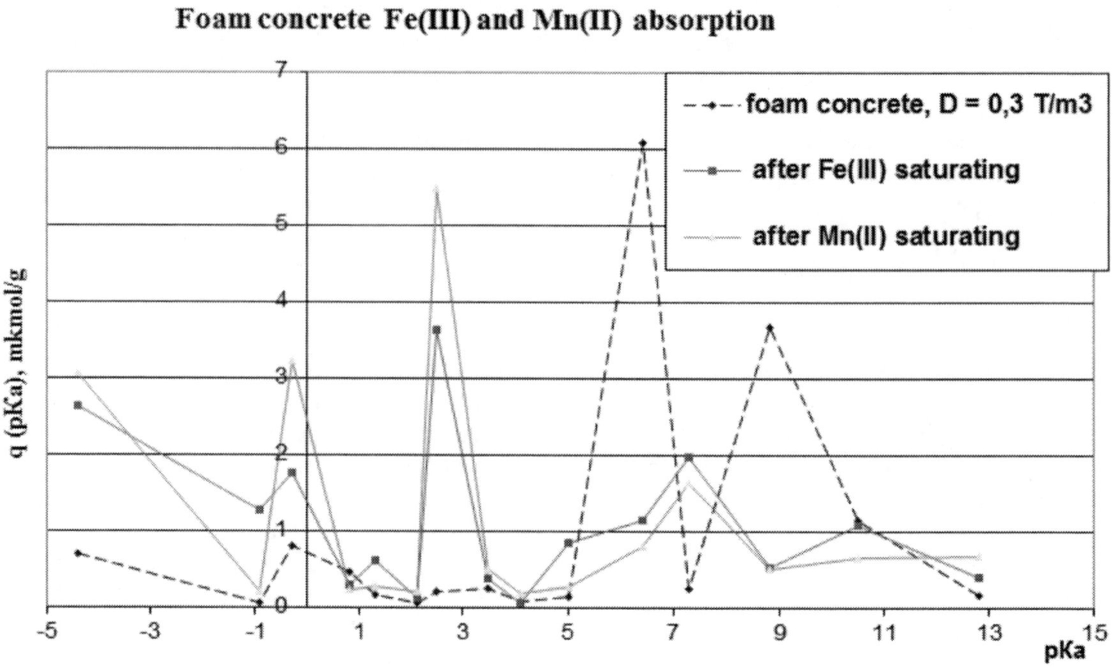

Figure 1. Absorption of heavy ions by foam concrete with density 0.3 T/m^3.

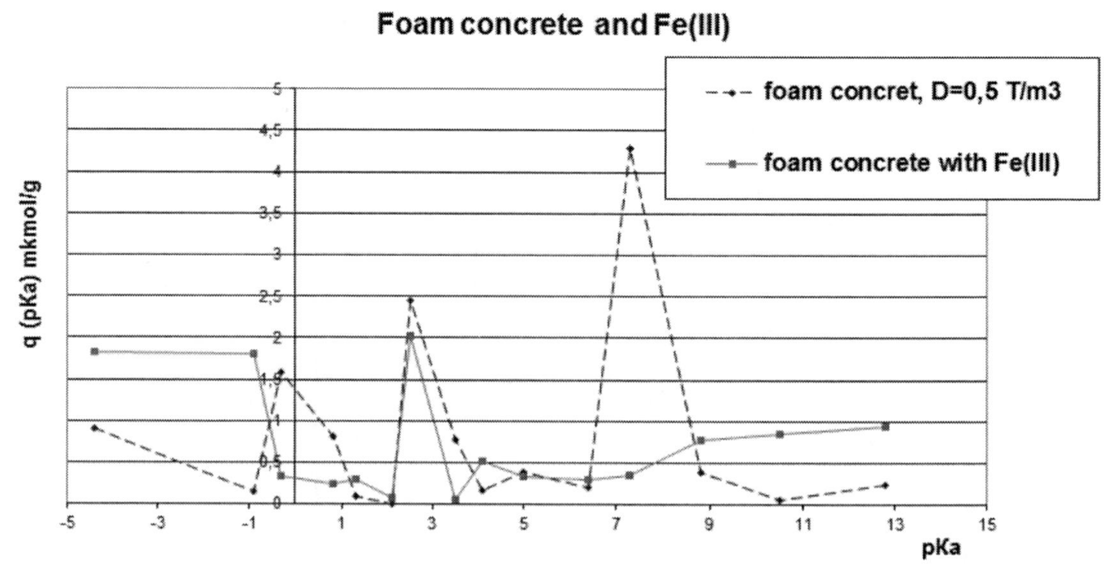

Figure 2. Absorption of heavy ions by foam concrete with density 0.5 T/m^3.

3 Conclusions

The new parameter of geoecological protective ability of construction materials and articles made of them is suggested. Parameter reflects ability of material to adsorb heavy metal ions, oil products and other pollution components. Its value depends on materials density and amount of surface adsorption centers capable to interact with pollutants.

It is shown that foam concrete is able to absorb and remove from the water toxic Fe(III) and Mn(II) ions and thus may be used in railway construction to protect environment form contamination with heavy metal ions originated from exploitation of railway transport.

References

1. L.B. Svatovskaya, M.M. Shershneva, M. Baidarashvily, A.M. Sychova, M. M. Sychov. Proceedia Engineering. 117 (2015), 350-354.
2. R. Kamnik, B. Kovačič, B. Pribicević, A. Đapo, Geodetski List, **69** (**3**), 171-188 (2015)
3. B. Kovačič, R. Kamnik, Z. Kapović, Geodetski List, **63** (4), 315-327 (2009)
4. N. Gubeljak, J. Predan, D. Kozak, J. Tuma, B. Kovačič, P. Konjatić, J. Sertić, Strojarstvo, **51** (4), 263-271 (2009)
5. B. Kovačič, R. Kamnik, International Journal for Engineering Modelling, **20** (1-4), 77-84 (2007)
6. M.M. Study of active surface centers in electroluminescent ZnS:Cu,Cl/ M.M. Sychov [et al.] // Applied Surface Science. – 2005. – №244. – P.461 – 464.
7. B . Kovačič, B. Geodetski Vestnik, **48** (1), 32-39 (2004)
8. S. Rutešić, J. Ćetković, M. Knežević, M. Žarcković, N. Vatin, Procedia Engineering, **117** (**1**), pp. 642-650 (2015)
9. S.V. Myakin , V.G. Korsakov, T.I. Panova, E.A. Sosnov, Yu.C. Fomchenkova, M. M. Sychov, O.A. Shilova. Effect of the Modification of Barium Titanate on the Permittivity of Its Composites with Cyanoethyl Ester of Polyvinyl Alcohol. Glass Physics and Chemistry, 2011, Vol. 37, No. 6, pp. 624–628.
10. Z. Popović, L. Lazarevic, L. Brajović, M. Vilotijević, Procedia Engineering, **117 (1)**, 601-613 (2015)
11. Z. Popović, L. Lazarevic, L.J. Brajović, P. Gladović, Metalurgija, **53 (4)**, 721-724 (2014)
12. Z. Popović, L. Brajović, , L. Lazarevic, L. Milosavljević, Tehnicki Vjesnik, **21 (1)**, 147-153 (2014)
13. B. Šćepanović, M. Knežević, D. Lučić, Journal of Applied Engineering Science, **10 (1)**, 15-20 (2012)
14. M. Lazarevska, J. Dražić, M. Knežević, T. Samardžioska, M. Cvetkovska, Journal of Applied Engineering Science, **10 (1)**, 27-30 (2012)
15. M. Knežević, Journal of Applied Engineering Science, **10 (1)**, pp. (2012)
16. D. Bojović, D. Jevtić, M. Knežević, Revista Romana de Materiale/ Romanian Journal of Materials, **42 (1)**, 16-22 (2012)
17. J. Cetkovic, S. Rutesic, T. Hanak, M. Knezevic, B. Melovic, Technics Technologies Education Management, **7 (1)**, 285-293 (2012)
18. M. Krgović, I. Bošković, M. Vukčević, R. Zejak, M. Knežević, R. Mitrović, B. Zlatičanin, N. Jaćimović, Materiali in Tehnologije, **45 (6)**, 609-612 (2011)
19. M. Krgović, M. Knežević, M. Ivanović, I. Bošković, M. Vukčević, R. Zejak, B. Zlatičanin, S. Durković, Materiali in Tehnologije, **43 (6)**, 327-331 (2009)
20. M. Knežević, B. Šćepanović, D. Lučić, R. Zejak, EASEC-11, Eleventh East Asia-Pacific Conference on Structural Engineering and Construction, (2008)
21. M. Krgović, N. Marstijepović, M. Ivanović, R. Zejak, M. Knežević, S. Durković, Materiali in Tehnologije, **41 (4)**, 189-192 (2007)
22. M. Knežević, B. Šćepanović, D. Lučić, Proceedings of the 10th East Asia-Pacific Conference on Structural Engineering and Construction, EASEC 2010, **2**, 269-274 (2006)
23. M.Krgović, M. Ivanović, Z. Blagojević, Ž. Jaćimović, R. Zejak, M. Knežević, InterCeram: International Ceramic Review, **55 (2)**, 104-106 (2006)

MATEC Web of Conferences **53**, 01025 (2016)
DOI: 10.1051/matecconf/20165301025

Features of Calculation of Consolidation Process of Weak Saturated Soils in View of Their Plastic Properties

Alexey Golubev[1], Alexey Savchenko[1], Luka Akimov[1], Artem Frolov[1,a] and Vladimir Lvov[1]

[1]Peter the Great Saint-Petersburg Polytechnic University, 29 Polytechnicheskaya st., St.Petersburg, 195251, Russian Federation

Abstract. Issue of construction of facilities in areas of weak saturated soils is becoming particularly relevant in recent years. This problem applies especially to objects of oil and gas industry. Saturated soils are characterized by a filtration anisotropy, an inhomogeneity and an ability to liquefy in violation of their structure because of water content. Therefore, the study of bearing capacity of such soils in the process of consolidation has significant practical importance in ensuring the reliability of structures. The formulation, the method and the algorithm for calculating the stress-strain state of weak saturated soils is described in this article. Solution of the consolidation task is based on the model of body forces of three-component soil considering the elastic-plastic deformation of its skeleton. Some results of calculations have been shown in this article.

1 Introduction

The Florin-Bio's model of body (volumetric) forces is used to solve the problem of consolidation which reduces to the definition of time-varying stress-strain state of saturated soils. This model considers the three-dimensional interaction forces between the solid, liquid and gaseous components of soil during their relative movement in compaction process. The Florin-Bio's theory postulates that soil is the continuous medium wherein all components are continuously distributed over the volume.

The complete system of equations of the consolidation task involves [1-4]:

The equilibrium equations

$$\sigma_{ij,i} + \delta_{ij} P_{ij} + F_i = 0, \tag{1}$$

The Cauchy's geometrical relationship

$$e_{ij} = \frac{1}{2}(u_{ij} + u_{ji}), \tag{2}$$

The physical equation for soil skeleton

$$\sigma_{ij} = D_{ijkl} \cdot e_{xl}, \tag{3}$$

[a] Corresponding author: artemfrolov007@mail.ru

The consolidation equation

$$-\dot{u}_{i,j} + \beta \cdot \dot{p} = (\frac{k_{ij}}{\gamma_b} p,i\ )i \tag{4}$$

In these equations e_{ij}, σ_{ij}, k_{ij} – components of tensors of deformations, tensions in the skeleton of the soil and the soil impermeability (filtration); F_i, U_i – components of vector body forces and displacements, p is the overpressure in the pore water, D_{ijkl} – components of physical matrix of the deformation of the soil skeleton, γ_b - unit weight of water, δ_{ij} – the Kronecker delta ($\delta_{ij} = 1$ when $i = j$, $\delta_{ij} = 0$ when $i \neq j$).

The coefficient β in Eq.4, interpreted by Florin as the coefficient of the volume compressibility of three-component soil owing to a presence of entrapped gas [1], is a variable which changes as pressure in the pore water p and volumetric soil deformation e_v vary:

$$\beta = \beta_0 \frac{p_0}{(p_0 + p)^2 (1 + e_1)}, \tag{5}$$

where p_0 - the initial pressure in the water, including atmospheric pressure; β_0 - the initial coefficient of volumetric compressibility, which is determined by the initial values of porosity of the soil n_0, the degree of soil saturation S_{r0} and the solubility coefficient of gas μ (for air $\mu = 0.0245$):

$$\beta = n_0(\mu S_{r_0} - S_{r_0} + 1). \tag{6}$$

2 Solution of a task

The analysis of experimental data and theoretical studies shows that the process of deformation of the soil skeleton could be satisfactorily described from the point of models of elastically perfectly plastic Prandtl body [5-7]. Using this model compared with other more complex elastoplastic models [8-12] greatly simplifies the solution of the task and does not requires additional experimental parameters which characterize the nonlinear deformation of soil.

Deformations are considered as elastic in the pre-limit stress state of soil and determined by the generalized Hooke's law using as characteristics the unit of the overall deformation and Poisson's ratio [13-15]. In the limiting condition, which is determined in accordance with Coulomb's yield condition [16], physical equations Eq.3 take the form of incremental ratio $d\sigma_{ij} = D^p{}_{ijkl} \cdot de_{kl}$, which represents a process of plastic deformation with the sequence of small deformations embroadening. The elements of $D^p{}_{ijkl}$ matrix have been obtained for the associated law of plastic flow [17].

The solution of the task of stress-strain state of saturated consolidating bases, based on the model of body forces in view of elastic-plastic deformation of the soil skeleton, is possible only by numerical methods involving computers to implement them [18-21]. In this article the finite element method has been used in the most common variant of the method of displacements [22]. The main provisions and the relations of the FEM allow to reduce the system of differential equations Eq.1 - Eq.4 to the elementary system of algebraic equations (represented in matrix form) which could be solved by the explicit difference scheme relative to the unknown vector of nodal displacements $U_{t+\Delta t}$ and the excess pore pressure $P_{t+\Delta t}$ at time point $t + \Delta t$:

$$K_{t+\Delta t} \cdot U_{t+\Delta t} + C \cdot P_{t+\Delta t} = K_t \cdot U_t + C \cdot P_t + \Delta F_{\Delta t}, \tag{7}$$

$$C^T \cdot U_{t+\Delta t} + (M - \frac{\Delta t}{2} \cdot H) P_{t+\Delta t} = C^T U_t + (M + \frac{\Delta t}{2} H) \cdot P_t. \tag{8}$$

The generalized matrices of stiffness K, gas deformation M, filtering H are formed from relevant matrices of each element of the system. A triangular simplex has been taken as a finite element. A triangular simplex is an element with nine degrees of freedom for which all the necessary matrices in finite element analysis have been obtained. The development of the plastic deformation region is defined with the transition of finite elements to the limit state because the stress state in such elements is homogeneous.

The programs which were designed for computers ES 1061and PCs of IBM type allow us to calculate the stress-strain state of inhomogeneous saturated soil masses under conditions of plane deformation during a random loading over time.

It is required to establish the input parameters of soils: the modulus of deformation E, the Poisson's ratio ν, the soil adhesion C, the angle of internal friction φ, the unit weight of soil γ, the coefficient of lateral pressure ξ (to determine the initial domestic stress state), the filtration coefficient K_f, the initial coefficient of volume compressibility β_0. The position of ground water level is given in addition to the boundaries of homogeneous soil areas.

The generalized matrix of Eq.8 could be built considering saturated and unsaturated stress state of the soil under the condition of Coulomb:

$$f = \sqrt{(\sigma_x - \sigma_y)^2 + 4\tau_{xy}{}^2} + (\sigma_x + \sigma_y) \cdot \sin\varphi - 2c \cdot \cos\varphi. \tag{9}$$

For $f < 0$ the soil is in the pre-limit (elastic) state and its deformability is determined by Eq.3 in which the physical matrix D_{ijkl} is equal to the elastic matrix $D^e{}_{ijkl}$:

$$D^e{}_{ijkl} = \frac{E}{(1+\nu)(1-\nu-2\nu^2)} \begin{vmatrix} 1-\nu^2 & \nu(1+\nu) & 0 \\ \nu(1+\nu) & 1-\nu^2 & 0 \\ 0 & 0 & (1-\nu-2\nu^2)/2 \end{vmatrix}. \tag{10}$$

For $f = 0$ the soil goes into limiting (plastic) state and elastic matrix is replaced by plastic $D^p{}_{ijkl}$ elements of which depend on the level of the achieved stress state:

$$D^p{}_{ijkl} = D^e{}_{ijkl} - D^e{}_{ijkl} \cdot \frac{\left(\frac{\partial f}{\partial \sigma}\right) \cdot \left(\frac{\partial f}{\partial \sigma}\right)^T \cdot D^e_{ijkl}}{\left(\frac{\partial f}{\partial \sigma}\right)^T \cdot D^e_{ijkl} \cdot \left(\frac{\partial f}{\partial \sigma}\right)}. \tag{11}$$

The calculation process at each time step is built on the following iterative scheme. On the first iteration the system Eq. 6 is solved at the time point $t + \Delta t$ with $K_{t+\Delta t} = K_t$. By the value of the displacement increment $(U_{t+\Delta t} - U_t)$ during the time Δt the increment of deformation Δe_{ij} is determined, which is divided into N sections. At each new n-th section the increment of stress is determined by $\Delta\sigma_{ij}^n = d_{ijkl} \cdot \Delta e_{kl}^n$. If the Coulomb yield function $f(\sigma_{ij}^{n-1})$ does not exceed a specified limit of transition to the plastic state (ε), then $d_{ijkl} = d^e{}_{ijkl}$; if $\varepsilon \le f(\sigma_{ij}^{n-1}) \le 0$, then $d_{ijkl} = d^e{}_{ijkl}(\sigma_{ij}^{n-1})$. In case of the transition to the "beyond" state, when $f(\sigma_{ij}^{n-1}) > 0$, the number of sections N is doubled and previous calculations are repeated. The matrix D_{ijkl}, connecting the increment of stress $\Delta\sigma_{ij}$ and the deformation Δe_{kl} during the time Δt is determined by the expression:

$$D_{ijkl} = \frac{1}{N} \cdot \sum_{n+1}^{N} d_{ijkl} \,. \tag{12}$$

Then the stiffness matrix of the system is calculated and Eq.8 is re-solved. Iterative process ends when the difference between the values of the angular displacement and the pore pressure, which were obtained in two successive iterations, is not greater then specified error.

Some results of calculations of the stress-strain state of 4 m. thick homogeneous soil base loaded with uniformly distributed vertical load q with width of 26.2 m. are given below. The calculation scheme and boundary conditions are presented in Fig.1.

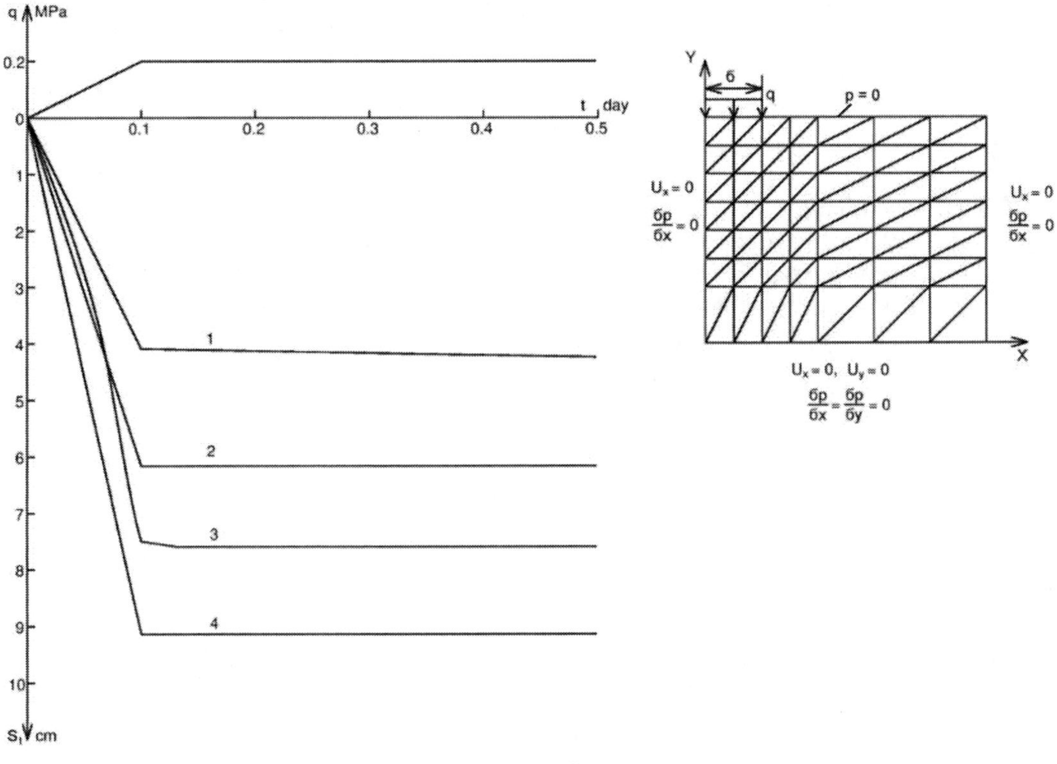

Figure. 1 Calculation scheme of the stress-strain state of homogeneous soil base and boundary conditions.

The soil has the following characteristics: $E = 500\ t/m^2$, $\nu = 0.3$, $C = 2.4\ t/m^2$, $\varphi = 15^\circ$, $\gamma = 1.0\ t/m^3$ (weighing of soil in water is considered), $\xi = 0.43$, $K_f = 0.001\ m/day$, $\beta_0 = 0$ (at full water saturation) or $\beta_0 = 0.03$ (at $n_0 = 0.41$ and $S_{r0} = 0.95$).

To analyze the effect of development of the plastic areas on the stress-strain state of soil consolidating the calculations were made using the elastic-plastic and the elastic model of the skeleton of the soil. In addition, calculations in these models were made for stable ground conditions when the overpressure in the pore water is equal to zero (no consolidation process). The figure shows the variation in time of the external load q and the precipitation of the center point of the band loaded by S_t. Curves 1 and 2 were obtained for the elastic model (respectively in the absence ($\beta_0 = 0$) and presence ($\beta_0 = 0.03$) of gas). Curves 3, 4 - for the elastic-plastic model (respectively in the absence and presence of gas). As it could be seen, the consideration of the development of the field of limit state of stress significantly affects on the amount of setting.

Thus, at $t = 0.1 day$, when the load has reached its maximum value, the ratio of the precipitate obtained by the elastic-plastic and the elastic models, 1.75 ($\beta_0 = 0$) and 1.43 ($\beta_0 = 0$); when $t = 0.5 day$, these values are respectively equal to 1.60 and 1.38. The values of stabilized sediment $S*$ for elastic model are 7.95, for elastic-plastic model this values are equal to 9.48. At time point $t = 0.5 day$ the degree of consolidation, estimated by the ratio $S_t / S*$, for fully saturated soil is equal to 0.65 (in elastic model), and 0.86 (in elastic-plastic model). In the presence of gas - 0.84 and 0.97, respectively. Despite the fact that the degree of consolidation of gas containing soil is higher, compaction process itself is slower than in a fully water-saturated soil (Fig.1): the rate of sediment in the range of time 0.1-0.5 is 0.58 cm/day (elastic model), 0.51 cm/day (elastic-plastic model). In water saturated state these values are equal to 2.08 and 1.65 cm/day, respectively.

3 Conclusions

The study of the consolidation process requires special attention in the calculation of bearing capacity of weak saturated soils. Presented results demonstrate that at calculations of stress-strain state of such soils it is necessary to consider their plastic properties. Designed calculation programs offer great opportunities to consider a nonlinear properties of stress-strain state of arrays of soils interacting with constructions. They could be used to predict the optimal mode of loads during the construction and operation of facilities.

References

1. V.A. Florin, *Osnovi mehaniki gruntov* (M. -L. Gosstroyizdat, St.Petersburg, 1961)
2. G. L. Martin, *The consolidation process in a partially saturated clay* (Abstract Ph.D, The University of Arizona, 1965)
3. L. D. Suits, T. C. Sheahan, B. Bai, Geotech. Test. J., **29**, 289-297 (2006)
4. D. T. C. Yao, W. L. De Oliveira-Filho, X. C. Cai, D. Znidarcic, Intern. J. for Numer. and Analyt. Meth. in Geomech., **26**, 139-161 (2002)
5. E.T.R. Dean, Intern. J. of Plast., **21**, 513-571 (2005)
6. I. F. Collins, Intern. J. of Mech. Sc., **47**, 493-508 (2005)
7. B. Loret, N. Khalili, Mech. of Mat., **34**, 97-116 (2002)
8. S. López-Querol, J. A. Fernandez-Merodo, P. Mira, M. Pastor, Intern. J. for Numer. and Analyt. Meth. in Geomech., **32**, 1431-1457 (2008)
9. Z. L. Wang, J. G. Wang, R. F. Shen, Comp. and Geosc., **86**, 849-859 (2008)
10. C. Menéndez, P.J.G. Nieto, F.A. Ortega, A. Bello, Appl. Math. and Comput., **216**, 458-476 (2010)
11. C. Menéndez, P.J.G. Nieto, F.A. Ortega, A. Bello, Appl. Math. and Comput., **49**, 2002-2018 (2009)
12. Soodeh Samimi, Ali Pak, Comp. and Geosc., **46**, 75-83 (2012)
13. P.G. Nicholson, *Soil Improvement and Ground Modification Methods* (Elsevier, Longford, 2015)
14. K. Soga, S.A. Jefferis, Contrib. to Geotech., **58**, 441-448 (2008)
15. R. I. Borja, Comp. Meth. in Appl. Mech. and Eng., **193**, 5301-5338 (2004)
16. A.M. Puzrin, Intern. J. of Solids and Struct., **38**, 6815-6827 (2001)

17. A. K. Bugrov, *O reshenii smeshannih zadach teorii uprugosti i teorii plastichnosti gruntov* (Base, foundation and soil mechanics. M., St.Petersburg, 1974)
18. H.W. Zhang, K.P. Wang, Z. Chen, Comp. Meth. in Appl. Mech. and Eng., **198**, 1456-1472 (2009)
19. Z.Y. Ai, Z.Y. Cheng, J. Han, Intern. J. of Eng. Sc., **46**, 486-498 (2008)
20. A. Ghassemi, A. Pak, H. Shahir, Comp. and Geotech., **37**, 10-24 (2010)
21. K.H. Park, P.K. Banerjee, Comp. Meth. in Appl. Mech. and Eng., **191**, 3233-3255 (2002)
22. L. Segerlind, *Primenenie metoda konechnih elementov* (M.: Mir, Moscow, 1979)

MATEC Web of Conferences **53**, 01026 (2016)
DOI: 10.1051/matecconf/20165301026

Problems and solutions of multi storey buildings: Okhta business center integration in historical center of Saint Petersburg

Daria Anishchenko[1,a], Egor Batkov[1], Maria Kukushkina[1] and Artem Korsun[1]

[1]*Peter the Great St. Petersburg Polytechnical University, Polytechnicheskaya st. 29, 195251, St. Petersburg, Russia*

Abstract. This research is an attempt to make the analysis of possible solutions to the problem of the land where the Okhta Center was to be constructed. The authors studied the case from the historical, practical and cultural sides. In order to find out the current public opinion on this matter two surveys were held both among St Petersburg residents and foreign guests of the city. The main aims of this article were to find the solutions, that would lead to improvement in the cultural and social level of St Petersburg citizens' life. The possible outcomes of this research are increasing incomes of the city budget, raising the status of the city and the development of Krasnogvardeisky region of St. Petersburg.

1 Introduction

The Nowadays the society needs to balance historic, cultural and modern urban aspects of the life of the city. Sometimes it means that every aspect could destroy another just because of the opportunity of the future profit from it. Of course, the profit is not only money matter, but also involves the reputation of the city and region in general, that also could reflect the economic stability of the region and nearby areas.

One of the best examples of the failure of the city to find the balance between culture, history and business is the Okhta Center. In 2010, Gazprom Company started the erection of a high skyscraper that was supposed to be the tallest tower in Europe, but citizen and archeological government funds stopped all works on the territory of the Okhta Cape. The main official reason was seven archeological layers discovered in the area and the possibility of their contribution to the development of Russian archeology and provide more information about life of ancient people in the Northern region.

As a result, the UNESCO, brought charges to the government of Saint Petersburg. According to them, the Okhta Center destroyed the exterior of Saint Petersburg skyline. They disapproved the idea, that the tower would be seen from all historical parts of the "Northern Venice" [1].

The main tendency in the architecture of Saint Petersburg is preserving the original look of the buildings and following the original plan. Saint Petersburg was the first thoroughly planned city in Russia. That is why even new buildings are sometimes constructed in baroque or neo classicism stiles to match in style with the surrounding buildings. This tendency is different from many other big cities like London, Paris, Madrid where contemporary architecture finds good balance with old architecture.

---

[a] Corresponding author : impurus@rambler.ru

All this cities have skyscrapers side by side with old architecture buildings; they find a solution in the balance of modern and old, economic and cultural requirements.

This paper analyze the reasons for the failure of large-scale project such as the Okhta Center and on the basis of public opinion, as well as taking into account the needs of the city, will offer few versions of the Okhta Cape. Therefore, an analysis of public opinion of citizens on the future destiny of the construction site Krasnogvardeisky region of St. Petersburg was held. In addition, it was reviewed and subjected to a thorough analysis of the arguments and submissions of the parties of the conflict, as well as reported and estimated accounting, after which the best option of the territory had been identified. The option includes taking into account features as well as archaeological value of the area, which was described in next paragraph.

2 Methods and materials

2.1 History of The Okhta Cape

The deepest archeological layer dates back to Neolithic Age, between IV and III millennium BC (the Baltic Sea was named Littorina sea at that time[2]). It represents life of ancient people in the North of Europe, whose main occupation was fishing. That is why fishing tools were found, such as plummet and trap for fish. The form of the plummet indicated the fact, that people on this territory used fishing bot and fishing nets. In addition, archeologist found a fish trap. It composed from wood sticks, which was thrust in the ground.

Together with fishing tools, ceramics, pendants made of amber and have trapezoidal or surround shape with V-form holes were found on this side

Besides the tools, such as special stone for wood processing, and ceramics, on Okhta Cape were excavated 40 building constructions: older living and industrial space and younger structures like houses, industrial and support structures [3].

Between Bronze and Rome Age only several single excavations were done.

In 1300, Royal governor Tergils Knutson came with 1100 warrior to build a temple. Sweden chronicler marked that Landskrona had moat and wall with eight towers. After ending of the construction, in the Landskrona based 300 warriors. Landskrona was destroyed the next year by Andrei, who was son of Alexander Nevskiy, the great Russian warrior and the patron saint of St Petersburg.

Today archeologists assume that under this fortification was another one, which could belong to Novgorod or Izhora, which was satellite of Novgorod. This layers include moat and earth wall.

This location couldn't stand along for a long time. So, in the 1500 chronicler noted about new commercial center in this place named Nevskoe Ustie. There were few settlement, where merchants fulfilled international transactions.

The excavations of this this period include ceramics made from red clay and kaolin [4].

The erection of Nyenschantz started during The Ingrian War, in 1611. The deficit of qualified workers and building material stopped construction, but in the end of 1611 the fortress could save 500 people and was almost end.

During next years Nyenschantz was mentioned in military correspondence.

In the 1644, planning of the fortress was presented. It looked like a square fortress with heptahedral towers on its corner and had three defensive front:

- moats and earth bank, which covered a city and fortress inside.
- fortification, including pentagonal fortress, four bastion and three ravelins
- old castle inside the fortress

Archeologists don't know the condition of Nyenschantz in 1661, but the fortification might have stayed unfinished.

After the ending of The Russo-Swedish War of 1656–1658, Nyenschantz still was the property of Sweden. And in 1659 a governor of the province got the order to strengthen Nyenschantz, named as the “significant and important” city, and the construction of the fortress continued. At that time, it had seven big “king” bastions and the earth bank.

The reconstruction and consolidation of the fortification had been continuing before the city was burnt in 1702.

The city near fortress was burnt in 1702 because the threat of the Russian attack, which started only years after [5].

After this day, the excavations lost their value for the Archeology, and this place has become important just for the Hisrory.

In 1727, dockyards were built on the territory of the Okhta Cape. It built rowing and sailing vessels such as schooners, centerboard boats etc. The Okhta Admiralty was created on the base of old dockyards. It built well-known sloop “Vostok”, which was include to The First Russian Antarctic Expedition.

In the Soviet time it was named Petrozavod. After 1931, it started building of tugs, passenger’s boats and liners.

During Great Patriotic War, the Petrozavod produced casing of mines, bombs, missiles, built pontoons for the evacuation from the besieged Leningrad and the delivering of the food to the city, repaired ships [6].

In 2006, Gazprom Company bought the Petrozavod. The industrial zone had been degraded by this time, and new owners had the idea of building a new the huge business center. It was to fulfill the requirements of the government to develop this region of Saint Petersburg [7].

Project of the Okhta Center had placed on 66.8 hectares. For its main part, which included a skyscraper and a nearby building, was disposed 4.6 hectares. According to the planning, the Gazprom Company and its sub companies had 16 percent of all the square, 35 % were planned to be placed for public use; other part of the square was used for the business center. The requirements of the modern business were fulfilled with this office area and made better conditions for attracting new taxpayers to the city.

The first zone of the project included: offices of "Gazprom" and its subsidiaries, the shopping and entertainment centre, the library, athletic facilities, the parking space.

The biggest Russian and World companies were supposed to hold space in the building that could provide workplace for more than 8500 specialists.

The building of the skyscraper should have been finished until the end of 2012. All project should have been finished until the end of 2016.

In December of 2010, works on the building site were stopped by the order of the Saint Petersburg’s governor.

The following part of this paragraph is a chronology of actions and events associated with the construction of the Okhta Center, which took place from 2005 to 2009.

November 15, 2005 - the competition for the development of the architectural concept of the administrative and business center "Gazprom-City" was announced. June 14, 2006 - the announcement of the results of the contest first stage for the preliminary concept of the center "Gazprom-City". From forty-six participants of the commission has selected the most promising projects. The British architectural bureau RMJM London won the Architectural design competition "Gazprom-City" project [8].

December 18, 2006 – the publication of the results of the sociological research "Social Information Agency". "According to the survey, the number of opponents and supporters of "Gazprom-City": 33% of peoples support the idea of the construction of St. Petersburg, 35% against the act. The relative given error of sociological research (it was 2.5%), we can conclude that with respect to the project "Gazprom-City" of the St. Petersburg population equally divided "(IA Regnum, 18.12.2006). At the same time, according to the study, 51% of the St. Petersburg young people supports the idea of the construction, while against the idea expressed only 25%. This is a good indicator expressing the needs of the city and the benefits of this project.

This information makes it possible to imagine how massive the project was and what aspects of the cultural and social life of the city it raised. However, one of the arguments of the opponents of the project was a huge price because it was planned to be fully financed from the city budget, but as we see, the first half, and then the entire construction cost of this project were carried out by Gazprom:

As a matter of fact, in 2007 St. Petersburg governor ,Valentina Matvienko, introduced the Legislative Assembly of the city with the draft Law of St. Petersburg "On Amendments to the Law of St. Petersburg "On Target Program of St. Petersburg "Construction of the administrative business center in St. Petersburg"'. According to the law, the main executor of the activities of the program becomes the project of "Cultural and Business Center Okhta ". It is worth mentioning that 49% of the share capital were to be owned by St. Petersburg and 51% - by JSC "Gazpromneft" [9].

However, the pressure on the government and criticism of the residents were conducted for a long time and in many ways. Especially important to mention is the letter from the Union of Architects, as well as the views of the head of the British architectural bureau Wilkinson Eyre Architect Chris Wilkinson and the opinion of the UNESCO about the importance of the project and its implementation in the whole the territory.

Back in July 2006, Valentina Matvienko was sent a collective appeal of the St. Petersburg Union of Architects, signed by its president, Vladimir Popov, alleging that the skyscraper would ruin the city's appearance. This idea was soon supported by the Union of Architects of Russia, in a letter which kept information about the competition broke all international and national rules and regulations. In protest Union boycotted the competition, which is why all the projects were foreign.

In January 2008, Chris Wilkinson, the head of the British architectural bureau Wilkinson Eyre Architect, during the interview to REGNUM correspondent assumed: "While developing any project, it is necessary to preserve as much as possible from the architectural heritage of the past and modern architecture needs to be very sensitive ... So in this case, in the case of the project skyscraper. Everything will depend on its location relative to the center of St. Petersburg. The farther the building, the better" [10].

In June 2007, New Zealand housed the 31th session of the World Heritage Committee of UNESCO. The decision of the session on the project of the World Heritage "Historic Centre of Saint Petersburg and Related Groups of Monuments" noted that: "The plans provided by a member state of 18 January 2007 and 5march 2007 do not comply with the committee and do not contain any clear boundaries and buffer zones of the objects in general, including the Leningrad Region". UNESCO adopted the strong recommendation to the Russian authorities: "To suspend implementation of the project, including the work permit until all relevant materials will not be considered"[11].

As mentioned, the entire cost of the project was now supposed to be covered by Gazprom.

The survey, conducted in 2008, showed increasing of the number of peoples supporting the construction. The Agency for Social Information, held a large survey (2,000 respondents). The study found that over the past two years the number of people supporting the construction of the public's business district "Okhta Center" in St. Petersburg had grown from 33 to 45%. This is not surprising, because during the 2008 crisis, the economic importance of the Okhta Center has only increased.

Vladimir Putin noted at a press conference, that on the one hand, he realized the cultural importance of the region where the center was being constructed, but on the other hand he believed that the city needed "fresh air" centers, which would give opportunities for the development of the business activity. Putin also said that he understood the concerns of people against the construction and he shared this concerns, he did not claim that this is the best solution, but he did not want to influence the decision-making. [12, 13, 14].

Retrospective could represent the value and social willingness to this construction. So, it`s easy to conclude, that profit from the Okhta Center and from saving the historical cityscape have the same value and importance for the city.

2.2 Interviewing

The building of the Okhta Center was closed in 2010. It is very interesting to know, how public opinion has been changing since the stopping of the erection.

In the special interview the respondents were offered the following options concerning the land where the center was to be constructed:

- Removing layer by layer of all archaeological evidences in a special museum, while constructing a business center on this place;
- Building a museum on this place;
- The combination of the museum and business center on this place
- No action
- Disinterest;

All this options have their own pros and cons, opportunities and threats (SWOT). The description of it is presented in the Table 1.

Table 1. SWOT analyzes

Option	Strength	Weakness	Opportunity	Threat
Removing layer by layer of all archaeological evidences in special museum, after it construct business center on this place	Saving of archeological layers, creation of the museum, creation of working places in the business center	High expenses for transportation and construction of detached building for the museum, disadvantages of using of the building square	The improvement of the social life of the individual district of Sankt-Petersburg and creation of big numbers of working places	High expenses for keeping the building as the museum , public organization critics because of the location of the business center in the historic area of Sankt-Petersburg
The museum on this place	Keeping the city reputation, availability for tourists	High expenses for construction of the museum, disadvantages of using building square, loosing possible working places	The improvement of cultural life of the individual district of Sankt-Petersburg, small number of workplaces	High expenses for keeping the building as a museum, stagnation in the social life of the region,
The combination of the museum and business center on this place	Saving of archeological layers, creation of the museum, creation of working places in business center, opportunity to upgrade the cultural level of office workers, and advantages of using the building square	High expenses, long- term building due to the developing of both projects	The improvement of cultural and social life of the district and city	A big traffic jam in this district, air pollution
No action	Saving possibility of block of flats construction , possibilities for the erection in the future	Disadvantages of the construction square using and losing interest for cultural values of Sankt-Petersburg,	Possibility of the community center foundation for improving social level of life in Saint Petersburg	Losing cultural value of great importance, regress of the district due to decreasing of interest of investors for this area as the potentially possible project

Disinterest – this option shows the degree of the concern of peoples for this problem

Other – this option indicates the degree of awareness of people about the condition of the construction site, their knowledge of city development programs, as well as a willingness to participate in public life

2.3 Theoretical formulas

More than 60 million rubles from the city budget were allocated in order to construct this building. Easy to prove that donation would recouped with more compact business-center and museum with a small cinema there would show films about history of Saint Petersburg, old Russian films etc.

The construction cost for 1m of area in 2013 was about 30.5 thousand of rubles [15]. Now it would be about 43.5 thousands. It is mean that we could build about 1380 m2 of area (Sgen) with using only the city budget. The janitorial service of the building [16]:

$$M_{aintance} = S_{gen} \cdot 22 \cdot f_{dollar\ rate} = 1380 \cdot 22 \cdot 62 \quad (1)$$

Some calculation for counting of a payback:

$$P_{rofit} = n \cdot 365_{days} \cdot k \cdot T_1 + m \cdot 365_{days} \cdot k_1 \cdot T_2 \quad (2)$$

n – the daily throughput of a museum. This coefficient depends on the size of the museum
m – the daily cinema throughput. This coefficient depends on numbers place in a movie hall
k – the average annual coefficient of the museum's room functioning capacity
k1 – the average annual coefficient of the cinema's room functioning capacity
T1 – the museum's ticket price
T2 – the cinema's ticket price
Souvenirs, excursions and DVD movie's sales in cinema could give:

$$\Delta P = 10\% P \quad (3)$$

All financial benefits could count such as:

$$F_{ull}P_{rofit} = P + \Delta P \quad (4)$$

Time will spent for the recoupment (in years):

$$N_{et}E_{aring} = F_{ull}P_{rofit} - M_{aintance} \quad (5)$$

$$T_{ime} = \frac{D_{onation}}{N_{et}E_{aring}} \quad (6)$$

3 Results and Discussion

3.1 Analyzes

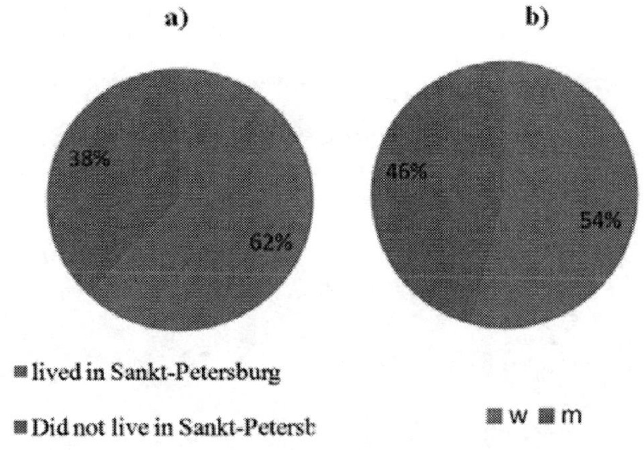

Figure1. (a) the location of respondents during Okhta proccess; (b) gender of interviewee

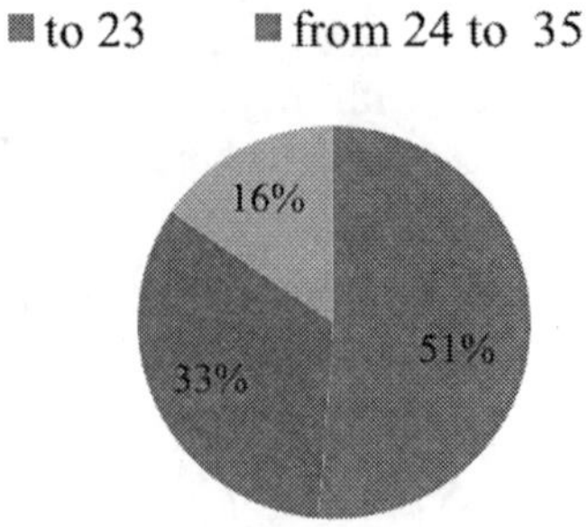

Figure 2. Diagram of the distribution answering of the respondents

The interview gave information about the gender of peoples and their location. A closer look at the data, which is presented on Figure 1, indicates that more than half interviewee didn't live on the territory of Saint Petersburg, so they couldn't participate in the social movement, so the decision about their integration in this problem during the conflict could be made.

That means that more than half people weren't involved in that problem at that time.

Gender of respondents just shows concern of peoples and their commitment.

The pie chart, demonstrated on Figure 2, shows the age of respondents, so, according this data, 51% of respondents was 18 and younger, 33 % was between 19 and 30 years old, and 16% was in the age between 31 and 50 years old during the Okhta conflict.

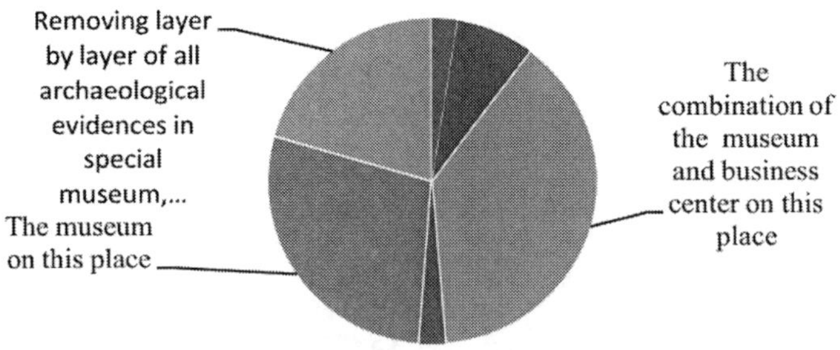

Figure 3. Diagram of the distribution answering of respondents/

The aim of this Figure 3 is to generalize the data and give the answer about the public estimation and involving in the proccess.

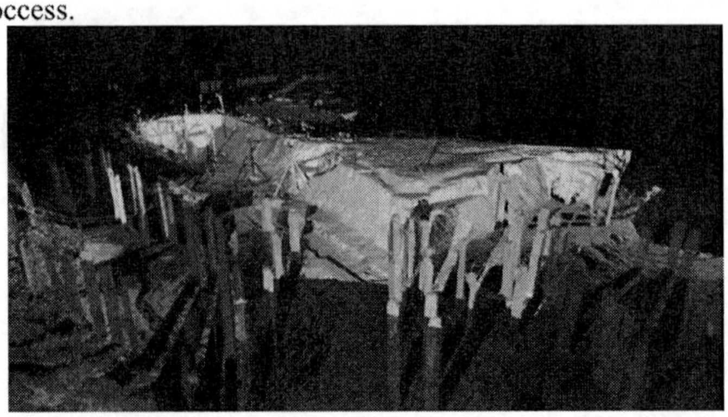

Figure 4. Results of the scanning [17, 18]

Respondents, who chose "other" usually say about the creation of the museum inside Nyenschantz. But it is impossible. Now Nyenschantz is just a huge hill of stones. It is easy to see on Figure 3, results of the scanning. [17, 18].

About three tenth of the respondents would like to see only museum on this place.

More than one third of the respondents want to see combination if museum and business center on this place.

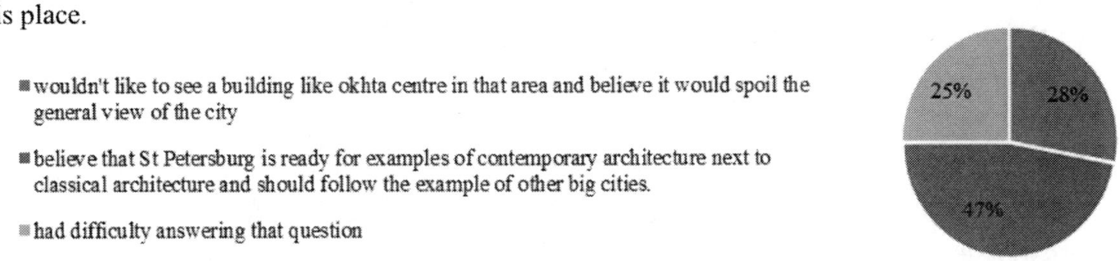

Figure5. Opinion of foreign guests of the city about the construction of a modern skyscraper in the historical center of city.

The last pie chart shows another view of this problem, from the side of foreign guests, who has the same project in their own countries.

3.2 Practical formulas

$$M_{aintance} = S_{gen} \cdot 22 \cdot f_{dollar\ rate} = 1380 \cdot 22 \cdot 62 = 1882320 \quad (7)$$

In Hermitage more than 12 thousands peoples, every day comes [19]. Of course, this museum one of the famous in the world, but it is real have more 600 people (n*T1) in day for the historical museum near the center, and 150 (m*T2) people that come to the cinema every day. The average costs of tickets in the museum is less than 250 rubles [20], the same in the cinema [21]. Try to find time with using formulas (2-6):

$$P_{rofit} = 600 \cdot 365_{days} \cdot 250 + 150 \cdot 365_{days} \cdot 250 = 68437500 \quad (8)$$

$$\Delta P = 0.1 \cdot 68437500 = 6843750 \quad (9)$$

$$F_{ull}P_{rofit} = 68437500 + 6843750 = 75281250 \quad (10)$$

$$N_{et}E_{aring} = F_{ull}P_{rofit} - M_{aintance} = 75281250 - 1882320 \quad (11)$$

$$T_{ime} = \frac{60000000}{73398930} \approx 0.8 \quad (12)$$

of year, or about 9 and half months.

3.3 Projects

One of the projects of the archeological museum on the Okhta Cape is represented as the already reconstructed model of Nyenschantz of XVII century. It would occupy all the area of the cape.

The museum includes few numbers of levels, one by one having expositions about the ancient times since Neolith Age to the times of Nyenschantz.

Figure 6. Visualization of the museum.

The layer, which was covered by the Yakornaya street, would be placed above the ground on special cantilevers. For free moving on Malookhtinskiy prospect could be built a podium with porte-cochere.

Figure7. The concept of the museum exterior.

The main question would be: what about the costs of building, and how it is possible without shutting off Malookhtinskij prospect and Yakornaya street [22]

Saint Petersburg needs to make workplaces and it couldn't spend a high volume of funds for it. The owner of this place wouldn't do it either, because of the low profit value of this place. In the same time the exterior is significant characteristic of the building, but Okhta Center had a playful name "corm". That factor shows the ratio of public to the appearance and understanding [23]. To such statements did not arise, it is necessary to clarify that: "Ergonomics is everywhere design. Aim of each architect should be the optimization and efficiency of the proposed design solutions " [24].

The idea of combination of the business center and the museum didn't need so much losses. Furthermore, it is possible to create also a big green zone closely to the center, because of using the shape of the cape, which looks like a spearhead. A huge dome made with long-span membrane helps to increase the solar affects [25] and to make sustainable [26].

After all, it will not only be beneficial to the environment. "Since A good planning of open space in the residential area will encourage people or the residents to spend more of their time at outdoor. Open space not only creates a meeting opportunities between the residents, but also can relief people from urban stresses and provided relaxation " [27].

The dome could be used as a museum, the main body of the building could be used as office area, first floor also could be used as a combination of the museum and the cinema with the small giftshop.

The erection of the building should start from the survey among the citizens, a big company that improves relationships between citizens and owners. We find it useful to make some briefings.

Charity events open great opportunities to attract additional investments, and work well for the city reputation.

The most important and historically significant object distinguished, with connection to archeological, historical and culturological spheres will be located inside the Okhta museum. Some of the less valuable items should be located in other museums. In that time also the dome of the building will be constructed at the same time because of its unusual shape, which is sure to provide high durability and high level of solarization.

Later the building site discovered during excavations, must be saved for further investigation and future exhibitions. Then it is possible to start the erection of the center.

This building could be unique in Russia as a building joining old and new architecture not only as aesthetic and historical exterior.

We realize the possibility of certain problems, connected to: the disagreement of citizens; troubles with the excavations on the site; long construction process in case of complications during the construction of the dome; limitations in the amount of space because for aesthetic reasons; given the length of the building, it couldn't be a zero-building type, but it is possible to reduce the energy consumption. In addition, it is possible to use architecture properties of Saint Petersburg and make facade looks like natural parts of Saint Petersburg history by stylizing them in eclecticism style.

4 Conclusion

The data yielded by this paper provides convincing evidence that erection of building composition of museum and business center on this territory might give good benefits for Gasprom and for the city in general. The result of the research is also based on the public opinion.

The result of our work, is the view of the need to build on the territory of Okhta Cape, a complex consisting of a business center and a museum, our opinion was confirmed by the survey conducted in 2008 and, accordingly, in 2015. But to have an alternative point of view, it was also taken for the analysis of the second most popular option (museum). "Ergonomics is everywhere design. Aim of each architect should be the optimization and efficiency of the proposed design solutions". Each of these projects has pros and cons. From the cultural point of view, both options meet the moral needs of the city. From the practical point of view, the second project can satisfy not only moral needs of the citizens, but also help to improve the economical situation in the east of St Petersburg. "Ergonomics is everywhere design. Aim of each architect should be the optimization and efficiency of the proposed design solutions"

References

1. *List of The World Heritage of UNECO*
2. *Littorina Sea. Brockhaus and Efron Encyclopedic Dictionary: to 86 m.* (1890-1907)
3. Mesolithic and Neolithic EASTERN EUROPE: HISTORY AND CULTURAL INTERACTION International Conference on the 100th anniversary of NN Gurina. 182-199 (2012)
4. P. E. Sorokin, Science in Russia, **3**, 19-25 (2011)
5. CORNERSTONE archeology, history, art and culture of Russia and neighboring countries. Institute of History of Material Culture RAS, 361-378 (2010)
6. L.N. Belova, G.N. Buldakov, A.J. Degtyarev *St. Petersburg. Petrograd. Leningrad: Encyclopedic Reference* (1992)
7. Target Program of Saint Petersburg "ADMINISTRATIVE BUILDING BUSINESS CENTER IN ST PETERSBURG",**14**, (2006)
8. «"Okhta Center "signed with the architects"' Business Petersburg »1606-1829 (Online) the press-service company "Gazpromneft", (2007)
9. Law "On Amendments to the Law of St. Petersburg "On Target Program of St. Petersburg "Construction of administrative business center in St. Petersburg"
10. Official conference, cities with IA REGNUM

11. Full text of the resolution of the 31st session of the World Heritage Committee of UNESCO on the convention dedicated to the preservation of world cultural and natural heritage, 116 (2007)
12. Official conference of Vladimir Putin, citing IA REGNUM
13. http://dic.academic.ru/dic.nsf/ruwiki/1081440
14. Online papers Fontanka: http://www.fontanka.ru/2007/01/29/115/
15. L. Lopatin, Vedomosti, **3457**, (2013)
16. http://pax-romana.ru/projects/competitions/museum-131?lc=ru\
17. http://trimetari.com/images/projects/20101125_okhta/okhta_points_intensity_shaded.jpg
18. http://trimetari.com/ru/proekty/ohranno-spasatelnye-arheologicheskie-issledovaniya-na-ohtinskom-mysu-v-sankt-peterburge
19. Rossiyskaya Gazeta , **5059**, (2007)
20. http://www.spb-guide.ru/page_13292.htm
21. According researching of NEVAFILM RESEARCH ™
22. E.V. Melnikova *The project of creating the historical and archaeological museum on Okhta Cape*
23. Bryan Lawson, Procedia - Social and Behavioral Sciences, **170**, 545–550 (2015)
24. Klaudiusz Fross, Procedia Manufacturing, **3**, 1625–1632 (2015)
25. Vera Murgul, Nikolai Vatin, Inna Zayats,Procedia Engineering, **117**, 819–824, (2015)
26. Volodymyr Mushchanov, Yevgen Gorokhov, Arusyak Vardanyan, Margarita Kashchenko, Darya Nemova, Procedia Engineering, **117**, 990–1000 (2015)
27. S. Yasmin Sofia Hussain, Ismail Said, Procedia - Social and Behavioral Sciences, **170**, 545–556 (2015)

MATEC Web of Conferences **53**, 01027 (2016)
DOI: 10.1051/matecconf/20165301027

Usage of Empirical-Statical-Dynamical (ESD) method for data extrapolation in Tunnel Construction

Zlatko Zafirovski[1,a]

[1]*University Ss. Cyril and Methodius in Skopje, Faculty of Civil Engineering, Partizanski odredi 24, 1000 Skopje, Republic of Macedonia*

Abstract. This article describes a methodology that shows how it is possible to integrate all these approaches in a problem for extrapolation of the parameters for hydrotechical tunnels. During the design process for tunnels in hydrotechics, one of the main problems is how to extrapolate the deformability and shear strentgh rock mass parameters from the zone of testing to the whole area (volume) of interes for interaction analyses between structure abd natural environments. Computers development in recent decades has contributed to the development of numerical calculation method in rock mechanics which enabled new and wider possibilities of stress and deformation calculation. This had significantly stimulated the development of rock mechanics and tunneling as scientific and technical discipline as well as the wider application of research results into practice.

1 Introduction

The investigation in rock masses in interaction with engineering structures is extremely important in a process of design of tunnels. The main problem is how to extrapolate the parameter from the zone of testing to the whole volume that is of interes for interaction analyses of the system rock mass-structure [1-3]. In this article Empirical-Statical-Dynamical (ESD) methodology of extrapolation is presented. The basis of the methodology lies in combination of the results from geotechnical and geophysical testings and rock mass classification, connected with definition of adequate regressive models [4-8].

2 Methods of analyses

Limitations in a process of investigation in rock masses comes from the fact that the whole tunnel length can not be completely covered with detailed geological and geotechnical investigations. So, it is necessary to find a way to extrapolate the necessary parameters from smaller volume of testing zone to the whole volume of the rock mass along the tunnel length [9, 10].

The given approach in a frame of this article can be defined as Empirical–Static-Dynamic (ESD) methodology of extrapolation [11-13]. The prerequisite for using this methodology is following:

- to have enough data for reliable rock mass classification.
- to have enough testing data for deformability with static tests.
- whole structural zone (in this case tunnel) to be covered with geophysical seismic tests.

---

[a] Corresponding author : zafirovskij@gmail.com

Such testing must be performed in a manner that will insure reliable data for geotechnical modeling of the whole area along the tunnel length.

One of the key problems here is to divide tunnel length in so called quasi-homogenous zones with relative uniformity of the deformability and shear strength rock mass properties as basic and constitutive elements of geological model. Inside such zone some conditions or properties are the similar in every point, and very different outside it. Each and every zone is determined by space limits and consists, in some way, properties which are important for the study [14-17].

It shall be noted, that the process of extrapolation is strictly connected and interrelated to the process of geotechnical modelling of the terrain. The complex geotechnical model is consisted of three basic models:

- model of natural geological environment; '
- model of engineering activity - geotechnical model in narrow sense (GM);
- model of interaction - model of stress-strain behavior.

A flowing-chart which shows the connections between each model is presented in Fig. 1.

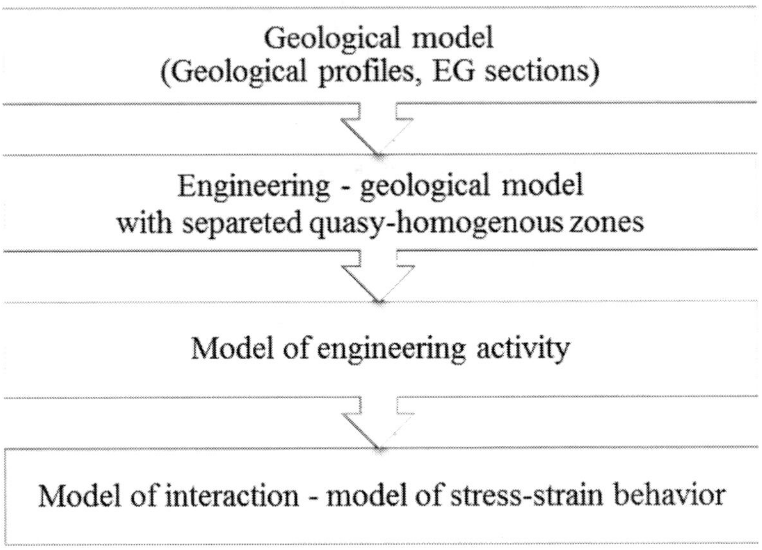

Figure 1. Steps for connection between different models.

It can be underlined that the model of engineering activity and the model of interaction are final phases of geotechnical modelling.

3 Results

To illustrate the methodology, one example is shown for the hydro technical tunnel constructed for the area of arch dam “Sveta Petka” on a river Treska in Republic Macedonia (Fig. 2, Tab. 1).

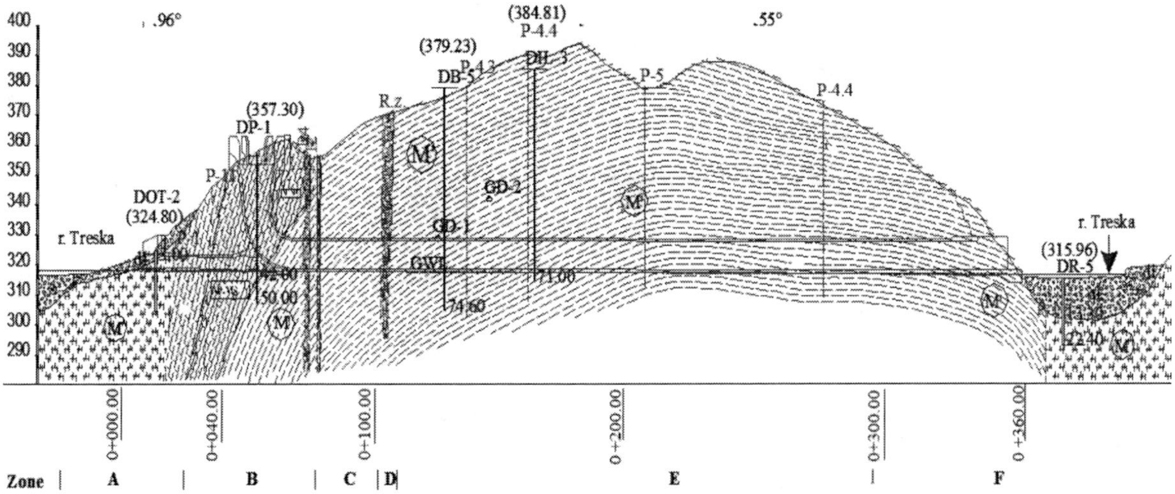

Figure 2. Engineering Geological Section for hydro technical tunnel for arch dam "Sveta Petka" in Republic of Macedonia.

In the Figure 2 the following notation:

M-marbles; M'-foliated marbles; al-alluvial deposits; GWL-groundwater level; DP-1-boreholes; GD-investigation galleries; Rz-fault zones; P-faults; D-deformation modulus; E-elasticity modulus; Vp-value of longitudinal seismic waves defined with geophysical tests; Q-value of rock mass quality after Barton et all; RMR-Rock Mass Rating after Bieniawski.

Table 1. Hydro technical tunnel for arch dam "Sveta Petka"

Zone	Deformation modulus, D [MPa]	Elasticity modulus, E[MPa]	Value of longitudinal seismic waves defined with geophysical tests, Vp [m/s]	Value of rock mass quality after Barton et all, Q	Rock Mass Rating after Bieniawski, RMR
A	3900-4500	8900-12000	4000-4600	1.46	49
B	4900-5800	11000-14000	4600-4700	1.58	52-56
C	4900	11000	4700	1.80	54
D	700-800	1600-1700	2600	0.40	21-25
E	6500-7500	15000-17000	4800-5000	1.66-2.0	57
F	3200-3500	7000-8000	3600	1.33	48

To define this Engineering-Geological Section, the following methodology of investigations is used:

- Collection of data for rock massif test results, particularly laboratory and field test results of strength, deformation, discontinuities and other parameters.
- Specific laboratory and field testing for a specific purposes.
- Statistical analysis of the tested parameters.

After that, all of the results from geological, geotechnical and geophysical investigations were used for establishing physical model through the RMR, Q and GSI classification.

Correlations between the quality of rock massif (RMR, GSI and Q indexes), dynamic (Vp) and static properties (D and E) of rock masses are expressed using results of the detailed classification of the rock massif around the measuring point with dilatometer testing's.

Typical deformability diagrams from dilatometer tests are given in Fig. 3.

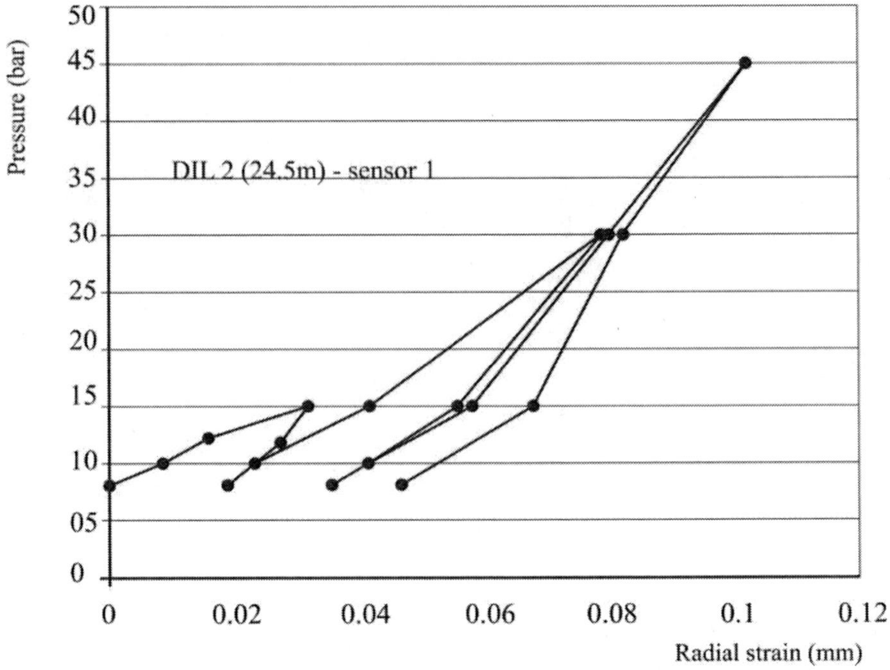

Figure 3. Typical diagram from dilatometer testing for a Rock Mass with low rating (RMR=20-25)

Diagrams shown on Figure 3 are basic for estimating of deformation modulus and elasticity modulus and imply not only to its' value than also to dependence of the modulus on pressure itself, so the point is rock mass „strenghthening“ or „softening“ regarding to pressure.

Based on detail analyses, a numerous regression models are obtained in order to fulfill the necessary criteria for extrapolation. Here, several type of regression models can be used. For example, some regressive dependances between quality of rock mass RMR and velocities of longitudinal elastic waves v and deformation modulus (D) for several locations (between them for dam „Sveta Petka" is presented in Fig. 4 and Fig. 5.

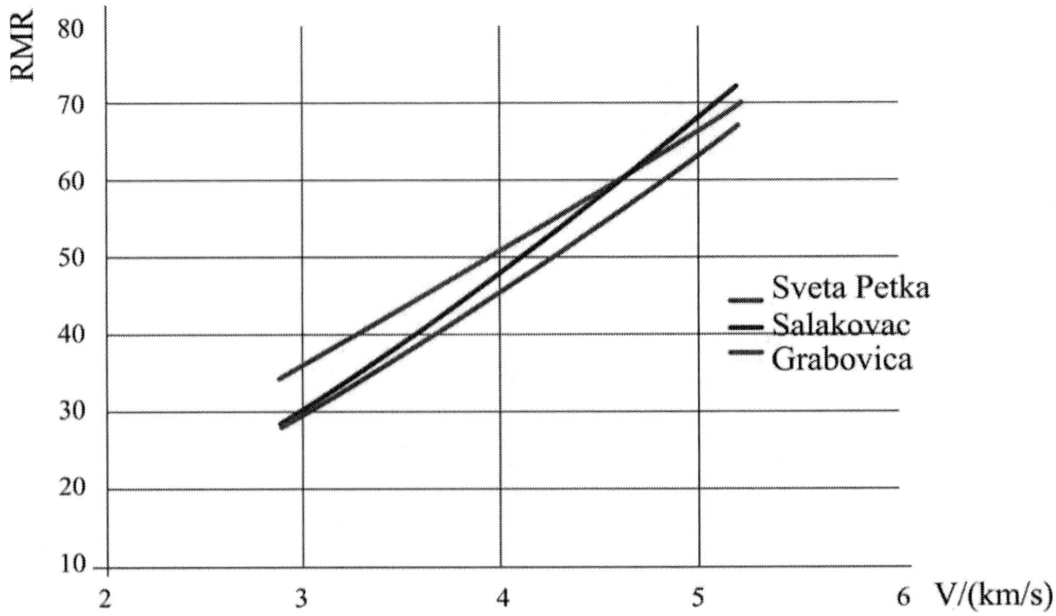

Figure 4. Regression curves between quality of rock mass RMR and velocities of longitudinal elastic waves vl from the location on "Salakovac" dam $RMR=9,8519xv_l^{1,1721}$ and "Grabovica" dam $RMR=9,4537\ xv_l^{1,2179}$ with correlative dependances fort the „Sveta Petka" dam $RMR=5,6848xv_l^{1,4979}$

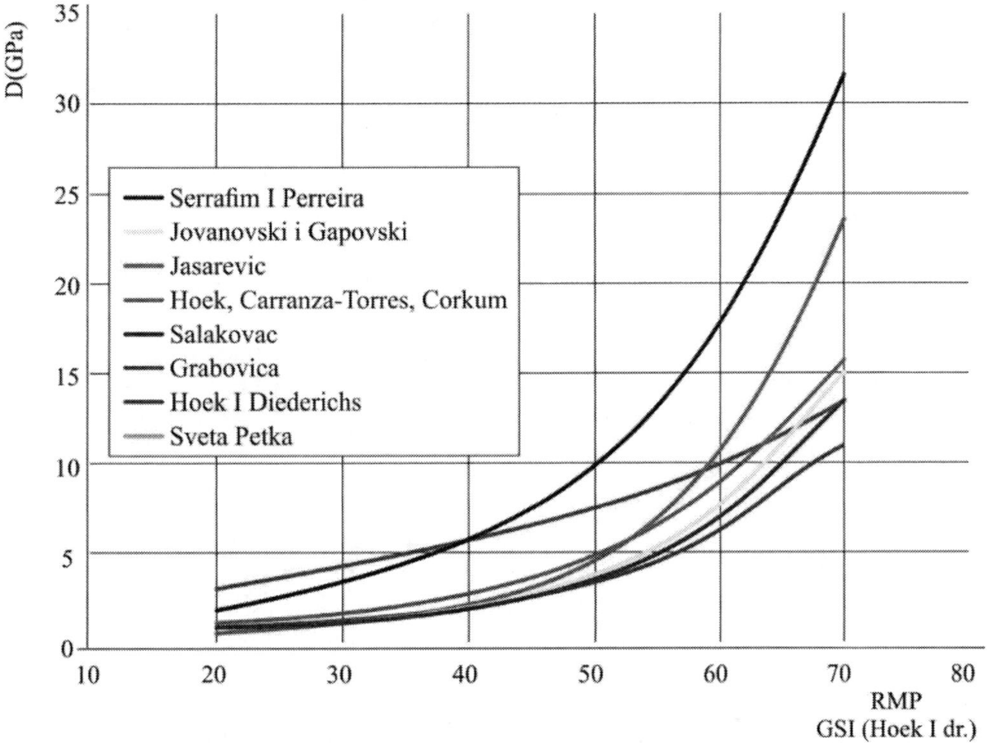

Figure 5. Correlative dependences between quality of rock mass RMR (GSI) and deformation modulus D from the location on "Salakovac" dam D =0,1369 x e 0,0657 RMR (GPa), „Grabovica" dam D =1,6963 x e 0,0295 RMR (GPa) and "Sveta Petka" D =0,1104 x e0,0703RMR

Advantage of analytical models showed in Figures 4 and 5, is that can make predictions and extrapolations relatively quickly and accurately and in that form is very appropriate for practical using. Whereat, when prediction of the parameters is made in this way, it needs to mention that different in situ test are made with different levels of vertical stress, for different strenghts, anisotropy and etc [18-22]. It is clear that if we combine empiric an field's methods, we can succesfully cover a lot of cases which are important for project analysis, but it is also clear that examples on the figures always have to be carefully used, reexamined and attentively engaged in geotechnical models.

4 Summary

The presented empirical–static–dynamic method for data extrapolation can be very useful tool in preparation of geotechnical models for numerical analyses in tunneling.

It can be mentioned, that aanalytical models for prognosis of possible intervals of deformation modulus D are useful as input data in numerical analysis for relatively shallow tunnels.

It is important to underline, that the process of modelling must correspondence with design phases. Results from initial models in first design phases for a level of Preliminary Design, can indicate the need for new data in next phases, and this, in the other hand, influences the improvement of models or leads to new ideas for new model types. We can conclude that there are many possibilities for further researches in this area. The purpose is to improve and confirm the methodologies suggested in this article, yet not only when it comes to tunnelling but also for other types of structures.

References

1. Hoek, E., Brown E.T. J. Geotech. Engng Div., ASCE 106 (GT9), pp 1013-1035 (1980).
2. Hoek, E., Marinos, P., Benissi, M. Bull Eng Geol Environ 57, pp 151- 160 (1998).

3. J. Ćetković, M. Knežević, M. Žarković, V. Murgul, N. Vatin, Applied Mechanics and Materials, 638-640, 2465-2470 (2014)
4. R. Kamnik, B. Kovačič, B. Pribicević, A. Đapo, Geodetski List, **69 (3)**, 171-188 (2015)
5. B. Kovačič, R. Kamnik, A. Štrukelj, N. Vatin, Procedia Engineering, **117 (1)**, 800-806 (2015)
6. S. Rutešić, J. Ćetković, M. Knežević, M. Žarcković, N. Vatin, Procedia Engineering, **117 (1)**, 642-650 (2015)
7. B. Kovačič, R. Kamnik, M. Premrov, Survey Review, 43 (320), 150-161 (2011)
8. B. Kovačič, R. Kamnik, A. Štrukelj, Promet - Traffic - Traffico, 22 (1), 15-22 (2010)
9. R. Usmanov, I. Mrdak, N. Vatin, V. Murgul, Applied Mechanics and Materials, **633-634**, 932-935 (2014)
10. R. Usmanov, V. Murgul, N. Vatin, Applied Mechanics and Materials, **633-634**, 1082-1085 (2014)
11. S. Ognjenovic, Z. Krakutovski, N. Vatin, Procedia Engineering, **117 (1)**, 564-572 (2015)
12. S. Ognjenovic, R. Donceva, N. Vatin, Procedia Engineering, **117 (1)**, 556-563 (2015)
13. B. Kovačič, R. Kamnik, Z. Kapović, Geodetski List, **63 (4)**, 315-327 (2009)
14. N. Gubeljak, J. Predan, D. Kozak, J. Tuma, B. Kovačič, P. Konjatić, J. Sertić, Strojarstvo, 51 (4), 263-271 (2009)
15. B. Kovačič, R. Kamnik, M. Premrov, N. Gubeljak, J. Predan, Z. Tišma, Strojniski Vestnik/Journal of Mechanical Engineering, **54 (5)**, 364-371 (2008)
16. B. Kovačič, R. Kamnik, International Journal for Engineering Modelling, **20 (1-4)**, 77-84 (2007).
17. B. Kovačič, B. Supej, (2004) Geodetski Vestnik, 48 (1), 40-49 (2004)
18. B . Kovačič, B. Geodetski Vestnik, **48 (1)**, 32-39 (2004).
19. Z.Popović, L. Lazarevic, N. Vatin, Procedia Engineering, **117 (1)**, 846-853 (2015)
20. S. Ognjenovic, R. Ristov, N. Vatin, (2015) Procedia Engineering, **117 (1)**, 575-578 (2015)
21. S. Rutešić, J. Ćetković, M. Žarcković, M. Knežević, N. Vatin, Procedia Engineering, **117 (1)**, 905-915 (2015)
22. S. Ognjenovic, R. Donceva, N. Vatin, Procedia Engineering, **117 (1)**, 549-555 (2015)

MATEC Web of Conferences **53**, 01028 (2016)
DOI: 10.1051/matecconf/20165301028

Analysis of stress-deformation state in the rock massif using Z-SOIL program

Zlatko Zafirovski[1,a]

[1]*University Ss. Cyril and Methodius in Skopje, Faculty of Civil Engineering, Partizanski odredi 24, 1000 Skopje, Republic of Macedonia*

Abstract. The Rock Mass and artificial constructions, during the changes of the stress-strain conditions are involved in specific types of interactions. The case explained in the article is a very specific, having in mind the closely spaced arch dam and existing intake tunnel. In order to insure the data for analyses, it is necessary to define the reliable constitutional model or analyses. According to the authors opinion, the combination of discontinual Rock Mass models and models for analyzing of quasi-continual media is a prerequisite for successful design and construction.

1 Introduction

The arch dam Matka on a river Treska in R.Macedonia is one of the oldest constructions in a Balkan region, constructed during the beginning of 20 century after First World War. (Fig. 1) It is still an example of well designed and constructed object. During the last years, there are some ideas to improve the efficiency of all system in a term of larger production of electricity. Beside other analyses, the design of larger surge tank is necessary, which means that new excavations shall be prepared very close to the slope surface, as well as to the zone of arch dam. Having in mind the specific surroundings, the main goal of this analysis is to introduce adequate methodology for excavation of surge tank enlargement of HEC MATKA1. In this analysis information's from the geological, geophysical and geotechnical investigations are used, and additional, some specific structural measurements are prepared in order to collect data about the joint set properties and orientation [1-4]. The project analyses includes slope stability of the pre-cut, stability analyses of individual blocks during excavation, and newly secondary stress – deformation state as a result from enlargement was analysed. Finally, an adequate support system is designed.

[a] zafirovskij@gmail.com

Figure1. The arch dam Matka on a river Treska.

2 Basic geological and geotechnical data

In a frame of Project solution, the input data for geological and geotechnical properties are analysed in order to insure reliable data for analytical and numerical analyses [5]. An approach where a combination of the methods for analysing of discontinued (jointed) and quasi- continual media is used. So, from the results of investigations, the following can be underlined:

The Rock Masses in an area of future excavation is built up of platy (foliated) gray to white marbles, locally with mica component concentrations [6, 7, 9].

All necessary joint properties are defined with structural-geological measurements, as a basis for analyze of structurally controlled instabilities [8-10]. The range of value for marbles expressed with Rock Quality Designation parameter is from RQD =66-82%.

Uniaxial compressive strength (for intact parts) is in a dependence of the direction of testing, and the anisotropy is evident (the strength normally to foliation planes is in a range from sc=78-86 MPa, while parallel to foliation in a range from sc=44-65 MPa. According to the Rock Mass Rating (Bieniawski,1989), the rock masses are in a class III with a value of RMR=43-49. The value of Geological Strength Index after Hoek and Brown classification is in a range GSI=50-59. The shear strength parameters and the deformability after the Hoek and Brown modified strength criteria, for different input parameters are in a following range:

- Modulus of deformation Em=3354-7152 MPa
- Cohesion C=166-379 kPa
- friction angle fm=43-50o

The look of the zone for future excavation, as well as the existing shape of the thank is given in a Fig.2

a) b)

Figure 2. a) A look of the Rock Masses at the zone of future pre-cut; b) a look inside theexcavation

3 Methods of stability and usedsoftware

In a first phase, the methods of stability analyses related to the discontinual Rock Mass media are analysed with a help of software for wedge and plane analyses (the software packages SWEDGE and ROCPLANE areused) [11-13].

For the structurally controlled instabilities in an underground, the program UNWEDGE is used. The software packages are design to incorporate different possible influences as simulation of water pressure along joints, earthquake influences etc, according to the principles of limit equilibrium methods [14, 15]. The additional loads as influences of blasting and earthquake are incorporated with a help of quasi-static analyses, where the dynamical forces are simulated with a help of horizontal c omponent as a results of coefficient of horizontal acceleration (a,) as a percent of ground accelerations (g). In some cases, the input parameters are used with some variations, in order to see the effect on the factor of safety (F). Just for an illustration, a stability conditions for the enlarged excavations are given on a Fig. 3

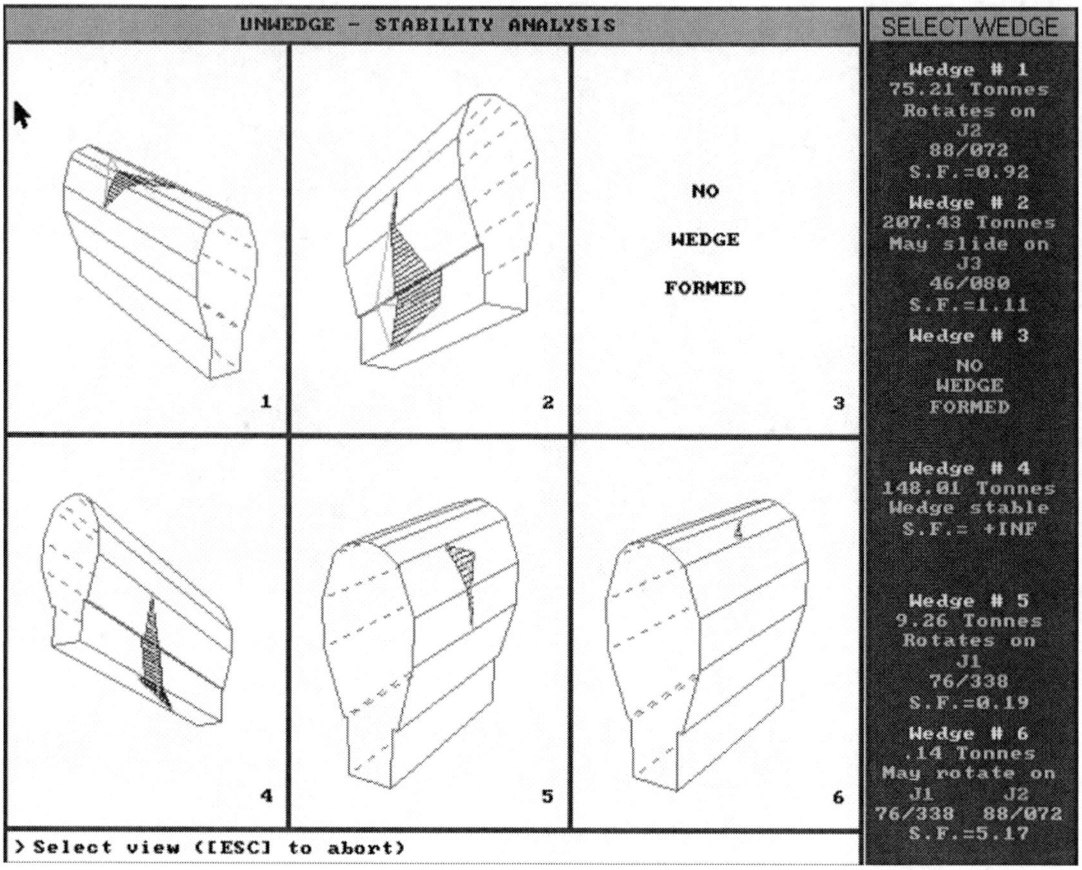

Figure 3. An overview of the stability conditions of individual wedges on a roof and sides for a case up to the final phase ofexcavation

4 Analysis of stress-deformation state in the rock massif and temporary lining with fem.

Based on detailed analyses, it was concluded that for secure excavation of surge tank enlargement, temporary lining will be used. This lining is consist of shotcrete with thickness d=8- 10cm, C-25 reinforced with welded mesh Q-335 500/560MPa, and adhesion anchors "SN" – Ø25mm, 4 meters long La=4m, at spacing ap/an=2.5/2.5m.

For numerical modelling of excavation and temporary lining of surge tank enlargement Finite Element Method (FEM) is used, incorporated in ZSOIL software produced by ZACE SERVICES Ltd (Switzerland).

The rock massif is modelled with quadrilateral FE with elastic-plastic behaviour and Mohr-Coulomb constitutive model, the existing intake tunnel with FE with elastic behaviour, and the temporary lining with beam elements [16, 17].

Following parts are incorporated in the numerical model:

- Shotcrete lining;
- Concrete lining of existing Intake Tunnel;
- Fresh (non-weathered) Rock Mass ate the lower zones bellow surface ;
- Zone of altered marbles with 3.5m thickness;
- Excavation disturbed zone with 1.5m thickness.

As critical section from aspect of load intensity, one section at the end of the surge tank is analyzed where the overburden is Hn≈23.5m. The horizontal distance from the top arch to the surface in this section is ≈7.7m. Numerical model is presented on the following Fig. 4:

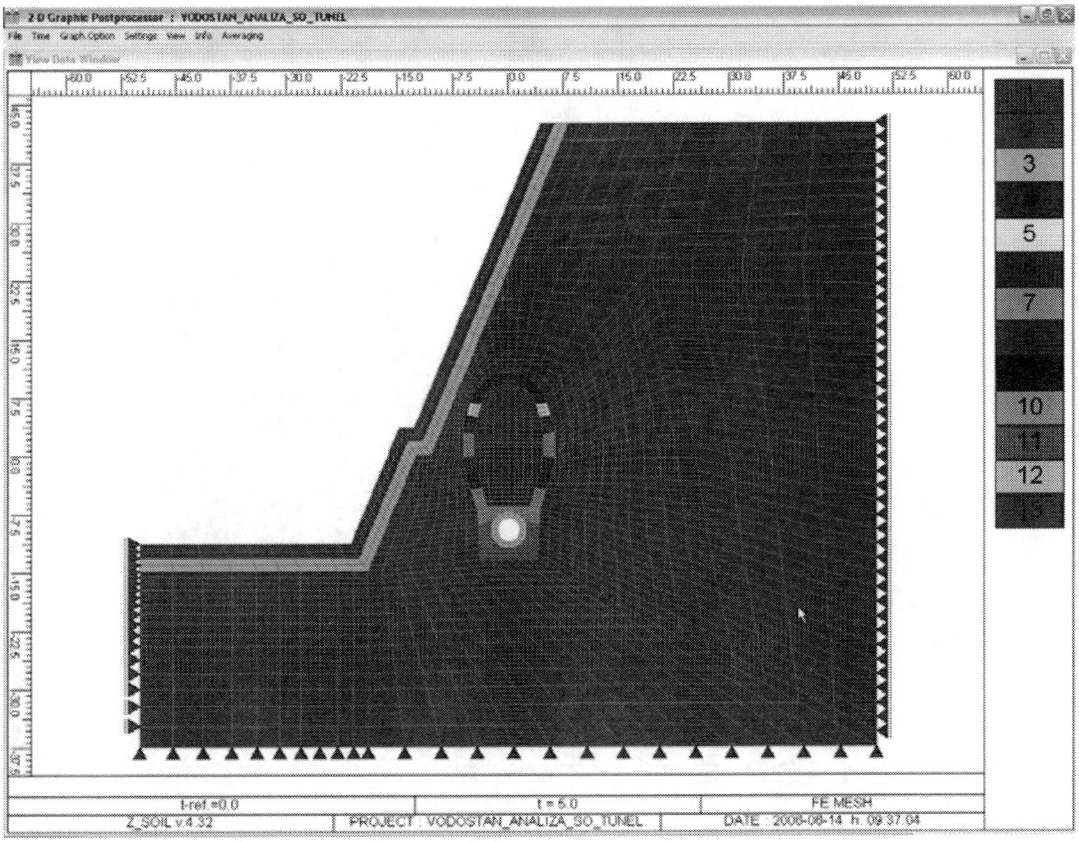

Figure 4 Numerical models for surge tankenlargement.

The excavation of surge tank enlargement will be made in 8 phases. As beginning (zero step) in the analysis, the primary stress state is taken into account which exists in the rock massif before any excavation was made. At this stage in the massif primary stresses exists, and deformation are equal to zero, because they are already execute. As a result of excavation of certain phase, disturbance of primary stress state arise, and secondary stress - deformation state appears, which acts as load in the moment of setting the primary lining.

Phase excavation is modelled with introduction of Existing Functions, and Unload Functions on the parts of the numerical model where excavation was made. From the moment of excavation to the time of primary lining setting some certain time passes, and during this time certain deformations are executed, which means that surrounding rock material take over some part of the loads. According to ZSOIL recommendations, 30% of the total deformations are executed when the primary lining start to take over the loads [18-20].

5 Results

In the part below in graphical form, only some of results from the analysis are given. Deformed state of the model as a total displacements due to excavation (without displacements from the primary stress state which are already executed), as well as internal forces in the primary lining of the surge tank are shown on the following figure.

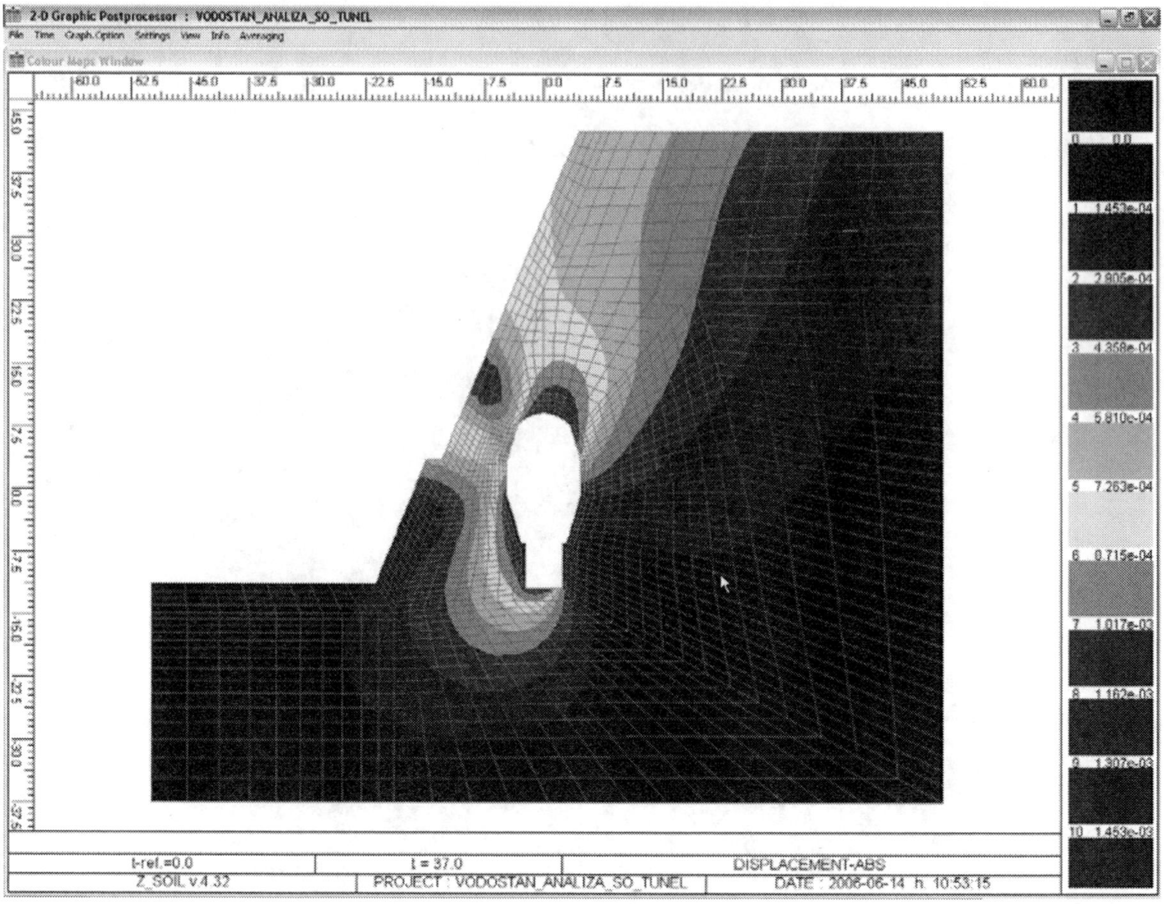

Figure 5. Total displacements of the model due to excavation of surge tank

Most important internal forces for dimensioning of shotcrete lining at the final stage of excavation are summarized in the following Tab. 1.

Table 1. Most inviolable internal forces in the shotcrete lining.

Section	N[KN]	M [KNm]
1	-221	0.7
2	90	0.2
3	86	1.7

In Tab. 2, the ultimate internal forces multiplied with partial safety coefficient (γu = 1.7), and dimensioning of shotcrete lining with thickness d=8cm, C-25MPa, reinforced with steel mesh MA 500/560 MPa, are given.

Table 2. Ultimate internal forces and dimensioning of shotcrete lining.

Section	Nul [KN]	Mul [KNm]	Necessary Reinforcement [cm2]
1	-376	1.2	Atop=0 ; Abot =0
2	153	0.3	Atop =1.53 ; Abot =1.53
3	146	2.9	Atop =2.55 ; Abot =0.64

6 Summary

The Rock Mass and artificial constructions, during the changes of the stress-strain conditions are involved in specific types of interactions. The case explained in the article is a very specific, having in mind the closely spaced arch dam and existing intake tunnel. In order to insure the data for analyses, it is necessary to define the reliable constitutional model or analyses. According to the authors opinion, the combination of discontinual Rock Mass models and models for analyzing of quasi-continual media is a prerequisite for successful design and construction.

Having this in mind, the works which will be performed for surge tank enlargement must be performed very carefully, in accordance with the proposed technical solutions and law regulations for safety during this kind of works

References

1. Z.T. Bieniawski Pergamon Press, **3**, 561-572 (1993).
2. E. Hoek, C. Carranza, B. Corcum, *Hoek-Brown failure criterion-2002 Edition*, (Rockscience Inc., Toronto, Canada, 2002)
3. J. Ćetković, M. Knežević, M. Žarković, V. Murgul, N. Vatin, Applied Mechanics and Materials, 638-640, 2465-2470 (2014)
4. R. Kamnik, B. Kovačič, B. Pribicević, A. Đapo, Geodetski List, **69 (3)**, 171-188 (2015)
5. B. Kovačič, R. Kamnik, A. Štrukelj, N. Vatin, Procedia Engineering, **117 (1)**, 800-806 (2015)
6. B. Kovačič, R. Kamnik, M. Premrov, Survey Review, 43 (320), 150-161 (2011)
7. B. Kovačič, R. Kamnik, A. Štrukelj, Promet - Traffic - Traffico, 22 (1), 15-22 (2010)
8. R. Usmanov, I. Mrdak, N. Vatin, V. Murgul, Applied Mechanics and Materials, **633-634**, 932-935 (2014)
9. R. Usmanov, V. Murgul, N. Vatin, Applied Mechanics and Materials, **633-634**, 1082-1085 (2014)
10. I. Mrdak, M. Rakočević, L. Žugić, R. Usmanov, V. Murgul, N. Vatin, Applied Mechanics and Materials, **633-634**, 1069-1076 (2014)
11. B. Kovačič, R. Kamnik, Z. Kapović, Geodetski List, **63 (4)**, 315-327 (2009)
12. N. Gubeljak, J. Predan, D. Kozak, J. Tuma, B. Kovačič, P. Konjatić, J. Sertić, Strojarstvo, 51 (4), 263-271 (2009)
13. B. Kovačič, R. Kamnik, M. Premrov, N. Gubeljak, J. Predan, Z. Tišma, Strojniski Vestnik/Journal of Mechanical Engineering, **54 (5)**, 364-371 (2008)
14. B. Kovačič, R. Kamnik, International Journal for Engineering Modelling, **20 (1-4)**, 77-84 (2007).
15. B. Kovačič, B. Supej, (2004) Geodetski Vestnik, 48 (1), 40-49 (2004)
16. B . Kovačič, B. Geodetski Vestnik, **48 (1)**, 32-39 (2004).
17. S. Ognjenovic, Z. Krakutovski, N. Vatin, Procedia Engineering, **117 (1)**, 564-572 (2015)
18. S. Ognjenovic, R. Donceva, N. Vatin, Procedia Engineering, **117 (1)**, 556-563 (2015)
19. S. Ognjenovic, R. Ristov, N. Vatin, (2015) Procedia Engineering, **117 (1)**, 575-578 (2015)
20. S. Ognjenovic, R. Donceva, N. Vatin, Procedia Engineering, **117 (1)**, 549-555 (2015)

MATEC Web of Conferences 53, 01029 (2016)
DOI: 10.1051/matecconf/20165301029

The Information Technology Project

Andrei Smirnov [1,a]

[1] St. Petersburg State University of Architecture and Civil Engineering, 2-Krasnoarmejskaja, 4, Saint-Petersburg, 190005, Russia

Abstract. The author raises the question of suitable use of different software in construction. Currently, there are several design techniques using computer technology. The various stages of the project require their parameters and software settings. Comparison of programs is fundamental in this work. The author also raises the question of the unification of all the software and the possible establishment of an absolute tool for the design in the future.

1 Introduction

All data modeling programs may be divided into two big groups: CAD (Computer-Aided Design) and BIM (Building Information Modeling). The first one is based on drawings, preparing, special standards application and software management and development. The second group of data modeling programs uses another way. BIM-software is used for full model making [1]. Ready model is very useful in each stage of construction and building: project, erection, maintenance, engineers' cooperation, etc.

BIM-prepared project presents structure in realistic conditions. The model considers all stresses and deformations of the structure, the effect of different parts of the building to each other and allows designing engineering systems with the building.

Nowadays there is a big amount of proposed BIM-software in the world. They are divided into 6 groups: architecture, sustainability, structures, MEP (Mechanical, Electrical, and Plumbing), construction and facility management programs. The most popular in the world are ArchiCAD and Revit. There are popular software in different regions, such as Nemetschek in Germany or Tekla in Finland [2].

Most of design and building companies cooperate, and there is a problem to choose the one program for joint working. There are a lot of reasons to use certain program: habit, personnel training, different agreements, etc. In other words there no absolute program used by all building industry members.

3D-visualization simplifies the understanding of the project by all interested parties, as well as reduces the risk of making mistakes during installation. However, even 85% of engineers in Europe still prefer to use a 2D-program. Partly this is explained by the fact that the standards for the documentation are created under this format. A legal document recognized as a traditional drawing and expensive 3D-model is a welcome addition to the status of the project. And there are no special

[a] Corresponding author: andrewsmirnov@inbox.ru

laws about mandatory use of specific software. In addition, not all allied partners for the development and equipment suppliers are willing to receive information in the form of 3D-models [3].

Thus, the choice of software begets a problem in cooperation between building industry members. The vantage of 2D works is the reason of slow development and promotion of 3D technologies in building design. Probably the absolute program will be the decision of these problems.

The first stage of each project is an architectural design. The architect has to explain his own idea accessibly and technically correct. Also the final project has to be applicable for fellow engineers' and builders' working. CAD-models are very useful, but other members of building process have to have imagination for full project understanding. By the way, CAD-model may be in 3D. But BIM-models allow understanding the idea very graphically. And further working with this model are more comfortable and effectively because of import and cooperation abilities.

Thus, to construct the base architectural clear need to choose suitable software. Possibly select multiple criteria such as: the speed of the proposed instruments, the possibility of importing into other programs, library facilities and so on. As a global provider of software spectrum is important to choose taking into account the specific characteristics of each company and the particular specialist.

2 CAD-software: AutoCAD and SketchUp

The absolute market leader for the design of the proposed program is AutoCAD. It is safe to assume that about 90% of the world's construction industry uses this software. The remaining 10% comes from similar programs, such as nanoCAD or DraftSight [4]. Analogous programs are usually offered for free or much cheaper than AutoCAD. Often, however, the functionality is much lower, and export opportunities are limited at best, IFC-format (Industry Foundation Classes), and sometimes non-existent [8].

AutoCAD offers a wide range of intuitive tools for creating drawings. All manner of plugins allow to automatically producing documentation to the required standards.

Newer versions of AutoCAD can support 3D-modeling. Creators from Autodesk are developing new technologies to accelerate the process [4].

Therefore, skill in AutoCAD is a prerequisite of successful work of the designer in our time. This program became the absolute software in the CAD-design, which is so lacking in BIM. All the necessary legislation and project drawings, specifications and cost estimates may be prepared using AutoCAD.

However, there are drawbacks. Despite extensive export opportunities in other software AutoCAD's formats (including IFC) are not supported by the majority of known visualization software. In other words, for the manufacture of the classic model for 3D-presentations AutoCAD is insufficient. It is possible to use a secondary program such as Revit or SketchUp to solve this problem.

Also, because AutoCAD is the CAD-software, it cannot edit user in the case of engineering error or offer the best solutions for a particular design.

SketchUp is a very progressive software design. Simple intuitive operation allows obtaining 3D-models of any complexity. It is widely used not only in construction but also in the production of furniture and interior planning. The great advantage of SketchUp is a huge library of ready components, where possible to find everything from small interior items and household appliances to aircraft and the world's interest. The user can create a project of any configuration with wider surroundings. The time spent on the production of realistic 3D-model via SketchUp is minimal [5].

However, it is advisable to use SketchUp in the initial stages of design: explanation of ideas, presentation, and marketing. This program does not manufacture finished drawings, sections, and plans. It is possible to estimate the volume of work required, compared with the existing facilities, but own experience and third-party software using. Also SketchUp cannot correct any errors or to offer solutions of design problems.

Own SketchUp graphics are "cartoonish"; not the quality for a presentation. But SketchUp format files are supported by most visualization programs. For example, SketchUp-Lumion ligament model allows producing the highest quality photorealistic render. It is also possible to make a video or panorama.

Both programs have the actual drawing tools: lines, shapes, polylines, etc; auxiliary editing copy, reflection, "squeezing" in 3D, zoom, etc. In other words, the user manually makes the necessary structural elements.

Thus, the CAD-software is an indispensable tool for the manufacture of the design documentation, calculation and presentation. However, the functionality of this technology is quite low as compared to BIM. In fact, a large number of works which can be performed automatically, is done by hand [8]. When producing 3D-model, the user should automatically receive all the necessary sections, plans and specifications, but it is not available for the CAD-software.

3 BIM-software: ArchiCAD and Revit

Undisputed leaders in the BIM-technologies are two competitors: ArchiCAD and Revit. Similar in functionality, they, however, have different audiences [8].

ArchiCAD was established in 1984 under the name of the company by Graphisoft Radar CH, working only with the water supply. The latest version is ArchiCAD 18 and it is already a complete solution for the design of buildings and structures. Despite the name, containing the product range CAD, this program is a classic example of BIM-software. The fact that this name was given before specific wording BIM, and in order to maintain the audience Graphisoft members decided to keep the brand ArchiCAD [6].

Revit is a product of Autodesk, the largest provider of software for architects, engineers and designers. For example, a well-known AutoCAD - is also a product of this company. And Revit was created specifically as a competitor for ArchiCAD. The first version was introduced in 2000 in the form of three programs: Revit Structure, for engineering calculations; Revit Architecture, for architects; and MEP-version for the design of communications structures. But in 2013, all three branches have been merged into one program as a result received a powerful tool conceived to fully integrated design of all building systems [7].

These two programs use similar construction elements that have counterparts in the real world: walls, columns, windows. To create shells have special manual tools. This is one of the fundamental differences between the CAD-software. Binding system is based on the preliminary establishment of the grid axes and elevation plans, on the basis of which all objects are built. Also, both programs have intelligent systems that allow the user to adjust in the case of engineering error or offer more optimal solutions of structural problems [8].

According to the survey, ArchiCAD used by about 30% of active users. Revit's audience is about 25% [8]. The spread is small, linked, rather, to the fact that ArchiCAD was introduced to the market for a much longer time to collect specific targeted clientele. Revit, however, the number of users reduces the gap each year [2].

ArchiCAD and Revit are categorized BIM-software. Therefore, their use in stages from design to operation is quite appropriate. However, Revit has a much wider field of application. This program not only prepares the complete drawings of all building elements, but some specifications and sheets for further operation and even dismantle the building. Also, as has been said, Revit combines opportunities for all designers: structural engineers, MEP software, engineers, architects, and so on. ArchiCAD cannot yet boast full internal integration. It requires separate programs for different sections of the design.

A very important issue is the creation time of the project. Models of ArchiCAD and Revit are much more time consuming. Even the creation of the elementary project in the first stages of the design must require a fairly large amount of time for the placement of axes and elevations, installation of all design elements, further imaging process. Therefore, if the initial stages of the project requires only to explain ideas and presentations, ArchiCAD and Revit - not always the best choice [8].

Exporting abilities of Revit is much higher than that of ArchiCAD. For example, visualization programs Artlantis, Lumion and V-Ray work with Revit format files. There is also an export from

AutoCAD. ArchiCAD in this plan falls behind. Artlantis has been specially created for it, with no problems exporting. But if the user wants to use the other software, then it is required to transfer finished ArchiCAD models in IFC-format, which is fraught with losses, bugs, time-consuming [1] [2] [6, 9-12].

Both programs have their own tools for visualization. Assign different texture layers model transforms the image into a realistic object. However, in comparison with similar methods render that offer Lumion or Artlantis, own visualization of ArchiCAD or Revit looks worse as they are simplified to the minimum necessary, so no settings of light and shade, reflection, reflections, transparency, global illumination, etc. For the presentation of the object, making of clips, panoramas and displays need to use third-party software [13-15].

Thus, these programs have their own advantages and disadvantages. The possible association of the experiments would create a product that is suitable for everyone [16, 17]. This is what Autodesk is currently seeking by perfecting Revit each year to obtain new customers and consumers. At the same time Graphisoft is releasing a new version, that does not much change the functionality and flexibility of the software, but appeals to a closed circle of conservative admirers of the software, valuing stability and not spending time learning new techniques and technologies on a broad scale [18-21] .

4 Conclusions

Thus, it is possible to draw the following conclusions:

- In the various design stages require different indicators. To produce a model of the facility as soon as possible it is better to use CAD-software. If do not needed presentation-quality 3D-models to explain all the ideas and proposals, it is logical to use AutoCAD. But in case of the need for photo-realistic images and objective indicators of exteriors and interiors of the building is better to use tools of SketchUp followed by visualizing.
- CAD-software can export to IFC-format and BIM. Such a course of action to accelerate the initial basic design with subsequent upgrades in complex project for further work during construction and operation. However, more time may be the cost of bringing objects from one program to the requirements of the other.
- In the case of the initial solutions for integrated project and construction work at several engineers or companies it is advisable to use a BIM-software. Parallel works of engineers of different specialties require the use of such technologies. BIM-program will take into account the impact of all the elements and structures of the building on each other.

However, the question remains about the absolute program, which will use the vast number of users, is still open. In the case of CAD-software the leader is AutoCAD, and already it is doubtful that any other product will be able to take its place. A possible candidate for the same position in BIM-technology is Revit, which is developing rapidly.

Thus, at the moment of virtual design technology are in a stage of rapid development. New ideas emerge every year, and the degree of optimization are becoming higher and higher. So perhaps, in the future there will be introduced a mandatory design through modeling or predictive calculation with attendant checks and correcting.

References

1. B. Hardin, D. McCool, *BIM and Construction Management: Proven Tools, Methods, and Workflows. 2nd Edition* (Indianapolis, 2015)
2. M. Hergunsel *Benefits of building information modeling for construction managers and BIM based scheduling* (Master's Thesis, Worcester, 2011)

3. M. Day, AEC Magazine, **1(47)**, 16-19 (2013)
4. L. Allen *AutoCAD Professional Tips and Technique* (San-Rafael, 2014)
5. A. V. Petelin *3D-modyelirovaniye v SketchUp 2015 in 2 volumes* (St. Petersburg, 2015)
6. N. I. Malova *ArchiCAD 18 v primerah* (St. Petersburg, 2015)
7. J. Vandezande, P. Read, E. Krygiel *Autodesk Revit Architecture 2013: Autodesk Official Training Guide* (San-Rafael, 2015)
8. A. V. Smirnov *The comparison of the architectural design softwares* (Bachelor's Thesis, Lappeenranta, 2015)
9. V.L. Kvint, V.V Okrepilov, Herald of the Russian Academy of Sciences, **84 (3)**, 188-200 (2014)
10. Z. Jakšic, D. Ladjinović, M. Trivunić, N. Harmati, N. Vatin, (2015) Procedia Engineering, **117 (1)**, 507-520 (2015)
11. K. Strelets, N. Vatin, (2015) Rocznik Ochrona Srodowiska, **17 (1)**, 104-112 (2015)
12. N. Harmati, Z. Jakšić, N. Vatin, (2015) Procedia Engineering, **117 (1)**, 791-799 (2015)
13. A. Milajić, D. Beljaković, D. Davidović, N. Vatin, V. Murgul, Procedia Engineering, **117**, 916-923 (2015)
14. B. Stamatovic, R. Upadhyay, Procedia Engineering, **117 (1)**, 660-667 (2015)
15. Z.Popović, L. Lazarevic, N. Vatin, Procedia Engineering, **117 (1)**, 846-853 (2015)
16. M. Jevrić, M. Knežević, J. Kalezić, N. Kopitović-Vuković, I. Ćipranić, Tehnicki Vjesnik, **21 (4)**, 873-879 (2014)
17. M. Lazarevska, M. Knežević, M. Cvetkovska, A. Trombeva-Gavriloska, Tehnicki Vjesnik, **21 (6)**, 1353-1359 (2014)
18. N. Pavličić, M. Perazić, D. Durić-Jocić, M. Knežević, Journal of Applied Engineering Science, **12 (1)**, 11-18 (2014)
19. B. Melović, S. Mitrović, A. Djokaj, N. Vatin, Procedia Engineering, **117 (1)**, 807-812 (2015)
20. B. Kovačič, R. Kamnik, A. Štrukelj, N. Vatin, Procedia Engineering, **117 (1)**, 800-806 (2015)
21. S. Rutešić, J. Ćetković, M. Knežević, M. Žarcković, N. Vatin, Procedia Engineering, **117 (1)**, 642-650 (2015)

MATEC Web of Conferences 53, 01030 (2016)
DOI: 10.1051/matecconf/20165301030

Complex Testing of the Bridges

Adrijana Savić[1,a]

[1]Direkcija za urbanizam i izgradnju grada Leskovca, 18000 Niš, Srbija

Abstract. This article describes the process of examining bridges. AB road bridge, founded on the columns by the Karpoš system and damaged due to erosive action of the river, is given as an example. The first tests of the bridge are conducted on the impact load of V 300, because of the appearance of longitudinal cracks. The results of the columns testing are presented in this paper, as well as the rehabilitation measures thus increasing the load capacity of the column.

1 Introduction

Systematic and complex testing of bridges allows timely notice of dangerous irregularities in their structure. Properly implemented, the rehabilitation measures are able to fully establish the performance of a bridge and thus to extend the term of their exploitation [1-4]

Traditionally, the testing involves the following steps [5-12]:

1) Introduction to the technical documentation.

2) Visual observation of the bridge structure.

3) Instrumental testing's. Conducted with the aim to assess the overlap with the project of geometrical proportions, also with the physical-mechanical properties of the applied materials. Evaluation of transport - exploitation characteristics of the pavement in the bridge is performed by geodetic surveying of the road surface [6, 7, 8].

Composition of the basic instrumental tests:

1) Concrete strength is determined under the following non-destructive testing methods: shearing, rebound hammer testing and ultrasonic tool. Fig. 1 (a, b, c)

2) Concrete shield layer determination. Fig. 2 (a)

3) Water resistance of concrete by air permeability degree. Fig. 2 (b)

4) Concrete moister determination. Fig. 2 (c)

5) Determination of strains in pre-stressed reinforcement. Fig. 3 (a)

6) Determination of strains is reinforcement wires using guy method. Fig. 3 (b)

7) Estimation of degree of corrosion according to the method electric potential measurement method. Fig. 3 (c)

8) Temperature determination of the bridge structure members using thermal imaging device. Fig. 4 (a)

9) Determination of adhesion of various surfaces and reinforcing materials to structure members. Fig. 4 (b)

10) Determination of paint-coat thickness under a non-destructive testing method. Fig. 4 (c)

[a] Corresponding author: sadrijana@bk.ru

11) Determination of concrete carbonization depth. Fig. 5 (a)

12) Determination of crack depth. Fig. 5 (b)

13) Land survey measurements execution (axial and cross-section profiles of traffic-bearing determination, deck surveying along bottom surfaces, checking whether actual marking points correspond to designed ones). Fig. 5 (c) [13- 18]

Figure 1. Concrete strength determination under a non-destructive testing method

Figure 2. Determination of concrete shield layer, water resistance and moisture

Figure 3. Determination reinforcement strains and estimation of corrosion degree

Figure 4. Thermal imaging survey, surfaces adhesion determination, paint-coat thickness determination

Figure 5. Determination of concrete carbonization depth, crack depth and land survey measurements.

2 Testing of the bridge, constructed in Predejane (Serbia), which suffered the corrosive effects of the river.

2.1 Disposition of the bridge.

AB Road Bridge is located in the village of Predejane over the South Morava river and consists of eight fields (8x10,0m). The width of the road lane is 4,0 m. The bridge is founded on the columns by the system Karposh [19, 21]. And under each support are the four columns. Due to the erosive action of the river bed, since the South Morava has torrential character, there was some damages in the 5th, 6th and 7th row, so the buckling length in those sections is far larger than the calculational, which is 5.0m (Fig. 6, 7, 8).

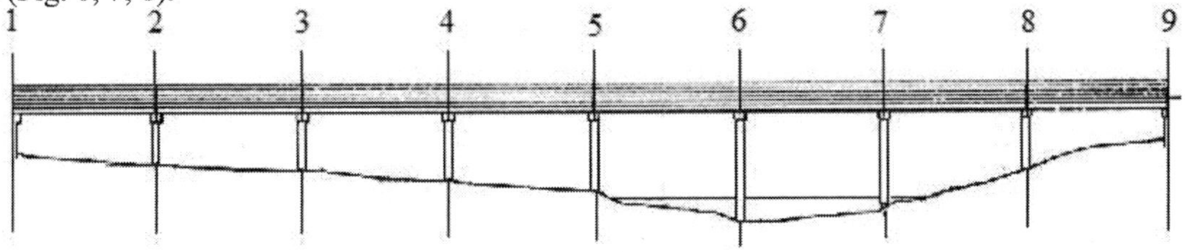

Figure 6. The scheme of the bridge

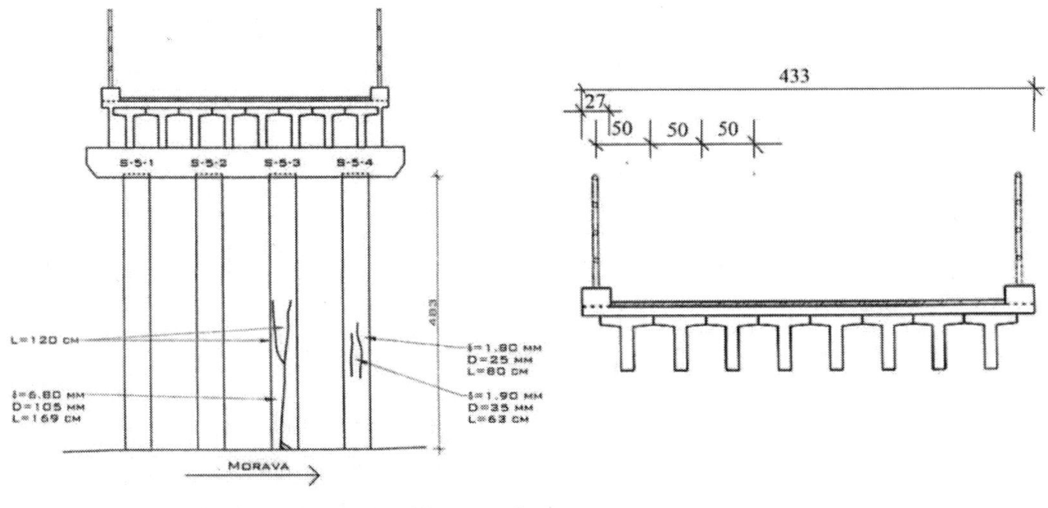

Figure 7. Cross-sectional and cracks on the pillars - Axis 5

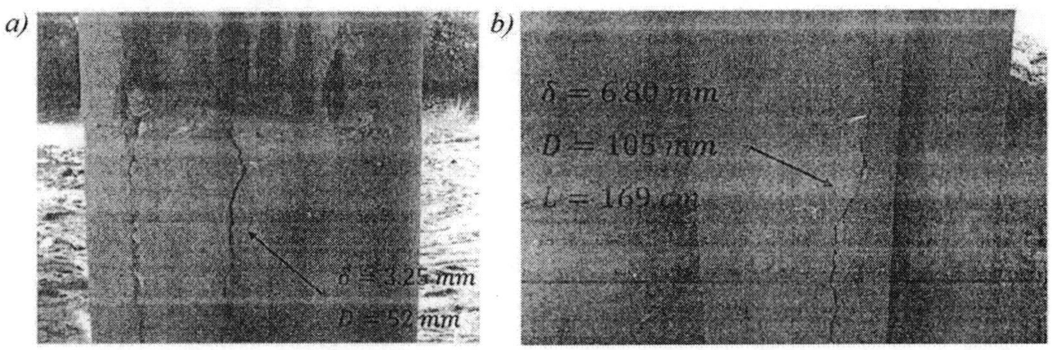

Figure 8. a) Corrosion of the reinforcement and cracks of the column S-4-1and b) Cracks on the column S-5-3

2.2 The conducted tests

Considering that large longitudinal cracks on some piles of the bridge had appeared, we started a pilot test of the V300 vehicle in a different mode of the vehicle motion and in a position which depended on the line of influential. The paper presents the results of columns testing in the row no.5 from the perspective of the vehicle motion and braking over the tested piles [22-29].

The measuring sensor V50, from Hottinger Baldwin company, was set in half of the range, in order to monitor the movements due to buckling and slenderness of approximately l = 6.50m.

The results of recorded diagrams in the program are by the system Spider - CatMan. When the force P acts, it is clear that the reactions of the vehicle at the point approximately correspond to half of the range of the pile, causing a change of the elastic line, which indicates that there is a second tone of oscillation [30-37]. The appearance of opening cracks in the investigated column, is also noticed, which further threatens the local stability of the bridge as a whole (Fig. 9).

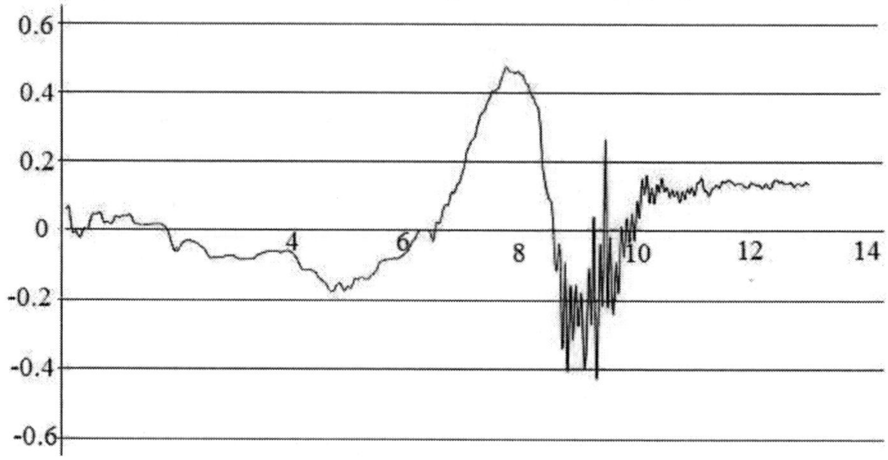

Figure 9. Diagram of the horizontal shift of the column

3 Static calculation of the columns with strengthens in 5th, 6th and 7th rows (Fig.10).

$l_k = 0.70 \cdot l_0$

$l_k = 0.70 \cdot 5.0$

$l_k = 3.5\ m$

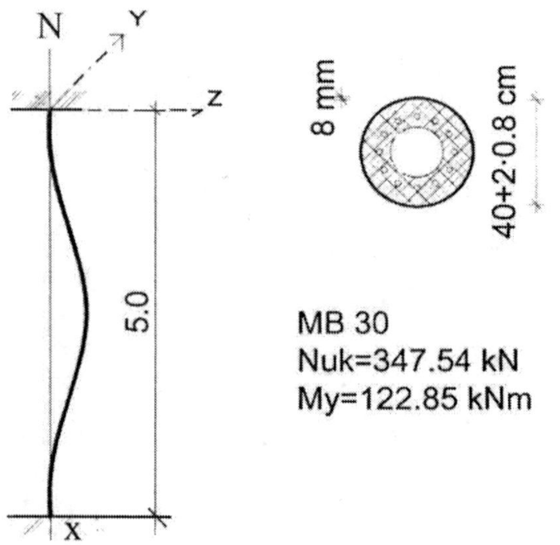

Figure 10. Static calculation of the columns

Other elements of the bridge superstructure weren't considered for possible rehabilitation measures, since they are in good condition, as such they can serve their purpose [38].

Adopting C.0361

Limits of stretching: β_{sa}=24 kN/cm^2

MB 30

Calculated value of concrete strength: β_r=0.7*3=2.1 kN/cm^2

GA 240/360 – 12 Ø22

$$A_b = \frac{40^2 \pi}{4} - \frac{19^2 \pi}{4} = 973.11 cm^2 \tag{1}$$

$$A_a^{profil} = \frac{41.6^2 \pi}{4} - \frac{40^2 \pi}{4} = 102.54 cm^2 \tag{2}$$

$$A_s = 12 \frac{2.2^2 \pi}{4} = 45.62 cm^2 \tag{3}$$

Coefficient of safety for the base load: γo=1.7

Active stiffness on bending of the composite column:

Ea=21 000 kN/cm2

$$(EI)w = E_a I_a + E_{bi} I_b + E_s I_s \tag{4}$$

$$I_a = \frac{(41.6^4 - 40^4)\pi}{64} = 21345.06 cm^4 \tag{5}$$

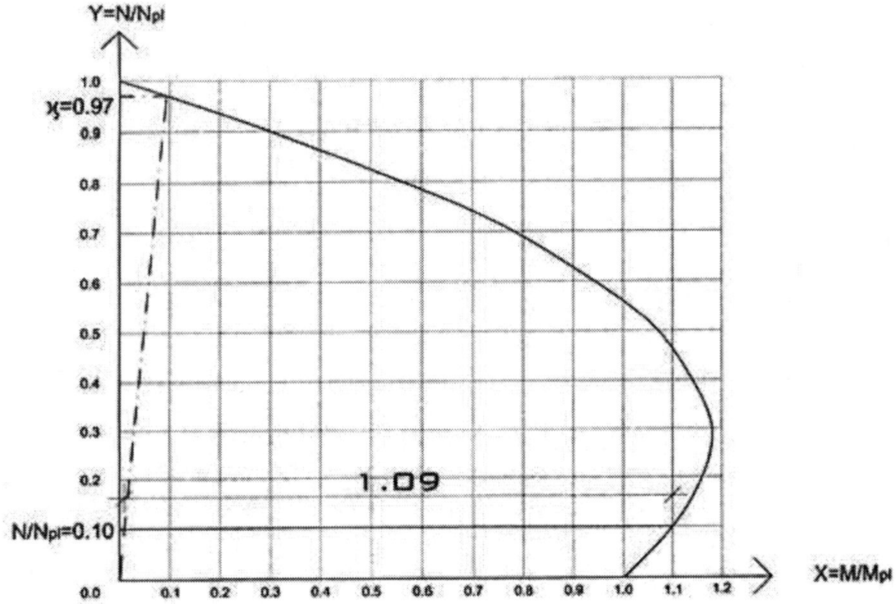

Figure 11. Interactional diagram of the composite column

4 Rehabilitation measures

Applied rehabilitation measures are the result of the test program with the aim of protecting the existing piles from adverse impacts of water and also to return the resulting buckling length to the projected norm of 5.0m. Protection of the piles was carried out by refuting Larsen talpe 604n.

Columns rehabilitation in 5th, 6th and 7th rows. The large number of cracks is present in these columns, so reinforcement is performed on the columns with steel having the thickness δ = 8,0mm, throughout the height of the columns. The columns are calculated as a couple, with the existing cross-section of MB 30. From the interactional curve above, it is clear that the performed reinforcement raised load capacity with very small changes in geometry as in width, and height [39-42].

Details of the foundation rate at the same time represents reinforcement of the piles as well as the nominal supports of the newly designed composite columns (Fig. 12).

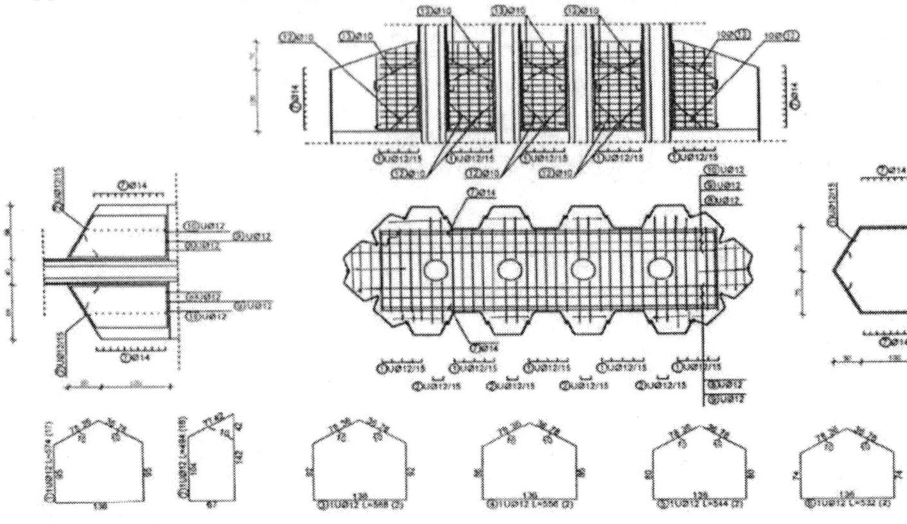

Figure 12. Details of the reinforced foundational rate

5 Summary

Systematic and complex testing of bridges allows timely notice of dangerous irregularities in their structure. Properly implemented, the rehabilitation measures are able to fully establish the performance of a bridge and thus to extend the term of their exploitation.

References

1. EN, *Design of composite steel and concrete structures*, (1994)
2. PBAB 87 Rules for concrete i reinforced concrete (2013)
3. Grinfeldi, G., A. S. Gorshkov, N. I. Vatin, Advanced Materials Research, **941-944**, 786-799, (2014)
4. Mandić, R., Hadži-Niković, G., Ćorić, S., Geofizika, **28 (1)**, 145-160 (2011)
5. R. Usmanov, V. Murgul, N. Vatin, Applied Mechanics and Materials, **633-634**, 1082-1085 (2014)
6. R. Kamnik, B. Kovačič, B. Pribicević, A. Đapo, Geodetski List, **69 (3)**, 171-188 (2015)
7. B. Kovačič, R. Kamnik, A. Štrukelj, N. Vatin, Procedia Engineering, **117** (1), 800-806 (2015)
8. R. Usmanov, M. Rakočević, V. Murgul, N. Vatin, Applied Mechanics and Materials, **633-634**, 927-931 (2014)
9. B. Kovačič, R. Kamnik, M. Premrov, Survey Review, **43** (320), 150-161 (2011)
10. Z. Radovanović, R. Sinđić Grebović, S. Dimovska, N. Serdar, N. Vatin, V. Murgul, Procedia Engineering, **117**, 870-878 (2015)
11. B. Kovačič, R. Kamnik, A. Štrukelj, Promet - Traffic - Traffico, **22** (1), 15-22 (2010)
12. B. Kovačič, R. Kamnik, Z. Kapović, Geodetski List, **63** (4), 315-327 (2009)

13. N. Gubeljak, J. Predan, D. Kozak, J. Tuma, B. Kovačič, P. Konjatić, J. Sertić, Strojarstvo, **51** (4), 263-271 (2009)
14. B. Kovačič, R. Kamnik, M. Premrov, N. Gubeljak, J. Predan, Z. Tišma, Strojniski Vestnik/Journal of Mechanical Engineering, **54** (5), 364-371 (2008)
15. B. Kovačič, R. Kamnik, International Journal for Engineering Modelling, **20** (1-4), 77-84 (2007).
16. B. Kovačič, B. Supej, (2004) Geodetski Vestnik, **48** (1), 40-49 (2004)
17. B . Kovačič, B. Geodetski Vestnik, **48** (1), 32-39 (2004).
18. V.V Okrepilov, A.Yu. Smirnov, Measurement Techniques, **56 (1)**, 54-60 (2013)
19. V.V Okrepilov, Studies on Russian Economic Development, **24(1)**, 35-42 (2013)
20. V.V Okrepilov, Standarty i Kachestvo, **10**, 52-55 (2005)
21. V.V Okrepilov, Ivanova, G. Standarty i Kachestvo, **12**, 62-68. (2004)
22. V.V Okrepilov, Standarty i Kachestvo, **3**, 94-96. (2003)
23. N. Vatin, N. Lavrov, A. Shipilov, Procedia Engineering, **117 (1)**, 374-380 (2015)
24. J. Ćetković, S. Rutešić, M. Zarković, M. Knežević, N. Vatin, Procedia Engineering, **117 (1)**, 780-790 (2015)
25. S.G. Popović, S. Vukanić, D.F. Komatina, E. Alihodzic-Jasarevic, N. Vatin, Procedia Engineering, **117 (1)**, 614-620 (2015)
26. S. Ognjenovic, Z. Krakutovski, N. Vatin, Procedia Engineering, **117 (1)**, 564-572 (2015)
27. S. Ognjenovic, R. Donceva, N. Vatin, Procedia Engineering, **117 (1)**, 556-563 (2015)
28. B. Stamatovic, R. Upadhyay, Procedia Engineering, **117 (1)**, 660-667 (2015)
29. Z.Popović, L. Lazarevic, N. Vatin, Procedia Engineering, **117 (1)**, 846-853 (2015)
30. S. Ognjenovic, R. Ristov, N. Vatin, (2015) Procedia Engineering, **117 (1)**, 575-578 (2015)
31. S. Rutešić, J. Ćetković, M. Žarcković, M. Knežević, N. Vatin, Procedia Engineering, **117 (1)**, 905-915 (2015)
32. V. Korsun, N. Vatin, A. Franchi, A. Korsun, P. Crespi, Mashtaler S. Procedia Engineering, **117 (1)**, 975-984 (2015)
33. S. Ognjenovic, R. Donceva, N. Vatin, Procedia Engineering, **117 (1)**, 549-555 (2015)
34. Procedia Engineering, **117 (1)**, 924-937 (2015)
35. S. Ognjenovic, Z. Zafirovski, N. Vatin, Procedia Engineering, **117 (1)**, 579-584 (2015)
36. B. Melović, S. Mitrović, A. Djokaj, N. Vatin, Procedia Engineering, **117 (1)**, 807-812 (2015)
37. S. Rutešić, J. Ćetković, M. Knežević, M. Žarcković, N. Vatin, Procedia Engineering, **117 (1)**, 642-650 (2015)
38. N. Vatin, J. Havula, L. Martikainen, A. Sinelnikov, A. Orlova, S. Salamakhin, Advanced Materials Research, **945-949**, 1211-1215 (2014)
39. N. Vatin, T. Nazmeeva, R. Guslinscky, Advanced Materials Research, **941-944**, 1871-1875 (2014)
40. N. Vatin, A. Sinelnikov, M. Garifullin, D. Trubina, Applied Mechanics and Materials, **633-634**, 1037-1041 (2014)
41. D.Arseniev, A. Rechinskiy, K.Shvetsov, N. Vatin, O. Gamayunova, Applied Mechanics and Materials, **635-637**, 2076-2080 (2014)
42. V.Chechevichkin, N. Vatin, Applied Mechanics and Materials, **641-642**, 409-415 (2014)

MATEC Web of Conferences **53**, 01031 (2016)
DOI: 10.1051/matecconf/20165301031

Classification of fire-technical characteristic of roofing materials in European and Russian regulation documents

Marina Gravit[1], Artem Vaititckii[1,a], Marina Imasheva[1], Diana Nigmatullina[1] and Alexandra Shpakova[1]

[1]Peter The Great Saint-Petersburg Polytechic University, Polytekhnicheskaya St., 29, St. Petersburg, 195251, Russia

Abstract. The authors analyze the fire-danger index of building materials (such as roofing materials) in European and Russian regulations. The parameters which were obtained during tests in accordance with standard of roofing materials in European and Russian methods of testing were analyzed. Some of the tests to determine the properties of the fire hazard of building materials are identical. Contributors show that the European standardization system has a much wider range of test methods for determining the properties of fire safety of construction materials. In Russia, for some tests is not unique (the test in a corner room, non-flammability tests, etc.). We concluded that the European standards for building materials are more stringent and also that the correlation between the various indicators missing. We demonstrate significant differences in the European and Russian fire hazard classification of identical roofing materials.

1 Introduction

The establishment of the uniform fire safety's requirements for products, rules and forms of confirmation's accordance, comparable tests' methods is aimed to the ensuring the country's integration into the world economy and, in particular, to the export-oriented "import substitution" (government strategy that emphasizes replacement of some agricultural or industrial imports to encourage local production for local consumption, rather than producing for export markets) in the construction industry. The harmonization of European and Russian standards for test methods and classification of building materials and structures are actively investigated the last five years [1-6].

All of the properties of fire danger of building materials (combustibility, flammability, smoke-forming ability, toxicity of products of combustion, flame spread on the surface), are divided into groups. Belonging to the certain group is determined by the regulated tests and the precise grading, in Russia [7-10]. The properties of fire risk are determined on the basis of the tests, and on the totality of these properties a building materials are classified into five classes of fire danger KM0, KM1, KM2, KM3, KM4, KM5 according to the requirements of Federal Law No. 123-FZ "Technical Regulation of fire safety" [11].

The method of classification of fire hazard of building products and materials, including the products and materials applicable in building structures established in the European standard EN 13501-1-2001 «Classification of building products and materials for fire safety" [12]. According to

[a] Corresponding author : twitwhite@gmail.com

[12], the building materials depending on the characteristics of their behavior on the fire are divided into two main groups: floor coverings and all other materials, excluding the floor materials in European classification.

The more differentiated approach is applied in the Russian classification. it establishes the following types of building materials: materials for covering walls and ceilings, flooring, including carpeting, roofing, waterproofing, and vapor barriers.

In the table 1 quoted comparative assessment of Russian and European classification principles for determining the fire hazard of building materials, taking into account the purpose of the building material, is shown. As we can see, despite that in the both classifications the ability to inflammation building materials from ignition source with different power is considered also as the possibility of the combustion of a variety of building materials fire hazards (high temperatures, smoke aerosol, etc.), experimental methods, which determine fire danger, and principles of classification of the building materials have absolutely different methodological approaches, as we know from the tests' results [13].

Classes of fire hazard of building materials in the Russian Federation are established according to the groups of fire risk. The assignment of the building material to a fire risk group (NG, G, B, RP, A, T) is carried the classification of parameters registered during the test according to standard experimental procedures. The degree of fire risk in the EU applies a methodological approach that is significantly different from the methods and principles of the classification adopted in the Russian Federation [14].

Table 1. Comparison of the principles of determining the classes of fire prevention class.

Fire prevention class (parameters according to the [8])	Purpose of building material	Parameters defining the fire prevention class	Notes
European (A1, A2, B, C, D, E, F)	All except flooring	ΔT, Δm, t_f, PCS, FIGRA, LFS, THR_{600s}, F_s, SMOGRA, TSP_{600s}, dropping point	
	Flooring	ΔT, Δm, t_f, PCS, smoke emission	
Russian (KMO, KM1, KM2, KM3, KM4, KM5)	For the walls and ceilings, including coatings of paints, enamels, varnishes and insulating	Г, В, Д, Т	The defining parameters for each mark: Г— T, S_m, S_L, $t_{c.r}$; В — q_{ign}; РП — q_{spr}; Д— D_m; $Т_n$— H_{CL50}
	Flooring and Roofing	В, РП, Д, Т	
	Roofing	Г, В, РП	
	Waterproofing and vapor barrier with a thickness of more than 0,2 mm	Г, В	

Designations of the classification indexes: Fs—length of flame spread, mm; FIGRA – coefficient of heat W/s; LFS – lateral flame spread, m; SMOGRA – fume evolution coefficient m2/s2; THR600s – total amount of heat evolved in 600 s, mJ; TSP600s – total amount of fume, evolved in 600 s, m2; PCS – high heat value, mJ/kg mJ/m2 critical heat flux achieved after the 30 min test (i.e. quantity keeping with flame spread).

Analyzing the criteria and indicators of fire danger in the European and Russian standards, it can be concluded that European standards for building materials to satisfy more stringent requirements [1, 2, 13, 15].

Thus, for the classification of building materials is necessary to adopt the European standard classification completely. In this case, the relevance is determined by the need for additional research methods and changes in the currently existing test methods for determining fire danger and fire resistance building structures [16].

2 Results and Discussion

Table 2 provides data on the parameters of fire danger according to the Russian and European classification of roofing. For roofing harmonized standard GOST R 56026-2014 "Building materials. The method of determining the group of fire danger roofing materials" [17] match to ENV 1187:2002 «Test methods for external fire exposure to roofs. Test 2: Method with burning brands and wind» [18]. Standards harmonized as a non-equivalent and European classification is not taken into account.

It will cause the acceptance of all the provisions of tests' ways. Some tests for determination the fire safety's properties of construction materials are similar according to the Russian and European standards. The European system of standardization has the much wider range of test methods for determining the fire safety's properties of construction materials which have no analogues in Russian regulations (table 3). The following table compares the roofing materials of different manufacturers. In connection with this direct correlation between the European and Russian parameters is difficult.

Table 2. Comparison of Russian and European classification of building materials on fire danger.

Classification parameter according to EN 13501-1	Test methods for fire danger	Classification parameter according to FZ-123	Test methods for fire danger
A1	EN ISO 1182:2010	KM 0	ГОСТ 30244-1994
A2 (s1,s2,s3)* (d0,d1,d2)**	ISO 5657:1997 ISO 9239- 1:2010 ISO 5659 – 2:2012 ISO 13344:2004	KM1 KM2 KM3 KM4 KM5	GOST 30402-96 GOST P 51032-97 GOST 12.1.044-89 p. 4.18 GOST 12.1.044-89 p. 4.20
B (s1,s2,s3) (d0,d1,d2)			
C (s1,s2,s3) (d0,d1,d2)			
D (s1,s2,s3) (d0,d1,d2)			
E/E-d2			
F	not rated		

Notes: *s1 – SMOGRA ≤ 30 m2/ sec2, TSP600 ≤ 50 m2, s2 - SMOGRA ≤ 180 m2/ sec2, TSP600s ≤ 200 m2, s3 – all other cases.
**d0 – burning melt drops are absent (according to EN 13823) during 600 sec, d1 – more than 10 sec, but no more than 600 sec, d3 – all other cases.

Table 3. Comparison of Russian and European classification of roofing materials on fire danger.

Name of material and number	Classification parameter according to [19]	Classification parameter according to [11]	Notes
1. Reinforced synthetic roof waterproofing sheet based on polyvinylchloride (PVC-P-NB), WOLFIN (thickness: 1,2- 3,0 mm.)	Class E	KM5, Г4, B3, РП4	

2. Roof waterproofing PVC-P membranes for mechanically fastened systems LOGICROOF V-RP (thickness: 1,2- 2,0 mm.)	Class E	КМ2*, Г1, В2, РП1 КМ3**, Г2, В2, РП1	*1,2 mm. **1,5 - 2,0 mm.
3. Plastic sheets for roof waterproofing covered sheets, gravel ballasted ALKORPLAN L 35177 (thickness: 1,2 mm., 1,5 mm.)	Class E	КМ5, Г4, В3, РП2	
4. Reinforced bitumen sheets for roof waterproofing Monarplan FM, Icopal (thickness: 1,2 mm., 1,5 mm.)	Class E	КМ3*, Г1, В2, РП1 КМ3**, Г2, В2, РП1	*1,2 mm. **1,5 mm.
5. Underlays for discontinuous roofing (thickness: 0,3-6 mm.) type Armourbase Pro, «IKO»	Class E	КМ5, Г4, В3, РП4	
6. Bitumen shingles with mineral reinforcement SHINGLAS type «Flamenco» «Jive» и «Foxtrot»	Class E	КМ5, Г4, В2, РП2	
7. Roof waterproofing membrane for covered flat roof systems with gravel ballast, roof garden or park deck Sikaplan-SGmA	Class E	КМ5, Г4, В3, РП2	
8. Roof waterproofing membrane for exposed mechanically fastened flat roof systems Sikaplan 18VG, Sikaplan 15VG	Class E	КМ3, Г2, В2, РП2	
9. Roof waterproofing membrane for exposed adhered flat roof systems Sarnafil G 410-12EL	Class E	КМ5, Г4, В3, РП1	

Felt			
10. Closed- cell polystyrene for thermal insulation of inverted roofs and floors Ravatherm XPS 500	Class E	КМ5, Г4, В2, Т3, Д3	
11. The wired mat for thermal acoustic insulation of industrial applications reaching high temperatures ProRox WM 950 ALU	Class A1	КМ1, Г1, В1, Д1, Т1	
12. Clay roofing tiles and fittings Koramic, «Wienerberger Ceramika Budowlana»	Class A1	КМ0, НГ	
13. Liquid applied roof waterproofing kit, based on polyurethane HYPERDESMO PB 2K	Class F	КМ3, Г2, В2, РП1	
14. Liquid applied roof waterproofing kit, based on polyurethane HYPERDESMO	Class F	КМ5, Г2, В2, РП3	
15. PVC membrane for waterproofing of roofing FLAGON SR	Class E	КМ5, Г1, В3, РП1	
16. Roof waterproofing membrane Protan SE	Class E	КМ5, Г2, В3, РП2	
17. Homogenous sheet waterproofing membrane with a signal layer, based on polyvinylchloride (PVC-P) (thickness: less 3 mm.) Sikaplan WP 1100-15HL, Sikaplan WP 1100-30HL, Sikaplan WP 2110-20HL	Class E	КМ5*, Г3, В3 КМ5**, Г4, В3 КМ5***, Г2, В3	* Sikaplan WP 1100-15HL ** Sikaplan WP 1100-30HL *** Sikaplan WP 2110-20HL

For instance, as we can see, the materials HYPERDESMO PB 2K and HYPERDESMO (numbers 13 and 14) have a class F. But these materials have respectively KM3 and KM5 class in Russian classification.

Materials ProRox WM 950 and Koramic (numbers 11, 12) have class A1 in European classification, while they have classification codes KM3 and KM5 in Russian classification. If we look in more detail, even if the same class, such as KM5, indicators flammability, flame propagation will also vary (numbers 14 - 17), etc.

3 Conclusions

As a result of analysis of the harmonization's process in European and Russian normative and technical documentation, it was found that currently in force in the Russian Federation regulations of fire safety in many ways contain different as from the European fire safety's requirements for products (buildings and structures, building materials and construction, fire -technical products, products of general purpose) [20, 21]. Research methods and measurements of fire safety's requirements which are comparable with European ones are also absent [17].

The scientific community thinks that after the harmonization of test methods will be necessary to come to uniformity in the classification of obtained parameters and their ranking on the principle of increasing the fire safety of construction materials and structures [22-26].

References

1. T. Eremina, N. Konstantinova, M. Grigorieva, Construction materials, the equipment, technologies of XXI century, 25 (2004)
2. M. Grigorieva, M. Gravit, VI All-Russia scientific-practical conference, 214 (2014)
3. T. Eremina, M. Gravit, Y. Dmitriev, Architecture and construction of Russia, **8**, 24 (2012)
4. M. Gravit, Coating materials and their application, **3**, 38 (2013)
5. M. Gravit, *Fire resistance of building structures in the European and Russian standards,* Standard and quality, **2**, 36 (2014)
6. I. Kxasanov, M. Gravit, A. Kosachev, A. Pekhotikova, V. Pavlov, Fire and explosion safety, **3**,. 49 (2014)
7. M. Grigoryeva, T. Eremina, N. Konstantinova, Fire and explosion safety, **8**, 34 (2015)
8. EN 13501-1-2001, *Reaction to Fire Classification for all products and building elements*
9. V.V Okrepilov, Standarty i Kachestvo, **3**, 94-96. (2003)
10. I. Khasanov, T. Eremina, M. Gravit, A. Makeev, Architecture and construction of Russia, **3**, 24 (2013)
11. Federal Law No. 123-FZ, *Technical Regulation of fire safety*, date of text: 22 July 2008
12. T. Eremina, M. Grigoryeva, Fire safety, 34 (2014)
13. D. Trushkin, Fire and explosion hazard substances and materials, **21** (2012)
14. T. Eremina, D. Minaylov, Fire and Emergency, 32 (2014)
15. A. Pestritsky, Technology building, **1**, 10 (2007)
16. L. Cooper, K. Steckler, NISTIR 5842, 45 (1996)
17. GOST R 56026-2014
18. ENV 1187:2002
19. EN 13501-1:2007+A1:2009
20. V.V Okrepilov, Ivanova, G. Standarty i Kachestvo, **12**, 62-68. (2004)
21. V.V Okrepilov, Standarty i Kachestvo, **10**, 52-55 (2005)
22. O. Artsybasheva, G. Vizgalova, A. Sivenkov, R. Aseeva, B. Serkov, Technology technospheric security, 3 (2014)
23. V.V Okrepilov, Studies on Russian Economic Development, **24(1)**, 35-42 (2013)
24. L. Gaile, A. Sprince, L. Pakrastinsh, Procedia Engineering, **57**, 294 (2013)
25. A. Khodakov, M. Tochenyy, S. Belyaeva, O. Nikonova, L. Pakrastinsh, Applied Mechanics and Materials, 997 (2014)
26. N. Vatin, S. Rutešić, J. Ćetković, M. Žarcković, M. Knežević, Procedia Engineering, **117**, 905 (2015)

MATEC Web of Conferences **53**, 01032 (2016)
DOI: 10.1051/matecconf/20165301032

Calculation of Limits of Fire Resistance for Structures with Fire Retardant Coating

Artem Krivtcov[1,a], Marina Gravit[1], Sergei Zimin[1] , Oleg Nedryshkin[1] and Valerii Pershakov[2]

[1]*Peter the Great St. Petersburg Polytechnic University, 195251, 29, Politekhnicheskaya st., Russia*
[2]*National Aviation University, Kosmonavta Komarova Ave, 1, Kiev, Ukraine, 03680*

Abstract. This article is devoted to fireproof processing of steel structures. The main task is to consider different types of sections of rod elements and to choose the most effective section for a steel column from the point of view of fire protection. For the solution of this task the steel columns with various cross sections working in identical entry conditions were considered. All necessary calculations for all types of sections were carried out. Results of calculations were presented in the summary table according to which the comparative analysis was made. At the end of work the conclusion that the compound section from four equal corners is the most effective from the point of view of fire protection.

1 Introduction

Building constructions, including steel are used widely because of the opportunities that they give to various objectives. Steel is often used in construction due to a wonderful combination of technological and operational properties, exemplified by many unique architectural constructions in the world. However, there is an important problem, with which humanity faced since its inception - fire protection. The main objective in this protection is to achieve security of people in the building. The second most important task is the preservation of wealth that can be lost in a fire. Many studies, standards, numerous tests of materials and structures are devoted to this problem. However, because of the importance of this problem and the development of technology, the relevance of the fire safety and search for new technical solutions in this area are retained [1, 2, 16].

Due to the numerous technical solutions and requirements for steel structures, it is difficult to expect the same solutions for different constructions. Firstly, the solution of fire protection of steel structures considers a number of factors the importance of which often depends on the function of construction. Therefore, it is important to determine the fundamental criteria that define the ways of fire protection of steel structures [6, 14].

The main task of this work was the calculation of the steel column in order to identify the most beneficial cross section in terms of fire protection. For this purpose in this paper was considered a fireproof material Knauf-Fireboard. The calculation was made according to the Russian regulations: Code Specification 16.13330.2011 "Steel Constructions", Federal Law of 22 July 2008 N 123-FZ "Technical Regulations on fire safety requirements".

---

[a] Corresponding author : artcomp2007@yandex.ru

The rod element was an object of this research was. For practical application of this work the steel vertical columns working for the central compression were considered. Entry conditions on loading, a class of steel, length and conditions of fixing for all columns were identical, only cross sections were various. All columns were faced with the fireproof plate material Knauf-Fireboard.

Main objective of this work was comparison of necessary amount of fireproof material for rod elements of various cross sections. Golovanov V. I. researches were taken as a basis of this work [3, 4, 6, 8, 11, 16]. Also the results received in works [9, 12, 18] were considered.

For the solution of an objective the following tasks were set:

- to carry out calculations of all considered sections on durability, stability and necessary thickness of fireproof material for these conditions;
- to carry out the comparative analysis and to draw conclusions on a consumption of metal and fireproof material depending on the cross section of a rod element.

2 Description of the calculation

2.1 Fire retardant material

The choice of materials for fire protection of steel structures is based on an understanding of their behaviour under high temperatures. It is important to know that some of the materials having good thermal properties can mechanically collapse from heat stress and thus not provide thermal protection. Others, however, having a flammability, but maintaining the mechanical stability under load during combustion (for example: wood), do not break down for some time and thus provide a barrier between the fire source and the structure. It is important to remember that the construction of the outer device fireproof cladding can significantly affect the fire resistance of a structure. The presence or appearance of fire-permeable holes can negate all the efforts of fireproofing [16].

In this paper material Knauf-Fireboard has been considered. It is a non-combustible plate materi-al developed by Knauf for fire safety of buildings.

Knauf-Fireboard is fire retardant facing material and is widely used in the finishing of technical premises, where, according to the fire safety requirements is necessary to use non-combustible materials, such as emergency exits of offices and shopping centers, evacuation routes. This product is also used in places where there is an increased risk of fire. Application of this material facilitates localization of fire in a room and preventing distribution of a flame outside [20, 22].

The difference between Knauf-Fireboard and other gypsum products is that the moisture evaporates from gypsum core after interaction with the flame, wherein the plate does not break or crack rather long time. Also plates Knauf-Fireboard have non-flammable fiberglass, which is a fire-resistant reinforcing frame product. Such reinforcing frame provides enhanced fire of entire structure [5, 17].

2.2 Classification of buildings and structures according to fire danger

The time of the fire cannot be infinite both because of the limited quantity of combustible material and due to the limited lifetime of constructions under fire exposure. Therefore, in Fire safety regula-tions for most constructions fire is normalized in the range of 0 to 6 hours [6].

Constructions separated into bearing and non-bearing have a significant difference in the required fire resistance. Therefore, different technical retardant solutions are used for them, although they can be made of a same steel assortment. For load-bearing elements an important factor is their responsibility for geometric variability of the structure under load during a fire, which can significantly reduce the critical temperature of steel. For non-bearing elements fire protection requirements are less stringent, as their role in ensuring the stability of the frame is negligible [12, 21].

Table 1 is a part of Table 21 from №123-FZ «Technical regulations for fire safety requirements» - the main Russian regulation in fire safety. There are five fire classes of building in Russian regulations. In this paper, we considered only load-bearing structures in the building of the first fire class.

Table 1. Compliance of fire class and fire-resistance of building constructions.

Fire class of building	Fire-resistance	
	load-bearing constructions	non-bearing constructions
I	R120	E30
II	R90	E15
III	R45	E15
IV	R15	E15
V	No regulations	No regulations

2.3 Calculation of the fire-retardant treatment

Steel for building structures has a high thermal conductivity, which significantly effects on structural fire protection engineering solutions. On one side elements of the steel structure heat up quickly, and the other - well distribute and dissipate heat. Balance of heat flows in heating and cooling determines the change in temperature of the structure and the time to reach the critical temperature [10, 13, 15].

To make decisions on fire protection of steel structures two criteria were introduced: the given thickness of the metal and the critical temperature. The given thickness of metal allows consider the cross section of the steel profile and the nature of the heat supply in case of fire. Like every criterion the given thickness of metal involves some simplification, but for many practical solutions it is very useful [12, 19].

The given metal thickness is calculated by the formula:

$$\delta = A_n/P \tag{1}$$

where:

A_n is cross sectional area, cm^2;

P is heated perimeter, cm.

The heated perimeter value depends on the geometry of the cross section and is calculated according to the formulas in Table 3 of the Instruction for calculation of the actual limits of fire resistance of steel structures with flame-retardant facing made of plates Knauf-Fireboard, approved with EMERCOM Russia [11, 14].

This criterion shows how effective heat removal is in relation to heating depending on the shape of the cross section under the conditions of heating [7].

Another important criterion is the critical temperature. The fire resistance of steel structures occurs as a result of warming up their sections or parts of it to the critical temperature. The critical temperature of steel structures under a load depends on the type of design scheme of its bearing, the metal stamp and the value of the load [8, 9, 14].

The critical temperature is defined as the lowest value of the two found in Table 1 of the Instruction for calculation of the actual limits of fire resistance of steel structures with flame-retardant facing made of plates Knauf-Fireboard, approved with EMERCOM Russia, depending on the values of the coefficients γ_t and γ_e. Values of the coefficients γ_t and γ_e, take into account changes in regulatory resistance and modulus of elasticity of steel.

Coefficient γ_t is calculated by the formula:

$$\gamma_t = N/A_nR_y \tag{2}$$

where:

N is longitudinal force, acting on the rod, kg;

R_y is calculated resistance of steel working on tensile, compression, bending in yield strength, kg/cm^2.

Coefficient γ_e is calculated by the formula:

$$\gamma_e = Nl_0^2/\pi^2 EI_{min} \tag{3}$$

where:
l_0 – calculated length of the rod, cm;
I_{min} – minimal moment of inertia, cm^4;
E - modulus of elasticity of steel, kg/cm^2.

3 Calculations

3.1 Input data

An object of research is a vertical steel column with applied longitudinal force. The input data for the task is:

Class of fire resistance of the building: I;
Type of rod work: compression;
Type of fixing: fixing the ends;
Type of cladding and heating conditions: facing a box-shaped with 4 sides;
Longitudinal force: 90000 kg;
Geometrical length of the rod: 300 cm;
Grade of steel: C235;
The initial modulus of elasticity of steel: 2100000 kg/cm2.

The task is to select cross sections satisfying the conditions of strength and stability and to calculate the necessary thickness of the plate material Knauf-Fireboard for each cross-section in the current input data.

3.2 Types of used cross-sections

In the calculation the most popular types of cross-sections were used: I-profiles, pipes, square profile, the composite section of the two channels, the composite sections of four equal corners.

3.3 Calculation example

3.3.1 Characteristic of construction:

Degree of fire resistance of the building: I.

In accordance with the Table 1: the fire resistance of structure is 120 min.

3.3.2 Characteristic of section:

Cross-sectional view: Hot-rolled steel I-profile (GOST 8239-89) number 33.

In accordance with the assortment of metal, used in GOST 8239-89, geometrical characteristics of the section are presented in the Table 2.

Table 2. Geometrical characteristics of the section.

h, cm	**b, cm**	**A_n, cm^2**	**I_{min}, cm^4**
33.0	14.0	53.8	419.0

3.3.3 Fire-retardant treatment

Type of cladding and heating conditions: facing a box-shaped with 4 sides. In accordance with Table 3 of the Instruction for calculation of the actual limits of fire resistance of steel structures with flame-retardant facing made of plates Knauf-Fireboard, approved with EMERCOM Russia, heated perimeter should be determined by the formula:

$$P = 2b+2h \tag{4}$$

$$P = 2\cdot 14+2\cdot 33$$

$$P = 94\ cm$$

Given metal thickness is calculated by the formula (1):

$$\delta = 53.8/94$$

$$\delta = 0.572\ cm$$

Coefficient γ_t is calculated by the formula (2):

$$\gamma_t = 90000/53.8\cdot 2250$$

$$\gamma_t = 0.74$$

In accordance with Table 1 of the Instruction for calculation of the actual limits of fire resistance of steel structures with flame-retardant facing made of plates Knauf-Fireboard, approved with EMERCOM Russia, critical temperature:

$$T_{cr} = 344.18°C$$

Coefficient γ_e is calculated by the formula (3):

$$\gamma_t = 90000\cdot 150^2/3.14\cdot 2100000\cdot 419$$

$$\gamma_t = 0.23$$

In accordance with Table 1 of the Instruction for calculation of the actual limits of fire resistance of steel structures with flame-retardant facing made of plates Knauf-Fireboard, approved with EMERCOM Russia, critical temperature:

$$T_{cr} = 898.28°C$$

So the minimal critical temperature:

$$T_{cr} = 344.18°C$$

In accordance with the Instruction for calculation of the actual limits of fire resistance of steel structures with flame-retardant facing, approved with EMERCOM Russia, the minimum allowable thickness of Knauf-Fireboard in these conditions is 32,5 mm.

4 Results

4.1 Result of calculation for different cross-sections

The results of calculation for different cross-sections are presented in the Table 3.

Table 3. Results of calculation for different cross-sections.

Cross-section	I-profile №33 (GOST* 8239-89)	Pipe E325x9 (GOST 10704-91)	Square profile 160x8 (GOST 30245-2003)	Composite section of the two channels № 20 (GOST 8240-89)	Composite section of four equal corners 75x8 (GOST 8509-93)
Picture	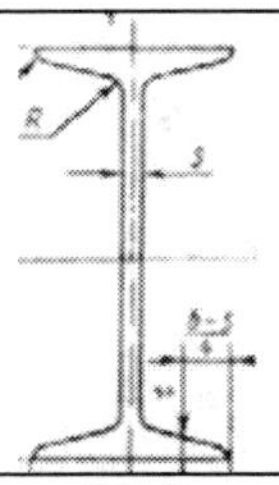	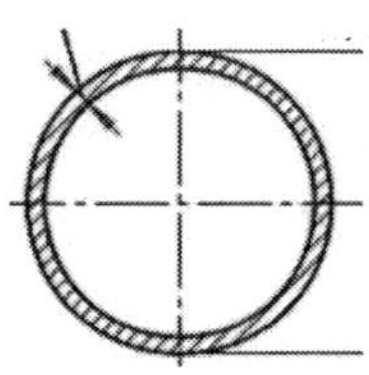	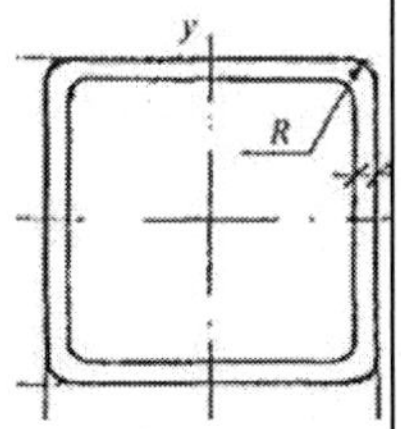	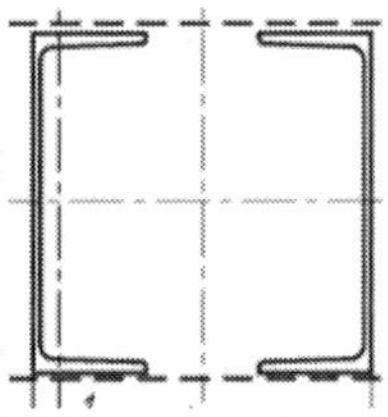	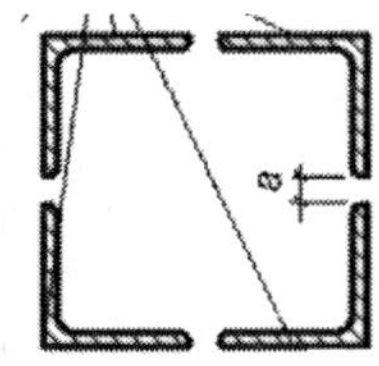
h, cm	33.0	32.5	16.0	20.0	15.0
b, cm	14.0	32.5	16.0	15.2	15.0
A_n, cm^2	53.8	45.31	46.44	46.8	46.0
I_{min}, cm^4	419.0	5818.9	1740.0	1657.2	1556.0
P, cm	94.0	130.0	64.0	70.4	60.0
δ, cm	0.572	0.349	0.726	0.665	0.767
T_{cr}, °C	344.18	178.75	193.99	198.69	188.15
Δ, mm	32.5	45.0	32.5	40.0	32.5
V, m^3	0.104	0.175	0.075	0.104	0.071
m, kg	126.6	210.4	109.4	110.4	108.24

where:

Δ, mm - minimum allowable thickness of Knauf-Fireboard;

V, m^3 – total volume of Knauf-Fireboard for a column;

m, kg – total weight of a column.

*GOST - National Standart of Russian Federation.

4.2 Conclusions by results of calculations

After the made calculation it is possible to draw the following conclusions:

- As a result of calculating a cross-section consisting of four equal corners proved the most cost-effective in terms of the consumption of fireproof material. This section of the column other conditions being equal has the least amount of material Knauf-Fireboard – 0.071 m^3. Also, this section is the most advantageous in terms of the consumption of metal in the column – 108.24 kg.
- Also very economical with both points of view turned cross-section of the square profile. Material consumption Knauf-Fireboard appeared 0.075 m^3, and steel consumption was slightly higher than that of the cross-section of four equal corners – 109.4 kg.
- The least advantageous in terms of consumption of steel and fire retardant turned cross-section of the pipe (0.175 m^3, 210.4 kg). Arranging conditions of stability and strength cross-section was too large, leading to a waste of material.

5 Conclusions

The purpose of this work was to reveal the most effective from the point of view of fire protection the cross section of a steel rod element. By the results of conducted research the following conclusions were drawn:

- calculation of the cross sections of steel rod elements which are the most used today was made;
- as a result of calculation by the most effective from the point of view of a consumption of fireproof material there were sections from four equal corners and a square profile.

References

1. M. Lazarevska, M. Cvetkovska, M. Knežević, A. Trombeva Gavriloska, M. Milanovic, V. Murgul, N. Vatin, Applied Mechanics and Materials, **627**, 276-282 (2014)
2. M. Lazarevska, M. Knežević, M. Cvetkovska, A. Trombeva-Gavriloska, Tehnicki Vjesnik, **21 (6)**, 1353-1359 (2014)
3. V. Golovanov, V. Pavlov, A. Pekhotikov. Pozharnaya bezopasnost, **3,** 48-58 (2002)
4. Yu. Morozov, L. Efron, O. Chevskaya, N. Shtychkov, P. Odesskiy, D. Solovyev, V. Moskalenko, A. Stepashin, I. Shabalov, D. Kulik. Stal, **9,** 48-53 (2004)
5. M. Jevrić, M. Knežević, J. Kalezić, N. Kopitović-Vuković, I. Ćipranić, Tehnicki Vjesnik, **21 (4)**, 873-879 (2014)
6. B. Parlour. Fire safety, **5,** 18-19 (2004)
7. O. Lamkin, M. Gravit, O. Nedryshkin. Construction of Unique Buildings and Structures, **11(38)**, 42-58 (2015)
8. V. Strakhov, A. Garashchenko, G. Kuznetsov, V. Rudzinskii. Combustion, explosion, and shock waves, **2**, 212-220 (2001)
9. M. Gravit. Fire and Explosion Safety, **11**, 42-45 (2014)
10. R. Kunze, B. Scharte, M. Bartholmai, D. Neubert, R. Schriever. Journal of thermal analysis and calorimetry, **3**, 901-913 (2002)
11. K. Langille, D. Nguyen, D. Veinot. Fire technology, **2**, 99-110 (1999)
12. M. Heinisuo, M. Laasonen, J. Outinen, J. Hietaniemi. *Application of Structural Fire Design*, 405-410 (2011)
13. M. Gravit, V. Gumenyuk, O. Nedryshkin. Procedia Engineering, **117**, 114 – 118 (2015)
14. A. Krivtcov, V. Kazakova, I. Mingalimov, P. Bogdanov, I. Nitsa. Construction of Unique Buildings and Structures, **6(33)**, 34-46, (2015)
15. M. Lazarevska, M. Knežević, M. Cvetkovska, A. Trombeva-Gavriloska, Tehnicki Vjesnik, **21 (6)**, 1353-1359 (2014)
16. O. Halyavin, M. Gravit, A. Pryanikova. *Science Week STU*, 22-25 (2015)
17. M. Lazarevska, M. Milanović, M. Knežević, M. Cvetkovska, A.T. Gavriloska, T. Samadzioska, Journal of Applied Engineering Science, **12 (1)**, 63-68 (2014)
18. Y. Dong, G. Wang, J. Yang. JCT Research, **2**, 231-237 (2014)
19. P. Kraus, M. Mensinger, F. Tabeling, P. Schaumann. Journal of Structural Fire Engineering, **6**, 237-246 (2015)
20. Z. Arabasadi, M. Khorasani, S. Akhlaghi, H. Fazilat, U. Gedde, M. Hedenqvist, M. Shiri. Fire Safety Journal, **61**, 193-199 (2013)
21. M. Salminen, M. Heinisuo. Journal of Constructional Steel Research, **97**, 105-113 (2014)
22. R. Puri, A. Khanna. Progress in Organic Coatings, **92**, 8-15 (2016)

MATEC Web of Conferences **53**, 01033 (2016)
DOI: 10.1051/matecconf/20165301033

Features of the Masonry Structure Calculation with Vertical Ring Beams based on the European Standards

Danijela Zejak

University of Montenegro, Faculty of Civil Engineering, Cetinjski put bb, 81000, Podgorica, Montenegro

Abstract. The design of masonry structures should remain a combination of traditional and modern knowledge. The Eurocodes in the field of the design of earthquake-resistant masonry structures, carry mostly acceptable and expected recommendations and requirements. This paper shows that they contain less acceptable principles, which require further experimental and analytical studies. Particularly it should be noted that Eurocode knows only a shearing mechanism by the horizontal coupling, although during the seismic load, the fracture usually occurs due to exceeding the main tensile stress, which manifests by opening diagonal cracks. The mechanism of the horizontal shearing by Eurocode is treated as critical for the seismic resistance of most masonry buildings, thus the masonry structure is attributed as sufficiently secure, although its actual seismic resistance would be far below from the required.

1 Introduction

The hazard of earthquakes in the Balkan region, situated in very complex geological and tectonic conditions, determine the relevance of this study (Fig. 1). Before the final adoption of European standards for the design of masonry structures on seismic areas, as much as they are considered advanced, it is necessary to perform detailed analysis and comparisons with our current regulations in order to obtain the synthesized experiences and accurate structural solutions, as well as eventually correcting and supplementing them [1-5]. In this regard, this paper analyzes classical three-story masonry structures with vertical ring beams applying valid (PZZ'81 and PIOVS'91) and European regulations (EC-6 and EC 8), for comparison of results.

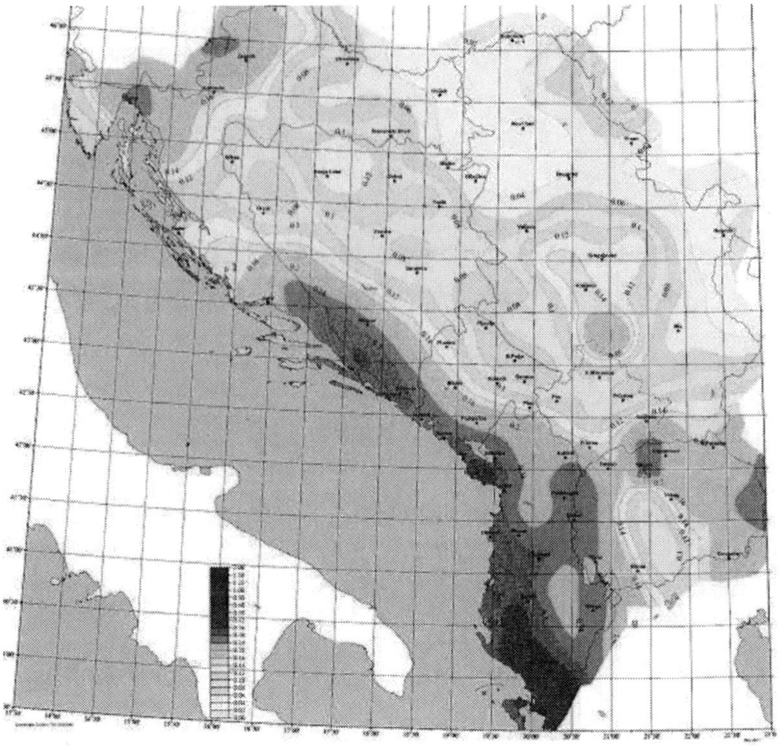

Figure 1. Seismic hazard maps, the Balkan region [7].

2 Materials and methods

We analyzed the masonry structure with vertical ring beams P+2S, floor height 2.8 m, with premises intend for household and housing. Supporting structure consists of walls of solid brick, 25 cm thick, built using a general-purpose mortar (Fig. 2).

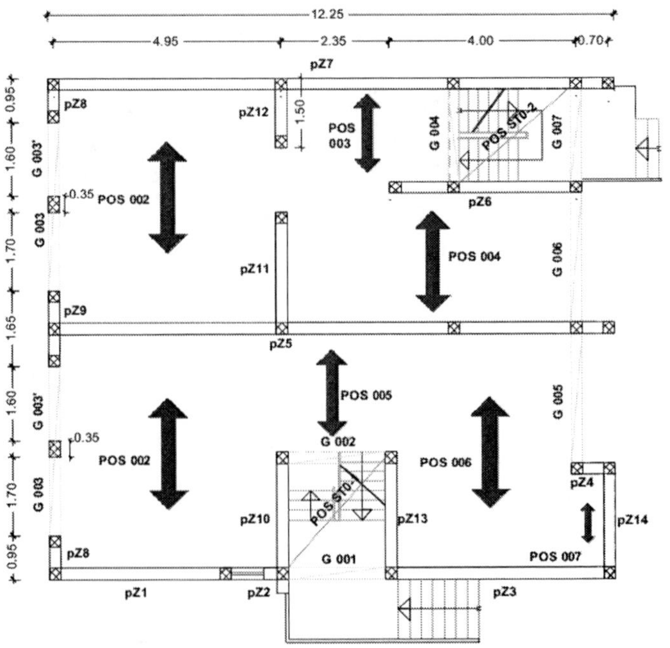

Figure 2. Ground floor basis [1].

Each wall is framed with vertical and horizontal reinforced concrete elements so they together act as a single structural element upon exposure of some influences [6]. When positioning horizontal and vertical ring beams, the requirements of EC 8 are adhered to, they are stricter than the applicable regulations. Ring beams are predicted at the level of each floor, on confrontment walls and on both sides of every orifice greater than $1.5m^2$ (along with PZZ, the length of orifice is greater than 2.5 m). Additional ring beams are placed in the walls where the maximum spacing between the existing, both horizontally and vertically, was 4 m (5 m). Semi-prefabricated mezzanine structure type FERT, was applied [7-9].

The structure is positioned in the VIII zone of seismicity. The structure is founding on the type B of soil (according to Eurocode, which is equivalent to the category I by PZZ) [10, 11].

2.1 Determination of seismic load of elements

In the calculation we have considered the following loads acting on the structure: net weight of the structure, imposed load and seismic load. The intensity of the loads is determined in accordance with European regulations. The vertical load of the walls is determined using the area of influence [12, 13].

Since the majority of masonry structures has regularity in arrangement of the walls in the base and the height, it appears that the seismic response of the structure is not affected by the higher tones of oscillation, which allows the use of simple linear static method, the method of lateral force (MBS) [14]. The project's transverse seismic force on the ground floor of the structure, F_b, for each direction of action of horizontal forces is calculated as follows:

$$F_b = S_d(T_1) \cdot m \cdot \lambda \tag{1}$$

where:

Sd (T_1) - project spectrum ordinate. Since the masonry structures are rigid with their own period of oscillation, which is generally in the range where project spectrum of response is by Evrocode 8, horizontal line, Sd (T_1) can be determined by the expression (2);

m - total mass of the structure above the foundation or above the rigid basement area. The total mass includes permanent and temporary loads (temporary loads are multiplied by the corresponding coefficients given in the Eurocode 1, which take into account the purpose of the structure and the fact that the total weight of the imposed load is not present at the time of the earthquake in its full strength);

λ - the correction factor which takes into account that the effective modal mass of the building, with at least three levels and translational degrees of vibration in both orthogonal directions, less than the entire weight of the building.

The horizontal components of the seismic action by Eurocode 8:

$$S_d(T_1) = a_g \cdot S \cdot \eta \cdot \frac{2.5}{q} \tag{2}$$

where: a_g - design acceleration of the ground ($a_g = 0.2g$), S - factor of soil, η - correction factor of attenuation, q - structural behavior factor.

The elastic seismic forces reduction with the behavior factor of structure q (masonry structures with ring beams have the value from 2 to 3) is shown as a pretty good solution, by preventing the demolition of buildings and also the requirement about extent of damage retains an acceptable level [6]. The resulting value of the seismic force, F_b = 1139 kN, is proportionally divided into elements by their stiffness, where the accidental eccentricity of the mass during an earthquake, is considered, in two normal directions. Eccentricities obtained as recommended by Eurocode 8 (EC 8) and specific values obtained by subtracting the corresponding coordinates of the center of stiffness (C_k) and of the center of mass (C_m), are quite different, and hence the difference in the torque. The higher value is taken, the one obtained by EC 8, where the accidental eccentricity is estimated on 5% of the span length. Comparison of the seismic forces obtained by the applicable regulations (the method of equivalent static loads - PIOVS'91) and European regulations makes sense if it is about marginal influences. Specifically for this example, the European regulations obtained influences about 45% higher than those obtained by PIOVS. It is important to note that the safety factor with the effects of the seismic load by EC is $\gamma = 1.0$. The seismic force is reduced by a factor of behavior $q = 2.0$, in order to provide adequate ductility to the structure, it would not make sense to subsequently increase the impact. To ensure bearing capacity of the building, the impacts are increased by 50% with the current regulations [15-18].

2.2 Modeling of the structure

The structure is modeled in the software package SAP2000 v14 (Fig. 2). The amount of the seismic load transferred to the individual elements is calculated. The 'spatial line model' is used, where each floor is modeled separately because of the change in length of the walls of the individual floors. The walls of the ground floor are wedged on the bottom side, and prevented from rotation of both horizontal axes on top. All elements are connected by a diaphragm, which is rigid in its plane. The first floor is modeled in the same way, except the adequate axial and seismic load are added. And, the walls of the top floor are modeled to represent the associated console, et. from the top side like the other floors, elements are stiffened by the diaphragm. It should be noted that every wall is originally determined by the effective height and by the solid part, in order to take into consideration the fact that the seismic load does not activate the entire height of the wall.

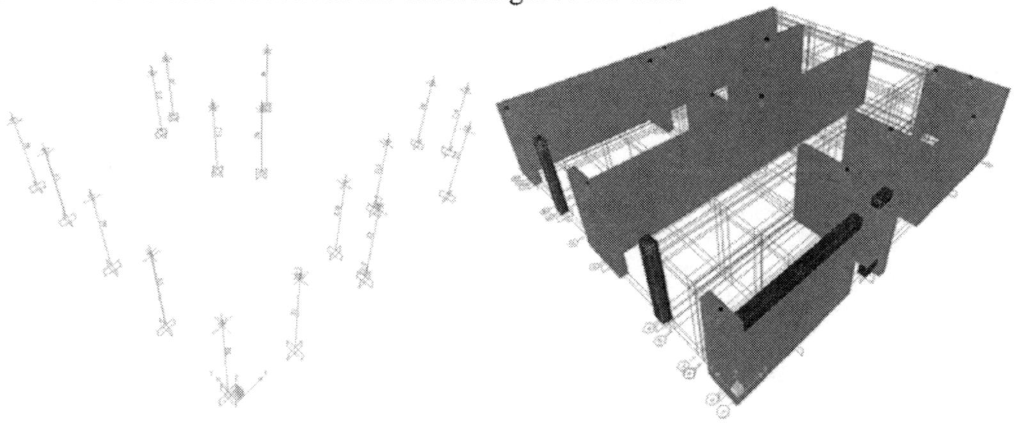

Figure 3. Spatial line model of the ground floor of the structure [1].

In modeling, special attention is given to proper allocation of material characteristics [19]. The values for the modulus of elasticity $E = 3660$ kN/m^2 and Poisson ratio $\upsilon = 0.3$ are inputted in the model, so the obtained value of the shear modulus is equal to the $G_{SAP} = 1407$ MPa, while the actual modulus of brick shear, because of the anisotropy and heterogeneity of the material which is obtained experimentally is actually $G_{real} = 400$ MPa. Therefore, the modification of this modulus by the reduction coefficient has been made (Fig. 3):

$$0.5\frac{G_{real}}{G_{SAP}} = 0.1423 \qquad (3)$$

where with this reduction coefficient of 0.5 is considered that during an earthquake the brick wall is going to enter the zone of plasticity.

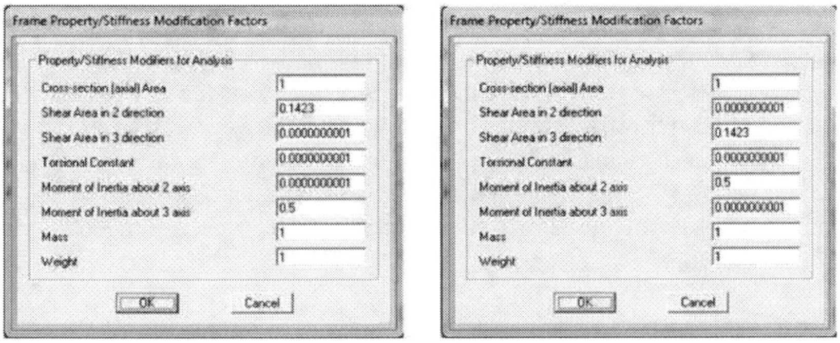

Figure 4. Material characteristics of the walls with the SAP2000 for the clay brick walls in the X and Y direction.

The period of oscillation obtained in SAP 2000 wasn't used as the relevant because of the nature of the model.

2.3 Dimensioning of the characteristic vertical structural element

Considering the geometry of the walls, their axial loads as well as horizontal, ie. seismic, the wall Z12 with critical direction is observed (Fig. 4). The conclusion was that this wall may be critical, due to its relatively low vertical load in relation to the horizontal, in terms of the resistance to the seismic loading.

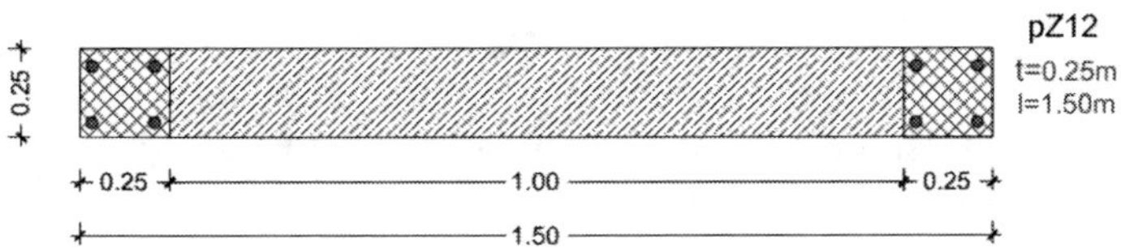

Figure 5. Dimensions of a typical wall at ground level.

The required material properties are determined in accordance with the provisions of Eurocode 6, except there wasn't basis to take the shear modulus, as previously noted, as well as the elastic materials, as recommended in this standard (40% of modulus of elasticity), because it's actual value in the best case reaches 10% of the modulus of elasticity [20-24].

2.4 Capacity of the wall exposed with predominantly vertical load

Bearing capacity of unreinforced walls in relation to the vertical load, must be based on the geometry of the wall, on the impact of present eccentricity and on the properties of used materials for building. In the calculation, the hypotheses of section flatness are valid, as well as the assumption that perpendicular tensile strength of the wall to the horizontal couplings is equal to zero [2]. Applicable and European standards consider the same influential factors when it comes to carrying capacity of the wall under pressure. It is believed that eccentricity is in the wall and is always at least 5% of its thickness. However, although the calculation principle is similar, the results are significantly different as can be seen in Tab. 1. With applicable regulations we are getting smaller capacity because the value of the buckling coefficient is less, and the value of the partial coefficient of the material is higher [25-27].

2.5 Capacity of the wall on shearing

Eurocode 6, which provides basic guidelines for the calculation of masonry structures, knows only the mechanism of sliding on a horizontal coupling (Fig. 5). This mechanism of the shear occurs if stresses on the pressure are low, so seismic forces can cause shearing of the upper part of the wall to the lowest on one of the horizontal couplings, in the long walls with poor quality of mortar. Sometimes walls behave like that in the upper part of buildings, under a rigid roof structure, where the accelerations are highest, and vertical loads are lowest [1].

To calculate the resistance of the wall on shearing, the boundary conditions as well as vertical and horizontal load acting on some walls, should be known. For the verification of the walls framed with the ring beams, capacity on the shearing (V_{Rd}) is taken as the sum of the capacity of the wall ($V_{Rd,z}$) and the concrete of the ring beams ($V_{Rd,c}$). Capacity on the shearing of the ring beam is obtained according to the rules in Eurocode 2, for the elements where is not necessary to calculate the reinforcement for the shearing:

$$V_{Rd} = V_{Rd,z} + V_{Rd,c} = f_{vd} \cdot t \cdot l_c \geq V_{Ed} \tag{4}$$

where: l_c is the length of the pressed part of the wall, neglecting any part of the wall that is exposed to tension, t the thickness of the wall, f_{vd} represents the value of the compressive strength of the wall on shearing, based on the average value of the vertical stresses of the pressed part of the wall that provides capacity on the shearing.

Figure 6. Mechanism of the shearing on horizontal coupling.

However, in a normal situation seismic forces cause the demolition of shearing, characterized by diagonal cracks (Fig. 6). It is, in fact, the mechanism of the destruction which in most cases determines the seismic resistance of masonry structures, and because of the cause of the cracks sometimes it is called demolition by diagonal tension.

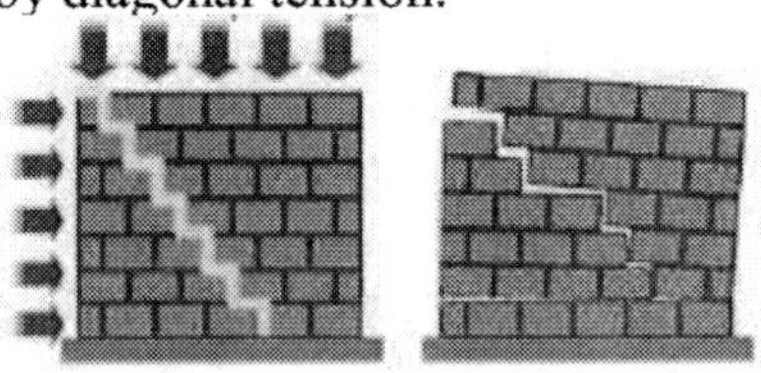

Figure 7. Mechanism of shearing with the appearance of diagonal cracks.

Applicable regulations, namely PIOVS'91, unlike European regulations, identified this mechanism of destruction ie. shearing of the wall, due to reaching the main tensile stresses. The average stress of the shearing in the wall element (τ_o), from the seismic action which the element is receiving, per unit length of the wall, is obtained by the formula:

$$\tau_0 = \frac{\sigma_{0,rus}}{1.5} \sqrt{1 + \frac{\sigma_0}{\sigma_{0,rus}}} \tag{5}$$

where:

σ_o - average stress in the wall element of the vertical load;

$\sigma_{o,rus}$ - major tightening stress in the wall element at the moment of demolition, for certain types of walls.

Regarding the calculation of the capacity of an unreinforced wall, Turncek and Cacovic proposed an equivalent equation to (5), except that factor b may vary from 1.5 to 1.1, but the equation above uses the constant value (1.5). The geometry of the wall and the actual ratio of vertical and horizontal load during demolition, on distribution of the shearing stress over the wall, are taken into account with this factor [28 – 30].

2.6 Capacity of the walls exposed to bending and / or axial load

In practice, when the wall is loaded by wind or earthquake mostly expressed stresses are shearing. Damages of the load-bearing walls that arise solely due to bending are rare [31].

In the section of control of the capacity on bending in its own plane, PZZ remained incomplete, while PIOVS'81 regulations don't mention this load [32-33].

By European standards, during the verification of the walls framed with the ring beams that are exposed to bending and / or axial load, valid assumptions for reinforced masonry elements, given in the EC 6, are approved. The highest calculated moment of bending the wall (M_{Rd}), loaded in its plane, with the neglected impact of press fitting of the ring beams, is equal to:

$$M_{Rd} = \frac{\sigma_d \cdot t \cdot l^2}{2} \cdot \left(1 - \frac{\sigma_d}{f_d}\right) + A_{rv} \cdot f_{yd} \cdot \frac{1-d}{2} \tag{6}$$

where: σ_d - average compressive stress in the wall of the vertical load; f_d - calculated value of the compressive strength of the wall; f_{yd} - calculated value of the strength of reinforcing steel; l, t, d - the dimensions of the wall; A_{rv} - surface of ring beam reinforcement.

3 Results and discussion

Discussed wall of thickness t = 25 cm, Z12, loaded with a normal force of constant load N_g =70.3 kN and with a horizontal force due to seismic impacts F_d = 83.8 kN, did not satisfy the stress control on the shearing, as well as the control of the main tensile stresses, an iterative procedure determined an adequate wall thickness (Tab. 1).

The wall with a thickness of 64 cm, according to the calculation, is sufficient, and satisfies the main control of tensile stress. However, the shear capacity is only satisfied with a wall thickness of 125 cm, leading to the formulas given in PZZ, or 86 cm, by European regulations. Approval of the wall of that thicknesses is absurd, because in practice, the fracture due to only shearing, is very rare [1].

Table 1. Values of the capacity of the wall Z12 on the pressure (N_{gr} / N_{Rd}), shearing (T_{gr} /V_{Rd}, bending (M_{gr} / M_{Rd}) and tension ($T_{gr,z}$ /$V_{Rd,diag}$)

	Applicable regulations				European standards			
T[cm]	N_{gr} [kN]	T_{gr} [kN]	M_{gr} [kN]	$T_{gr,z}$ [kN]	N_{Rd} [kN]	V_{Rd} [kN]	M_{Rd} [kNm]	$V_{Rd,diag}$* [kN]
25	228.8	125.7	254.9	64.3	658.8	35,0	214.8	42.4
38	347.7	47.4	283.4	88.7	974.4	45.4	216.8	56.8
51	466.7	60.4	310.9	112.8	1331.3	55.8	217.5	70.6
64	585.6	73.4	323.3	136.6	1688.1	66.2	218.1	84.2
Note*: Taken from the Slovenian regulations								

4 Summary

It is undisputed that the design of masonry structures should remain a combination of traditional and modern knowledge. Analysis of masonry constructions is done simultaneously by applying the Rules on technical standards for masonry walls (PZZ'81), Regulation on technical standards for the

constructions in seismic areas (PIOVS'91) and Eurocodes (EC 6 and EC 8). Eurocode presupposes shear force to be a reliable mechanism form calculating shear wall resistance force, which conflicts to the real behavior of constructions during earthquakes. It is therefore recommended determining the seismic masonry resistance according to the current rules that require verification tensile strain, whose possible exceeding leads to the appearance diagonal wall cracks.
The Eurocodes in the field of the design of earthquake-resistant masonry structures, carry mostly acceptable and expected recommendations and requirements. However, the paper shows that they contain less acceptable principles, which require further experimental and analytical studies. Particularly it should be noted that Eurocode knows only a shearing mechanism by the horizontal coupling, although during the seismic load, the fracture usually occurs due to exceeding the main tensile stress, which manifests by opening diagonal cracks. So the mechanism of the horizontal shearing by Eurocode is treated as critical for the seismic resistance of most masonry buildings, thus the masonry structure is attributed as sufficiently secure, although its actual seismic resistance would be far below from the required.

References

1. Zejak, D. Specialist Thesis, Faculty of Civil Engineering University of Montenegro, (2012)
2. Evrokod 6. EN 1996-1-1(2005)
3. Evrokod 8. EN 1998-1(2004)
4. Regulation on technical norms for masonry walls ("Fig. SFRY". No. 87/9) and the following standards JUS
5. Regulation on technical norms for the construction of buildings in seismic areas ("Fig. SFRY" No 31/81, 49/82, 29/83, 21/88 and 52/90.)
6. Tomažević, M. *Introducing Eurocodes and the calculation of seismic resistance of the masonry constructions,* Ljubljana 263 p. (2008)
7. Information on: http://www.wbalkanseismicmaps.org/index.php/seismic-hazard-maps
8. Zejak, D. 5th Internacional Conference of Civil Engineering Science and Practice, GNP 2014, 387-394 (2014)
9. Mrdak, I., Rakočević, M., Žugić, L., Usmanov, R.., Murgul, V., Vatin, N. Applied Mechanics and Materials. **633-634**, 1069-1076 (2014)
10. Kovačič, B., Kamnik, R., Kapović, Z. Geodetic Journal, **63 (4)**, 315-327 (2009)
11. Brcic, S., Zugic-Zornija, Lj. International Conference on Structures Under Shock and Impact, SUSI, 375-384 (1998)
12. Kovačič, B., Kamnik, R., Premrov, M. Survey Review, **43 (320)**, 150-161 (2011)
13. Kovačič, B., Kamnik, R., Premrov, M., Gubeljak, N., Predan, J., Tišma, Z. Strojniski Vestnik / Journal of Mechanical Engineering, **54 (5)**, 364-371 (2008)
14. Ber, B., Premrov, M., Strukelj, A., Sustersic, I., Dujic, B. Challenging Glass 4 and COST Action TU0905 Final Conference - Proceedings of the Challenging Glass 4 and Cost Action TU0905 Final Conference, 219-226 (2014)
15. Z. Radovanović, R. Sinđić Grebović, S. Dimovska, N. Serdar, N. Vatin, V. Murgul, Procedia Engineering, **117**, 870-878 (2015)
16. V. Pukhkal, V. Murgul, M. Garifullin, Procedia Engineering, 117, 624-627 (2015)
17. V. Murgul, Procedia Engineering, **117**, 808-818 (2015)
18. Premrov, M., Dujic, B., Ber, B. COST Action TU0905 Mid-Term Conference on Structural Glass - Proceedings of COST Action TU0905 Mid-Term Conference on Structural Glass, 25-32 (2013)
19. Jancar, J., Dujic, B., Premrov, M. World Conference on Timber Engineering 2012, WCTE 2012, **1**, 504-510 (2012)
20. Premrov, M., Dobrila, P. WIT Transactions on the Built Environment, **104**, 409-418 (2009)
21. S. Popović, N. Djurovic, V. Murgul, Procedia Engineering, **117**, 832-840 (2015)
22. R. Usmanov, M. Rakočević, V. Murgul, N. Vatin, Applied Mechanics and Materials, **633-634**, 927-931 (2014)

23. S. Nikolskiy, N. Vatin, O. Pertseva (2015) Journal of Applied Engineering Science, **13 (1)**, 11-18 (2015)
24. Z. Bonić, G.T. Ćurčć, M. Trivunić, N. Davidović, N. Vatin, (2015) Procedia Engineering, **117 (1)**, 424-435 (2015)
25. Z. Jakšic, D. Ladjinović, M. Trivunić, N. Harmati, N. Vatin, (2015) Procedia Engineering, **117 (1)**, 507-520 (2015)
26. J. Ćetković, S. Rutešić, M. Zarković, M. Knežević, N. Vatin, Procedia Engineering, **117 (1)**, 780-790 (2015)
27. S. Rutešić, J. Ćetković, M. Žarcković, M. Knežević, N. Vatin, Procedia Engineering, **117 (1)**, 905-915 (2015)
28. S. Rutešić, J. Ćetković, M. Knežević, M. Žarcković, N. Vatin, Procedia Engineering, **117 (1)**, 642-650 (2015)
29. I. Garanzha, N. Vatin, Applied Mechanics and Materials, **633-634**, 965-971 (2014)
30. B. Melović, S. Mitrović, A. Djokaj, N. Vatin, Procedia Engineering, **117 (1)**, 807-812 (2015)
31. N. Gubeljak, J. Predan, D. Kozak, J. Tuma, B. Kovačič, P. Konjatić, J. Sertić, Strojarstvo, **51** (4), 263-271 (2009)
32. L. Muravyeva, N. Vatin, Applied Mechanics and Materials, **635-637**, 468-471 (2014)
33. L. Muravyeva, N. Vatin, Applied Mechanics and Materials, **635-637**, 451-456 (2014)

MATEC Web of Conferences **53**, 01034 (2016)
DOI: 10.1051/matecconf/20165301034

Calculation the options of cracks in the earth dam

Yulia Ibraeva [1, a], Georgiy Bulatov [2] and Philipp Tarasevskii [3]

[1]*Peter the Great St.Petersburg Polytechnic University, 195251,St. Petersburg, Russian Federation*

Abstract. The article presents the calculation of the limit depth of cracks in the dam, the calculation of step cracks in the dam, the calculation of full disclosure, the calculation of the direction of the cracks in the dam, the position of the first crack and their natural clogging. As well in this article has been shown all the necessary formulas for these calculations.

1 Introduction

Despite the considerable progress made in the soil construction of dams, reserves to reduce the cost of construction of earth dams and increase their strength filtration completely exhausted. Mainly, in our opinion, is to increase reserves crack resistance ground shields, cores, and ensuring the principle of equal strength of all elements of the dam by further improving the design and technological solutions.

2 Calculation of limiting the depth of cracks in the dam

The limiting depth the propagation of vertical crack [1–3]

$$h_{c.lim}=\frac{2\cdot(C+C_w)}{\gamma_{gr}\cdot tg\,\beta_a} \tag{1}$$

Where: C – clutch of soil in shear, C_w– additional clutch which equivalent to support actions of filtration forces and called overpressure of water in the cavity of cracks (at the phenomenon of hydraulic fracturing), γ_{gr}– unit weight the soil of dam, averaged over the height of the crack, β_a– the angle of the active soil collapse.

$$\beta_a=\frac{\pi}{4}-\frac{\varphi}{2} \tag{2}$$

Where: φ – the angle of internal friction of soil.

$$C_w=\eta_c\cdot p_w \tag{3}$$

Where: η_c– average value of water pressure in the cavity cracks, p_w– the coefficient of additional clutch from water pressure (in the first approximation, equal to 0.1).

[a]Corresponding author: ulchi412@mail.ru

By separating at the crestal area of the dam on the surface and underwater, and taking the water pressure in the cavity of the crack below the water averaged hydrostatic, we write

$$h_{c.lim}=\frac{2C\text{-}h_f\left[\left(\gamma_f\text{-}\gamma_{sus}\right)\cdot tg\beta_a+\eta_c\cdot\gamma_w\right]}{\gamma_{sus}\cdot tg\beta_a\text{-}\eta_c\cdot\gamma_w} \tag{4}$$

Where: h_f – freeboard of the dam, γ_f и γ_{sus}– densities and soil respectively on the surface and in suspension in the water condition, γ_w–specific gravity of water.

Eq. 1 for C_w=0 is recommended by many authors to determine the depth of cracks as well as the height of the maximum sustainable vertical ground slope [1–17].

It can also be obtained from the condition of soil strength in compression at the base of the vertical slope in the form of

$$h_{vert}=\frac{R_c}{\gamma_{gr}} \tag{5}$$

Where: R_c– strength of the soil compression by the limit equilibrium theory [11–19].

The height of the vertical slope is defined as

$$h_{vert}=\frac{3.8\cdot C}{\gamma_{gr}} \tag{6}$$

Were considered above options for the conditions in the plane problem. But when the transverse cracks in the dam equilibrium problem of vertical slope is voluminous, and for its, taking into account increased risk of fractures, we write

$$h'_{c.lim}=\frac{2\cdot(C+C_w)}{\gamma_{gr}\cdot tg\beta_a}\cdot K_v \tag{7}$$

Where: K_v– coefficient taking into account the spatial tasks.

Assuming that there exist flat surfaces collapse obtain

$$K_v\cong\frac{l}{l_0}=\frac{B+m\cdot h_{c.lim}}{B+2\cdot m\cdot\frac{h_{c.lim}}{3}}\cdot\frac{sin\beta_a}{sin\beta_{a0}} \tag{8}$$

Where: l и l_0– calculated lengths of the surface of collapse equal to the ratio of body volume of collapse toward to area surface collapse, respectively for flat and volume tasks, B and m –width of the crest of the dam and the average inception of its slope in the direction of the axis of the crack, β_{a0}– the angle of active caving ground in the volumetric conditions task (may be taken as approximately β_a).

3 Calculation step cracks in the dam

Average cracking step in analogy with the width of prism the active collapsing soil is determined based on the data [20] by one of the following expressions

$$a_c=h_{c.lim}\cdot\eta_a\cdot\xi_{rg}\ , \tag{9}$$

$$\psi_m\cdot h_{c.lim}\ at\ h_l < h_{c.lim} \tag{10}$$

Where: $h_{c.lim}$ –maximum depth of crack propagation, η_a– coefficient which takes into account the widely observed random factor (scatter) in the formation of cracks, expressed in a significant natural irregularities of the distance between them (approximately 1 ... 2), ξ_{rg}– function unloading solid ground in the event of a crack in it, taken as a

$$\xi_{rg} = tg\beta_a \qquad (11)$$

Where: β_a – active caving angle of soil, ψ_m– coefficient which reflecting layer monolithic taken equal to 1, 2 ... at the transition from unconsolidated soil to little connected dense soils, given that individual rocks to approximate an average of the cubic form, h_l– thickness of the layer of soil (for example dams or soil impervious element) [21–24].

The place of potential cracks will be any weakened vertical section in the dam or in its base, defined throughout their hydrogeological history at the material time: cracks in the layer, ground interlayer, thinning, chink, and conjugation of elements structures, technology seams and others.

Note. At the junction of dirt and concrete parts of the building or in the sectional seams of deformation extension of the earth surface will be compensated (absorbed) by the appropriate disclosure of cracks, which calculated similarly fissures.

Often in the base of dams under a layer unconsolidated sediments overlain layer of solid rock. In this case it is useful to check the parameters of cracks on the stratum, as it can make a decisive contribution to the character of crack in the dam body. In one of the options, this calculation should be equal to the pitch of cracks at the base formation of the hard rock. In the presence in the base laminate thickness desired verification options cracking and of other layers.

A large range of properties of natural soils and including their strength characteristics, can also introduce the concept of the minimum step of cracks.

For example, the minimum step of cracks on the conditions of fracture cracks dirt console

$$a_{t.k.} = h_c \cdot \sqrt{\frac{R_p}{3 \cdot \gamma_{gr} \cdot h_c}} \qquad (12)$$

Or conditions shear of console

$$a_{t.k.} = \frac{C}{\gamma_{gr}} \qquad (13)$$

Where: h_c – crack depth, R_p и C –strength of the soil, respectively, and the tensile shear, γ_{gr}– unit weight of soil.

Natural to assume that the extent value of deformations of the step cracks will be changes accordingly. Approximate regularity of marked changes proposed below. By analogy with the expression permanent deformation in the form of CES

$$R^* \cdot \varepsilon^* = C_R^* \qquad (14)$$

We can write the expression for constant cracking

$$a_c^* \text{-} \varepsilon^* = C_a^* \qquad (15)$$

From a comparison of these relations we obtain

$$a_c^* = R^* \cdot C_c^* \geq a_{c\ min} \qquad (16)$$

At a constant pitch of cracks

$$C_c^* = \frac{C_a^*}{C_R^*} \qquad (17)$$

Where: a_{Tmin} – minimum step of cracks.

4 Calculation of the total crack opening

In accordance with the entire disclosure of cracks

$$\delta=\delta_0+\Delta\delta\ , \tag{18}$$

If

$$\delta_0=\delta_p+\delta_{man}+\delta_{oth} \tag{19}$$

Where: δ_0– crack opening before erosion, $\Delta\ \delta$ –the same additional, in the process of erosion, δ_p, δ_{man}, δ_{oth}– the same against part–time, managers and other factors [24–27].

Crack opening of part–time work

$$\delta_p=(\varepsilon^*+\frac{\Delta R}{R^*}+\varepsilon_{man}-\varepsilon_k)\cdot a_c\ , \tag{20}$$

If

$$\varepsilon_{man}=\frac{R_p}{E_{pz.p}}\cdot\lambda(z) \tag{21}$$

Where: ε^*–the expected horizontal deformation of elongation, ε_{man}– deformation of elastic recovery during unloading from stretching, $E_{pr.p}$– the modulus of deformation of soil which is corresponding to it, $\lambda(z)$ – the function of restore the elastic deformations.

In a first approximation

$$\lambda(z)=\frac{1}{3}\cdot(1-\frac{z}{h_c})\cdot(1-\mu^2)\geq 0 \tag{22}$$

Where: h_c – the intermediate depth of the crack propagation.

Entering values into Eq. 20 and equating the left side to zero, and of assuming that $z = h_c$, taking into account other influences obtain

$$h_T=\frac{\varepsilon^*+\dfrac{\Delta h}{R^*}-\varepsilon_l(1-\mu^2)+\varepsilon_{oth}+\varepsilon_{man}}{\dfrac{1}{R^*}+\dfrac{\xi\cdot\gamma_{gr}\cdot K_a}{E_{pz.c\ (T)}}(1-\mu^2)}\leq h_{c.lim} \tag{23}$$

With the notation introduced earlier. In practice, part–time work has been cases where a single point at the beginning of CES undergoes compressive strain ε_1^*, and then lengthening ε_2^*. In this case it is recommended to take

$$\varepsilon^*=[\varepsilon_2^*-\varepsilon_1^*\left(1-\frac{E_c}{E_{pz.c}}\right)] \tag{24}$$

Where: E_c, $E_{pr.c}$– the modulus of soil deformation respectively in compression and unloading of the compressive stress which are caused through compressive strain.

5 The calculation of the direction of cracks in the dam

Dams can change the direction of cracks developing on the Earth's surface (CES). Deflection angle χ line breaks in the dam plan (Fig. 1) from the normal to the major principal elongation ε^* (i.e. in terms of the crack deviation) is determined from the condition

$$\frac{\varepsilon}{F}=max \tag{25}$$

Where: ε – component strain elongation in the direction normal to the line break, F – cross–sectional area of the dam along the line break

$$\varepsilon=\varepsilon^{*}\cdot cos^{2}\chi \,, \tag{26}$$

$$F=\frac{H_d}{2}(B_{gr}+\frac{B_{gr}+H_d(m_B+m_H)}{cos(\theta-\chi)}) \tag{27}$$

Where: ε^*– most importantly elongation, χ – deviation in terms of a crack, H_d–height of the dam,

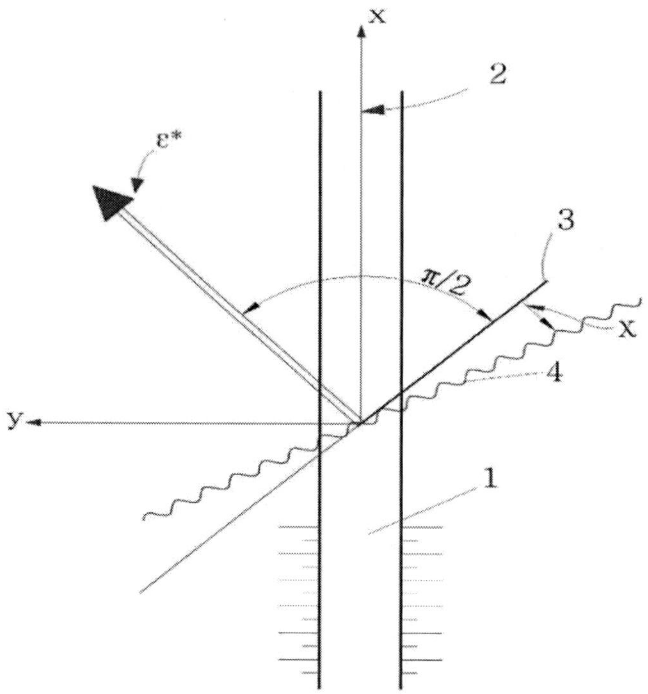

Figure 1. The scheme of deformation of the dam in terms of: 1 – crest of the dam; 2 – axis of the dam; 3 – normal to the main elongation; 4 – axle cracks

B_{gr}– width of the dam, m_B и m_H– foundations, the upper and lower slope, θ – the angle of deflection of the main extension at longitudinal axis of the dam [18–25].

Of the conditions for the crack deviation we obtained the equation

$$\frac{B_{gr}}{B_{gr}+H_d(m_B+m_H)}\cdot cos(\theta-\chi)+1-\frac{tg(\theta-\chi)}{2tg\chi}=0 \tag{28}$$

The first term may be taken as approximately 0. From equation that

$$\chi\cong\frac{\theta}{3} \tag{29}$$

In the vertical plane, we can assume that in the side prisms of dam the cracks will be deviate from the vertical on the direction by an angle of slope (deviation in a vertical sectional view of a crack) [24–26].

$$\psi\leq\beta_a \tag{30}$$

6 Status of the first cracks

Location and direction of the crack is determined by well–known law of nature "where the thin, there and tear." Initially, there is only one (the first) crack – and this is also the law. And the feature of it is that the appearance of subsequent fractures is essentially determined by the first of them.

The first crack may not coincide with the position of the maximum elongation CES, and deviate in the direction of various kinds weakened regions due to multiple causes of natural and man–made, non–ideal creations of both nature and human hands.

For example, the place of appearance the cracks on the slope of the dam may be incoming (internal) angle berm, etc. Both obvious and subtle at first glance factors can significantly alter the picture of fracture. Sometimes arise and unexpected phenomena, such as the complete cracking dam – with the rapid emptying of the repository. Of course, as elsewhere, in this case also can help only a closer analysis of the system.

7 The natural cracks clogged

Rainwater flows periodically come to life on the crest of dam mainly in the ruts of transport, at a meeting with cracks fell through in them and spread out over the slopes. These flows are accompanied by phenomena of transport (erosion–per wash) soil particles, which are the analogous phenomena of clogging of cracks in the pavement slabs or parquet floor cracks.

The phenomenon of clogging reminiscent the central alluvium with the axis predominantly on the middle ridge width. The density of the material filling of the cavity thin cracks will be small due to the appearance the effects of silage and loose appearance at elevated heterogeneity filling.

At the beginning washed away the small particles that fill up the tip of crack. Then wash up in the cavity of the new layers of soil, washed from the crest of the dam. In heavy rain on the slopes are possible the formation of micro ravines eroded cracks in the upper part (the rivers is usually confined to the geological fractures).

In spring can be partially flood of cracks due to water–logged soil. At the end of the dry season, the crack may become more pronounced. Reliability natural clogging cracks will be determined by the ratio of the rate of formation of the tampon to the speed of progress the cavity of crack. Generalized criterion of adequacy clogging represented as

$$\Pi_i(T) \geq [\Pi_i(T)] \tag{31}$$

Where: $\Pi_i(T)$– expected parameters tamponing cracks (in brackets and their allowed values of the terms of reliability), i= 1 – thickness of the tampon, i=2 – height of the tampon, i= 3– length, i= 4 – permeability and density, i= 5 – its strength of filtration.

Expected parameters clogging cracks are determined based on the theories of climate predictions water erosion and wash. The valid options are evaluated according to the criteria of reliability, below.

8 Results and Discussion

The result of the article is the formula by which, in the earth dam are determined: the depth of the cracks, the value of disclosure, presumably the direction of opening and a step appearance. Using this methodology can be compute required constructively-technologic protection measures.

9 Conclusion

Thus, we can calculate the parameters of cracks appearing in the earth dam and to predict the appearance of the first crack. Due to the calculation of crack size and step their formation we can introduce the event name for the specified concrete fracture parameters, and to avoid their occurrence in groundwater dam and as a consequence of further negative impact on the body of a subsurface dam.

References

1. R. Usmanov, M. Rakočević, V. Murgul, N. Vatin, Applied Mechanics and Materials, **633–634**, 927–931 (2014)
2. R. Usmanov, I. Mrdak, N. Vatin, V. Murgul, Applied Mechanics and Materials, **633–634**, 932–935 (2014)
3. M. Jocovic, B. Melovic, N. Vatin, V. Murgul, Applied Mechanics and Materials, **678**, 644–647 (2014)
4. Y. Sonoda, Applied Mechanics and Materials, **566**, 10–25 (2014)
5. E. Panulinová, S. Harabinová, Advanced Materials Research, **969**, 245–248 (2014)
6. S. Harabinová, E. Panulinová, Advanced Materials Research, **969**, 208–211 (2014)
7. M. Sharp, Y. Seda–Sanabria, E. Matheu, Applied Mechanics and Materials, **82**, 428–433 (2011)
8. I. Mrdak, M. Rakočević, L. Žugić, R. Usmanov, V. Murgul, N. Vatin , Applied Mechanics and Materials, **633 – 634**, 1069–1076 (2014)
9. D. Kubečková, Advanced Materials Research, **1020**, 883–887 (2014)
10. G. Bulatov, V. Teleshev, V. Leonov, International Water Power & Dam Construction, **117**, 111–115 (2003)
11. D. Hilyard, International Water Power & Dam Construction, **5**, 20–22 (2010)
12. S. Bortkevich, Power Technology and Engineering, **5**, 287–292 (2009)
13. M. Jocovic, B. Melovic, N. Vatin, V. Murgul, Applied Mechanics and Materials, **678**, 644–647 (2014)
14. N. Vatin, N. Lavrov, N. Korzhavin, Applied Mechanics and Materials, **641–642**, 353–358 (2014)
15. V. Vladimirov, Yu. Zaretskii, V. Orekhov, Power Technology and Engineering, **3**, 161–166 (2003)
16. S. Shkol'nikov, I. Sekisova, Power Technology and Engineering, **5**, 331–339 (2008)
17. V. Bukhartsev, M. Petrichenko, Power Technology and Engineering, **46 (3)**, 185–189 (2012)
18. G. Bulatov, Trudy LPI, **375**, 91–94 (1981)
19. V. Istomina, V. Burenkova, G. Mishurova, *The filter strength of clay soils* (Stroyizdat, 220, 1975)
20. A. Bougrov, V. Markevich, Gidrotekhnicheskoye stroitel'stvo, **4**, 24–26 (1980)
21. G. Bulatov, N. Vatin, D. Nemova, Y. Ibraeva, P. Tarasevskii, Applied Mechanics and Materials, **725–726**, 342–349 (2014)
22. V. Deković, A. Andelković, N. Milošević, G. Gajić, M. Janić, Carpathian Journal of Earth and Environmental Sciences, **8 (2)**, 107–112 (2013)
23. L. Rasskazov, M. Smirnova, Power Technology and Engineering, **48 (2)**, 85–88 (2014)
24. A. Bauer, S. Haug, WasserWirtschaft, **104 (4)**, 28–33 (2014)
25. J. Guo, I. Hasan, P. Graeber, Springer Series in Geomechanics and Geoengineering, **2015**, 195–209 (2014)
26. M. Balzannikov, M. Rodionov, International Journal on Hydropower and Dams, **20 (6)**, 60–63 (2013)
27. A. Feringer, N. Kabanov, Power Technology and Engineering, **46 (5)**, 380–383 (2013)

MATEC Web of Conferences **53**, 01035 (2016)
DOI: 10.1051/matecconf/20165301035

Measures of struggle against appearance of cracks in earth dams

Yulia Ibraeva [1,a], Georgiy Bulatov[1] and Philipp Tarasevskii[1]

[1]*Peter the Great St.Petersburg Polytechnic University, 195251,St. Petersburg, Russian Federation*

Abstract. The article describes a method calculation of the basic parameters of the transverse rows of pile of simple printed or precast dam. As well, in this article have been shown all the necessary formulas for this calculation and have been proposed solutions to prevent cracking in the dams.

1 Introduction

Despite the considerable progress made in the soil construction of dams, reserves to reduce the cost of construction of earth dams and increase their strength filtration completely exhausted. Mainly, in our opinion, is to increase reserves crack resistance ground shields, cores, and ensuring the principle of equal strength of all elements of the dam by further improving the design and technological solutions. Works in this direction have are many research in institutes and laboratories of the Russian Federation (in VNIIG, Waters GEO Gidroproekt and SPbSTU).

2 Materials and Methods

Known that nuclei in ground cracks occur due to insufficient compressive stresses. The application in earth dam the nucleus of dome shape [1] (Fig. 1.a) are reduce the negative impact of "suspension" nucleus and increase the level of its compressive stresses [2]. In this case, the core is been performed with a double curvature (in the form of sickle or bow) in cross section and in plan view.

Also envisaged the option of slope the axis O–O of the nucleus in the downstream direction at an angle of 45°, wherein the core takes the form of half bow. Due to the curved shape the nucleus is been achieved the increasing the stability how in the upper slope (compared with a flat–screen) so in the lower slope (compared with a flat core). For a given level of reliability of the dam, its slopes can be make steeper, and the kernel – spanning more thinly than the flat anti filtration elements. The weak point of the dam is a seam conjugation with the base kernel. However, the filtration resistance of the seam can be significantly enhanced if the kernel into the bottom section divided groove [3] filled with drainage fine–grained material into several branches, each of which independently interfaces with base [2].

[a] Corresponding author: ulchi412@mail.ru

In the grooves can be installed drainage galleries, which are connected the piping system. These devices allow you to optimize the distribution of pressure between the individual branches of the nucleus.

In case of an occurrence of cracks it is will not pass–through but its will be cross in several places of internal drainage, which provides the effect of "keyway" and contributes to making the reliability of conjugation with the base kernel [4].The possibility of a significant reduction in the thickness of the main part of the nucleus with subsequent benefits in addition provide the cost–effectiveness.

The destructive effect of cracking can be greatly weakened by the creation in a body of the dam (or unpaved impervious element) of artificial seam–incision, which filled with drainage material, the direction of which coincides with the direction of the expected cracks [3].

Fig. 1 shows a homogeneous dam of cohesive soil on the upper slope of the network of which is made of vertical grooves [2].

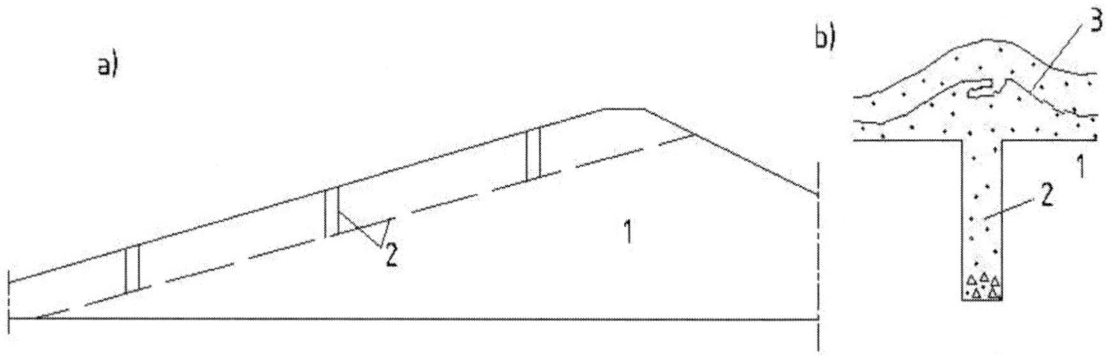

Figure 1.a Homogeneous dam of cohesive soil **Figure 1.b** Transverse section

One of the structural layouts of incisions (transverse section) is be presented in Fig. 2. Cavity which developed excavator are be filled a bulk material with formation a reverse filter and for improve tightness overlapped ribbon membrane with pleat, which compensates deformations. In another embodiment, the hermetic member can been formed as a V – shaped width which inserted in the cavity. Pressurization seam– notch may also been provided at the expense of installing it in the cavity of the balloon of elastic material, filled with pressurized fluid [3].

In the latest version is also will be achieved an additional compression of the array (block) the soil which located between the seams – incision. Application of seams – incisions, as well as use of the known solutions proposed V.N. Zhilenkov [5], eliminates the possibility of emergence destructive for the dam through cracks in its anti–filtration elements.

Eliminate cracking about–crest part of the dam can be due to hammer in cross pile rows [2], which will provide wedging dam, in the longitudinal direction [6]. Pile series (Fig. 3) may be precast, stamped and placed along the axis of alleged cracks in areas of increased deformation elongation dam crest [7].

To improve the crack resistance of the dam piles may been formed of two halves between which placed elastic chambers that filled with liquid at a pressure greater than the water pressure on the pile tip. For fully exclude the occurrence of any additional cracks in spacing between pile rows the anchor rods with prestressing connect the halves of piles in adjacent rows.

Below is shows a method of calculating the basic parameters of the transverse pile rows of simple printed or precast dam in accordance with the notation given in Fig. 2 (a–cross–section of the dam, b – its Area).

The length of piles row could been determined by the expression:

$$Lp = \frac{Hp}{[Jp]} \quad (1)$$

Where Hp – estimated pressure at around the bottom of the pile row, Jp – allowable seepage gradient material for a dam on his contact with material piles, which could been taken on the recommendations of [8].

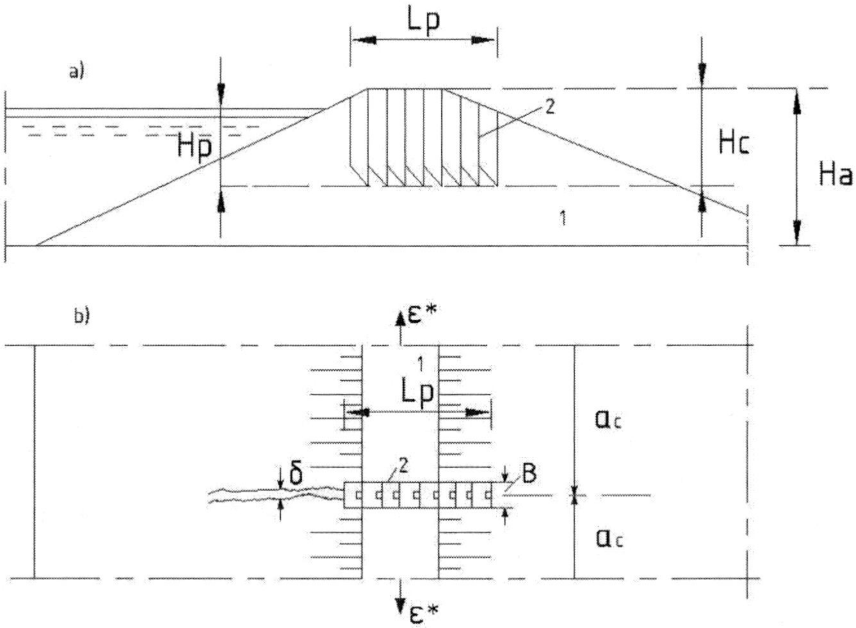

Figure 2. Transverse rows of pile

$$B = \delta(1+n) \quad (2)$$

Where δ – the maximum anticipated thickness of the crack, n – degree propping fractures (in first approximation n=2) [9].

The depth of pile series is be defined by the formula

$$H_p = h_c \cdot K_o \quad (3)$$

Where h_c – expected depth of the crack, K_o – factor over deepened pile of ($K_n \cong 1.2$).

Expected thickness of the crack recommended to define the expression similar to the one in the [10].

$$\partial = (\varepsilon^* - [\varepsilon_m]) \cdot a_p \quad (4)$$

Where ε^*– the expected relative deformation elongation crest of the dam, ε_m – permissible relative deformation elongation array (block) of soil between adjacent rows of pile, a_p – step along the rows of pile rowing dam [11].

Step rows defined in the form

$$a_p = \frac{a_c}{m} \quad (5)$$

Where a_c – expected step cracks along the crest of the dam, m – frequency series ($m \cong 0.5 ... 1.0$). The greatest depth of the crack propagation is be determined by the expression [12]

$$h_t = \frac{2 \cdot (C + C_w)}{\gamma_{gr} \cdot \tan\left(\frac{\pi}{4} - \frac{\varphi}{2}\right)} \qquad (6)$$

Where φ and C – shear strength parameters, γ_{gr} – volumetric weight of soil dam (generally in a suspended state in water), C_w – extra grip equivalent to support the actions of seepage forces caused by the water pressure in the plane of the crack ρ_w [13].

In the first approximation we can take

$$C_w \approx 0.1 \cdot \rho_w \qquad (7)$$

Step cracks determined by the smallest of the following conditions:

$$a_c \leq (1 ... 3) \cdot h_c \qquad (8)$$

$$a_c \leq (1 ... 2) \cdot H_D \qquad (9)$$

Where H_D – height of the homogeneous dam or thickness of impervious element of cohesive soil layered dam [14].

In the case of the impact of undermining the base of the dam works in the above conditions accepted the largest value of numerical coefficients; the smallest of these values are be taken in the absence of undermining and in the case of seismic effects [15].

Since the individual parameters repeatedly associated with each other calculation to carried out using the method of selection followed by selection of the optimal variant [16].

As expected the strain ε^*evolve over time T_i gradually, the most logical wedging body of the dam by the introduction of additional compensating deformation elongation, volume administered in the form of piles in batches at certain stages of deformation of the dam.

5. The sample of conventional calculation the piling (excluding stages of protection). For subsurface dam of loam at $\varphi = 31^0$; $C = 0.12\ kg/cm^2$; $\gamma_{gr} = 1.1\ t/m^3$; $\varepsilon^* = 7 \cdot 10^{-3}$; $[\varepsilon_m] = 1 \cdot 10^{-3}$ were obtain, in accordance with the above procedure the following results: $L_p = 13.2\ m;\ a_p = 15\ m;\ H_p = 6.6\ m\ ;\ \delta = 0.09\ m;\ h_c = 5.5\ m;\ H_p = 6.6\ m;\ B = 0.27\ m.$

The calculation results show the feasibility and effectiveness of the implementation of the pile rows of padded fine sand, as well as precast wood or concrete.

6. Each of stage there comes provided if

$$\varepsilon(T_i) = [\varepsilon_m] \cdot \eta_0 \qquad (10)$$

Where $\varepsilon(T_i)$– increasing with time deformation of the dam or its anti–filtration element. The feature of this protection is continuous monitoring of deformation elongation, η_0 – degree of efficiency of the implementation phase of the protective measures, which depends both on the rate of increase of the deformation elongation, and the time required for installation of piles of this phase of protection against cracks [17].

Thus, depending on the thickness B_i piling will be not only extinguished tensile stress but also achieve compressive stress an optimal level on the protected portion of the dam. Moreover, continues to the total deformation of elongation ε^* a dam at the beginning should extinguish compressive stresses in the soil between the rows of pile and then create a tensile stress corresponding to the condition [18].

3 Results and Discussion

The article represented by four activities to enhance equal strength of the various elements of earth dams and for prevent cracking. Provided the methodology of calculation of the basic parameters for the event the installation of transverse rows of pile.

4 Conclusion

Considered above measures allow increase level equal strength of the various elements of earth dams intended for operation in difficult conditions (with increased risk of fracture), and reduce the cost of material resources to build structures while providing a specified level of reliability.

References

1. R. Usmanov, M. Rakočević, V. Murgul, N. Vatin, Applied Mechanics and Materials, **633–634**, 927–931 (2014)
2. R. Usmanov, I. Mrdak, N. Vatin, V. Murgul, Applied Mechanics and Materials, **633–634**, 932–935 (2014)
3. M. Jocovic, B. Melovic, N. Vatin, V. Murgul, Applied Mechanics and Materials, **678**, 644–647 (2014)
4. Y. Sonoda, Applied Mechanics and Materials, **566**, 10–25 (2014)
5. E. Panulinová, S. Harabinová, Advanced Materials Research, **969**, 245–248 (2014)
6. S. Harabinová, E. Panulinová, Advanced Materials Research, **969**, 208–211 (2014)
7. M. Sharp, Y. Seda–Sanabria, E. Matheu, Applied Mechanics and Materials, **82**, 428–433 (2011)
8. I. Mrdak, M. Rakočević, L. Žugić, R. Usmanov, V. Murgul, N. Vatin , Applied Mechanics and Materials, **633 – 634**, 1069–1076 (2014)
9. D. Kubečková, Advanced Materials Research, **1020**, 883–887 (2014)
10. G. Bulatov, V. Teleshev, V. Leonov, International Water Power & Dam Construction, **117**, 111–115 (2003)
11. D. Hilyard, International Water Power & Dam Construction, **5**, 20–22 (2010)
12. S. Bortkevich, Power Technology and Engineering, **5**, 287–292 (2009)
13. M. Jocovic, B. Melovic, N. Vatin, V. Murgul, Applied Mechanics and Materials, **678**, 644–647 (2014)
14. N. Vatin, N. Lavrov, N. Korzhavin, Applied Mechanics and Materials, **641–642**, 353–358 (2014)
15. V. Vladimirov, Yu. Zaretskii, V. Orekhov, Power Technology and Engineering, **3**, 161–166 (2003)
16. S. Shkol'nikov, I. Sekisova, Power Technology and Engineering, **5**, 331–339 (2008)
17. V. Bukhartsev, M. Petrichenko, Power Technology and Engineering, **46 (3)**, 185–189 (2012)
18. G. Bulatov, Trudy LPI, **375**, 91–94 (1981)
19. V. Istomina, V. Burenkova, G. Mishurova, *The filter strength of clay soils* (Stroyizdat, 220, 1975)
20. A. Bougrov, V. Markevich, Gidrotekhnicheskoye stroitel'stvo, **4**, 24–26 (1980)
21. G. Bulatov, N. Vatin, D. Nemova, Y. Ibraeva, P. Tarasevskii, Applied Mechanics and Materials, **725–726**, 342–349 (2014)
22. *V. Deković, A. Anđelković, N. Milošević, G. Gajić, M. Janić, Carpathian Journal of Earth and Environmental Sciences, **8 (2)**, 107–112 (2013)*
23. *L. Rasskazov, M. Smirnova, Power Technology and Engineering, **48 (2)**, 85–88 (2014)*
24. *A. Bauer, S. Haug, WasserWirtschaft, **104 (4)**, 28–33 (2014)*
25. *J. Guo, I. Hasan, P. Graeber, Springer Series in Geomechanics and Geoengineering, **2015**, 195–209 (2014)*
26. *M. Balzannikov, M. Rodionov, International Journal on Hydropower and Dams, **20 (6)**, 60–63 (2013)*
27. *A. Feringer, N. Kabanov, Power Technology and Engineering, **46 (5)**, 380–383 (2013)*

MATEC Web of Conferences **53**, 01036 (2016)
DOI: 10.1051/matecconf/20165301036

Functionality of Ventilated Facades: Protection of Insulation

Mikhail Petrichenko[1], Tatiana Musorina[1,a], Elena Statsenko[1], Anastasia Ostrovaia[1] and Vladimir Tarasov[1]

[1]*Peter the Great St.Petersburg Polytechnic University, Polytechnicheskaya 29, 195251, St.Petersburg, Russia*

Abstract: This article discusses about methods of construction of the ventilated facades. The ventilated facade is not only the element of facing, it is the supporting structure. Their main objective - creation of air ventilating space between a facade and an external wall of the building. Moving of air in this gap protects a heater from destruction, interfering with a moisture congestion. In addition, the ventilated facade protect the building from aggressive influence of external environment, have a sound and thermal insulation properties. There are several problems of systems of the ventilated facades connected with an application of a heater. For more effective using it is necessary to minimize contact of a heater with environment.

1 Introduction

HVF - is a complex system consisting of a substructure, insulation, air gap and a protective screen, which is mounted on the outside of the main walls of the building. [1-11] They allow you to quickly and easily reconstruct the existing buildings and to increase its resistance to heat walling. They also comply with the requirements of the recently adopted heat and can be used for the construction of buildings, improve the appearance of the building, sound insulation, moisture conditions enclosing walls, resistance to weathering, and others. [13]. The cost of hinged ventilated facade substructure depends on the thickness of the insulation [5].

Mineral wool insulation in facade and roof structures subject to strong operational influences. Inside the fiber structure are constantly alternating processes: humidification - drying, freeze - thaw, mechanical vibration. In the Russian climate the most devastating are the processes of freezing and thawing damp mineral wool. The rate of destruction of the insulation is determined by the volume of incoming and freezing rain. The combination of effects after prolonged use leads to the fact that break down the organic binder is subjected to vibration fatigue fracture mineral fiber [12]. It should be borne in mind that most of the insulation in our conditions have become wet during transportation and installation. Insulation used in curtain wall systems must meet a number of requirements: it must be fire-resistant, durable, have good sound insulation qualities. It should have a sufficient density - from 80 to 100 kg / m3. Since the air flow is separated small particle material - going emission fiber. When installing an illegal armed formation at a certain wind speed they can start to whistle and hoot, most often it occurs in places eddies of wind flows. Describes the phenomenon caused more long brackets for attaching external components and ductility of cotton. The system of ventilated facades require high quality components and, as mentioned earlier, a very high-quality installation, as corrosion of one or two fasteners can result in collapse of the entire façade. [9, 23-25]

[a] Corresponding author : flamingo-93@mail.ru

One possible solution, the application of ventilated facades may be the development and introduction of new energy-saving technologies for these systems [10, 17-19]

The hinged ventilated facades of is the most multifunctional data element systems - a layer of air or an air gap. It is this layer performs both heat shield and blocking function. Additionally, when properly projected structure in this layer HVF moisture is removed, which is a waste product of people staying in the building and allocated in various household and industrial processes.

But it is also necessary to determine how much Ventfasad affect the thermal conductivity of the insulation. The main feature of insulation consisting of bonded mineral fibers - high air permeability. System through interfiber pores in a material permeable to air, gas, and liquid water, irrespective density of the material insulator. Some manufacturers believe most breathable wool her dignity: fibrous insulation materials do not interfere with the movement of vapor out through the exterior walls. But none SNIP does not allow the use of insulation and walling to dehumidify indoor air exist for this ventilation system. Upon contact with cold external sites wet insulation inside air condenses, wetting the insulation. Limit moisture condensation in the insulation can only design solutions that reduce the diffusion and convection currents flowing water [2-3]

Hydration wool comes from different sides, and the water-repellency fiber insulation does not reduce the volume of condensation water.

1. Moisturize from the outside. Facade cladding ventilated facade, consisting of separate elements, there is a gap (rusty), through which the rain and the cold air penetrate and moisturize the insulation system. From moisture completely protected even solid lining, as is always the possibility of defects installation, mechanical damage, the number of which increases with the square facing, the number of window frames and various frames.

2. Moisturize from the inside. In cold weather, a heater threatened moisture from the warm side. When carrying fencing have enhanced vapor permeability (porous concrete wall brick wall with a poor grouting, walling bulkhead slots, poorly executed, steam, attics), vaporous moisture from the premises condenses in cold regions insulation. Tall buildings have a high partial pressure of steam on the upper floors. In this case, the vapor barrier properties of the fence, even made of cast concrete, may not be enough, need more steam, [1,15]

With regard to the ventilated facade weight loss of mineral wool installed without wind and waterproof membrane, leads not only to a decrease in strength, thermal insulation properties, but also to a serious violation of the ecology of the environment and living space. For example, a nine-storey building with insulation Series 90, with an area of 1498 square meters of insulation. m requires 135 cubic meters. m of modern mineral wool density 74 kg / cu. m. In the 25 years of operation of conventional building ventilation air flow can endure because of the skin venfasada 1875 kg fiber dust. These data convince appropriateness of the design of ventilated facades wind-hydroprotective membranes on the outer surface of the mineral wool. A more expensive solution - the use of panels of high density (above 150 kg / cu. M), does not provide similar protection of insulation from external moisture and filtering, [4,20-22].

Warranty service life of 50 years, mineral wool can be obtained subject to the limitation (rationing) destructive factors. To do this, the system should be regulated to protect the insulation, working in the air gap of the facade:

1. Protection of the inner surface Minplita - severe restriction of moisture from the premises provided a high level of vapor barrier fence;

2. protection of outer surface Minplita - strict limitation of atmospheric moisture, air filtration, the loss of mineral fibers is ensured by the presence of wind-hydroprotective diffusion membrane.

The traditional objection to the use of wind-hydroprotective ventilated facade membrane systems is that "the presence of water vapor permeability of the membrane reduces the insulation system." Vapor-permeable membrane slightly, by 0.5%, reduces the diffusion of water vapor through the multilayer structure of the outer walls with a ventilated air gap. Afraid to install a vapor barrier in a residential area due to lack of ventilation. At the same time over the fence in the heater can do a number of internal moisture from the premises, which can not cope with a membrane having a water vapor permeability of 1,000 g / sq. m per day. As a result, a "cost-effective" solution as the lack of ventilation in the room and the membrane in a ventilated gap leads to the inner insulation and moisture to eliminate its external protection. By limiting the admission of moisture from the premises membrane will display only the gaseous moisture which passes itself under adverse weather conditions - high humidity, fog and rain. Water vapor permeability of 1,000 g / sq. m per day maintains a heater in the equilibrium dry in all weather conditions. [14]

Application of wind - hydroprotective vapor barrier membrane and insulation systems with a ventilated gap will exclude from the selection criteria of insulation characteristics such as water absorption, air permeability, density, leaving only the thermal conductivity, dimensional stability, mechanical strength. [27]

Goals and objectives

The aim of the article is - taking into account the impact of the analysis of a ventilated facade on the insulating properties of the insulation.

To achieve this goal it is necessary to solve the following problems:

- Investigate the thermal insulation properties of walls with hinged ventilated facades with a shield and without protective screen;
- An analysis of studies on the thermal insulation of building materials walling considering shield;

Shown, that:

- There is a dependence of the density of the air permeability of the fibrous insulation; With this, the influence of a number of design features and climatic conditions on the longitudinal filtration of air in the insulation and heat-shielding properties of ventilated facades. [6]

2 Materials and Methods

2.1 Types of mineral wool

Thermal insulation for ventilated facade should have the following main functions: waterproof, high water vapor permeability, lack of strain in the winter, the most tightly against the wall. The main materials used for facade insulation - mineral wool, expanded polystyrene, extruded polystyrene foam, fiberglass materials.

For example, a heater as IZORUF is a semi-rigid mineral boards, based on - the rock. Because of this, the material can withstand temperatures up to 750 degrees without losing their acoustic and thermal insulating properties. Plates thus not deform or subside. The high degree of vapor permeability prevents the accumulation of water in the heat-insulating layer, ensuring its rapid evaporation. Among the materials produced on the basis of extruded polystyrene foam, it is possible to allocate Tepleks 35. The low humidity of this material is provided by closed cell structure, which distinguishes it from other moisture-absorbent insulation. Tepleks 35 is not affected by temperature, it has a minimal deformation during thermal exposure. Antibacterial properties (resistance to rot) and chemical resistance allows the use of polystyrene in a corrosive environment. [14, 26]

In current practice as insulation material in the territory of Russia are widespread mineral wool based on basalt binder, as well as boards based on expanded polystyrene. In some parts of the building, such as caps and place of constant accumulation and retention of water used slabs of polystyrene, obtained by extrusion. At this historic moment, the current regulatory documentation did not reflect all the nuances of the use and production of such materials. For example, it is completely absent standards for durability of such products, the terms of service and operational requirements in different climatic conditions in Russia. [7, 8]

3 Impact of ventilated facades in the mineral wool

We find the thermal resistance of the wall

1. Consider the case where no shield

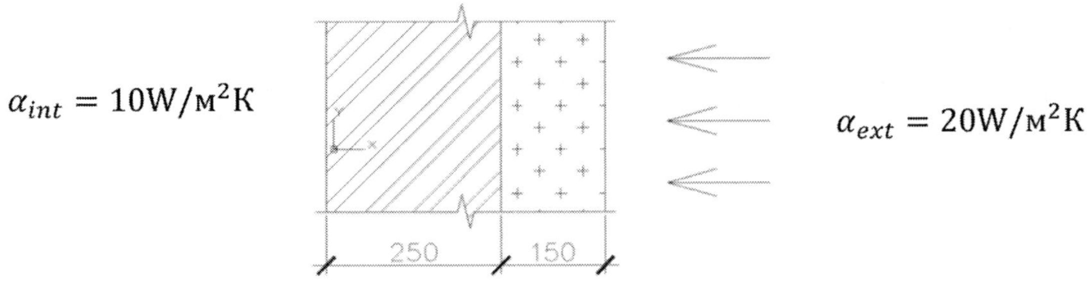

Figure 1. Cut the wall without a protective screen

Find the thermal resistance of the wall.
The resistance is given by:

$$R = \frac{1}{\alpha_{int}} + \frac{1}{\alpha_{ext}} + \frac{\delta_К}{\gamma_К} + \frac{\delta_y}{\gamma_y} \tag{1}$$

Initial data:
The temperature inside the premises $\alpha_{int} = 10W/(м^2K)$
The temperature of the outside $\alpha_{ext} = 20Вт/(м^2K)$

Table 1. The original data for calculating

	Thermal conductivity	Thickness
silicate bricks	$\gamma_К = 0{,}7$	$\delta_К$=0,25 м
Mineral wool	$\gamma_y = 0{,}07$	δ_y ==0,15 м

$$R = \frac{1}{10} + \frac{1}{20} + \frac{0.25}{0.7} + \frac{0.15}{0.07} = 2.65 \ м^2K/W \tag{2}$$

Conclusion: due to the fact that the external (cold) wall surface exposed to weathering, low thermal resistance of the fence. The construction of the wall does not provide the minimization of heat flow

2. In the presence of the screen (double façade)

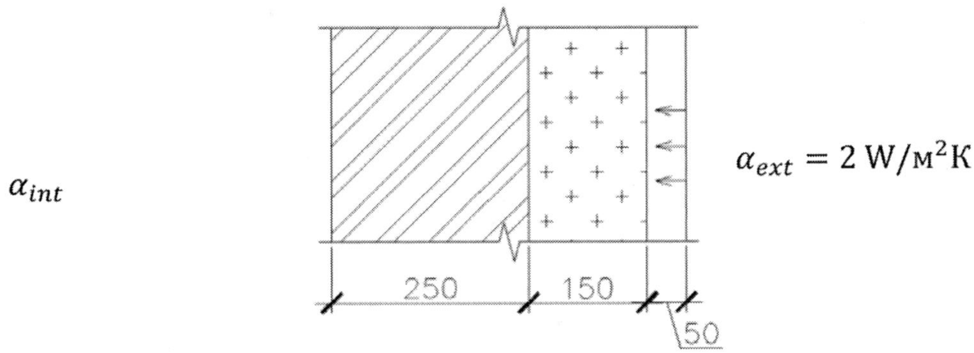

Fig 2. Incision wall based shield

The temperature inside the air gap will be equal to:

$$\alpha_{ext} = \frac{\alpha_{air}}{\delta_{air\ layer}} = \frac{0{,}1}{0{,}05} = 2\ W/м^2K$$

$$R = \frac{1}{10} + \frac{1}{2} + \frac{0.25}{0.7} + \frac{0.15}{0.07} = 3{,}1\ м^2K/W$$

Conclusion: the introduction of HVF, thermal resistance increases.

4 Results and Discussion

Consider the mass transfer of moisture in the insulation. The air gap as a pump pulls moisture from the insulation. It is necessary to move the dew point in the air gap is necessary that the air temperature is greater than the saturation temperature ($t_в > t'$), wherein, the water vapor: t' at this pressure $p' = (\frac{t'}{100})^4$

Take the extreme case where the saturation temperature of the of insulation is on the edge, taking into account the protective screen
Heat flow stream is determined by:

$$t_{ext2} = t_c + \frac{r_2}{r_1}(t_{ext1} - t_c) \quad (3)$$

The temperature difference is equal to:

$$\Delta t = \Delta t_{ext} = t_{ext2} - t_{ext1} \quad (4)$$

Conclusion: The facade retains heat output in the wall

The evaporation of moisture from the surface of a heater in a free-convective flow in a viscosity-gravity flow regime described by the system of equations Eckert. In dimensionless form this system of equations is of the form:

$$\begin{aligned} &f''' + 3ff'' - 2f'^2 + \theta = 0, \\ &\theta'' + 3\sigma f\theta' = 0, \end{aligned} \quad (5)$$

Where $f=f(y/\delta)$ – the dimensionless function of the current, y- transverse coordinate (direction - from a hot surface into a stream),
δ - thickness of the boundary layer lifting force,
θ - the concentration of a conservative and passive impurity,
σ - the number of Prandtl for the diffusion of impurities, primes denote derivatives with respect y/δ.
Limit conditions for the system Eckert put this way:

$$f(0) = f''(0) = f'(\infty) = 0, \theta(0) - 1 = \theta(\infty) = 0. \quad (6)$$

In this paper, for simplification replace the differential equation of pollutant transport algebraic condition "barotropic". Namely, let

$$\theta = \beta f'^2, \quad (7)$$

Where β - coefficient of proportionality.
Then the first equation Eckert takes the form:

$$f''' + 3ff'' - (2 - \beta)f'^2 = 0, \quad (8)$$

wherein.

Nont $f(0) = f''(0) = f'(\infty) = 0$ rivial solution of this equation is:

$$f' = u_0(1 - F^2), F := f / f_0, f_0 := f(\infty). \quad (9)$$

proved by: $\beta=5$,

u_0 – the speed of free-convection flow on the hot surface (as a fraction of the average speed), $f_0 = \sqrt{2u_0/3}$ - dimensionless stream function (as a fraction of the flow).

Then, finally:

$$f = f_0 th(xu_0 / y_0),$$
$$u = u_0 \operatorname{sech}^2(xu_0 / y_0), \tag{10}$$
$$\theta = \beta u_0^2 \operatorname{sech}^4(xu_0 / y_0)$$

5 Conclusions

1. In the current practice as insulation material in the territory of Russia are widespread mineral wool based on basalt binder, as well as boards based on expanded polystyrene. Thermal protection is not dependent on the thickness of the insulation. All it related to the properties of the material itself.
2. As shown in the calculations based on a ventilated facade thermal resistance increases.

References

1. V.G Gagarin, AVOK, **2** (2005);
2. B.V. Gusev, V.A. Yezerskiy, P.V. Monasteries Industrial and civil construction, **1** (2005)
3. V.G. Gagarin V.V. Kozlov E.Y.Tsykanovsky AVOK, **2**, 20-26 (2004)
4. S.N. Ovsyannikov T.S. Vyazova, **6**, 20-28 (2013)
5. G.Radovic, , V.Murgul, N.Vatin, Applied Mechanics and Materials, **584-586**, 564-569 (2014)
6. W. Guoa, X. Qiaoa, Y. Huanga, M. Fanga, X. Hanb. Energy and Buildings Volume In Press, Corrected Proof, 43-53 (2012).
7. Z. Xinhong, M Congyu., G. Pingdao, Energy Energy Procedia, **14**, 1523–1527 (2012)
8. M. Petrichenko, X .Rakova , M. Vyatkin, , T. Musorina, , D. Kuznetsova, Applied Mechanics and Materials, **725-726**, 1101-1106 (2015)
9. N.V. Nemov, V.Y.Olszewski, D.N. Tseitin Scientific and technical statements STU **4** (183), 295-301 (2003)
10. G. Radovic, , V. Murgul, , N .Vatin, Applied Mechanics and Materials, **584-586**, 564-569 (2014)
11. A.S. Gorshkov Civil Engineering magazine, **1** (2010)
12. A.S Gorshkov., D.Y Popov. A.V. Glumov Civil Engineering magazine, **8 (18)** (2010)
13. M.V. Knatko M.N.Efimenko, A.S. Gorshkov Civil Engineering magazine **2** (2008)
14. S.D. Maloedov, V.N. Vyguzov Building materials **5** (2001)
15. I. Ptuhina, T. Spiridonova, T. Musorina, Applied Mechanics and Materials, **725-726**, 153-159 (2015)
16. A.F. Ostrovaja, M.R. Petrichenko, E.A. Stacenko, Applied Mechanics and Materials, **725-726** (2015).
17. T. Smirnova Architecture and Construction Requirements, **5** (2009)
18. V. Starodubtsev, S.V. Povetkin Industrial and Civil Engineering, **5** (2009)
19. A.A. Stepanov Building materials, 5 (2009)
20. E.T. Tazeeva, A.S. Gorshkov Vseros works. scientific and engineering. Conf. - SPb .: Publishing House of the Polytechnic University in 2011 (2011)
21. N.P. Umnyakova Architecture and Construction, 5 (2009)
22. D. Nemova, , V. Murgul, , V. Pukhkal, , A. Golik, , E. Chizhov, N. Vatin, Journal of Applied Engineering Science **12**, 37-44 (2014)
23. N.I. Vatin D.V. Nemova The world of construction and real estate, **36**, 1-4 (2010)
24. N. Mingottia, T. Chenvidyakarn, A.W. Woods, Energy and Buildings, 58, 237–249 (2013)
25. F. Marques da Silva, M. Gloria Gomes, A. Moret Rodrigues, Building and Environment, **87**, 292-301 (2015)
26. N. Mingotti, T. Chenvidyakarn, A. W. Woods, Building and Environment, **46,** 807-823 (2011)
27. A.S. Gorshkov, X.M. Rakova, T.A. Musorina, D.N. Tseytin, K.N. Agishev Construction of Unique Buildings and Structures, **4(31),** 232-247 (2015)

MATEC Web of Conferences **53**, 01037 (2016)
DOI: 10.1051/matecconf/20165301037

Changing of Cations Concentrations in Waters of Polluted Urban River

Maria Andrianova[1,a] and Ekaterina Bondarenko[1]

[1]*Peter The Great Saint-Petersburg Polytechnic University, 195251, St.Petersburg, Russian Federation*

Abstract. Water from urban river Okhta polluted with domestic and industrial wastewaters was investigated. Specific electric conductivity (k), molar concentrations of ions Na+, K+, Mg++ and Ca++, concentration of total nitrogen (TN) were measured in water samples. Increasing of k happened together with increasing of molar fraction of sodium-ion (R_{Na}) among all studied cations (and correspondingly decreasing of molar fractions of other cations). Good correlations were found between R_{Na} and TN ($r = 0.67$), k and TN ($r = 0.84$). The results support the idea of the leading role of wastewaters in changing of k and cations concentrations. Electric conductivity and R_{Na} could be used to distinguish between polluted and not polluted waters in the Okhta.

1 Introduction

Urban rivers are usually influenced by industrial and domestic wastewaters. In 2013 in St. Petersburg about 98.4% of wastewaters were accepted by city canalization system and delivered for treatment to waste water treatment plants. Amount of treated wastewaters was esteemed as 2.18 mln m^3/day [1]. However, water pollution remains possible because of untreated or poorly treated wastewaters discharged directly to rivers of the city in the frames of allowed limits. There is some evidence of underestimation of pollutants amount discharged by industrial enterprises in their annual official reports [2, 3]. Also emissions over allowed discharge limits and illicit disposal of liquid wastes from unknown sources through storm water or drainage canalization (which outlets open directly to rivers) still occur [3, 4]. Such emissions are harmful for fish and other aquatic organisms [5]. They also can influence on water quality at intakes of waterworks and demand additional treatment procedures for drinking water treatment [6 – 9].

It is important to use fast methods of water quality monitoring in order to detect pollution as soon as possible. This will help to find the source of pollution and prevent illicit discharge of wastewaters in future. However, it is difficult to control pollution of river Neva because of its high ability to dilute impurities from wastewaters. Its average flow rate is 80 km^3/year (about 220 mln m^3/day) [10]. Tributaries of the Neva are more vulnerable to pollution and can be more easily controlled because their total flow rate is rather lower (it forms about 2% of the Neva flow rate) [10].

The Okhta is a left-bank tributary of the Neva. It is considered to be the most polluted river in St.Petersburg [3] due to impurities from agricultural, municipal and industrial objects. According to

[a] Corresponding author : maandrianova@yandex.ru

official reports of enterprises city wastewaters contribute for about 20% of its flow rate [11]. This river was chosen for the current investigation.

In order to control water quality measurement of water specific electric conductivity (k) can be used. It takes very little time and can be done on site or by on-line monitoring devices. Electric conductivity has been investigated as marker of contamination from wastewaters discharges [12 – 16]. Specific electric conductivity is considered to be useful in providing screening of the pollution level [15, 16].

For natural surface waters in the North-Western part of Russia (to which the Okhta belongs) small or moderate concentrations of inorganic salts are usual [17]. This is explained by feeding of rivers mainly from snow and rain. Salts from the soil in their basins had been washed out during thousands of years with often rains and meltwaters. Also the significant part of rocks in that region has hardly washable crystal structure [10].

Calcium ions and hydrocarbonates are prevailing over other ions [17, 18] in this region. Sum of concentrations for the main ions responsible for electric conductivity of water (hydrocarbonates, sulfates, chlorides, nitrates, calcium, magnesium, sodium and potassium) in the Okhta vary in the range 60 –140 mg/L [17]. Concentration of dissolved solids (that include ions) in untreated domestic wastewaters is higher and is esteemed as 200 – 1000 mg/L [19, 20]. In St.Petersburg limits for total dissolved solids permitted for discharge to municipal canalization are 1000 mg/L [21]. These values suppose applying conductometry for revealing pollution with domestic wastewaters in the studied water objects.

Previous investigations of the Murinsky creek and the Okhta showed good correlation of water electric conductivity with concentrations of other pollutants characteristic for domestic wastewaters – total nitrogen, inorganic carbon, fluorimetric parameters [22, 23]. Aim of the present work was to study the role of several cations in formation of electric conductivity in polluted urban river Okhta.

2 Materials and Methods

2.1 Water Samples

Water samples were collected from the Okhta. Average flow rate of the Okhta was 5.18 m^3/s and sewerage inflow was esteemed as 1.36 m^3/s [11].

Sampling points were marked as O1 – O17. Outlets of wastewaters were localized according to map of canalization pipelines: several outlets between O10 and O11, O12 and O14. Presence of several outlets was confirmed visually during water collection (O11 and O12, O14 and O15). Scheme of sampling points is given at Fig. 1.

Water samples of about 1 liter were collected from July 2013 to March 2014. Three series of sampling were done. The bottles with the samples were stored in laboratory at +4...+6°C for several days in vertical position while suspended matter was sedimenting. The upper part of water was analyzed.

2.2 Chemical Analysis

Specific electric conductivity of water (k) was measured at the day of sampling by conductometer “HI 8713” (HANNA Instruments) at 20 °C with measurement error ± 5%. Other parameters were determined after storage of samples in the supernatant. Concentrations of inorganic carbon (IC) and total nitrogen (TN) were determined by analyzer “TOC Lvpn-TNM” (Shimadzu, Japan) with measurement error ± 10%. Concentrations of inorganic cations (potassium, sodium, magnesium, calcium) were determined by capillary electrophoresis by “Capel-103R” (Lumex, Russia) with measurement error ± 14%.

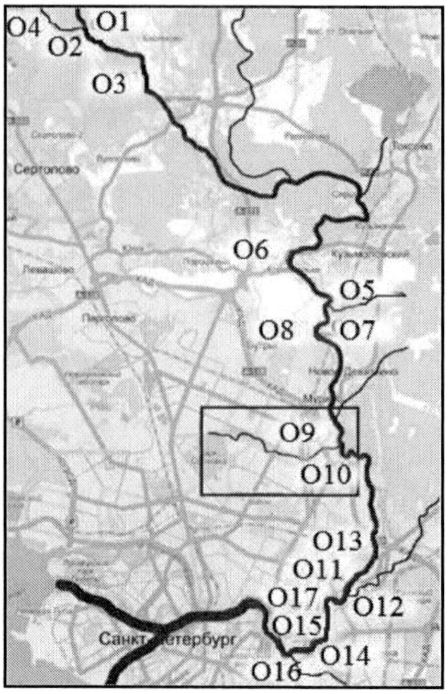

Figure 1. Scheme of sampling points at the Okhta. The Murinsky creek is shown in rectangular frame.

3 Results and Discussion

In order to esteem background values for natural (not polluted) waters of the Okhta data from [17] could be used. They include results of regular chemical analysis of water collected at various seasons from the Okhta before (upstream) St.Petersburg at observation station near village Novoye Devyatkino. Sampling point O7 of the present research is located near that station. Data in Table 1 show that during periods of low and high water flow rate concentrations of main ions changed not more than one order of magnitude [17].

Table 1. Ranges of main ions concentrations in the Okhta at various seasons in 1946-1965 [17].

Concentration	Ca^{2+}, (mg/L)	Mg^{2+}, (mg/L)	K^{+}+Na^{+}, (mg/L)	HCO_3^-, (mg/L)	Cl^-, (mg/L)	SO_4^{2-}, (mg/L)	NO_3^-, (mg/L)	NO_2^-, (mg/L)
Okhta	4.6 – 9.6	0.3 – 4.0	4.0 – 29.0	5.5 – 56.7	4.3 – 16.3	10.9 – 22.9	1.0 – 3.0	0.02 – 0.47

Table 2. Ranges of parameters of water samples from the Okhta in 2013-2014.

Concentration	Ca^{2+}, (mg/L)	Mg^{2+}, (mg/L)	K^{+}, (mg/L)	Na^{+}, (mg/L)	IC, (mg/L)	HCO_3^-, (mg/L)	TN, (mg/L)	k, (mkSm/cm)
Okhta	14 – 196	12 – 857	3 – 32	14 – 289	3 – 38	14 – 192	0.3 – 4.5	40 – 725

Chemical analysis of samples from the Okhta in the present work showed that concentrations of cations and specific electric conductivity of water (k) varied 1–2 orders of magnitude with time and place. Minimal and maximal values are given in Table 2. Maximal values in Table 2 were significantly higher than in Table 1, demonstrating influence of pollution from the city on water content. Selected data for the Okhta are shown at Fig. 2. It can be seen from Fig. 2 and other data that molar concentration of sodium was maximal among other inorganic cations.

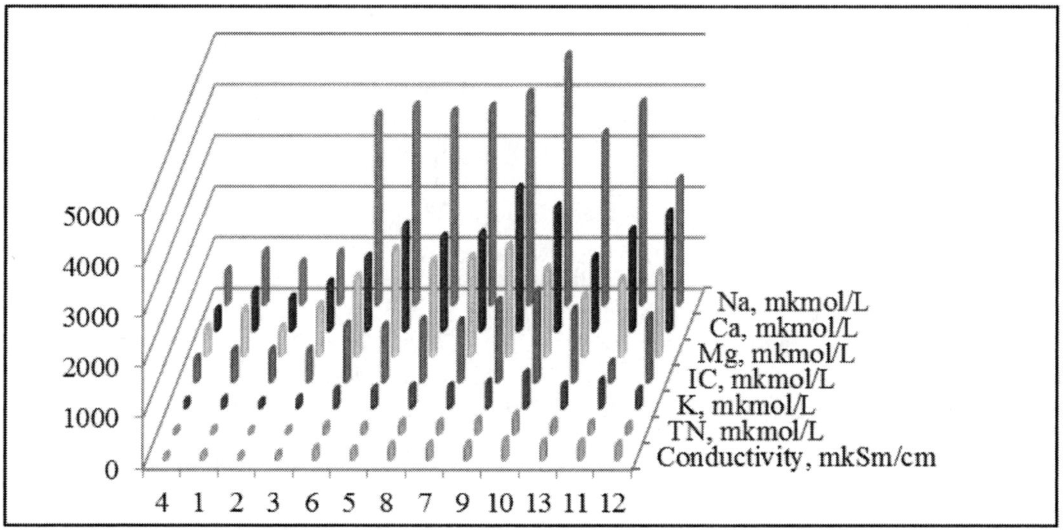

Figure 2. Parameters of the Okhta water samples from one series. Numbers of sampling points (on x axis) are placed in order of water flow.

In Table 3 molar ratios of four inorganic cations are shown for selected points of sampling. The ratios were calculated according to the formula

$$R_i = (C_i \cdot 100\%) / (C_K + C_{Na} + C_{Mg} + C_{Ca}) \quad (1)$$

where C_i - concentration of i-ion (mol/L), C_K , C_{Na} , C_{Mg} , C_{Ca} - concentrations of corresponding ions (mol/L).

Table 3. Ratios of inorganic cations (K, Na, Mg or Ca) molar concentration to sum of inorganic cations molar concentrations (in %) and specific electric conductivity k (in mkSm/cm) in water samples from the Okhta

Sample	September 2013					November 2013					March 2014				
	R_K	R_{Na}	R_{Mg}	R_{Ca}	k	R_K	R_{Na}	R_{Mg}	R_{Ca}	k	R_K	R_{Na}	R_{Mg}	R_{Ca}	k
3	4	33	33	30	68	-	-	-	-	-	3	27	37	33	54
6	4	53	22	20	175	-	-	-	-	-	4	47	24	25	141
8	5	48	24	23	238	7	41	24	29	122	-	-	-	-	-
7	5	48	24	23	244	7	41	23	29	123	-	-	-	-	-
9	5	44	23	29	252	6	33	23	38	143	4	37	25	34	186
10	7	51	18	25	334	4	58	16	22	725	4	53	18	24	263
11	6	50	19	25	296	5	39	25	32	229	4	42	21	32	252
12	5	37	24	34	254	6	46	20	28	231	4	40	25	31	239
15	-	-	-	-	-	5	44	22	29	295	5	44	24	27	281

It can be seen from Table 3 that increasing of water conductivity led to increasing of molar fraction of sodium and in most cases decreasing of molar fractions of other cations. This effect is supposed to be a result of pollution with sewage in which sodium usually prevails over other inorganic cations [24]. The most notable effect was between samples O3 and O6, O9 and O10. Sampling point O3 is placed in a forest and sampling points O6, O8, O7 – after big settlements. Sampling points O9 and O10 are situated near the boundary of St.Petersburg before and after inflow of the Murinsky creek, which is heavily polluted with domestic wastewaters. However, increasing of water conductivity could be explained also by changing of soils and ground water properties as it was done for river Volga in [25]. But for river Okhta we also observed decreasing of parameters downstream the river flow. It could be explained by dilution with less polluted water from other tributaries. In general the received data support the idea of the leading role of wastewaters in changing of water electric conductivity and cations concentrations for the studied river. Further analysis of anions concentrations could give more proof to this idea because chlorides are known to be dominating in domestic wastewaters [15].

In order to find another proof of this idea Pearson correlation coefficients (r) were calculated between parameters connected with cations concentrations and TN – a parameter characteristic for wastewaters. The values were the following: 0.84 for k and TN; 0.67 for R_{Na} and TN; 0.24 for R_K and TN; (–0.78) for R_{Mg} and TN; (–0.28) for R_{Ca} and TN. High positive values for k and R_{Na} demonstrate that electric conductivity and R_{Na} could be used to distinguish between polluted and not polluted waters in the Okhta.

4 Conclusions

Specific electric conductivity (k) and concentrations of four inorganic cations (potassium, sodium, magnesium, calcium) and total nitrogen (TN) were determined in water samples of urban river Okhta polluted with domestic and industrial wastewaters from St.Petersburg. Correlation coefficients (r) were calculated between parameters connected with cations concentrations and TN – a parameter characteristic for wastewaters.

The data obtained showed that increasing of k in samples happened together with increasing of molar fraction of sodium-ion (R_{Na}) and in most cases decreasing of molar fractions of other cations. Good correlations were found between R_{Na} and TN (r = 0.67), k and TN (r = 0.84).

In general the data support the idea of the leading role of wastewaters in changing of water electric conductivity and cations concentrations for the studied river. Electric conductivity and R_{Na} could be used to distinguish between polluted and not polluted waters in the Okhta.

References

1. *Annual report-2014 of Vodokanal St.Petersburg*, 250p. (2015)2
2. E.V Kolesnikova, V.A Shelutko, Vestnik of Saint Petersburg university. Series 7. Geology, Geography, **Vol. 3**, pp. 81-88 (2008)
3. D.A Golubev, N.D Sorokin *Environmental protection, natural resources and environmental safety in St. Petersburg in 2010*, 434 p. (2011)
4. V.N.Chechevichkin, N.I.Vatin, Applied Mechanics and Materials, 641-642, pp. 409-415 (2014)
5. M.Daniel, A.Monteneblo, M.Bernardes, J.Ometto. P.DeCamargo, A.Krusche, M.Ballester, R.Victoria, L. Martinelli, Water, Air, and Soil Pollution, **136,** pp. 189–206 (2002)
6. *Water supply and wastewater discharge in St.Petersburg* 318p. (2008)
7. L.M.Molodkina, D.D.Kolosova, E.I.Leonova, M.F.Kudoyarov, M.Y.Patrova, Y.V.Vedmetskii, Petroleum Chemistry, **52 (7)**, pp. 487-493 (2012)
8. A.Kim, N.Chernikov, Applied Mechanics and Materials, 725-726, pp. 1338-1334 (2015)
9. N.I.Vatin, V.N.Chechevichkin, A.V.Chechevichkin, E.S.Shilova Magazine of Civil Engineering **2(37),** pp.81-88. (2013)
10. R.A Nezhihovskij, *River Neva and Nevskaya Guba* Gidrometeoizdat, p. 112 (1981)
11. *Wastwater discharge and treatment in St.Petersburg* Stroyisdat, 418 p. (2002)
12. D. Chalupová, P.Havlíková, B. Janský, Environmental Monitoring and Assessment, **184**, pp.6283–6295 (2012)
13. M.Y.Thompson, D.Brandes, A.D.Kney, Journal of Environmental Management, **104**, pp. 152–157 (2012)
14. A.J. Stewart, Environmental Management, **27**, pp. 37–46 (2001)

15. D.N.R. De Sousa, A.A. Mozeto, R.L. Carneiro, P.S. Fadini, Science of The Total Environment, **484**, pp. 19-25 (2014)
16. F.Bonvin, R.Rutler, N.Chèvre, J. Halder, T. Kohn, Environmental Science and Technology, **45 (11)**, pp. 4702–4709 (2011)
17. *Resursy poverkhostnyh vod SSSR.* **Tom 2**, Gidrometeoizdat, pp. 463-518 (1972)
18. E.Ya.Yakhnin, S.G.Gumen, E.L.Proletarskaya, A.V. Shubin, I.V.Sycheva, Ecologicheskaya Khimiya, **8 (3)**, pp. 145-154 (1999)
19. M.L. Davis, *Water and wastewater engineering*, McGraw-Hill, pp. 18-1– 18-9 (2010)
20. E. Gogina, N. Makisha, Applied Mechanics and Materials, 361-363, 2013, pp. 628-631(2013)
21. *Govermnent of St.Petersburg. Regulation of committee on energy and engineering from 4 June 2014 # 81*, 5 p. (2014)
22. A.N. Chusov, E.A. Bondarenko, M. Ju. Andrianova, Applied Mechanics and Materials, 641-642, pp. 1172-1175 (2014)
23. M. Ju.Andrianova, E.A.Bondarenko, E.O.Krotova, A.N. Chusov, Proceedings of IEEE Workshop on Environmental, Energy and Structural Monitoring Systems, September 17-18, 2014, Naples, Italy, pp. 198-202 (2014)
24. G.Tchobanoglous, F.L. Burton, *Wastewater Engineering. Treatment, Disposal, Reuse*, McGraw-Hill, 1820 pp.. (1991)
25. I.A. Nemirovskaya, Water Resources, **39 (5)**, pp. 533–545 (2012)

MATEC Web of Conferences **53**, 01038 (2016)
DOI: 10.1051/matecconf/20165301038

Sandwich Panel as a Structural Element of Overlap

Maxim Novikov[1a], Vyacheslav Polushin[2] and Sergey Zimin[3]

[1, 3] *St. Petersburg State Polytechnical University, Politekhnicheskaya, 29, Saint-Petersburg, 195251, Russia*
[2] *OOO "Yoris Ide", Kooperativnaya, 1A, Leningrad region, Gatchinsky district, 188349, Russia*

Abstract. This paper considers the issue of sandwich panels using as load-bearing structural elements. The comparison of deflections and critical failure loads were obtained by the results of the full-scale roof sandwich panels tests conducted by the company "Joris Ide" and the theoretical design, according to the calculation method described in Euronorms. Based on these results it was concluded that sandwich panels can be treated as a load-bearing structure only with more taught manufacturing requirements. Thus, the reduced spread of critical loads can be achieved.

1 Introduction

Ceiling is very important constructive element in building, that's why it will be always actual, to find optimal solution in this sphere.

In our days reinforced concrete is the most popular material for construction ceiling. It is strong enough, durable and fire-resistant. But it has some shortcomings, the main shortcoming is own weight load comparing with effective load, that constriction perceive. In addition, when the premises are presented the individual requirements of the thermal insulation to the own weight of overlap added weight envelopes. That's why it is not rationally to *use* it when requirements to ceiling aren't high, otherwise it could cause increase of load on bearing elements, and hence cause increasing of cross-section of columns, beams, size and price of foundation. All this greatly affects the cost of the building as a whole.

Eventu2 ally arise a number of cases, when it must be taken another material, which also meet the requirements of strength, fire-resistant, thermal conductivity (if such requirements imposed), but it would had significantly lower weight. Sandwich panels could be used as such material, which can combine functions of both constructive elements and walling.

Sandwich panels - frameless three-layered product, in which the individual layers are glued together. For external and internal layers used galvanized steel sheet coated, profiled or smooth, and for the middle layer - insulation of expanded polystyrene or mineral fiber [1].

The question of the sandwich panels, most often comes up when discussing the construction of prefabricated buildings. Usually it is the building for production purposes, the construction of which is carried out continuously. But despite the obvious relevance of use of this material is limited. This is due to the fact that the thermal engineering calculation is widely studied [2-11], which allows us to use sandwich panels as walling. However issues of technological and regulatory plan and strength calculation remain open and require active discussion [12].

In Russian regulatory literature, there are no calculation method by which with reasonable certainty can be assessed strength properties of sandwich panels. But if you try to start using this material as the load-bearing structural elements it is necessary to enter of such a method. According to EA, Petrova [13] in the

[a] Corresponding author: maximnn1720@gmail.com

European Union for over 20 years, there are regulations dedicated to this issue, but before the application of these instruments on the territory of Russia, they need to be adapted to the realities of our country.

Nomenclature

v – the deflection of the beam
q – the uniformly distributed load
L – the length beam
E – the Young's modulus of face
I – the bending stiffness
x- the coordinate cross-section of longitudinal axis
S – the shear stiffness
M – the bending moment
γ – the slip
σ_{F1}, σ_{F2} – the normal stress of face
e – the distance between the centroid of the faces
A_f- the square of face's cross-section
τ_c – the shear stress in the core
V – the shear force of the beam
B – the width (depth) of the beam
M_{max}- the maximum bending moment of length beam
V_{max} - the maximum shear force of length beam
R_A, R_B- support reactions
q_1- the limit load of a sandwich panel, from the calculation of strength and wrinkling of faces
q_2 – the limit load on the sandwich panel, the calculation of shear strength of the core material
q_3 – the limit load on the sandwich panel, the calculation of compressive strength of the core material
$[\sigma]$, $[\tau]$ – the limit stress
l_{sup} - the width of the support
q_{cr} - the limit load is minimum of q_1, q_2, q_3
γ_M – the relevant material factor

2 Review of the literature

Review of the national regulatory literature showed no documents, allowing to make the calculation of a "sandwich" - panels. The only exception is the recommendation [14], which are not mandatory and that not all options are taken into account fracture mechanisms, in particular shear failure of the middle layer, the middle layer of compressive, shear strength in the long-term load.

However, in the literature can be found works, dedicated to search for optimal methods of calculating both the "sandwich" - panels, and nodes of their interface with other design elements, such as OA Tusnina in his articles [15,16] described the method of calculation of conjugation z-shaped run with a "sandwich" - panel. Mircea Georgescu [17] and Marcin Kujawa [18] are also exploring this problem. Kahina Saad Saoud [19] considers the problem of general and local loss of compressed facing. Robert Studziński [20] explores the impact of microprofiling of facing on the bearing capacity of sandwich panels. Bronisław Kosowski [21] in his article offers calculation of multispan sandwich panels with intermediate supports elastic pliability under the influence of uniformly distributed load and stress caused by the difference in temperature.

The presence of scientific works confirms the relevance of the question.

S. Petrov in his article [22] gives the experimental results which show that the work of sandwich panels can be represented as a beam, working on a bend of the uniformly distributed load with uniform distribution of the normal stresses in the facing to the width of the panel, considering that the average a layer of insulation is works only for shear and due to local deformation of the facing of the reaction steps, it is necessary to set the support is elastic pliability. In [23] he described a theoretical method for determining the stress-strained state of sandwich panels using these assumptions. The given experimental data confirm the validity of calculations. However, the described method time consuming and takes quite a lot of time, for this reason, the author was forced to write a program [24], allowing to make the calculation more quickly. But this approach also has drawbacks, such as the program describes a finite number of individual cases, so it does not turn out to apply in every situation. Therefore European standards [25, 26], offer a method of calculation for the simplified design scheme. It is assumed that you can ignore the impact of compliance supports, if you enter a check for a deformation of the metal supports. This approach greatly simplifies the calculations while the results are still quite accurately coincide with the full-scale tests.

3 Statement of the Problem.

The purpose of this paper is to analyze the proposed European standards methodology for assessing stress-strained state of sandwich panels and determine the characteristics that should be taken into account when considering the material as a carrier component.

4 Materials and Methods

4.1 Experimental studies.

Tests of wall sandwich panels MEGA made by order of the company "Joris Ide", were conducted in accordance with [27], the purpose of which was to determine the strength of the sample panels at lateral bending. According to [23]. The problem can be reduced to the calculation of the beam pattern, as the stress across the width in the same section does not vary significantly, so the model was a one span slab-beam with a uniformly distributed load applied over the entire width of the panel. The load was applied gradually. After each stage of the application load, the testing product is kept under load till full stabilization of the deflection for 5 minutes. During this time, the value of the load was kept constant.

Were tested sandwich panels MEGA brands SP-Z-100-MV and MEGA SP-Z-150-MV. Three samples of each brand were tested for spans of 4 and 6 meters. The thickness of the metal facings 0.5 mm, insulation of mineral wool brand "Tehnosendvich C", a two-component polyurethane glue Daltofoam 44203-Suprasek 2026. To measure the deflection in the middle of the panels and deformation of the supports and displacement in the longitudinal direction of the facings six deflection meters PAO 6 were install, one at mid-span, two devices in the immediate vicinity of the supports and one on the end panel.

The samples were brought to the exhaustion of the panels bearing capacity (to failure). According to data of tests plotted dependence of deflection at mid-span of the applied load.

4.2 Theoretical studies.

In [26] proposed a method for calculation of sandwich panels as a result of which can be obtained analytically plots deflection at mid-span of the applied load and compare with the data of full-scale tests.

According to this method, the middle layer materials and facings for the given range of deformation are linear elastic. Rigidity in tension of the middle layer as compared with the rigidity of the tensile facings so small that the influence of the normal stress in the middle layer can be disregarded. Because of these assumptions, the bearing capacity of sandwich panels with thin and smooth or slightly profiled facings running on pure bending can be divided into two components - the carrying capacity of facings, sensing compressive and tensile forces from the bending moment and the carrying capacity of the middle layer, the sensing stress shift. As a result, according to [26] equation of the bent-axis will be by formula 1:

$$-EIv'' = M - EI\gamma' \tag{1}$$

In the case of a uniformly loaded one span beam, after integration, we obtain an expression for calculating the values of the movement:

$$v = \frac{1}{24}\frac{qL^4}{EI}\left[\left(\frac{x}{L}\right)^4 - 2\cdot\left(\frac{x}{L}\right)^3 + \left(\frac{x}{L}\right)\right] + \frac{q}{2S}\left(Lx - x^2\right) \tag{2}$$

The characteristic values of resistance for sandwich panels are shown in Table 1

Table 1. Normative values of resistance.

Normative values of resistance	Values
Yield strength of the faces	230 MPa
Compressive strength of the core materials	0.06 MPa
Wrinkling strength of the faces	90.51 MPa

The normal stress in the faces can be determined from the expression 3, according [25]

$$\sigma_{F1} = \sigma_{F2} = \frac{M}{e\cdot A_f} \tag{3}$$

The shear stress in the core is calculated by formula 4, according [25]

$$\tau_c = \frac{V}{e\cdot B} \tag{4}$$

Maximum bending moment, shear force and values support reactions for case one-span sandwich-panel with uniformly distributed load can be determined from the expressions 5, 6

$$M_{\max} = \frac{qL^2}{8} \tag{5}$$

$$V_{mqx} = R_A = R_B = \frac{qL}{2} \tag{6}$$

Therefore, it is possible to obtain the values of critical load:

1. For case of normal stress in the faces by formula 7

$$q_1 = \frac{[\sigma]\cdot 8eA_f}{L^2} \tag{7}$$

For case of shear stress in the core by formula 8

$$q_2 = \frac{[\tau]\cdot 2eB}{L} \tag{8}$$

2. For case of compressive stress in the core by formula 9

$$q_3 = \frac{2 \cdot l_{\text{sup}} \cdot B \cdot [\sigma]}{L} \tag{9}$$

5 Results and Discussion

The values of critical loads, taking into account the relevant material factor shown in Table 2.

The dependences deflection of the beam from uniformly distributed load, according to data of test sandwich-panels and theoretical calculation are presented at figure 1, 2, 3.

Table 2. The value of critical loads

	The length of span	Limit load, q_{cr}	The damage mechanisms
MEGA SP-Z-100-MV	L = 4 m	2.274 $kN/m2$ (1.819 $kN/m2$)	Wrinkling of compressed face in the span (γM=1.25)
	L = 6 m	1.011 $kN/m2$ (0.809 $kN/m2$)	Wrinkling of compressed face in the span (γM=1.25)
MEGA SP-Z-150-MV	L = 4 m	3.00 $kN/m2$ (2.40 $kN/m2$)	Compression of the core over a support (γM=1.4)
	L = 6 m	1.514 $kN/m2$ (1.211 $kN/m2$)	Wrinkling of compressed face in the span (γM=1.25)

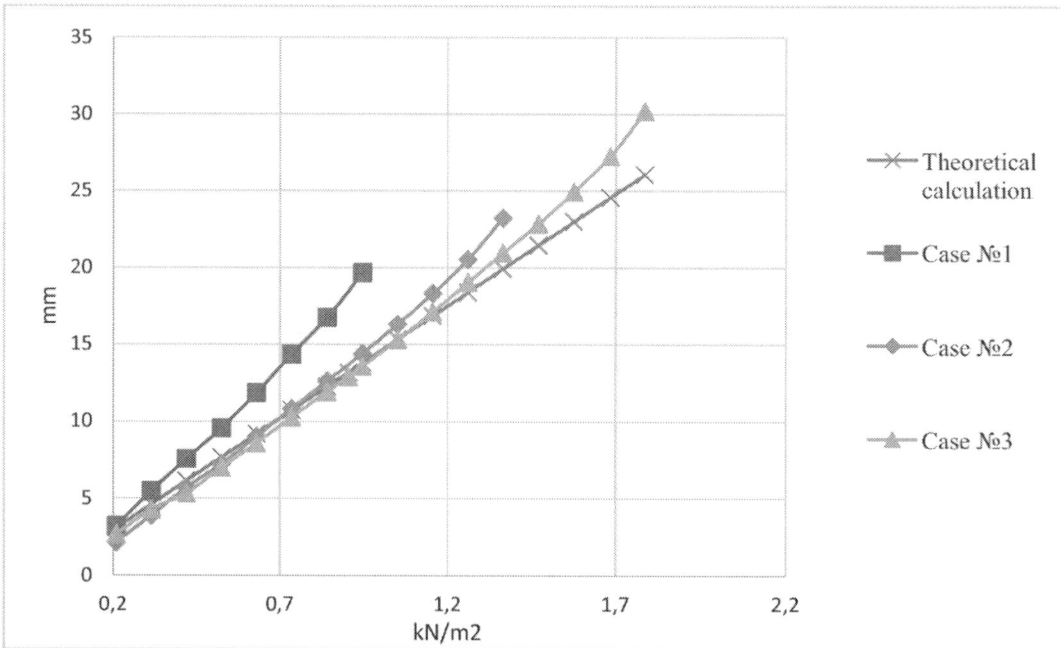

Figure 1. The dependence deflection of the beam from uniformly distributed load, according to data of test sandwich-panels MEGA SP-Z-100-MV with 4 m span and theoretical calculation

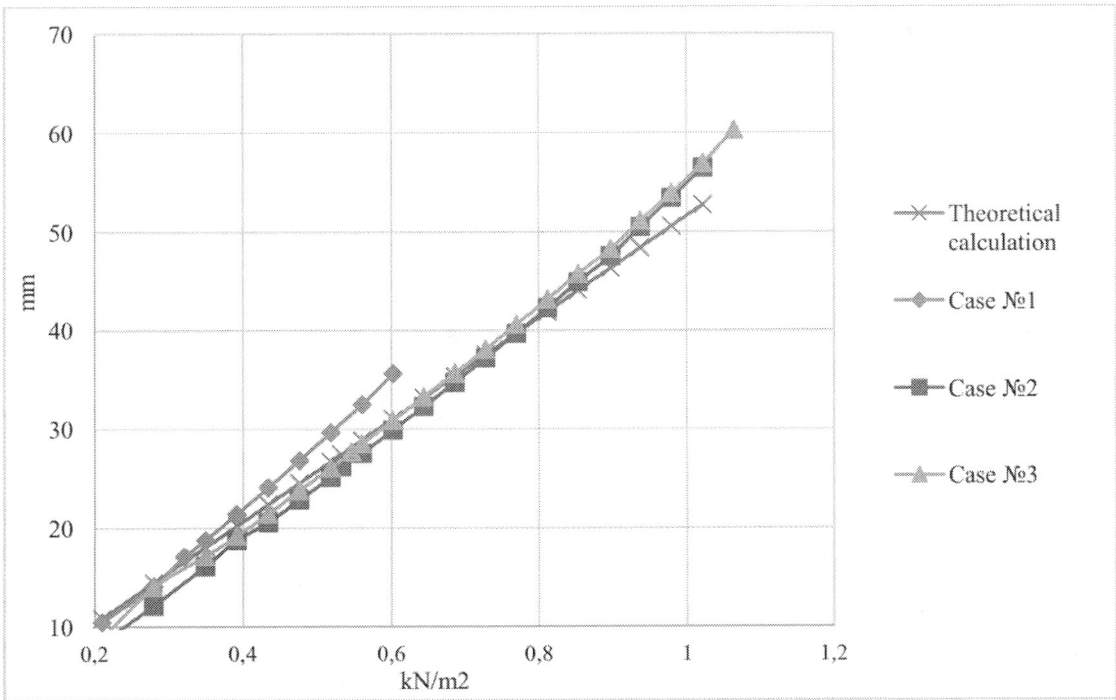

Figure 2.. The dependence deflection of the beam from uniformly distributed load, according to data of test sandwich-panels MEGA SP-Z-100-MV with 6 m span and theoretical calculation

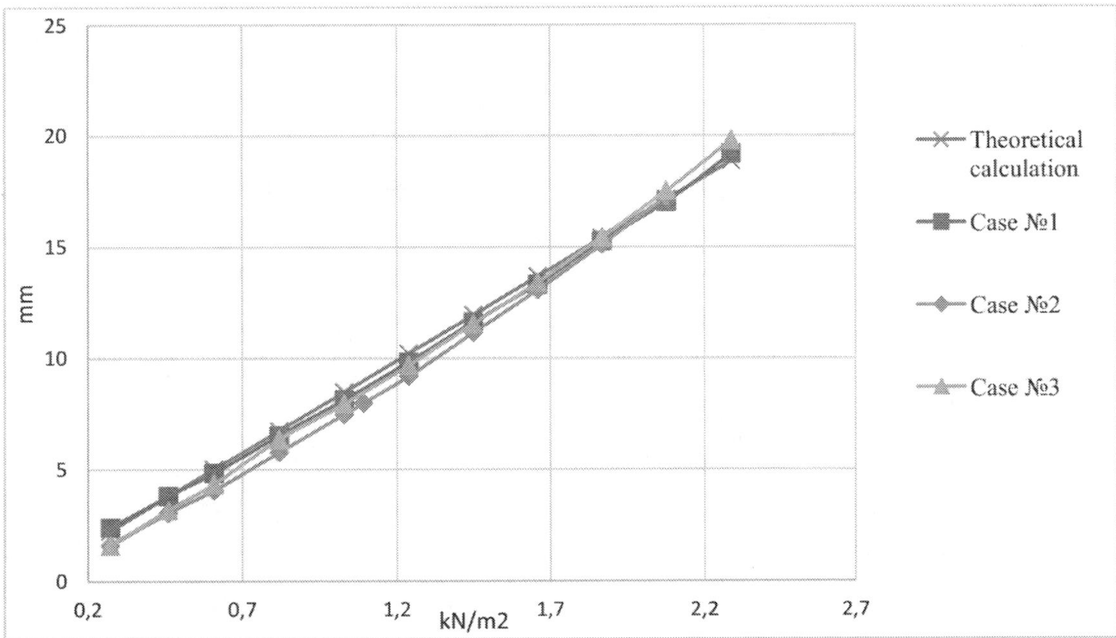

Figure 3. The dependence deflection of the beam from uniformly distributed load, according to data of test sandwich-panels MEGA SP-Z-150-MV with 4 m span and theoretical calculation

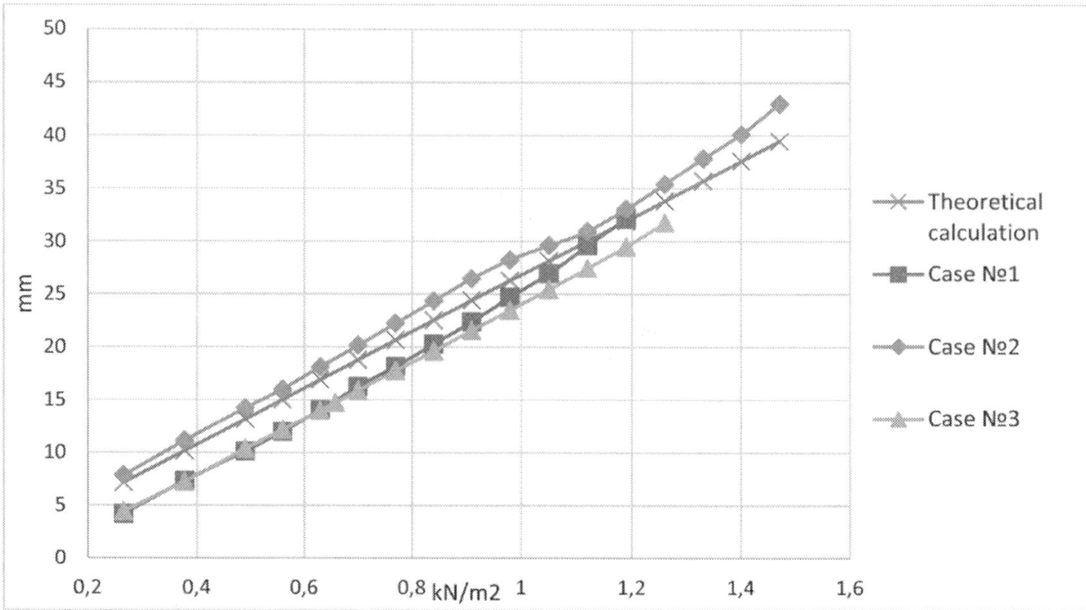

Figure 4. The dependence deflection of the beam from uniformly distributed load, according to data of test sandwich-panels MEGA SP-Z-150-MV with 6 m span and theoretical calculation

Table 3. The value of critical loads and damage mechanisms made by theoretical and actual cases

	The length of span	Theoretical critical load, q_{cr}	The damage mechanisms	Actual critical load, q_{cr}	The damage mechanisms
MEGA SP-Z-100-MV	L = 4 m	2.27 $kN/m2$ (1.82 $kN/m2$)	Wrinkling of compressed face in the span (γm=1.25)	1.05 $kN/m2$	Wrinkling of compressed face in the span at a distance of 3/8 of the support
				1.47 $kN/m2$	Wrinkling of compressed face in the span at a distance of 300 mm of the support
				1.81 $kN/m2$	Wrinkling of compressed face in the span at a distance of 500 mm of the support
	L = 6 m	1.01 $kN/m2$ (0.81 $kN/m2$)	Wrinkling of compressed face in the span (γm=1.25)	0.64 $kN/m2$	Wrinkling of compressed face in the span at a distance of 3/8 of the support
				1.06 $kN/m2$	Wrinkling of compressed face in the middle part in the span
				1.09 $kN/m2$	Wrinkling of compressed face in the span at a distance of 3/8 of the support
MEGA SP-Z-150-MV	L = 4 m	3.00 $kN/m2$ (2.40 $kN/m2$)	Compression of the core over a support (γm=1.4)	2.48 $kN/m2$	Compression of the core over a support
				2.23 $kN/m2$	Compression of the core over a support
				2.48 $kN/m2$	Compression of the core over a support
	L = 6 m	1.51 $kN/m2$ (1.21 $kN/m2$)	Wrinkling of compressed face in the span (γm=1.25)	1.26 $kN/m2$	Wrinkling of compressed face in the middle part in the span
				1.54 $kN/m2$	Wrinkling of compressed face in the middle part in the span
				1.32 $kN/m2$	Wrinkling of compressed face in the middle part in the span

The results of the experiment it is possible to create a table with the actual breaking load for each case and compare it with the results obtained by means of calculation (Table 3). It should be noted that in each of these tests delamination facing from insulation was observed, whereby facing lost it's resistance and failure occurs. Also the table shows that, the data theoretically obtained on the strength are overstated and failure occurs at lower loads, but the load carrying capacity "sandwich" - panels enough to use them as building envelopes. To use of sandwich panels as constructive bearing elements, it is necessary to reduce the variation of critical loads. The table shows that most of the panels can bear load that stated on the calculation, but in order to ensure the strength of the panels is necessary to understand why the calculated bearing capacity is overstated. To do this, pay attention to the feature of fracture, namely, the peeling of insulation from the facing. This may mean that the glue does not provide sufficient compound of components of panel or insulation is not sufficient shear strength in the longitudinal plane. When making more stringent requirements at the manufacturing stage, in particular, can be used methods of nondestructive testing [18] to control of mechanical damage actually achieve stability in the destruction, then should adjust the calculated method of reliability coefficients of material and tests to confirm the results. After these measures, it will be possible to recommend the use of "sandwich" - panels as bearing structures in practice.

On the other hand, it is possible to approach the problem from the new constructive solutions, in particular as an example of technological solutions of this problem can be leaded work of H. Kim [29] or Petrova EA [30], which proposed to strengthen the construction of sandwich panels by reinforcing elements.

6 The conclusion

As a result, were analyzed the method of estimating the stress-strain state calculation of sandwich panels proposed by source [26]. It can be concluded that the method of sufficiently accurately describes the behavior of this material, however due to volatility characteristics of each individual panel is currently not possible to ensure sufficient carrying capacity of this material for using as a structural load-bearing elements. It should be the introduction of more stringent requirements for the production of sandwich-panels in the case of their application as load-bearing structures to solve the problem.

References

[1] TY 5284-003-79850813 (2010)
[2] A. Gorshkov, N. Vatin, D. Nemova, A. Shabaldin, L. Melnikova, P. Kirill, Procedia Engineering, **117 (1)**, 1085-1094 (2015)
[3] N. Harmati, Z. Jakšić, N. Vatin, Procedia Engineering, **117 (1)**, pp. 791-799 (2015)
[4] N. Vatin, A. Gorshkov, P. Rymkevich, D. Nemova, D. Tarasova, Applied Mechanics and Materials, **670-671**, 365-369 (2014)
[5] N. Vatin, A. Gorshkov, A. Kazimirova, K. Gureev, D. Nemova, Journal of Applied Engineering Science, **12 (3)**, 207-216 (2014)
[6] M. Penić, N. Vatin, V. Murgul, Applied Mechanics and Materials, **680**, 534-538 (2014)
[7] V. Pukhkal, M. Tanić, N. Vatin, V. Murgul, Procedia Engineering, **117**, 864-869 (2015)
[8] E. Aronova, N. Vatin, V. Murgul, Procedia Engineering, **117 (1)**, 771-779 (2015)
[9] M. Tanic, D. Stankovic, V. Nikolic, M. Nikolic, D. Kostic, A. Milojkovic, S. Spasic, N. Vatin, 2015, Procedia Engineering, **117 (1)**, 924-937 (2015)
[10] N.I. Vatin, D.V. Nemova, D.S. Tarasova, A.A. Staritcyna, Advanced Materials Research, **953-954**, 854-870 (2014)
[11] N.I. Vatin, D.V. Nemova, A.S. Kazimirova, K.N. Gureev, Advanced Materials Research, **953-954**, 1537-1544 (2014)
[12] V. Ermoshkin, S. Dymovskih, I. Vojlokov, S. Yakubov, StrojProfil, **2(72)**, 33-34 (2009)
[13] E. Petrova, O. Kalmykov, Gorodskoe kommunalnoe hozyajstvo, 38-44 (2015)
[14] Recommendations on determination, CNIIPSK them Mel'nikova, 1-44 (2013)
[15] O. Tusnina, M. Hejnisuo, Promyshlennoe i grazhdanskoe stroitel'stvo, 67-70 (2012)
[16] O. Tusnina, A. Danilov, Vestnik MGSU, 26-36 (2014)
[17] M. Georgescu, V. Ungureanu, Thin-Walled Structures, 242-249 (2014)
[18] M. Kujawa, C. Szymczak, Thin-Walled Structures, 43-52 (2014)
[19] K. Sad Saoud , P. Le Grognec, Thin-Walled Structures, 13-23 (2014)
[20] R. Studziński, Z. Pozorski, A. Garstecki , Journal of Constructional Steel Research, 227-234 (2014)
[21] B. Gosowski, M. Gosowski, Journal of Constructional Steel Research, 53-60 (2014)
[22] S. Petrov, E. Il'diyarov, N. Popkov, I. Holopov, M. Mosesov, A. Solov'yov, 44-47 (2009)
[23] S. Petrov, Stroitel'naya mekhanika i raschyot sooruzhenij, 27-33 (2010)
[24] I. Holopov, S. Petrov, Promyshlennoe i grazhdanskoe stroitel'stvo, 36-40 (2013)
[25] Self-supporting double skin metal faced insulating panels, EN 14509, (2013)
[26] M. Davies, CIB Working Commission, 195-204 (2001)
[27] Metal panels with insulation foam. General specifications, GOST 32603, (2012)
[28] L. Pieczonkaa, F. Aymerichb, W. J. Staszewskia, International Symposium on Dynamic Response and Failure of Composite Materials, 216–221 (2014)
[29] H. Kim, B. Hyun Cho, H. Hur, K. Kang, Materials & Design, 231-242 (2015)
[30] E. Petrova, Collection of scientific articles, 26-43 (2015)

MATEC Web of Conferences **53**, 01039 (2016)
DOI: 10.1051/matecconf/20165301039

Use of CAE-Systems at Evaluation of Shock Absorbers for Metallurgical Equipment

Viktor Artiukh[1,a], Talzhan Raimberdiyev[2] and Vladlen Mazur[3]

[1]*St. Petersburg State Polytechnical University, Polytechnicheskaya, 29, Saint-Petersburg, 195251, Russia*
[2]*Hoja Akhmet Yasawi International Kazakh-Turkish University, Sattarkhanova, 29, Turkestan, 161200, Kazakhstan*
[3]*LLC "Saint-Petersburg Electrotechnical Company", Pushkin, Parkovaya, 56, Saint-Petersburg, 196603, Russia*

Abstract. Work presents results of strength calculations of absorber elastic element that were done by CAE-system and standard method of calculating. It is shown, that use of FEM software allows studying thoroughly the distribution of appropriate stresses on surfaces of elastic element.

1 Introduction

At current economic conditions it is necessary to design new equipment during short terms in order to increase its competitiveness. There are a lot of different features of FEM calculations done for big complex and sometimes unique details [1–5]. Mathematical model (MM) which is a basis for any technical task in CAE-system (Computer – Aided Engineering system) should be universal with possibility to change numerous parameters. The most popular MM is one which is created by means of CAD-system (Computer – Aided Design system) with FEM [6–9].

2 Materials and Methods

According to authors of this article, preferred CAE-system for making design and stress calculations of metallurgical equipment might be SolidWorks – COSMOSWorks. In addition, some key producers of metallurgical heavy machinery use SolidWorks – COSMOSWorks [10, 11]. Also, it is well known about stress calculations of details from different steel grades [12–14] but there is a lack of information about use of FEM for stress calculations of absorbers elastic elements of metallurgical equipment [15–18].

Example of elastic elements FEM stress calculation is given below. It is done for elastic strip (polyurethane grade adiprene L-167), which is a part of absorber for equipment base installed on foundation. Sketch of elastic strip is shown on fig. 1.

[a] Corresponding author : artiukh@mail.ru

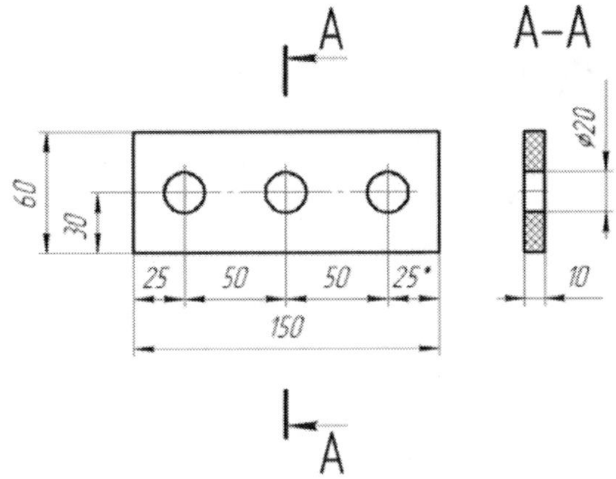

Figure 1. Sketch of the elastic strip from polyurethane grade adiprene L-167.

MM was created in SolidWorks – COSMOSWorks for making stress calculation of the elastic strip from polyurethane grade adiprene L-167. The MM strip has particular design and material properties (elastic modulus E = 60 MPa, Poisson's ratio $\mu = 0{,}495$). The elastic strip undergone by compressive force.

Maximum loading of the strip during usage of absorber is equal to its uniform shrink by 2 mm. Scheme of the strip loading is shown on fig. 2.

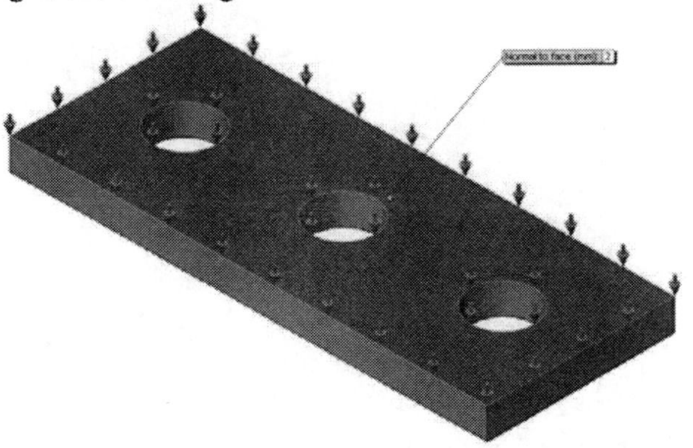

Figure 2. Scheme of the absorber strip loading by deformation in MM.

Boundary conditions were done according to recommendations [12] such as zero movement of surfaces normal to their face (refer to fig. 3).

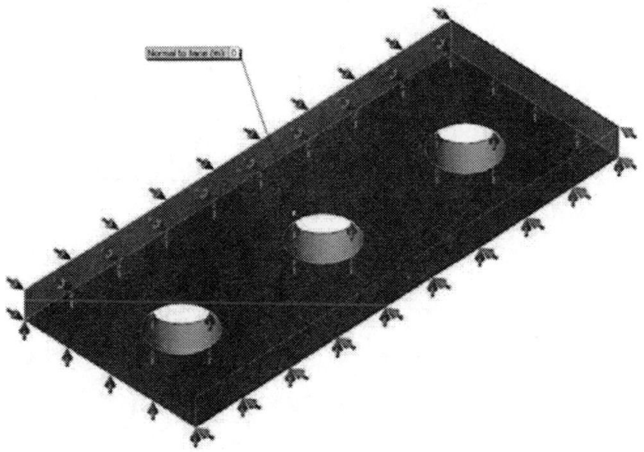

Figure 3. Loading scheme of the strip.

Stress calculation sequence of the absorber strip was done according to recommendations [7, 11], that made it possible to clarify dimensions of elements equal to 1mm with tolerance of 0.05mm, amount of mesh nodes is 703049; amount of mesh elements is 494837.

Calculation results of normal stress in X-axis (SX direction) are shown on fig. 4, where maximum value of normal stress $\sigma_{SX\ \max\ MM}$ is in node number 693380 and $\sigma_{SX\ MM}^{693380} = -22,65$ MPa. It means that $\sigma_{SX\ \max\ MM} = \sigma_{SX\ MM}^{693380}$.

It is clear that nodes number 693380 and number 702954 are positioned symmetrical in reference to the central hole of the strip. Hence, accuracy of the stress calculation of MM is

$$\Delta_{MM} = \frac{\sigma_{SX\ MM}^{693380} - \sigma_{SX\ MM}^{702954}}{\sigma_{SX\ MM}^{693380}} \cdot 100, \tag{1}$$

where $\sigma_{SX\ MM}^{702954} = -22,02$ MPa – value of normal stress in node number 702954.

Among shown stress calculation results on fig.4 the biggest value of normal stresses in X-axis on lateral surface (LS) of the strip $\sigma_{SX\ \max\ MM}^{LS}$ is in node number 1451 and it is equal to $\sigma_{SX\ MM}^{LS\ 1451} = -7,797$ MPa. It means that $\sigma_{SX\ \max\ MM}^{LS} = \sigma_{SX\ MM}^{LS\ 1451}$.

Maximum value of normal stress $\sigma_{SX\ MM}^{693380}$ is possible to compare with allowable stress of polyurethane [19] which can be equal to:

$$\left[\sigma_{SX}\right] \approx 0,4 \cdot E = 0,4 \cdot 60 = 24 \text{ MPa}, \tag{2}$$

It confirms fulfillment of strength stipulation $\left|\sigma_{SX\ MM}^{693380}\right| < \left[\sigma_{SX}\right]$, |-22,65| MPa < 24 MPa.

Shown stress calculation of the elastic strip can be compared to theoretical/standard calculation shown in [20]. Geometrical characteristics of the strip with three equal round holes are taken into account. Scheme that describes interconnection of FEM and theoretical calculations is shown on fig. 5, where $\sigma_{SX\ T}^{A}$ is normal stress in point A acting in X-axis on surface of the strip contour, i.e. geometrical position of node number 693380 of MM and point A of theoretical calculation are coincide. It means they have the same stress $\sigma_{SX\ T}^{A} = \sigma_{SX\ MM}^{693380} = -22,65$ MPa.

Deviation of results of FEM and theoretical calculations is

$$\Delta = \frac{\sigma_{SX\ MM}^{LS\ 1451} - \sigma_{SX\ \max T}^{i}}{\sigma_{SX\ MM}^{LS\ 1451}} \cdot 100, \tag{3}$$

where $\sigma_{SX\max T}^{i}$ – maximum value of normal stress in i-point of lateral surface acting in particular direction.

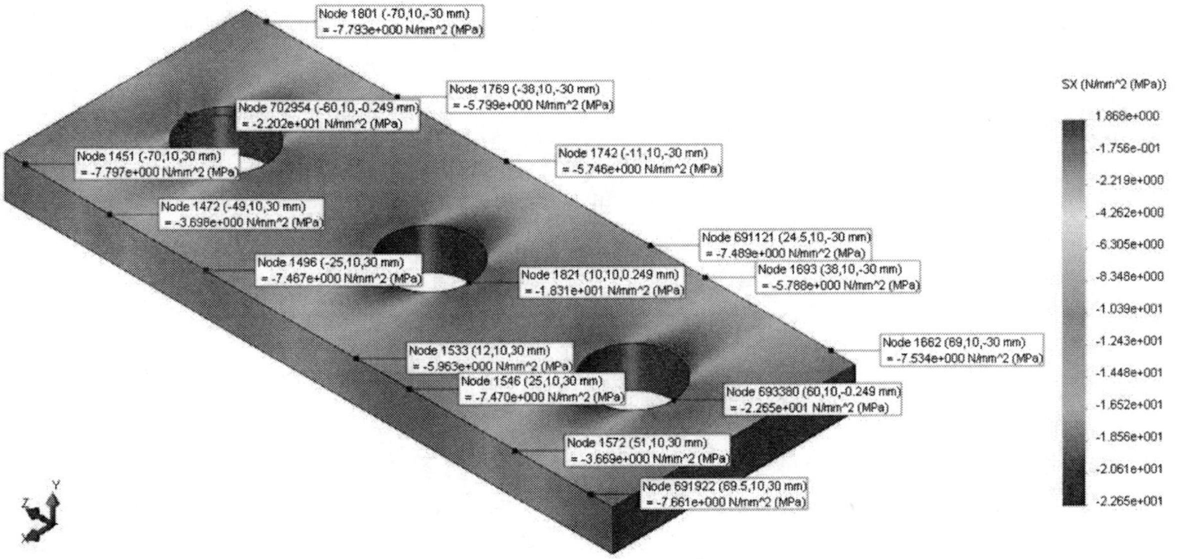

Figure 4. Distribution of normal stress acting in X-axis in the absorber strip.

Maximum stress in i-point of lateral surface:

$$\sigma^{\mathrm{i}}_{SX\,\max\,\mathrm{T}} = \frac{\sigma^{\mathrm{A}}_{SX\,\mathrm{T}}}{K^{\mathrm{A}}_{\mathrm{T}}}, \tag{4}$$

where $K^{\mathrm{A}}_{\mathrm{T}} = 2{,}975$ – coefficient when taking into account dimensions of strip and amount of uniform round holes [20].

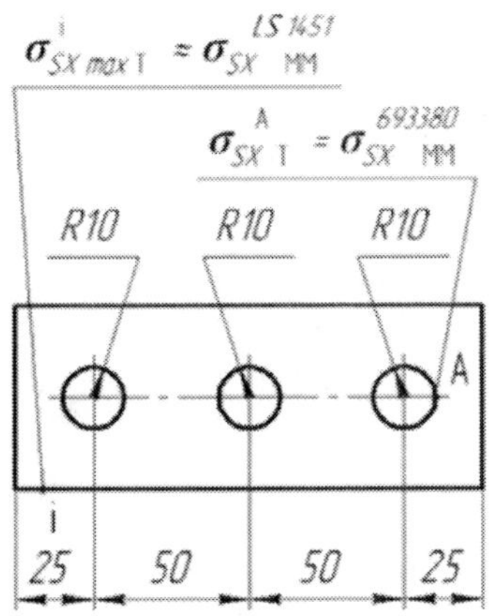

Figure 5. Interconnection scheme of FEM and theoretical calculations.

Results of FEM and theoretical stress calculations of absorber strip are given in table 1.

Table 1. Analyses results of MM and theoretical stress calculation.

Parameter	Δ_{MM}, %	Δ, %
Value	2,78	2,36

3. Conclusions

According to described above and values in table 1 it is clear that use of CAD-systems allows to make relationships between parameters of elasticity theory and set of design features of developed elastic elements for absorbers of metallurgical equipment.

Acknowledgments

This article is written under grant «Verification and development of models of inelastic deformation at the passive loading» of Russian Foundation for Basic Research.

References

[1] B.N. Polyakov, *Optimization and improvement of billet and sheet rolling mills* (Machines builder, **11**, 39–48, 2005)
[2] V.G. Artiukh, *Precision of Safety Devices for Iron and Steel Machines* (Mariupol: PSTU, 177, 2000)
[3] V.G. Artiukh, *Loads and Overloads of Iron and Steel Machines* (Mariupol: PSTU, 246, 2008)
[4] V.G. Artiukh, *Basis of Protection of Metallurgical Machines from Breakdown* (Mariupol: Publishing Group "University", 288, 2015)
[5] Mohammed I. Matarneh, Nabeel S. Gharaibeh, V.G. Artyukh, IJESIT, **4(2)**, 1–7, (2015)
[6] Y.I. Nyashin, A.N. Skorokhodov, etc, *Matrix method of the solution of elastoplastic problems in the theory of metal forming* (Theory of rolling, M: Metallurgy, 75–77, 1980)
[7] A.V. Chigaryov, A.S. Kravchyuk, etc, *ANSYS for engineers* (Mechanical engineering, 512, 2004)
[8] E.A. Nekliudova, A.S. Semenov, S.G. Semenov, B.E. Melnikov, Applied Mechanics and Materials, **725–726**, 648–653, (2015)
[9] I. Maniak, B. Melnikov, A. Semenov, S. Saikin, Applied Mechanics and Materials, **725–726**, 943–948, (2015)
[10] V. Sevastyanov, A. Volvach, Computer design and technical flow of documents, **2**, 28–31, (2007)
[11] A.V. Belevich, A.V. Goncharuk, etc, Universities News: Nonferrous Metallurgy, **9**, 32–35, (2002)
[12] A.A. Alyamovskiy, *SolidWorks/COSMOSWorks. Engineering FEM analysis* (DMK Press, 432, 2004)
[13] H. Magnus, *Prediction of Roll Torque in Hot Rod and Wire Rolling* (Örebro University, 77, 2004)
[14] A.B. Kaplun, etc, *ANSYS in Engineer's Hands. Practical issue* (Editorial URSS, 272, 2003)
[15] V. Mazur, V. Artyukh, G. Artyukh, M. Takadzhi, Engineering Designer, V.**37**, No.**1**, 26–29, (2012)
[16] Firas M.F. Al-Quran, M.E. Matarneh, V.G. Artukh, Research Journal of Applied Sciences, Engineering and Technology, **4(11)**, 1585–1589, (2012)
[17] Nabeel S. Gharaibeh, Mohammed I. Matarneh, V. G. Artyukh, Research Journal of Applied Sciences, Engineering and Technology, **8(12)**, 1461–1464, (2014)
[18] V.G. Artiukh, S.Yu. Karlushin, E.N. Sorochan, Procedia Engineering, **117**, 938–944, (2015)
[19] G.V. Artiukh, Protection of Metallurgical Machines from Breakdowns, **2**, 91–105, (1997)
[20] G.N. Savin, V.I. Tul'chiy, *Handbook of stress concentration* (K.: High School, 412, 1976)

MATEC Web of Conferences **53**, 01040 (2016)
DOI: 10.1051/matecconf/20165301040

Modification of Wastewater Treatment Technology at Cottonseed Oil Plant

Mary Shick Alshabab[1], Maria Andrianova[1,a], and Dergham Alsalloum[2]

[1]*Peter the Great Saint-Petersburg State Polytechnic University, 195251, St.Petersburg, Polytechnicheskaya street, 29, Russian Federation*

[2]*Al-Baath University, Homs, Syrian Arab Republic*

Abstract. Wastewaters from cottonseed oil producing plant in Syria were studied in laboratory experiments. Aim of the study was to suggest modification of wastewater treatment technology in order to increase its efficiency. Concentration of pollutants in wastewaters was controlled by measurement of COD. According to the results of experiments it was suggested to decrease significantly (8-20 times) dosages of reagents (acidifier, coagulant, flocculant) in several actual stages of treatment (acidification, separation, coagulation and sedimentation) and add stage of dispersed air flotation before coagulation treatment. The modified wastewater treatment technology would reduce COD to the values allowed for irrigation waters by Syrian National Standard.

1 Introduction

Application of wastewaters for crops irrigation is important in Eastern Mediterranean where agricultural sector remains the largest water consumer [1, 2, 3]. Among discharges of various industries wastewaters from food production seem to be more suitable for application in agriculture. Due to the natural origin of main components, low toxicity and good biodegradability could be expected for impurities in such wastewaters. According to national water quality standards, wastewaters need treatment before irrigation [1, 2].

There is a long tradition in Syria to use wastewater for irrigation. In general, the reuse of wastewater in Syria is largely unplanned and mainly untreated wastewater is used. Wastewater is often taken uncontrolled from sewage channels or polluted rivers [3].

Wastewaters of a vegetable oil company in Syria are studied in this paper. The study was conducted to evaluate efficiency of wastewater treatment technologies and suggest means to improve them in order to meet the requirements of the Syrian National Standard 2752/2003 for crops irrigation waters [5].

The studied company produces mainly cottonseed oil. Water as steam and liquid is used at many stages of oil production [4]. The main impurities in wastewater are organic substances from crude cottonseed oil such as proteins, phospholipids, free fatty acids, hydrocarbons, sterols, pigments and substances that cause unwanted taste and odor [4]. Wastewaters of vegetable oil production have low ratio of biological oxygen demand in 5 days (BOD_5) to chemical oxygen demand (COD) [6, 7, 8].

[a] Corresponding author : maandrianova@yandex.ru

Organic matter is present in dissolved, colloidal and suspended form. Suspended matter also includes parts of seeds and inorganic matter such as sand, clay, insoluble products formed as a result of reagent treatment [8]. According to these characteristics of impurities wastewater quality at various stages of its treatment can be controlled mainly by measuring concentration of total suspended solids (TSS) and COD. It is also necessary to control pH because it is an important indicator in different stages of physicochemical treatment.

Several types of technologies and constructions are used in treatment of wastewater from vegetable oil plants. Many of them are also widely applied in treatment of other types of waters. For example, coagulation and filtration are used in treatment of drinking water [9, 10], municipal wastewater [11, 14], drainage water [15], separation and flotation are used in treatment of industrial wastewaters polluted with petroproducts [10, 16, 17].

In oil wastewaters treatment processes driven by gravity force are applied to remove settling materials (sand, clay, fine gravel) and isolation of free floating oils in separators and settling tanks. Physical chemical methods (coagulation, flotation, etc.) are used to remove finely dispersed material, emulsified oils, and dissolved pollutants. Biological treatment allows removing dissolved pollutants, which can be biologically degraded, such as phenols, aldehydes and others. But it is rarely used in vegetable oils plant wastewaters due to the low value of their BOD_5/COD ratio [6 - 8]. Additional technologies of treatment (filtration through sand, coal, membrane filtration) are applied in some cases [6, 7, 13, 18, 19].

2 Objects of investigation

2.1 Oil production and wastewater generation at the plant

Al-Ahliah Company for Vegetable Oil Production was studied. It is one of the biggest nutritional oils companies in Syria. The company produces a group of vegetable oils products under several brand names. The most used raw material is cottonseeds. The company is located about 5 km to the north of the city of Hama. The company occupies area of 200000 m^2 where the technological parts of the company occupy 60000 m^2, the rest of the area is agricultural land that belongs to the company. The wastewater is treated in order to irrigate the neighbouring lands, which are usually planted with industrial crops.

Process of oil production starts with seeds preparation. Seeds are screened on vibrating sieves to separate sticks and stones. The cotton fluff is also removed. Then cottonseeds are cracked to separate the kernel from the external shell. The kernels are converted to thin flakes which are cooked in boilers equipped with sprinklers. The boilers have double walls to heat the flakes using steam and are equipped with a motor to flip the contents. The purpose of this stage is to eliminate the toxic effects of gossypol [20-21] and to make the mechanical pressing easier.

Cooked flakes release oil and water under mechanical pressing. The resulting liquid mixture is then moved to separation tanks, where the oil floats on surface of water and later is transported to crude oil tanks. The remaining solid material (seed cake) contains about 5–10% oil, and a solvent (hexane) is added to extract oil from it. The oil-solvent mixture is heated using steam in an extraction device (a double-walled boiler). The mixture is then moved to a distillation tower to separate the oil from the solvent. The solvent is then condensed and collected in tanks, where it floats on the surface of the water and can be separated and used again.

Crude oil treatment is done to remove the maximum amount of undesirable substances from oil. It includes stages of degumming, refining, colour and odour removal. Degumming removes resins, proteins, phosphides which form 1-3 % of crude oil. For this purpose, the process of rehydration is used, by spraying water, steam or acid solutions (often phosphoric acid) on a limited quantity of crude oil at a temperature of 35-50 °C for several minutes along with slow stirring. A turbid layer is formed on the interphase boundary between the oil and water. Gums precipitate in the water layer. The wastewater resulting from this process contains most of the impurities existed in crude oil such as free fatty acids, phosphides (lecithin), sterols, saturated and unsaturated hydrocarbons, glycerine.

During refining alkaline solution (such as sodium hydroxide solution) is added according to the acidity of the oil. Then with stirring and raising the temperature to 70°C free fatty acids, phosphides, resins and materials are deposited in the form of sodium salts, in addition absorbing a portion of sterols, waxes and gums in the formed soap stock. Wastewater resulting from this process contains some oil, soda, soap and glycerine.

Colour removal (bleaching) is designed to remove the largest part of oil pigments using adsorbents. The most used substance is a clay material that consists mainly of hydrated aluminium silicate, in the form of a fine powder. The bleaching process takes place in cylindrical devices with a conical bottom, equipped with mixers for flipping the content, under a cooking temperature of 80-90 °C and with a stirring period between 20-40 min under normal atmospheric pressure. After that oil is pulled to compressed refineries to get rid of depleted clay material.

Deodorization is done to remove substances that cause the unwanted smell and taste of oil. The oil is placed in cylindrical device, all the air in the devices is vacuumed, steam is passed through the oil under a low pressure (2-5 mmHg) with a working temperature of 200-250°C. The steam bubbles go through the oil, carrying with it water-soluble chemicals. Wastewater resulting from deodorization carries aldehydes, ketones and free fatty acids with a low molecular weight.

Mechanical filtering is done by compressed refineries to remove the solid impurities suspended in oil. Often the oil goes through two filtering processes: the first after the bleaching process, and the second after the end of all operations. After filtering oil is sent to the last producing section - filling and packing.

The wastewaters from oil production are transported to wastewater treatment plant.

2.2 Wastewater treatment procedure

During treatment wastewaters from the cottonseed oil plant underwent the following stages. In the primary separation tank oil fraction was isolated from water fraction. Big drops of oil floated to the surface of separator, some suspended solids settled to its bottom. Oil fraction was sucked by floating pumps and taken to special tanks.

The water after that passed through fine screens (diameter 1.5 mm). Wastewater at this stage had alkalic reaction (pH between 8.3 and 11.2). After that water was mixed with sulfuric acid to reach pH 3–3.5. In separation tank for acidified water the oily acidified material floated on the surface where it was scraped and pumped to collection tanks.

Before the next stages of treatment water was neutralized with slaked lime in mixing and sedimentation tank. After that reagents for coagulation treatment (polyacrylamide and aluminum sulfate) were added to water and it was then transferred to lamellar settler, where sediments were removed. The final treatment of water was done by sand filters and coal filters. Filtered water was chlorinated and transported to treated water tank.

The problems with wastewater treatment at the studied plant were the following. The process of dosing was not properly controlled. The main operating problem at the plant was the removal of sludge and sediments from the bottom of the tanks, which caused the need to stop the work of the treatment plant from once to while to clean the tanks and sometimes to use air pumps to help in the cleaning operation. The efficiency of the final lamella clarifier was good, but soon it reduced because of the sticky sediments and the difficulty of finding an effective technology to clean the surfaces of the plates. The sludge showed high viscosity which might be imputed to the uncontrolled dosage of polymer.

2.3 Wastewaters sampling and analysis

Wastewater samples were collected from the last section of the primary separation tank (after separating the oil from the surface of alkaline wastewater). Two samples of 20 L each were collected in amber glass bottles in June, 2009. The samples were stored in refrigerator for about a week prior to experiments and analyses.

COD was determined in wastewater samples diluted 50-100 times by distilled water. Concentration of potassium dichromate that remained in solution after incubation at high temperature in presence of sulfuric acid was measures by titration with Mohr`s salt with indicator ferroin. Measurement error was ± 15%. Water pH was measured by a laboratory pH-meter. Measurement error was ± 0.05 pH units.

3 Results and discussion

Parameters of wastewater from the studied oil production plant and requirements of Syrian National Standard 2752/2003 are shown in Table 1. The data demonstrates that wastewaters cannot be used for irrigation because of high values of COD (at least decreasing of about 50-200 times is needed), TSS (decreasing of 5-16 times needed) and high pH value. However the wastewater treatment technology that had been applied at the plant at the period of these experiments also failed to give water enough good quality for irrigation, especially in reduction of COD (data not shown). According to sets of laboratory data from the company BOD_5/COD was about 0.2 and therefore COD can be used as a main parameter for wastewater quality control.

Table 1. Selected requirements of Syrian Standard No. 2752/2003 for irrigation and parameters for wastewater from the plant.

Parameter	Category A (Cooked vegetables, playgrounds, parks and roadsides in residential areas)	Category B (Fruit trees, grain, fodder crops, green areas and roadsides outside residential areas)	Category C (Forest trees and industrial crops)	Wastewater sample from plant
COD (mg/l)	75	200	300	14000
TSS (mg/l)	50	150	150	800
pH	6-9	6-9	6-9	11

In order to modify the existing treatment procedures laboratory experiments were fulfilled with samples of wastewater. Fig. 1 depicts selected results of wastewater acidification (with 98% H_2SO_4) and subsequent removal of free oil in separator. The best treatment was observed at pH 2. For future treatment it was decided to add acid up to pH 4 because the results in the range pH 2-4 were close and within the measurement accuracy. Also this decision was done to save reagents because the dosage of sulfuric acid at low pH was 20 times more than at pH = 4. In general at this stage COD of water decreased by 40-50%.

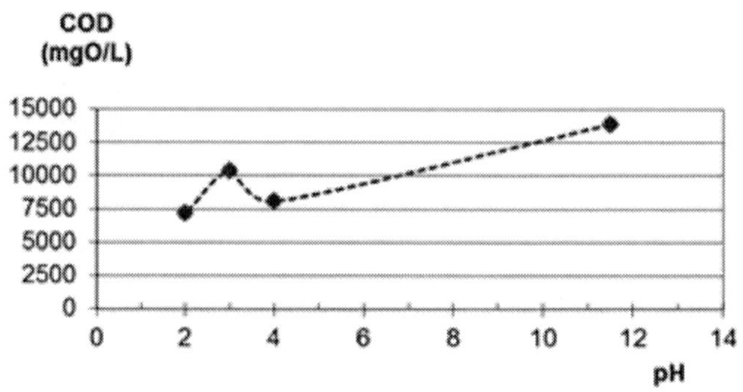

Figure 1. Changing of COD in wastewater after acidification and separation.

In wastewater treatment, oil-water separators need to be followed by subsequent treatment to remove other forms of oils: mechanical dispersions, chemical emulsions and dissolved oils [18]. After choosing the suitable pH for acidification it was suggested to use additional stage of treatment – flotation. Acidic conditions could be favourable for it due to reduction in electrostatic repulsive forces between dispersed fine oil droplets [20].

In the laboratory experiment, portative aerator producing 1.5–2 L of air per minute was used. The efficiency of flotation of acidified water (without other chemical additives) was tested for different periods. As it is shown in Fig. 2b, flotation successfully decreased concentration of COD in wastewater after collection of froth away from the surface. A flotation period of 15 min provided the best removal efficiency. However, experiments at site should be recommended in order to choose the most effective regime and construction of flotator.

a) b)

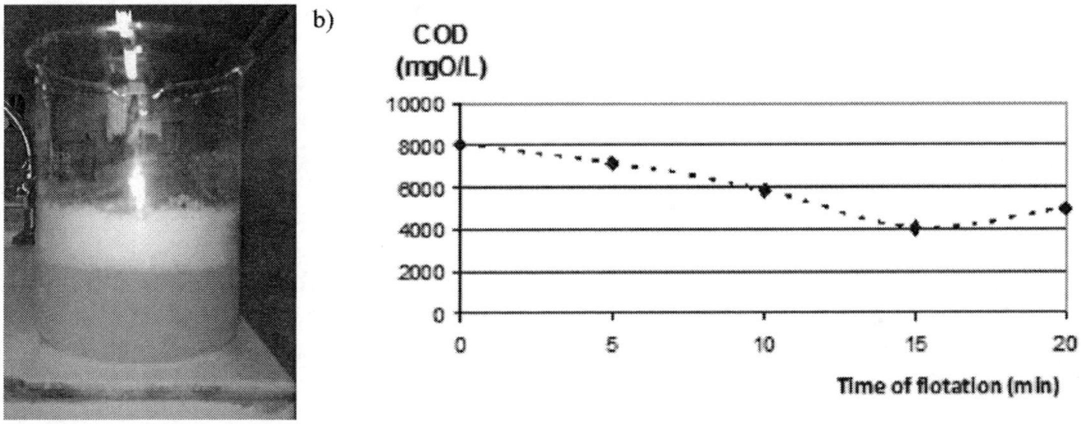

Figure 2. (a) Froth on the surface of wastewater after aeration; (b) Changing of COD in wastewater after various periods of flotation.

The next stage at the plant was coagulation treatment. For effective use of aluminum salts neutral reaction of water is preferable [7, 9]. Therefore after flotation 140 – 200 mg/L of $Ca(OH)_2$ was added to wastewater up to pH 7–8. Then coagulant was added at dosage of 4 – 15 mg/L. After that samples were left for 30 minutes for flocculation and sedimentation. COD was measured in water above the settled sediments. The results in Fig. 3a show that dosage over optimal value could decrease efficiency of treatment. Optimal dosage obtained in laboratory experiments (4-5 mg/L of aluminium sulfate) was one order of magnitude lower than dosage used at the plant (35 mg/L).

a) b)

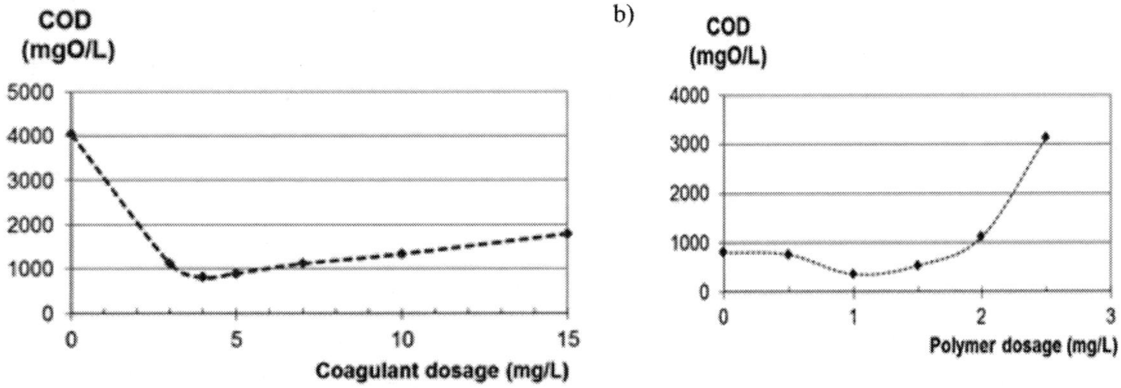

Figure 3. (a) Changing of COD in wastewater after neutralization and treatment with various dosages of aluminium sulfate; (b) Changing of COD in wastewater after neutralization and treatment with various dosages of polymeric flocculant and dosage of coagulant 4 mg/L.

In the last part of experiment, the influence of flocculant on the efficiency of coagulation treatment was studied. Polymer polyacrylamide was added to neutralized wastewater at dosage 0.5-2.5 mg/L. After that aluminum sulfate was added at dosage of 4 mg/L, and the samples were left for 30 min. The results are shown at Fig. 3b. The optimal dosage of flocculant is about 1 mg/L and higher values decrease treatment efficiency significantly. Such effect is typical for coagulation treatment, for high concentrations of coagulant and flocculant can stabilize colloidal and suspended matter and hinder water treatment process [13]. It can also be seen from Fig. 3a and Fig. 3b that overdosage of coagulant is less harmful to the treatment process than overdosage of flocculant. Optimal flocculant dose found from experiment was several times lower than dosage actually used at the plant (20 mg/L).

After all the stages of treatment modelled in the laboratory experiments, COD of wastewater decreased from 14000 mgO/L to 350 mgO/L. Concentration of chemically oxidized matter in water decreased 40 times and the reached values of COD were very close to the value allowed for irrigation of industrial crops by Syrian National Standard. It can be supposed that filtration of such water through sand and coal filters as it was done at the plant would decrease COD to a level allowed for irrigation.

4 Conclusions

Treatment of wastewater at vegetable oil producing plant in Syria was studied. Preliminary monitoring of the wastewater treatment plant showed that dosages of chemical additives were high compared to the recommended values in industrial wastewater treatment. The efficiency of treatment, measured mainly by the COD removal, was not always enough to use the treated water in irrigating the neighbouring agricultural lands.

In order to achieve a COD removal efficiency of about 97%, a series of laboratory experiments was conducted to define the proper dosages of reagents and check efficiency of dispersed air flotation suggested as a new stage of treatment. Based on the results of experiments the following wastewater treatment stages were recommended.

1. alkaline wastewaters (pH 11) are treated in primary separator where oil floats up and is removed from water fraction,
2. waters are acidified with sulfuric acid to pH 4 and free oil is removed in separator (pH 3-3.5 was used in actual treatment at the plant),
3. part of remained organic matter is removed from water as froth during dispersed air flotation (this stage is suggested instead of in separation actually used at the plant),
4. water is neutralized by $Ca(OH)_2$ to pH 7-8,
5. water undergoes coagulation treatment with the following dosages: 4 mg/L of aluminium sulfate (35 mg/L was actually used at the plant) and 0.5-1 mg/L of polyacrylamide (20 mg/l was actually used at the plant).
6. subsequent filtration and disinfection.

The suggested modifications of wastewater treatment allow to reduce COD 40 times and reach the values of COD allowed for irrigation by Syrian national Standard 2752/2003.

References

1. M.Y. Saidam, *A regional overview of wastewater management and reuse in the Eastern Mediterranean Region* (Regional CEHA, Cairo, Egypt, 2005)
2. A.N. Angelakis, *Wastewater Reclamation and Reuse Practices in Mediterranean Region* (WHO regional Office for Europe, EURO Project Office, MAP, Athens, Greece, 2002)
3. S. Rothenberger. *ACWUA Summary report - Wastewater Reuse in Arab Countries* (ACWUA, 2010)
4. O. B. Adelola, E. A. Ndudi. Iinternational Journal of Basic and Applied Science, **01**, 398-402 (2012)
5. *Syrian National Standard No. 2752/2003, Treated wastewater used for irrigation* (2003)

6. N. Azbar, T. Yonar. Process biochemistry **39**, 869–875 (2004)
7. K. B. Chipasa. Polish Journal of Environmental Studies **10**, 141-147 (2001)
8. S. Aslan, B. Alyüz, Z. Bozkurt, M. Bakaoğlu. Polish Journal of Environtal Studies, **18**, 533-538, (2009)
9. Andrianova M. Ju., Molodkina L.M., Chusov A.N. Applied Mechanics and Materials, **587-589**, 573-577 (2014)
10. N.I.Vatin, D. Nemova, N. Kharkov. Applied Mechanics and Materials, **680**, 354-358 (2014)
11. D.D. Kolosova, A.A. Maslak, N.N. Rolle, L.M. Molodkina. Magazine of Civil Engineering. **6 (41)**, 22-30 (2013)
12. N. I. Vatin, V. N. Chechevichkin, A. V. Chechevichkin, Ye. S. Shilova, L. A. Yakunin. Applied Mechanics and Materials, **587-589**, 565-572 (2014)
13. Tchobanoglous, G., Burton, F.L. *Wastewater Engineering : Treatment, Disposal, Reuse* (New York: McGraw-Hill, 1991)
14. L.M. Molodkina, D.D. Kolosova, E.I. Leonova, M.F. Kudoyarov, M.Y. Patrova, Y.V. Vedmetskii. Petroleum Chemistry, **52(7)**, 487-493 (2012)
15. V.N. Chechevichkin, N.I. Vatin. Applied Mechanics and Materials, **641-642**, 409-415 (2014)
16. M.Ju. Andrianova, L.M. Molodkina, E.A. Bondarenko, A.N. Chusov, M.P. Fedorov. *Proceedings of CRETE 2014 The International Conference on Industrial and Hazardous Waste Management*, 79-80 (2014)
17. M. R. Beychok. *Aqueous Wastes from Petroleum and Petrochemical Plants,* (John Wiley & Sons, 1967)
18. A.I. Zouboulis, A. Avranas, Colloids and Surfaces A: Physicochemical and Engineering Aspects **172**, 153–161 (2000)
19. M.L.S. Welz, N. Baloyi, D.A. Deglon, Water SA **33(4)**, 453-458 (2007)
20. W.-S. Ye, J.C. Liang, T.C. Hsu, In Vitro **19(1)**, 53-57 (1983)
21. M. Schwab, *Encyclopedia of Cancer* (Springer-Verlag Berlin Heidelberg, 2012)

MATEC Web of Conferences **53**, 01041 (2016)
DOI: 10.1051/matecconf/20165301041

Effective methods of parameter identification for creep models with account of III stage

Alexey I. Grishchenko[1,a] and Artem S. Semenov[1]

[1] St.Petersburg State Polytechnical University, Polytechnicheskaya st.29,195251,Saint-Petersburg,Russia

Abstract. The account of tertiary (third) stage of creep is important for the failure analysis of structures. The correct description of third stage of creep is based on taking into account of damage accumulation process. An identification of the parameters of coupled creep-damage model leads to the optimization problem of finding the global minimum of a functional, having a complex, non-linear structure and a large number of local minimums. The Nelder–Mead method is applied to the solution to the multiparameter identification problem. The effectiveness and robustness of the applied approach are illustrated on examples of mechanical and civil engineering.

1 Introduction

The creep process is described by the so-called creep curve, which represents the time dependence of the strain at constant temperature and applied load (or stress). It is conventionally divided into three sections or stages (Fig. 1): initial stage, or primary creep, when the strain rate slows with increasing time, secondary or steady state of creep, when the strain rate eventually reaches a minimum and becomes near constant and in tertiary creep, the strain rate exponentially increases with stress. Fracture always occurs at the tertiary stage.

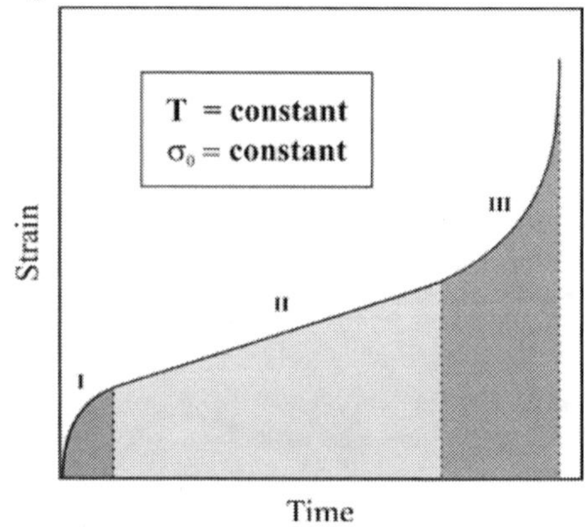

Figure 1. Schematic representation of creep curve showing primary (I), secondary (II) and tertiary (III) deformation regimes.

Corresponding author: gai-gr@yandex.ru

For most materials, the main time before the failure takes the second stage of the creep. Therefore, to describe the creep curves of these materials is sufficient to use the model, taking into account only the second (steady state) stage of the creep. However, some materials, such as single crystal nickel base superalloys [1,2], heat-resistant steel [3,4] or asphalt mastic [5], demonstrate domination of the third stage of creep. To describe the creep of such materials, it is necessary to use models that describe correctly a third stage of creep. The correct description of third stage of creep is based on taking into account of damage accumulation process.

2 Materials and Methods

2.1 Constitutive equations of creep

In this work the Kachanov-Rabotnov model [3] is used for description a third stage of creep:

$$\frac{d\varepsilon^c}{dt} = B\frac{\sigma^n}{(1-D)^l}\left(\varepsilon^c\right)^m, \tag{1}$$

$$\frac{dD}{dt} = K\frac{\sigma^p}{(1-D)^q}, \tag{2}$$

where D, is scalar damage variable ($0 \le D \le 1$).

After integration (1), (2) obtain:

$$\varepsilon^c = \left\{\frac{1-m}{1-\frac{l}{1+q}}\frac{B}{K(q+1)}\sigma^{n-p}\left[1-\left(1-K(q+1)\sigma^p t\right)^{1-\frac{l}{q+1}}\right]\right\}^{\frac{1}{1-m}}. \tag{3}$$

Thus, it is generally necessary to determine the seven constants: *B, n, K, q, l, p, m.*

Checking of the correctness of determining the coefficients K and p equation (2) may be carried out by means of conditions for the destruction $D(t = t_f) = 1$, where D is found by integrating (2) with zero initial conditions $(D(t = 0) = 0)$ and constant stress. The result of the integration of (2) may be expressed as follows:

$$D = 1-\left(1-t/t_f\right)^{1/q+1}, \tag{4}$$

where t_f is time before the failure

$$t_f = \frac{1}{(q+1)K\sigma^p}. \tag{5}$$

Taking into account equation (4) and (5) the equation (3) can be rewritten as:

$$\varepsilon^c = \left\{ \frac{1-m}{1-\dfrac{l}{1+q}} B\sigma^n t_f \left[1 - \left(1 - \frac{t}{t_f} \right)^{1-\frac{l}{q+1}} \right] \right\}^{\frac{1}{1-m}} \quad (6)$$

In this paper $n=l$ is used.

For simplification the definition of material parameters (6) can be rewritten as:

$$\varepsilon^c = \left\{ A\sigma^x \left[1 - \left(1 - K_1 \sigma^p t \right)^v \right] \right\}^{m_1}, \quad (7)$$

where

$$A = \frac{1-m}{1-\dfrac{l}{1+q}} \frac{B}{K(q+1)}; \; x = l - p; \; K_1 = K(1+q); \; v = 1 - \frac{l}{q+1}; \; m_1 = \frac{1}{1-m}, \quad (8)$$

Units of measurement are follows: $[A] = \dfrac{1}{[MPa]^x}; [K_1] = \dfrac{1}{s \cdot [MPa]^p}; [x] = [l] = [v] = [m_1] = [1]$.

Defining the parameters of the material was carried out by means of minimization of the objective function:

$$L = \sum_{i=1}^{n} \sum_{j=1}^{k} \left[\left\{ A\sigma_i^x \left[1 - \left(1 - K_1 \sigma_i^p t_j \right)^v \right] \right\}^{m_1} - \varepsilon(\sigma_i, t_j) \right]^2 \rightarrow min. \quad (9)$$

This functional has a complex, non-linear structure and a large number of local extremums. The solution to this problem is not trivial. When we using the Newton's method, the Gradient descent method we have not received the result. In this works for solution of this problem was use Nelder–Mead method.

2.2 Nelder–Mead method

The Nelder–Mead method is a numerical method used to find the minimum or maximum of an objective function in a many-dimensional space. It is applied to nonlinear optimization problems for which derivatives may not be known. This method is described in detail in literature [6,7], and we are shown only principal scheme of this method.

1. Initially selected n + 1 point $x_i^{(k)} = x_i\left(x_{i_1}^{(k)}, x_{i_2}^{(k)}, \ldots\ldots, x_{i_n}^{(k)} \right), i = 1, n+1$, where k is iteration number.
2. Ordering. Determine three points:
   $x_h^{(k)} : f\left(x_h^{(k)}\right) > f\left(x_i^{(k)}\right), i = 1, n; x_l^{(k)} : f\left(x_h^{(k)}\right) < f\left(x_i^{(k)}\right), i = 1, n;$
   $x_g^{(k)} : f\left(x_h^{(k)}\right) > f\left(x_i^{(k)}\right), i = 1, n, i \neq h$
3. Calculate the mass center $x_0^{(k)}$ of all points. $x_0^{(k)} = \sum_{i=1}^{n+1} x_i^{(k)} - x_n^{(k)} \Big/ n$
4. Transformation. Compute the new working simplex from the current one.

5. Reflect. Compute reflected point $x_r^{(k)} = x_0^{(k)} + \alpha\left(x_0^{(k)} - x_{n+1}^{(k)}\right)$. If, $f\left(x_l^{(k)}\right) \leq f\left(x_r^{(k)}\right) < f\left(x_g^{(k)}\right)$accept $x_r^{(k)}$ and go to 7.
6. Expand. If $f\left(x_r^{(k)}\right) < f\left(x_l^{(k)}\right)$ compute the expansion point $x_e^{(k)} = x_0^{(k)} + \gamma\left(x_0^{(k)} - x_{n+1}^{(k)}\right)$. If $f\left(x_e^{(k)}\right) < f\left(x_r^{(k)}\right)$, accept $x_e^{(k)}$ and go to 7. Else, if $f\left(x_e^{(k)}\right) \geq f\left(x_r^{(k)}\right)$, accept $x_r^{(k)}$ and go to 7.
7. Contact. It is certain that $f\left(x_r^{(k)}\right) \geq f\left(x_g^{(k)}\right)$, compute the contraction point:
8. If $\left(x_g^{(k)}\right) \leq f\left(x_r^{(k)}\right) < f\left(x_h^{(k)}\right)$, compute $x_c^{(k)} : x_c^{(k)} = x_0^{(k)} + \rho\left(x_r^{(k)} - x_0^{(k)}\right)$. If $f\left(x_c^{(k)}\right) < f\left(x_r^{(k)}\right)$, accept $x_c^{(k)}$ and go to 7.
9. If $\left(x_r^{(k)}\right) \geq f\left(x_h^{(k)}\right)$, compute $x_c^{(k)} : x_c^{(k)} = x_0^{(k)} + \rho\left(x_h^{(k)} - x_0^{(k)}\right)$. If $f\left(x_c^{(k)}\right) < f\left(x_h^{(k)}\right)$, accept $x_c^{(k)}$ and go to 7.
10. For all but the best point, replace the point with $x_i^{(k)} = x_1^{(k)} + \mu\left(x_0^{(k)} - x_1^{(k)}\right), i = 2, n+1$. Go to step 7.
11. End the search criterion: $\left\{\frac{1}{n+1}\sum_{i=1}^{n+1}\left[f\left(x_i^{(k)}\right) - f\left(x_0^{(k)}\right)\right]^2\right\} \leq \varepsilon$, where ε is the arbitrary small number, $f\left(x_0^{(k)}\right)$ – is the value of the objective function at the center of mass of all points.

3 Results and Discussion

3.1 Result of parameter identification for creep models of single crystal nickel base superalloys

The constants of the creep model (1)-(2) for single crystal alloys VZhM4 and VZhM5Y with crystallographic orientation <001> at 1050 °C were determined. Experimental creep curves were obtained in the laboratory of Strength of materials [8-10]. Creep curves for the following stress: 250, 280 and 320 MPa were investigated. The experimental creep curves and their approximation by (7) are shown in Fig. 2. Mean square deviation between the experimental points and approximation was 1%. The creep model parameters for single crystal alloys VZhM4 and VZh5MY are given in Tab. 1.

Table 1. The creep model parameters for single crystal alloy VZhM4 and VZhM5Y

	A	x	K_I	p	v	m
VZhM4	2010	$1{,}0\cdot10^{-6}$	$1{,}58\cdot10^{-13}$	5,63	$9{,}6\cdot10^{-4}$	1,83
VZhM5Y	929	$5{,}0\cdot10^{-5}$	$1{,}12\cdot10^{-12}$	5,01	$2{,}66\cdot10^{-3}$	1,51

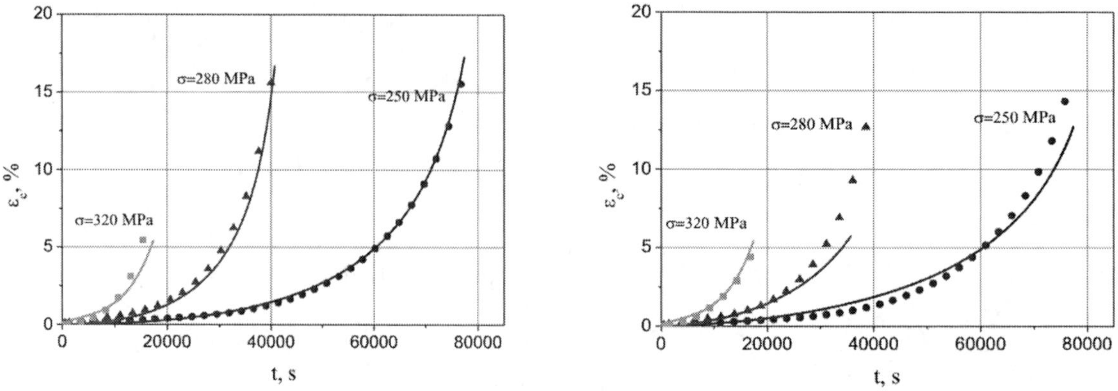

Figure 1. Experimental creep curves (points) [8-10] and their approximation with using creep model (solid lines) for single crystal nickel base superalloys VZhM4 (a) and VZhM5Y (b).

3.2 Result of parameter identification for creep models of heat-resistant steel

The constants of the creep model (1)-(2) for Type304H austenitic stainless steel at 650°C are determined. We use experimental creep curves from [4]. Creep curves for the 250, 280 and 320 MPa are considered. The experimental creep curves and their approximation by (7) are shown in Fig. 2. Mean square deviation between the experimental points and approximation was 0.4%. The creep model parameters for Type304H austenitic stainless steel were shown in Tab. 2.

Table 2. The creep model parameters for Type304H austenitic stainless steel

	A	x	K_1	p	ν	m
HR steel	$2{,}31\cdot10^{-7}$	3,01	$1{,}63\cdot10^{-15}$	6,91	$3{,}09\cdot10^{-7}$	1

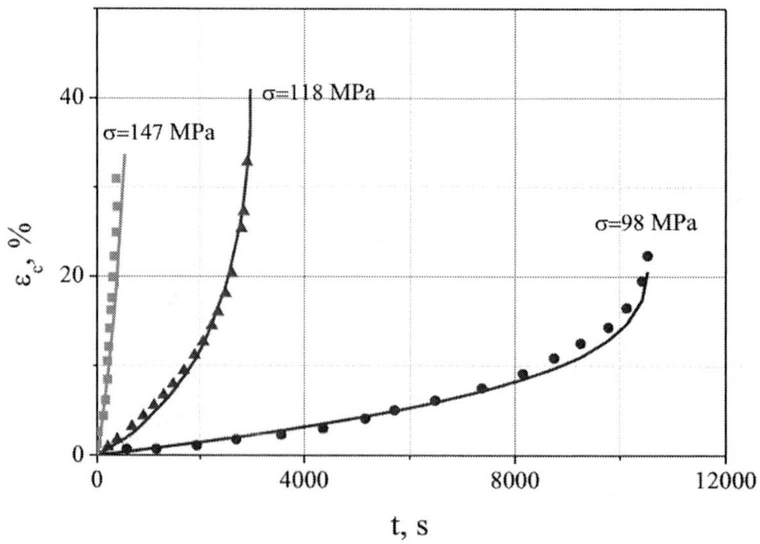

Figure 2. Experimental creep curves and their approximation with using creep model (1)-(2) for Type304H austenitic stainless steel at 650°C. The lines show the approximation of creep curves, points shows the experimental data [4].

3.3 Result of parameter identification for creep models of asphalt mastic

The constants of creep model for asphalt mastic at 40°C were determined. We use experimental creep curves from [5]. Creep curves for the following stress: 0.1, 0.15, 0.2, 0.25 and 0.3 MPa was investigated. The experimental creep curves and their approximation by (7) are shown in Fig. 3. Mean square deviation between the experimental points and approximation was 2%. The creep model parameters for asphalt mastic was shown in table 3.

Table 3. The creep model parameters for asphalt mastic

	A	x	K_I	p	v	m
Asphalt mastic	46,6	1,33	$2{,}61\cdot10^{-2}$	2,19	$2{,}21\cdot10^{-5}$	0,49

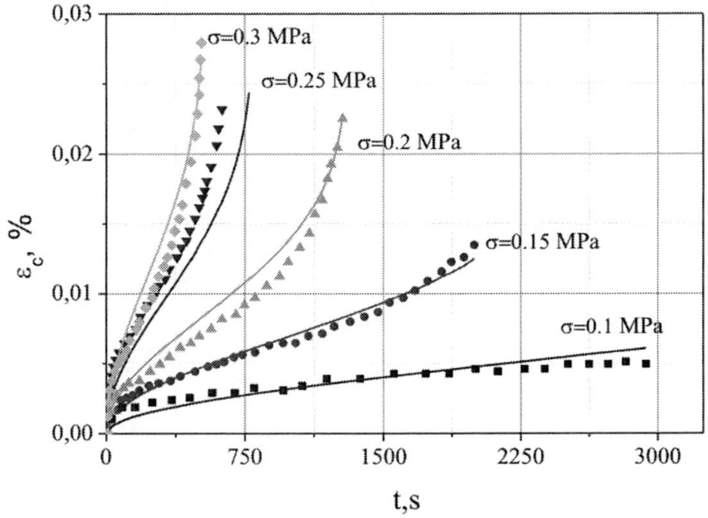

Figure 3. Experimental creep curves and their approximation with using creep model (1)-(2) for asphalt mastic. The lines show the approximation of creep curves, points show the experimental data [5].

4 Conclusions

The modeling of creep process has been analyzed for three different class of materials:

- single-crystal nickel-based alloys VZhM4 and VZhM5Y,
- 304H austenitic stainless steel,
- asphalt mastic.

The obtained results of creep analysis allow to find out:

- the creep model in the form of the hardening theory considering damage accumulation is applicable for creep behavior prediction with account of III stage for all considered materials,

Acknowledgments

This research was supported by the Russian Foundation for Basic Research (Grant No. 15-08-08779). The authors are grateful to O.A. Fedorov for help in writing program of creep curves approximation.

References

1. R.E. Shalin, I.L.Svetlov, E.B. Kachanov, Mechanical Engineering, Moscow, (1997)
2. R.C. Reed, Cambridge University Press, (2006)
3. S.R. Holdsworth, Woodhead Publishing; CRC Press, pp 403–420, (2008)
4. M. Igarashi, Woodhead Publishing; CRC Press, 539–596 (2008)
5. Y. Ye, X. Yang, C. Chen, Construction and Building Materials, **23**, 3161–3165 (2009)
6. J.A. Nelder, R. Mead, Comput. J., **7**, 308–313 (1965)
7. D.M. Himmelblau, McGraw-Hill, New York (1972)
8. L.B. Getsov, A.S. Semenov, E.A. Tikhomirova, A.I. Rybnikov, Materiali in Tehnologije, **48(2)**, 255-260 (2014)
9. S.G. Semenov, L.B. Getsov, E.A. Tikhomirova, A.S. Semenov, Metal Science and Heat Treatment, 2016 (to be published)
10. S.G. Semenov, L.B. Getsov, A.S. Semenov, N.V. Petrushin, O.G. Ospennikova, A.A. Zhivushkin, Problems of mechanical engineering and reliability of the machines, **2,** 2016 (to be published).

MATEC Web of Conferences **53**, 01042 (2016)
DOI: 10.1051/matecconf/20165301042

Modified finite element analysis for exterior boundary problems in infinite medium

Natalia Chernysheva[1,a] and Leonid Rozin[1]

[1]*St.Petersburg State Polytechnic University, 195221 St.Petersburg, Russian Federation*

Abstract. Modified algorithm of the 3d finite element analysis developed for solution of external boundary problem is considered. The method in question is based on incorporating the FEM and Somigliana's integral formula. Efficiency of software implementations of the algorithm has been tested. A stress-strain analysis of inhomogeneous medium with a cavity has been carried out to display the approach.

1 Introduction

The underground non-reinforced caverns subjected to internal pressure or any other loads are often used in construction and mining as the gas or oil storages or for the waste dumping (see Fig.1, a). Large-sized cavern is often located in an inhomogeneous medium because of complicated structure of the rock formation. Analysis of stability of such caverns should be made with account for specifics of the layered rock structure [1-5]. Also for the land surface subsidence prediction and risk analysis of buildings it is necessary to estimate deformation of the rock mass caused by the presence of underground karst caverns (see Fig.1, b). Consequently, the ground subsidence and horizontal displacement under building foundation are the important topics for the analysis [6-11].

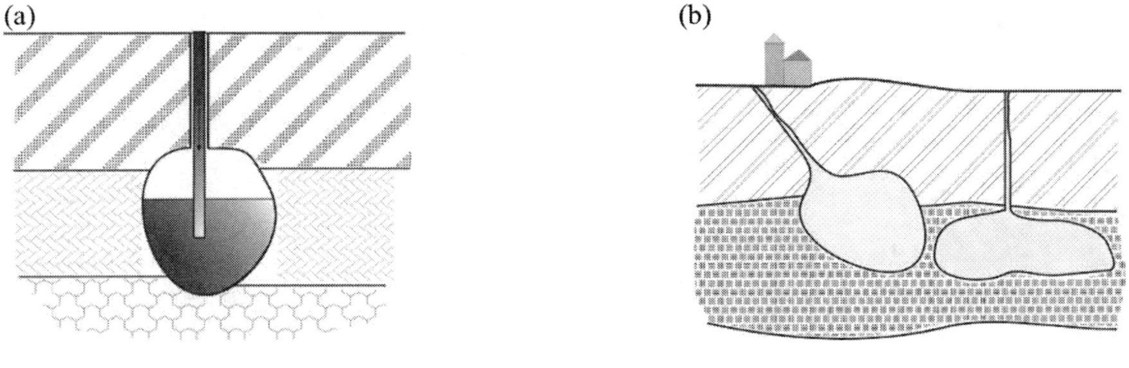

Figure1. (a) Underground oil storage in a cavern (b) Underground karst caverns (i.e. solution cavities caused by ablation with underground water)

[a] Corresponding author : chernat0000@mail.ru

In many cases of practical importance the analysis of structures considering their interaction with infinite foundation or the surrounding elastic medium is based on numerical solution of the three-dimensional external boundary problem [12-14]. Application of algorithms of the finite element method (FEM) or the boundary element method (BEM) reveals some peculiar properties of solution in infinite domain [15-17]. The FEM analysis is performed assuming a reduced domain of finite size, or by application of special "infinite" elements [18, 19]. The problems inevitably arise in using this approach, such as validation of the reduced domain size and selection of boundary conditions on the external boundary in the first case, or complications encountered in solution of the system of equations in the second case. When the BEM approach application is considered the analysis of interaction of a structure and of inhomogeneous medium occurs inconvenient, as well as the case of domains with several cavities, meanwhile these problems are not rare in practice.

These specifics of the boundary problem in the case of infinite domain cause some computational complications and increase the computing cost [20, 21]. Various known algorithms offer to avoid such problems at the expense of combining different methods and equations [22-27]. In the following, the algorithm based on combining the FEM facilities and Somigliana's integral formula, combined method (CM), is considered and its efficiency displayed [28-30].

2 Somigliana's integral formula

The Somigliana's integral formula allows determining components of the displacement vector $\mathbf{u}(\xi) = (u_1(\xi), u_2(\xi), u_3(\xi))^T$ at arbitrary point ξ in a space Ω bounded by a surface S (see Fig.2); in the absence of the volume forces:

$$u_j(\xi)=\int_S (t_i(\eta)G_{ij}(\eta,\xi) - F_{ij}(\eta,\xi)u_i(\eta))dS(\eta), \tag{1}$$

where $\xi\in\Omega$, $\eta\in S$, $i, j = 1, 2, 3$ are indexes of Cartesian coordinate axes x_i, $t_i(\eta)$ is a component of the actual force vector and $u_i(\eta)$ is a component of the actual displacement vector at the point η at the surface S; $G_{ij}(\eta,\xi)$ and $F_{ij}(\eta,\xi)$ are the components of vectors of force and displacement caused by the unit force acting in direction x_j at the point ξ (the fundamental solution of Navier's equilibrium equations).

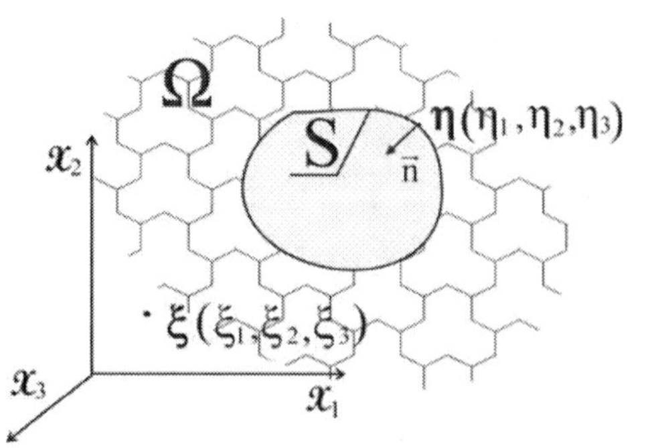

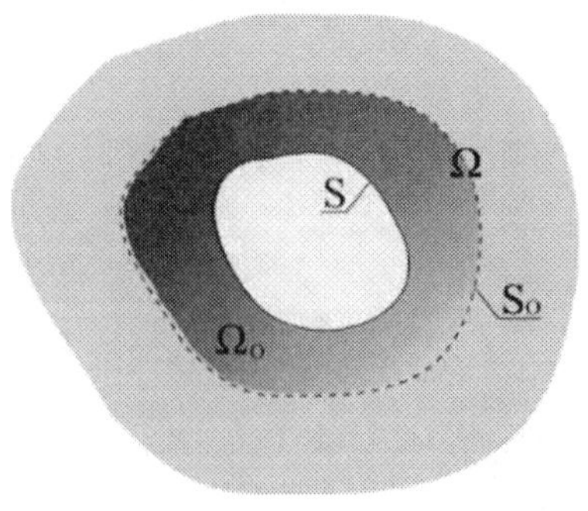

Figure 2. Exterior domain bounded by a surface S **Figure 3.** Selection of analyzed domain in a space Ω

3 Algorithm of combined method (CM)

In the following, the algorithm of the 3d analysis for external boundary problem by combined method (CM) is presented.

Let the displacement vector $\mathbf{u}(\xi)$ should found in elastic space Ω bounded by a surface S, satisfying at $\xi\in\Omega$ the Navier's equilibrium equations and boundary conditions. The boundary conditions are

$$t_i(\xi) = \sigma_{ij}(\xi)\, n_i(\xi) = p_i(\xi),\ \xi\in S_1\ ;$$

$$u_i(\xi) = u_{iS}\,(\xi), \xi\in S_2\ ;\ S_1\cup S_2 = S, \tag{2}$$

where $n_i(\xi)$ are components of the unit outer normal at the surface S, $\sigma_{ij}(\xi)$ are the components of stress tensor, $p_i(\xi)$ and $u_{iS}\,(\xi)$ are given components of force and displacement vectors.

A sub-space Ω_0 bounded by a surface S_0 is selected in the semi-infinite space Ω (see Fig. 3). The iteration procedure can be performed as follows. The first step in process is establishing of boundary condition $u_i^{(0)}$ at the surface S_0 (assumed usually $u_i^{(0)} = 0$). Then the boundary problem in the sub-space Ω_0 can be solved by FEM to complete boundary condition (Eq. 2) on the surface S. This allows using the Somigliana's integral formula (Eq. 1) to determine the first approximation, $u_i^{(1)}$.

Generally, having (k-1)-th approximation

$$u_i(\xi) = u_i^{(k\text{-}1)}\ ,\ \xi\in S_0\ , \tag{3}$$

the k-th approximation $u_i^{(k)}(\xi)$, $\xi\in S_0$ can be found by applying the Somigliana's integral formula (Eq. 1), using $u_i^{(k\text{-}1)}(\xi)$, $\xi\in S_1$ and $t_i^{(k\text{-}1)}(\xi)$, $\xi\in S_2$ determined by the FEM analysis of boundary problem in the sub-space Ω_0 with boundary conditions (Eq. 2,3). The process can be terminated when the difference between the successive approximations $u_i^{(k)}(\xi)$ and $u_i^{(k+1)}(\xi)$ ($\xi\in S$) achieves the required accuracy ε

$$max\ \left|\left(u_i^{(k+1)}(\xi) - u_i^{(k)}(\xi)\right)/\, u_i^{(k)}(\xi)\right| < \varepsilon\ ,\ \xi\in S,\ i = 1, 2, 3. \tag{4}$$

The following approach is used for integrating the first term in Eq. 1 which contains $t_i(\eta)$, $\eta\in S$. The nodal forces P_i in the nodes $\xi^{\,l}$ of a finite element ΔS in the right-hand side of the FEM equations can be treated as some integral characteristics of surface forces $t_i(\eta)$:

$$\Sigma\, P_i(\xi^{\,l}) = \int t_i(\xi)\, dS(\xi),\ \xi\in\Delta S,\ l = 1, 2, 3. \tag{5}$$

An approximation is assumed:

$$\int t_i(\xi)\, G_{ij}(\xi,\eta)\, dS(\xi) \approx \Sigma\, P_i(\xi^{\,l})\, G_{ij}(\xi^{\,l},\eta),\ \xi\in\Delta S,\ i = 1, 2, 3. \tag{6}$$

Strictly, this formula (Eq. 6) is not a quadrature formula but it provides reasonable accuracy as an alternative of numerical differentiation in the FEM solution.

The software developed based on implementation of this algorithm has been tested in a series of the model problems. These were external boundary problems with self-balanced and non-self-balanced external loads applied to a sphere-shaped concavity, such as uniform pressure, two self-balanced point forces and one point force applied symmetrically or non-symmetrically. These problems were also analyzed by the FEM for the purpose of comparing the results. The FE meshes were different by the fineness and the number of finite element rows. In general, the accuracy of the CM results even at the number of elements far less that in the FEM modeling was higher than for the FEM. Apart from that, the displacements obtained by the CM for the case of non-self-balanced external loads include the rigid-body components unlike those determined by the FEM. This feature of the CM may be considered as providing better accuracy of results in such cases.

4 Double-boundary algorithm of combined method (DCM)

The algorithm of combined method was modified for the case of inhomogeneous medium on the assumption that the inhomogeneous medium may have a substantial influence on the strain-stress state of the structure within a limited surface L. The remaining space Ω_L can be considered as a homogeneous one. A surface S_0 is selected in the space Ω_L (see Fig. 4) and a closed sub-space Ω_0 bounded by surfaces S and S_0 is considered.

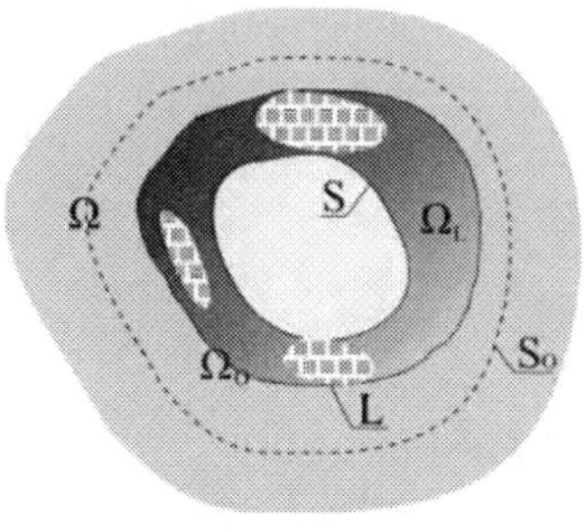

Figure 4. Scheme of the inhomogeneous space selection for analysis

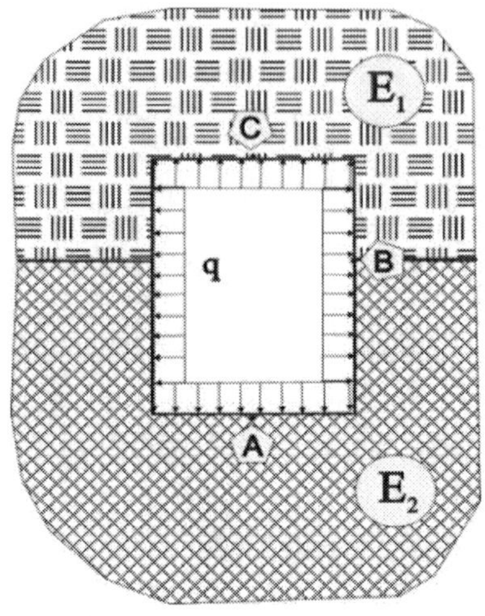

Figure 5. Underground cavern in inhomogeneous rock massif

The first step in the process is establishing of boundary condition $u_i^{(0)}$ at the surface S_0 (usually $u_i^{(0)} = 0$). Then the boundary problem in the sub-space Ω_0 can be solved by FEM to define boundary condition (Eq. 2) on the surface L. The domain of integration in Eq. 1 is the boundary of homogeneous region, consequently it will be the surface L instead of the surface S. Then the boundary problem in the sub-space Ω_0 is solved by FEM to define the second approximation $u_i^{(2)}$ on the surface L and so far. It is a double-boundary algorithm of combined method (DCM).

The values of stress and strain in Eq. 1 in the modified algorithm in question are calculated by the FEM approximately, while in the original algorithm part of them is defined by Eq. 2. By this reason designing of the FE mesh has to be more careful to decrease the error.

The developed software based on implementation of the DCM algorithm has been tested in stress-strain analysis of sphere-shaped cavity surrounded by a spherical layer with physical properties different from those in the surrounding space under the internal uniform pressure. The dimension of the analyzed domain for the CM analysis was 3 times less than that for the FEM analysis. The accuracy of the CM and the FEM results was comparable. The both methods resulted in partly underestimated stresses and strains compared to the results of exact solution due to insufficient dimensions of the considered domain.

5 Strain analysis of rock massif with a cavity

For example selected is an underground closed cylindrical cavern subjected to internal pressure on the boundary of two rock layers; a complete cohesion of the layers on the boundary was assumed (see Fig. 5). The layers' properties are characterized by the Young's modules, E_1 and E_2, with E_2 varied from E_1 to 10 E_1. The variety may be caused by variability of rock crumbling even in a homogeneous massif. The influence of relationship of Young's modules E_1 / E_2 on the stress-strain state of massif was evaluated. Calculations were carried out by applying the DCM as well as FEM to verify results.

The modeled area for the DCM was two times smaller than that for FEM. The DCM FE model had two layers where the first one Ω_L consisted of two different materials and the second one $(\Omega_0|\Omega_L)$ had assumed average Young's modulus E_3. The FEM model had four layers consisting of two different materials.
The intrinsic stress state of massif caused by the weight of upper rock was accounted for by the method of deleting loads. For this purpose the FEM analysis of massif without cavern under the action of vertical and horizontal mass forces corresponding to depth of location was performed. By this way the loads on the surface S to be deleted were determined as nodal forces and then equivalent forces opposite in sign to above deleted forces were used to define the boundary conditions.

Safety factors $K_\sigma = \sigma_u^*/\sigma$ were calculated in the inner layers of finite elements around the cavern. There $\sigma_u^* = 0.75\sigma_u$ is the weakened rock compressive strength diminished with respect to the standard rock compressive strength σ_u. As might be expected, the tangential stress σ_θ was the highest at the side surface of cavern ($K\sigma_\theta = 1.09 \div 1.43$) as well as the vertical stress σ_z at the upper and the bottom surfaces ($K\sigma_z = 1.15 \div 1.71$ for the upper surface and $K\sigma_z = 1.05 \div 1.15$ for the bottom one). Values of the safety factor for the sound layer with modulus E_2 were smaller than those for the weak one; they decreased accordingly the ratio E_1/E_2. Therefore, it is the reasonable conclusion that the volumes where $K\sigma < 1.2$ should be reinforced.

Fig. 6, 7 illustrate displacements u_z at the points A, C and u_ρ at the point B compared with corresponding displacements in homogeneous massif with Young's modulus E_1 for point A or with modulus E_2 for point C as functions of the ratio E_1 / E_2. Fig. 4 shows that displacement u_{zC} of the point C in the weak rock volume is almost unaffected by the increasing modulus E_2, while

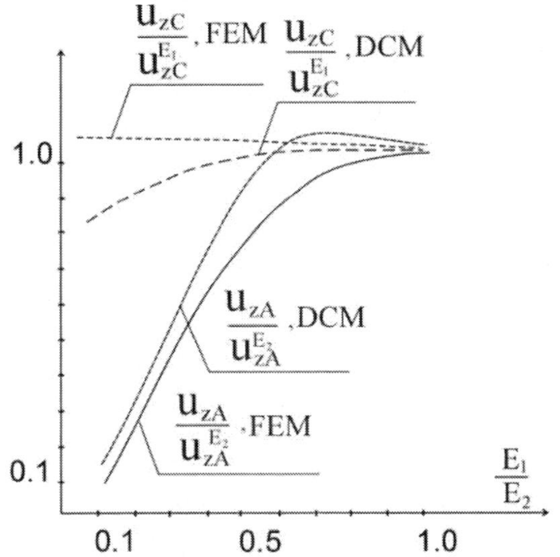

Figure 6. The vertical displacements of the upper and bottom surfaces of the cavern

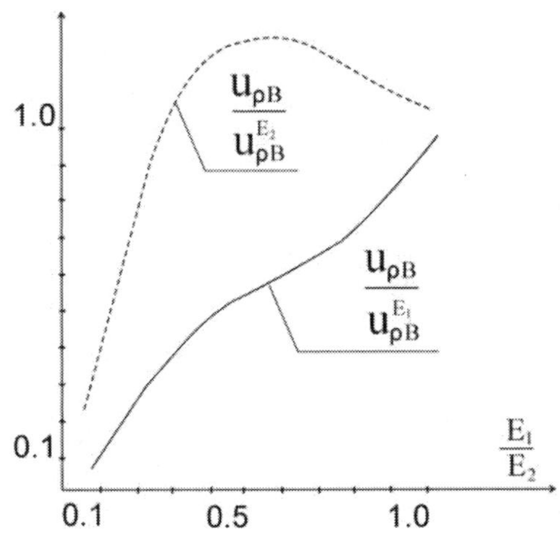

Figure 7. The radial displacement of point on the cohesion surface

displacement u_{zA} of the point A in the sound rock volume decreases irrespective the changes of modulus E_2. The radial displacement $u_{\rho B}$ of the point B on the contact surface is smaller than that in homogeneous massif with Young's modulus E_1. The displacement $u_{\rho B}$ is larger than that in homogeneous massif with Young's modulus E_2 if $E_1 / E_2 > 0.3$ because of the weak rock dominant influence. In case of $E_1/E_2 < 0.3$ displacement $u_{\rho B}$ decreases because of the strong rock influence.

As is also evident from the above (see Fig. 6, 7) all the results of the FEM and DCM analyses are mostly the same. So far, the stress-strain state of the massif can be estimated by applying the DCM. Besides, the considered domain and the number of finite elements necessary for the DCM analysis were far smaller than those for the FEM analysis.

6 Concluding remarks

Results of the analysis performed present as evidence the practicality of the method under study. Accuracy of the results obtained is mostly the same as accuracy of the FEM' results even though a reduced computational space and number of finite elements are smaller by comparison.

References

1. X. Hong, K. Yong, Z. Mou, Advanced Materials Research, **308-310**, 1683–1687 (2011)
2. C. S. Desai, J. T. Christian, *Numerical Methods in Geotechnical Engineering* (McGraw-Hill Book Company, New York, 1973)
3. W. Jing, C. Yang, J. Kong, W. Ji, Rock and Soil Mechanics, **32, No. S2**, 544–550 (2011)
4. A.V. Vorobev, G.G. Kashevarova, Magazine of Civil Engineering, **9 (44)**, 10-22 (2013)
5. K.A. Issen, J.W. Rudnicki, Journal of Geophysical Reserach, **105, No. 9**, 21529—21536 (2000)
6. S.N. Eigenson, N.V. Korikhin, A.I. Golovin, Magazine of Civil Engineering, **1 (45)**, 59-70 (2014)
7. A.A. Moisja, N.I. Vatin, Magazine of Civil Engineering, **3 (5)**, 7–10 (2009)
8. N.I. Vatin, D.V. Nemova, V. Murgul, V. Pukhkal, A. Golikm, E. Chizhov, Journal of Applied Engineering Science, Istrazivanja i Projektovanja za Privredu, **1**, 37-44 (2014)
9. V.V Okrepilov, A.Yu. Smirnov, Measurement Techniques, **56 (1)**, 54-60 (2013)
10. R. Kamnik, B. Kovačič, B. Pribicević, A. Đapo, Geodetski List, **69** (**3**), 171-188 (2015)
11. B. Kovačič, R. Kamnik, N. Vatin, Applied Mechanics and Materials, **725-726**, 913-921 (2015)
12. O.C. Zienkiewicz, R.L. Taylor, *The Finite Element Method, Volume 2: Solid Mechanics, fifth ed.* (Butterworth-Heinemann, 2000)
13. A.A. Shabana, *Computational continuum mechanics* (CU Press, Cambridge, 2008)
14. B.I. Wohlmuth, *Discretization Methods and Iterative Solvers Based on Domain Decomposition* (Springer, 2001)
15. E.H. Dill, *The Finite Element Method for Mechanics of Solids with ANSYS Applications* (CRC Press, 2012)
16. A. Levandovskiy, B. Melnikov, Applied Mechanics and Materials, **725-726**, 928-936 (2015)
17. I. Maniak, B. Melnikov, A. Semenov, S. Saikin, Applied Mechanics and Materials, **725-726**, 943-948 (2015)
18. A.S. Semenov, S. Sähn, B.E. Melnikov, Proceeding of SPIE, **3687**, 427-436 (1999)
19. B.E. Melnikov, A.S. Semenov, Applied Mechanics and Materials **831**, 187-192 (2014)
20. A.S. Semenov, B.E. Melnikov, M.Yu. Gorokhov, Proceeding of SPIE, **5831**, 167-173 (2005)
21. B.E. Melnikov, A.S. Semenov, Z. angew. Math. Mech (ZAMM), **7, s. 2**, 615-616 (1996)
22. V.V. Lalin, L.A. Rozin, D.A. Kushova, Magazine of Civil Engineering, **1 (36)**, 87-96 (2013)
23. G.S. Kolosova, V.V. Lalin, A.V. Kolosova, Magazine of Civil Engineering, **5 (40)**, 76–85 (2013)
24. E.A. Nekliudova, A.S. Semenov, B.E. Melnikov, S.G. Semenov, Magazine of Civil Engineering, **3(47)**, 25–39 (2014)
25. B.E. Melnikov, N.V. Chernysheva, N.V. Podgornaya, I.N. Chigareva, Construction of Unique Buildings and Structures, **2(7)**, 71-77 (2013)
26. N. Vatin, Ja. Havula, L. Martikainen, A. Sinelnikov, A. Orlova, S. Salamakhin, Advanced Materials Research, **945-949**, 1211-1215 (2014)
27. A.O. Gordeeva, N.I. Vatin, Magazine of Civil Engineering, **3 (21)**, 36-46 (2011)
28. B. Kovačič, R. Kamnik, A. Štrukelj, N. Vatin, Procedia Engineering, **117** (1), 800-806 (2015)
29. B. Kovačič, R. Kamnik, M. Premrov, Survey Review, **43** (320), 150-161 (2011)
30. B. Kovačič, R. Kamnik, A. Štrukelj, Promet - Traffic - Traffico, **22** (1), 15-22 (2010)

MATEC Web of Conferences 53, 01043 (2016)
DOI: 10.1051/matecconf/20165301043

Modeling of nonlinear multiaxial deformation of concrete on the base of hyperelastic orthotropic model

Kirill Lavrov[1,a], Artem Semenov[1] and Andrey Benin[2]

[1]Institute of Applied Mathematics and Mechanics, Peter the Great St. Petersburg Polytechnic University, St. Petersburg 195251, Russia
[2]PSTU reseach division, Peterburg State Transport University, St. Petersburg 190031, Russia

Abstract. The hyperelastic orthotropic material model is proposed to describe the nonlinear behavior of concrete under monotonic multiaxial loading with taking into account the tension-compression anisotropy. The orthotropy is introduced for the correct description of concrete cracking. The hyperelasticity provides unconditional thermodynamical consistency and advantages in numerical solving of boundary value problems. Identification of model parameters is based on four experimental deformation diagrams of concrete: axial stress - axial strain and axial stress - transverse strain under uniaxial tension and compression. The results of the hyperelastic orthotropic model are compared with Karpenko's orthotropic model and experimental data for multiaxial loading.

1 Introduction

Nonlinear structural behavior of concrete is mainly caused by the processes of microcracking under tension and crushing under compression. The modeling of three dimensional nonlinear anisotropic concrete behavior is still unsolved problem. The nonlinear models of concrete are discussed in many various articles and books [1-11 and other]. A main attention is focused on the plasticity models [1-12]. However, in some cases, the use of such complex evolutionary models may be excessive. Under monotonic loading the elastic models may be sufficient that can be a lot easier to take into account the properties of the anisotropy and multiaxiality.

In present study there is shown an attempt of an account of nonlinear, multi-axial, tension-compression anisotropic concrete characteristics based on using of a locally orthotropic hyperelastic material with axes of orthotropy that are coincident with principal stresses directions. The model due to the potential existence has advantages in numerical solution of boundary value problems [13-16]. The advantage of the proposed model is also a simplicity of identification of the model parameters.

[a] Corresponding author: kirill@lavrov.spb.ru

2 Materials and Methods

2.1 Constitutive equations

The formulation of elastic potential (free energy) totally determines the constitutive equations and tensor of elastic moduli for the orthotropic hyperelastic material. Symmetry group of the orthotropic material includes reflections of three orthotropic planes, characterized by normals n_1, n_2, n_3. The axes of orthotropy n_1, n_2, n_3 coincide with principal stresses directions for concrete and can vary (rotate) under complex loading program. It is assumed that the elastic potential ψ is a function of invariants of strain tensor and tensors, which characterize anisotropic material properties:$M_1=n_1\times n_1$, $M_2=n_2\times n_2$ and $M_3=n_3\times n_3$, only two of which are independent due to the equality $M_1+M_2+M_3=1$, where 1 is a unit tensor. In general case there are 7 invariants the for orthotropic material [17]: I_1=trε $I_1 = tr\boldsymbol{\varepsilon}$ $I_2 = \frac{1}{2}\left(I_1^2 - tr\boldsymbol{\varepsilon}^2\right)$, $I_3 = \det\boldsymbol{\varepsilon}$, $I_4 = \mathbf{M}_1 \cdot\cdot\, \boldsymbol{\varepsilon} = tr(\mathbf{M}_1 \cdot \boldsymbol{\varepsilon})$, $I_5 = \mathbf{M}_1 \cdot\cdot\, \boldsymbol{\varepsilon}^2 = tr\left(\mathbf{M}_1 \cdot \boldsymbol{\varepsilon}^2\right)$, $I_6 = \mathbf{M}_3 \cdot\cdot\, \boldsymbol{\varepsilon} = tr(\mathbf{M}_3 \cdot \boldsymbol{\varepsilon})$, $I_7 = \mathbf{M}_3 \cdot\cdot\, \boldsymbol{\varepsilon}^2 = tr\left(\mathbf{M}_3 \cdot \boldsymbol{\varepsilon}^2\right)$.

A second-order Taylor series expansion of the potential $\psi(I_1, I_2, I_4, I_6)$ can be written as seven-term formulation:

$$\psi = \tfrac{1}{2}\lambda J_1^2 + \mu J_2 + \tfrac{1}{2}\alpha I_4^2 + \tfrac{1}{2}\beta I_6^2 + \gamma I_1 I_4 + \delta I_1 I_4 + \varepsilon I_4 I_6 \tag{1}$$

where λ and μ are Lame parameters; α, β, γ, δ and ε are constants of material. In formulation (1) the only liner invariants in respect to strain and I_2 are used. The quadratic and cubic invariants I_3, I_5, I_7 are not used in this formulation for simplicity. The potential (2) is able to describe only the linear material behavior taking into account difference in tension/compression.

The possible generalization of (1) is proposed with aim to take into account the nonlinear behavior of concrete:

$$\psi = \tfrac{1}{2}\lambda J_1^2 + \mu J_2 + a(I_4) + b(I_6) + c(I_1 I_4) + d(I_1 I_4) + e(I_4 I_6)\,, \tag{2}$$

where a, b, c, d and e are piecewise twice continuously differentiable functions determined from experiments under tension and compression.

Expression (2) is a generalization of the formulation [18], which using only the four-term presentation ψ (without last three members in (2)). This simplified four-term formulation allows to describe different resistance of the concrete in tension/compression under uniaxial loading. Introduction of the functions $c(I_1 I_4), d(I_1 I_4), e(I_4 I_6)$ allow to more accurately describe the transverse strain-stress behaviour in a row with longitudinal strain-stress dependence.

For the hyperelastic material the stress tensor is defined by the equation:

$$\boldsymbol{\sigma} = \frac{\partial \psi(\boldsymbol{\varepsilon})}{\partial \boldsymbol{\varepsilon}}\,. \tag{3}$$

A substitution of the potential (1) into the equation (3) leads to the relation:

$$\boldsymbol{\sigma} = \lambda I_1 \mathbf{1} + 2\mu\,\boldsymbol{\varepsilon} + a'\mathbf{M}_1 + b'\mathbf{M}_3 + c'\left(I_1\mathbf{M}_1 + I_4\mathbf{1}\right) + d'\left(I_1\mathbf{M}_3 + I_4\mathbf{1}\right) + e'\left(I_4\mathbf{M}_3 + I_6\mathbf{M}_1\right). \tag{4}$$

Differentiation of (3) leads to a quasi-linear relationship between the rate of stress and rate of strain

$$\dot{\boldsymbol{\sigma}} = {}^4\mathbf{C}(\boldsymbol{\varepsilon}, \mathbf{M}_1, \mathbf{M}_3) \cdot\cdot\, \dot{\boldsymbol{\varepsilon}}\,, \tag{5}$$

where the fourth rank tensor of elastic moduli ${}^4\mathbf{C}=\frac{\partial\boldsymbol{\sigma}}{\partial\boldsymbol{\varepsilon}}=\frac{\partial^2\psi(\boldsymbol{\varepsilon})}{\partial\boldsymbol{\varepsilon}^2}$ can be presented as follows:

$$
\begin{aligned}
{}^4\mathbf{C}=\lambda\mathbf{1}\otimes\mathbf{1}+\mu\left(\mathbf{1}\overline{\otimes}\mathbf{1}+\mathbf{1}\underline{\otimes}\mathbf{1}\right)+a''\mathbf{M}_1\otimes\mathbf{M}_1+b''\mathbf{M}_2\otimes\mathbf{M}_2+c''\left(\mathbf{M}_1\mathbf{1}+\mathbf{1}\mathbf{M}_1\right)+\\
+d''\left(\mathbf{M}_3\mathbf{1}+\mathbf{1}\mathbf{M}_3\right)+e''\left(\mathbf{M}_1\mathbf{M}_3+\mathbf{M}_3\mathbf{M}_1\right)
\end{aligned}
\tag{6}
$$

where the symbols of direct and indirect diadic product $(\mathbf{A}\otimes\mathbf{B})_{ijkl}=A_{ij}B_{kl}$, $(\mathbf{A}\overline{\otimes}\mathbf{B})_{ijkl}=A_{ik}B_{jl}$, $(\mathbf{A}\underline{\otimes}\mathbf{B})_{ijkl}=A_{il}B_{jk}$ are used, a'' is denotes second derivation $\frac{d^2a}{dI_4^2}$, similarly for all other functions.

Equation (5) with account of (6) can be expressed in the vector basis $\mathbf{n}_1,\mathbf{n}_2,\mathbf{n}_3$ in the following matrix form:

$$
\begin{Bmatrix}\dot\sigma_{11}\\ \dot\sigma_{22}\\ \dot\sigma_{33}\\ \dot\sigma_{12}\\ \dot\sigma_{23}\\ \dot\sigma_{31}\end{Bmatrix}=
\begin{bmatrix}
\lambda+2\mu+a''+2c'' & \lambda+c'' & \lambda+c''+d''+e'' & 0 & 0 & 0\\
\lambda+c'' & \lambda+2\mu & \lambda+d'' & 0 & 0 & 0\\
\lambda+c''+d''+e'' & \lambda+d'' & \lambda+2\mu+b''+2d'' & 0 & 0 & 0\\
0 & 0 & 0 & \mu & 0 & 0\\
0 & 0 & 0 & 0 & \mu & 0\\
0 & 0 & 0 & 0 & 0 & \mu
\end{bmatrix}
\begin{Bmatrix}\dot\varepsilon_{11}\\ \dot\varepsilon_{22}\\ \dot\varepsilon_{33}\\ \dot\varepsilon_{12}\\ \dot\varepsilon_{23}\\ \dot\varepsilon_{31}\end{Bmatrix}.
\tag{7}
$$

The inversion of the equation (7) can be rewritten in the principal axes of the stress tensor (assumed that the inequalities $\sigma_1\geq\sigma_2\geq\sigma_3$ fulfilled), as follows:

$$
\begin{Bmatrix}\dot\varepsilon_1\\ \dot\varepsilon_2\\ \dot\varepsilon_3\end{Bmatrix}=
\begin{bmatrix}
\lambda+2\mu+a''+2c'' & \lambda+c'' & \lambda+c''+d''+e''\\
\lambda+c'' & \lambda+2\mu & \lambda+d''\\
\lambda+c''+d''+e'' & \lambda+d'' & \lambda+2\mu+b''+2d''
\end{bmatrix}^{-1}
\begin{Bmatrix}\dot\sigma_1\\ \dot\sigma_2\\ \dot\sigma_3\end{Bmatrix}.
\tag{8}
$$

For comparison let consider the orthotropic elastic model N.I. Karpenko [1], which also takes into account the orthotropy property, but does not possess the property of having the potential:

$$
\begin{Bmatrix}\varepsilon_1\\ \varepsilon_2\\ \varepsilon_3\end{Bmatrix}=\frac{1}{E}
\begin{vmatrix}
\frac{1}{\nu_1} & \frac{-\nu_{t12}}{\nu_2^-} & \frac{-\nu_{t13}}{\nu_3^-}\\
\frac{-\nu_{t21}}{\nu_1^-} & \frac{1}{\nu_2} & \frac{-\nu_{t23}}{\nu_3^-}\\
\frac{-\nu_{t31}}{\nu_1^-} & \frac{-\nu_{t32}}{\nu_2^-} & \frac{1}{\nu_3}
\end{vmatrix}
\begin{Bmatrix}\sigma_1\\ \sigma_2\\ \sigma_3\end{Bmatrix},
\tag{9}
$$

where E is initial Young's modulus of concrete; ν_i is the coefficient characterizing the change of secant modulus under uniaxial loading by $\nu_i=\sigma_i/E\varepsilon_i(i=1,...,3)$; $\nu_{t_{ji}}$ is a secant coefficient of transverse strain defined as $\nu_{t_{ji}}=-\varepsilon_i/\varepsilon_j(j=1,2,3;j\neq i)$. The approximation for ν_i and $\nu_{t_{ji}}$ are proposed in [1, 19, 20]:

$$\nu_i = \nu_i^{\max} \pm (\nu_{0i} - \nu_i^{\max})\sqrt{1 - e_{1i}\eta_i - e_{2i}\eta_i^2} \ , \tag{10}$$

$$\nu_{tji} = \nu_{tji}^{\max} \pm (\nu_{0tji} - \nu_{tji}^{\max})\sqrt{1 - \eta_i^2} \ (i = 3, ji = 13,23) \ , \tag{11}$$

where the positive sign agrees with rising diagram branch and negative sign to descending branch, $\eta_i = \sigma_i / \sigma_i^{\max}$, $\nu_i^{\max}$ is the maximum of secant modulus variation coefficient (top of the diagram), ν_{0i} is the initial value of ν_i (for rising branch $\nu_{0i} = 1$, for descending branch $\nu_{0i} \approx 2.05\nu_i^{\max}$), $e_{1i} = 2 - 2.05\nu_i^{\max}$, $e_{1i} + e_{2i} = 1$.

2.1.1 Identification of material functions.

The four deformation diagrams for concrete are used for the material functions identification:

- axial stress - axial strain under uniaxial tension $\sigma_{xx} = h_t(\varepsilon_{xx})$;
- axial stress - transverse strain under uniaxial tension $\sigma_{xx} = q_t(\varepsilon_{zz})$;
- axial stress - axial strain under uniaxial compression $\sigma_{xx} = h_c(\varepsilon_{xx})$;
- axial stress - transverse strain under uniaxial compression $\sigma_{xx} = q_c(\varepsilon_{zz})$.

Under the uniaxial *tension* along axis x the eigenvalues of stress tensor are $\sigma_1 = \sigma_{xx} > 0$, $\sigma_2 = \sigma_3 = \sigma_{yy} = \sigma_{zz} = 0$ (assumed $\sigma_1 \geq \sigma_2 \geq \sigma_3$), therefore $\mathbf{M}_1 = \mathbf{i} \otimes \mathbf{i}$, $\mathbf{M}_3 = \mathbf{k} \otimes \mathbf{k}$. Under natural supposition $\varepsilon_{yy} = \varepsilon_{zz}$ we have $I_1 = \varepsilon_{xx} + 2\varepsilon_{zz}$, $I_4 = \varepsilon_{xx}$, $I_6 = \varepsilon_{zz}$. The equation (7) for the considering loading case leads to the equations:

$$\begin{cases} \dot{\sigma}_{xx} = (\lambda + 2\mu + a'' + 2c'')\dot{\varepsilon}_{xx} + (2\lambda + 2c'' + d'' + e'')\dot{\varepsilon}_{zz} \\ \dot{\sigma}_{zz} = (\lambda + c'' + d'' + e'')\dot{\varepsilon}_{xx} + (2\lambda + 2\mu + b'' + 3d'')\dot{\varepsilon}_{zz} = 0 \end{cases} \tag{12}$$

The tension deformation diagram $\sigma_{xx} = h_t(\varepsilon_{xx})$ is determined after integration of $\dot{\sigma}_{xx} = h_t'\dot{\varepsilon}_{xx}$, which can be obtained by the substitution of $\dot{\varepsilon}_{zz}$ as a function $\dot{\varepsilon}_{xx}$ from the second equation of the system (12) to the first equation of the system. Similarly the diagram $\sigma_{xx} = q_t(\varepsilon_{zz})$ can be found by the substitution of $\dot{\varepsilon}_{xx}$ as a function $\dot{\varepsilon}_{zz}$ from the second equation of the system (12) to the first equation.

Under the uniaxial *compression* along axis x the eigenvalues of stress tensor are $\sigma_1 = \sigma_2 = \sigma_{zz} = \sigma_{yy} = 0$, $\sigma_3 = \sigma_{xx} < 0$, therefore $\mathbf{M}_1 = \mathbf{k} \otimes \mathbf{k}$, $\mathbf{M}_3 = \mathbf{i} \otimes \mathbf{i}$. Taking into account $\varepsilon_{yy} = \varepsilon_{zz}$ we have $I_1 = \varepsilon_{xx} + 2\varepsilon_{zz}$, $I_4 = \varepsilon_{zz}$, $I_6 = \varepsilon_{xx}$. The equation (7) for the considering loading case and invariants concretization leads to the equations:

$$\begin{cases} \dot{\sigma}_{xx} = (\lambda + 2\mu + b'' + 2d'')\dot{\varepsilon}_{xx} + (2\lambda + c'' + 2d'' + e'')\dot{\varepsilon}_{zz} \\ \dot{\sigma}_{zz} = (\lambda + c'' + d'' + e'')\dot{\varepsilon}_{xx} + (2\lambda + 2\mu + a'' + 3c'')\dot{\varepsilon}_{zz} = 0 \end{cases} \tag{13}$$

The compression deformation diagram $\sigma_{xx} = h_c(\varepsilon_{xx})$ is determined after integration of $\dot{\sigma}_{xx} = h_c'\dot{\varepsilon}_{xx}$, which can be obtained by the substitution of $\dot{\varepsilon}_{zz}$ as a function $\dot{\varepsilon}_{xx}$ from the second equation of the system (13) to the first equation of the system. Similarly the diagram $\sigma_{xx} = q_c(\varepsilon_{zz})$ can be found by the substitution of $\dot{\varepsilon}_{xx}$ as a function $\dot{\varepsilon}_{zz}$ from the second equation of the system (13) to the first equation.

Elastic constants λ, μ we assume known from the initial slope of stress-strain diagram. For definition of five functions a, b, c, d, e we have four equations (12)-(13) therefore to eliminate the uncertainties, we can introduce an additional constraint. To simplify the presentation we assume (there are other options):

$$c''+d''+e''=0\,. \tag{14}$$

In view of (14) we get four equations for the four functions a, b, c, d :

$$\begin{cases} \dot{\sigma}_{xx}=\left(\lambda+2\mu+a''+2c''-(2\lambda+c'')\dfrac{\lambda}{2\lambda+2\mu+b''+3d''}\right)\dot{\varepsilon}_{xx}=h_t^{'}\dot{\varepsilon}_{xx} \\ \dot{\sigma}_{xx}=\left(2\lambda+c''-(\lambda+2\mu+a''+2c'')\dfrac{2\lambda+2\mu+b''+3d''}{\lambda}\right)\dot{\varepsilon}_{zz}=-q_t^{'}\dot{\varepsilon}_{zz} \\ \dot{\sigma}_{xx}=\left(\lambda+2\mu+b''+2d''-(2\lambda+c''+2d''+e'')\dfrac{\lambda}{2\lambda+2\mu+a''+3c''}\right)\dot{\varepsilon}_{xx}=h_c^{'}\dot{\varepsilon}_{xx} \\ \dot{\sigma}_{xx}=\left(2\lambda+c''+2d''+e''-(\lambda+2\mu+b''+2d'')\dfrac{2\lambda+2\mu+a''+3c''}{\lambda}\right)\dot{\varepsilon}_{zz}=-q_c^{'}\dot{\varepsilon}_{zz} \end{cases} \tag{15}$$

where $h_t^{'}$ and $h_c^{'}$ - the inclinations of stress and strain diagrams that are found experimentally.

After the transformation of (15) we obtain

$$\begin{cases} (\lambda+2\mu+a''+2c''-h_t^{'})(2\lambda+2\mu+b''+3d'')-(2\lambda+c'')\lambda=0 \\ (\lambda+2\mu+a''+2c'')(2\lambda+2\mu+b''+3d'')-(2\lambda+c''+q_t^{'})\lambda=0 \\ (\lambda+2\mu+b''+2d''-h_c^{'})(2\lambda+2\mu+a''+3c'')-(2\lambda+d'')\lambda=0 \\ (\lambda+2\mu+b''+2d'')(2\lambda+2\mu+a''+3c'')-(2\lambda+d''+q_c^{'})\lambda=0 \end{cases} \tag{16}$$

The solution of (16) is

$$\begin{cases} c''=\left\{\lambda\left[\dfrac{q_t^{'}}{h_t^{'}}\left(\dfrac{q_c^{'}}{h_c^{'}}-1\right)-2\right]-q_t^{'}\right\}\Big/\left(\dfrac{q_t^{'}}{h_t^{'}}+1\right) \\ d''=\left\{\lambda\left[\dfrac{q_c^{'}}{h_c^{'}}\left(\dfrac{q_t^{'}}{h_t^{'}}-1\right)-2\right]-q_c^{'}\right\}\Big/\left(\dfrac{q_c^{'}}{h_c^{'}}+1\right) \\ a''=\dfrac{q_c^{'}}{h_c^{'}}\lambda-3c''-2\lambda-2\mu \\ b''=\dfrac{q_t^{'}}{h_t^{'}}\lambda-3d''-2\lambda-2\mu \end{cases} \tag{17}$$

The relations $\dfrac{q_t^{'}}{h_t^{'}}$ and $\dfrac{q_c^{'}}{h_c^{'}}$ represent the tangential coefficients of transverse deformation (analog of Poisson's ratio in a nonlinear case) under tension and compression respectively.

Integrating of (17) is complicated in a general case, however in some cases it could be possible. For instance, piecewise linear approximation of concrete deformation diagrams is obtained by using the piecewise quadratic functions:

$$a(x) = \tfrac{1}{2}\sum_i A_i x^2 \left[H(x - x_{i+1}) - H(x - x_i)\right], \tag{18}$$

$$b(x) = \tfrac{1}{2}\sum_i B_i x^2 \left[H(x - x_{i+1}) - H(x - x_i)\right] \tag{19}$$

and linear functions

$$c(x) = \sum_i C_i x\left[H(x - x_{i+1}) - H(x - x_i)\right], \tag{20}$$

$$d(x) = \sum_i D_i x\left[H(x - x_{i+1}) - H(x - x_i)\right], \tag{21}$$

where $H(x)$ is a Heaviside function. The coefficients $A_i = a'', B_i = b'', C_i = c'', D_i = d''$ are defined by (17) within each interval.

Differing from piecewise linear approximation of concrete deformation diagrams (polynomial, exponential and power-type functions) can be also used [18], but integration of (17) is not easy in these cases and therefore variant with (18)-(19) is preferable.

3 Results and Discussion

The various monotonic three-axial loadings are considered. In the *first* example the stresses components are defined as $\sigma_3 : \sigma_1 : \sigma_2 = -1 : -1/10 : -1/10$. The dependences $\varepsilon_3(\sigma_3)$ and $\varepsilon_1(\sigma_3)$ can be found after inversion and integration of Eq. (8) for for hyperelastic orthotropic model and on the base of (9) for the orthotropic elastic model of N.I. Karpenko. In computations the material parameters corresponding to the concrete are used: $R_b = 41.8$ MPa, $R_{bt} = 2.6$ MPa, ε_R=0.000215, E_0=30400 MPa. Stress-strain diagrams for the concrete have been taken from [1] and [21]. The results of comparison of the hyperelastic orthotropic model with Karpenko's orthotropic model and experimental data [24] are shown in Fig. 1. Both models show a good prediction accuracy compared with experiment.

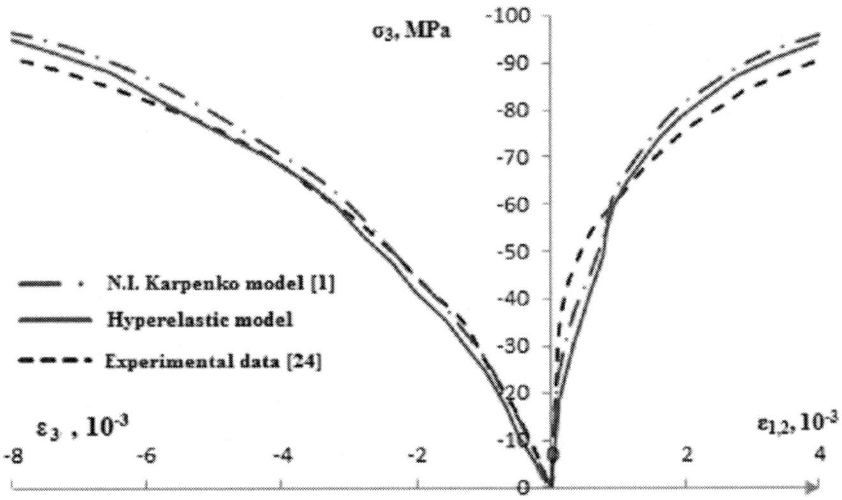

Figure 1. Comparison of the hyperelastic orthotropic model prediction with N.I. Karpenko's orthotropic model results and experimental data for multiaxial loading $\sigma_3 : \sigma_1 : \sigma_2 = -1 : -1/10 : -1/10$.

In the *second* example the stress tensor components are defined as $\sigma_3 : \sigma_1 : \sigma_2 = -1 : -1/14.5 : -1/14.5$. The dependences $\varepsilon_3(\sigma_3)$ and $\varepsilon_1(\sigma_3)$ can be found in the same manner as in example 1. In computations the material parameters corresponding to the concrete are used. The results of comparison of the hyperelastic orthotropic model with Karpenko's orthotropic model and experimental data [24] are shown in Fig. 2. Both models demonstrates a good prediction accuracy in comparison with experiment.

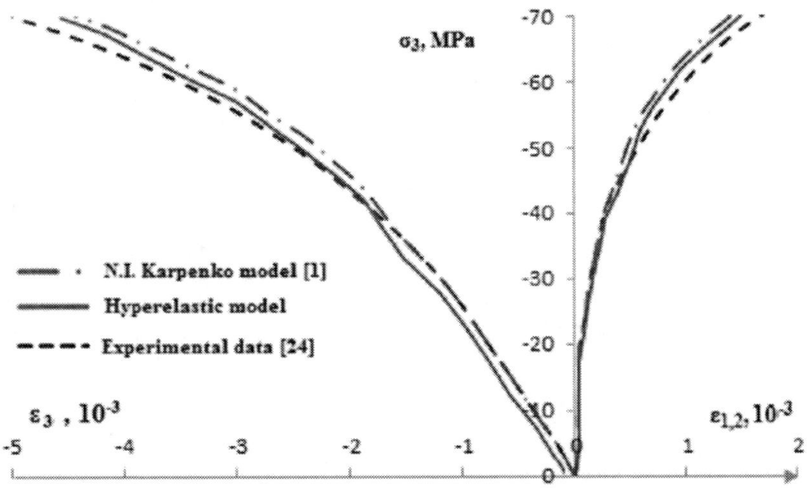

Figure 2. Comparison of the hyperelastic orthotropic model prediction with N.I. Karpenko's orthotropic model results and experimental data for multiaxial loading $\sigma_3 : \sigma_1 : \sigma_2 = -1 : -1/14.5 : -1/14.5$.

4 Conclusions

The seven-term formulation of hyperelastic orthotropic material model is proposed to describe the nonlinear behavior of concrete under monotonic multiaxial loading with taking into account the tension-compression anisotropy. The method of material parameters identification is suggested based on four experimental deformation diagrams: axial stress - axial strain and axial stress - transverse strain under uniaxial tension and compression. The responses of the hyperelastic orthotropic model are compared with Karpenko's orthotropic model and experimental data for three-axial loading.

In the next steps of investigations there are the comparison with plastic models of concrete and finite-element implementation [22] of the proposed model and solutions of nonlinear boundary value problems [13-16, 23].

References

1. N.I. Karpenko, General solution of armored conrete mechnics, 416 p. (1996)
2. G.A. Geniev, V.N. Kissyuk, G.A. Tupin, The theory of plasticity of concrete and reinforced concrete, 316 p. (1974)
3. T.M. Petsold, Reinforced concrete structures, 308 p. (2003)
4. J. Lubliner, J. Oliver, S. Oller, E. Onate, Int. J. Solids Struct., **25(3)**, 299-326 (1989)
5. Ortiz, M. Mech. Mater., **4,** 67-93 (1985)
6. Z.P. Bazant, Y. Xiang, P.C. Prat, J. Eng. Mech., **122,** 245-254 (1996)
7. K.J. Willam, E.P. Warnke, International Association for Bridges and Structural Engineering, Bergamo (1974)
8. H. Kupfer, H.K. Hilsdorf, H. Rush, ACI Journal, **66,** 656-666 (1969)
9. S.P. Shah, Mech. Engrg. Mat., 579-590 (1984)
10. M.A. Taylor, A.K. Jain, M.R. Ramey, ACI. J, **69**, 758-764 (1972)
11. Bangash, Elsevier Publications, (1989).
12. B.E. Melnikov, A.S. Semenov, S.G. Semenov, Proc. of the A.N. Krylov Shipbuilding Research Institute, **53**, 85-92 (2010)
13. A.V. Benin, A.S. Semenov, S.G. Semenov, B.E. Melnilov, Magazine of Civil Engineering, **7 (33),** 32-42 (2012)
14. A.V. Benin, A.S. Semenov, S.G. Semenov, Advances in Civil Engineering and Building Materials, 233-237 (2012)
15. A.V. Benin, A.S. Semenov, S.G. Semenov, B.E. Melnilov, Magazine of Civil Engineering. **5 (40)**, 86-99 (2013)
16. A.V. Benin, A.S. Semenov, S.G. Semenov, B.E. Melnilov, Magazine of Civil Engineering. **1 (45)**, 23-40 (2013)
17. A.J.M. Spencer, Theory of invariants, 239–353 (1971)
18. K.I. Lavrov, A.S. Semenov, A.V. Benin, Proc. of Conf. XLIII Science Week of St. Petersburg state polytechnical University. Institute of Applied Mathematics and Mechanics, 21-23 (2014)
19. N. I. Karpenko, Structural Mechanics and Engineering. **2**, 31-36 (1982)
20. N. I. Karpenko, Concrete stress-deformation diagram master curve composition, 164-173, (1983)
21. A.V. Benin, Straining and destruction of armored concrete: analytical, numerical and experimental research, 127 p. (2006)
22. A.S. Semenov, Proc. of V Int. Conf. "Scientific and technical problems of forecasting the reliability and durability of the structures and methods for their solution", 466-480, (2003)
23. A.V. Benin, A.S. Semenov, S.G. Semenov, Advanced Materials Research, **831,** 364-369 (2014)
24. Y. N. Malashkin, Deformation and destruction of a concrete under multiaxial stress condition 28p. (1984)

MATEC Web of Conferences **53**, 01044 (2016)
DOI: 10.1051/matecconf/20165301044

Improved energy efficiency and optimization of microclimate in buildings

Vesna Trifunovič Dragišić [a],

College of Applied Studies in Civil Engineering and Geodesy, University of Belgrade, Hajduk Stankova 2, 11000 Belgrade, Serbia

Abstract. Nowadays it is possible to reduce energy consumption without losing comfort as a result of using efficient energy saving technologies and advanced environment control methods for buildings. One of the measures to improve energy performance of buildings can be installation of decentralized air intake and exhaust mechanical ventilation systems with plate heat exchangers in apartments making it possible to «return» up to 85% of thermal energy. The article deals with the decentralized system controlled ventilation with heat recovery and alternative solutions heating supply air in residential buildings.

1 Introduction

Improved energy efficiency of buildings is one of the most urgent problems of construction science nowadays. Lower energy consumption and energy saving technologies make it possible to comply with the standards of sustainable development of the society and provide comfort to every household. Energy consumption of the municipal sector accounts for about 40 % of the total amount. Reduced energy consumption and improving the quality of the microclimate inside the buildings are the main goals of energy-efficient renovation of old buildings [1-5]. Besides, energy efficient upgrading of old buildings will improve the indoor climate, reduce the cost of electricity and heat, decrease carbon dioxide emissions, increase present value of the building and improve its condition and durability. Schematic diagrams of the ventilation systems of historic buildings are shown in Figure 1 [3-5]

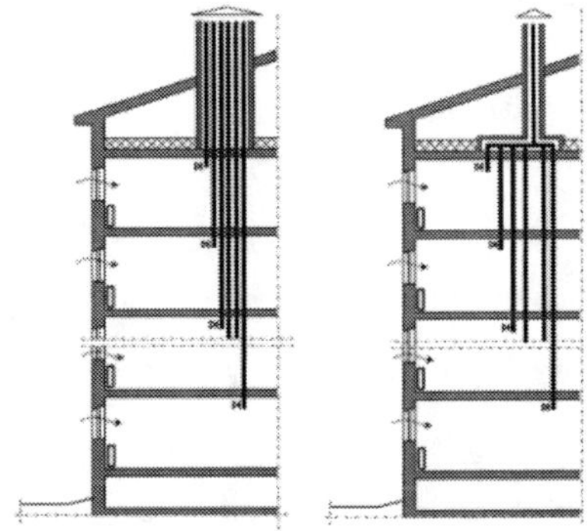

Figure 1. Schematic diagrams of the ventilation systems of historic buildings.

[a] Corresponding author : vesna.tg@bk.ru

A set of measures should be taken to achieve reduction of heat energy needed for residential buildings. These could be the following measures: thermal insulation of the building's envelope, recovery of heat from ventilation emissions wastewater disposal, heating systems optimization. Heat energy can be saved to a significant extent due to heat recovery in building ventilation system . The «Passive house» concept has appeared to be the most valuable one with respect to energy efficiency. The rate of 15 kW / (m^2 per year) is the typical one in matter of energy consumption needed to heat a «passive house» under weather conditions of Central Europe. In Stockholm it can reach 20 kW / (m^2 per year), and in Rome it can't be over 10 kW / (m^2 per year) [5]. The concept is based on the statement to reduce heating costs to zero and achieve constant comfortable temperatures owing to efficient thermal insulation and impermeability of a building's envelope, any home heat recovery and passive solar heating (Figure 2).

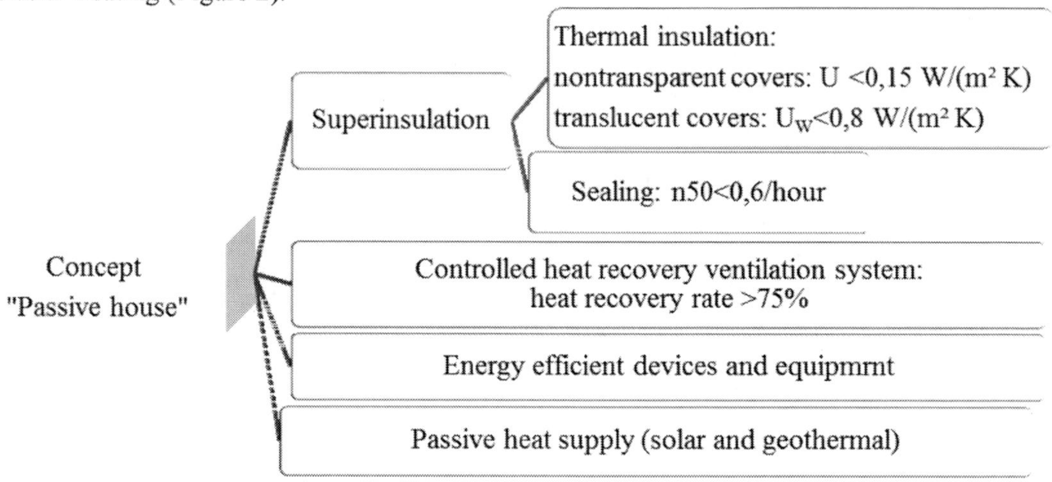

Figure 2. Basic components of the «Passive house» concept.

2 Decentralized ventilation system in residential buildings and apartments

One of the measures to improve energy performance of buildings can be installation of decentralized air intake and exhaust mechanical ventilation systems with plate heat exchangers in apartments making it possible to «return» up to 85% of thermal energy [7, 8]. In this case separate individual single-block intake and exhaust ventilation systems with cross-flow plate-based heat exchangers are installed in every apartment (Figure 3, 4). An air ductwork is arranged throughout every apartment. Air intake into apartments is supposed to be arranged through air diffusers, and air recovery is supposed to be arranged in kitchens, toilets and bathrooms. Outer air is heated in a heat exchanger by the air recovered from an apartment.

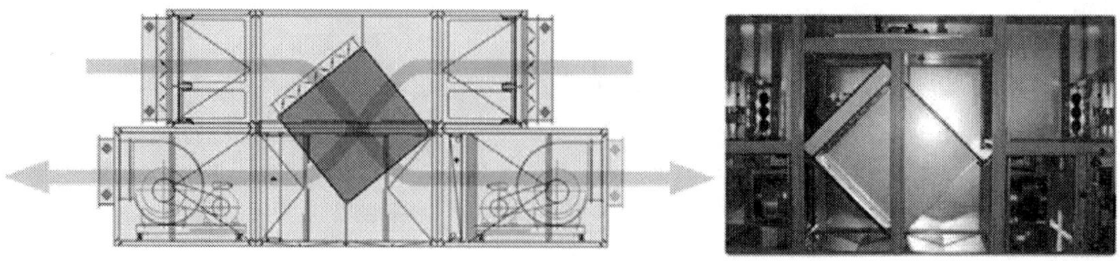

Figure 3. Plate heat exchanger designed to recover heat from air in ventilation systems and general view of a heat exchanger [6]

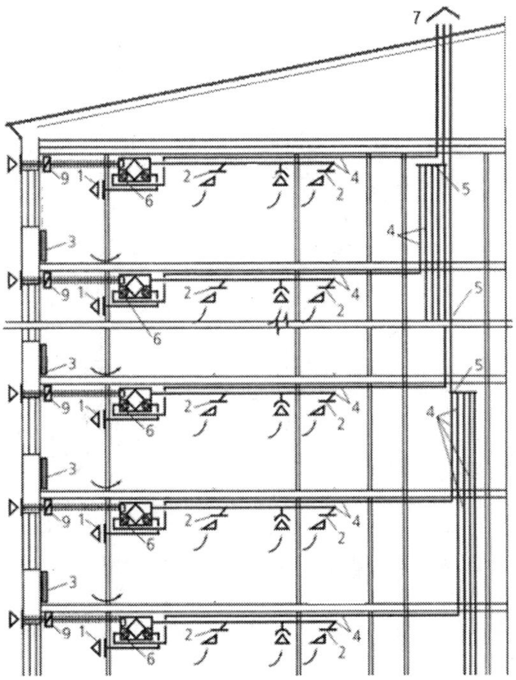

Figure 4. Scheme decentralized mechanical supply and exhaust ventilation systems with heat recovery boiler plate type [4].

1 – air intake device; 2 – exhaust device; 3 – heating device; 4 – satellite-duct; 5 – assembly exhaust duct; 6 - heat exchangers (individual); 7– exhaust shaft

Fans have a possibility of 9-stage digital regulation. Residents may change air exchange levels from 0 to 9 (the value '0' implies that the system is switched off; the values 1, 2, 3 are optimal for all the types of apartments and ensure the standard level of air exchange depending on the area of apartments, the values above 3 can level air exchange up but increase the level of noise).

It is necessary to meet all the regulatory requirements for air exchange before to put energy efficient building into operation. To achieve this aim the following intake and exhaust air consumption rates have been regulated: 110 m3/h is for 1-2 room apartments, 130 m3/h is for 3- room apartments and 180 m3/h is for 4- room apartments. The balance of air consumption in intake and exhaust channels has been regulated with due account for standard exhaust air volumes in kitchens and toilets.

Dispatching systems ensure that information on the parameters of controlled air exchange modes in every apartment is received. The figure 5 shows general automated survey data on system performance in apartments in reference to actual operation quality of electric regulators meant for setting appropriate parameters by residents.

The major parameter in respect to thermal energy balance available for residents is an air exchange level. It can have impact on both the value of heat losses in the building and thermal loads between separate apartments of the building. A controlled air exchange system makes it possible to save energy due to management of ventilation at different times of the day. Air exchange level can be limited up to 50% of the standard rate for those residents who work over 70 hours a week and are outside the building (at work, in a shop, or absent for a walk) [9-13].

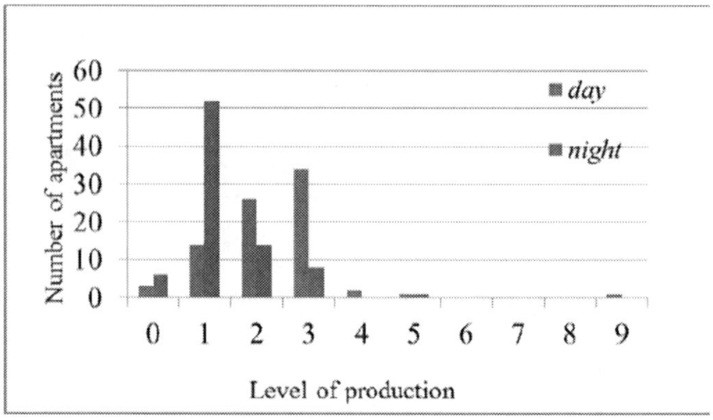

Figure 5. Air exchange level set by residents during daytime and nighttime.

The notion of comfort is subjective, often we take cold for freshness and superheated air for stuffiness. In addition, as a rule, family oriented to those to whom "cold". Mechanical ventilation system with individual control may lead to a lack of ventilation in the premises relatively normalized values. Also, for many residents an opportunity to open and close windows is a necessary component of personal comfort.

The tightness is one of the main principles energy efficiency design and construction.

In our time, "in the pursuit of energy efficiency" designed and advertised times sealed windows, do not let the air, and thus turned the main advantage of windows - ensure supply of fresh air - in a threatening health disadvantage.

Requirements of air permeability for windows in Russian regulations about air permeability changed in the following way [14]:

- 1971 - 18 kg/ (m^2•h);
- 1979 - 10 kg/ (m^2•h);
- 1998 - 5 kg/ (m^2•h).

The natural ventilation system of multi-storey buildings is still traditionally based on air infiltration through cracks and gaps of windows. Therefore we can conclude that imposing tougher requirements of air permeability for windows has been the first step to draught-proofing, which has led to worsening microclimate with inappropriate air interchange.

In older buildings energy efficiency reconstruction is usually started by occupants who replace their old windows, providing natural ventilation, with new tightness windows. Thereby air interchange reduces, microclimate worsens and damp rises.

It is essential to switch from energy saving at all cost to microclimate improvement. The priority target of the contemporary construction is to form the appropriate climate in living spaces.

The use of solar energy for power supply of residential buildings is a must-have for the "green design". Solar energy has been used successfully in the reconstruction and restoration of historic buildings [15 - 19].

Reception heating the incoming air in the glazed space of the building facade without the tightness of the building envelope is implemented in a number of projects (Figure 6, 7, 8)

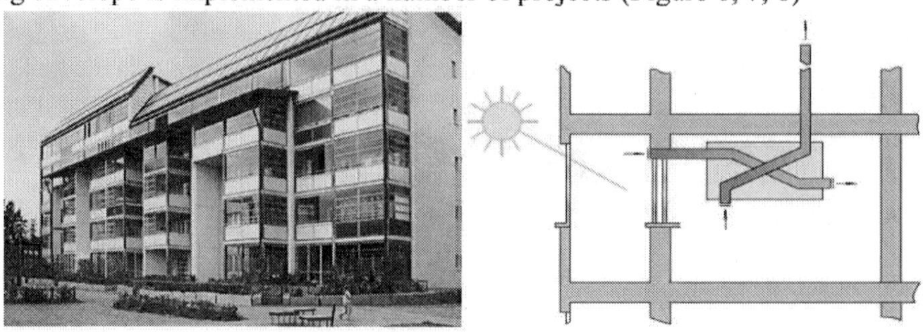

Figure 6. "Ecological housing" area Viikki, Helsinki (Finland).

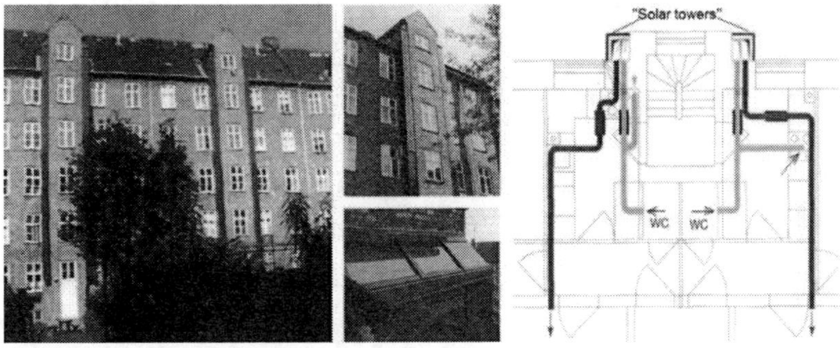

Figure 7. Reconstruction of a residential building, Frederiksberg (Denmark) [20].

Figure 8. Reconstruction of a residential building, Østerbro, Copenhagen (Denmark) [20].

Commerzbank famous building is an example of energy-efficient construction with natural and controlled ventilation. Effective management of heat flow in a double glass façade provides natural air conditioning of the building and significant energy savings (Fig. 9)

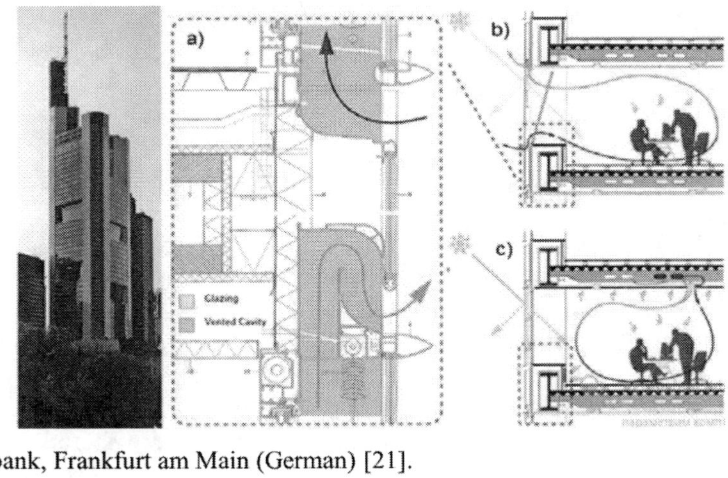

Figure 9. Commerzbank, Frankfurt am Main (German) [21].

a) Double skin façade (vented cavity); b) passive heating/cooling/ventilation system; c) active heating/cooling/ventilation system

"Plan, 'sky' gardens, and a full-height atrium that allows for every office in the tower to have operable windows for views, natural ventilation, and daylight.

The four-storey 'sky' gardens provide fresh air and allow for passive solar gain, while the central atrium space acts like a natural ventilation chimney for the inward-facing offices. Despite the effectiveness of passive systems in Frankfurt's temperate climate zone, Foster recognizes the potential for technical (active) systems to maximize the building's efficiency. A computer controlled building management system decides whether passive or active systems are most appropriate for use at a given time and adjusts openings and shading devices accordingly. During the design of the glazing, ventilation slots are included allowing air to enter through the bottom sill of the insulated glass and exit at the head. The fixed exterior glass within the operable window system slows the flow of air avoiding the creation of a draught or the entry of water when the interior portion of the window is opened" [21].

3 Summary

Provide the necessary ventilation rate and avoid heat losses due to cooling of you can seal the building as the unit mechanical ventilation with heat recovery exhaust air. Using mechanical air intake and exhaust ventilation system in multi-storey residential buildings requires creating new solutions for air ventilation arrangement. Accordingly the following actions should be taken: to choose an appropriate air intake and recovery scheme, define output of intake and exhaust air fans, determine duct sections values required. But can also be used passive solar heating of the incoming air in the glazed facade. But certainly not always architectural solution allows dominance of glass on the facade. In fact, in this case we are talking about managing heat flows by means of architectural design.

References

1. Y. Nikitin, V. Goryunov, V. Murgul, N. Vatin, Applied Mechanics and Materials, **680**, 504-509 (2014)
2. N. Harmati, Z. Jakšić, N. Vatin, (2015) Procedia Engineering, **117 (1)**, 791-799 (2015)
3. M. Penića, S. Golovina, V. Murgul, Procedia Engineering, **117**, 883-890 (2015)
4. Z. Jakšic, D. Ladjinović, M. Trivunić, N. Harmati, N. Vatin, (2015) Procedia Engineering, **117 (1)**, 507-520 (2015)
5. Feist, W., Schnieders, J., Dorer, V., Haas, A. Re-inventing air heating: Convenient and comfortable within the frame of the Passive House concept (2005) Energy and Buildings, Vol. 37(11), pp. 1186–1203.
6. Gabriyel, I., Ladener, Kh. *Rekonstruktsiya zdaniy po standartam energoeffektivnogo doma* (St. Peterburg: BKhV-Peterburg, 2011).
7. V. Murgul, D. Vuksanovic, V. Pukhkal, N. Vatin, Applied Mechanics and Materials, **633-634**, 977-981 (2014)
8. V. Murgul, D. Vuksanovic, N. Vatin, V. Pukhkal, Applied Mechanics and Materials, **635-637**, 370-376 (2014)
9. V. Murgul, D. Vuksanovic, N. Vatin, V. Pukhkal, Applied Mechanics and Materials, **680**, 524-528 (2014)
10. V. Pukhkal, N. Vatin, V. Murgul, Applied Mechanics and Materials, **680**, 529-533 (2014)
11. D. Vuksanovic, Y. Nikitin, V. Murgul, N. Vatin, V. Pukhkal, Applied Mechanics and Materials, **680**, 499-503 (2014)
12. V. Pukhkal, V. Murgul, M. Garifullin, Procedia Engineering, 117, 624-627 (2015)
13. V. Pukhkal, N. Vatin, V. Murgul, Applied Mechanics and Materials, **633-634**, 1077-1081 (2014)
14. V. Murgul, Journal of Applied Engineering Science, **12 (1)**, 1-10 (2014)
15. R. Alihodzic, V. Murgul, N. Vatin, E. Aronova, V. Nikolić, M. Tanić, D. Stanković, Applied Mechanics and Materials, **624**, 604-612 (2014)
16. G. Radovic, V. Murgul, N. Vatin, E. Aronova, Applied Mechanics and Materials, **627**, 357-364 (2014)
17. V. Murgul, N. Vatin, I. Zayats, Procedia Engineering, **117**, 824-829 (2015)
18. G. Radović, V. Murgul, M. Cvetkovska, E. Aronova, N. Vatin, Journal of Applied Engineering Science, **12 (4)**, 277 – 284 (2014)
19. M. Penić, N. Vatin, V. Murgul, Applied Mechanics and Materials, **680**, 534-538 (2014)
20. Y. Tabunschikov, M. Brodach, *Energy efficient buildings* (Moscow: AVOK-PRESS, 2003)
21. http://www.mbenkert.com/arend.benkert.defilippis.tillmaand.pdf (12.12.2015)

MATEC Web of Conferences **53**, 01045 (2016)
DOI: 10.1051/matecconf/20165301045

Forecasted Payback Period in the Case of Energy-Efficient Activities

Alexander Gorshkov[1], Vera Murgul[1,a] and Olena Oliynyk[2]

[1]*St. Petersburg State Polytechnical University, Polytechnicheskaya st. 29, St. Petersburg, 195251, Russia*
[2]*National Aviation University, Kosmonavta Komarova, 1, Kiev, 03680, Ukraine*

Abstract. There is a very important governmental issue, which is to secure energy efficiency of buildings under design. This issue is stated in the federal law, governmental regulations and orders of the subject Ministries. However, issues related to energy-efficiency are appropriate for both existing buildings and the ones under design. Total volume of buildings constructed according to new heat protection rules and regulations valid since 2000, accounts for 5% of the total volume of all the housing stock in the Russian Federation. Consequently, reduction of heat energy consumption in houses built before 2000 is getting more crucial in the nearest future. Nowadays there is a national programme to execute major repair works in the existing buildings. It is important to have this programme correlated with the energy-saving program. However, the economy of investment projects shouldn't be disregarded when running this programme. Energy-efficient activities should contribute to a better return on investment, and reduction of the energy consumed.

1 Introduction

One of the ways to reduce heat losses for heating can be additional insulation of exterior building envelopes (walls, slabs, attic slabs, exterior doors, etc.). Increase in insulation in building envelopes leads to reduction of heat losses due to transmission. The less there are heat losses in a building the less amount of heat energy is required to supply to the building from a heat supply source in order to recover heat losses due to transmission (provided the requirements for microclimate are met) [1-8]. Thus, insulation leads to reduction of the energy consumed in the building, and, consequently, to cut-down of heating costs [8]. The economic effect due to this energy-saving activity is based on this principle. However, it will require additional capital investments. Economic efficiency of the energy-saving activity to be introduced can be featured by its payback period. In the case of the payback period less than estimated service life and maintenance duration of the technological solution to be applied [9-12], it should be stated as economically viable, i.e. energy-efficient.

As to insulation in existing buildings it should be noted that buildings constructed according to regulations valid before the introduction of the Federal Law of the Russian Federation "On energy-efficiency …" (FL № 261 dated 23 November 2009) are outdated (but not from physical standpoint). They became outdated due to insufficient level of heat protection of exterior envelopes. With the energy-saving programme coming into existence, standard requirements for heating protection in

[a] Corresponding author: october6@list.ru

buildings were strengthened. For this reason existing buildings do not comply with the new standards. And there is only one opportunity to make them conform to new rules and regulations, which is insulation of existing buildings at up-to-date or even higher level.

The method to calculate the payback period of energy-efficiency investments is proposed in the works [13, 14]. The method to estimate the level of heat protection efficiency in the case of exterior envelopes using net present value (NPV), obtained as a result of the energy-saving activity given, is demonstrated in the work [15]. The calculation method to estimate economic efficiency of energy-saving translucent structures is stated in the industry standard IS UMPP (Union of Manufacturers of Polymer Profiles) 4.5-2012 [16].

This survey deals with the calculation method of the payback period in the case of investments meant for façade insulation in the existing building.

2 Calculation model

A calculation method of specific costs has been used herein to calculate the payback period of investments meant for façade insulation in the existing block of flats. The following basic equation is given to calculate a simple payback period [17, 18].

$$T = \frac{K_2}{Э_1 - Э_2} = \frac{\Delta K}{\Delta Э}. \tag{1}$$

where

ΔK – difference of capital costs meant for renovation per 1 m2 of exterior walls, Rub/m2 (ΔK= K2, as K1=0);

Э1 – operation and maintenance costs allowing for heat losses through 1 m2 of an exterior wall during one heating season before insulation works of exterior walls are executed, Rub/ m2·per year;

Э2 – operation and maintenance costs allowing for heat losses through 1 m2 of an exterior wall during one heating season after insulation works of exterior walls, Rub/ m2·per year;

ΔЭ – the difference between heat losses through 1 m2 of an exterior wall before and after insulation works of facades in the existing building $Э_1$-$Э_2$ are executed.

In the formula (1) annual savings per year ΔЭ , Rub/ m2 to be achieved due to renovation of façades in the existing building and reducing transmission heat losses in the building can be determined as follows [14]:

$$\Delta Э = (U_1 - U_2)\cdot\frac{0{,}024\cdot HSDD}{1163}\cdot c_h, \tag{2}$$

where

U1 – heat transfer coefficient (U-value) of exterior walls in the existing building before renovation (insulation) works of the facades, W/(m2·K);

U2 – heat transfer coefficient (U-value) of exterior walls in the existing building after renovation (insulation) works of the facades, W/(m2·K);

HSDD – heating season degree-day, K·day;

0.024, 1163 – conversion factors;

Ch– cost of heat energy for heating in the dwelling area given, Rub/Gcal.

Let us remark here that the payback period calculated under the formula (1) does not allow for:

- the growth of rates for heating energy;
- loan interest (provided a loan for exterior walls insulation is taken);
- future discounted cash flows obtained as a result of the energy saving activity considered herein such as reducing of heat losses for heating..

Due to this reason the value of a non-discounted payback period, calculated under the formula (1), can be considered as an estimated one.

Provided a construction company or private individual implements insulation works using own (non-borrowed) resources then capital costs ΔK will be equal to cost estimates. If an executive company uses a loan (a bank credit) to perform works on the basis of equal monthly installments (annuities) total investments into energy efficiency $\Delta \widetilde{K}$ should be determined under the formula:

$$\Delta \widetilde{K} = m \cdot A \cdot \Delta K, \qquad (3)$$

where

m – a number of redemption periods (e.g., if there is a 1-year loan: m=12, if there is a 2-years loan: m=24 and etc.);
A – annuity factor;
ΔK– own resources of a performing company (investments without regard to loan payments).
Annuity factor A can be determined according to the formula:

$$A = \frac{p_l \cdot (1+p_l)^m}{(1+p_l)^m - 1}, \qquad (4)$$

where

p_l – monthly interest of a bank loan rounded to two decimal places per payment frequency (e.g., in the case of 12 % per annum and monthly payments (12 times per year): p_l=0.12/12=0.01);

m – the same as in the formula (3).

Coming from the analysis of the formulas (1) and (2), in particular, it follows that with the facades insulation project given herein (ΔU=U1-U2) under climatic conditions of the area given HSDD, rate of return on investment depends just on the value of heat energy C_h and its dynamics of change through time (rate growth for heat energy).

Rates for heating energy rise annually. It means that with every year (heating season) annual savings of monetary funds $\Delta Э_i$ will increase.

When considering the model it should be taken into account that money funds saved in subsequent years must be calculated on the basis of actual value of money in n years, i.e. future cash flows must be discounted.

Allowing for all the factors mentioned above, the forecasted payback period of investments meant for additional facades insulation should be determined using the following equation [17, 20]:

$$T = \frac{\ln\left[1 + \frac{\Delta \widetilde{K}}{\Delta Э} \cdot \frac{(r-i)}{(1+i)}\right]}{\ln\left[\frac{1+r}{1+i}\right]}, \qquad (5)$$

where

$\Delta \widetilde{K}$ – the same as in the formula (3), Rub/m2;

$\Delta Э$ – the same as in the formula (2), Rub/m2;

r – average annual growth of rates for heating energy;
i – discount rate.

The equation (5) makes it possible to calculate the payback period T of the energy-saving activity under consideration with due account for total capital costs to have it performed $\Delta \widetilde{K}$, loan payments (P_l), growth of the rate for heat energy (r), discounting of future cash flows (i), which can be achieved using fund savings due to implementation of such an energy-saving activity.

Let us remark that the equation (5) makes it possible to estimate the payback period of any energy-saving activity or technical solution, including engineering one. It is highly important to evaluate its energy-saving potential capacity ΔЭ and capital costs $\Delta \widetilde{K}$ for its implementation.

As to the measure for future cash flows to be discounted, the following factors, which can make an impact on the value of future cash flows, can be considered: average inflation rate taken for a certain period (for instance, for 5 or 10 previous years), interest rate of the Russian Central Bank, return on investments (such alternative investments as a deposit) and others.

It should be pointed out that the equation (5) considers a number of time that denote time, in particular dynamics of rate growth for heat energy (expressed as r) and the rate (i), according to which discounting of future cash flows accumulated due to the energy-saving activity is estimated. Nowadays it's impossible to determine for sure how the variables may vary through time. Consequently, with the aim to estimate the forecasted payback period of energy-efficient investments, there are some probable scenarios how the variables in the equation (5) may change, and then choose the most probable one from the list of the data obtained.

3 Calculation example

Let's take the case of exterior walls insulation in the house built in Saint-Petersburg before 2000. Let's assume that a resistance to heat transfer of exterior walls complies with the requirements assigned for the envelopes before 2000, and is equal to R_0^{bas} =1.0 м2·K/W.

Notes: an actual resistance to heat transfer of exterior walls in the building can be determined as the result of field measurements.

Having R_0^{bas} = 1.0 m2·K/W, we will find that U1 (see the formula (2)) is equal to 1.0 W/(m2·K). The amount of heat season grades in Saint-Petersburg is shown in the Table 1 and accounts for 4796 °C. Heating in the building is a centralized (supply from a city HPP). The cost of heat energy C_t= 1408.01 Rub/Gcal including VAT (see Order of the Committee on Tariffs of Saint-Petersburg dated 18.12.2013 state. № 527-p).

It is required to insulate exterior walls of the existing building to comply with the up-to-date specifications for the level of thermal protection (according to CR R23-02-2003 (Construction Rules and Regulations) «Thermal protection of buildings») and calculate the payback period of additional insulation activities. As an insulator let's take mineral wool products with a thin plaster layer as an insulator, which are assigned for facade works.

Table 1. Estimated climatic conditions for dwelling houses located in Saint-Petersburg.

Indicator	Parameter	Units of measure	Calculated value
Estimated outside air temperature	t_{out}	°C	- 26
Average outside air temperature during the heating period	t_{hp}	°C	- 1.8
Heating season duration	z_{hp}	days/year	220
Heating season degree day	HSDD	°C·days/year	4796
Estimated inside air temperature	tв	°C	20

Let us calculate the necessary thickness of the insulator, m. Let's use the following formula [9]:

$$\delta_{req} = \left(R_0^{new} - R_0^{bas}\right) \cdot \frac{\lambda_{ins}}{r_{wb}}, \tag{6}$$

where

R_0^{new} – required (rated) value of the heat transfer resistance of exterior walls in the building, m2·K/W;

R_0^{bas} – original (actual) value of the heat transfer resistance of exterior walls in the building before additional insulation works are made, m2·K/W;

λ_{ins} – thermal conductivity of the insulator, W/(m·K);

r_{wb} – heat transfer performance uniformity factor for additional insulation of the additional insulation layer.

Let's assume the coefficient of thermal conductivity for the mineral wool insulator under operation conditions (equal to 0.045 W/(m·K)), and the heat transfer performance uniformity factor r_{wb} equal to

0.8.

The required (rated) resistance to heat transfer for exterior walls in dwelling buildings under climatic conditions of Saint-Petersburg (HSDD=4796 °C·days) 3.08 m2·K/W. The value of the resistance to heat transfer 3.08 m2·K/W corresponds to the coefficient of thermal conductivity U2=0.325 W/(m2·K). Taking into account the actual resistance to heat transfer in the building, which is not insulated, assumed as 1.0 m2·K/W, the required value of the thickness of the insulation layer, will be:

$$\delta_{req} = \left(R_0^{new} - R_0^{bas}\right)\cdot\frac{\lambda_{ins}}{r_{wb}} = (3.08-1.0)\cdot\frac{0.045}{0.8} = 0.117\,(m). \quad (7)$$

Consequently, to implement further calculations, let us assume that the required thickness value of the insulator made of mineral wool is equal to 120 mm.

Capital costs for additional insulation of the exterior wall in the existing building with the thickness of the insulator equal to 120 mm, which is subject to be covered with the thin plaster layer, ΔK are assumed equal to 1950 Rub/m2, and they are as follows:

- 300 Rub/m2 is the cost of dry mixtures;
- 550 Rub/m2 is the cost of heat insulation (with an estimated price of the mineral wool insulator 4560 Rub/m3);
- 60 Rub/m2 – the cost dish-shaped studs;
- 40 Rub/m2 – the cost of profiles;
- 1000 Rub/m2 – the cost of total construction and assembly works (service preparation, insulator installation, fixing to a lower layer, plasterwork).

As it is shown in the structure given herein the capital costs for additional insulation of the exterior walls in the existing building, insulation costs (550 Rub/m2) amount to less than 30 % taken from the total costs (1950 Rub/m2).

Let us assume that a company has taken a loan at 12 % annual interest rate for 3 years (m=36) to finance insulation works for the existing building.

In this case the annuity factor will be:

$$A = \frac{p_l\cdot(1+p_l)^m}{(1+p_l)^m-1} = \frac{0,012\cdot(1+0.012)^{36}}{(1+0.012)^{36}-1} = 0.034. \quad (8)$$

Then total investments $\Delta\widetilde{K}$ to implement energy-saving project taking into account interest loan payments will account for (with annuity monthly payments):

$$\Delta\widetilde{K} = m\cdot A\cdot\Delta K = 36\cdot 0.034\cdot 1950 = 2386.8\left(Rub/m^2\right).$$

Let us calculate reduction value of the maintenance and operation costs during the first heating season due to energy saving activity introduction according to the formula (2):

$$\Delta Э = (U_1-U_2)\cdot\frac{0.024\cdot HSDD}{1163}\cdot c_h = (1-0.325)\cdot\frac{0.024\cdot 4796}{1163}\cdot 1408.01 = 94.1\left(Rub/m^2\right) \quad (9)$$

Rate-growth dynamics for heat energy in Saint-Petersburg is shown in the Table 2.

Table 2. Rate -growth dynamics for heat energy in Saint-Petersburg since 2006 till 2011 in the case of central heating supplies.

Year	Rate value, Rub/Gcal (incl. VAT)	Grounds
2006	500.40	Administrative order of the Regional energy commission issued by the Saint-Petersburg Government dated 16 November 2005 N 100-p
2007	575.46	Administrative order on the tariffs issued by the Saint-Petersburg Government dated 15 November 2006 N 123-p

2008	650.00	Administrative order on the tariffs issued by the Saint-Petersburg Government dated 31 October 200. N 139-p
2009	795.73	Administrative order on the tariffs issued by the Saint-Petersburg Government dated 19 November 2008 N 141-p
2010	931.00	Administrative order on the tariffs issued by the Saint-Petersburg Government dated 14 December 2009 N 199-p
2011	1050.00	Administrative order on the tariffs issued by the Saint-Petersburg Government dated 13 December 2010 N 334-p

From the data, shown in the Table 2, it follows that during the period under consideration (from 2006 to 2011) an average value of the relative tariff growth for heat energy per year ΔC_h accounted to 16 %. Thus let's take an average annual tariff growth for heating energy r equal to 0.16.

Let's assume discounting of future cash flows (i) at the key interest rate CB RF (11 %), i.e. the parameter i when calculating the payback period of investments will be 0.11.

On the grounds of the data obtained let us calculate the payback period of investments into façade insulation in the existing building. And we will obtain:

$$T = \frac{\ln\left[1+\frac{\Delta\tilde{K}}{\Delta Э}\cdot\frac{(r-i)}{(1+i)}\right]}{\ln\left[\frac{1+r}{1+i}\right]} = \frac{\ln\left[1+\frac{2386.8}{94.1}\cdot\frac{(0.16-0.11)}{(1+0.11)}\right]}{\ln\left[\frac{1+0.16}{1+0.11}\right]} = 17.3(\text{years}). \quad (10)$$

Provided a construction company uses own (non-borrowed) resources for façade renovation works with subsequent insulation, the payback period will be:

$$T = \frac{\ln\left[1+\frac{\Delta K}{\Delta Э}\cdot\frac{(r-i)}{(1+i)}\right]}{\ln\left[\frac{1+r}{1+i}\right]} = \frac{\ln\left[1+\frac{1950}{94,1}\cdot\frac{(0.16-0.11)}{(1+0.11)}\right]}{\ln\left[\frac{1+0.16}{1+0.11}\right]} = 15.0(\text{years}). \quad (11)$$

4 Conclusion

Thus with the current economic situation and tariffs in Russia the payback period of investments into insulation of the facades in the existing building under Saint-Petersburg climatic conditions will account for less than 18 years. Notes. The calculations and conclusions given above are true provided there is renovation (insulation) of the facades and simultaneous setting up of AHSCU (automated heat supply control unit) upon putting the building in operation. Otherwise, insulation facades may result in increase of the inside air temperature in the buildings maintained and failure of the estimated rebound effect for energy efficiency (actual reduction of operational and maintenance costs $\Delta Э$ may appear to be less than estimations).

The factors, which positively impact on the reduction of the payback period of the investments into renovation of the facades, are as follows:

- advance rate-growth of tariffs for heat energy (r);
- reduction of bank loan interest rates (pl);
- inflation and risks inflation (i);
- increase of the parameter $\Delta Э$, which shows the difference between heat losses through 1 m2 of an exterior wall before and after insulation works of facades in the existing building (increase of $\Delta Э$ may be achieved only due to magnifying the thickness of the insulation layer, and this will automatically lead to increase in capital costs $\Delta\tilde{K}$, and it is unknown which of these parameters will be increasing faster);
- reduction of the capital costs for heating $\Delta\tilde{K}$ (though it may lead to impairing quality of

construction and assembly works).

For this reason the actual objective factors, which contribute to reduction of the forecasted return on investments according to the formula (5), will be the difference (r-i) between annual relative rate growth and the coefficient, which shows discounting of future cash flows (inflation, risks, alternative investments and others), and reduction of bank loan interest rates (pl) provided the company uses the loan with the purpose to implement this energy-saving activity as well.

For this very reason a number of EU countries have grants aimed at reconstruction of buildings. In Germany, there is a state programmer «Energy reconstruction», according to which a set of reconstruction measures was approved, with the purpose to reach the level of energy saving such as Effizienzhaus 100 and Effizienzhaus 85 [20]. The group of German state banks KfW (Kreditanstalt für Wiederafbau) provide gainful loans and grants to implement energy-efficient reconstruction making it possible to reduce return on investment, as it follows from the analysis of the equation (5) and calculation example mentioned above.

References

1. N. Harmati, Z. Jakšić, N. Vatin, (2015) Procedia Engineering, **117 (1)**, 791-799 (2015)
2. Y.A. Tabunschikov, M.M. Brodatch, ASHRAE Journal, **48 (11)**, 26-31 (2006)
3. Y. Tabunschikov, M. Brodatch, Supplement, **14** (SUPPL. 7), 168-174 (2004)
4. N. I. Vatin, D. V. Nemova, P. P. Rymkevich, A. S. Gorshkov, Inzhenerno-stroitel'nyj zhurnal, **34**, 4-14 (2012)
5. V.V. Okrepilov, V.N. Krutikov, G.I. El'kin, Measurement Techniques, **57 (2)**,109-116 (2014)
6. V.V. Okrepilov, A.Yu. Smirnov, Measurement Techniques, **56 (1)**, 54-60 (2013)
7. V.V. Okrepilov, Studies on Russian Economic Development, **24(1)**, 35-42 (2013)
8. A. S. Gorshkov, Stroitel'nye materialy, **8**, 12-17 (2009)
9. A. S. Gorshkov, D. Y. Popov., A. V. Glumov, Inzhenerno-stroitel'nyj zhurnal, **8**, 5-8 (2010)
10. A. S. Gorshkov, M. V. Knat'ko, P. P. Rymkevich, Inzhenerno-stroitel'nyj zhurnal, **8,** 20-26 (2010)
11. A. S. Gorshkov, P. P. Rymkevich, I. I. Pestrjakov, M. V. Knat'ko, Stroitel'nye materialy, **9**, 49-53 (2010)
12. A. N. Dmitriev, Y. A. Tabunshhikov, I. N. Kovalev, N. Shilkin, AVOK-PRESS, **9**, 8-13 (2005)
13. Y. A. Tabunshhikov, N. V. Shilkin, AVOK, **7**, 10-21 (2005)
14. G. P. Vasil'ev, Energosberezhenie, **6**, 14-23 (2011)
15. *STO SPPP 4.5-2012 Raschetnyj metod ocenki jekonomicheskoj jeffektivnosti jenergosberegajushhih svetoprozrachnyh konstrukcij* (SPbPU, St. Petersburg (2012)
16. A. S. Gorshkov, Energosberezhenie, **4**, 12-19 (2014)
17. A. S. Gorshkov, P. P. Rymkevich, Stroitel'nye materialy, oborudovanie, tehnologii XXI veka, **188**, 40-45 (2014)
18. A. S. Gorshkov. *Inzhenernye sistemy. Rukovodstvo po proektirovaniju, stroitel'stvu i rekonstrukcii zdanij s nizkim potrebleniem jenergii* (SPbPU, St. Petersburg (2013)
19. A. S. Gorshkov, P. P. Rymkevich, D. V. Nemova, N. I. Vatin, Inzhenernye sistemy. AVOK - Severo-Zapad, **3**, 32-36 (2014)
20. Gabriyel, I., Ladener, Kh. *Rekonstruktsiya zdaniy po standartam energoeffektivnogo doma* (St. Peterburg: BKhV-Peterburg, 2011)

MATEC Web of Conferences **53**, 01046 (2016)
DOI: 10.1051/matecconf/20165301046

Methodology to Improve Energy Efficiency of Residential Historic Buildings in St. Petersburg

Vera Murgul[1,a]

[1]*St. Petersburg State Polytechnical University, Politekhnicheskaya, 29, Saint-Petersburg, 195251, Russia*

Abstract. The paper contains the analysis of the goals and objectives of improving energy efficiency of residential buildings, as well as the methodology of selecting energy-efficient modernization measures for historic buildings. The priority objective of this study was selected as a residential housing energy efficiency of historic buildings as a tool to improve the quality of the human environment. The development of renewable energy technologies is presented as an alternative for energy saving. If we take the purpose of improving energy efficiency as an improvement of the quality of the human environment (from the living quarters level to global environmental sustainability), the alternative to energy saving of traditional energy resources can be the replacement of them by energy from renewable sources, even lavish spending of which does not harm the environment . All energy saving should be focused primarily in man-made environments (industrial processes, heating systems and etc.), the anthropogenic environment (living environment) should get the maximum energy for the stable provision of quality microclimate.

1 Introduction

It is well-known that energy efficiency is the goal to reduce the amount of energy required to provide products and services. Improvements in energy efficiency achieved by adopting a more efficient technology or by application of commonly accepted methods to reduce energy losses [1].

Energy efficiency can be seen as:

- reducing energy consumption tool in all sectors of the economy and one of the main conditions for improving competitiveness. Main methodology is energy saving;
- a tool to fight climate change; main methodology - energy saving and switching to renewable energy sources
- ensuring energy security tool (reduction of dependence on imports of energy resources from donor countries and reducing the risks of possible depletion of traditional energy resources).

Thus, we see that the ultimate goals pursued by the desire to improve energy efficiency are fundamentally different.

Russian Energy Strategy by 2035 considers improving energy efficiency as the main way to improve the efficiency of the economy [2]. ES-2035 involves the reduction of GDP electric capacity by 40% and energy consumption - by 50% by 2035 (comparing to the level of 2010), without it the energy sector will inevitably hold back the economic development. Missed potential of organizational

[a] Corresponding author: october6@list.ru

and technological energy saving is estimated to be 1/3 of the total consumption of energy resources in the country.

Reducing the energy consumption of the economy is one of the main conditions for improving the competitiveness of any country. Fig. 1 shows the energy use per unit of GDP of different world regions.

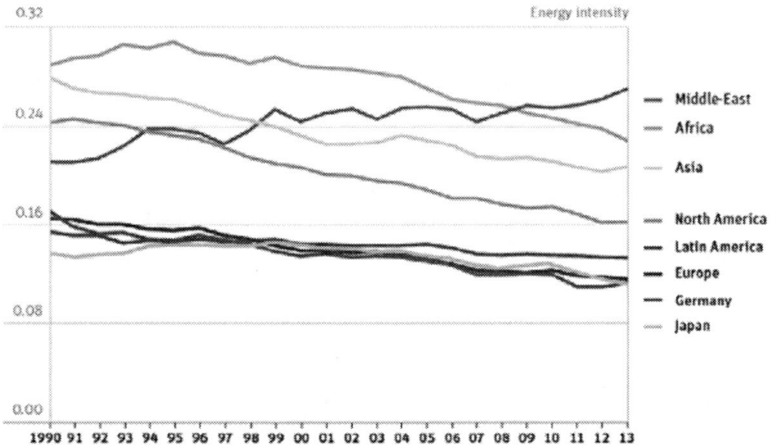

Figure 1. Energy intensity (= energy use per unit of GDP) of different world regions [3]

The improvement of energy efficiency of housing and communal services becomes particularly important as those are the largest consumers of energy resources [4-8].

Reducing energy consumption in housing for causing its percent of greenhouse gas emissions is possible in three key ways:

1) reducing the amount of energy consumed by reducing the wasteful loss (energy saving potential disclosure)
2) reducing the amount of energy consumed (reduced comfort). This option is clearly impossible for a dynamically developing city
3) reducing the amount of energy consumed, 3hich causes the emission of greenhouse gases, with the replacement of them with similar amounts of energy from renewable sources, the use of which is environmentally friendly.

In fact, the transition to renewable energy is an alternative to energy saving if the improvement of the quality of the human environment is in priority

The goal of reducing the energy consumption of traditional, non-renewable energy sources in different countries tries to resolve various problems. For Russia, energy efficiency is primarily a tool to reduce energy consumption in all sectors of the economy and is considered as a major condition for competitiveness increasing [2]. According to it, the key to success is represented by energy saving. Besides, slowdown in the growth of CO2 emissions becomes a natural attendant effect.

If we consider the foreign programs of improving energy efficiency, the aims are primarily claimed as the fight against climate change and also energy security (reducing dependence on energy imports from donor countries). Sometimes one can also hear quite a controversial statement about the imminent exhaustion of traditional energy resources. The key to success in a foreign world is said to be not only energy efficiency, but also the switch to the usage of renewable energy sources (RES). Moreover, the development of renewable energy technologies receives strong state support.

Considering residential buildings as an object of study, authors chose residential energy efficiency improvement of historic buildings as a tool for improvement of the quality of human environment, both at the microclimate level of living quarters and at the level of global environmental sustainability.

Overview of energy strategies of the largest countries in the world shows aiming at a significant increase in energy efficiency and diversification of the fuel and energy balance structure through the development of renewable energy sources. According to Russian energy policy, though, the problem of increasing energy efficiency is included in the priority directions of science development, technology and engineering development of Russian Federation, however, renewable energy in conjunction with energy efficiency are not even mentioned, but noted the development of nuclear energy [9].

However, the potential of renewable energy should not be neglected. In fact, the transition to renewable energy is an alternative to energy savings, if the improvement of the quality of the human environment at the global level is in priority.

Thus, the main "duet" of this study: energy savings plus the replacement of traditional energy sources to renewable sources of energy in the living environment of a large city.

All energy saving should be focused primarily on man-made environments (industrial processes, heating systems and etc.), the anthropogenic environment (living environment) should get the maximum energy for the stable provision of quality microclimate and adequate level of comfort. It is necessary to shift the focus from the energy saving "at any price" to the improvement of the quality of the microclimate in the buildings.

An alternative to economy can and should be the development of technologies. This applies primarily to renewable energy sources. If the main purpose of improving energy efficiency will become the improvement of human's environment quality (from the quarters level to global environmental sustainability), the alternative to energy saving traditional energy resources can be their replacement by energy from renewable sources, even lavish spending of which will not harm the environment .

2 Risks of energy-efficient modernization of historic buildings

Specific objects of energy efficient modernization of residential buildings are historic buildings built during pre-industrial times. Modern policy of protection of historical heritage is various in different countries. But there is something in common - concern for the safety of architectural heritage.
The pursuit of compliance with current regulations of energy consumption is often the impulse for an inconsiderate reconstruction of facades, the result of which is the complete loss of historic appearance of buildings. In Fig. 2, you can see examples of facades, which led to the complete loss of the original appearance of the buildings.

The pillar is partially covered by insulation (Berlin)

Facades before and after thermal insulation, Nuremberg Germany

Figure 2. Negative examples of warming of facades and historical buildings [10]

And it's not just the loss of the external appearance of the historic facades and loss of visual historic environment of urban areas in general, but also the loss of the history of the constructing are. Figure 3 presents the history of the building's reincarnation.

Figure 3. The sad story of transformation of a historic building Luitpold Haus in Nürnberg: a) bis ca. 1945, b) bis ca. 2007, c) today [10]

3 Speciality of energy-efficient modernization of historic buildings of St. Petersburg

St. Petersburg - a fast-paced metropolis. Conflict of interests of the urban environment with the interests of preserving the historical authenticity of the city center requires the development of a balanced strategy, not harmful to the cultural and historical heritage of the city and allow it to evolve rapidly. The problem of preserving the historical environment of St. Petersburg is solving in the conditions of major changes taking place in the socio-economic sphere, as well as on the background of significant technological and climate change.

Security experience in St. Petersburg demonstrates a reasonable approach to the protection of buildings in historic center. Interests of preserving the authenticity of the historic center are put above the interests of energy conservation. And it is absolutely correct in the authors' opinion.

Residential historic housing of St. Petersburg is a unique object of study. Despite the fact that the level of heat energy consumption doesn't correspond to the current date requirements, there are no restrict for their energy modernization.

"Code of practice CP 50.13330.2012 does not apply to a thermal protection of buildings, structures and facilities that are in accordance with the legislation referred to objects of cultural heritage (monuments of history and culture)." "Code of Practice CP 50.13330.2012 applies individually during the construction and renovation of existing buildings considering their architectural and historical importance, taking into account their historical value based on the decisions of the authorities and coordination with the bodies of state control in the sphere of protection of monuments of history and culture." [11]

This is primarily due to the inability or complexity of application of standard measures for the modernization of energy efficiency for historic buildings because of the presence of protection objects (front facades, interiors and other securities.).

However, the currently available methods of improving energy efficiency make it possible to reduce the consumption of thermal energy without harming the historical authenticity of the residential area, this applies not only to the facades, but also to the historical building system.

Therefore, the problem of adaptation of existing methods and techniques to improve energy efficiency of buildings to buildings of historic buildings that have architectural and historical significance is important and urgent is, and adaptation to the buildings, which are objects of cultural heritage in accordance with their main specificity - presence of objects of protection and the constraints imposed by the current guarding regulations.

4 Methods of improving the energy efficiency of residential historic buildings of St. Petersburg

The method was developed for the historic buildings of the background construction and for cultural heritage buildings in historical central districts of St. Petersburg. The technique is a sequence of actions, following which heat energy costs could be reduced to a normalized level without compromising the external appearance of historic buildings and historical construction system, without a damage to the subjects of protection. The technique is based on the updated regulations on thermal protection of buildings design: RD 23-16-2012, CP 50.13330.2012. Regarding the buildings, which are objects of cultural heritage, the technique is based on the Federal Law N 73-FL (Ed. of 03.08.2015.) "On the objects of cultural heritage (monuments of history and culture) of the peoples of the Russian Federation" and the St. Petersburg Law "On protection of cultural heritage of St. Petersburg (Ed. of 18 December 2014). With regard to historic buildings of background construction, the technique is based on the limits set out in the Appendix called "Modes of land use within the boundaries of protection areas" of the law of Saint-Petersburg "On the boundaries of protection areas of objects of cultural heritage and on the conditions of using the land in the above boundaries and on the introduction of changes into the law of Saint-Petersburg "on the master plan of Saint-Petersburg and on the boundaries of the protection areas of cultural heritage in the region of Saint-Petersburg". (Ed. of 26 June 2014).

Table 1. Methodology of improving energy efficiency of residential historic buildings in St. Petersburg

Methods of energy efficiency modernization				Normable parameter of energy efficiency (if there is one), based on RD 23-16-2012	Protection area										Area of construction to be regulated (RA)								
					Protection area of cultural heritage-listed object PA 1					Protection area of cultural heritage-listed object PA 2					RA 1					RA 2			RA 3
					Street-faced buildings				Intra-district buildings	Street-faced buildings				Intra-district buildings	Street-faced buildings				Intra-district buildings	Street-faced buildings		Intra-district buildings	All buildings
					On the territory of open urban spaces		Buildings in other areas			On the territory of open urban spaces		Buildings in other areas			On the territory of open urban spaces		Buildings in other areas						
					Front facade	Courtyard facade	Front facade	Courtyard facade		Front facade	Courtyard facade	Front facade	Courtyard facade		Front facade	Courtyard facade	Front facade	Courtyard facade		Front facade	Courtyard facade		
Energy saving	Improved thermal protection of individual building external walls	Wall insulation, R_{req} [m²·°C/W]	external	3.08																			
			internal																				
		Thermal insulation of attic floor, R_{req} [m²·°C/W]		4.06																			
		Thermal insulation of ceilings above unheated basements and cellars, R_{req} [m²·°C/W]		4.06																			
		Thermal insulation of floors over passages R_{req} [m²·°C/W]		4.60																			
		Thermal insulation of floors over the ground R_{req} [m²·°C/W]		4.50																			
		Thermal insulation of windows and balcony doors with a rate of glazing:																					

		- $k_g \leq 18\%$; - $18 < k_g \leq 36\%$; - $36 < k_g \leq 54\%$; - $k_g > 54\%$/ R_{req} [m^2·°C/W]	0.51 0.64 0.81 1.02																			
		Thermal insulation of showcases and stained-glass windows R_{req} [m^2·°C/W]	0.51																			
		Thermal insulation of bay windows with vertical glazing R_{req} [m^2·°C/W]	0.37																			
		Thermal insulation of entrance doors R_{req} [m^2·°C/W]	0.79																			
		Sealing of building envelope. 1. The air permeability of building envelopes: - exterior walls; - windows and balcony doors (at 10 Pa pressure difference); - the overlap of the first floor; - entrance doors to the apartments G_{req} [kg/m^2·h] 2. The air permeability of apartments (with closed supply and exhaust vents, and the pressure difference between indoor and outdoor air 50 Pa) with ventilation: - With a natural impulse; - With mechanical drive.	1.0 5.0 0.5 1.5 Air exchange rate $n_{50} \leq 4$ h^{-1}; $n_{50} \leq 2$ h^{-1}																			
	Modernization of services and equipment	Automatic exhaust ventilation with mechanical drive and recovery of heat in exhaust air and natural inflow through vents																				
		Energy-efficient appliances and facilities																				
		Regulation and registration of energy consumption																				
		Smart building management systems																				
The spatial transformation		Construction of superstructure floor with high heat-insulating properties	Without an increase of thermal load on the heating system			*		**			*		**			*		**	*		**	
		Construction of attic floors with high heat-insulating properties		***										***					***	***		
		Overlapping closed court																				
New energy-efficient construction		Reconstruction of historical urban environment by recreating the lost historical structures - With the number of storeys of buildings: - 1-3 storeys; - 4 to 6 storeys. q_h^{reg}, [kWh/(m^2·year)]	135-191 126-178																			
Additional energy from renewable sources	Solar energy	Solar PV-systems: BIPV																				
		Solar PV-systems: BAPV																				
		Solar PV-systems: BITPVC																				
		Thermal solar collectors																				
		Technologies of passive solar design																				

Adopted conventions
Allowed
Allowed with restrictions
Forbidden
q_h^{reg} - the specific annual heat consumption for heating, ventilation and hot water supply;
R_{req} - normable thermal resistance, m2·°C/W;
q_h^{reg} – required breathability of building envelope, kg / m2 • h;

In accordance with the land use regulations within the boundaries of protected areas of cultural heritage located in the historical central districts of St. Petersburg:

For PA 1

*- height to the eaves according to the characteristics of the historical environment of the present environmental zone, height to the roof ridge is limited by the roof angle of 300.

** For intra-district development: not higher than street-faced housing, located directly in front of reconstructed object;

*** - without changing the roof configuration (rising the level of the ridge at 1 m is allowed)

For PA 2

*- height to the eaves according to the characteristics of the historical environment of the present environmental zone, height to the roof ridge is limited by the roof angle of 300.

** For intra-district development: not higher than street-faced housing, located directly in front of reconstructed object;

For RA 1

* height to the eaves according to the characteristics of the historical environment of the present environmental zone, height to the roof ridge is limited by the normable roof angle;

** For intra-district development: the object must not be visible from the public urban spaces;

*** without changing general architectural solutions

For RA 2

* formation of the street front development with height to the eaves not exceeding 28 m and 33 m to the ridge;

** providing visual excess of the street front over the intra-district development

*** without changing general architectural solutions, rising the level of the ridge at 1 m is allowed

All changes of the architectural solution of building and structure facades are produced in accordance with the conclusion of a public authority of cultural heritage protection. This methodology can be applied to the historic buildings on the territory of the historically formed central districts and does not apply to buildings - monuments and memorials of history and culture.

5 Conclusion

A technique developed for improvement of energy efficiency of residential historic buildings in St. Petersburg on the basis of complex usage of energy-saving measures and solar energy is a versatile tool to select the methods of energy efficiency modernization for the historic buildings of protection areas and zones of reguable construction and buildings of cultural heritage, taking into account the limitations of protection imposed by the regulations.

The main attention during the energy modernization of historic buildings should be focused on the engineering retrofitting, common house, as well as inside apartment (energy-efficient appliances and equipment, management, registration and accounting of consumed energy, intelligent building management systems, local energy production from renewable sources, etc.). Historic building system, as well as the historic building and finishing materials should be regarded as a potential object of protection and not to be partially or completely replaced with modern materials and designs.

A detailed study of historical methods of ventilation in the 3-4 storey buildings showed the presence of heat impulses (heat kiln channels), and the organization of draught from all dwellings. For example, a modern 9-storey building does not have a motive in the ventilation system. Features of historical methods of ventilation system's organization should be taken into account and maximally preserved during the process of modernization of the building.

Modern ventilation systems with heat recovery of exhaust air "return" up to 80% of the heat vent discharges. However, the installation of such systems is a controversial solution, in addition to the imposing cost of the installation and the inevitable presence of a network of air ducts, it is necessary not to open the window. Heat recovery is possible only in the absence of the so-called "salvo airing," ie, during constantly closed windows. As a result, a person loses a certain connection with the natural environment: smells and sounds of the open space. In fact, the sealing of buildings - the next step for a "house is a machine to live in", created by Le Corbusier. "I want to make sure that the bird having flown into the office, did not even notice that it is not outside the building, but inside it" wrote James Winsome. But this already can not be considered. Just as organic architecture once came into conflict with the needs of modern urbanism, also today, pollution of the urban environment has placed new priorities: unique advantage of mechanical ventilation is the ability to clean the incoming air with wet filters.

Warming of facades of historical buildings better be abandoned. The threat of losing the authenticity of the historic environment (even in the case of common background development) is incommensurable in importance with insignificant energy savings. If the decision on the insulation of facades will still be accepted, a reasonable scheme can be external insulation of courtyard facades and internal warming of front facades.

There is a need for development of a special set of energy efficiency measures for historical buildings, taking into account their specific construction system. It should be the revival of traditional techniques and traditional materials, first of all in order to not to disturb the building and energy balance, which was reached by the art of construction of the 19th century. Invasion by foreign (modern) materials in general should not be allowed.

The historic building system with unique environmental qualities and a proven experience of two centuries of exploitation can and should be the subject of protection itself, along with valuable protected facades, interiors and so on.

The current practice of protection of historical background development is focused on the visuals. Protection subject includes facial facades, silhouettes of buildings, video of open urban spaces. For buildings referred to the cultural heritage, individual items are isolatedly protected (for example, valuable interior elements). Based on the study of the unique qualities of the historic building system it seems necessary to protect building construction themselves as well. And also historical methods of heating and ventilation of buildings. Under a layer of insulation facades and styles disappear. The total reconstruction with the demolition of the walls and ceilings makes historical design disappear too. One of the problems of modern practice of reconstruction and restoration of historic buildings is the disappearance of the building's history.

Reference

1. https://en.wikipedia.org/wiki/Efficient_energy_use (13.11.2015)
2. Russian Energy Strategy to 2035 http://www.energosovet.ru/stat835.html (13.11.2015)
3. http://energytransition.de/2012/09/efficiency/ (13.11.2015)
4. Y.A. Tabunschikov, M.M. Brodatch, ASHRAE Journal, **48** (**11**), 26-31 (2006)
5. Y. Tabunschikov, M. Brodatch, Supplement, **14** (SUPPL. 7), 168-174 (2004)
6. W. Feist, J. Schnieders, V. Dorer, A. Haas, Energy and Buildings, 37, Issue 11, 1186-1203 (2005)
7. F. Ochs, G. Dermentzis, W. Feist, Energy Procedia, **48**, 1124-1133 (2014)
8. Leskovar, V.Ž., Premrov, M. Energy and Buildings, **43** (**12**), 3410-3418 (2001)
9. Presidential Decree of 07.07.2011, № 899 On approval of the priority directions of development of science, technology and engineering in the Russian Federation
10. https://stadtbildberlin.wordpress.com/schwerpunktthema-energetische-sanierungen/ (13.11.2015)
11. Code of practice CP 50.13330.2012

MATEC Web of Conferences **53**, 01047 (2016)
DOI: 10.1051/matecconf/20165301047

The Experimental Study of Concrete Beams Reinforced with Different Types of Bars Carrying Capacity

Andrey Benin[1,a] , Sergey Semenov[1] and Bogdanova Ekaterina[1]

[1]*Petersburg State Transport University, 190031, 9 Moskovsky pr., Saint Petersburg*

Abstract: The results of experimental study on concrete beams reinforced with glass fiber reinforced plastic (GFRP) bars are presented and compared with steel reinforced concrete beams and beams reinforced with steel and GFRP bars together. Three series of reinforced beams were tested in the flexure. The experimental data are showed that possible area in which GFRP bar possesses potential to employ is secondary reinforcement in concrete structures.

1 Introduction

Fiber reinforced plastic (FRP) bars are widely used construction material in abroad. This type of a bar has been successfully used as reinforcement for road construction, in bridge decks and for concrete structures in severely aggressive environments. FRP bars have several advantages compared to steel reinforcement such as high strength, low weight, electromagnetic neutrality, resistance to corrosion and low cost. All of these advantages could lead to increased durability of concrete structures, and reduced maintenance costs.

A large-scale program on the application of FRP bars for concrete structures has been initiated in USSR: FRP bars have been used for construction without insulation crossmembers for power line supports, as a reinforcement for an electrolysis bath, for construction of a warehouse for fertilizers, in bridge decks [1, 2]. But all scientific developments were stopped in 1990, due to the high production cost of this construction material. That is why the use of FRP bars have been widely spread in abroad in comparison with Russia.

Today the use of FRP bars for concrete structures in Russia leads to the necessary of experimental and theoretical researches of structures reinforced with FRP bars in order to develop the national normative documents for the design and construction of concrete structures using FRP bars [3-8]. It may be worth noting that today a lot of experimental programs are carried out in Russian research laboratories and institutes to gain an insight into the behavior of concrete structures reinforced with FRP bars [9-15].

The Mechanical laboratory of Petersburg State Transport University with the collaboration of St. Petersburg State Polytechnical University conduct experimental and numerical investigation FRP bars bond behavior with concrete and FRP bars behavior in the flexure [16].

The aim of the current study is to experimentally investigate the flexural behavior of concrete beams reinforced with FRP bars and to compare of carrying capacity and deflection of beams reinforced with steel bars and beams with combined reinforcement.

2 Experimental program

Three series of reinforced concrete beams were tested in four-point-flexure. The geometry of the beams, the diameter of reinforcement and loading regime were identical for all specimens.

Three series included beams reinforced with four steel bars A500 (12 mm diameter), beams with four glass fiber reinforced plastic (GFRP) bars (the bar is made of E-glass fibers in a polyester matrix Epikote 828, using a pultrusion process) and beams with two steel and two GFRP bars. Geometry of the beams and reinforcement pattern are given in Table 1. The beams were made from a ready-mixed concrete B45. The beams were loaded by Instron machine model SATEC 1200 KN. Overall view of experimental setup is shown in Fig. 1, a).

[a] Corresponding author: benin.andrey@mail.ru

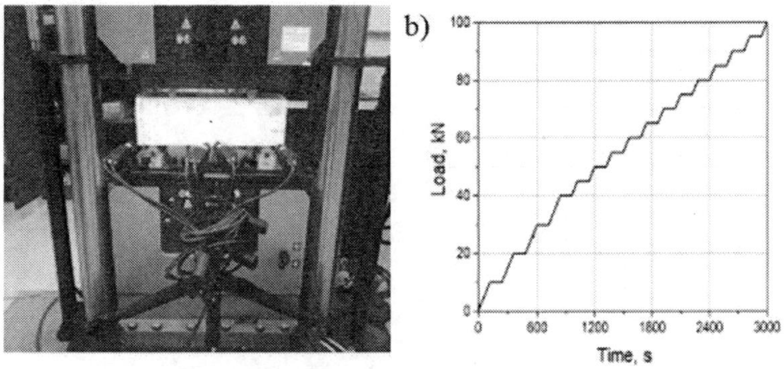

Figure 1. a) Overall view of experimental setup, b) Loading regime

Loading was applied stepped to the beams at a rate of 10 kN per step with intermediate exposures until fracture as shown in Figure 1, b). The load was measured with the machine's load cell. The beam deflections were measured with a linear variable differential transformer (LVDT) at mid-span and under loading points. Strain fields were measured with two digital image correlation Vic3D systems on side of the beam painted by speckle pattern (Table 2).

3 The experimental results

The experimental dates are given in Table 1. Typical evolution of strain fields for experimental load at first crack, maximum load capacity and mode of failure of the concrete beams are presented in Table 1.

Table 1. The experimental dates

Reinforcement pattern (b=150 mm, h=150 mm, a_1=35 mm, a_2=35 mm)	Maximum load capacity, Pmax, kN	Deflection, mm			Mode of failure (inclined crack fracture)
		Left point	Mid-span	Right point	
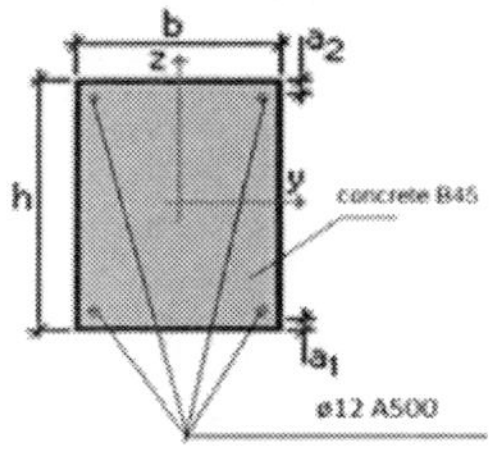	310	1,37	1,64	1,36	
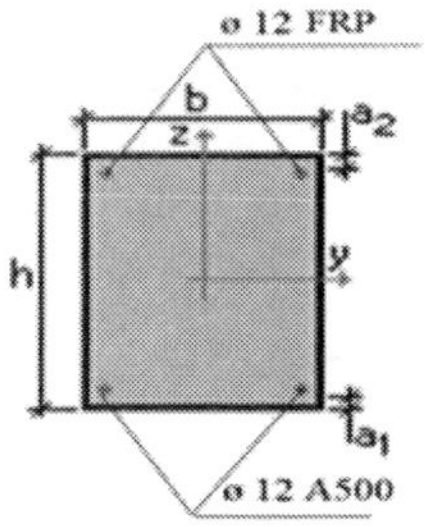	300	1,31	1,77	1,47	
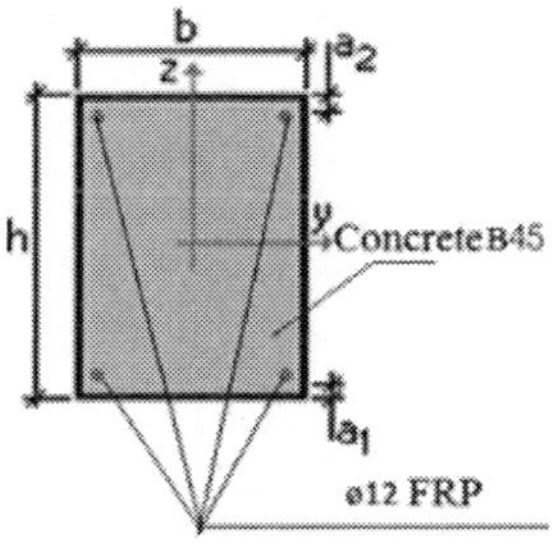	220	2,75	3,76	3,31	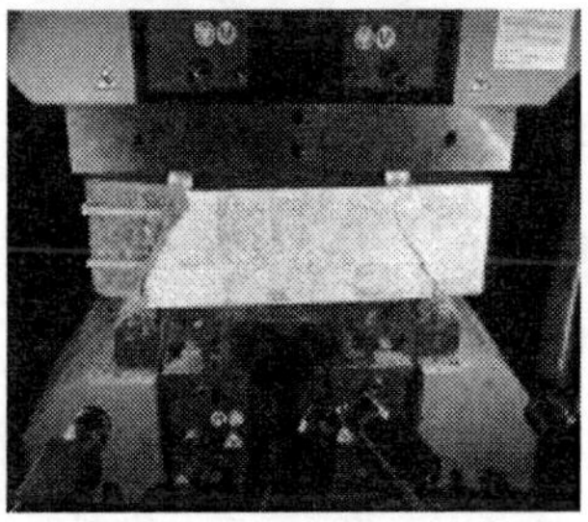

Table 2. Displacement and strain fields

Reinforcement pattern	Maximum principal strain fields	Experimental load, P_{max}, kN
Beams reinforced with four steel bars A500	First crack 	40,6
	Crack growth 	310
	Fracture of the sample 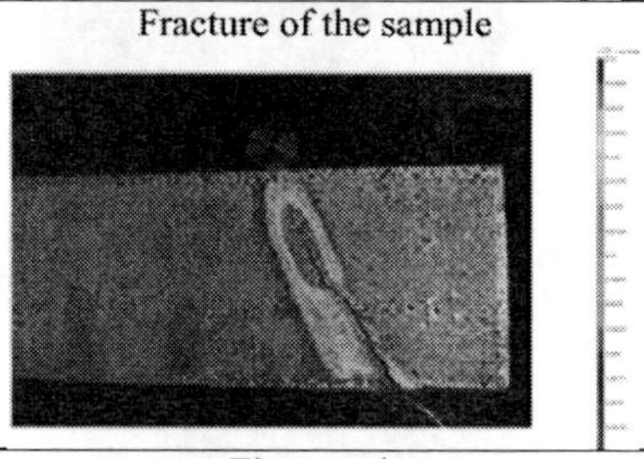	310
Beams with four glass fibre reinforced plastic (GFRP) bars	First crack 	40
	Crack growth 	211
	Fracture of the sample 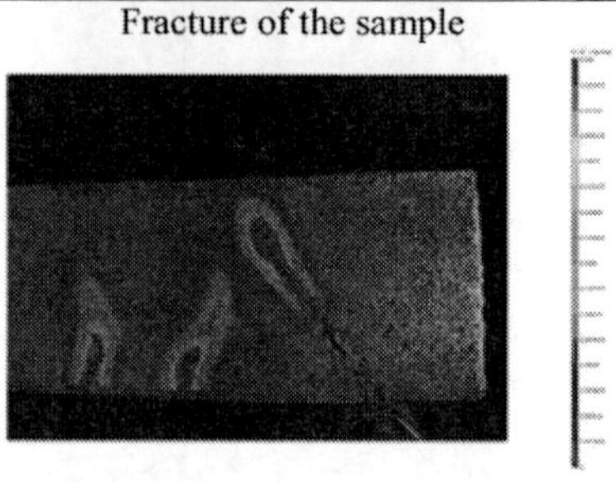	220

Beams with two steel and two FRP bars	First crack 	50
	Crack growth 	290
	Fracture of the sample 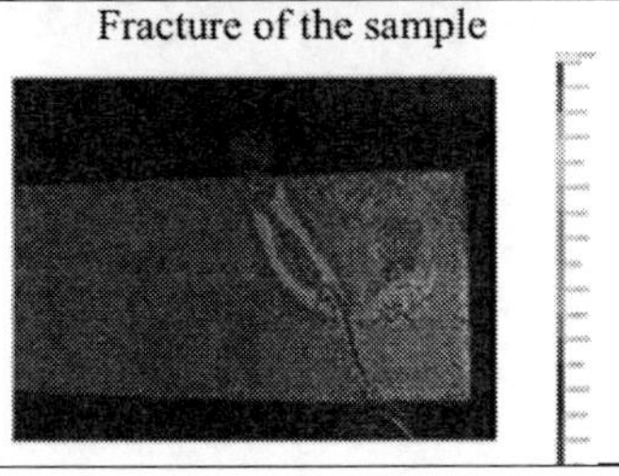	300

4 Conclusions

The experimental results have showed:
1. Crack initiation was started in bending zone. This cracks were stopped by the lower rebars. Later on inclined cracks was initiated and propagated from the bearing area. The main beam fracture reason is inclined cracks started from the bearing area.
2. Using of fiberglass rebars as main reinforcement leads to decreasing of the load capacity of reinforced concrete specimens in comparison with steel reinforced beam up to 29%.
3. Concrete specimens reinforced by fiberglass rebars demonstrate high deformability in comparison with steel reinforced specimens which leads to appearance of meaningful deflections and as a consequence increasing of crack openings.
4. Using of hybrid reinforcement (fiberglass bars in compression zone and steel bars in tension zone) was practically not changed load capacity.
5. Thus, due to increased deformability of concrete elements reinforced with fiberglass rebars, the most appropriate area of using such type of rebars is secondary reinforcement, placed in nonleaded zones for decreasing amount of steel using for reinforcement.

References

1. V. Stepanova, A. Stepanov, E. Jirkov, *Fibre reinforced plastic bars* (2013)
2. ICI 440.1R-06 Guide for the Design and Construction of Structural Concrete Reinforced with FRP bars
3. V. Stepanova, Building materials, equipment, technologies of XXI century, **10,** 12-14 (2014)
4. A. Benin, G. Bogdanova, S. Semenov, Applied Mechanics and Materials, **617,** 215-220 (2014)
5. T. Zinnurov, A. Piskunov, L. Safiulina, O. Petropavlovskikh, D. Yakovlev, Internet-journal Science of science, **4,** 1-12 (2015)
6. A. Benin, S. Semenov, Industrial and Civil Engineering, **9,** 74-76 (2013).
7. V. Khozin , R. Gizatullin, A. Kuklin, A. Piskunov, News of the KSUAE, **1,** 211-213 (2013)
8. Y. Klimov, O. Soldatenk, D. Oreshkin // URL: http://www.ekibar.org/files/frp_rebar_test_adhesion_concrete.pdf
9. S. Nikolskiy, N. Vatin, O. Pertseva (2015) Journal of Applied Engineering Science, **13 (1)**, 11-18 (2015).
10. Z. Bonić, G.T. Ćurčć, M. Trivunić, N. Davidović, N. Vatin, (2015) Procedia Engineering, **117 (1)**, 424-435 (2015)
11. V. Korsun, N. Vatin, A. Franchi, A. Korsun, P. Crespi, Mashtaler S. Procedia Engineering, **117 (1)**, 975-984 (2015).
12. I. Garanzha, N. Vatin, Applied Mechanics and Materials, **633-634**, 965-971 (2014)
13. N. Vatin, I. Pestryakov, S.Kiski, Z. Teplova, Applied Mechanics and Materials, **584-586**, 1381-1387 (2014)
14. B. Kovačič, R. Kamnik, M. Premrov, N. Gubeljak, J. Predan, Z. Tišma, Strojniski Vestnik/Journal of Mechanical Engineering, **54 (5)**, 364-371 (2008)
15. B. Kovačič, R. Kamnik, International Journal for Engineering Modelling, **20 (1-4)**, 77-84 (2007)
16. E. Bogdanova, A. Benin, M. Belyaev, S. Semenov., A. Semenov, Collection of The XLII theoretical and practical conference with international participation «Week of Science in SPbSPU», 6-9 (2014)

MATEC Web of Conferences **53**, 01048 (2016)
DOI: 10.1051/matecconf/20165301048

The Different Methods of Displacement Monitoring at Loading Tests of Bridges or Different Structures

Boštjan Kovačič[1,a], Rok Kamnik[1] and Andrii Bieliatynskyi[2]

[1]*University of Maribor, Faculty of civil engineering, transportation engineering and architecture, Smetanova 17, SI 2000 Maribor, Slovenia*
[2]*National Aviation University, Kosmonavta Komarova, 1, Kyiv, 03058, Ukraine*

Abstract: By measuring the displacements and deformations at different structures we deal in the Faculty of Civil Engineering, transportation Engineering and Architecture in University of Maribor for about 20 years. At that time we measured over 600 structures. Most loading tests of bridges and Viaducts were made. The measurements of movements needed to be as precise and accurate as possible. To do that laboratory test of instruments were made to see which instrument gives us reliable results. Displacements can be determined by geodetic and physical methods, depends of the construction. The use of geodetic methods are still preferable. In the paper the measurements with the total station, the level and rotation level, photogrammetry and solutions on the field by physical methods with inductive transducers are presented. We need to measure displacements as quick as possible but efficiently because we can not repeat the measurements under the same conditions. Also the surveying on the bridge and in the lab with the comparison of methods is presented under the different hard terrain conditions - water beneath the construction, big height of the structure, unapproachability, large span structures.

1 Introduction

Scientific discipline test structures and buildings was formed somewhat later than other disciplines in the construction industry. However, some special cases testing facilities exist in very ancient times. The first testing of structures dating back to the XV. century. At the time, Leonardo da Vinci made the first examination of the various forms of construction. The first works and forms of security testing beams were carried out by Galileo in the sixteenth century. He tried to discover the relationship between the impact force to the structure and its budget. At that time was not yet known law of elasticity of the structure, because the dependence set by Galileo Galilei, although it was principally set properly, contained many inaccuracies.

Experimental testing methods of construction in the beginning were based on a theoretical calculation that describes the process of deformation and fracture under the action of the load as described in the book by T. Thomas,1855 [1]. With the development of numerical science in the last 30 years with the rapid development of information technology, analysis of movements and deformations have dealt with a number of experts who have described their achievements in a number of articles [2-7], and books [8].

[a] Corresponding author : bostjan.kovacic@um.si

2 Displacements and displacements

Displacements and strains are due to external and internal influences such as wind force, effect of temperature changes, tectonic and seismic effects, changes in the amount of floor (ground) water, static and dynamic loading of objects, say in the form of folds, tilt, turn, distortion of the building, and possible damage in the form of cracks and even fractures [9-12].

Examination of the structure (Fig. 1) or object as a measurement of displacement and strain to determine the change in the position and shape of the building, considering the environment, depending on the time.

Basic indicators for the interpretation of measurement results are character, the direction and magnitude of displacement and strain, and we share in regard to the character in uniform and uneven shifts and the elastic and plastic deformation, in the light of changes in the direction of the horizontal and vertical and considering the size of the non-hazardous, tollerant , serious and critical [13-18].

The measurement of displacement and strain so necessary for assessment of the state of construction, assessment of the state of the object associated with the process of construction or production, or to repair or reinforcement structure and verify the accuracy of theoretical assumptions. The need for experimental testing occurs in the following cases [9]:

- the possibility of using the structure in case of exceeding the useful load,
- testing the durability of construction elements of mass production,
- tests of large and special facilities at the impact of useful load,
- tests of complex systems construction, and testing of its safety.

Figure 1. The pilot testing of the bridge

3 Materials and methods

3.1 Methods for displacements measuring

Vertical movements can measure in different ways and with several methods. What method will be used according to the terms of structures, provided accuracy, planned developments, courts around the structure, external conditions, the number of staff. Displacement measurements usually performed using methods and leveling at higher and higher buildings this method more we can give you results, so we have to use any other surveying method (trigonometry, GPS, photogrammetry ....) or non-geodetic method (measuring inductive the senses, the senses of voltage ...) [19, 20]

To know which method we should use to know the characteristics of the structure, ie. construction height, length range, which is under construction, scheduled shifts ...

At the Faculty of Civil Engineering, Transportation Engineering and Architecture, University of Maribor we are sample surveys of bridges and viaducts from 2003. At that time, we conducted more than 300 test trials in Slovenia and abroad. That progress measured the highest accuracy in order to test the method on different structures in the lab and test the bridge [21].

This year we made the test result on the bridge where we used different methods. We used the methods of surveying, leveling method, the method of GPS - RTK method and photogrammetry method trigonometry with three different instruments (traditional total station, station-automatic and robotic cells with built-in scanner), and non-geodetic methods with built-in senses voltage (strain gauge) [22-24].

3.2 Levelling method

This method is the oldest method of monitoring progress. So we can get only vertical movement and high accuracy.

Leveling is so well-known methods and the most used. In this method needs to be mindful that we have a point that is used for comparison to the field where we do not have a fixed displacement and strain so most times stabilize the bridge where there is no cargo and no oscillations (Figure 2).

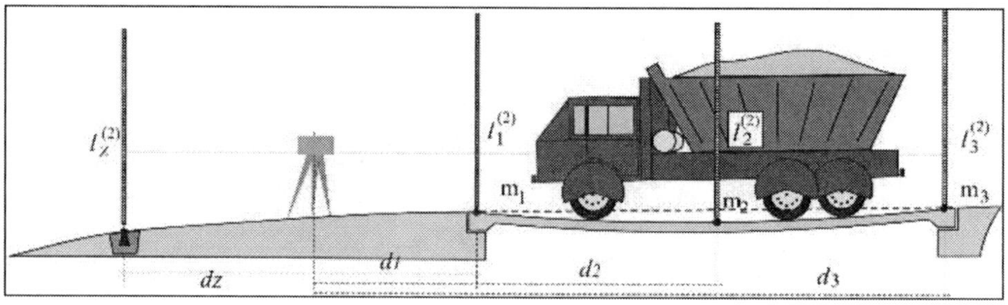

Figure 2. An example of leveling the reference points outside the structure

The problem of this method is the construction of a large range. Thus, the method should be leveled leveling the middle there and back. It takes us a long time. It's the same problem with the construction of a large harbor. So that this method has meaningfully used in smaller structures (bridges over highways or smaller rivers). In Figure 3 we can see a small overpass Pekel near Maribor who has been tested and leveling.

Figure 3: Levelling the overpass Pekel near Maribor.

In practice, with small bridges can be used and leveling rotary laser with sensors that are attached to your computer. This method can be used and the dynamic tests.

3.3 Method of trigonometry

The method of trigonometry are the most useful. With this method, we can measure progress in greater lengths and heights of the structure. To obtain high precision measurements should be made in several gyms that are in traditional total station takes too much time. Recently, using the robotic total station where during one gyrus measuring classical instruments make 3 gyrus. In this way we get more data from which we can calculate the precise measurement and an ellipse of errors. It is important at the micro-shift, where shifts in the range of accuracy of the instrument because using robotic cells with 0.5 sec possibility of measuring angles [25, 26].

Practice has shown us, that the best method to use trigonometry using electronic tachymeter with which we can measure the length and vertical angle. With the import of meteorological conditions, the measured length and angle calculations with coordinates or new item or change in terms of height. Based on the height of trigonometry and import of all parameters and error can be calculated using the change in H-known formula [10][12]:

$$\Delta H = S \cdot ctgZ_A + i_A - l_B + \left(\frac{1-k_a}{2}\right) \cdot \left(\frac{S^2}{R}\right) + \omega \qquad (1)$$

where is:
S - Length
Z_A - vertical angle
i_A - the height of the instrument at point A
l_B - Reflector height at point B
k_a - the coefficient of refraction (for Slovenia to f = 0.13)
R - radius of the Earth
ω- correction coefficient

The method of trigonometric leveling, which is used in pilot testing of structures, that is what affects the results. For each measure it is necessary to pre-calculate the accuracy of the measurement, which is based on the classic geodetic works where the height difference obtained by the above formula.

In the pilot tests and the measurement of vertical movement can of formula 1 drop height of the instrument and A, as measured in the relative coordinate system and the local system. From the formula we can drop and reflector height b, because we are, lock the retro brand and using the height of 0.0 mm. The measured lengths were relatively short (S <150 m) and we dropped a correction due to the curvature of the Earth.

Thus, formula 1 can be written in another form [11],[12]:

$$\Delta H_a = S \cdot ctgZ_A + \frac{S^2}{2R} - k_a \frac{S^2}{2R} \qquad (2)$$

In our example replaces the zenith length of the vertical angle and the length of the hair, so the formula is:

$$\Delta H_a = D_p \cdot \sin\alpha + \frac{D_p^{\;2}}{2R} - k_a \frac{D_p^{\;2}}{2R} \qquad (3)$$

For variable size can take the length Dp, vertical angle and refraction coefficient k.

With this formula, we can get very precise vertical movement. Also this method gives us the ability to measure shifts in the direction of x, y, so that we get shifts in the 3D direction of what we used in the monitoring of movements and deformations of the stairs and dams and most of geodetic monitoring of the 3D direction. In Figure 4, we see two positive tools for proof testing Pekel overpass near Maribor.

Figure 4: Two total station next to the overpass Pekel

In our example, the testing instrument for monitoring the progress we used a total station Nikon NPL series 352 and robotic station LEICA MS50.

Table 1 shows the resulting shifts in the point T1 in the middle of the range 1 and T2 in the middle of the range 2 with both instruments. With a classical station we made a survey in one gyrus with a robotic station in 5 gyrus. From the results it can be seen and the accuracy of the Leica total station with classical instrument from one gyrus, we can not get accurate measurements so that in this example uses the accuracy of the instrument.

Table 1. Comparison of the measurement results

	Vertical displacements [mm]			
Point	Phase 1		Phase 2	
	NPL	MS50	NPL	MS50
T1	-2.0	-1.8	0.5	0.3
T2	1.0	1.8	2.0	-2.7

3.4 Scanning Method

Lately there is much talk about field scanners. By now we know the classic scanner and the company made a LEICA instrument Multi Station, which is a combination of robotic cells, and scanners with the precision of sight 0.5 ". With this instrument, we also measured the progress in the construction of the test 3 gyrus and in the same time and make a scan at the time of construction, without load and when we put the load as seen in Figures 5, 6. With the structure we obtain the image scan to the coordinate system and gain a complete model of the structure that we used for different simulations.

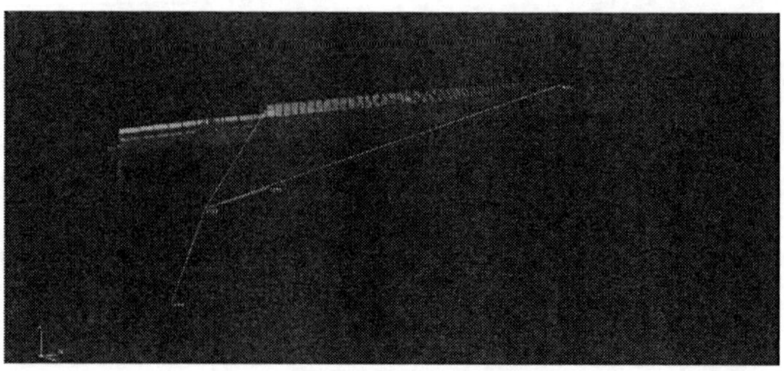

Figure 5. Overpass Pekel and measurement points - an empty object (scan)

Figure 6. Overpass Pekel - loaded object (scan)

We can also get the mathematic comparison image and shifts for each point of the structure. It gives us a whole picture of structural behavior during load. Certainly this method is accurate and effective. Just to make structural model in office it would take several days and there we get in a few minutes the entire model.

3.5 GPS method

Global Positioning System (GPS) is recently used increasingly for testing. Surveyors use it for recording terrain, positioning objects ... We also tested the GPS to track vertical movement. We used the method of RTK and every 5 seconds made a survey (Figure 7). I need to know, that the accuracy of GPS in the range from a few millimeters to a few centimeters so that this method can not get a shift rank 1-3 mm. After analyzing the results, we get the shift level 12 mm which is more than all the other results obtained with other instruments. GPS method is mostly used for geodetic monitoring of landslide where the antenna is left in the same place for 24 hours, and we use the statics readings every 15 seconds.

Figure 7. GPS device on Puchov bridge in Ptuj

3.6 Method of photogrammetry

For comparison measurement method shifts we used and the method of stereo-photogrammetry. In this method we use brand - points that are fixed in the vicinity of geodetic mark, so the comparisons shift (Figure 8).

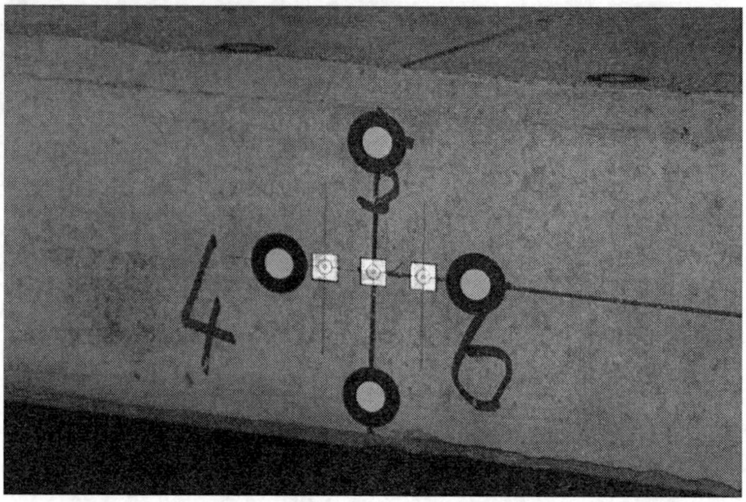

Figure 8. Self-adhesive retro brand

Mark - point is the white circle with a diameter of 18 mm on a black background. It is used a system for stereo-photogrammetry TRIPOP manufacturer GmbH (Braunschweig, Germany), in which the CCD camera Fuji S3 Pro with a resolution of 4256 x 2848 pixel and a software package for the analysis of the object coordinates. In every step of the load was made at least 8 shots at different angles and positions and, most importantly, the images should mind all brands. Photos of individual phase loads are processed with software. The results of processing the motion vector in the vertical direction (Figure 9). The maximum standard deviation results in each phase was 0.08 mm.

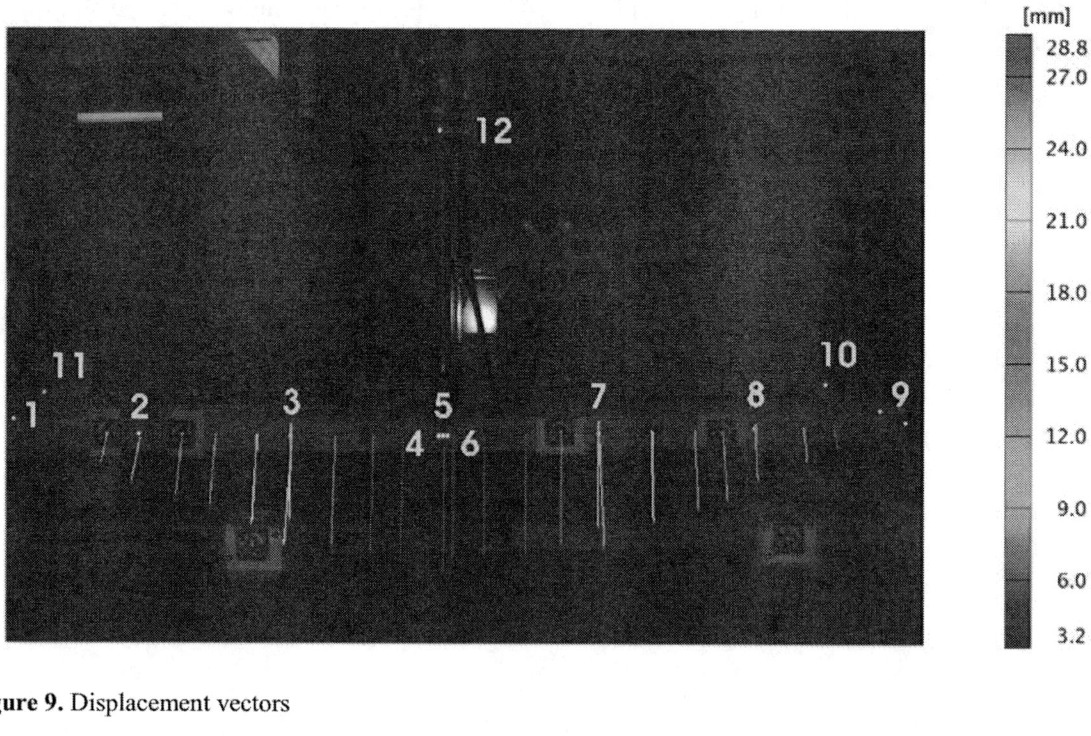

Figure 9. Displacement vectors

Photogrammetry method is used for smaller construction and testing in the laboratory. For large structural problem resolution photos and positions with which we make shots, because we do not see all stamps from the point.

3.7 The relative methods

By these methods, most use inductive felt we put under construction. These sensors are highly sensitive and provide us with the most accurate results. We use them in the dynamic tests where the geodetic methods can not be so quick to make the recording points.

Gauging - strain gauge can be put on the reinforcement of the bridge already at the time of construction or glued to the structure when it is already built. Gauging give us the results of the specific deformations in terms of voltage change in the material. When building the largest viaduct in Slovenia - Crni Kal - we placed our over 100 papers, and at the time of load track and structural behavior change specific strain.

Figure 10. Measurements of voltage below the overpass Pekel

4 Comparison of results and discussion

In preliminary tests structure we used different methods. Before you do the test should see what is the structure, which has beneath it, which rank shift predict and range structure. Based on that we decide which method we use. In practice, it's always good to use two methods at the same time so that we can make analysis and comparison. You need to know that with geodetic methods can not make repeated measurements under the same conditions, so you should be very careful in measuring and concentrated.

Results of the test tests gave us some answers about the accuracy of the measurements. From the results it can be seen that the improvements obtained with different methods on the same level with a margin of a few tens of mm. For such a small structure is sufficient to use inductive sense for the dynamic portion of the test and the method of leveling, for larger structures and robotic cell with or without scanner and inductive measuring senses and leaves.

5 Conclusion

Observation of the structure under load can be performed in various ways, as described by many authors. Which method will be used mainly depends on the range of construction, planned developments, access to the construction and weather conditions. In practice, it makes sense to use at least two independent methods which we partially increases costs and increases measurement time, but can provide us with more reliable results. On the basis of multiple measurements can be performed by a variety of statistical analyzes and comparisons.

In our case, we used seven different methods. The aim of the measurements was to compare and determine the most appropriate method. If you can already in the construction phase set Gauging it would be most appropriate as we get some amount (as happens with the structure, temperature, stretch material ...) during construction. Other surveying methods are used for the static portion of the test. From the test examination we see that without surveying can not carry out trial tests.

References

1. T. Thomas. *Principles of geometry, mensuration, trigonometry, land-surveying and levelling* (Longman, London, (1855)
2. M. Tatin, M. Briffaut, F. Dufour, A. Simon, J.-P. Fabre, Engineering Structures, **91**, 26-39 (2015).
3. J. Yu, X. Meng, X. Shao, B. Yan, L. Yang. Engineering Structures, **81**, 432-443 (2014)
4. N. Vatin, N. Lavrov, A. Shipilov, Procedia Engineering, **117 (1)**, 374-380 (2015)
5. C. Almeida Santos, C. Oliveira Costa, J. Batista. Mechanical Systems and Signal Processing, In Press, (2015)
6. S. Ognjenovic, Z. Krakutovski, N. Vatin, Procedia Engineering, **117 (1)**, 564-572 (2015)
7. S. Ognjenovic, R. Donceva, N. Vatin, Procedia Engineering, **117 (1)**, 556-563 (2015)
8. M. Dilena, M.P. Limongelli, A. Morassi, Mechanical Systems and Signal Processing, **52–53**, 162-180 (2015)
9. C. Venkatramaiah, *Textbook of Surveying* (Orient Blackswan, London, (2011)
10. B. Kovačič, R. Kamnik., M. Premrov. Survey Review, 43 (320), pp. 150-161 (2011)
11. Vodopivec, F. The precise leveling (Ljubljana, FUJ (1988)
12. B. Kovačič, R. Kamnik. AVN. Basic Vermessungs messages, **113**, 10, p. 322-329 (2006)
13. Marendic A., Z. Kapovic, R. Paar, Geodetski list, **3**, 175-190 (2013)
14. Z.Popović, L. Lazarevic, N. Vatin, Procedia Engineering, **117 (1)**, 846-853 (2015)
15. S. Ognjenovic, R. Ristov, N. Vatin, (2015) Procedia Engineering, **117 (1)**, 575-578 (2015)
16. S. Rutešić, J. Ćetković, M. Žarcković, M. Knežević, N. Vatin, Procedia Engineering, **117 (1)**, 905-915 (2015)
17. S. Ognjenovic, Z. Zafirovski, N. Vatin, Procedia Engineering, **117 (1)**, 579-584 (2015)
18. S. Popović, S. Vlahović, N. Vatin, Procedia Engineering, **117 (1)**, 854-863 (2015)
19. S. Rutešić, J. Ćetković, M. Knežević, M. Žarcković, N. Vatin, Procedia Engineering, **117 (1)**, 642-650 (2015)
20. R. Usmanov, I. Mrdak, N. Vatin, V. Murgul, Applied Mechanics and Materials, **633-634**, 932-935 (2014)
21. R. Usmanov, V. Murgul, N. Vatin, Applied Mechanics and Materials, **633-634**, 1082-1085 (2014)
22. R. Usmanov, M. Rakočević, V. Murgul, N. Vatin, Applied Mechanics and Materials, **633-634**, 927-931 (2014)
23. V.V Okrepilov, M.V. Leonidovich, Asian Social Science, **11 (7)**, 312-325 (2015)
24. I. Mrdak, M. Rakočević, L. Žugić, R. Usmanov, V. Murgul, N. Vatin, Applied Mechanics and Materials, **633-634**, 1069-1076 (2014)
25. V.V Okrepilov, V.L. Makarov, A.R. Bakhtizin, S.N Kuzmina, Economy of Region, **2**, 301-313 (2015)
26. V.L. Kvint, V.V Okrepilov, Herald of the Russian Academy of Sciences, **84 (3)**, 188-200 (2014)

MATEC Web of Conferences **53**, 01049 (2016)
DOI: 10.1051/matecconf/20165301049

Efficiency of Urban Development under High-rise Construction of Districts

Irina Ptuhina[1], Tatiana Spiridonova[1], Tatiana Musorina[1,a], Svetlana Kanyukova[1] and Anna Rezvaia[2]

[1]*Peter the Great St.Petersburg Polytechnic University, Polytechnicheskaya 29, 195251, St.Petersburg, Russia*
[2]*Magadan Institute of Economics - Branch Office of the Private Educational Establishment of Higher Professional Education Saint-Peterburg, Universi ty of Management and Economics, Magadan, Novaya ul. 27, 685000, Russia*

Abstract. In article the problem of irrationally organized massive shabby, including emergency quarters on the example of St. Petersburg reveals. The goal of research: to offer a method of development (renovation) of urban areas by construction of quarters from exclusively high-rise buildings. Because of building of high-rise objects, considered quarters can assume more modern air, increase in the housing areas by 10 times not less in comparison with the previous indicators of this territory, the improved infrastructure of the area, and as a result, emergence of new workplaces. Authors made the analysis of economic indicators of the investment project. The minimum term, for which saved-up income exceeds initial investments (a project payback period) made 2.3 years. The analysis of risks is presented in this article

1 Introduction

Creation of living conditions, which meeting requirements of Russian society of the XXI century has to be one of the priority directions of social and economic policy of the State. In article the problem of irrationally organized massive shabby, including emergency quarters on the example of St. Petersburg reveals. The main attention in article is focused on a way of renovation of quarters by demolition of outdated buildings and constructions and construction on their place of the high-rise units combining modern requirements to quality of inhabited massifs, infrastructures, production and other capacities. High-rise construction is meant as construction of buildings and structures more than 100 meters. The functional thought-over high-rise buildings concentrate in themselves the high density of infrastructure, or for example, housing stock, releasing a useful area of city lands within the existing megalopolis borders.

There are 18 districts including "painful" quarters at the beginning of 2015 in St. Petersburg, which area makes about 1439 sq.km.

At the first stage of selection of the suitable district for research areas in which according to High-rise regulations it isn't allowed to create high-rise buildings (table 1) were excluded.

Table 1. Characteristic of the districts of St. Petersburg by Russian Regional Information Agency

No.	Title of district	Square, [km2]	Background height
1	Admiraltejskiy	13.82	----

[a] Corresponding author : flamingo-93@mail.ru

2	Vasileostrovskiy	14.64	to 70 [m]
3	Vyborgskiy	115.38	to 100 [m]
4	Kalninskiy	40.12	to 100 [m]
5	Kirovskiy	48.00	to 80 [m]
6	Kolpinskiy	105.8	to 80 [m]
7	Krasnogvardeyskiy	56.83	to 80 [m]
8	Krasnoselskiy	115.00	to 120 [m]
9	Kronshtadskiy	19.35	to 50 [m]
10	Kurortniy	267.92	to 30 [m]
11	Moscowskiy	71.07	to 90 [m]
12	Nevskiy	61.79	to 140 [m]
13	Petrogradskiy	24.00	----
14	Petrodvortcovyj	109.00	to 36 [m]
15	Primorskiy	109.87	to 100 [m]
16	Pushkinskiy	239.95	to 50 [m]
17	Frunzenskiy	37.47	to 100 [m]
18	Thsentralnyj	17.12	----

So, for consideration 6 districts of St. Petersburg in the territory of which construction of high- rise buildings is allocated:

- Vyborgskiy,
- Kalninskiy,
- Krasnoselskiy,
- Nevskiy,
- Primorskiy,
- Frunzenskiy.

On the basis of the analysis of the offered way of renovation of quarters an attempt is made is shown that the method will allow enabling the realization more effectively:

1. Social needs of the population of considered quarters for educational and entertaining resources, sports resources, medical resources, thereby having created the new centers of social and economic development of St. Petersburg.

2. Commercial requirements, by growth of appeal of peripheral territories to placement of industrial and public and business high-rise complexes, and also ensuring high-rise profitability of investments owing to uniqueness of objects.

3. The budgetary requirements – creation of new workplaces and sources of replenishment of the local budget at the expense of taxes [1].

1.1 Goals and objectives

Object of research is the extensive method of renovation [2] of the residential quarter by demolition of shabby housing and construction of high-rise buildings in the freed territory.

Subject of research is the assessment of efficiency of carrying out renovation by an extensive method.

The goal of research: to offer a method of development (renovation) of urban areas by construction of quarters from exclusively high-rise buildings.

Objectives:

- research of problems of reproduction of the dwelling and analysis of a condition of housing stock,
- drawing up an economic assessment before renovation of quarter and after,
- identification and analysis of financial and non-financial forms of the state (municipal) participation in process of renovation of housing stock.

Scientific novelty of work comes to light in use only of high-rise unique buildings and constructions in a method of development (renovation).

2 Materials and Methods

2.1 International experience of urban high-rise construction of districts

30% of a world's population lived in the cities in 1950. It was 47% in 2000. The sum can reach even 60% by 2030 of also quickly growing urbanization, especially in developing countries, creates a lot of opportunity and problems [23].

Considerable distribution was gained by construction of panel buildings, as well as in Russia, after World war II in Western and Eastern Europe. A lot of buildings considerably became outdated and demanded capital repairs by 1980-2000th years. Panel buildings in Europe, as well as in Russia, were also inefficient, because considerable costs of repair and operation, low popularity among tenants and buyers from a commercial point of view. So, in 2000 renovation of a mass housing estate became one of priority tasks in Europe.

Many realized projects of renovation it is possible to note in such countries as Germany, Poland, Lithuania, the Netherlands, Romania, France and England. Development of projects of renovation of a housing estate happens due to reconstruction of houses or due to partial or full demolition of buildings and construction new. The significant role at their realization is played by public opinion of researches, public hearings, presentations of projects are conducted. Also through mass media it is told residents about the course of works.

Building of quarters can be both multifunctional, and single-function.

The only example of single-function high-rise office building in the city Frankfurt am Main one-functionality is determined by historically developed city role as all-European financial center. Today in the city offices of 400 largest banks are concentrated. According to the general function their compact placement in the uniform business high-rise downtown could be represented expedient.

Successfully projects in Paris are "The Front of Seine", and then a multipurpose complex La Défense on the western suburb of the city which replaced invaluable chaotic suburban building (Nanterre, Pyuto, Kurbevua). Building of the area La Défense became the largest and effective example of realization of the European principles of the integrated urbanism. Multifunctional building provided the round-the-clock bright rich life of the area. Concentration of high-rise office building on the suburb of the city relieved historical central building of Paris of inevitable reconstruction at placement in it new office objects and brought economic revival: in offices of La Défense about 80% of agreements and transactions in economy of France are annually concluded.

Large-scale town-planning reconstruction of an outdated industrial zone of the city was in London. There are an area of docks (Doklend) stretched on both coast of the Thames on more than 10 km to the east from the Tower was made. The State investments were pulled for this project and then investments of private investors for which the government planned essential tax privileges and big state subsidies for 10-year term went. The concept of reconstruction provided creation of a multipurpose urban environment with office center, various cultural and manufacturing enterprises, housing estate for people with various income, schools, medical and other institutions of infrastructure, parks which are arranged well by foot embankments, reservoirs and their coast. The first built in 1991 at that time the highest in Europe tower office Kanari-Verf (height of 245 m). Here it is supposed to build not less than 104 thousand new dwellings and to provide to 249 thousand new workplaces till 2016.

Russian builders can adopt practical experience of consecutive concentration of forces on the extremely limited number of sites from the European builders, as, for example, La Défense in Paris or Doklend in London, and also ensuring complexity of building at the expense of a combination of objects of a different functional purpose in buildings which space-planning decision most harmoniously answers its function that means creation of multifunctional skyscrapers, and also combination in construction of different buildings of different function [24].

2.2 Concept of renovation of housing stock, justification of a choice of renovation method

Renovation of housing stock – is a process of replacement and restoration of the elements of housing stock (residential buildings and their parts) which left as a result of functional and physical wear with new elements or objects [2].

Renovation of housing stock is a cyclic process including generally sequence of transformations of housing renovation methods, which are a new construction, overhaul, modernization, restoration and reconstruction, demolition of residential buildings and construction of new objects [2]. The last way is considered in this article, and it is carried out by means of wrecking of the existing building and repeated construction of buildings on the released land plots.

The author of the thesis Khokhlov, O.B [2] specified in the research that among different ways of renovation demolition and new construction is the most perspective method (table 2).

Table 2. Main economic results of realization of methods of renovation of housing stock

Renovation methods	Results of renovation process		
	Growth of the income from operation of object	Decrease in costs of operation of object	Increase in service life of structures and object in general
Demolition and new construction	+	+	+
Strengthening of bearing structures			+
Modernization of engineering communications		+	+
Installation of systems of the account and regulation of a water consumption and energy resources		+	
Re-planning and reshaping of rooms	+		
Strengthening of thermal insulation		+	+
Extension, superstructure of rooms	+		

Secondary building of the territory by high-rise buildings is a convenient way of renovation of quarter by means of which conditions of accommodation of the population improve, the quantity of shabby housing in the area decreases, there are large volumes of new inhabited space, and, therefore, and security of the population with housing increases, the infrastructure and improvement of the territory improves. Thus it isn't required development of new territories, because the site of building remains within the available borders [3].

2.3 The characteristic of the considered quarter and the description of a condition of housing stock

Authors of this article made an attempt to consider a possibility of renovation of the quarter which is situated in Nevskiy district in the territory limited to streets Karavayevsky and Teplovoznoy, Slepushkin and Skachkov's lanes and makes 141129.6 sq.m (see fig. 1). The distance to historic center is 18.7 m.

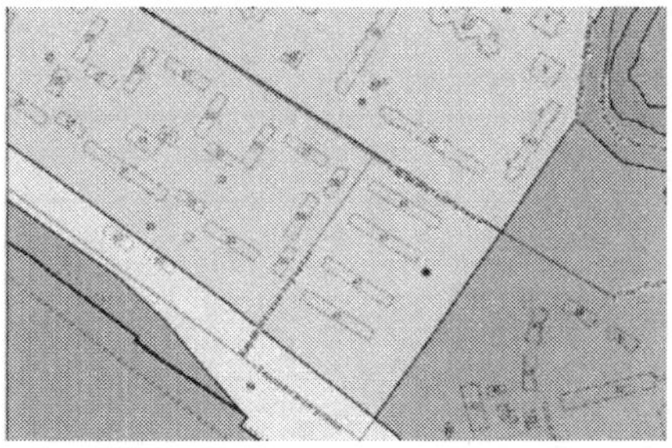

Figure 1. Fragment of quarter

The maximum background height is 110 m on the considered territory. There are 25 houses are situated now in the territory of the quarter.

Zone is inhabited for middle- and multi-storey multi-flats houses located out of the territory of historically developed districts of St. Petersburg's center with inclusion of objects of welfare and household appointment. Quarter borders from Karavayevskaya st. on a fire brigade.

Figure 2. Karavayevskaya st. – a) panorama of the street. b) and c) courtyard

All houses, which are situated in the territory of this quarter (see fig. 2, table 3), were made in accordance with project of Series 1-528KP. Usually, people calls such houses "Brezhnevky". A brick series 1-528KP, actively was under construction in suburbs in Leningrad. There are exist more than two tens modifications of different number of storey from two to five, and also quantity of front doors from two to eight in the nomenclature. Features of these houses are qualitative parquet, panel framed doors and well-adjusted window frames. There are enough high ceilings (2.5 m), small kitchens and halls, existence of several adjacent rooms [4-5].

General characteristics of a Series 1-528KP are:

- houses are made from bricks,
- there are about 2-5 storeys,
- height of premises is 250 cm
- the number of rooms are 1.2, and 3 room in a one flat,
- it usually makes with local building materials,
- years of construction are 1959-1975.
- cities of distribution are St. Petersburg and Nizhny Novgorod Novgorod [22]. Table 3. General analysis of infrastructure of the district

Table 3. The situation in the quarter

Type	Number and distance	Additional
Restaurants and cafe	8 (1-1.5 [km])	"McDonald's", "KFC" ets.
shopping mall	2 (1 [km])	"Parade" mall, "Port Nakhodka" mall
Hotels	2 (5 [km])	Hotel in Metallostroy
Leasure		-
Main sightseeings	many in historical center, a lot of in district	
Health-improving zone	2 fitness clubs	
Educational centres	7	
Medicine	Government's Clinique for children, more than 5 private clinics for adults	
Suway (underground)	Rybatskoye 1	
Buses	nearby subway, more than 10 routes	
Fixed-run taxis	nearby subway, more than 10 routes	
Railway station	Rybatskoye	

2.4 Sequence of implementation of renovation

Participation of the State in renovation, research of the processes, which are connected with resettlement of residents.

It is supposed that responsibility for demolition of available buildings, cleaning of the territory of construction garbage and sale of a site are assumed by the state. Therefore, the municipality of the area "prepares" a site for sale and enables its realization at auction specifically for construction of a new residential building.

Possible ways of resettlement of residents at demolition of a residential building are shown in table 4.

Table 4. Ways of resettlement of residents at demolition of a residential building is possible in…

1. …in the new housing acquired in primary market, also in the territory of Nevskiy district	2. …in the housing acquired in the secondary market, also in the territory of Nevsky district	3. …in the new housing constructed on a place of demolished objects

Authors choose the third way in this article, because it allows to make:

1. often renovation for many inhabitants is a social stress, citizens could get used to a residence and have difficulties with process of resettlement on a constant basis,

2. use the aspect of quality improvement of inhabited environment for residents, who live in

"Brezhnevka" for confirmation of effectiveness of considered method of renovation allocating.

This way has additional difficulty, which is the need of providing residents with the lodging.

Authors of the thesis [2] claim that nevertheless "even with a growth of average security with housing, the share of the people having housing of "low" quality doesn't decrease". Really, such problem has a wide circulation in the territory of St. Petersburg, but by sight authors of this article can be found the ways allowing "to go to a meeting" to the population, having reduced housing cost, for example, by means of cooperation of government bodies and commercial banks to present the most favorable conditions for receiving a mortgage in the new building, for example, to young families [6-9].

Thus, it is necessary to move temporarily previously before dismantle of buildings for the residents living in the territory of the quarter captured by renovation processes.

Let renovation of the I-IV buildings 41 on Karavayevskaya st. will become the first step. Therefore, it will be required to settle in removable housing of 400 families (100 families live in each house). And here, the first difficulties connected with requirement of receiving at the initial stage of rather large volume of investments are possible. By authors of article it was calculated that as of January, 2015 the total amount allocated for temporary resettlement of residents for 3 years (for estimated time of duration of construction of the new high-rise building), has to make about 3,5 billion rubles in case of moving on leased squares. It is important to note that in case of moving on the squares relating to the state funds of reserve housing, the sum considerably will decrease depending on reduction of a rent [10-12].

According to Russian Housing Codes the maintenance responsibility and repair of residential buildings of the privatized areas lies on owners of housing. But it is impossible to carry out renovation, because in a year on routine maintenance and overhaul repairs residents of four cases of the forty first house pay in the sum about 50000 rubles. These means will be enough only for payment of cargo transport for implementation of transportation of personal belongings of residents of three of four houses on a temporary residence for comparison [13-15].

2.5. Participation of commercial subjects of renovation

The project of renovation of quarter, first of all, is the investment project, where can be used by investors not only the State sources of financing, but also sources from other subjects of the market relations, in other words, non-budgetary sources.

Thus, who can invest in renovation of the considered quarter?

- the population, due to accumulative effect from monthly payment under the article "capital repairs",
- commercial subjects, at the expense of an investment of own and borrowed funds, for example, the bank credits.
- public funds, for example, budgetary funds and means of trust off-budget funds [16].

3 Results and Discussion

3.1 Efficiency of the project of renovation by construction on a place of demolished buildings of high-rise buildings

Economic effect demonstrates the difference between results of economic activity (for example, a product in value terms) and the expenses made for their receiving and use.

The project of renovation of quarter, as it has already written in paragraph 3, is investment project and consequently it is necessary to prove to investors that the project investment and attractive therefore it is necessary to calculate efficiency of implementation of the project [19-21].

Let from 25 available in the territory of the considered quarter of 4 buildings (41, I-IV parts) will be accepted as the first step (table 5). Construction of one high-rise building is planned in the territory of these buildings. It is possible to consider social and economic efficiency of such project on this example [17-18].

Table 5. Characteristic of objects of renovation (1st step)

Indicators	Values for one house	Values for four houses
1. Number of flats Including in private property, in government property	100 60% 40%	400 60% 40%
2. General square of flaats, m2	4000	16 000
3. Number of families	100	400
4. Number of floors	5	
5. Number of entrances	5	25
6. Management company	„Ryb.-Stroyservice“	
7. District	Nevskiy	
8. Year of construction	1975	
9. Physical deterioration	60%*	
10. Volume of building, m3	58 600	234 400
11. Market cost of 1 m2 of general square, rub.	83,000	
12. Market cost of all general square, rub.	334 500 000	1 338 000 000
Contributions to capital repairs, rub per year	11364 – per month	45 456
13. The average cost of rent is 1 sq.m of housing, rub/month **	496 (without utilities)	

Note: * Physical deterioration is accepted conditionally for calculation, isn't the valid wear of the considered building

** The average cost of rent of cubic meter of housing without utilities is accepted.

Before an assessment of efficiency it is necessary to define what finantial streams will be considered in calculations [2]. The interval of time of investment into the project is T = 3 years, it was accepted as the horizon of an assessment of financial streams (table 6, table 7, table 8).

Table 6. Project of renovation of the residential building 41, building of I.

Calculation horizon period	
t = 0-1st year	t = 1-3 years
Demolition, resettlement of residents and building of new object	End of investment, settling of residents

Table 7. Basic data on the project of new construction

Calculation of outcoming money	Square, [m2]	Cost of 1 [m2]	The total cost
Construction cost of 1 m2 of the city		80 000.00	
building area	196000		
The cost of construction			15 680 000 000.00
built-up area	16000		
The cost of land 1m2		15 000.00	
Value of the land			240 000 000.00
Design and operation izyskat 3% of the cost of building.	0.03		470 400 000.00
Field supervision of 0.2%	0.002		31 360 000.00
Technical supervision 0.1%	0.001		15 680 000.00
Non-investment costs costs 2% of the value	0.02		313 600 000.00
The total cost total			16 751 040 000.00
The cost of construction of 1m2			85 464.49
Calculation of incoming money			
selling price of 1m2 to 25 thousand more than the actual		110 464.49	
Estimated inflow from sales			21 651 040 000.00
The distribution of data of outcoming money			
0 year (cost of land, design and izyskatelkie work)			710 400 000.00
Costs in future years with 1 to 3 (total price minus costs year 0) are			16 040 640 000.00
1 year 35%	0.35		5 614 224 000.00
2 year 40%	0.4		6 416 256 000.00
3 year 25%	0.25		4 010 160 000.00
Total otok the period from 0 to 3 years			16 751 040 000.00
Money distribution by years (10% risk)			
0 year 0%	0		0.00
1 year 10%	0.1		2 165 104 000.00
2 year 50%	0.5		10 825 520 000.00
3 year 30%	0.3		6 495 312 000.00
			19 485 936 000.00

Table 8. Calculation of NPV

			Characteristics of capital investments (outflow)			Features inflow			Features income		
Year inve stme nt (t)	The interes t rate discou nt (E)	Coeffi cient. discou nt (η)	Capex data (Kt)	Discounted . capex data Ktη	Discounted cumulative total capex Ktη	The influx of data (Rt)	Discounted inflow data Rtη	Discounted pritokpo years Rtη accrual basis	Income data without discount	Income data at a discount	Income data on an accrual basis (NPV)
0	0.2	1.000	578 400.00	578 400,00	578 400.00	0.00	0.00	0.00	- 578 400.00	- 578 400.00	- 578 400.00
1	0.2	0.833	5 614 224.00	4 678 520.00	5 256 920.00	2 151 904.00	1 793 253.33	1 793 253.33	-3 462 320,00	-2 885 266.67	-3 463 666.67
2	0.2	0.694	6 416 256.00	4 455 733.33	9 712 653.33	10 759 620.00	7 471 958.33	9 265 211.67	4 343 364.00	3 016 225.00	- 447 441.67
3	0.2	0.579	4 010 160.00	2 320 694.44	12 033 347.78	6 455 712.00	3 735 944.44	13 001 156.11	2 445 552.00	1 415 250.00	967 808.33
			16 619 040.00	12 033 347.78		19 367 236.00	13 001 156.11		2 748 196.00	967 808.33	

Result:

NPV – 967 808.33 rub, Index of profitability – 1.08, Payback period – 2.3 years. Calculation results are presented in Figure 3 and 4.

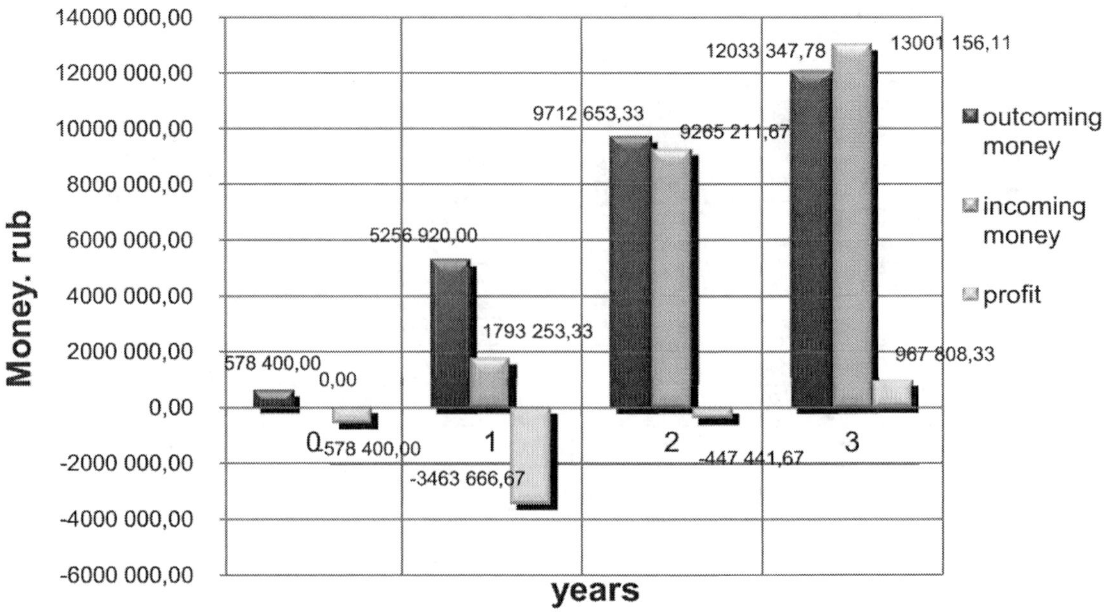

Figure 3. Characteristics of payback period

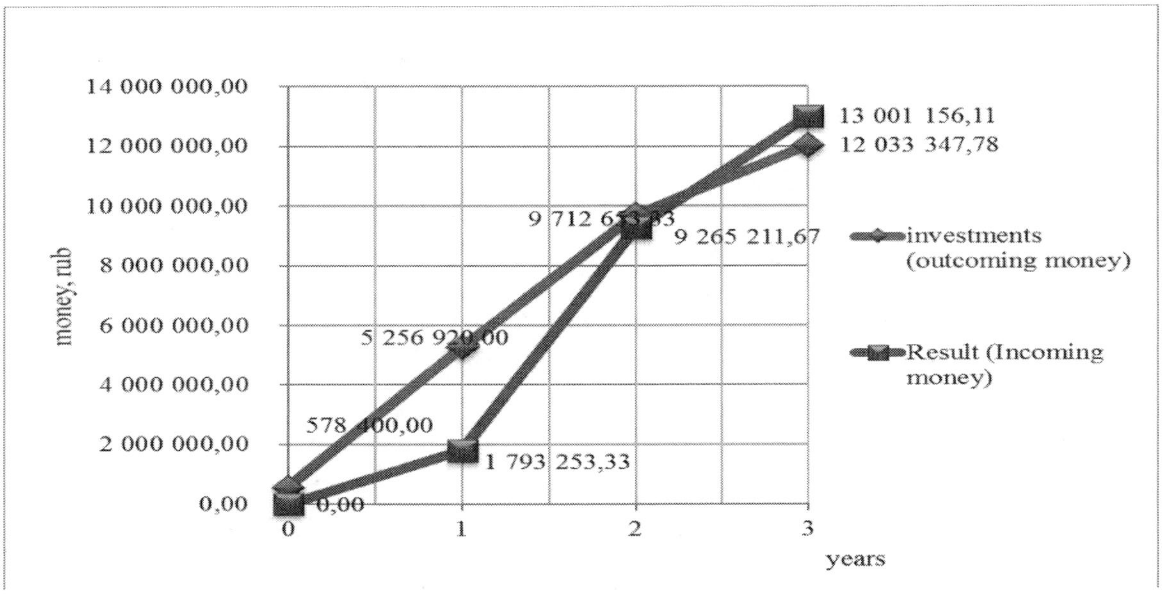

Figure 4. Characteristics of incoming, out coming money and profit

Let us assume that renovation would be carried out construction of a new building, but not the high-rise and ldf 15-17 storey. Then, on the basis of a similar calculation we take the values of NPV – 300 716.38 rub, Index of profitability – 1.18 and payback period – 1.875 years. Thus, there is an opportunity approximately to estimate and compare possible technical and economic risks.

Factors of economic risk are change in discount rate E=18%, change in discount rate E=20%, errors in budgeting and appreciation of resources.

Factors of technological risk are errors in project documentation, violation of technology of performance construction installation works, low professional level and violation of the schedule diagram of performance of installation and construction works.(Results are seen in Table 8.1, 9, 10)

Table 8.1. Supposed characteristics of projects

(i)	Project of high-rise construction (1)		Project of building with 15-17 floors (2)	
	The probability of getting result (pi)	Value of of the result (ci)	The probability of getting result (pi)	Value of of the result (ci)
1	0.40	967.80	0.35	360.84
2	0.25	483.90	0.45	0.00
3	0.35	871.02	0.20	300.70

Table 9. Risk assessments of investment project

Indicators	Number of investment projects	
	P	Projec
The mathematical expectation of profit	812.95	186.43
The dispersion of earnings	37840.13	28898.39
The standard deviation of earnings	194.53	170.00
The coefficient of variation	0.24	0.91

Table 10. Assessment of borders of the most possible profit on projects

Indicators	Number of investment projects	
	P	Project
The upper limit of the possible profit	1396.5	696.4
The lower limit of the possible profit	229.4	-323.6

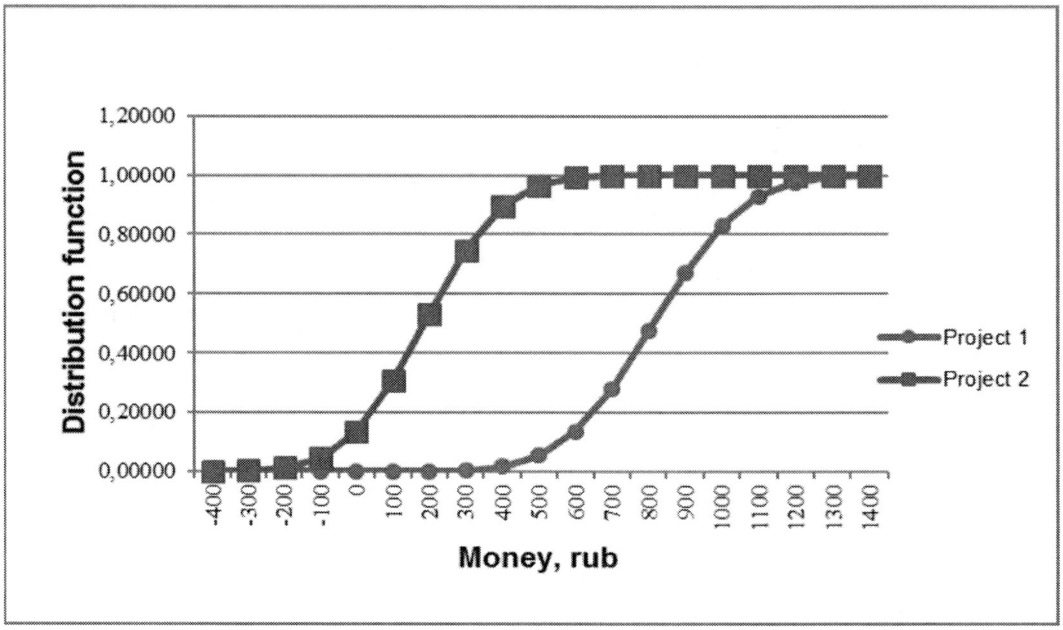

Figure 5. Distribution function of possible profit

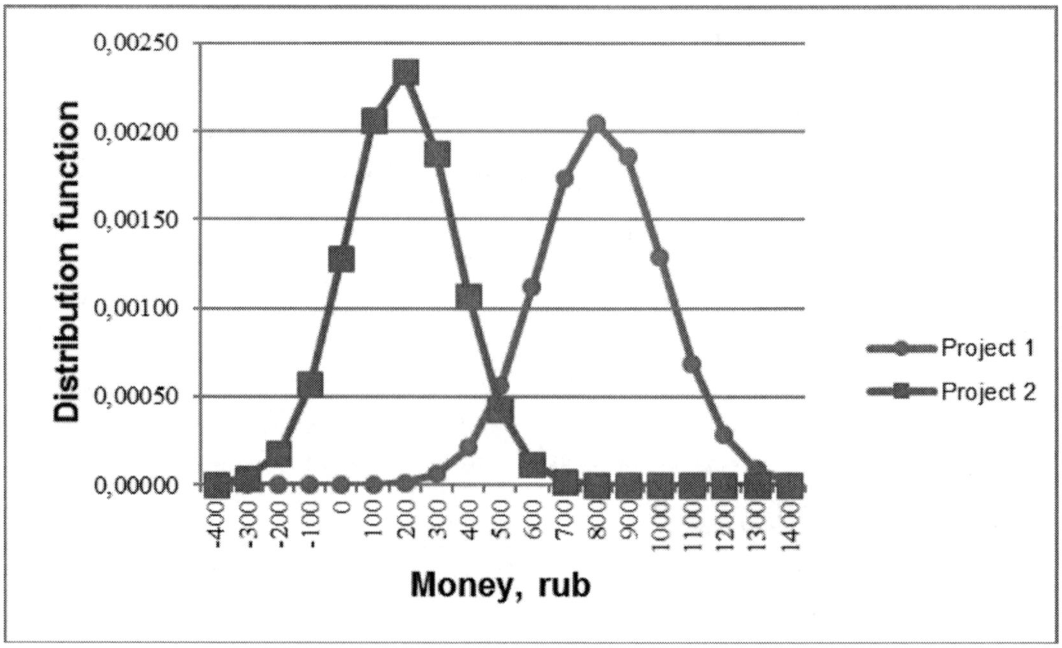

Figure 6. Distribution of density of possible profit

The possible profit makes: from 229.4 million rubles to 1396.5 million rubles at economic risks and from -323.6 million rubles to 696.4 million rubles at technological hazards by results of calculation of economic and technological risks for the first project (see Figure 5 and 6). Cooperation with insurance companies will allow reducing probability of economic risks [22].

4 Conclusions

1. The problem of reconstruction of shabby quarters can be solved with the help of demolition of housing which doesn't meet modern requirements of quality, and construction on a place of the demolished buildings of skyscrapers. Because of building of high-rise objects, considered quarters can assume more modern air, increase in the housing areas by 10 times not less in comparison with the previous indicators of this territory, the improved infrastructure of the area, and as a result, emergence of new workplaces.
2. Authors made the analysis of economic indicators of the investment project. The project 35 of a floor skyscraper is considered economically effective: in three years of investment the house will bring in the income in 967808.33 thousand rubles. Efficiency of investments is more than unit, namely the index of profitability is equal 1.08. The minimum term, for which saved-up income exceeds initial investments (a project payback period) made 2.3 years.
Renovation by demolition of shabby housing and construction of new buildings on a place of the taken down is the most favorable way of renovation owing to which advantages are got by both the state, and commercial investors.
3. The first project of high-rise renovation is more "safety" than the second as even taking into account economic and technological risks as the minimum border of possible profit can be positive.
4. Russian builders can adopt practical experience of consecutive concentration of forces on the extremely limited number of sites from the European builders and ensuring complexity of building at the expense of a combination of objects of a different functional purpose in buildings which space-planning decision most harmoniously answers its function that means creation of multifunctional skyscrapers, and also combination in construction of different buildings of different function

References

1. A.A. Uversky, O.G. Nikonov, Construction of Unique Buildings and Structures, **4** (19), 139-150 (2014).
2. O.B. Khokhlov, Performance evaluation of projects and programs renovation housing (Thesis, Tomsk, 2006)
3. Ptuhina, T. Spiridonova, T. Musorina, Applied Mechanics and Materials, **725- 726**, 153-159 (2015).
4. M. Petrichenko, X. Rakova, M. Vyatkin, T. Musorina, D. Kuznetsova, Applied Mechanics and Materials, **725-726**, 1101-1106 (2015)
5. N. Lipovac, A.M. Jandriček, Prostor, **19** (2), 443-455 (2011)
6. Z.A. Gaevskaya, X.M. Rakova, Advanced Materials Research, **941-944**, 825-830 (2014)
7. N.S. Bolshakov, S.A. Krivoy, X.M. Rakova, Advanced Materials Research, **941- 944**, 895-900 (2014)
8. Ptuhina, M. Viatkin, T. Musorina, Construction of Unique Buildings and Structures, **9** (14), 46-56 (2013)
9. O.N. Popova, T.L. Simankina, Magazine of Civil Engineering, **7** (42), 40-50 (2013)
10. I. Čerpes, N. Pandol, A. Fikfak, European Countryside, **6** (3), 225-242 (2014)
11. C. Zappulla, C. Suau, A. Fikfak, Journal of Architecture and Urbanism, **38** (4), 247-264 (2014)
12. A. Fikfak, Journal of Architecture and Urbanism, **37 (4)**, 257-267 (2013)
13. T. Teslya, T. Spiridonova, Construction of Unique Buildings and Structures, 7 (12), 5-9 (2013)
14. J. Tamošaitienė, J. Šipalis, A. Banaitis, E. Gaudutis, International Journal of Strategic property and Management, **17 (1)**, 92-108 (2013)

15. Ptuhina, M. Viatkin, T. Musorina, Construction of Unique Buildings and Structures, **5 (10)**, 58-67 (2013)
16. J. Tamošaitienėa, E. Gaudutis, M. Kračkaa, Procedia Engineering, **57**, 1151–1155 (2013)
17. K. Al-Kodmany, Journal of Urban Technology, **18** (3), 115-140 (2011)
18. http://gov.spb.ru/gov/terr/ (date of reference: 11.01.2015).
19. http://spbren.ru/proekti/9/34968.html (date of reference: 11.01.2015).
20. D. Nemova, V. Murgul, V. Pukhkal, A. Golik, E. Chizhov, N. Vatin, Journal of Applied Engineering Science, **12** (**1**), 37-44 (2014)
21. G. Radovic, V. Murgul, N.I. Vatin, Applied Mechanics and Materials, **584-586**, 564-569 (2014)
22. http://domavspb.narod.ru/index/0-32 (date of reference: 14.01.2015).
23. http://database.cobraman- ce.eu/abstracts/abstract_BR_17.pdf (date of reference: 15.01.2015).
24. http://library.stroit.ru/articles/verhstroy/index.html (date of reference: 15.01.2015).
25. N. Vatin, D. Nemova, A. Staritcyna, D. Tarasova, Advanced Materials Research, **953-954**, 854-870 (2014).
26. N. Vatin, M. Petrichenko, D. Nemova, A. Staritcyna, D. Tarasova, Applied Mechanics and Materials, **633-634**, 1023-1028 (2014)

MATEC Web of Conferences **53**, 01050 (2016)
DOI: 10.1051/matecconf/20165301050

Organization of Business Processes of the Company on the Basis of the Systems Approach Teners

Valentina Vaganova[1], Lubov Vorona-Slivinskaya[1], Anna Medvedeva[1] and Svetlana Kuzmina[1,a]

[1]*Saint-Petersburg University of State Fire Service of the Ministry of the Russian Federation for Civil Defense, Emergencies and Elimination of Consequences of Natural Disasters, Moskovsky pr. 149,Saint-Petersburg, 196105, Russia*

Abstract. The article considers the management specificity on the basis of the systems approach tenets and description of business processes of the industrial enterprises. As the Company is a service-provider, its functional features are taken into consideration when modeling the business processes. The authors highlight challenges the Company faces in performance management because the existing system doesn't allow to predict the financial results at the stage of formation of orders portfolio, to evaluate adequacy of financial resources to objectives set and to operate cash flows. All these issues are considered in sufficient detail when analyzing a range of problems and are taken as a basis of the project on implementation of the budgeting system based on the process approach.

1 Introduction

The enterprise (further the Company), whose design of the systems approach to business-process management is taken as a basis of the study, renders the wide-ranging environmental impact assessment services [1].

Indispensable condition for work of any organization operating hazardous production facilities, generating waste, releasing contaminants into the atmosphere or water objects is compliance with requirements of the industrial ecology [2-4].

The concept of industrial ecology includes assessment and monitoring of chemical and biological pollution of the environment by the industrial enterprises, regulation of economic efficiency of actions on protection from pollutions and compliance of the industrial processes with the legal requirements. To identify, monitor and standardize harmful impacts on environment, the law provides a list of mandatory documents to be developed by legal entities and individuals, whose activity have direct or indirect impacts on environment (emission and/or dumping of the polluting substances, generation of production and consumption waste) [5-8].

Taking into account extensiveness of the environmental legislation, complication and diversity of the requirements in this field, one should have profound knowledge of laws, responsibility for legal and engineering decision-making, ability to predict the consequences. These demand the expert knowledge in the fields of ecology and industrial safety [9, 10].

[a] Corresponding author : kuzmina2003@bk.ru

2 Problem statement

Today it is safe to say that even the large industrial enterprises and organizations mostly prefer to outsource solving of these problems to the specialized expert organizations, rather than to train and employ stuff experts. This, in turn, is related to the concepts "environmental works and services", "market of environmental works and services", their development and application [12].
The environmental services provided by the Company are:

- inventory of pollution sources (air, water, waste);
- development of environmental projects for any type of industrial enterprises;
- consulting services in the fields of environmental security and environmental protection;
- environment-related information services;
- environmental audit, independent environmental assessment (including public assessment).

The Company is registered as a small business entity. Its core activity is research and development in the fields of natural and engineering sciences.
The income of the Company consists of two components:

- revenue from engineering/ technological projects in the fields of development and coordination of documentation in the sphere of industrial ecology and elimination of oil spills;
- revenue from expertise in the international and regional projects.

The core activity of the Company is development of the project documentation in the field of industrial ecology, approval and examination of projects. The list of the projects developed by the Company includes:

1. Plans on Prevention and Elimination of Oil Spills (PEOS);
2. Plans on Emergencies Elimination (PEE);
3. Plans on Localization and Elimination of Accident (PLA);
4. Material Safety Data Sheet of Dangerous Object (MSDSDO);
5. Draft Standards on Waste Formation and Limits on their Placement and Passports of Dangerous Waste (DSWFLP);
6. Draft Standards of Maximum-Permissible Emissions of the Polluting Substances in the Atmosphere (MPE);
7. Standards on Admissible Dumpings of the Polluting Substances in Environment (SAD);
8. Actions on Assessment of Impact on Environment (AAIE);
9. Projects of Justification of the Size of the Sanitary Protection Zone (SPZ).

The advancing activity of the Company is rendering environmental support to the enterprises (outsourcing).
The Company also renders expert services, participating in the international and regional environmental projects.
However the existing performance management system doesn't allow the Company to predict the financial results at the stage of order portfolio building, assess adequacy of financial resources to the targeted objectives and operate cash flows.
This range of problems was has been taken as a basis for the project on implementation of the budgeting system based on the process approach [12, 13].

2.1 Stage of the Company's life cycle and setting objectives in course of implementation of the budgeting system

For today the Company has passed the stage of formation and search for its market niche and mastering the survival skills. The long-lasting formation stage is characterized by a simple and informal organizational structure without clear division of functions and financial responsibility centers, and also by concentration of powers in hands of owners. There was no budgeting system, the managerial decisions were generally made by the Company's owners on the intuition basis. From time to time there were disagreements among founders concerning the accepted managerial decisions.

Nevertheless, the Company managed to occupy its market position, achieve competitive edge and create positive reputation.

Now the challenge for the Company is to stabilize its position, fully realizing its available advantages. Today it is necessary to model its business processes, form the organizational-and-financial structure, using a linearly-functional principle, namely to establish the financial responsibility centers, bring powers down to managers, formalize procedures, focus on sales growth. It is also necessary to form a reserve fund and establish tight control over expenses [15].

The model of budgeting system being developed on the basis of the process approach consists of the following stages:

1. Task definition and collection of information, necessary for budgeting.
2. Examination and analysis of the available information, generalization and analysis of the past achievements, calculation of performance indicators, draft budgeting.
3. Expert examination and evaluation of the draft budget.
4. Adoption of the budget.

2.2 Results

The business processes of the Company are relatively simple due to its specificity and operation in the services sector.

When modeling, all the business processes of the Company were divided into 3 groups (Fig. 1):

1. Core business processes.
2. Supporting business processes.
3. Management business processes.

This modeling has resulted in formation of the financial responsibility centers, including: profit centers, revenue center, cost centers and investment center.

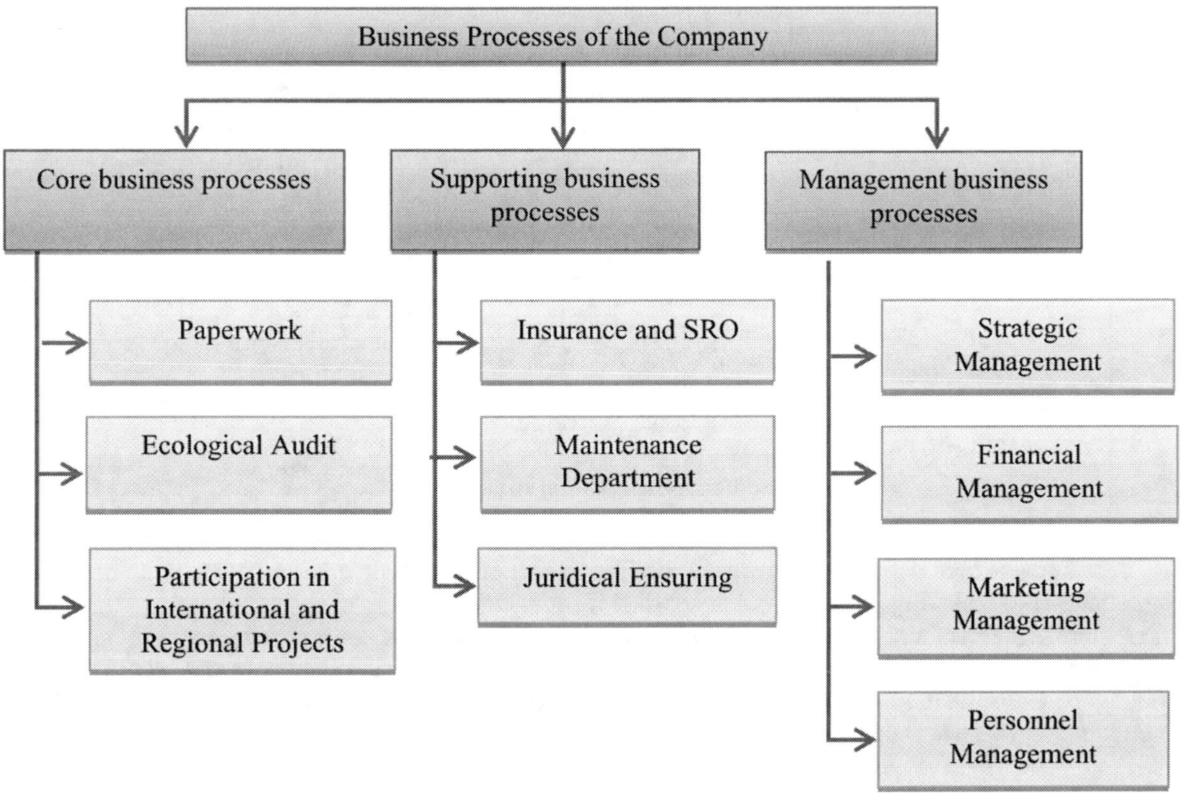

Figure 1. Business processes (BP) of the Company

Development of the budgeting model is based on the following assumptions. As the financial status of the Company is displayed in a format of three main statements (balance sheet, profit-and-loss statement, cash flow statement) forming the system of economic assessment of the Company's performance, the most effective financial management is possible when the budgeting of the Company's economic activity allows to link all these statements [12, 23-25].

It is expedient to organize the financial planning according to the following scheme:

REVENUE-AND-EXPENDITURE BUDGET =>; CASH FLOW BUDGET (DIRECT METHOD) =>; BALANCE SHEET BUDGET

Suggested is the following budgeting procedure:

The Financial responsibility centers of the enterprise draw up revenue-and-expenditure budgets and BCF. The costs centers draw up budgets of labour remuneration, overheads, purchases, business and managerial expenses.

Allocation of the Company's services among the profit centers is presented in Table 1 [17].

Table 1. Allocation of the Company's services among the profit centers

No	Profit center	Services	Customers	Financial responsibility
1	Department	Development of PEOS, AAIE, PLA, MSDSDO	Enterprises of the oil and gas industry, Large and medium-sized industrial enterprises, environment-related public organizations	Director – 1 Experts – 2
2	Department of environmental design	Development and coordination of the DSWFLP, MPE, SAD projects, and so forth. Environmental maintenance of activity of the enterprises (outsourcing) Environmental audit	All enterprises and the organizations having the status of nature users	Director – 1 Experts – 2
3	Department of international projects	Expert, organizational, coordination services with participation in the international and regional projects in the environmental field	The interested commercial, semi-state, public organizations, funds, state bodies of Russia, Norway, Finland	Director – 1 Manager – 1

These budgets being consolidated within the framework of the whole enterprise enable to draw up the revenue-and-expenditure budget and the cash flow budget (Fig. 2).

The profit centers draw up sales budgets, direct costs budgets, labour remuneration budget and overheads budget. The costs centers draw up a labour remuneration budget, overheads and marketing budgets. Consolidation of the above-mentioned budgets enable to draw up the enterprise-wide master budget.

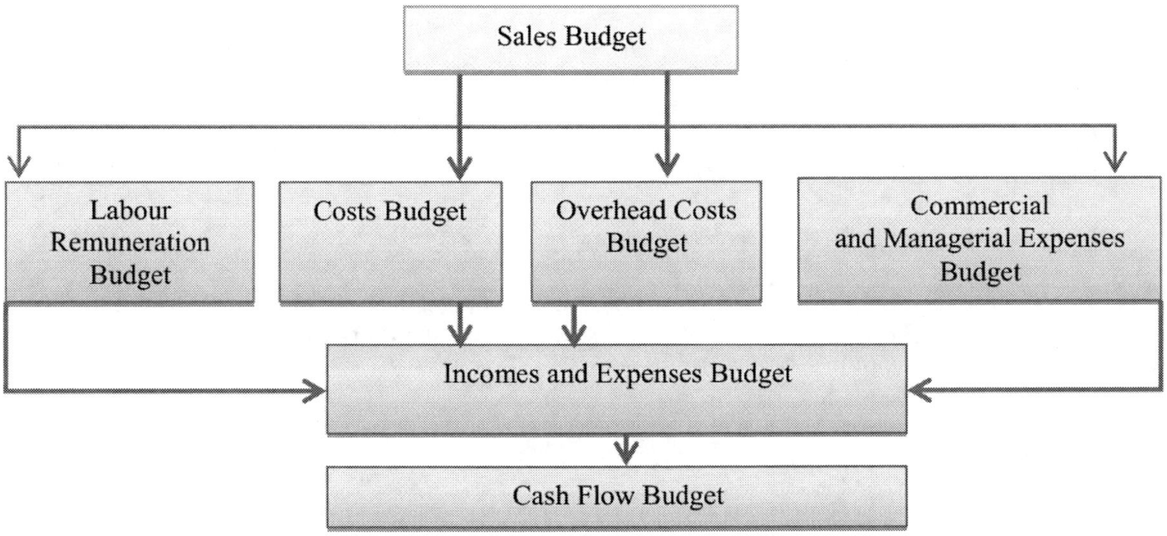

Figure 2. Model of the Company's budget

It is usual for the budget period to be a year, broken down into a month or a quarter control periods.

The general financial plan is formed by consolidation and coordination of the financial plans for business lines and includes the revenue-and-expenditure budget and expected balance, and simplified taxation plans:

– sales budget;
– cost budget;
– overheads budget;
– commercial and managerial expenses budget;
– revenue-and-expenditure budget, cash flow budget and break-even point.

The new budgeting model was used for drawing up the master budget of the Company: the revenue-and-expenditure budget, the cash flow budget and calculating of the break-even point.

3 Conclusions

Calculation of the break-even point is one of the major tasks when planning sales of products and services.

The break-even point is a point at which total cost and total revenue are equal. There is no loss or profit. It represents the sales amount that is required to cover total costs. Sometimes it is also called the critical volume: the minimum output that must be exceeded in order to make profit.

To calculate the break-even point, total costs are disaggregated into fixed and variable costs. In theory, variable costs are increasing in proportion to increase in production (sales) of products. Fixed costs are not dependent on volume of the produced (sold) products and services and business activity.

Break-even point = Revenue x Fixed costs / (Revenue – Variable costs)

Thus, to cover the Company's total costs in 2015, its annual revenue shall be at least 13 117 069 rubles

Table 2 –Data for the break-even chart

% of services provided	Fixed costs, ruble	Variable costs, ruble	Revenue, ruble
0%	8 149 975	8 149 975	-
10%	8 149 975	8 207 821	2 305 000
20%	8 149 975	8 265 668	4 610 000

30%	8 149 975	8 323 514	6 915 000
40%	8 149 975	8 381 360	9 220 000
50%	8 149 975	8 439 206	11 525 000
60%	8 149 975	8 497 053	13 830 000
70%	8 149 975	8 554 899	16 135 000
80%	8 149 975	8 612 745	18 440 000
90%	8 149 975	8 670 591	20 745 000
100%	8 149 975	8 728 438	23 050 000

The cash-flow forecast for 2015 (Table 2) has shown annual cash revenue of 25 164 400 rubles, that gives reason to believe that, in case of successful implementation of all the projects, the revenues will allow not only to avoid losses, but also to get profit (Fig. 3).

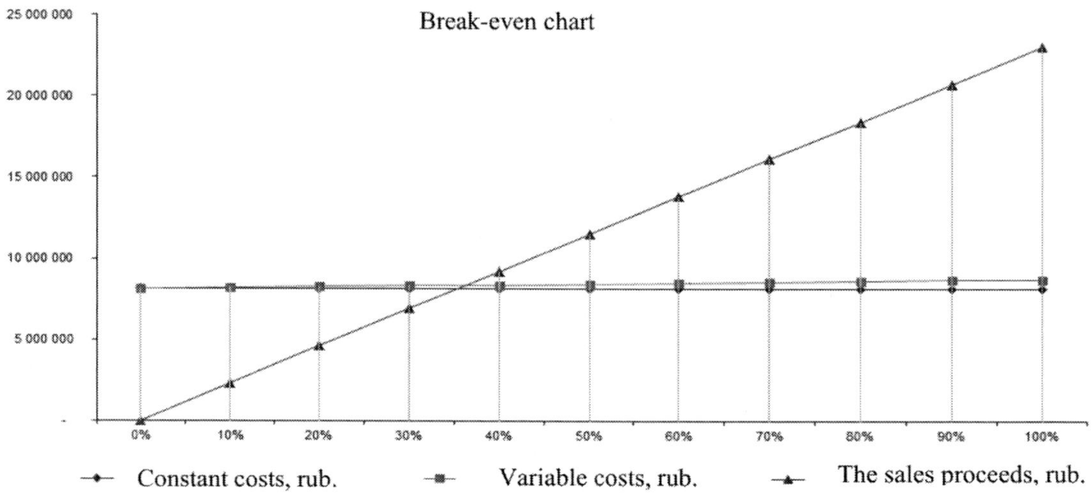

Figure 3. Break-even chart

References

1. *Scientific and practical portal "Production Ecology". Power and environmental contradictions in national security system* http://www.ecoindustry.ru (23.11.2015)

2. J. Ćetković, S. Rutešić, M. Zarković, M. Knežević, N. Vatin, Procedia Engineering, **117 (1)**, 780-790 (2015)

3. V. Urbanavičienė, A. Kaklauskas, E.K. Zavadskas, J. Šliogerienė, J. Naimavičienė, N. Vatin, International Journal of Strategic Property Management, **18 (3)**, 213-224 (2014)

4. B. Melović, S. Mitrović, A. Djokaj, N. Vatin, Procedia Engineering, **117 (1)**, 807-812 (2015)

5. D.Arseniev, A. Rechinskiy, K.Shvetsov, N. Vatin, O. Gamayunova, Applied Mechanics and Materials, **635-637**, 2076-2080 (2014

6. S. Isaev, P. Baranov, N. Vatin, Y. Zhukova, A. Sudakov, Technical Physics Letters, **40 (8)**, 653-656 (2014)

7. K. Strelets, N. Vatin, (2015) Rocznik Ochrona Srodowiska, **17 (1)**, 104-112 (2015)

8. S. Rutešić, J. Ćetković, M. Knežević, M. Žarcković, N. Vatin, Procedia Engineering, **117 (1)**, 642-650 (2015)

9. S. Rutešić, J. Ćetković, M. Žarcković, M. Knežević, N. Vatin, Procedia Engineering, **117 (1)**, 905-915 (2015)

10. V.Chechevichkin, N. Vatin, Applied Mechanics and Materials, **641-642**, 409-415 (2014)

11. V. Korsun, N. Vatin, A. Franchi, A. Korsun, P. Crespi, Mashtaler S. Procedia Engineering, **117 (1)**, 975-984 (2015)

12. A. G. Konstantinov, A.V. Berestenko, V. Ya Ushakov, *Collected papers of the International Youth Conference "Energy Saving Technologies"*, **I**, 6-7 (2011).

13. S.N. Kuzmina, Journal of Economic Regulation, **3,** 3 (2012)

14. V. V. Okrepilov, Innovations, **12**, 88-91 (2008).

15. N.I. Babkina, A.V. Babkin, Approaches and methods of an innovative potential assessment of an industrial cluster. Innovative economy and industrial policy of the region (EKOPROM-2014) (St. Petersburg: Publishing House of Peter the Great St. Petersburg Polytechnic University 2014).

16. V. A. Vaganova, *Management system improvement of capital repairs projects in the power generation companies* (St. Petersburg: Publishing House of Peter the Great St. Petersburg Polytechnic University 2012).

17. V. A. Plotnikov, *The Belgorod State University Scientific Bulletin*, "History. Political science. Information science" series, **vol. 23, no 13-1**, 5-10 (2012).

18. S. N. Kuzmina, Naukovedenie (electronic journal), **4** (7), 1-9 (2012).

19. L. G. Vorona-Slivinskaya, *Strategic planning of industrial production development: features of implementation in the conditions of unstable macro-microenvironments* (FINEC), St Petersburg, 2007)

20. S. V. Kuznecov, S. S. Neustroev, *Program and target approach in management of innovative development of economics* (Saint Petersburg: IPRE RAN, 140, 2012).

21. V. L. Kvint, V. V. Okrepilov, Innovations, **9 (191)**, 41-51 (2014).

22. V.L. Kvint, V.V Okrepilov, Herald of the Russian Academy of Sciences, **84 (3)**, 188-200 (2014)

23. V. V. Okrepilov, Problemy prognozirovaniya, **1**, 52-62 (2013)

24. V.V Okrepilov, M.V. Leonidovich, Asian Social Science, **11 (7)**, 312-325 (2015)

25. V.V Okrepilov, V.L. Makarov, A.R. Bakhtizin, S.N Kuzmina, Economy of Region, **2**, 301-313 (2015)

MATEC Web of Conferences **53**, 01051 (2016)
DOI: 10.1051/matecconf/20165301051

Methods of Parametric Optimization of Thin-Walled Structures and Parameters which Influence on it

Anton Kibkalo[1,a], Maria Lebedeva[2] and Mikhail Volkov[1]

[1]*Peter the Great St. Petersburg Polytechnical University, Polytechnicheskaya st. 29, 195251, St. Petersburg, Russia*
[2]*St. Petersburg State University of Architecture and Civil Engineering, 2-nd Krasnoarmeiskaya St. 4, 190005, St. Petersburg, Russia*

Abstract. The question of efficiency of thin-walled structures contains a number of contradictions. You need to select the best from all the existing structures on the criteria of optimization options. The search is conducted by varying of the parameters at parametric optimization. As a rule the aim of building structure optimization is reducing of material consumption, the labor input and cost. The costs of a particular variant of construction most full describes the given cost. There are two types of optimization parameters - immutable and varying. The result of the optimization of thin-walled beams will be a combination of parameters for each design situation in which provides the required strength and the minimum of the objective function - factory cost of production.

1 Introduction

The question of efficiency of thin-walled structures contains a number of contradictions. For example there is a saving of steel at a flexible wall of beam, but in the same time the high of the beam is increased. It leads to reduction of useful workshop volume or to the increasing amount of the wall panels. The erection of corrugated wall more cumbersome: there are additional process steps, increasing the amount of weld metal needed additional equipment.

Search of the best of all possible options, in terms of a given optimality criterion is called optimization. Depending on the nature of the differences of one embodiment (alternatives) from the other there are two types of optimization [1,2].

At parametric optimization search is conducted by varying the process parameters that affect its state. This is the geometric dimensions and physical properties of the materials of construction components. Often when a parametric search of the best option there is a situation, when all options of combination of parameters have been exhausted, and the optimal solution is not found. In such cases we speak of improper structure of the object. Thus, structural optimization objective is to achieve the object of optimality due to the choice of its structure. In the case of building structure it is its constructive solutions (type of elements, their number, the principle of mutual location, connection type).

[a] Corresponding author : chelsi-1993@mail.ru

2 Methods and materials

To turn on the contradictory nature of the optimization problem in mathematical analysis handy tool, it requires a clear statement of the problem, which traditionally consists of three components:

1. The objective function determines the optimal solution, while there are three kinds of purpose of the objective function: maximizing, minimizing or permanent limit (assignment setpoint).
2. Restrictions set immutable relationship between the permanent parameters and variable parameters, and of these dependencies following limits of varied parameters.
3. The boundary conditions indicate the extent to which may be the value of optimization parameters.

The formulated optimization problem will have a number of possible solutions that satisfy all constraints and boundary conditions. Type of objective function sets a certain number of optimal solutions.

The optimization problem arises in situations where there is no possibility to make a decision without a quantitative assessment of options (alternatives). For this assessment in the first place need to identify the purpose of optimization.

As a rule the aim of building structure optimization is reducing of material consumption, the labor input and cost. There is may be another target in depending on the specific conditions. For example, in the case of the construction of a long distance from the factory it could be reduce the cost of transportation design to on-site storage.

To formulate an optimality criterion - means to identify the signs, which should be carried out a comparative evaluation of alternatives and select the best among them. From the criterion depends on the form of the objective function, and hence the method for solving the optimization problem. Also optimality criterion affects the objectivity of the results, depending on the completeness of test parameters in the objective function.

The most accurate optimization results can be obtained by considering the problem of multi-criteria, which could be taken into account several parameters. However, the transition to this task seriously complicates the solution.

To avoid such complications, complex criteria can be selected. It would cover most of the indicators characterizing the optimization goal. That criterion is the price, the basic terms of which are always the mass and the complexity of the structure of production.

The costs of a particular variant of construction most full describes the given cost [2], which include the cost of "in erection" and operating costs. In turn, the cost of "in erection" consists of the production cost (the cost of materials, implementation of manufacturing operations for the production, the costs in the factory and costs out of production) and technological installation cost (the sum of the costs of implementing operations on installation design in the construction of the building).

It is important that the optimality criterion is formed of such costs, which are amenable to fairly precise definition at the stage variant design. The complexity of the procedure of calculating the cost should be justified by the reliability of the results.

Costs of transportation and installation of construction, especially in the operation, it is difficult to define precisely because of their dependence on a number of factors not related to the merits of the problem (distance from the construction site of the manufacturer, methods of transportation, conditions and methods of installation, power construction company, the projected production process, etc.)

From this perspective, the objective results can be achieved by taking the manufacturing cost structure of production as optimality criterion.

Results of parametric optimization allow us to describe the nature of the influence of each parameter on the objective function.

Methods for calculating the technical and economic indicators, including production cost, are divided into two main groups [3, 4, 5]:

- accrued methods based on a detailed calculation of the required rental at list prices, the calculation of operational labor input;

- analytical (enlarged) methods with the use of formulas, tables and nomograms, approximating statistics.

In the present study to cover different design situations is preferred analytical method that allows more compact to formulate an objective function. This approach corresponds to the method proposed by JM Likhtarnikov [6] in the 70-ies of XX century. Factory structure manufacturing costs are determined according to the complexity of the design weight of PI by the formula:

$$C_u = \left[G_k \cdot C_{\text{ОМ}} + T \cdot a \cdot \left(1 + K_{\text{НР}}\right) + G_k \cdot H\right] \cdot K_u \cdot K_{\text{р}}, \quad (1)$$

where G_k – structural mass;

$C_{\text{ОМ}}$ – cost of materials;

T – the total complexity of design;

a – hourly wages of workers;

$K_{\text{НР}}$ – coefficient taking into account of overheads;

H – conditionally fixed costs for maintenance and production management;

K_u, K_p – coefficients taking into account the profitability of the construction and deductions for NCI, bonuses, etc.

Object parameter optimization called all the characteristics of an object, describing his condition. The total number of parameters can be divided into two groups: the variable parameters that change with a certain pitch in a certain range; unchangeable parameters that describe the source data for the design. The true solution of the problem of optimization is such a solution, in which the combination of the optimal parameters does not conflict with the embodiment of the project in practice. To do this, formulated the boundary conditions (border regions of varying parameters), set iteration step, which provides product mix parameters.

3 Optimization parameters

3.1 Immutable parameters

3.1.1 Span

Beams with a flexible or corrugated wall coverings used as a truss for buildings or roof purlins. Span length, which should be overlapped by beam depends on the intended manufacturing process. But in any case the span length should be a multiple of the module of 3 m or 6 m [7]. Length of the beam is determined by step of bearing trusses and beams of coverage (transverse frames), which usually equals the modular size of wall panels, covering panels, that 6 m or 12 m. If the pitch of bearing frames exceeds the specified values for overlapping used secondary structures usually through the section. From the variety of buildings is emited large-span buildings with 36-120 meters span length: industrial buildings for manufacturing or assembly of large products - aircraft, ships and other equipment, the storage of various equipment: aircraft hangars, depots for electric, garages, etc. Already during the flight 40 ... 60 m in the middle of the span large internal forces develop and the use of continuous wall structures is inefficient compared with through structures (farms).

3.1.2 Load

To limit an extensive range of possible values of the load, consider some design situations. The use of thin-walled beams is most likely in the projects with the use of modern lightweight materials. Usually galvanized corrugated decking steel plays the role of a light carrier flooring, which is laid on the girders, made of galvanized cold-formed channel section profiles. The use of non-combustible materials contributes to reduction of mass insulation; these materials has small volume weight and the minimum value of thermal conductivity, such as extruded polystyrene foam. The need for insulation, type and thickness depends on the process that determines the required indoor temperature and climatic region. From climatic region also depends on the snow load. Weight of sheeting and runs depends from steps transverse frames. Boundary conditions "load" parameter defined extreme combinations of input data.

3.2 Variable parameters

3.2.1 Steel grade

As the mechanical properties of steel is conventionally divided into three groups: normal strength (C235 ... C285, Ry <Ry 40 kN / cm2), high strength (C345 S390 ... 29 kN / cm2 <Ry 40 kN / cm2). The last class of steels ineffective for thin-walled beams due to their low plasticity.

3.2.2 The relative height of the beam and wall flexibility

The smallest conditional flexibility is limited by the rules [8] $\overline{\lambda_w} \geq 6$, which corresponds to the physical flexibility of the wall $\lambda_w = \frac{h_w}{t_w} = 250$.

Wall height h_w - the width of rolled sheet, which is determined by the length of the barrel of the working rolls of sheet rolling mills. Limiting the length and width of the sheets depends on the thickness of rolled [20] (to avoid death and curvatures at the handling and storage). In the case of steel coils thinnest limited only by the width of the rolled products.

3.2.3 The relative length of the compartment

The minimum length of the compartment limited by the condition of using supercritical reserve bearing capacity of the thin wall. Stress which corresponds to the emergence the diagonal band in the wall (2):

$$\tau_{cr} = 10{,}3 \cdot \left(1 + \frac{0{,}76}{\mu^2}\right) \cdot \frac{R_s}{\overline{\lambda}_w^2}, \tag{2}$$

If the length of the compartment is less than half the height of the wall (μ <0,5), the value of the critical shear stresses close to the ultimate stress. In other words, supercritical bearing capacity of the flexible wall significantly (up to three times) reduced when μ <0,5.

3.2.4 Crimping parameters

The complexity of the search for the optimum size of the corrugations: variation of the relative dimensions of the corrugations $(\frac{f}{a}; \frac{f}{t})$ has a contradictory effect on the shear strength of the wall. Increase of the rigidity of the corrugation doesn't always contribute to hardening of the walls. Moreover, the size of the different shape of the corrugations has a different effect on the strength of the wall.

Optimization of parameters of corrugation is facilitated by the lack of influence of height of the wall on the critical shear stresses.

Maximum lateral force of the corrugated wall bounded on one hand of loss of stability; on the other hand - the development of plastic deformation in the wall (shear strength steel). This means that the corrugations of the walls need not be very gentle, low rigidity, but at the same time extremely rigid to the critical stress not exceed the yield strength of steel [9]. High intensity crimping results in excessive overspray, wear plate-bending equipment and welding tractors.

To make the best decisions it makes no sense to consider all the values of the objective function corresponding to each of the optimization parameters. Only those combinations of variable parameters are interesting for the analysis that are able to provide a building structure the required strength and stiffness without overspending material (excessive safety margin).

Of course, along with these requirements to the designed device and imposed other requirements: aesthetics, compactness, manufacturability, and so on. However, such requests are usually difficult to formalize and submit to the mathematical notation, a qualitative assessment of an expert required to meet them [10-14].

The result of the optimization of thin-walled beams will be a combination of parameters for each design situation in which provides the required strength and the minimum of the objective function - factory cost of production.

4 Conclusions

Results of parametric optimization will describe the nature of the influence of each parameter on the objective function.

1. The use of high-strength steel leads to a rise in the cost of thin-walled beams at 18-30%. Meanwhile, the mass of construction can be reduced by 30-60%, that in some cases may be a very valuable asset. However, the use of higher strength steel helps reduce the cross-sectional area zones, thereby increasing the maximum deflection and leads to some overspray.
2. Increasing of relative height of beam leads to economies of steel until to a certain limit, after which the mass is increasing due to the necessity of more powerful devices belts.
3. Increasing the flexibility of the wall helps to alleviate the beam. However, the range of values for which the weight is reduced effectively, not so great.
4. Relative compartment length affects its shear strength: the weaker the compartment, the greater its length. As the reduction in the number of compartments (increasing their length) beam becomes easier and less time-consuming to manufacture.
5. Changing crimping parameters affect the labor and cost of construction. Tightening of corrugations makes sense only in cases of unstable form of exhaustion bearing capacity - at a flexible wall with gentle corrugations ultimate shear force is determined by the critical stress. In other cases, the shear strength is determined by the tensile strength of steel and material consumption to improve the stability of the corrugations is irrational.
6. In most cases, the cost and complexity grow proportional to the mass.

References

1. I. P. Norenkov, *Fundamentals of computer-aided design* (2002)
2. M. R. Garifullin, N. I. Vatin, Construction of Unique Buildings and Structures,**21**, 32-57 (2014)
3. V. V. Kuznetsov *The steel structures of buildings and structures: a handbook designer* (1998)
4. A. I. Kolesov, A. A. Lapshin, A. V. Valov, Privolzhsky Scientific Journal, **1**, 28-33 (2007)
5. N. I. Vatin, V. A. Rybakov, StroyPROFIL, **56**, 60-63 (2007)
6. Y. M. Likhtarnikov *Variant design and optimization of steel structures* (1979)
7. *GOST 28984-91. Modular coordination of dimensions in construction* (1991)
8. *Steel construction. The updated edition of SNiP II-23-81: SP 16.13330.2011* (2011)
9. Y. I. Olkov, A. S. Poltoradnev, Proceedings of the International Congress "Science and Innovation in modern construction - 2012" International scientific-practical conference on the scientific direction "Actual problems of modern construction and ways to effectively address the" Ph. - St. Petersburg, (2012)
10. V. Z. Vlasov, *Thin elastic rods* (1959)
11. R. Kamnik, B. Kovačič, B. Pribicević, A. Đapo, Geodetski List, **69** **(3)**, 171-188 (2015)
12. B. Kovačič, R. Kamnik, A. Štrukelj, N. Vatin, Procedia Engineering, **117** (1), 800-806 (2015)
13. L. Lazarević, M. Kovačević, Z. Popović, Facta Universitatis, Series: Mechanical Engineering, **13** **(2)**, 133-141 (2015)
14. B. Kovačič, R. Kamnik, M. Premrov, Survey Review, **43** (320), 150-161 (2011)

MATEC Web of Conferences **53**, 01052 (2016)
DOI: 10.1051/matecconf/20165301052

Methods of determining the optimal project of reconstruction of The Petrovsky Dock in Kronstadt

Marina Romanovich [1] and Anna Vilinskaya[2, a]

[1] *State unitary enterprise "Municipal government of inventory and assessment of real estate", Russia, 191104, St. Petersburg, Belinskogo St., 13*
[2] *Peter the Great Saint-Petersburg Polytechnic University, Russia, 195251, St. Petersburg, Polytechnicheskaya St., 29*

Abstract. Today in Russia there are many historical monuments, which are in a derelict and not the operational state. So, this is actual question about their reconstruction with preservation of historical significance, with current technology, and innovation. One of these abandoned objects is The Petrovsky Dock in Kronstadt. Priority idea of its reconstruction it is museum. The Russian and foreign experience in development of the modern sea museums is analysed. In article the existing options of reconstruction of Dock are considered. The method of calculation of total optimality project coefficient of reconstruction is offered. Survey of experts showing what functions are most interesting to the museum is conducted. As a result, we determine the optimal reconstruction project of Dock with unique functionality.

1 Introduction

Kronstadt – naval capital of Russia. There are the unique marine structures, historical and cultural monuments. Unfortunately, not all structures have survived to our time, some are in a derelict state, for example, Petrovsky Dock, an object of cultural heritage of federal importance.

The Petrovasky Dock – In the 18th century, it was the first dry dock in Russia, designed by Peter the Great. Dock was intended to repair the underwater parts of ships and vessels. It was a grand structure. Over time, Dock stopped using for its intended purpose, up to 2008 part of the Petrovsky Dock was used for repair of ships, but now it is not used and is in a derelict state.

In 2011, Petrovsky Dock was transferred to the ownership of St. Petersburg and was transmitted to the administration of Kronstadt district. With the view of restoration with preservation the historical and cultural features, with adaptation for modern purposes and museum like unique monument, the Administration is developing options of involving the object in economic circulation.

For consideration of the constantly offered different options for Reconstruction and the use of Petrovsky Dock. But so far no project is not implemented. In order to find the best reconstruction project it's necessary to analyze the proposed options, compare all pluses and minuses, as well as refer to successful projects both abroad and in Russia. So, to collect a complete set of information needed for a successful reconstruction project of The Petrovsky Dock.

[a] Corresponding author : willi-ann@yandex.ru

1.1 Literature review

Since Petrovsky Dock building of the 18th century, it is most often mentioned in the literature devoted to the history of Kronstadt. For example, there is N. Durov's article [1], which is fully dedicated to Peter the Great Canal. F.A. Timofeevsky in his brief historical essay [2] not only tells the story of Peter's building docks and canals, but also shows a detailed plan of Kronstadt with an indication of the southern coast in 1723 and designed by Peter the Great harbors and channels. In N.A. Rykachev's article "Kronstadt old" [3] there is a separate chapter "Channel description", which described in detail the structure and the materials used in the construction of a cross-channel. The historical notes S.I. Elagina and A. Treskina [4-5] there you can also find information about construction of the Petrovsky canal. And other writers, historians have not left such a great facility without attention [6-11].

1.2 Problem statement

The purposes of article are:

1. An overview of the Russian and foreign experience in the field of modern museums that popularizing maritime culture;
2. Analysis and comparison of the already proposed reconstruction projects and the use of The Petrovsky Dock;
3. Determination of criteria that show the degree of success and acceptability of the reconstruction project Petrovsky Dock;
4. Choosing the best reconstruction option, taking into account the analysis and preservation of the historic value of the monument.

2 The review of the existing sea museums

In Russia, a lot of naval museums, but the biggest of them is located in St. Petersburg - Central Naval Museum. This is one of the oldest museums in Russia and one of the largest maritime museums in the world. In 2013 the museum moved to a renovated complex Kryukov (Marine) barracks, but, despite this, the museum didn't become modern. In our modern times visiting museums is the most interesting for people, where in addition to the study of museum exhibits can spend time interactively, immerse themselves in some of the periods of history, to do something with their own hands, to feel part of the story that keeps the museum.

From foreign experience [15-22], for example, Maritime Centre "Vellamo" in Kotka (Finland). It is a multifunctional complex that includes several museums. In addition, there is a separate information center (books, magazines, video and audio materials, pictures, and more). The museum used 3D-simulator, allowing visitors to make a virtual dive to the "Frau Maria" on the bottom. On the roof of the building maritime center is a viewing platform.

In Portsmouth (UK) is a maritime museum – Mary Rose Museum. The ship, had lain on the seabed more than four years, was raised in 1982 and was built around it the historical museum, which tells of an entire era of British history. This is really a high-tech museum, because there is the biggest museum showcase in the world. The museum building is also unique, it has the form of jewelry box, inside which is itself the renovated flagship and more than 19 thousand artifacts raised with him.

An interesting museum is located in Stockholm (Sweden) "Vasa" Museum. The museum opened in 1990 around the Swedish «Vasa» warship that sank at the entrance to the bay of Stockholm in 1628. The building allows you to inspect the hull of a wooden ship from all sides, as well as to get some room. The museum has a cinema where is shown a film about the Vasa on sixteen languages.

The modern "living" museum is Lennusadam Maritime Museum in Estonia - one of the biggest museums in the country, located in Tallinn. The museum expositions are located in three-dimensional space. In addition, the museum has a restaurant and a shop MARU, cinemas, simulators, children's area and a lot of other interesting things. There is also an outdoor roof terrace with a beautiful view of the passenger port, the city center and the Old Town. Comparing the Russian and foreign experience, we can say that Russia doesn't have enough good, interesting and modern maritime museum. Kronstadt – as the sea capital of Russia, a great place to create such a museum. A Petrovsky Dock can be really high-tech, multi-functional museum complex [23-26].

3 Data description and methology

In 2011, a competition was announced for the best reconstruction project of Petrovsky Dock, but none of them has been implemented. Offers a variety of construction companies for the reconstruction of Dock also didn't find its real application.

Consider and compare all renovation options previously proposed and use of Dock Peter the Great.

Table 1. Comparative characteristics of projects of reconstruction The Petrovsky Dock.

№	Description of project	Construction costs, [million rubles]	Construction period, [Months]	Functions	Pluses	Minuses
1	Floating fountain complex	500	12	1. Fountain 2. Concert hall 3. Beautification the waterfront pool 4. Overview terraces	- Strengthening the attractiveness of the city as a tourist center - Concert hall - Panoramic view	- Use only as a concert hall - Focus more on tourists - Not saved historical value - Not attractive to the local people, and especially for tourists (in terms of functionality)
2	The park area	150	10	1. Walking and cycling paths 2. City Park	- Place for walks - Promotion of healthy lifestyles - Clean air	- Not saved historical value - Seasonality (in will not fall and winter a popular place) - No income
3	The creative space	275	20	1. Shops 2. Restaurants 3. Offices 4. Galleries 5. Hostels 6. The loft housing 7. City Park	- A large number of services - Place for walks - Multifunctionality	- Not saved historical value - Not attractive to the tourists

4	The re-construc-tion-conserva-tion	290	11	1. Installation of the vessel, with restaurant and mu-seum exposition 2. Performance 3. Reconstruction of craft workshops	- Preservation a cultural monument - Its transformation into a commercially successful project - Re-creation histor-ical atmosphere	- A little functional - Not enough modern technologies
5	Indoor Museum of Un-derwater Archaeol-ogy	130	18	1. Demonstration of the ships in the water through the glass 2. Training pool 3. Exhibition and conference Hall 4. Restoration workshops 5. Office and aux-iliary rooms	- Combining build-ings in Marine glo-ry's historical com-plex - income from diving - Multifunctionality	- glass sarcophagus - Not saved historical value
6	The Air-ship	210	10	1. Airship	- Airship - a new symbol of Kronstadt	- dome of the adminis-trative building resem-bles the silhouette of the dome of the Naval Ca-thedral - Not saved historical value
7	Multi-functional complex	395	18	1. Exhibition of warships of various types 2. Commercial and entertainment area 3. The objects of public catering 4. Warehouses - gallery and official-ly-business zone	- Multifunctionality - Income from ser-vices	- Not preserved unique set of hydraulic and urban landscape at a location nearby
8	Building in the form of nuclear submarine "Ty-phoon"	255	15	1. Museum, exhi-bition 2. Restaurant with a rotating figure of St. Andrew on the roof 3. Shops	- Multifunctionality - Income from ser-vices	- Not saved historical value,because in Kronstadt never built submarines

9	Multi-functional center	280	16	1. Museum 2. Diving Centre 3. Sea Club for children	- Multifunctionality - Income from services	- Not saved historical value
10	Museum complex	620	19	1. Underwater excursions 2. Exposition of real ships 3. Diving Centre	- Preserving history - Multifunctionality	- relatively expensive

3.1 Definition of the optimum project

To conduct a comprehensive assessment of each of the above applications projects, we calculate weights for each project, using widely applicable-employed in the construction method of expert evaluations [20]. Each of the ten projects evaluated by five experts on a scale from 1 to 10. Matrix of weighted estimates constructed after affixing estimates [21, 22], presented in Table 2.

Table 2. Matrix of weighted estimates.

Experts	Coefficients of optimality										Total score
	1	2	3	4	5	6	7	8	9	10	$\sum_{k=1}^{m} h_{ik}$
1	3	4	4	5	5	4	4	5	5	5	44
2	4	3	1	4	5	2	2	3	3	4	31
3	4	5	2	3	4	4	3	4	5	5	39
4	3	3	2	5	5	4	5	4	4	5	40
5	1	3	0	5	3	0	4	0	2	0	18

Next, the total score is determined for each criterion, obtained by summing the values of each column of the matrix [21], presented in Table 3.

Table 3. Matrix of weighted estimates.

Experts	Coefficients of optimality										Total score
	1	2	3	4	5	6	7	8	9	10	$\sum_{k=1}^{m} h_{ik}$
1	3/44	4/44	4/44	5/44	5/44	4/44	4/44	5/44	5/44	5/44	44
2	4/31	3/31	1/31	4/31	5/31	2/31	2/31	3/31	3/31	4/31	31
3	4/39	5/39	2/39	3/39	4/39	4/39	3/39	4/39	5/39	5/39	39
4	3/40	3/40	2/40	5/40	5/40	4/40	5/40	4/40	4/40	5/40	40
5	1/18	3/18	0/18	5/18	3/18	0/18	4/18	0/18	2/18	0/18	18
Total mark r_i	0,43	0,56	0,22	0,72	0,67	0,36	0,58	0,41	0,57	0,50	

The values of weighting coefficients calculated by the formula (1), at the same time the condition presented in a formula 2 has to be satisfied [21]:

$$\sigma_i = \frac{r_i}{\sum_{i=1}^{m} r_i} \tag{1}$$

$$\sum_{i=1}^{n} \sigma_i = 1 \tag{2}$$

$$\sum_{i=1}^{m} r_i = 0{,}43 + 0{,}56 + 0{,}22 + 0{,}72 + 0{,}67 + 0{,}36 + 0{,}58 + 0{,}41 + 0{,}57 + 0{,}50 = 5{,}02$$

The weighting coefficients for each project are presented in Table 4.

To determine the total optimality project coefficient following indicators have been introduced:

K_{cost} – construction cost coefficient, calculated based on the average cost of construction = 313,5 million rubles, corresponding to 5 points on 10-point scale;

K_{period} – construction period coefficient, calculated based on the average period of construction = 14,9 months corresponding to 5 points on 10-point scale.

K_{multi} – multifunctionality coefficient corresponds to the number of functions, which performs one or another product (table 1).

The total optimality project coefficient calculated by the formula (3):

$$\sum K_{optimal} = (K_{cost} + K_{period} + K_{multi}) \times K_{weight} \tag{3}$$

The values of all coefficients are given in Table 4.

Table 4. The calculation of the total optimality project coefficient.

№	K_{cost}	K_{period}	K_{multi}	K_{weight}	$\sum K_{optimal}$
1	**2**	**3**	**4**	**5**	**6**
1	7,98	4,03	4,00	0,09	1,44
2	2,39	3,36	2,00	0,11	0,85
3	4,39	6,71	7,00	0,04	0,72
4	4,63	3,69	3,00	0,14	1,58
5	**2,07**	**6,04**	**5,00**	**0,13**	**1,70**
6	3,35	3,36	1,00	0,07	0,54
7	**6,30**	**6,04**	**4,00**	**0,12**	**1,96**
8	4,07	5,03	3,00	0,08	0,97
9	4,46	5,37	3,00	0,11	1,41
10	**9,89**	**6,38**	**3,00**	**0,10**	**1,93**

3.2 Topical museum functions

For definition of optimum functions of the museum 7 experts have been interrogated. The list of functions for the museum has been made. Experts had to choose the most pleasant options. Results are presented in Table 5.

Table 5. Determination of the optimal project functions.

		Experts							
№	**Function name**	**1**	**2**	**3**	**4**	**5**	**6**	**7**	**∑opinion**
1	Fountain								0
2	Concert hall								0
3	**Viewing platforms**	+	+	+	+	+		+	**6**
4	Shops			+					1
5	Restaurants	+	+					+	3
6	Offices	+							1
7	Galleries			+					1
8	**Exposition of real ships**	+	+	+	+		+	+	**6**
9	Craft workshops		+		+		+		3
10	**Aquariums**	+	+			+	+	+	**5**
11	Airship		+						1
12	Sea Club (for children)		+	+	+	+			4
13	Diving	+	+					+	3
14	Underwater excursions		+						1
15	**Quest (on the marine theme)**	+		+	+	+		+	**5**
16	**The visualization system, interactive (visitors can lay on the levels, rotate, zoom in and almost make the cut image and reveal hidden details of exhibits)**	+	+	+	+	+	+	+	**7**
17	Multimedia books (living books)	+			+		+	+	4
18	Excursions from St. Petersburg (and other cities)	+						+	2

4 Conclusion

After review and comparison of the Russian and foreign experience in the creation of maritime museums, we can conclude that Russia needs an interesting and cotemporary Maritime Museum. As the sea capital Kronstadt is a great place to create such a museum. As a variant – the reconstruction of Petrovsky Dock for high-tech museum complex.

It was previously suggested options for the restoration of Petrovsky Dock. In the course of this work were analyzed polls, the results of which determined the most optimal, according to experts, projects. As a result of the leading position is occupied Multifunctional complex (project №7 table 1), which includes an outdoor exhibition of military vehicles of different types, shopping and entertainment area, catering facilities, parkgauzy gallery and officially-business zone, construction cost 395 million rubles, the construction period – 18 months. Construction costs slightly higher than the average figure for all projects, if we add the functional, based on the interests of the experts (table 5), the cost increase that immediately makes the project not optimal.

In second place The museum complex (project №10 table 1), which offers excursions under water, exposure of real ships, diving center. Construction period 19 months, the cost - 620 million rubles. Minus of this project - relatively expensive, which immediately raises the question of its implementation. And on the third place Indoor Museum of Underwater Archaeology (project №5 table 1), which is a demonstration the ships in the water through the glass, a training pool, exhibition and conference halls, restoration workshops, office and auxiliary premises. Construction period – 18 months, the cost – 130 million rubles. Even if you increase the functional components of the project, the cost of the construction will not exceed the average for projects.

Comparing and analyzing these three projects, we conclude that the last project is optimal for the reconstruction of Petrovsky Dock. It is a modern Indoor Museum of Underwater Archaeology, where you can see the exposure of real ships in the water through the glass, scuba diving training in the pool, spend some time in the restoration workshops, as well as to conduct business meetings in the exhibition and conference halls, offices and auxiliary rooms. Based on the interest of visitors to the innovative technologies, interactive visualization system is ideal for the exhibition, visitors can lay out on the levels, rotate, zoom in and almost make the cut image and reveal hidden details of artifacts. Today the quests become more and more topical on the marine theme could be especially interesting. Finally, the viewing platforms to enjoy the view around Petrovsky Dock, because the location is really unique.

References

1. N. Durov *Materialy dlya istorii stroitel'nogo dela v Rossii* (Saint-Petersburg, Spb, 1862).
2. F. A. Timofeevskii *Kratkiy istoricheskiy ocherk dvuhsotletiya goroda Kronshtadta* (Kronshtadt, Spb, 1913).
3. N. A. Ryikachyov *Kronshtadtskaya starina* (Kronshtadtskiy vestnik, Kronshtadt, 1885).
4. S. I. Elagin *Nachalo Kronshtadta* (Kronshtadtskiy vestnik, Kronshtadt, 1866).
5. A. Treskin *Istoricheskiy ocherk Kronshtadta* (Kronshtadtskiy vestnik, Kronshtadt, 1862).
6. A. V. Shelov *Istoricheskiy ocherk kreposti Kronshtadt* (Kotlin, Spb, 1904).
7. P. I. Shalikov *Puteshestvie v Kronshtadt 1805 goda* (Moscow, 1817).
8. A. A. Inostrantsev *Geologicheskoe stroenie ostrova Kotlina* (Kronshtadt, Spb, 1912).
9. P. N. Stolpyanskiy *Istoriko-obschestvennyiy putevoditel po Kronshtadtu* (Petrograd, Spb, 1923).
10. L. I. Tokareva *Ekskursiya po gorodu s Lidiey Tokarevoy i «Kronshtadtskim vestnikom»* (Kronshtadtskiy vestnik, Kronshtadt, 2004).
11. M. A. Shlendova *Oni rodilis v Kronshtadte* (MKS, Spb, 2004).
12. P. Karon, InfoWorld, **17(45)**, 68 (1995).
13. British Heritage, **21(3)**, 13-14 (2000).
14. S. Kenderdine, Archives and Museum Informatics, **12(1)**, 17-38 (1998).
15. National Geographic, No **179(5)**, 1 (1991).
16. L.A. Kvarning, Scientific American, **269(4)**, 84-91 (1993).
17. M. Mändel, O. Orro, Construction History, **27**, 65-85 (2012).
18. P.B. Jones, Architectural Review, **213(1271)**, 48 (2003).
19. J. Weirick, Architecture Australia, **4**, 14-18 (2001).
20. M. A. Romanovich T. L. Simankina, D. P. Ilchenko, Promyishlennoe i grazhdanskoe stroitelstvo, **5**, 73-77 (2015).
21. O. A . Grebneva *Teoriya prinyatiya resheniy: Ucheb. Posobie* (IrGTU, Irkutsk, 2012).
22. V. V. Fedoseev , A.N.Garmash ,.D. M. *Dayitbegov Ekonomiko-matematicheskie meto-dyi i prikladnyie modeli* (UNITI, Moscow, 1999)
23. N. Pavličić, M. Perazić, D. Durić-Jocić, M. Knežević, Journal of Applied Engineering Science, **12 (1)**, 11-18 (2014)
24. M. Ćetković, D. Vuksanović, M. Knežević, T. Samardžioska, M. Cvetkovska, Journal of Applied Engineering Science, **10 (1)**, 31-36 (2012)
25. J. Cetkovic, M. Knežević, I. Vujovic, J. Cerovic, Technics Technologies Education Management, **7 (4)**, 1646-1654 (2012)
26. J. Ćetković, S. Rutešić, M. Zarković, M. Knežević, N. Vatin, Procedia Engineering, **117 (1)**, 780-790 (2015)

MATEC Web of Conferences **53**, 01053 (2016)
DOI: 10.1051/matecconf/20165301053

Causes of formation of cracks in the outer masonry of multilayer walls in the zone interwindow belts and partition

Sergey Zimin[1], Raushan Alzhanova[1,a], Vadim Litovchenko[1,b]

[1]*Peter the Great St.Petersburg Polytechnic University, 29 Polytechnicheskaya st., St.Petersburg, 195251, Russian Federation*

Abstract. This article examine the causes of crack formation in the exterior layer multilayer enclosing structures, on early terms of operation (from 3 to 5 years). The problem is investigated on a concrete example of the existing residential building to St. Petersburg. We explore the cracks in the zone interwindow belts and partition. For explore this question used methods such as: analysis of the experience, visual inspection, the experimental methods in laboratories, numerical modeling in the software package ANSYS. As a result of visual examination of sediment soils are not detected. According to the results tests of a selected samples bricklaying, compliance of its durability to existing rules - is defined. Identified a number of causes of crack formation. Also, a number of recommendations is developed for the design, to avoid similar problems on early terms of operation.

1 Introduction

Currently, in the Russian Federation is actively building constructed using multilayer structures. Because of search of economic options of the protecting designs with the increased heat transfer resistance. Technology sandwich structures approached for this role better than others.

This type of walling consists of the following layers:

- bearing structure - it can be made of brick, reinforced concrete, aerocrete blocks, etc.;
- insulation - is not regulated, taken depending on the heat engineering calculation;
- a obverse layer - facing brick.

However, this technology was borrowed from European developers, but not adapted to Russian conditions and design methods. Over a short period of time (three to five years) after commissioning, began to arise problems such as cracks in the zone interwindow belts and piers in the obverse layer (bricklaying) and to moisture ingress.

In a consequence of cyclic freezing and thawing of moisture in cracks, they start expanding and finally fragments of a brick and/or a laying collapse. In turn, this leads to the ingress of moisture on the insulation to deteriorate its thermal insulation properties.

Premature repair which is expensive, big flow of complaints, loss of cost of apartments in houses with collapses, damage of an aesthetic look, etc. All these effects, after which in some regions of the Russian Federation, it was decided to prohibit the use of this technology.

[a] Corresponding author : alzhanova03@gmail.com
[b] Corresponding author : litovchenko.v93@gmail.com

With regard to existing differences our climate from the European, in the domestic housing construction should be more stringent requirements for the material and the design method of an obverse layer [1].

Great contribution to the study of bricks made by Russian scientists M.K. Ischuk [2-11].

Just research the problem have dealt with foreign scientists, such as K.D. Beysli [20], V. Jager, G. Pfeiffer [21], M.E. Phillips, T.A. Montagu [22] - who wrote many works to anchor system in the lining. N.A. Romani [23] was engaged in research pre intense designs in the lining. G. Simudik, A.V. Paige [24] research the development of cracks and crack resistance in the masonry.

N.P. Umnyakova conducted a research about the durability of sandwich walls lined with bricks. Her work explores the causes of the destruction of the sandwich construction of exterior walls with an effective insulation.

Also a great contribution to the explore of the problem made the following researchers: Prof. R.B. Orlovich [12,13], a graduate student N.M. Rubtsov [13], a graduate student S.S. Zimin [12-14], the purpose of their study was to determine the bearing capacity and deformability of the anchoring of the compounds used in the building envelope. They conducted laboratory experiments which resulted in the following conclusions:

1. The bearing ability of anchors pull-out is almost identical to layings of voidness and corpulent bricks in case of use of solutions of bigger brand. At the smaller durability mortar joints of seams characteristics of anchors have paramount value: flexibility, diameter and roughness of a surface.

2. Increased strength and rigidity of anchor links does not guarantee the ability to increase their anchors.

3. Section anchors should be chosen in such a way that their increase is not destroyed mortar joints, while at the same time reducing the gap did not occur from stretching. [13]

Application in the obverse layer of a hollow bricks/stone high voidness, but not corpulent, or a brick/stone with small voidness (at their identical frost resistance), can adversely affect the strength of brickwork. For the combined laying with a relatively thick obverse layer, this fact largely relates to the outside (obverse) exterior tier, which is directly contacting to the atmospheric environment (in given to the combined obverse laying exterior tier it is executed from a brick by voidness of 40%). Voidness brick is a kind of structure formed from intersecting two directions thin walls, break the connection between them is relatively easy, which is especially affected:

- at local action of loading;
- at action of loading perpendicular to voids (to thin partitions between them);
- at alternate freezing-thawing of a brick, in the state sated with moisture (water when freezing increases in volume, presses on walls of a time and destroying them, leads to destruction and reduction of durability of a brick). Under adverse conditions when filling with moisture of emptiness (for example, at condensate loss; at active passing of steam through an external wall; through cracks, etc.), water when freezing presses, except walls of a time, on walls of emptiness that leads to the progressing destruction of a brick and, as a result, – layings.

Here it is worth noting that the project is put into the brand of frost resistance of brick characterizes its ability to withstand alternate freezing-thawing, when the latter, as well as moisture saturation brick, wear volume character. In the real conditions atmospheric moisture exposed to outer bulkhead voidness bricks (deep inside the walls of it does not penetrate because of the direction the partial pressure "from inside the building to the outside"). Effects of negative and positive temperatures more susceptible to the same outer bulkhead brick (deep wall temperature aligned). All this leads to emergence of tangent tension τ_y, on border of an external bulkhead with intervoid (under adverse conditions there is an office of an external bulkhead from intervoid).

In most projects, fixing the outer layer of laying carried to the inner layers or reinforced concrete frame with flexible connections.

The following are the most common errors when the device flexible linkages that have led or may lead after some time for emergencies:

• insufficient corrosion resistance;
• excessive pliability of the plane of the wall;

• excessive shear stiffness;
• unsatisfactory anchoring the laying obverse and internal layers;
• large distance between connections.

Connection should be rigid enough when working out of the plane of the wall.

The distance between connection in many projects is appointed without any justification. During the construction of the distance between communications and the places of their binding specified in the project on many of the buildings surveyed are not met. [8]

Despite the large amount of work on this subject, the question of how to prevent the formation of cracks at the design stage is not solved.

2 Research objective

The purpose of this work is identification of the reasons of formation of damages (cracks) of a obverse layer of external walls to zones of interwindow belts and piers, and also a number of recommendations about design for prevention of formation of cracks is developed.

It will cause the acceptance of all the provisions of tests' ways. Some tests for determination the fire safety's properties of construction materials are similar according to the Russian and European standards. The European system of standardization has the much wider range of test methods for determining the fire safety's properties of construction materials which have no analogues in Russian regulations (table 3). The following table compares the roofing materials of different manufacturers. In connection with this direct correlation between the European and Russian parameters is difficult.

The objectives of the work are:

1. Conduct a visual inspection of the exterior walls of the building.
2. Identify the structural features of the building, including outer walls.
3. Conduct a local autopsies and sensing external walls with subsequent analysis of their design decisions, as well as pairing conditions to the supporting structure of the building.
4. Collect samples of brick and mortar of the laying obverse layer, to conduct laboratory tests to determine the mechanical properties of laying.
5. Perform simulation of fragments of the walls and supporting structures.
6. As a result of the survey and calculations to identify the causes of cracking.

To issue recommendations for prevention of damage at the design stage.

3 Object of research

Object of research is the residential building to the address St. Petersburg, Kultury Ave., 19, the case 3.

4 Inspection of a residential building

Visual and constructive examination of external walls of object of research was conducted. With the application of sensing and local autopsies exterior walls.

5 Description of object of research

The object of the survey is a residential building (Figure 1), constructed and commissioned in 2009.

The building consists of three sections different storey, separated by expansion joints.

The constructive scheme of building frameless; the load-bearing structure made of reinforced concrete and are mostly walls, and matching them with plates of overlappings.

Elbow, arranged around the perimeter of the building, constructively formed internal transverse reinforced concrete walls; in places where the transverse wall are not provided - reinforced concrete columns (plates of overlapping are monolithic interfaced with columns of beamless option). As the transverse walls and columns arranged between aerated concrete laying, forming elbow are thereby inner layer outer end walls formed from facades of the Z- and P-shaped portions.

Plates of the balconies/loggias noted above are constructive continuation of interfloor plates of overlappings.

External walls of the building – multilayered (a detailed description of the external walls and their pairing with reinforced concrete structures, including plates, is presented in section 5.2).

Interior walls are made of aerated concrete. Blocks of elevator shafts are made of precast reinforced concrete. Ventilation units - precast reinforced concrete, arranged on the entire height of the building with the output of gases, vapours, etc. outward directly through the coating.

The foundations of the building – pile; piles unite a tape grillage.

To determine if uneven sediment of the building foundations and understanding the distribution pattern of defects and damage to building structures of the building as a whole - a general inspection of the building from the inside out. Based on the results of the inspection of the damage in the form of typical cracks indicating a substantial residue of irregular foundations, while inspecting the premises of the basement and underground parking, technical floor, rooms of ladder and lift unit, as well as sample dwellings - have been discovered.

Figure 1. General view of the building

6 Pair of external walls with reinforced concrete structures

Determining the actual composition and structural features of the outer walls of the building is made on the basis of the results of autopsies and local sensing obverse layer, followed by a comparison of the actual design decisions. Also, the analysis of constructive solutions of exterior walls was carried out visually from the balconies / loggias and internal rooms.

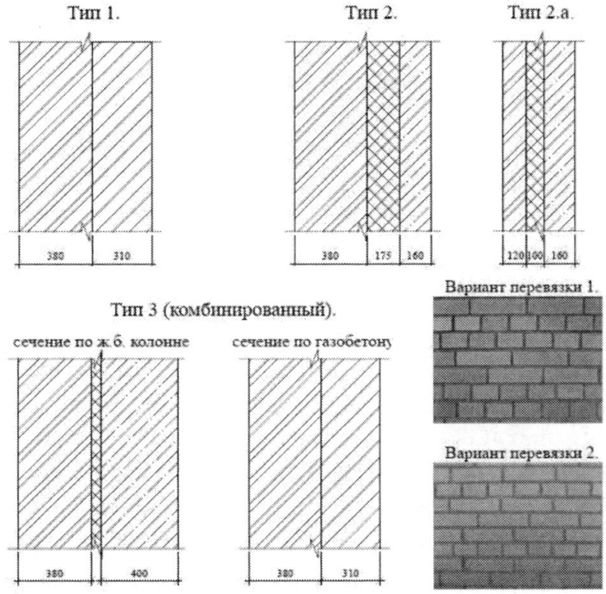

Figure 2. The decision on the external walls

There are the following types of external walls (Figure 2), differing primarily on the composition of the inner layer (outer (obverse) layer is made of a single solution for all types of walls):

Type 1. The outer walls of the rectilinear sections, i.e. sections, going along the facades - double layer: inner layer is presented laying of aerated concrete blocks with a thickness 310 mm, external (obverse) - combined masonry thickness of 380 mm (decisions laying of an obverse layer, see below).

Type 2. Outside end walls of Z-and P-shaped sections formed by internal cross-walls - three-ply: the inner layer of reinforced concrete thick wall represented 160 mm, external (obverse) - combined masonry thickness of 380 mm (decisions laying of an obverse layer, see below). Between the internal and obverse layers provides a layer of insulation from min 175 mm mineral wool.

Type 3 or combined. Outside end walls of Z-and P-shaped sections formed by the inner columns - double layer: inner layer is represented by two reinforced concrete columnsm, forming an end face and the laying from aerated concrete blocks arranged between columns. Between the columns and an obverse layer provides insulation.

The inner layer is presented laying of aerated concrete blocks of thickness 380 mm (types 1, 3) is made with floor-by-floor support on a reinforced concrete plates of overlappings (overhang layer of plates is not provided) and is arranged between the internal transverse reinforced concrete walls (Type 1) or columns (type 3). The deformation seam between plate of overlappings, columns, cross-walls, on the one hand, laying or aerated concrete - the other, is not provided. Thus, the inner layer included in the total work support structures.

The inner layers provided reinforced concrete walls with a thickness of 160 mm (type 2, 2.a), as the main load-bearing structures of the building, performed at full height with support (pairing) on a base grillage. Between the end walls and columns on one side and a obverse layer - on the other hand, there is provides insulation.

By the nature of the work it is possible to distinguish two kinds of an obverse layer depending on whether or not foreseen at sites of facades, representing this view, balconies/loggias:

Type 1 is typical for areas of facades, where there is provided a device of balconies / loggias. This type of an obverse layer can be described as layer arranged on the entire height of the building and connected to the inner layers of aerated concrete and reinforced concrete with flexible links submitted by braces, studs and t. p.

Type 2 is characterized by areas of facades, where balconies/loggias device (section 5.1 stated that balcony plates are the constructive continuation of interfloor overlappings slabs that are going outwards, as if "cut" obverse layer). An obverse layer of this type in relation to balkonnym plates can be described as follows:

- plate, "cutting" an obverse layer share his height into separate sections with a height equal to the distance between the plates;
- lack of horizontal deformation seam between a plate and a laying of an obverse layer;
- the floor-by-floor jamming of a laying between plates (or plates in a laying) allows you to talk about the association of individual sections in a single obverse layer, wherein the reinforced concrete plates are the kind of rigid links connecting the obverse layer of the building.

In addition to the above, it is worth mentioning:

- two variants of bandaging of a laying of an obverse layer (Figure 2); and consistency in their application is missing, that provokes concentration of tension in a laying;
- for architectural reasons laying of interwindow belts on a certain length by height not bandaged with laying piers. This circumstance provokes a warping of sections on border of an interwindow belt and piers in the next zone on height where bandaging is provided;
- lintel in a laying of a obverse layer are presented by combined reinforced concrete crosspiece and the steel corner arranged in obverse exterior tier 120 mm thick and unprotected outside. Considering an essential difference of coefficients of linear temperature expansion of a laying and steel, it is possible to note that their combined use in one design provokes emergence of cracks and destruction of a laying in a contact zone;
- in deformation seams of the building the laying of obverse layers of bandaging has no.

7 Description of deformations

Development of cracks begins in an interwindow belt with the bottom and/or top corner of apertures with a tendency to their merge, including with cracks above - and the below-located interwindow belts, cross-cutting and the trunk (Figure 3).

In bottom corners of apertures of a crack have local sites of an exit to piers; the cracks beginning in the top corners bypassing crossing points, come also to piers. Nevertheless in general, cracks have an obvious vertical focus.

Figure 3. Cracks of vertical orientation on border of interwindow belts and piersu

The width of cracks not more disclosure 2-3 mm.

It should be noted, that cracks of this type don't prove in interwindow belts which laying isn't bandaged with a laying of piers for architectural reasons.

8 The results of strength tests

To determine the actual strength characteristics of a bricklaying from a obverse layer made sampling of brick and mortar, followed by tests in accredited scientific testing laboratory "POLYTECH-SKiM-TEST" of the St. Petersburg State Polytechnic University, according to the procedures of state standards (testing brick compressive and bending in accordance with GOST 8462-85, the solution - compression in accordance with GOST 5802-86). Results of testing samples of bricks are given in Table 1. The test results of the solution shown in Table 2.

Table 1. Results of tensile strength tests definition bricks flexural and compressive.

Marking	Compression			Bending		
	Breaking load in compression, kN	**Durability limit at compression, MPa**	**Average value of durability at compression, MPa**	**Breaking load at a bend, kN**	**Durability limit at a bend, MPa**	**Average value of durability at bend, MPa**
1	172,24	12,4	12,8	4,16	2,46	2,41
2	176,12	12,7		4,00	2,37	
3	122,75	13,4				

Table 2. Test Results Determination of compressive strength solution.

Sample number	Breaking load, kN	Durability limit at compression, MPa	Average value of durability at compression, MPa
1	24,95	4,99	4,87
2	27,92	5,58	
3	20,20	4,04	

It may be noted low rates of laying tensile strength, shear and flexural bandaged section. These low levels are not associated with this specific laying, and is characterized in principle for masonry structures (good resistance to compression and low tensile, shear and bending). It is also worth noting the thickness of an obverse layer in 380 mm and a high voidness brick / stone, which is reflected in the area of cutting and stretching (tensile strength and shear is directly related to the durability characteristics, as well as the cross-sectional area where the efforts of developing countries).

9 Modeling of structures in the software package Ansys

In section 5.2 noted that the value of the given plot height of an obverse layer type 2 hred.2 substantially less value given the height of the plot of an obverse layer type 1 hred.1 (hred.2 << hred.1). An obverse layer pier can also be attributed to the type 2 (with values listed heights respectively hred.1 and hred.2). Considering only the difference of the actual heights of a pier and an interwindow belt, it may be noted that the value of the given height is hred.3 differ from hred.1 и hred.2. Difference given heights (hred.1 – hred.3) and (hred.2 – hred.3) leads to a difference of longitudinal deformation and . Bandaging of layings obverse layer of a pier and interwindow beltconstrains noted differences that causes tangent tension , which, given the low resistance of laying shear, leads to warping sections on the boundary layers and cracks (obverse layers are moved relative to each other by their height). Capping model with floor-by-floor support on plates of overlapping in Figure 4, the full height of the building in Figure 5.

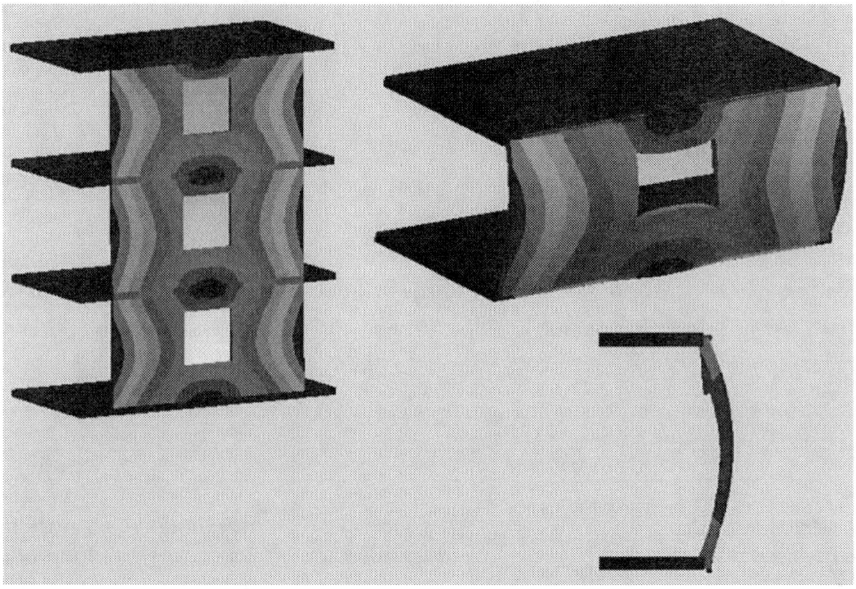

Figure 4. The device of an obverse layer with a floor-by-floor support on overlapping plates

Lack of a horizontal deformation seam between a plate and a laying of an obverse layer, led to that plates put an obverse layer into the general operation of reinforced concrete building structures. At the same time, the corresponding loadings are transmitted through plates on a laying of an obverse layer (the weight of plates, payloads, the load on the balcony fences and etc.). Also we will note that vertical efforts in a laying can arise as from both its own weight of a laying and weight of the designs leaning on it, and from control of temperature influences that finally provoke emergence of cracks.

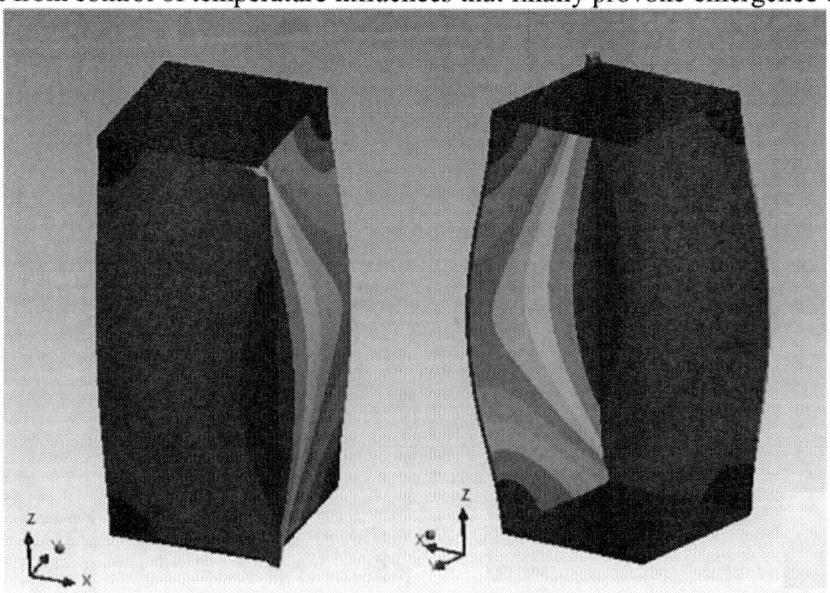

Figure 5. The device of an obverse layer on all height of the building

Thus, these mechanisms also describe the shear adjustment layers, caused by the difference of the longitudinal strain.

However, in this case, it is impossible to call lack of a vertical deformation seam between obverse layers of piers and interwindow belts a background on which these mechanisms prove. Manifestation of mechanisms happens rather against imbalance the rigid of characteristics of communications of interwindow belts and piers concerning deformations of obverse layers. These circumstances caused emergence of vertical cracks on border of interwindow belts and piers.

10 Conclusions

1. The analysis revealed that the main cause of damage to the obverse layer protecting designs is the wrong design.

2. For operational efficiency and durability of the obverse layer sandwich walling should make calculations of a bricklaying shear deformation and temperature, as well as to take into account the difference of temperature deformation of laying and steel lintel.

3. It should be laid in the expansion project of horizontal seams in places of a floor-by-floor support of an obverse layer. It will promote that there was no tension from not projected loadings and flexural deformations.

4. It is necessary to carry out a special calculation of anchors for each floor in places interwindow belts and piers. As a result which does not necessarily have to be identical anchor. The purpose of this calculation must have the same shear strain at each height level.

5. It is not advisable to use a voidness bricks for lining. Since they help to reduce the work anchors due to decrease in the area of friction.

References

1. A.Y. Navchuk. V.N. Derkach. R.B. Orlovich. Architecture and Constr., **5 (216)**, 66-68 (2010).
2. M.K. Ishchuk. Construction materials. **10**. 87-90 (2008).
3. M.K. Ishchuk. M.V. Shirav. Construction materials. **5**. 93-95 (2012).
4. M.K. Ishchuk. A.V. Zuveva. Structural Mechanics and Analysis of Constructions, **4**, 71-73 (2006).
5. M.K. Ishchuk. A.V. Zuveva. Industrial and Civil Engineering, **3**, 40-43 (2007).
6. M.K. Ishchuk. Housing Construction. **5**. 15-19 (2008).
7. M.K. Ishchuk. Housing Construction. **3**. 28-31 (2008).
8. M.K. Ishchuk. Construction materials: architecture. **4**. 101-105 (2008).
9. M.K. Ishchuk. Housing Construction. **4**. 23-27 (2008).
10. M.K. Ishchuk. Integral. **1**. 20-22 (2001).
11. M.K. Ishchuk. Technology Building. **2 (64)**. 28-37 (2009).
12. S.B. Serikkhaliyev. S.S. Zimin, R.B. Orlovich, Construction of Unique Buildings and Structures, **5 (20)**. 28-38 (2014).
13. R.B. Orlovich. N.M. Rubtsov. S.S. Zimin. Magazine of Civil Engineering. **1 (36)**. 3-11 (2013).
14. S.S. Zimin. N.P. Romanov, O.V. Romanova, Construction of Unique Buildings and Structures, **9 (36)**. 33-43 (2014).
15. N. Vatin. J. Havula. L. Martikainen. A. Sinelnikov, A. Orlova, S. Salamakhin, Advanced Materials Research. **945–949**. 1211–1215 (2014).
16. N. Vatin. J. Havula. L. Martikainen, A. Sinelnikov, L. L. Shurovkina, Applied Mechanics and Materials. **725–726**. 722–727 (2015).
17. J.G. Rots. Heron. **3 (2)**. 49-63 (1991).
18. M. Gilbert. B. Hobbs. C.K. Molyneaux. International journal of impact eng., **3**, 253-275 (2002).
19. R. Jasinski. Przeglad Budowlany. **9**. 28-36 (2009).
20. K.J. Beasley. The Construction specifier. **2 (51)**. 25–28 (1998).
21. W. Jager. G. Pfeifer. Mauerwerk Kalender Berlin. 233-264 (2005).
22. M.E. Phipps. T.I. Montague. 7th North American Masonry Conference. **2**. 789-798 (1996).
23. N.A. Roumani. *The shear strength of prestressed brichwork sections* (Ph.D thesis, University of Manchester. 1985).
24. G. Simudic, A.W. Page, 7th North American Masonry Conference, **2**,1007-1018 (1996).
25. A. Ponomarev, M. Knežević, N. Vatin, S. Kiski, I. Ageev, Journal of Applied Engineering Science, **12 (3)**, 227-231 (2014).
26. M. Lazarevska, M. Milanović, M. Knežević, M. Cvetkovska, A.T. Gavriloska, T. Samadzioska, Journal of Applied Engineering Science, **12 (1)**, 63-68 (2014).
27. B. Šćepanović, M. Knežević, D. Lučić, O. Mijušković, Proceedings of the 13th East Asia-Pacific Conference on Structural Engineering and Construction, EASEC 2013 (2013).
28. J. Cetkovic, M. Knežević, I. Vujovic, J. Cerovic, Technics Technologies Education Management, **7 (4)**, 1646-1654 (2012).
29. Z. Radovanović, R. Sinđić Grebović, S. Dimovska, N. Serdar, N. Vatin, V. Murgul, Procedia Engineering, **117**, 870-878 (2015).

MATEC Web of Conferences **53**, 01054 (2016)
DOI: 10.1051/matecconf/20165301054

Wavelet analysis function of changing average work amounts in monolithic construction

Tatiana Simankina[1], Marina Romanovich[2, a] and Oleg Tsvetkov[3]

[1] *Peter the Great Saint-Petersburg Polytechnic University, Russia, 195251, St. Petersburg, Polytechnicheskaya St., 29*
[2] *State unitary enterprise "Municipal government of inventory and assessment of real estate", Russia, 191104, St. Petersburg, Belinskogo St., 13*
[3] *Federal state budgetary educational institution of higher professional education "Komsomolsk-on-Amur state technical University" ("KnASTU"), Russia, 681013, Khabarovsk, Lenin prospect, 27*

Abstract. In work periodicity of impact of various external and internal factors on function of changing average work amounts and extent of various factors influence on the process of monolithic construction are defined by wavelet transform. Basic data are data on daily amount of the concreted structures (CBM) in 4.5 months. Using the continuous wavelet transform wavelet spectrum was constructed, which showed the periods of impact of external and internal factors on the construction process. Using the discrete wavelet transform the algorithm of an exception of critical deviations from a function of changing average work amounts was developed. This algorithm allows to plan future works taking into account the revealed factors.

1 Introduction

Many scientific papers are dedicated to the analysis of various phenomena and processes using wavelet transforms [1-14]. For example, in the field of research of social processes [6], in the study of patterns of signals in radio electronics [7, 8] in the context of manufacturing processes for prediction of injuries [9], in the prediction of climate and weather conditions, studies of seismic effects [4, 10, 12], to predict the technological characteristics of industrial production [11], and also in the medical field to detect various abnormalities occurring in the human body, for the processing of digital images, in the fields of physics, chemistry, etc. However, in the field of analysis and prediction of different characteristics of the construction process, the wavelet transform is not widely used.

In a general view the formula of the wavelet transform:

$$\psi_{ab}(t) = \frac{1}{\sqrt{a}}\psi\left(\frac{t-b}{a}\right) \tag{1}$$

where $\psi_{ab}(t)$ – wavelet, defined for the set parameters a and b, generated by a maternal initial wavelet $\psi(t)$;

b – shift parameter in time;

a – parameter of time scale change (levels of detail);

1/a^1/2– the multiplier providing independence of norm of functions from the value a.

[a] Corresponding author : p198320@yandex.ru

Small values of a correspond to small scale $\psi ab(t)$ or at higher frequencies where $\omega \sim 1/a$. Large values of parameter a correspond to large scale $\psi ab(t)$ and low frequencies. Due to the change of scale, wavelets are able to identify the difference in characteristics at different frequencies. By shifting wavelets are able to analyze the properties of the signal at different points throughout the studied interval. Therefore in the analysis of the signals created by a number of data, the wavelet transform is a significant advantage over Fourier transform. The Fourier transform gives a global information about the frequencies of the analyzed signal when used at system of functions defined on infinite intervals.

2 Data description and methology

The analysis of data on daily average amounts of work performed in the construction of monolithic house was carried out. Basic data are data on daily amount of the concreted structures (CBM) in 4.5 months (October, November, December, January, February).

Average daily amounts are calculated by the formula:

$$V_{averagei} = V_{di} + \frac{V_{st}}{N_t} \tag{2}$$

where $V_{averagei}$ is the average amount of work performed in the *i-th* day, CBM;

V_{di} – the volume of the concrete designs (walls, columns), etc. in the *i-th* day, cubic meters, CBM;

V_{st} – the volume of concrete section of slab for the period of time t;

N_t – number of the working days in the period of time t.

Time period t is the number of days between the concreting sections of floor slabs. The function of changing average work amounts was made using the calculated values of the average daily amount of work performed in each day of the reporting period (4.5 months).

The function of change in average amounts was investigated using continuous wavelet transform and discrete wavelet transform.

Continuous wavelet transform may be direct and reverse. Direct continuous wavelet transform is defined by a formula (3), reverse continuous wavelet transform is defined by a formula (4).

$$W_S(a,b) = (f(t), \psi_{ab}(t)) = \frac{1}{\sqrt{a}} \int_{-\infty}^{+\infty} f(t) \psi\left(\frac{t-b}{a}\right) dt \tag{3}$$

$$f(t) = \frac{1}{C_\psi} \int_{-\infty}^{+\infty} \int_{-\infty}^{+\infty} W_S(a,b) \psi_{ab}(t) \frac{dadb}{a^2} \tag{4}$$

where C_Ψ – the normalizing coefficient:

$$C_\Psi = \int_{-\infty}^{+\infty} |\Psi(\omega)|^2 |\omega|^{-1} d\omega \angle \infty \tag{5}$$

where $(\cdot,\cdot)$ – scalar product of the corresponding factors,

$\Psi(\omega)$ – Fourier transform of a wavelet $\psi(t)$.

The wavelet spectrum $Ws(a,b)$ or scale-time spectrum is a function of two arguments – parameter of time scale change a and shift parameter in time b. Spectrum $Ws(a,b)$ is a surface in three-dimensional space. However, as a rule, instead of the image of a surface spectrum represents it's projection to the *ab* plane. It is also possible to analyze pictures of lines of local extrema (maxima) of these surfaces, so-called skeleton which helps to reveal structures of the analyzed initial signal [15].

It is necessary to perform many computing operations for calculation of a wavelet spectrum at continuous change of parameters a and b. Many functions $\psi_{ab}(t)$ are redundant. It is necessary to conduct the discretization of these parameters at preservation of possibility of restoration of a signal from its transformation. Discretization is usually carried out for constructions of number two in degree [17-19].

Let a = 2m, b = k·2m, then the expression (1) takes the following form:

$$\psi_{mk}(t) = \frac{1}{\sqrt{a}}\psi\left(\frac{t-b}{a}\right) = \frac{1}{\sqrt{2^m}}\psi\left(2^{-m}t-k\right) \tag{6}$$

where m и k – integers.

Thus earlier considered ab plane turns into the corresponding grid of m, k. Direct and reverse wavelet transform will be determined by formulas:

$$c_{mk} = (f(t), \psi_{mk}(t)) = \int_{-\infty}^{+\infty} f(t)\psi_{mk}(t)dt, \tag{7}$$

$$f(t) = \sum_{m,k} c_{mk}\psi_{mk}(t) \tag{8}$$

Drawing an analogy with the Fourier transform, the coefficients c_{mk} can be determined using continuous wavelet transform $Ws(a,b)$ [15]:

$$c_{mk} = W(2^m, k \cdot 2^m) \tag{9}$$

Thus, from the formula (8) shows that the basic signal $f(t)$ can be represented as the sum of «wavelet waves» with defined coefficients c_{mk}.

3 The results of wavelet analysis

To study the function of changes in average amounts of work discrete approximation of Meyer wavelet ("dmey") was used, and 14-th approximation of the studied function [16]. In Figure 1 some schedules are submitted:

a) 14-th approximation of function of changes in average amounts of work;

b) the wavelet spectrum $W(a,b)$coefficients as projections on the ab plane (parameter of time scale change, shift parameter in time). In any point of the plane it is possible to determine the frequency corresponding to a certain value of coefficients a and b. Color shows the value of the spectrum $W(a,b)$ on the module. Light areas correspond to great values coefficients, dark tone fixes transitions of a signal through zero. In our case average value of a signal is accepted to zero. Shades of gray-brown flowers in each area allocated ranges of transitional values of the wavelet spectrum $W(a,b)$;

c) for example values of the wavelet spectrum $W(a,b)$ which correspond to scale a=32 are shown;

d) shows the wavelet spectrum $W(a,b)$ with well-defined lines of local maxima, passing approximately in the middle of light areas of the plane ab graphics b).

The larger the scale of the wavelet, the wider area of the signal will affect the result [9]. Each area of the wavelet spectrum $W(a,b)$ is formed as a result of impact on function of changing average work amounts of these or those factors.

The wavelet spectrum reveals not only the hierarchical structure of the analyzed time series, but also carries information about the processes taking place during the period of construction, in particular about influence of external and internal factors on the progress of construction. Using the analysis of the wavelet spectrum it is possible to find global relationships between different parameters and the influence of some parameters on others, in a certain period of time. Analyzing the alternation of areas with different frequency and how function of changes in average amounts of work is influenced by external and internal factors, it is possible to predict behavior of function. And most importantly, the analysis allows to diagnose and predict unstable conditions and the moments of approach of critical deviations of function.

In the analysis of the wavelet spectrum $W(a,b)$ the lines of local maxima of surfaces presented on graphics of d) Figure 1 were also analyzed.

In table 1 results of the analysis of the wavelet spectrum $W(a,b)$ for the 14th approximation function of changes in average amounts of work and for basic function are given in a period which equals 4.5 months (128 days). Levels of detail a of wavelet -decomposition are specified in the second column of the table, average values of the periods T are shown in the third column, average standard errors of the periods σT are given in the fourth column.

By comparison of statistical data about construction of the monolithic house to the data presented in table 1 the periods of emergence of various factors influencing the course of construction were determined (table 2).

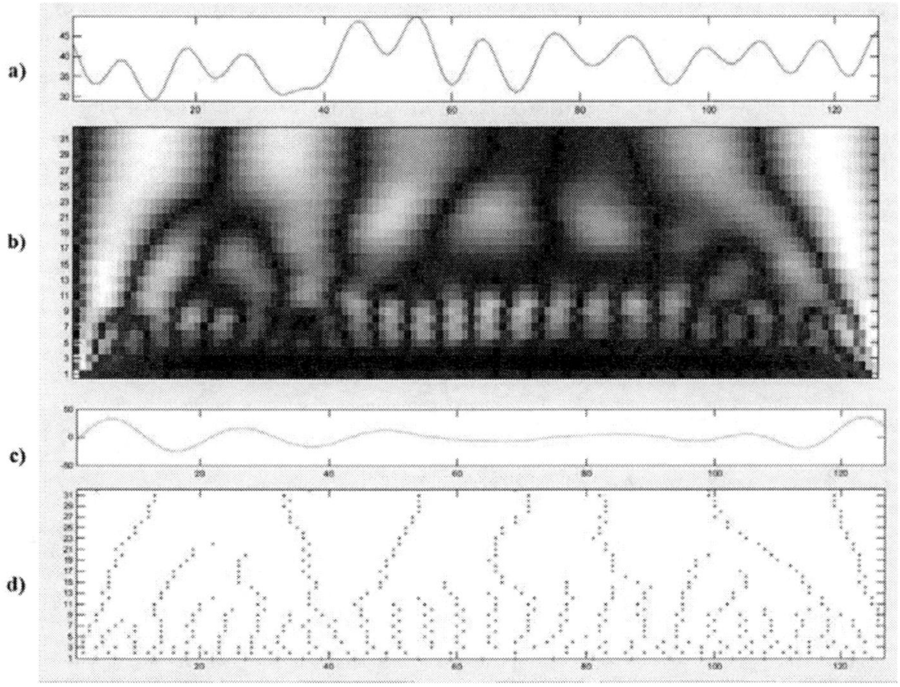

Figure 1. Continuous wavelet transform 14th approximation function of changing average work amounts.

Table 1. Analysis of the wavelet spectrum *W (a,b)*.

№	*a* (levels of detail)	T , days	σT , days
14th approximation function of changing average work amounts			
1	7	4	±1.5
2	15	10	±3.5
3	25	19	±4.5
4	from 5 to 25	70	±5.5
Basic function of changing average work amounts			
5	40	25	±4.5
6	61	45	±3

Table 2. The periods of impact of external and internal factors on the progress of construction.

№	Factors	T , days	σT , days
1	Failures in work of construction crew	4	±1.5
2	The elimination of defects revealed in the work	10	±3.5
3	De-energization of a building site	19	±4.5
4	Delays in supply of construction materials	70	±5.5
5	Repair of cranes and mechanisms	25	±4.5
6	Non-working condition of cranes and mechanisms	45	±3

In this work comparison of the latent features of function of changing average work amounts Figure of works was carried out with factors data which were managed to be collected in 4.5 months of construction of the monolithic house. However there could be more factors. Studying of their hidden influences on the course of construction can become separate area for further studying.

Figure 2 shows the discrete wavelet transformation of function of changing average work amounts for one month of construction works. As the base functions in the decomposition it is used Daubechie's wavelet (db) of the 1st order. Figure 2 shows the graphic of the function of changing average work amounts (*s/ds*) and a graphics of the components of the function, restored by the approximating coefficient a_4 and to the detailing coefficients d_1, d_2, d_3, d_4. The detailing coefficients show dynamics details at the allocated levels. The coefficient d_1 is the closest approximation to basic function, d_4 coefficient – the farthest approximation which shows details of the main trends of the high-frequency component of the time series. The low-frequency filter a_4 reflects the main tendencies of low-frequency components. The schedule a_4 shows that the function of changing average work amounts at deep 5-level decomposition reveals two main trends. The first trend – the average amount of completed work during the first 16 days of month equals 44 CBM a day. Further until the end of a month of construction works the main trend performed amount was 40 CBM a day.

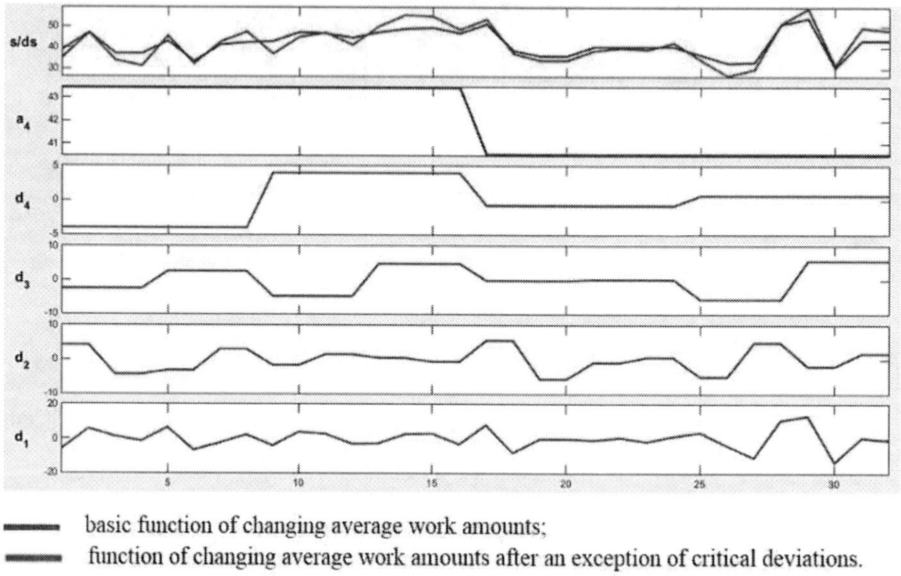

Figure 2. Continuous wavelet transform 14th approximation function of changing average work amounts.

Thus the general regularity of change of the performed works for construction of the monolithic house within one month was revealed. This regularity can't be revealed from basic function.

The algorithm of an exception from function of changing average work amounts critical deviations developed. Deviations arise as a result of the impact on the function of various factors leading to deviations of the function from the given trend. In our case, the given trend is the average value of the daily executed amount of works (CBM of the concreted designs). Deviations from the trend should be minimal, especially in the direction of decreasing values of the average amount of work performed. The algorithm is given below.

1.*Formation of a data number for basic function of changing average work amounts.* For a start it is necessary to define value of a number of data of function and the studied period (schedule s in Figure 2).

2.*Processing of data series functions.* At this stage it is necessary to carry out the analysis of a number of data of function by means of wavelet transforms, to reveal the maximum and minimum deviations of values, to define the factors influencing function and the periods of emergence of factors.

3.*Decomposition of basic function of changing average work amounts.* It is necessary to choose wavelet, its order and number of levels of decomposition of function N.

4.*Threshold processing.* For each level from 1 to N it is necessary to choose thresholds (depending on objectives). After that it is necessary to execute threshold processing of the detailing coefficients on those sites of function where it is necessary to minimize or at all to clean the critical deviations of function at all from the set trend resulting from impact on function of various factors.

5. *Reconstruction*. At the fifth stage of algorithm it is necessary to carry out restoration of values of function by means of the reverse wavelet transform after threshold processing. The result is a function without critical deviations or with the minimum deviations (schedule *ds* in Figure 2).

6. *Scheduling*. After an exception or minimization of critical deviations we get a function of changing average work amounts without impact on it of various factors. If we compare basic function and the function after an exception of critical deviations it is possible to define extent of impact of this or that factor on construction process at a specific point in time. If we know frequency and extent of influence of factors on construction process it is possible to change number of workers in this or that day, to correct the schedule of work of crews, to enter additional days of check and setup of machines and mechanisms. These actions will allow to optimize control of daily amount of the performed works and construction.

4 Conclusion

Thus frequency and extent of impact of various factors on function of changing average work amounts and the general course of construction can be defined by using the wavelet transforms. The developed algorithm allows to plan future works taking into account the revealed dependences.

References

1. N. Astaf'eva, Uspekhi fizicheskikh nauk, **11,** 1145–1170 (1998).
2. V. G. Zakharov, *Razrabotka i primenenie metodov vejvlet-analiza k nelinejnym gidrodinamicheskim sistemam* (Istitute of mechanics of continuous environments, Perm', 1997).
3. L. Novikov, Nauchnoe priborostroenie, **3**, 70–76 (2000).
4. V. Alekseev, Vestnik Yurskogo gosudarstvennogo universiteta, **3 (30)**, 3–10 (2013).
5. I. Dremin, O. Ivanov, V. Nechitajlo, Uspekhi fizicheskikh nauk, **5**, 465–501 (2001).
6. A. Davydov, Sotsiologicheskie issledovaniya, **11**, 89–102 (2003).
7. O. Lazorenko, S. Lazorenko, L. Chernogor, Radiofizika i radioastronomiya, **2**, 182–204 (2007).
8. O. Lazorenko, S. Lazorenko, L. Chernogor, Radiofizika i radioastronomiya, **3**, 278–294 (2007).
9. N. Muller, S. Seryj, Upravlenie v sotsial'no-ehkonomicheskikh sistemakh, **3 (21)**, 52 – 60 (2009).
10. V. Mkrtychev, A. Reshetov, Vestnik MGSU, **7**, 59 – 67 (2013).
11. A. P. Lushavin, *Prognozirovanie kharakteristik tekhnologicheskikh protsessov promyshlennogo proizvodstva s ispol'zovaniem iskusstvennykh nejronnykh setej i vejvlet-preobrazovanij* (FGBOU VPO Samarskij gosudarstvennyj tekhnicheskij universitet, Samara, 2012).
12. A. Y. Nikul'nikov, *Tekhnologiya interpretatsii rezul'tatov vejvlet-preobrazovaniya sejsmicheskoj zapisi* (Rossijskij Gosudarstvennyj Geologorazvedochnyj universitet im. Sergo Ordzhonikidze, Moscow, 2012).
13. O. Amosov, N. Muller, Internet-zhurnal Naukovedenie, **3**, 1 – 14 (2014).
14. O. Mandrikova, A. Tiristanov, trudy Vserossijsk. nauchn. konf. Proektirovanie nauchnykh i inzhenernykh prilozhenij v srede Matlab, 242-261 (2002).
15. A. N. Yakovlev, *Vvedenie v vejvlet-preobrazovaniy* (NGTU , Novosibirsk , 2003).
16. M. Romanovich, Sovremennye problemy nauki i obrazovaniya, **1**, URL: http://www.science-education.ru/121-19053 (2015).
17. O. V. Nagornov, V. G. Nikitaev, S. A. Tyuflin, A. N. Pronichev, T. I. Bukharova, K. S. Chistov, R. Z. Kashafutdinov, V. A. Khorkin, *Vejvlet-analiz v primerakh* (NIYAU MIFI , Moscow, 2010).
18. N. K. Smolentsev, *Osnovy teorii vejvletov. Vejvlety v MATLAB* (DMK Press, Moscow, 2005).
19. I. Dobeshi, *Desyat' lektsij po vejvletam* (NITS Regulyarnaya i khaoticheskaya dinamika, Izhevsk, 2001).

MATEC Web of Conferences 53, 01055 (2016)
DOI: 10.1051/matecconf/20165301055

The use of Transaction Costs Theory in Interorganizational Design

Jasmina Ćetković[1,a], Slobodan Lakić [1], Miloš Knežević[2], Miloš Žarković[3] and Tatiana Sazonova[4]

[1]Faculty of Economics, University of Montenegro, 81000 Podgorica, Montenegro
[2]Faculty of Civil Engineering, University of Montenegro, 81000 Podgorica, Montenegro
[3]Erste Bank AD Podgorica, 81000 Podgorica, Montenegro
[4]St. Petersburg State University of Architecture and Civil Engineering, 2-Krasnoarmejskaja, 4, Saint-Petersburg, 190005, Russia

Abstract. This paper analyzes the basics of economics of transaction costs, which grew into a suitable and practical tool to explore the wide problems in the domain of different sciences. The emphasis in this paper is placed on the use of economics of transaction costs in the determination of the growth limits and optimization of its organizational structure. This paper further explores the implications of the economics of transaction costs on contemporary forms of inter organizational design.

1 Introduction

For a particular product or a service to cross the boundaries of an organizational system or any part thereof it is necessary to perform an activity or set of activities that are called transactions.

The agreed price of the goods or services is not the only cost, because transactions conducted in the market are not free. Realization of transaction activities has an effect on the occurrence of transaction costs, which some authors call "transition costs" and represent the cost of preparing and controlling the movement of goods and services in the market or within the organizational system. Williamson defined transaction costs as the costs of planning, adapting and monitoring achievement of objectives in the case of alternative governance structures [1]. Rougher definitions of transaction costs under transaction costs include all costs except production costs – associated with the transactions.

Size of transaction costs depends on the type of transaction and its complexity. In the modern world, transaction costs increase as a result of costly information that are asymmetrically distributed among the groups that exchanged them. The limitations of certain cognitive capacity and the lack of information are the basic assumptions of the new institutional theory on which grounds the theory of transaction costs has been created. New institutional economics has begun to engage in defining research questions related to the structure of t he economic systems and research how to adapt traditional economic methods in order to explore these questions. Coase's historical essay from 1937 [2] is a classic example of this approach. Coase questioned why some transactions take place on the market, and others within the company, and what factors determine the allocation of transactions between these two spheres. Thus, the idea of costs of transactions was born which is then transformed into the concept of transactional costs which is now a key part of the economics of institutions.

2 The theory of transactions costs as the basis of the modern economics

The theory of transaction costs is the foundation of modern economics. Until the 70s of the last century many, more or less well-known economists considered this theory as controversial or wrong. The term "economics" was first used in the 20s of last century in the A. Marshall's book "Principles of Economics" [3] who is known as the originator of the Cambridge School and one of the opponents of unlimited effects of laissez-faire. He has under this term marked the science that has studied the economic problems which were previously dealt by the political economy. Neoclassics have defined economics as the science that studies the allocation of resources by human societies, mutual concurrence to each other.

The basis of modern economics presents a series of works synthesized in the theory of transaction costs, which is the basis of economics of transaction costs as well as contemporary economics. The economics of transaction costs studies the transaction costs of goods and services among market participants, as well as Demsetz noticed there is a cost of using "price mechanism" [4]. This theory has grown on the foundations of a new trend of economic thought in the 70s of the last century – the "new institutionalism" which founder is the Nobel laureate R.H. Coase. This economic thought was created as a response to the old "American institutional school."

The economics of transaction costs has grown into usable practical tool, suitable to explore the broad types of problems in different sciences. At the same time, it is also used in the study of the size of the company, to establish the limits of growth of the

[a] Corresponding author : jasmina@ac.me

company, and the boundaries of the market where the company performs.

R. H. Coase wrote his groundbreaking work entitled The Nature of the Firm back in 1937 in England. Later, transferring to the United States, his dilemmas are shared with the Chicago School, which was strongly influenced by Marshall's views. This is particularly evident in 1960 when he published a paper entitled The Problem of Social Costs in which he made a significant step forward compared to Pigoua (after Marshall's, the spiritual leader of the Cambridge school). Pigoua's understanding of solving problems through government interventions and legal system was enhanced with a new understanding of resolving relationship among market players in terms of opportunity cost, legal liability and property rights.

The term "economics of the new institutionalism" or shorter "new institutionalism" is associated with the science of the R.H. Coase, A.A. Alchian, H. Demsetz, O.E. Williamson and others who deal with the problem of transaction costs in the market, as well as the rights of ownership and responsibility in economic relations (such as H.A. Simon, E.F. Fama, M.C. Jensen, S.G. Winter, et al.). As a leading representative of the school of "new institutionalism" Coase was awarded with the Nobel Prize in 1991 for his contribution. However, Coase's noticed that when making a decision on the question of "make or buy" market price is not the only deciding factor. Important, often the deciding factor are significant transaction costs, which at that time were high and large companies rather decided to produce by themselves than to indulge in the purchase of products that have followed the high transaction costs of the procurement coordination.

The theory of transaction costs is essentially a theory of the firm which is constantly trying to respond to the question – why the company exists? The essence of this theory is the replacement of the price mechanism with more efficient performance of the transaction of goods or services under alternative institutions, under which the company is understood and any other structure (vertical integration, auction etc.) in order to reduce transaction costs. Coase's initial assumption is the assumption of the existence of the company due to the existence of transaction and marketing costs which they cannot reduce to a minimum. If these costs are equal to the cost of using the price mechanism, the existence of the company would lose the purpose.

Shift in his perception, Coase gave in his work The Problem of Social Cost in which he presented a new argument – the right of ownership of a particular property will not affect the efficiency of the allocation of goods, unless there are transaction costs of negotiating – this postulate was later called the Coase Theorem.

A further contribution to the development of the theory of transaction costs was given by O. Williamson. He perceives transaction costs as the costs of planning, adapting and monitoring the execution of the objectives in the case of alternative governance structures [1]. Williamson noted that the company has its own role in the economic system if the transactions are organized into companies at lower costs than those at which the transactions took place over the market. Market imperfections, the complexity of the organizational structure and the number of participants in the market are the key factors of the size of transaction costs. The dimension of market imperfections is a consequence of bounded rationality and opportunistic behavior of participants. He found a clear link between the three dimensions of the transaction – specific assets, uncertainty and frequency of transactions. The higher the degree of specificity is, the greater the need to organize transactions internally, within the hierarchy/firm. For an organization it would be more profitable to meet the requirement within its borders, than to provide it through the purchase on the market. Identification of appropriate methods for objective decisions about what should be internal and external transactions is the primary task of the analysis of transaction costs. Furthermore, due to the high degree of uncertainty in the behavior of parties involved in the transaction, the transactions are often internalized. Also, the greater the degree of frequency of transactions, the lower transaction costs is within the hierarchy/firms than the costs of market transactions.

The theory of transaction cost has defined the basis for determining the company growth limits. Whether a transaction is carried out within the firm or on the market, it depends on the relationship between costs incurred as a result of the transaction on the market and the cost of carrying out the same transactions within an organizational form. Organization size limit is determined by the volume of those transactions whose overspending leads to higher transaction costs within the company in relation to the costs of carrying out the same transactions in the market. With this kind of analysis we are able to determine the economic limits of the company growth, or economically feasible solution of vertical and horizontal expansion of activities within a single organizational system. The main objective of the theory of transaction costs is to minimize them through the establishment of any management structure that leads to the realization of the basic objective.

Use-value theory of transaction costs does not decrease the indisputable practical situations in which firms carry out transactions within the hierarchy even if the transaction costs of market transactions are zero, guided with the interests of greater reliability, dependence reduce of unpredictable market, profitability rate increase by exploiting effects of economies of scale and etc.

3 The use of Transaction costs theory in interorganizational design

One of the theoretical foundations of modern forms of interorganizational relations is the theory of transaction costs. Start of the theory of transaction costs use in the issue of organizational restructuring is linked to the works of Commons in 1934. In the transaction costs, he recognized factor and measure of the efficiency of the organizational structure [5]. Next, Coase concluded that the organizations are basically an alternative to the market and variable organization boundaries depend on the analysis of transaction costs which indicates the economic rational horizontal and vertical expansion activities. Essentially, in this analysis decisions about opening new businesses or retaining and developing certain parts of the company should be based. That solves one of the major issues, such as the question of optimization of the organizational structure.

J.C. Jarillo in his work On Strategic Networks [6] pointed on the use of the theory of transaction costs in organizations interorganizational design. Specifically, in addition to the extremes "produce or buy", he identifies the third option which is realized through various forms of cooperation between organizations in the field interorganizational design. Jarillo recognizes relations of cooperation between the companies as a source of competitive advantage. He pointed out to the greater efficiency of networks or alliances in relation to the hierarchies/firm and the market in terms of minimizing transaction costs. In this sense competitive and cooperative aspects of the behavior of companies are identified as complementary aspects of the behavior of a single reality.

Modern and intensified forms of cooperation between organizations relativize the importance of the opportunistic behavior of actors' phenomenon and the concept of the high cost of monitoring market transactions. Through modern and different types of networks between organizations a new mechanism for coordinating activities – trust and predictability is establishing, while conflicts prevention is resolved by balancing the interests of network elements [7]. The network forms of organization according to Powell are characterized by reciprocal communication patterns and the exchange representing sustainable economic organization. In Tab. 1 below, Powell presented a comparative overview of the differences between the basic characteristics of the markets, networks and hierarchy/company.

Table 1. A comparative overview of the differences between the basic characteristics of the markets, networks and hierarchy/company

Key Features	Market	Hierarchy	Network
Normative Basis	Contract – property Rights	Employment Relationship	Complementary Strengths
Means of Communication	Prices	Routines	Relational
Methods of Conflict Resolution	Haggling – resort to courts for enforcement	Administrative fiat - Supervision	Norm of reciprocity – Reputational concerns
Degree of Flexibility	High	Low	Medium
Amount of Commitment Among the Parties	Low	Medium to High	Medium to High
Tone or Climate	Precision and/or Suspicion	Formal, bureaucratic	Open-ended, mutual benefits
Actor Preferences or Choices	Independent	Dependent	Interdependent
Maxing of Forms	Repeat transaction (Geertz, 1978) Contracts as hierarchical documents (Stinchcombe, 1985)	Informal organization (Dalton, 1957) Market-like features: profit centers, transfer pricing (Eccles, 1985)	Status Hierarchies Multiple Partners Formal rules

The modern practice of interorganizational relations has overcome the limitations of traditional transactional arrangements of supplying and servicing, opening up a space for more complex forms of interorganizational cooperation and relations. Through modern ways of networking organizations are trying to achieve synergetic effects compiling internal, hierarchical and external coordination of market relations. Equity incentives under various forms of interorganizational relationships contribute to minimization of the opportunity costs.

However, the limits of growth are not something that cannot be reached. With the growing size of the organization, costs of coordinating activities and managing tempered relations are growing as well as the price of the size of the organizational system. One of the ways to counteract the increasing costs of coordination and management in relation to the use of internal production is the relocation of activities that are not of primary (non-core) and are called outsourcing. However, outsourcing companies must have access to a sufficiently large and quality market supplier which cannot always be safely counted.

In this regard, the hybrid contract forms between organizations occur in response to the transaction which marks the specificity of assets, while vertical integration remains in force with transactions that are characterized by strong bilateral dependence. Classic contracts represent a form of market coordination relationships in simple market transactions while relational contracts are the basis for many forms of interorganizational forms that are between all extremes (markets and hierarchy/firm).

The cooperation forms that facilitate cooperation and integration between different levels of organization are called the strategic alliances. The forms of cooperation that have characteristics of network and hierarchical organizations are the franchise arrangements and joint ventures. The highest degree of vertical integration and the internal value creation is achieved by forming mergers, acquisitions and hierarchically organized independent organizations. Typical forms of inter-organizational networks are based on vertical integration as mergers and acquisitions. The highest degree of vertical integration is achieved through mergers and acquisitions. With the growth of the transaction which subject is complex, non-standard product, organization development will be directed towards the vertical differentiation or integration in order to minimize management costs as a result of bounded rationality as a kind of transaction costs. In this way vertical integration has primacy over contracting.

Williamson has analyzed a number of benefits from an integration organizations trends such as better control over the assets, less obvious problem of informational imperfections, increasing the sense of belonging to the organization which leads decreasing the need for opportunistic behavior and so on.

On the other hand, Coase analyzed the factors of limiting the growth boundaries of the organization - increasing the cost of planning activities of the company due to an increase in the number of parts of the company, the likelihood of poor allocation of production resources with the growth of the number of transactions between the company and the possibility of monopoly of the market due to competition of choking firms growth [8].

3.1 Contemporary theoretical consensus – outsourcing as a form of interorganizational relation

Unexpectedly low rates on return that coorproations have begun to realize at the end of the last century sent a signal to organizations in order to review the organizational design. The positive effects of economies of scale, expanding the product range, high market share, and greater control over the sources of resources and distribution channels through large integration among organizations were absent compared to the expected. Contemporary theoretical consensus has given birth to the opposite recommendation – reducing the size of the "real" measure (rightsizing) company focus on small core activities.

One of the basic organizational trends becomes transmission of peripheral activities to third parties, except for the activities that are most important and unique and provide the company an edge over competitors (competitive advantage) [9]. Unlike traditional agreements on procurement and outosourcing are characterized by long-term business relationship and a high level of shared risk. In practice, there are various forms of outsourcing, the forms that include targeted, short-term forms of business cooperation, to forms of business cooperation in which they share responsibility and risks in the business. There are forms of outsourcing in which the tasks are delegated to a third party within the borders of one country but the forms which delegate abroad (offshore outsourcing).

Contemporary forms of outsourcing are not orientated to improve the efficiency of a single function or activity but on the reconfiguration of the entire business process which enhances the value of the organization as a whole as well. Thus outsourcing is recognized as a way for the implementation of structural changes in the organization that includes tactical (short-term) and long term (strategic) benefits. Short-term effects might be usage of the effects of economies of scale and resolution of financial problems and the long-term strategic are showing focus of the organization on key competences [10].

However, besides the positive effects, outsourcing produces certain drawbacks such as increased dependence on suppliers, loss control, employee redundancy, increased costs due to the "hidden" costs of outsourcing, etc.

Furthermore, the literature points to the other traps of outsourcing: the lack of managament commitment, lack of communication plan, the minimum knowledge and experience of suppliers, non-recognition of outsourcing risk, ignoring cultural differences and so on [11]. Therefore, the organization needs, during the contract design contract for outsourcing, in the best possible way to protect itself against all possible risks [12].

Due to a lack of outsourcing in the literature, the term backsourcing is introduced implying a "return" in the organization of activities that were the subject of outsourcing as a result of various factors [13].

It is certain that the company will opt for outsourcing when other organizations can perform the same activity with lower costs due to the use of the effects of economies of scale [14]. Some authors note that today outsourcing is used as a tactical way out of financial difficulties [15]. Although the biggest beneficiaries of outsourcing are larger companies, some authors state that the effects of outsourcing are biggest in small companies [16], because small companies cannot reveal greater effects of economies of scale.

The basic premise of successful outsourcing is correct identification of core competencies. Displacement, due to errors, core competencies "outside" of the competitive advantage of the organization can be permanently lost [17].

3.2 Continued use of the theory of transaction costs at interorganizational design – strategic alliances, joint ventures, mergers and acquisitions

Avoiding opportunity costs characteristic for transactions conducted through the market happens with establishing of interorganizational relationships or partnerships. One of the current forms of partnerships is strategic alliances.

On the possibility of applying the theory of transaction costs on the strategic alliance Carlos Jurrilo was the first to point out. He noticed that the cooperative behavior of an organization becomes a modern success factor. He recognizes the theoretical framework for the study of competitive/market and cooperative aspects of the company as compatible aspects of a single reality. In addition, cooperative relationships become the source of its competitive strength. At that point, Jarillo developed the concept of strategic networks as a tool for understanding cooperative relations among firms and analyze their role in creating the strategy of the company. Analyzing networks in relation to the market and hierarchy (as extreme models of internal and external relations) Jurrilo defined several characteristics of strategic networks [6]: hybrid mechanisms, trust and predictability as coordination mechanisms, balancing of interests as a mechanism to overcome the conflicts, mutual dependence as a preference or choice of actors, semiformal degree of formalization, mutual adaptation as a model of integration, etc. The economic limits of firm size are determined by the scope of activities that may be exceeded to overcome the costs of market transactions or in cooperation with another company. Often, when it reaches the limits of the economic size of the company, the company expands its business in a manner that is pooled with other companies.Strategic alliances are in the center of the spectrum of different organizational forms.

The role of strategic alliances is in cooperation facilitation and different degrees of integration between the companies which often leads to overgrowth in mergers and acquisitions. Competitive pressures have encouraged firms to seek additional sources in cooperation with others. Today participation in a number of strategic alliances is inevitable.

These cooperative arrangements, focused on the realization of common strategic objectives, which may lead to significant and lasting exchange, sharing and joining of new development of knowledge, products, services or technologies, allow access to complementary competencies that are expensive to develop within a company. As the deepest form of cooperation, the strategic alliance would have to deduct a portion of the autonomy from the partners and to leave a significant part of the autonomy to business partners [18] which are often direct competitors to each other. The famous and successful examples of such alliances are Sony and Ericsson (area of mobile telephony) or Star Alliance and United Airlines (area of air travel).

The structure of strategic alliances varies from medium and long term aspects (capital joint ventures and minority equity alliances) to the short and medium term structure (bilateral and unilateral alliance). The degree of organizational integration is the highest at the first structure and significantly lower in other structures of the alliance which is primarily a result of its ownership structure. In the capital there is a common joint venture equity and minority equity alliances with one-way or reciprocal capital ownership. In the capital joint ventures there is a common equity and minority equity alliances with one-way or reciprocal capital ownership. In bilateral and unilateral alliances there is no joint ownership partner. Some authors [19] are ranking rigorous strategic alliances, from transaction, over medium to permanent relations, such as cartels.

The joint venture involves a type of strategic partnership in which merger of a portion of resources between two or more organizations occurs in order to create a separate organization with common ownership.

The share of resources is often equal but different from a purely financial participation through participation license or brand. Studies show that the rate of degradation of the joint venture, as a high-risk forms of partnership, declined in the late 20th century to less than 50%. The main motives of making the joint venture, by many authors, are recognized as risk sharing in the implementation of a business idea and the realization of access to foreign markets. Partnership motives may be usage of the effects of economies of scale (production and distribution) and economy of wide.

Mergers involve a combination of the two organizations whereby one organization ceases to exist, while acquisitions mark to take control of another organization, where the merger organization becomes majority owner and mergee organization does not necessarily exist. Much of the research that was done in the world show a higher number of motives for the formation of mergers and acquisitions ranging from increased market share and its preservation, reduce competition and increasing revenues, purchasing technology to the rapid growth and use of multiple effects of synergy [20].

In the early 20th century, Bower, based on research conducted in the United States, provided a fairly comprehensive range of motivational factors for the case company acquisition. Results of this deep research showed that acquisitions occur for five reasons:

- to deal with overcapacity through consolidation in mature industries;
- to roll-up competitors in geographically fragmented industries;
- to extend into new products or markets;
- as a substitute for R&D and
- the use of eroded industrial boundaries to create new industries [21].

Results arising as a result of M&A, according to Robert F. Bruner can be measured over at least seven elements [22]:

- Creation of market value;
- Financial stability,
- Improved strategic position;
- Organizational strength;
- Enhanced „brand";
- Observance of the letter and spirit of ethical norms and laws and
- Improved process.

Mania of mergers and acquisitions was strongly expressed in 80s and 90s of the twentieth century [23] as shown in Table 2 and 3.

Table 2. Completed acquisitions in Western Europe, 1984-1989

Year	Total No	Value [£m]
1984	596	7.050
1985	1.138	13.414
1986	1.565	29.045
1987	2.246	36.274
1988	3.361	52.659
1989	5.138	85.759

Table 3. Acquisitions and mergers by industrial and commercial companies within the UK from 1984 to 1994

Year	No. of companies acquired	Total expenditure, [£m]
1984	568	5.474
1985	474	7.090
1986	842	15.367
1987	1.528	16.539
1988	1.499	22.839
1989	1.337	27.250
1990	779	8.329
1991	506	10.434
1992	432	5.941
1993	526	7.063
1994	-	-
1st quarter	183	2.720
2nd quarter	184	1.652
3rd quarter	154	2.109

The next wave that swept the business world began in 1998 and marks the internationalization wave of mergers, tackling Europe, Asia (Japan) and America (Canada). Entering in the 21st century and the current state characterizes weakening of this type of activity which was not significantly jeopardized by the previous trends in this area. Last weakening of these activities is primarily a result of a current financial crisis that has significantly affected the functioning of the global financial system. All processes of intense merger activity are characterized by a state of lower interest rates, increase in stock prices and economic expansion. Throughout the history, the experience with the global mergers and acquisitions has not been positive only. At the beginning of the 21st century, Marks and Mirvis analyzing the trends in the establishment and survival of mergers and acquisitions practice stated that three out of four mergers and acquisitions do not achieve the assigned strategic and financial goals [24] because as noted by psychologist Harry Levinson - "when companies merge and acquire, It's a lot like falling and love" [25] In order to improve these processes Marke and Mirvis analyzed the strategic and psychological challenges that organizations have to have and which should solve on the best way in the preparatory phase of this process in order to improve their performance [26-29]. Soon the factor analysis by which the theory tried to explain failures of a large number of mergers and acquisitions evaluated from traditional and rational factors to the soft factors of what the human factors and organizational behavior are [30-33].

4 Summary

The basis for determination of the company growth limits, established by the transaction costs theory, served as a guideline and a challenge to the use of the theory of transaction costs in modern interorganizational company design. As the limits of companies' growth can be reached, due to rising costs of coordinating activities and managing tempered relations of growing organizational system, the modern practice of interorganizational relations has opened the possibility of complex forms of interorganizational cooperation and relations. The current financial crisis has opened the issue of business uncertainty, which also revived the need for establishing an adequate vision of firms businesses. In these conditions, it's likely, modern forms of interorganizational relations and cooperation will represent an important means for their realization.

References

1. O. Williamson, Am. J. Soc., **87**, pp. 287-296 (1981)
2. R. H. Coase, Econom. New Ser., **4(16)**, pp. 386-405 (1937)
3. A. Marshall, *Principles of Economics London*: Macmillan; Books, p. 376 (1920)
4. H. Demsetz, *The Economics of the Business Firm: Seven Critical Commentaries* Cambridge University Press, UK, p. 285 (1997)
5. J. R. Commons, *Institutional Economics* McMillan, p. 385 (1934)
6. J. C. Jurillo, Str. Manag. J., **9**, pp. 31-41 (1988)
7. W. Powell, *Neither Market Nor Hierarchy: Network Forms of Organization Research in Organizational Behavior*, **12**, p. 300 (1990)
8. O. Williamson, S. Winter, Oxf. Uni. Pre. p. 327 (1991)
9. M.E. Porter, *Competitive Strategy* Free Press, New York, p. 361 (1980)
10. A. Douglas, J. Meehan, *Outsourcing for business excellence,* Tot. Qual. Manag., pp.278-293 (2001)
11. M. Power, C. Bonifazi, K.C. Desouza, J. Bus. Str., **25 (2)**, pp.37 - 42. (2004)
12. K. Andrade, R.B. Chapman, *Insourcing After the Outsourcing,* Guide Amacon New York, p. 287 (1998)
13. S.F. Wong, *Drivers for IT backsourcing decision IBIMA*, **2 (14)**, pp. 765-779 (2008)
14. A.P. Kakabadse, N. Kakabadse, Eur. Manag. J., **20 (2)**, pp. 756-764. (2002)
15. M. Johnson, *Outsourcing in Brief.* London: Butterworth- Heinemann Oxford, p. 186 (1997)
16. A.H. Jumah, D. Wood, UK Comp. W. St., **49 (7)**, pp. 265 – 275 (2000)
17. C.K.. Prahalad, G. Hamel, Har. Bus. Rev., pp. 765-773 (1990)
18. D. Jolly, Eur. Bus. For. **9**, pp. 567-584 (2002)
19. J.R. Harbison, P.P. Pekar, *Smart Alliances: A Practical Guide to Repeatable Success,* Wiley, p. 176 (1998)
20. P. Shrivastava, J. Bus. Str., **7**, pp. 65-76 (1986)
21. J. L. Bower, Not All M&As Are Alike-and That Matter Har. Bus. Rev., p. 98 (2001)
22. R.F. Bruner, *Applied Margers and Acquisititon* New York: John Wiley & Sons, p. 243 (2004)
23. S. Cartwright, C.L. Cooper, *Managing Mergers, Acquisition and Strategic Alliances: Integrating People and Cultures* Oxford: Butterworth Heinemann, p. 325 (1996)
24. M.L. Marks, P.H. Mirvis, Acad. Manag. Exec., **15(2)**, pp.745-756 (2001)
25. H. Levinson, Har. Bus. Rev., pp. 81-101 (1970)
26. J. Ćetković, S. Rutešić, M. Zarković, M. Knežević, N. Vatin, Procedia Engineering, **117 (1)**, 780-790 (2015)
27. B. Stamatovic, R. Upadhyay, Procedia Engineering, **117 (1)**, 660-667 (2015)
28. S. Rutešić, J. Ćetković, M. Žarcković, M. Knežević, N. Vatin, Procedia Engineering, **117 (1)**, 905-915 (2015)
29. S. Rutešić, J. Ćetković, M. Knežević, M. Žarcković, N. Vatin, Procedia Engineering, **117 (1)**, 642-650 (2015)
30. B. Melović, S. Mitrović, A. Djokaj, N. Vatin, Procedia Engineering, **117 (1)**, 807-812 (2015)
31. K. Strelets, N. Vatin, (2015) Rocznik Ochrona Srodowiska, **17 (1)**, 104-112 (2015)
32. L. Muravyeva, N. Vatin, Applied Mechanics and Materials, **635-637**, 451-456 (2014)
33. S. Ognjenovic, R. Donceva, N. Vatin, Procedia Engineering, **117 (1)**, 549-555 (2015)

MATEC Web of Conferences 53, 01056 (2016)
DOI: 10.1051/matecconf/20165301056

Mechatronic Device for Elbow Rehabilitation

Davide Fausti[a] and Gianluigi Petrogalli

[1]*Polibrixia, R&D Department, 25123 Brescia, Italy*

Abstract. This work proposes a mechatronic device for elbow rehabilitation. The realized therapy is based on continuous passive motion. A prototype was realized and design and dimensioning is here presented. Preliminary tests will be performed on non-pathological subjects to allow clinical experiments.

1 Submitting the manuscript

The advantages obtained by using automatic devices for rehabilitation include the ability to gain quantitative information about therapies in order to adapt it to the particular needs of patients [1].
Thanks to these mechatronic devices [2, 3], there is also the possibility of performing the therapy and the execution of exercises remaining autonomously at home or under a physiotherapist's supervision, which controls the precision of movements performed by patients [4].
Although the technology of communication systems and computer equipment are widely spread among the people of all ages, they are still subject to strong psychological resistance, especially by therapists [5].
The proposed device is classified as an automatic system suitable for both rehabilitation in passive mode (CPM, the person is totally "accompanied" by the machine), and in the active mode for patients [3, 6].

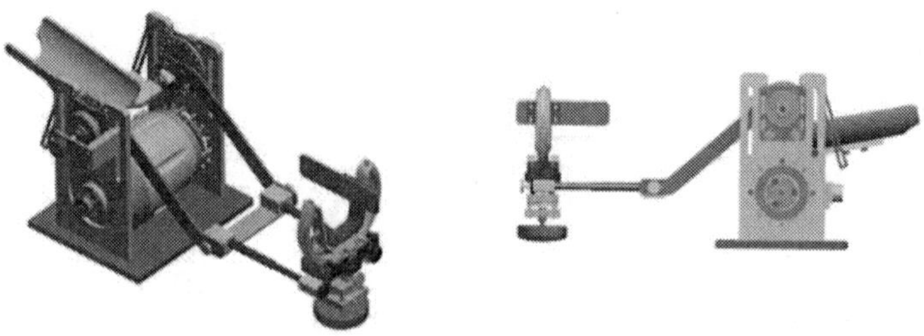

Figure 1. The mechanical device: CAD model.

Thanks to results obtained from preliminary tests (shown in paragraph 3) functional requirements have been defined for the device's control [7]. The definition of motion algorithms and their implementation are under development [8-12].

[a] Corresponding author : davide.fausti@polibrixia.it

The objective is to provide the physiotherapist with a device that, in addition to reproducing the most used types of exercise in a precise [13-16] manner, allows to test new rehabilitation techniques and to quantify the results [17].

2 Description of the mechanical structure

The device has an architecture consisting of exoskeleton structure [18]. There are two vertical side plates fixed to a base support, between which, in the lower part, there is a direct drive brushless motor. In the upper part of the plates are hinged two parallel rods, which have the function to link with the appendix for the support of the wrist.
The patient puts the upper arm into the support positioned at the rear of the device (see the following Fig.2, number 1) and, using a special orthopedic glove, engages the wrist to the support positioned in front of the device (number 2, in figure).

Figure 2. The mechanical device: CAD model.

The wrist support has a semicircular guide, that can rotate around its own axis, so as to allow the pronation-supination hand's movement.

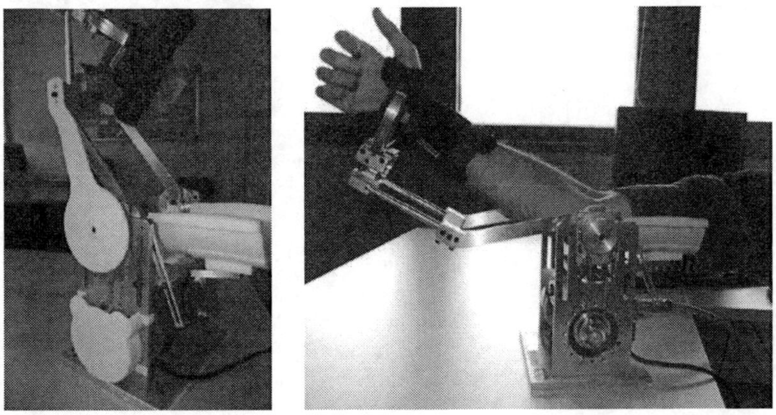

Figure 3. First prototype of the device.

The flexion-extension elbow's movement, as shown in following figure, is obtained thanks to a synchronous belt drive, having a τ=0.2 transmission ratio.

Figure 4. The transmission for flexion-extension.

Among different types of actuation: piezoelectric [19, 20], electric actuators [21-24], shape memory actuators [2, 16, 25-29] and different other technologies, the choice was for a DC motor.
Pronation-supination movement is realized by a transmission with cables, sliding inside sheaths, and fixed to the ends of circular guide. They are driven by a little winch with DC motor.
This type of transmission [30] gives a good degree of compliance [31, 32] to the system and allows to isolate motor from driven elements. Furthermore, with this architecture, is maintained as low as possible the masses connected to patient's arm, getting the advantage of a small inertia.
The presence of mechanical stops guarantees limitations to strokes of flexion-extension and pronation-supination movements. Moreover, in order to prevent dangerous situations [33, 34], in case of pronation-supination of the wrist, the device is equipped with a torque limiting clutch.
The transmission for elbow's flexion-extension is reversible, so as to allow the device to operate even in active mode.
The mechanical design has been performed according to ergonomic criteria and in accordance with the anthropometric parameters of the limb. There is a wide adaptation possibility to different articular dimensions.
The following image shows the different device's settings in order to adapt it to patient sizes.

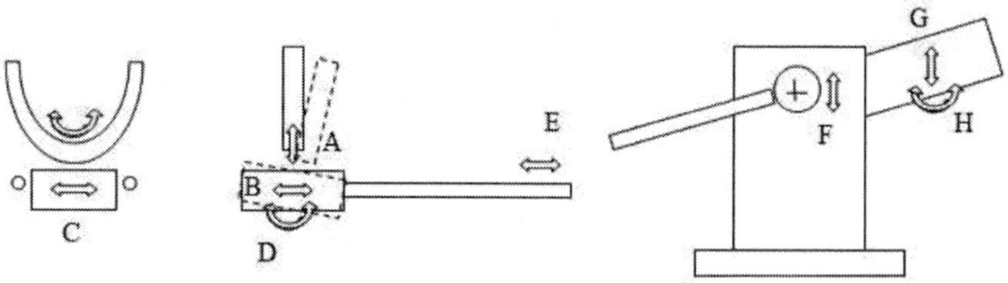

Figure 5. The transmission for flexion-extension.

They are:

A - vertical adjustment on the axis of pronation- supination (when device is stationary);
B - sliding wrist support to prevent a possible compression of the arm (either when the device is stationary or in motion);

C - Transverse scrolling of the wrist support to accommodate the lateral movement of the hand during both flexion- extension and pronation - supination (to use as a setting when the device is stationary or, as a degree of freedom, when the device is working);
D - rotation of the wrist support (degree of freedom);
E - arm length adjustment (at stationary device);
F - axis adjustment of the drive pulley - tensioning system belt (the device is stopped);
G and H - adjustable positioning arm (the device is stopped).
Kinematical [35-41] an dynamical [42-44] aspects of the proposed device are not described in this work, where we prefer put the attention on the possibilities of use. Dimensioning is realized according with EU standard [45-48].

3 Conclusions

After a study of pathologies and associated treatments, we proposed a device for elbow rehabilitation. A prototype was dimensioned and realized according with biomedical standards. Some preliminary tests was realized to verify the possibility of applications on different class of patients exhibiting sufficient results for further researches.

References

[1] T. K. Hong and G. S. Kim, *Journal of Institute of Control, Robotics and Systems,* **19**, 1011-1016, 2013
[2] F. Aggogeri, A. Borboni, and R. Faglia, *Applied Mechanics and Materials,* **373-375**, 130-133, 2013
[3] A. Borboni, R. Faglia, and M. Mor, *ASME 2014 12th Biennial Conference on Engineering Systems Design and Analysis ESDA 2014,* **1**, 1-10, 2014
[4] A. Karime, M. Eid, H. Dong, M. Halimi, W. Gueaieb, and A. E. Saddik, *IEEE Transactions on Instrumentation and Measurement,* **64**, 427-438, 2015
[5] G. Taveggia, J. H. Villafañe, F. Vavassori, C. Lecchi, A. Borboni, and S. Negrini, *Journal of Manipulative and Physiological Therapeutics,* **37**, 242-252, 2014
[6] C. Amici, A. Borboni, R. Faglia, D. Fausti, and P. L. Magnani, *IEEE/RSJ International Conference on Intelligent Robots and Systems IROS,* **2008**, 735-740, 2008
[7] J. Adams, J. Roberts, K. Simms, D. Cheng, J. Hartman, and C. Bartlett, *American Journal of Cardiology,* **103**, 762-765, 2009
[8] A. Borboni, M. Lancini, and R. Faglia, *ASME 2014 12th Biennial Conference on Engineering Systems Design and Analysis ESDA 2014,* **2**, 1-7, 2014
[9] A. Borboni, R. Bussola, R. Faglia, P. L. Magnani, and A. Menegolo, *Journal of Mechanical Design, Transactions of the ASME,* **130**, 0823011-0823016, 2008
[10] N. Vatin, N. Lavrov, A. Shipilov, Procedia Engineering, **117 (1)**, 374-380 (2015)
[11] B. Stamatovic, R. Upadhyay, and N. Vatin, *Procedia Engineering*, 2015, 660-667
[12] B. Melović, S. Mitrović, A. Djokaj, and N. Vatin, *Procedia Engineering*, 2015, 807-812
[13] F. Aggogeri, A. Borboni, A. Merlo, and N. Pellegrini, *Advanced Materials Research,* **590**, 252-257, 2012
[14] A. Borboni, F. Aggogeri, N. Pellegrini, and R. Faglia, *Advanced Materials Research,* **590**, 399-404, 2012
[15] F. Aggogeri, A. Borboni, R. Faglia, A. Merlo, and S. De Cristofaro, *Applied Mechanics and Materials,* **336-338**, 1170-1173, 2013
[16] A. Borboni and R. Faglia, *Procedia Engineering*, 2014, 1378-1381
[17] G. C. Burdea and C. Defais, *Proceedings - IEEE Virtual Reality*, 2013, 99-100
[18] S. K. Manna and S. Bhaumik, *Journal of Robotics,* **2013**, 2013
[19] A. Borboni, D. De Santis, and R. Faglia, *Proceedings of 8th Biennial ASME Conference on Engineering Systems Design and Analysis ESDA2006,* **2006**, 1-8, 2006

[20] A. Borboni and R. Faglia, *Journal of Applied Mechanics Transactions ASME,* **80**, 1-7, 2013

[21] S. F. Bart, T. A. Lober, R. T. Howe, J. H. Lang, and M. F. Schlecht, *Sensors and Actuators,* **14**, 269-292, 1988

[22] I. Boldea and S. A. Nasar, *IEEE Transactions on Energy Conversion,* **14**, 712-717, 1999

[23] K. Nakanoa, Y. Suda, and S. Nakadai, *Journal of Sound and Vibration,* **260**, 213-235, 2003

[24] C. Q. Ru, X. Mao, and M. Epstein, *Journal of the Mechanics and Physics of Solids,* **46**, 1301-1318, 1998

[25] A. Borboni, R. Faglia, and M. Palpacelli, *MESA 2014 - 10th IEEE/ASME International Conference on Mechatronic and Embedded Systems and Applications,* **1**, 1-7, 2014

[26] S. Pandini, A. Borboni, I. Bodini, D. Vetturi, D. Cambiaghi, K. Paderni, *et al., AIP Conference Proceedings,* **1599**, 306-309, 2014

[27] S. Pandini, T. Riccò, A. Borboni, I. Bodini, D. Vetturi, D. Cambiaghi, *et al., Journal of Materials Engineering and Performance,* **23**, 2545-2552, 2014

[28] M. Tiboni, A. Borboni, M. Mor, and D. Pomi, *Proceedings of the Institution of Mechanical Engineers. Part I: Journal of Systems and Control Engineering,* **225**, 443-451, 2011

[29] A. Borboni, F. Aggogeri, N. Pellegrini, and R. Faglia, *Advanced Materials Research,* **590**, 405-410, 2012

[30] Z. L. Sun, L. Y. Chen, Y. Zhang, and J. Y. Ding, *Dongbei Daxue Xuebao/Journal of Northeastern University,* **24**, 548-551, 2003

[31] A. Borboni, D. De Santis, and R. Faglia, *ASME 2010 10th Biennial Conference on Engineering Systems Design and Analysis ESDA2010,* **2**, 99-106, 2010

[32] A. Borboni and D. De Santis, *Meccanica,* **49**, 1327-1336, 2014

[33] S. Nikolskiy, N. Vatin, O. Pertseva (2015) Journal of Applied Engineering Science, **13 (1)**, 11-18 (2015

[34] Z.Popović, L. Lazarevic, N. Vatin, Procedia Engineering, **117 (1)**, 846-853 (2015)

[35] A. Borboni, *Proc of the IEEE International Conference on Fuzzy Systems,* **1**, 336-339, 2001

[36] G. Resconi, A. Borboni, R. Faglia, and M. Tiboni, *Lecture Notes in Computer Science,* **2178 LNCS**, 352-368, 2001

[37] C. Amici, A. Borboni, P. L. Magnani, and D. Pomi, *Proceedings of EUCOMES 2008 - The 2nd European Conference on Mechanism Science,* **2008**, 479-485, 2009

[38] C. Amici, A. Borboni, and R. Faglia, *Advances in Mechanical Engineering,* **2**, 1-9, 2010

[39] A. Borboni, F. Aggogeri, and R. Faglia, *ASME 2014 12th Biennial Conference on Engineering Systems Design and Analysis ESDA 2014,* **3**, 1-10, 2014

[40] A. Borboni, S. Pandini, D. Cambiaghi, M. Lancini, R. Adamini, R. Faglia, *et al., ASME 2014 12th Biennial Conference on Engineering Systems Design and Analysis, ESDA 2014,* **3**, 1-6, 2014

[41] G. Palmieri, M. Callegari, L. Carbonari, and M. C. Palpacelli, *MESA 2014 - 10th IEEE/ASME International Conference on Mechatronic and Embedded Systems and Applications, Conference Proceedings*, 2014

[42] A. Borboni, R. Bussola, R. Faglia, and M. Tiboni, *Proceedings of the 2004 Eleventh World Congress in Mechanism and Machine Science,* **1**, 1-5, 2004

[43] C. Amici, A. Borboni, P. L. Magnani, and D. Pomi, *Proceedings of EUCOMES 2008 - The 2nd European Conference on Mechanism Science,* **2008**, 487-493, 2009

[44] J. H. Villafañe, P. Pillastrini, and A. Borboni, *Journal of Chiropractic Medicine,* **12**, 176-181, 2013

[45] L. Solazzi, R. Scalmana, R. Adamini, R. Faglia, and A. Borboni, *Agricultural Engineering International: CIGR Journal,* **16**, 102-112, 2014

[46] T. Kandikjan, J. J. Shah, and J. K. Davidson, *CAD Computer Aided Design,* **33**, 721-737, 2001

[47] B. W. Olesen, M. Koschenz, and C. Johansson, *ASHRAE Transactions*, 2003, 656-668

[48] V.V Okrepilov, V.N. Krutikov, G.I. El'kin, Measurement Techniques, **57 (2)**,109-116 (2014)

MATEC Web of Conferences 53, 01057 (2016)
DOI: 10.1051/matecconf/20165301057

Test Bench Simulator for Heart Valves

Gianluigi Petrogalli[1,a] and Maurizio Mor[1]

[1]*Polibrixia, R&D Department, 25123 Brescia, Italy*

Abstract. The objective of this research is the creation of a testing bench prototype for comparing different types of heart valves, taking into account the variability of some important parameters involved in cardiovascular events. Thanks to performed tests and to obtained data, the comparison between the different valves allows physicians to make the most appropriate choice of heart valve for each patient. The study is still in preliminary stage: it is necessary to do many tests in order to obtain a large amount of data with which it is possible to proceed with the development and implementation of a testing bench stable and easily programmable.

1 Introduction

The objective of this research is the creation of a testing bench prototype for comparing different types of heart valves, taking into account the variability of some important parameters involved in cardiovascular events [1].
The high simplicity of the device allows a direct pressure and flow acquisition and processing , thanks to an appropriate software.
A linear motor is used to simulate the heartbeat and, thanks to changes in the parameters of a PID controller [2], is calibrated to reproduce the law of motion [3-7] taken from the literature [1]: this law was made more suitable for our use eliminating discontinuities and traits not connected.
For acquiring pressures and flows is adopted a measurement system that complies with UNI EN ISO 5840:2006 "Cardiovascular implants - Cardiac valve prostheses" [8].

2 The blood model

The blood is a connective tissue composed of both liquid part and particles that are in suspension. The first, called emulsion part, is represented by the plasma, and the second, called suspension part, includes red and white blood cells.
The plasma is commonly treated as a homogeneous and incompressible fluid with newtonian behavior. It has a dynamic viscosity of 0.0012 - 0.0016 kg / ms at 37 ° C and a specific weight equal to 1025-1028 times compared to water [9].
On the other hand the blood, with all its elements, cannot be considered as absolutely newtonian: from literature [10-12] it is known that its behavior depends on both the section of the duct in which it flows and heartbeat frequency.

[a] Corresponding author : gianluigi.petrogalli@polibrixia.it

At low frequencies (<10 s-1) it is non-newtonian, following the laws of Bingham [9] at high frequencies (60-100s-1) is typically newtonian and for midrange frequencies is thixotropic (liquid-gel transitions may occur).
During normal function of cardiovascular system, we can consider the blood as essentially newtonian with a viscosity of approximately four times larger than water.

Figure 1. The artificial aortic valve installed in the test bench: the section has a diameter of 23 mm. The valve is manufactured from SulzerMedica ®.

For the development of testing bench and for the design of various components is used the mathematical model of Walburn - Schneck (1), whose constitutive law of viscosity of the blood is shown below.

$$\mu = c_1 \cdot e^{c_2 \cdot Ht} \left[e^{c_4 \cdot \left(\frac{TPMA}{Ht^2} \right)} \right] \cdot \dot{\gamma}^{-c_3 \cdot Ht} \tag{1}$$

Where:

-μ is the blood's viscosity;
-TPMA represents the content of proteins, except albumin;
-Ht is the percentage of hematocrit;
-c1, c2, c3 and c4 are characteristic constants (tabulated);
-ϒ typical value depending on pressure, flow and geometric dimensions of the duct.

The normal Ht values for a man must be included in the range 42% -53%, while for a woman between 37% and 52%: we chose a Ht value 0.399. The parameter TPMA was set equal to 0.0032, selecting 8% of blood's protein of the plasma constituents and considering that albumin represents 60% of total proteins [13].
The great advantage of this law is that allows to compare physiologic parameters [14], such as the hematocrit or TPMA [1], with mechanical parameters as the viscosity μ of the blood, the pressure and

flow. The test bench is calibrated on the basis of parameters that are easily recognizable by doctors: we can also implement a programming mode using the instruments for the measurement of blood parameters.

3 Testbench

The primary target of testing bench is to acquire the variables that describe the dynamic behavior of the fluid during the heartbeat [8].
In order to optimize the functionality of the device, in the design phase, was considered the real flow dynamics blood pumped by the heart [1]; with this approach, the attention is focused only on the most interesting aspects.
The tests were performed on a mechanical aortic valve, which is usually located in the left part of the heart and more precisely in the left ventricle.
The left ventricle, isolated from the rest of the organ, can be modeled as a simple effect pump with inlet and outlet of the fluid controlled by two one-way valves [15-22].
As shown in Fig. 1 and Fig. 2, the heartbeat is reproduced using a linear motor [23-28] connected to a hydraulic cylinder with double effect: in this way, with a suitable law of motion is possible to flow the fluid within the circuit, and to adjust the heart's frequency.

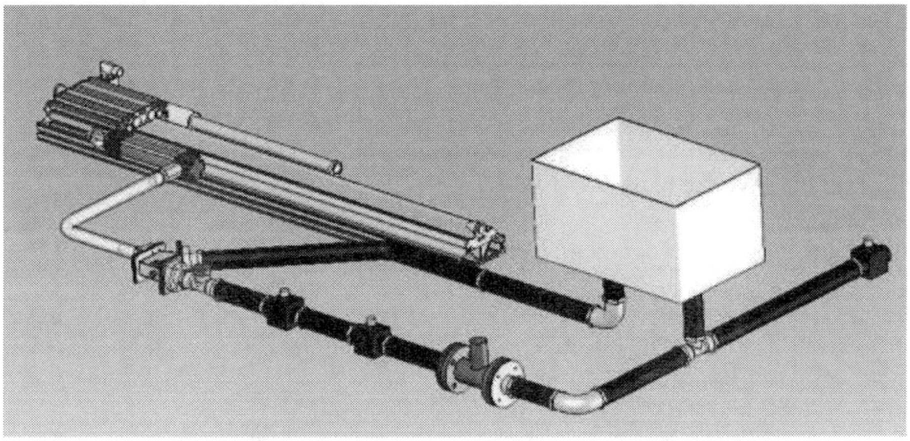

Figure 2. The design of testing bench consists of a linear motor, "heart" group and sensors.

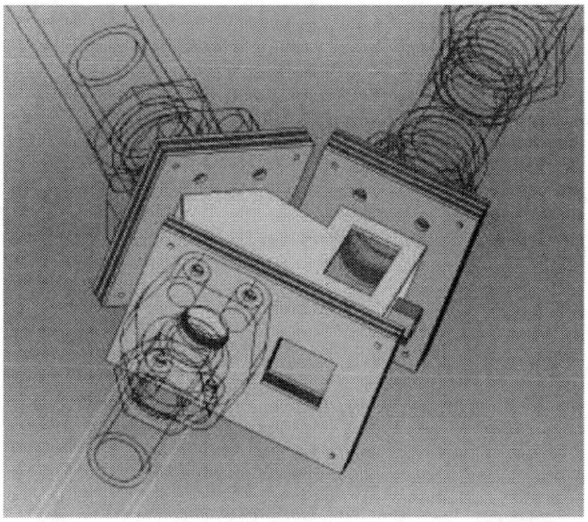

Figure 3. Detail of "heart" group, made of aluminum or plastic, in which the commercial heart valve is positioned.

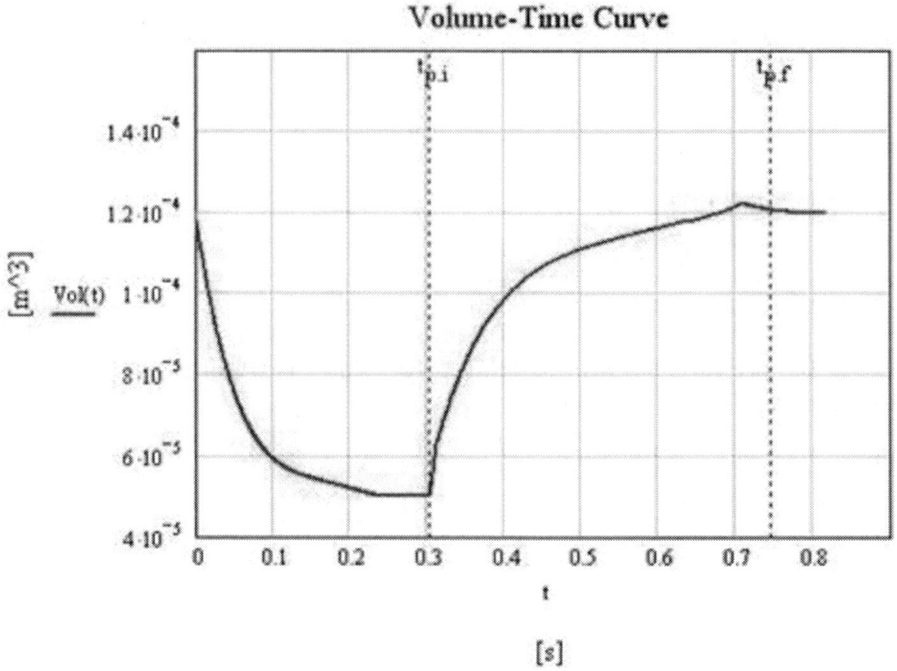

Figure 4. Curve of volume of fluid through the left ventricle experimentally acquired.

The parameters considered for linear motor choice are:
-the force required to move the blood's mass;
-the speed of fluid;
-the necessary acceleration for movement.

Fig. 4 shows a typical time-fluid volume curve for the left ventricle of a human heart [1]: it's easy to observe that there are discontinuities in transition from suction phase to sending phase and therefore it cannot be set directly to linear motor [29-32].
It was necessary to interpolate it through fourth order polynomials.
Under the literature advice [33-43], the linear motor should run easily with a law of motion as smooth as possible in order to avoid vibrations and other undesired effects associated with the presence of compliances and inertias [24, 44-46].
For measuring the pressure and flow were used respectively the sensor SEN-3284 B045 by KOBOLD ® and the flowmeter FLOMAG 3000 (model A3); signals were acquired with NI USB-6009 by National Instrument and signal processing was performed with a dedicated program developed with Labview [47].

4 Experimental results and conclusions

Kinematical [48-55] and dynamical [56-58] aspects of the system are related to the linear motor's calibration [59, 60] that is performed by changing PID parameters and by overlapping the obtained law of motion on the desired one, for a heart rate of 60 beats per minute.

When the deviation between two laws is acceptable, durability tests are performed by varying heart's frequency or by setting a random frequency and then by analyzing the heart valve's behaviour.

Since the obtained law of motion was composed of five parts connected by means of a polynomial curve, the algorithm for automatic calibration was not easily obtainable. For these reasons it was decided to proceed experimentally by adjusting first of all the proportional and integrative components and then the derivative part.

A not correct viscosity of the fluid, or an inadequate law of motion can cause undesired effects that generate instability within the device, until the engine failure [61].

In executed tests a motor's stroke deviation of 1mm was considered acceptable: in fact, a greater error could create vibrations, abnormal openings of the valve and errors in measuring instruments.

The objective of this research is to carry out a testing bench for various commercial heart valves , by simulating the heartbeat through a linear motor and by adjusting the law of motion with respect to the fluid properties.

Thanks to performed tests and to obtained data, the comparison between the different valves allows the doctors to make the most appropriate choice of heart valve for each patient.

The study is still in preliminary stage: it is necessary to do many tests in order to obtain a large amount of data with which it is possible to proceed with the development and implementation of a testing bench stable and easily programmable.

References

[1] J. C. Shryock and L. Belardinelli, *American Journal of Cardiology,* **79**, 2-10, 1997

[2] A. Emadi, Y. J. Lee, and K. Rajashekara, *IEEE Transactions on Industrial Electronics,* **55**, 2237-2245, 2008

[3] A. Borboni, R. Bussola, R. Faglia, and M. Tiboni, *Proceedings of the 2004 Eleventh World Congress in Mechanism and Machine Science,* **1**, 1-5, 2004

[4] M. Antonini, A. Borboni, R. Bussola, and R. Faglia, *Proceedings of 8th Biennial ASME Conference on Engineering Systems Design and Analysis ESDA2006,* **2006**, 1-8, 2006

[5] M. Antonini, A. Borboni, R. Bussola, and R. Faglia, *Proceedings of 8th Biennial ASME Conference on Engineering Systems Design and Analysis ESDA2006,* **2006**, 1-8, 2006

[6] A. Borboni, R. Bussola, R. Faglia, P. L. Magnani, and A. Menegolo, *Journal of Mechanical Design, Transactions of the ASME,* **130**, 0823011-0823016, 2008

[7] A. Borboni, M. Lancini, and R. Faglia, *ASME 2014 12th Biennial Conference on Engineering Systems Design and Analysis ESDA 2014,* **2**, 1-7, 2014

[8] B. Heublein, R. Rohde, V. Kaese, M. Niemeyer, W. Hartung, and A. Haverich, *Heart,* **89**, 651-656, 2003

[9] J. Shih, H. J. Kim, and C. A. Taylor, *Proceedings of the ASME Summer Bioengineering Conference, SBC2008*, 2009, 317-318

[10] G. Gerosa, V. Tarzia, G. Rizzoli, and T. Bottio, *Journal of Thoracic and Cardiovascular Surgery,* **131**, 1058-1064.e2, 2006

[11] A. Giovannelli, G. Alessandri, A. Rama, I. Gandjbakhch, G. Rondoni, and N. Alessandri, *European Review for Medical and Pharmacological Sciences,* **12**, 303-308, 2008

[12] M. K. Sharp and R. K. Dharmalingam, *ASAIO Journal,* **45**, 535-540, 1999

[13] M. Koeppen, T. Eckle, and H. K. Eltzschig, *Anesthesiology,* 2015

[14] G. Taveggia, J. H. Villafañe, F. Vavassori, C. Lecchi, A. Borboni, and S. Negrini, *Journal of Manipulative and Physiological Therapeutics,* **37**, 242-252, 2014

[15] M. Tiboni, A. Borboni, M. Mor, and D. Pomi, *Proceedings of the Institution of Mechanical Engineers. Part I: Journal of Systems and Control Engineering,* **225**, 443-451, 2011

[16] G. H. Feng and E. S. Kim, *Journal of Micromechanics and Microengineering,* **14**, 429-435, 2004

[17] A. Borboni, F. Aggogeri, and R. Faglia, *International Journal of Advanced Robotic Systems,* **10**, 1-10, 2013

[18] P. Q. Su and Z. H. Deng, *Applied Mechanics and Materials* **318**, 15-20, 2013

[19] V.V Okrepilov, V.N. Krutikov, G.I. El'kin, Measurement Techniques, **57 (2)**,109-116 (2014)

[20] V.V Okrepilov, Studies on Russian Economic Development, **24(1)**, 35-42 (2013)

[21] V.V Okrepilov, Standarty i Kachestvo, **10**, 52-55 (2005)

[22] G. I. Grinfeldi, A. S. Gorshkov, and N. I. Vatin, *Advanced Materials Research* **941-944**, 7

[23] A. Borboni, F. Aggogeri, N. Pellegrini, and R. Faglia, *Advanced Materials Research,* **590**, 405-410, 201286-799, 2014

[24] F. Aggogeri, A. Borboni, and R. Faglia, *Applied Mechanics and Materials,* **373-375**, 130-133, 2013

[25] A. Borboni, R. Faglia, and M. Palpacelli, *MESA 2014 - 10th IEEE/ASME International Conference on Mechatronic and Embedded Systems and Applications,* **1**, 1-7, 2014

[26] A. Borboni, S. Pandini, D. Cambiaghi, M. Lancini, R. Adamini, R. Faglia, *et al., ASME 2014 12th Biennial Conference on Engineering Systems Design and Analysis, ESDA 2014,* **3**, 1-6, 2014

[27] S. Pandini, A. Borboni, I. Bodini, D. Vetturi, D. Cambiaghi, K. Paderni, *et al., AIP Conference Proceedings,* **1599**, 306-309, 2014

[28] S. Pandini, T. Riccò, A. Borboni, I. Bodini, D. Vetturi, D. Cambiaghi, *et al., Journal of Materials Engineering and Performance,* **23**, 2545-2552, 2014

[29] G. D. Buckberg, M. Castellá, M. Gharib, and S. Saleh, *European Journal of Cardio-thoracic Surgery,* **29**, S98-S106, 2006

[30] C. Dragan, A. Samuila, S. Das, D. Iancu, M. Bilici, and L. Dascalescu, *Journal of Electrostatics,* **67**, 184-188, 2009

[31] H. Mamori, K. Fukagata, and J. Hoepffner, *Physical Review E - Statistical, Nonlinear, and Soft Matter Physics,* **81**, 2010

[32] H. Rau, S. Brody, C. Droste, and A. Kardos, *European Journal of Applied Physiology and Occupational Physiology,* **67**, 26-29, 1993

[33] A. Borboni, D. De Santis, and R. Faglia, *Proceedings of 8th Biennial ASME Conference on Engineering Systems Design and Analysis ESDA2006,* **2006**, 1-8, 2006

[34] A. Borboni, F. Aggogeri, N. Pellegrini, and R. Faglia, *Advanced Materials Research,* **590**, 399-404, 2012

[35] F. Aggogeri, A. Borboni, R. Faglia, A. Merlo, and S. De Cristofaro, *Applied Mechanics and Materials,* **336-338**, 1170-1173, 2013

[36] A. Borboni and R. Faglia, *Journal of Applied Mechanics Transactions ASME,* **80**, 1-7, 2013

[37] A. Borboni and R. Faglia, *Procedia Engineering*, 2014, 1378-1381

[38] Z. Guoyong, Z. Yugang, and H. Ronggguo, *Applied Mechanics and Materials* **29-32**, 2002-2007, 2010

[39] M. H. Korayem, M. Nazemizadeh, and H. R. Nohooji, *International Journal of Advanced Robotic Systems,* **9**, 2012

[40] T. S. Lee and Y. J. Lin, *International Journal of Advanced Manufacturing Technology,* **15**, 640-648, 1999

[41] J. Mattmüller and D. Gisler, *International Journal of Advanced Manufacturing Technology,* **45**, 1007-1016, 2009

[42] I. Storey, A. Bourmistrova, and A. Subic, *Proceedings of the Institution of Mechanical Engineers, Part C: Journal of Mechanical Engineering Science,* **226**, 2656-2673, 2012

[43] A. Borboni and M. Lancini, *Journal of Vibration and Acoustics, Transactions of the ASME,* **137**, 1-9, 2015

[44] A. Borboni, D. De Santis, and R. Faglia, *ASME 2010 10th Biennial Conference on Engineering Systems Design and Analysis ESDA2010,* **2**, 99-106, 2010

[45] A. Borboni and D. De Santis, *Meccanica,* **49**, 1327-1336, 2014

[46] A. Borboni, R. Faglia, and M. Mor, *ASME 2014 12th Biennial Conference on Engineering Systems Design and Analysis ESDA 2014,* **1**, 1-10, 2014

[47] M. Tsukiji, N. Watanabe, Y. Yamaura, N. Okahashi, K. Obase, Y. Neishi, *et al., Journal of the American Society of Echocardiography,* **21**, 43-46, 2008

[48] A. Borboni, *Proc of the IEEE International Conference on Fuzzy Systems,* **1**, 336-339, 2001

[49] G. Resconi, A. Borboni, R. Faglia, and M. Tiboni, *Lecture Notes in Computer Science,* **2178 LNCS**, 352-368, 2001

[50] C. Amici, A. Borboni, P. L. Magnani, and D. Pomi, *Proceedings of EUCOMES 2008 - The 2nd European Conference on Mechanism Science,* **2008**, 479-485, 2009

[51] A. Borboni, F. Aggogeri, and R. Faglia, *ASME 2014 12th Biennial Conference on Engineering Systems Design and Analysis ESDA 2014,* **3**, 1-10, 2014

[52] J. Denton and J. B. Cronin, *Journal of Strength and Conditioning Research,* **20**, 528-534, 2006

[53] A. C. Fisher, *Clinical Hemorheology,* **5**, 61-69, 1985

[54] G. Guidoboni, R. Glowinski, N. Cavallini, S. Canic, and S. Lapin, *Applied Mathematics Letters,* **22**, 684-688, 2009

[55] T. T. Nguyen, Y. Biadillah, R. Mongrain, J. Brunette, J. C. Tardif, and O. F. Bertrand, *Journal of Biomechanical Engineering,* **126**, 529-535, 2004

[56] C. Amici, A. Borboni, R. Faglia, D. Fausti, and P. L. Magnani, *IEEE/RSJ International Conference on Intelligent Robots and Systems IROS,* **2008**, 735-740, 2008

[57] C. Amici, A. Borboni, P. L. Magnani, and D. Pomi, *Proceedings of EUCOMES 2008 - The 2nd European Conference on Mechanism Science,* **2008**, 487-493, 2009

[58] C. Amici, A. Borboni, and R. Faglia, *Advances in Mechanical Engineering,* **2**, 1-9, 2010

[59] H. H. Mu, Y. F. Zhou, X. Wen, and Y. H. Zhou, *Mechatronics,* **19**, 577-585, 2009

[60] F. Aggogeri, A. Borboni, A. Merlo, and N. Pellegrini, *Advanced Materials Research,* **590**, 252-257, 2012

MATEC Web of Conferences **53**, 01058 (2016)
DOI: 10.1051/matecconf/20165301058

Review on Land-Based Wheeled Robots

Himanshu R. Rasam [1,a]

[1] *IIT Bhubaneswar, India*

Abstract. This paper presents an overview of the state of art of mobile robots presenting aspects of their current level of sophistication, fields mobile robots are being presently used, the future application and scope of mobile robots. Classifying the primary research topics of multi-robot systems into seven major categories (Biological inspirations, architecture, localization/mapping/exploration, object transport and manipulation, motion co-ordination, reconfigurable and learning robot systems, vision), we discuss the current state of research in these areas. An attempt to cover the points for which a general agreement has emerged within the scientific domain, are then presented, followed by a discussion of alternative approaches and currently unsolved problems. This paper also reviews the most relevant literature and previous research activity regarding mobile robotics. We conclude by identifying more topics of scientific consensus, open research issues and a discussion of some application areas where mobile robots could be used with existing technology.

1 Introduction

Literature [1] defines Mobile Robots as robots that can move from one place to another autonomously, that is, without assistance from external human operators. The field of autonomous mobile robots can be considered immature from research point of view. Unlike a fix based industrial robot [2-5], a mobile robot has a greater freedom for movement. The perks of mobile robots over conventional fixed/industrial robots are
i) Ability to operate in a larger workspace.
ii) Ability to explore unknown environments.
iii) Ability to co-ordinate with other robot/ group of robots to perform a single integrated task.
Due to such abilities mobile robots can substitute humans in many fields by carrying out the functions of exploration, patrolling, surveillance, home appliances (e.g. lawn mower and floor cleaner), butler etc. [1].There has been an explosive growth in the research of mobile robots in the last decade. Such mobile robots can be broadly classified into three categories based on environments they operate in a] Land/ ground based robots b] aquatic/ underwater robots c] aerial robots [6].
We would be discussing only about the type(a)- 'Land based mobile robots' in this paper. As they are designed to operate in unknown environments, mobile robots demand a much higher level of intelligence than the conventional industrial robots. This is now possible due to advancements in silicon technology and computing power. A field defined as cooperative mobile robotics has also emerged in the recent years which uses a combination of a number of mobile robots in coordination to

[a] Corresponding author : hrr10@iitbbs.ac.in

execute various tasks. Most of the work in cooperative mobile robotics began after the introduction of the new robotics paradigm of behavior based control [4]. Many cooperative robotics researchers find examining the social characteristics of insects and animals helpful while designing such cooperative mobile robots. Mobile robots find a huge application in military and defense. They are used for various purposes such as surveillance in enemy territory, mine detection and defusing, hostage rescue scenarios and as a substitute for soldiers.

A great segment of research in mobile robotics is dedicated to development of architectures, task planning capabilities and control. Developing an architectural structure for multi-robot systems has is being focused upon greatly [7].

Localization, Mapping and Exploration are classified as the major primary operations of a mobile robot (Land). The single and multi- robot approach to localization, mapping and exploration can be described using categories based on use of landmarks, scan-matching, use of graphs, range sensors and vision sensors [8]. One widely discussed question about single and multiple mobile robot systems is the effectiveness of multi-robot teams over single mobile robot. This question is currently being taken into study but much remains to be determined for the variety of approaches and methods for localization, mapping and exploration.

Some of the earlier work in robotics which was focused on reconfigurable distributed systems is now being researched further with an aim to achieve function from shape, allowing individual modules, or robots, to connect and re-connect in various ways to generate a desired shape to a particular operation. Theoretically speaking, these systems have a capability of showing great robustness, versatility, and even self-repair [8].

Each and every aspect of the topic would be broadly analyzed in the upcoming sections.

2 Types of Mobile Robots

The types of robots possible are unlimited, though we classify the mobile robots broadly into these categories:

1. Land-based wheeled robot
2. Land-based tracked robot
3. Land-based legged robot
4. Air-based: Plane, helicopter, quadcopter, drones
5. Water-based: Boat, submarine
6. Hybrid/combination Robot
7. Stationary robot (arm/manipulator)

Since we are only discussing about the Land-based robots, we will study only the first three categories in detail.

3 Land-based wheeled robots

The Wheeled mobile robots which operate on land/surface [9, 10] can be broadly classified into two categories

i) robots with locally restricted mobility- (i.e. with Non-holonomic constraints)

ii) robots with full mobility - (Omni-directional robots)

Mobile robot subsystems are: Locomotion [11] (concerning how the robot moves through the environment), Sensing (concerning how the robot measures properties of itself and its environment), Control (concerning how a robot generates physical actions), Reasoning (concerning how the robot maps measurements into action), and Communications (concerning how the robot communicates with other robots or with an outside operator).

Operating Environments: An operating environment for a robot falls into one of these three categories.

- •Structured Environments: The robot has a complete knowledge of the environment allowing an offline planning of the motion (due to availability of a map possibly acquired by the robot sensors in an exploratory phase).
- •Semi-structured Environments: The robot has incomplete knowledge of the environment thus allowing motion global planning but local flexibility has to be provided due to the presence of dynamic obstacles.
- •Unstructured Environments: The robot has a very limited environmental knowledge and has to face a wide variety of obstacles/situations [12].

The structured/semi-structured environments are also termed as Indoor (2-D) environments while the unstructured environments are termed as Outdoor (3-D) environments.

A basic motion problem for a land based wheeled robot is to plan its path through static and dynamic objects and to reach its destination [13]. A user may program the robot to travel via various paths such as the least time path or the least cost path. Many times there are multiple mobile robots involved in the task to be performed and they need to co-ordinate and work in unison to complete the task with a higher efficiency [14, 15].

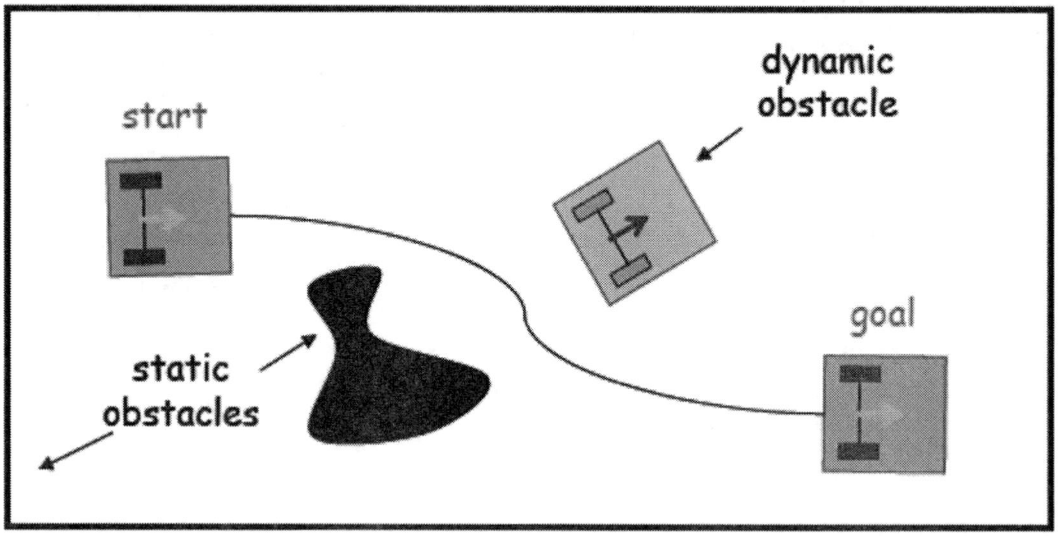

Figure 1. Basic motion problem.

The following are the elementary motion tasks executed by a wheeled robot

- ➢ Point to Point motion
- ➢ Path following
- ➢ Trajectory tracking
- ➢ Purely reactive (local) motion - [Obstacle avoidance, wall following, target tracking]

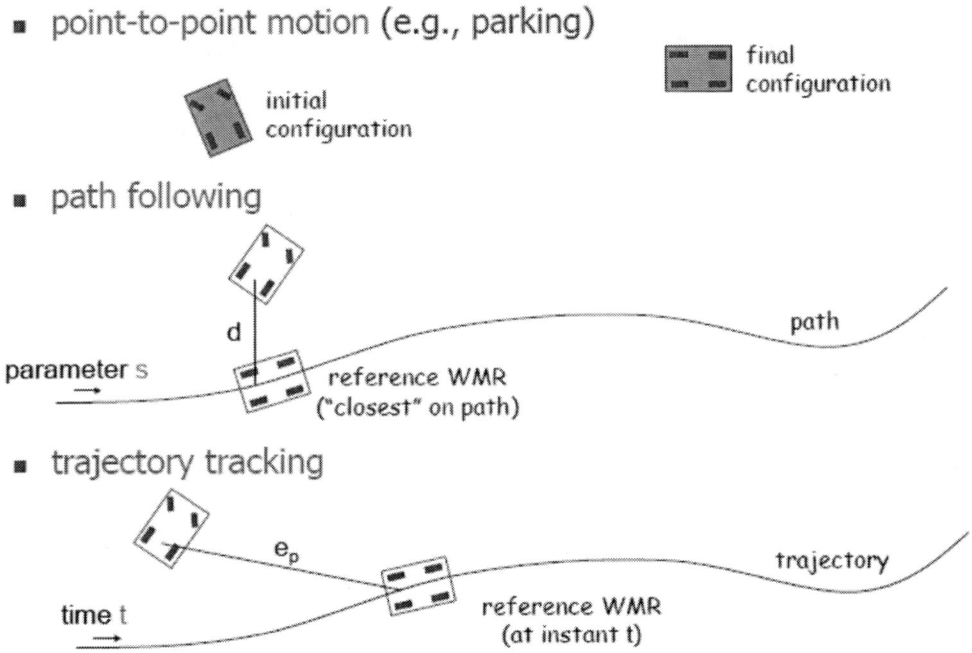

Figure 2. Types of motion.

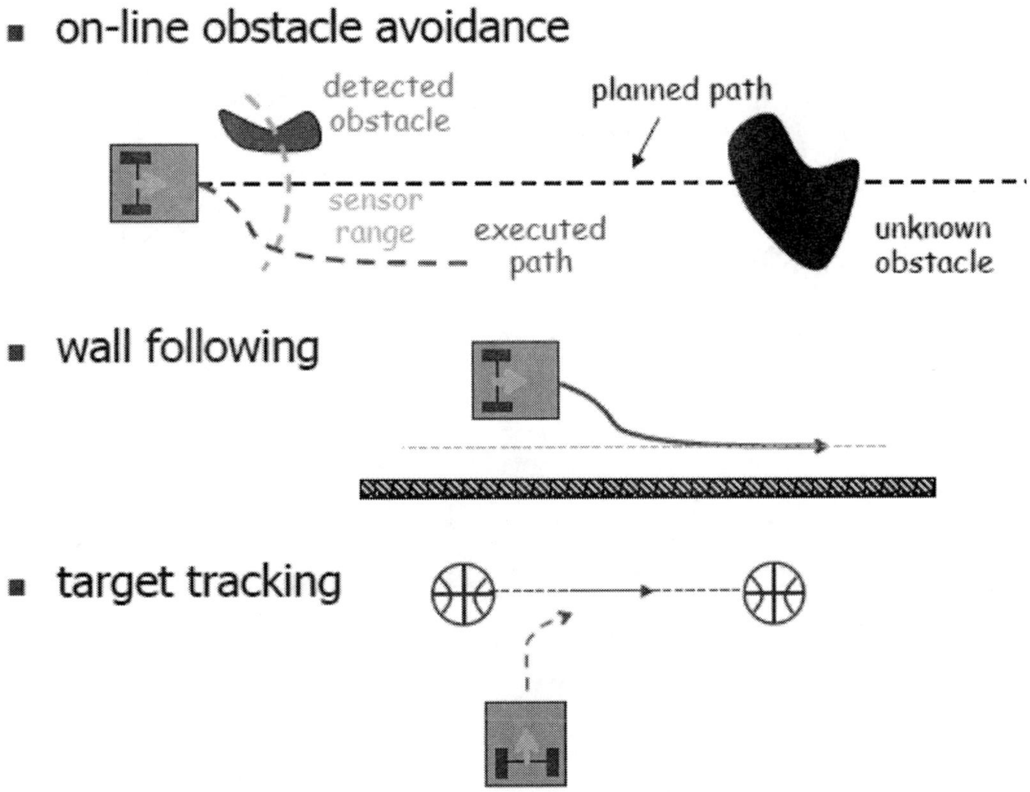

Figure 3. Examples of Reactive Motion.

3.1 Wheels of a WMR

The wheels of a WMR (Wheeled Mobile Robot) can be classified as conventional (directional) and omnidirectional [16].

Conventional wheels include

- fixed wheel (fixed axis and support)
- centered orient able wheel
- off centered orient able wheel (castor wheel)

Omni-directional wheels include

- Swedish wheel/ Mechanism
- Spherical wheel

Conventional robots have a non-holonomic constraint i.e. they can move in a certain direction (forwards and backwards), but not others (side to side). Generally, the conventional wheel only has a tangential velocity in direction of the desired direction. Omnidirectional wheel on the other hand has a normal velocity in direction perpendicular to the tangential velocity to facilitate the side by side motion of the wheel.

3.2 Modelling of the Wheeled Mobile Robots

i) Kinematics Constraints [17-22]:

We consider a mechanical system with n generalized coordinate's 'q' subject to 'm' kinematic constraints:

$$A(q)\dot{q} = 0 \tag{1}$$

where: $A \in R^{m \times n}$ is a full rank matrix [1]. Wheeled vehicle and mobile robots involve kinematics constraints like many other kinematic systems.

4 Kinematic Model

Let's consider the kinematics model [23-28] for an autonomous vehicle. The position of the mobile robot in the plane is shown below [29].

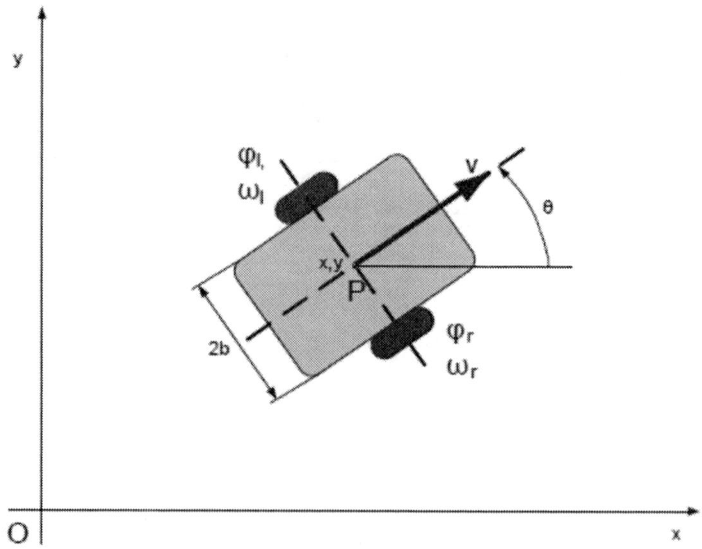

Figure 4. Position of mobile robot in a plane.

The inertial-based frame (Oxy) is fixed in the plane of motion and the moving frame is attached to the mobile robot. We will assume that the mobile robots are rigid cart equipped, with non-deformable conventional wheels, and they are moving on a non-deformable horizontal plane. During the motion: the contact between the wheel and the horizontal plane is reduced to a single point, the wheels are fixed, the plane of each wheel remains vertical, the wheel rotates about its horizontal axle and the orientation of the horizontal axle with respect to the cart can be fixed. The contact between the wheel of the mobile robots and the non-deformable horizontal plane supposes both the conditions of pure rolling and non-slipping during the motion. This means that the velocity of the contact point between each wheel and the horizontal plane is equal to zero. For low rolling velocities this is a reasonable wheel moving model. The center of the fixed wheel is a fixed point of the cart and b is the distance of the center of the wheel from P. The rotation angle of the wheel about its horizontal axle is denoted by $\Phi(t)$ and the radius of the wheel by R. Hence, the position of the wheel is characterized by two constants: b and R and its motion by a time-varying angle: $\Phi_r(t)$ – the rotation angle of the right wheel and $\Phi_l(t)$ – the rotation angle of the left wheel. The configuration of the mobile robot can be described by five generalized coordinates such as:

$$q = [x, y, \theta, \Phi_r, \Phi_l]^T \quad (2)$$

where:

x and y are the two coordinates of the origin P of the moving frame (the geometric center of the mobile robot),

θ is the orientation angle of the mobile robot (of the moving frame),

$\Phi_r(t)$ – the rotation angle of the right driving wheel,

$\Phi_l(t)$ – the rotation angle of the left driving wheel.

The vehicle velocity v can be found in equation (3):

$$v = R(\omega_r + \omega_l)/2 \quad (3)$$

where:

$\omega_r = \frac{d\Phi_r}{dt}$ - angular velocity of the right wheel,

$\omega_l = \frac{d\Phi_l}{dt}$ - angular velocity of the left wheel,

The position and the orientation of the mobile vehicle are determined by a set of differential equations (4-6) in the following form:

$$\dot{x} = (R\cos\theta\,(\omega_r + \omega_l))/2 \quad (4)$$

$$\dot{y} = (R\sin\theta\,(\omega_r + \omega_l))/2 \quad (5)$$

$$\dot{\theta} = R\,(\omega_r - \omega_l)/2b \quad (6)$$

Here,

$$\dot{x} = v\cos\theta \quad (7)$$

$$\dot{y} = v\sin\theta \quad (8)$$

Then the matrix form is:

$$\begin{bmatrix}\dot{x}\\ \dot{y}\\ \dot{\theta}\end{bmatrix} = \begin{bmatrix}\cos\theta & 0\\ \sin\theta & 0\\ 0 & 1\end{bmatrix}\begin{bmatrix}v\\ \dot{\theta}\end{bmatrix} \quad (9)$$

Finally the kinematics model of the vehicle velocity v and the angular velocity $\dot{\theta}$ can be represented by the matrix as follows:

$$\begin{bmatrix}v\\ \dot{\theta}\end{bmatrix} = \begin{bmatrix}R/2 & R/2\\ R/2b & -R/2b\end{bmatrix}\begin{bmatrix}\omega_r\\ \omega_l\end{bmatrix} \quad (10)$$

In this case, we now consider the mobile robot motion as a non-holonomic mechanical system, where three kinematics constraints exist:

$$\begin{aligned}\dot{x}\sin\theta - \dot{y}\cos\theta &= 0\\ \dot{x}\cos\theta - \dot{y}\sin\theta &= R\omega_r - b\dot{\theta}\\ \dot{x}\cos\theta + \dot{y}\sin\theta &= R\omega_l + b\dot{\theta}\end{aligned} \quad (11)$$

The constraints can be written in the form (1), where matrix A $\in R^{m \times n}$ (m=3, n=5) can be described as:

$$A = \begin{bmatrix}\sin\theta & -\cos\theta & 0 & 0 & 0\\ \cos\theta & \sin\theta & b & -R & 0\\ \cos\theta & \sin\theta & -b & 0 & -R\end{bmatrix} \quad (12)$$

In this paper the angular velocities of the two wheels of the mobile robot are independently controlled. In this review, we neglect dynamic matters [30-33]: contacts [34, 35], vibrations [36-42], deflections [43-49] and accuracy [50-59] of robot realization and mounting [60-65].

References

[1] C. X.Q., C. Y.Q., and C. J.G., "Mobiles Robots - Past Present and Future," *Mobile Robots - State of the Art in Land, Sea, Air, and Collaborative Missions*, C. X.Q., Ed.: InTech, 2009

[2] G. Resconi, A. Borboni, R. Faglia, and M. Tiboni, *Lecture Notes in Computer Science,* **2178 LNCS**, 352-368, 2001

[3] M. Antonini, A. Borboni, R. Bussola, and R. Faglia, *Proceedings of 8th Biennial ASME Conference on Engineering Systems Design and Analysis ESDA2006,* **2006**, 1-8, 2006

[4] A. Borboni, R. Bussola, R. Faglia, P. L. Magnani, and A. Menegolo, *Journal of Mechanical Design, Transactions of the ASME,* **130**, 0823011-0823016, 2008

[5] F. Aggogeri, A. Borboni, and R. Faglia, *Applied Mechanics and Materials,* **373-375**, 130-133, 2013

[6] R. Fernández, C. Salinas, H. Montes, P. G. De Santos, and M. Armada, *Adaptive Mobile Robotics - Proceedings of the 15th International Conference on Climbing and Walking Robots and the Support Technologies for Mobile Machines, CLAWAR 2012*, 2012, 723-730

[7] T. M. Knasel, *Robotics,* **2**, 149-155, 1986

[8] T. Arai, E. Pagello, and L. E. Parker, *IEEE Transactions on Robotics and Automation,* **18**, 655-661, 2002

[9] N. Gubeljak, J. Predan, D. Kozak, J. Tuma, B. Kovačič, P. Konjatić, J. Sertić, Strojarstvo, **51** (4), 263-271 (2009)

[10] B. Stamatovic, R. Upadhyay, N. Vatin, Procedia Engineering, **117 (1)**, 660-667 (2015)

[11] G. Taveggia, J. H. Villafañe, F. Vavassori, C. Lecchi, A. Borboni, and S. Negrini, *Journal of Manipulative and Physiological Therapeutics,* **37**, 242-252, 2014

[12] J. Xiao, A. Sadegh, M. Elliott, A. Calle, A. Persad, and H. M. Chiu, *IEEE/ASME International Conference on Advanced Intelligent Mechatronics, AIM*, 2005, 438-443

[13] B. Melović, S. Mitrović, A. Djokaj, and N. Vatin, *Procedia Engineering*, 2015, 807-812

[14] V.V Okrepilov, Standarty i Kachestvo, **3**, 94-96. (2003)

[15] V.V Okrepilov, Ivanova, G. Standarty i Kachestvo, **12**, 62-68. (2004)

[16] N. Silva, A. Vale, and L. Baglivo, *ICINCO 2013 - Proceedings of the 10th International Conference on Informatics in Control, Automation and Robotics*, 2013, 48-57

[17] P. Boucher, *Robotica,* **89**, 2014

[18] A. Borboni, *Proc of the IEEE International Conference on Fuzzy Systems,* **1**, 336-339, 2001

[19] C. Amici, A. Borboni, R. Faglia, D. Fausti, and P. L. Magnani, *IEEE/RSJ International Conference on Intelligent Robots and Systems IROS,* **2008**, 735-740, 2008

[20] A. S. Tubaileh, *International Journal of Advanced Manufacturing Technology,* **74**, 1521-1537, 2014

[21] A. Yazici, *International Journal of Advanced Robotic Systems,* **10**, 2013

[22] C. Amici, A. Borboni, P. L. Magnani, and D. Pomi, *Proceedings of EUCOMES 2008 - The 2nd European Conference on Mechanism Science,* **2008**, 479-485, 2009

[23] D. Coman and A. Ionescu, *Boundary Value Problems,* **2013**, 2013

[24] C. Amici, A. Borboni, and R. Faglia, *Advances in Mechanical Engineering,* **2**, 1-9, 2010

[25] J. Jezný, *Applied Mechanics and Materials* **611**, 107-114, 2014

[26] A. Borboni, F. Aggogeri, and R. Faglia, *ASME 2014 12th Biennial Conference on Engineering Systems Design and Analysis ESDA 2014,* **3**, 1-10, 2014

[27] J. Płaskonka, *2012 17th International Conference on Methods and Models in Automation and Robotics, MMAR 2012*, 2012, 617-620a

[28] A. Borboni, S. Pandini, D. Cambiaghi, M. Lancini, R. Adamini, R. Faglia, *et al.*, *ASME 2014 12th Biennial Conference on Engineering Systems Design and Analysis, ESDA 2014,* **3**, 1-6, 2014

[29] A. Rodic and G. Mester, *Acta Polytechnica Hungarica,* **10**, 113-133, 2013

[30] L. Ding, Haibogao, YuankaiLi, GuangjunLiu, and ZongquanDeng, *Mechanism and Machine Theory,* **86**, 235-264, 2015

[31] C. Amici, A. Borboni, P. L. Magnani, and D. Pomi, *Proceedings of EUCOMES 2008 - The 2nd European Conference on Mechanism Science,* **2008**, 487-493, 2009

[32] T. T. Tsung and N. Hoai, *Key Engineering Materials* **625**, 683-688, 2015

[33] M. Tiboni, A. Borboni, M. Mor, and D. Pomi, *Proceedings of the Institution of Mechanical Engineers. Part I: Journal of Systems and Control Engineering,* **225**, 443-451, 2011

[34] M. Brunner, T. Fiolka, D. Schulz, and C. M. Schlick, *Robotics and Autonomous Systems,* **63**, 89-107, 2014

[35] H. Xu, X. Liu, Z. Zhang, and H. Fu, *Jiqiren/Robot,* **35**, 208-217, 2013

[36] A. Borboni, D. De Santis, and R. Faglia, *Proceedings of 8th Biennial ASME Conference on Engineering Systems Design and Analysis ESDA2006,* **2006**, 1-8, 2006

[37] A. Akbarimajd and N. Sotoudeh, *Advanced Robotics,* **28**, 105-117, 2014

[38] A. Borboni and R. Faglia, *Journal of Applied Mechanics Transactions ASME,* **80**, 1-7, 2013

[39] J. Al-Azzeh, S. F. Yatsun, A. A. Cherepanov, I. V. Lupehina, and V. S. Dichenko, *Research Journal of Applied Sciences,* **9**, 597-602, 2014

[40] A. Borboni, M. Lancini, and R. Faglia, *ASME 2014 12th Biennial Conference on Engineering Systems Design and Analysis ESDA 2014,* **2**, 1-7, 2014

[41] G. Zhong, Y. Kobayashi, T. Emaru, and Y. Hoshino, *JVC/Journal of Vibration and Control,* **20**, 3-23, 2014
[42] A. Borboni and M. Lancini, *Journal of Vibration and Acoustics, Transactions of the ASME,* **137**, 1-9, 2015
[43] I. A. Gravagne, I. D. Walker, and C. D. Rahn, *Proceedings of the ASME Design Engineering Technical Conference*, 2001, 1943-1952
[44] A. Borboni, D. De Santis, and R. Faglia, *ASME 2010 10th Biennial Conference on Engineering Systems Design and Analysis ESDA2010,* **2**, 99-106, 2010
[45] T. Chanthasopeephan, A. Jarakorn, P. Polchankajorn, and T. Maneewarn, *Robotics and Autonomous Systems,* **62**, 38-45, 2014
[46] J. Frémy, F. Ferland, M. Lauria, and F. Michaud, *Robotics and Autonomous Systems,* **62**, 579-590, 2014
[47] A. Borboni and D. De Santis, *Meccanica,* **49**, 1327-1336, 2014
[48] A. Borboni, R. Faglia, and M. Mor, *ASME 2014 12th Biennial Conference on Engineering Systems Design and Analysis ESDA 2014,* **1**, 1-10, 2014
[49] S. Nam, J. Oh, G. Lee, J. Kim, and T. Seo, *Journal of Mechanical Science and Technology,* **28**, 5175-5187, 2014
[50] J. Moravec and P. Pošík, *Evolutionary Intelligence,* **6**, 171-191, 2014
[51] F. Aggogeri, A. Borboni, A. Merlo, and N. Pellegrini, *Advanced Materials Research,* **590**, 252-257, 2012
[52] A. Borboni, F. Aggogeri, N. Pellegrini, and R. Faglia, *Advanced Materials Research,* **590**, 399-404, 2012
[53] F. Aggogeri, A. Borboni, R. Faglia, A. Merlo, and S. De Cristofaro, *Applied Mechanics and Materials,* **336-338**, 1170-1173, 2013
[54] A. Borboni and R. Faglia, *Procedia Engineering*, 2014, 1378-1381
[55] W. K. Tey, C. F. Yeong, Y. L. Seow, E. L. M. Su, and S. H. Tang, *Applied Mechanics and Materials* **607**, 791-794, 2014
[56] D. Zhang, L. Mao, X. N. Tang, Z. B. Li, and J. P. Chen, *Applied Mechanics and Materials* **347-350**, 143-147, 2013
[57] A. Borboni, F. Aggogeri, and R. Faglia, *International Journal of Advanced Robotic Systems,* **10**, 1-10, 2013
[58] A. Borboni, F. Aggogeri, N. Pellegrini, and R. Faglia, *Advanced Materials Research,* **590**, 405-410, 2012
[59] A. Borboni, R. Faglia, and M. Palpacelli, *MESA 2014 - 10th IEEE/ASME International Conference on Mechatronic and Embedded Systems and Applications,* **1**, 1-7, 2014
[60] V. I. Guzeev and D. Y. Pimenov, *Russian Engineering Research,* **31**, 989-993, 2011
[61] D. Y. Pimenov, *Journal of Friction and Wear,* **34**, 290-293, 2013
[62] D. Y. Pimenov, *Journal of Friction and Wear,* **34**, 156-159, 2013
[63] D. Y. Pimenov, *Journal of Friction and Wear,* **35**, 250-254, 2014
[64] D. Y. Pimenov, V. I. Guzeev, and A. A. Koshin, *Russian Engineering Research,* **31**, 1151-1155, 2011
[65] D. Y. Pimenov, V. I. Guzeev, and A. A. Koshin, *Russian Engineering Research,* **31**, 1105-1109, 2011

MATEC Web of Conferences **53**, 01059 (2016)
DOI: 10.1051/matecconf/20165301059

An Innovative Shape Memory Actuator

Valter Cappellini[1,a] and Cinzia Amici[1]

[1] *Università degli Studi di Brescia, Mechanical and Industrial Engineering Department, 25123 Brescia, Italy*

Abstract. The work describes a NiTi linear actuator. This material is able to realize a contraction with heating produced through Joule effect. Then a cooling of the active device is realized with forced air. Finally the lengthening is realized with another active element. The particular structure of the geometry allows for an increment of reliability, because the electrical connections are mechanically stabilized and the active elements are compelled to avoid undesired electrical contacts through an insulated cylindrical core.

1 Introduction

One of the main requirements that an actuator needs to satisfy is to guarantee high mechanical performances in a limited space. In this way, SMA actuators play an important role to provide the highest power-to-weight ratio among light-weight technologies [1-3]. They have become a very attractive option for uses in actuators and they are implemented in a broad range of commercial applications (consumer electronics, fashion, fasteners, ...). SMA actuators can be actuated remotely with a low actuation voltage [4-6] guaranteeing reliable performances and ease in installation.

SMA actuators are widely used in wire or spring configurations, but upcoming applications (e.g. medical instrumentations or microsystems) also demand more complex shapes [7-12]. Designing a SMA actuator is a challenging task due to the complexity of the material selection and behavior. In this manner, the electrical, thermal and mechanical aspects have to be considered simultaneously [13]. SMAs are intermetallic compounds able to recover, in a continuous and reversible way, a previously defined shape when subjected to an appropriate thermomechanical load [14-16]. From a microscopic point of view, this transformation consists in a transition from a crystallographic stable phase at low temperature (e.g.. martensite) to a different crystallographic stable phase at high temperature (e.g. austenite). When the SMA is deformed by an external force, instead of breaking crystallographic bonds and damaging its structure, it starts a progressively arrangement of planes which closes the deformation without achieving significant atomic displacements. During this process, atoms have moved only slightly from original positions. When an imposed stimulus occurs, atoms move to restore the previous crystal structure before the deformation and then the recovery of macroscopic original shape.

2 SMAs application to linear motors

[a] Corresponding author : v.cappellini@unibs.it

This paper presents the study and realization of a SMA actuator able to achieve linear movements in a coil spring configuration. SMA material has an high electrical resistance, it can be heated to its transition temperature simply by passing current through it. This approach opens up many possibilities for providing mechanical actuation (movement) without any moving parts. The purpose of this research is the realization of SMA actuators which perform in defined operating temperature a specified movement. Temperatures have to meet limits due to actuator structure and application. The duty cycle for current application should be in agreement with user's needs, so as to allow a movement in a short time. For these reasons, authors have as primary objective the shortest working cycle with imposed temperature constraint, especially during the cooling phase. A robust approach (Design of Experiment) is accepted for performing tests in order to achieve a required output for thermal [17], voltages, stroke and time point of view [18, 19]. The developed experimental tests permitted a performance comparison between different operating conditions.
This paper focuses on two main issues in SMAs assessment and selection. The first one is the characterization of the thermo-mechanical behavior of SMA materials [20-22] and the other is the actuator application of one-way shape memory effect (SME). The selection of the operative temperature range is a crucial point in SMA selection [23]. All memory alloys have temperatures that characterize the internal structure [24]. In this study, the main temperatures of phase transformations for the adopted alloy (NiTiCu5) include the temperature at end of martensitic transformation (MF) 26°C, the martensitic transformation start temperature (MS) 37°C, the austenite transformation start temperature (AS) 48°C and the end temperature austenitic transformation (AF) 59°C (Hysteresis 24°C) [25-28]. The temperature of 59°C indicates the maximum operative temperature for SMA material, in fact for higher temperature no transformation phase are pointed out. The SMA element remains in the same position and all exceeded power is dissipated. Temperature of 26°C indicates the lower limit of the resting phase of SMA element and internal structure of alloy is all in the martensitic phase.
Generally, the one-way effect does not provide suitable mechanism of SMA actuator due to the host structure that is not able to return to initial shape after cooling process. However, different researches [29] proposed some designs of SMAs applied to muscles of robot hands demonstrating that it is possible to realize a hand actuated by SMAs. In this case, a model based controller does not perform in a satisfying way due to the open loop configuration. The control problem (e.g. hysteresis) can be partially solved restricting the displacements to a specific range simplifying the problem to a linear one. This behavior is linearized using two counteracting shape memory elements. This solution is further improved by a temperature sensor on the shape memory element. Electrical resistance position feedback is also used in common practice [30, 31]. An example of this application has been developed in the proposed actuator system.

3 Design of SMA actuator module

The actuator is designed to obtain a linear movement along horizontal axis of SMA elements. The shortest duty cycle was the primary objective in designing the device. This assumption highlights that the importance to meet specific functional requirements, severely affected the designers' choices [32, 33] regarding the shapes and configuration that actuators should possess in order to produce predefined strokes and output forces [34-41]. Cartesian geometry (Fig. 1) develops electrical and mechanical function on several layers [42, 43]. This configuration is a result of movement optimization as shown in the followings.
The actuator needs to moves along an horizontal axis or a plane to perform a linear movement. The intuitive geometry for this type of movement is the cylinder, where all the components are positioned around the main axis of movement. In this case functional axes (mechanical constraints, electrical contacts, movements, thermal heating and cooling) are coincident with the movement axis and the model is very compact [44-46]. A cylindrical geometry does not permit the geometrical diversification of the contact points that characterize the different functions.

These functions need of different materials and physical connections. In order to design properly the single elements of the actuator, it is useful to separate functions with Concentric cylinder geometry (separate elements, on concentric surfaces) or Cartesian geometry (different functions on different parallel plans). The physical model is more complex, but the design phase and maintenance activities are simplified by isolating geometrical features, with independent functional behaviors.

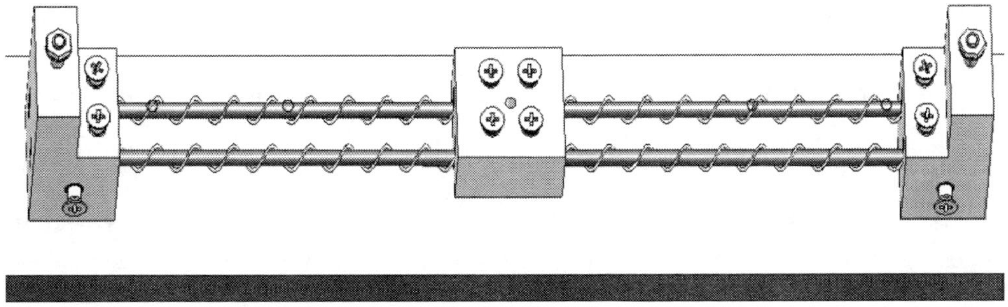

Figure 1. The modular actuation system.

From a Cartesian geometry is possible to develop a modular model (Fig. 2). SMA model is designed to have more actuator elements in parallel without changing the conformation of the individual components, but increasing overall dimensions, developed force and energy consumption [47, 48]. This approach represents an important benefit compared with concentric cylinder geometry. In particular, this configuration permits an increment of the nominal force of the actuator as shown in Fig.1 [49-51].

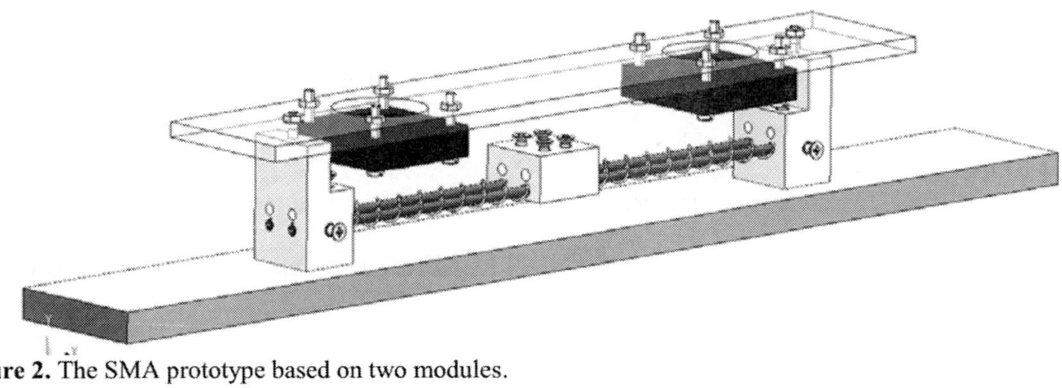

Figure 2. The SMA prototype based on two modules.

Fig. 2 describes the developed prototype, electrical and mechanical functional plans are separate. Electrical function is located in the upper part of the SMA actuator (Fig. 2) while the plane of the mechanical function is located on the axis of SMA elements. The number of springs has been chosen in order to obtain the best control and stability in the movement. In this study four SMA elements are used, positioned on each side between central slider and wall. The spiral diameter is equal to 4.00 mm and the wire thickness is 0.40 mm.

4 Conclusions

This paper presents the development of a modular actuator based on shape memory alloy material. Starting from theoretical observations, this study wants to highlight the main relations between the functional and geometrical properties of this device. The next steps include to improve some theoretical aspects and develop new prototypes to optimize the required perfomances.

References

[1] A. Borboni, R. Faglia, and M. Mor, ASME 2014 12th Biennial Conference on Engineering Systems Design and Analysis ESDA 2014, **1**, 1-10, 2014

[2] F. Aggogeri, A. Borboni, R. Faglia, A. Merlo, and S. De Cristofaro, Applied Mechanics and Materials, **336-338**, 1170-1173, 2013

[3] F. Aggogeri, A. Borboni, and R. Faglia, Applied Mechanics and Materials, **373-375**, 130-133, 2013

[4] A. Borboni, R. Faglia, and M. Palpacelli, MESA 2014 - 10th IEEE/ASME International Conference on Mechatronic and Embedded Systems and Applications, **1**, 1-7, 2014

[5] A. Borboni and R. Faglia, Procedia Engineering, 2014, 1378-1381

[6] A. Borboni, F. Aggogeri, N. Pellegrini, and R. Faglia, Advanced Materials Research, **590**, 405-410, 2012

[7] M. Tiboni, A. Borboni, M. Mor, and D. Pomi, Proceedings of the Institution of Mechanical Engineers. Part I: Journal of Systems and Control Engineering, **225**, 443-451, 2011

[8] G. Taveggia, J. H. Villafañe, F. Vavassori, C. Lecchi, A. Borboni, and S. Negrini, Journal of Manipulative and Physiological Therapeutics, **37**, 242-252, 2014

[9] C. Amici, A. Borboni, P. L. Magnani, and D. Pomi, Proceedings of EUCOMES 2008 - The 2nd European Conference on Mechanism Science, **2008**, 487-493, 2009

[10] C. Amici, A. Borboni, P. L. Magnani, and D. Pomi, Proceedings of EUCOMES 2008 - The 2nd European Conference on Mechanism Science, **2008**, 479-485, 2009

[11] C. Amici, A. Borboni, R. Faglia, D. Fausti, and P. L. Magnani, IEEE/RSJ International Conference on Intelligent Robots and Systems IROS, **2008**, 735-740, 2008

[12] C. Amici, A. Borboni, and R. Faglia, Advances in Mechanical Engineering, **2**, 1-9, 2010

[13] G. I. Grinfeldi, A. S. Gorshkov, and N. I. Vatin, Advanced Materials Research **941-944**, 786-799, 2014

[14] J. Čapek, D. Vojtěch, and P. Novák, METAL 2012 - Conference Proceedings, 21st International Conference on Metallurgy and Materials, 2012, 1323-1328

[15] V. Kučera, J. Čapek, A. Michalcová, and D. Vojtěch, Manufacturing Technology, **14**, 342-347, 2014

[16] P. Novák, H. Moravec, J. Vystrčil, J. Adámek, J. Kopeček, and T. F. Kubatík, Key Engineering Materials **647**, 95-101, 2015

[17] F. Aggogeri, A. Borboni, A. Merlo, and N. Pellegrini, Advanced Materials Research, **590**, 252-257, 2012

[18] A. Borboni and M. Lancini, Journal of Vibration and Acoustics, Transactions of the ASME, **137**, 1-9, 2015

[19] A. Borboni, R. Bussola, R. Faglia, P. L. Magnani, and A. Menegolo, Journal of Mechanical Design, Transactions of the ASME, **130**, 0823011-0823016, 2008

[20] B. O. Agboola, T. Baxevanis, and D. C. Lagoudas, ASME 2014 Conference on Smart Materials, Adaptive Structures and Intelligent Systems, SMASIS 2014, 2014

[21] J. De Brito Sirnões, F. F. R. Pereira, J. Otubo, and C. J. De Araújo, Materials Research Society Symposium Proceedings, 2014, 31-36

[22] D. Depriester, A. Maynadier, K. Lavernhe-Taillard, and O. Hubert, International Journal of Solids and Structures, **51**, 1901-1922, 2014

[23] A. Borboni, F. Aggogeri, and R. Faglia, International Journal of Advanced Robotic Systems, **10**, 1-10, 2013

[24] M. G. Suru, N. M. Lohan, B. Pricop, I. P. Spiridon, E. Mihalache, R. I. Comaneci, et al., International Journal of Materials and Product Technology, **50**, 276-288, 2015

[25] M. Abedini, H. M. Ghasemi, and M. Nili Ahmadabadi, Materials and Design, **30**, 4493-4497, 2009
[26] C. Gurau, G. Gurau, L. G. Bujoreanu, and F. M. B. Fernandes, Materials Today: Proceedings, **2**, S905-S908, 2015
[27] D. Mutter and P. Nielaba, European Physical Journal B, **84**, 109-113, 2011
[28] K. W. K. Yeung, K. M. C. Cheung, W. W. Lu, and C. Y. Chung, Materials Science and Engineering A, **383**, 213-218, 2004
[29] J. Peña, E. Solano, A. Mendoza, J. Casals, J. A. Planell, and F. J. Gil, Bio-Medical Materials and Engineering, **15**, 289-293, 2005
[30] N. Ma, G. Song, and H. J. Lee, Proceedings of SPIE - The International Society for Optical Engineering, 2003, 46-55
[31] G. Song, N. Ma, and H. J. Lee, Smart Structures and Systems, **3**, 189-200, 2007
[32] A. Borboni, D. De Santis, and R. Faglia, ASME 2010 10th Biennial Conference on Engineering Systems Design and Analysis ESDA2010, **2**, 99-106, 2010
[33] A. Borboni and D. De Santis, Meccanica, **49**, 1327-1336, 2014
[34] R. J. Chang, C. C. Shiu, and C. Y. Cheng, International Journal of Advanced Robotic Systems, **10**, 1-12, 2013
[35] F. Furgiuele, C. Maletta, and E. Sgambitterra, ASME 2012 Conference on Smart Materials, Adaptive Structures and Intelligent Systems, SMASIS 2012, 2012, 545-551
[36] A. Hadi, M. Elahinia, A. Yousefi-Koma, M. M. Moghadam, and C. Chapman, ASME 2010 Conference on Smart Materials, Adaptive Structures and Intelligent Systems, SMASIS 2010, 2010, 555-564
[37] A. Khodayari, B. Zarefar, M. M. Kheirikhah, and E. Mirsadeghi, Proceedings - 2011 IEEE International Conference on Control System, Computing and Engineering, ICCSCE 2011, 2011, 312-316
[38] A. Rezaeian and A. Yousefi-Koma, ASME International Mechanical Engineering Congress and Exposition, Proceedings (IMECE), 2014
[39] L. Xin, B. Xu, and J. He, Proceedings of SPIE - The International Society for Optical Engineering, 2012
[40] Y. G. Yang, G. B. Han, and C. W. Kim, Transactions of the Korean Society of Mechanical Engineers, B, **35**, 1301-1307, 2011
[41] P. Zhuang and S. D. Xue, Zhendong yu Chongji/Journal of Vibration and Shock, **32**, 204-212, 2013
[42] A. Borboni and R. Faglia, Journal of Applied Mechanics Transactions ASME, **80**, 1-7, 2013
[43] A. Borboni, F. Aggogeri, N. Pellegrini, and R. Faglia, Advanced Materials Research, **590**, 399-404, 2012
[44] V.L. Kvint, V.V Okrepilov, Herald of the Russian Academy of Sciences, **84 (3)**, 188-200 (2014)
[45] V.V Okrepilov, V.L. Makarov, A.R. Bakhtizin, S.N Kuzmina, Economy of Region, **2**, 301-313 (2015)
[46] V.V Okrepilov, M.V. Leonidovich, Asian Social Science, **11 (7)**, 312-325 (2015)
[47] B. Šćepanović, M. Knežević, D. Lučić, Journal of Applied Engineering Science, **10 (1)**, 15-20 (2012)
[48] B. Šćepanović, M. Knežević, D. Lučić, Informes de la Construccion, **66 (1)**, 1-13 (2014)
[49] B. Descamps and R. Filomeno Coelho, International Journal of Solids and Structures, **51**, 2390-2399, 2014
[50] S. G. Hoffman, M. P. Reed, and D. B. Chaffin, Ergonomics, **54**, 47-59, 2011
[51] A. Schäfer, 21st Conference on Measurement of Force, Mass and Torque Together with HARDMEKO 2010 and 2nd Meeting on Vibration Measurement, IMEKO TC3, TC5 and TC22 Conferences, 2010, 53-56

MATEC Web of Conferences **53**, 01060 (2016)
DOI: 10.1051/matecconf/20165301060

Inverse Kinematics of a Serial Robot

Cinzia Amici[1,a] and Valter Cappellini[1]

[1]*Università degli Studi di Brescia, Mechanical and Industrial Engineering Department, 25123 Brescia, Italy*

Abstract. This work describes a technique to treat the inverse kinematics of a serial manipulator. The inverse kinematics is obtained through the numerical inversion of the Jacobian matrix, that represents the equation of motion of the manipulator. The inversion is affected by numerical errors and, in different conditions, due to the numerical nature of the solver, it does not converge to a reasonable solution. Thus a soft computing approach is adopted to mix different traditional methods to obtain an increment of algorithmic convergence.

1 Introduction

Inverse kinematics problem is the computation needed to find the joint angles for a given Cartesian position and orientation of the end effectors. There is no unique solution for the inverse kinematics thus necessitating application of numerical iterative models. In the case of a redundant manipulator, the number of inverse kinematic solutions may become infinite and closed form solutions are impossible to find in general.

The methods available for solving these problems may be broadly grouped into two major classes: classical techniques based on the differential kinematic relationship between the end-effector velocities and the corresponding joint angle velocities [1, 2], and learning based techniques based on biologically inspired approaches like neural networks, fuzzy logic, machine learning [3-5].

The classical techniques require previous knowledge of the manipulator geometry represented by the Jacobian matrix. Pseudoinverse, Jacobian transpose, damped least square methods and their variants fall into this category. On the other hand, learning based approaches can be used to learn the kinematic relationship of the manipulators during action–perception cycles without requiring explicit knowledge of their geometry.

A robot manipulator is composed of a serial chain of rigid links connected to each other by revolute or prismatic joints. A revolute joint rotates about a motion axis and a prismatic joint slide along a motion axis. Each robot joint location is usually defined relative to neighboring joint [6-11]. The relation between successive joints is described by 4 × 4 homogeneous transformation matrices that have orientation and position data of robots. The number of those transformation matrices determines the degrees of freedom of robots. The product of these transformation matrices produces final orientation and position data of a n degrees of freedom robot manipulator. Robot control actions are executed in the joint coordinates while robot motions are specified in the Cartesian coordinates. Conversion of the position and orientation of a robot manipulator end-effector from Cartesian space to joint space, called

[a] Corresponding author : cinzia.amici@unibs.it

as inverse kinematics problem, which is of fundamental importance in calculating desired joint angles for robot manipulator design and control. If the joints of the manipulator are more complex, the inverse kinematics solution by using these traditional methods is time consuming [12-23]. In recent years, there have been increasing research interest of artificial neural networks and many efforts have been made on applications of neural networks to various control problems [24-27].

2 Problem definition

The study of precision positioning systems plays a relevant role in the design phase to obtain high performance in robot and mechatronic system accuracy [28-35]. This paper focuses on a serial manipulator modelled as a serial chain of rigid bodies (links) connected by joints, Fig .1[36, 37].
The joint J_i connects links l_i and l_{i-1} providing a relative movement. This movement is described by the scalar joint variable q_i. In this study only rotary and prismatic joints are considered. The first type of joints permits circular movements around the joint axis, while the second one guarantees only rectilinear movements between the connected links along the joint axis. In this structure two links are fundamental in the kinematic chain:

- l_1: it is the base and it is considered motionless;
- ln: it is the end-effector, the position and orientation are important since it used to act on the external world.

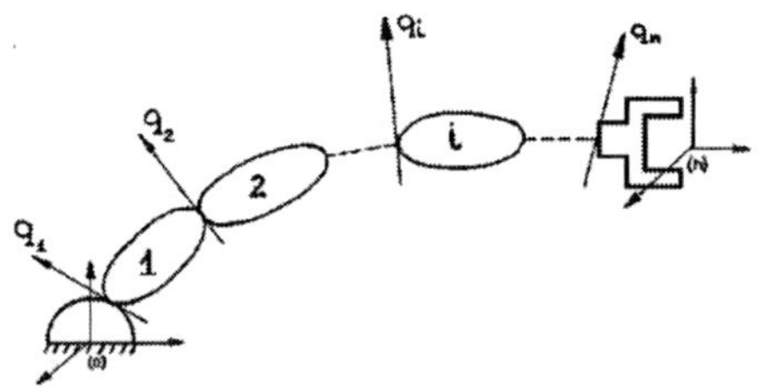

Figure 1. Open-loop kinematic chain scheme.

The position and orientation of the end-effector is described by a set of variables S (work space variables) or a set of joint variables Q. In a serial manipulator scheme, the relation between the joint and the work space is defined by:

$$S = F(Q) \tag{1}$$

Eq. 1 describes the positional kinematics of a serial robot [38-40]. The inversion of Eq. 1 can be usually difficult and computationally expensive and sometimes it is impossible to obtain a closed form. This task is called as positional inverse kinematic problem [41, 42]. In order to avoid these issues the robot manufacturers usually select a kinematic chain that solves the inverse kinematic problem in a closed form. These choices are focused to guarantee the required performance and reliability of robots and mechatronic systems [43-46]. This study aims to propose an alternative approach focusing on the numerical solution overcoming difficulties without losing in computational performances.
As case of study, a simple SCARA robot, is considered, as schematically shown in Fig. 2.

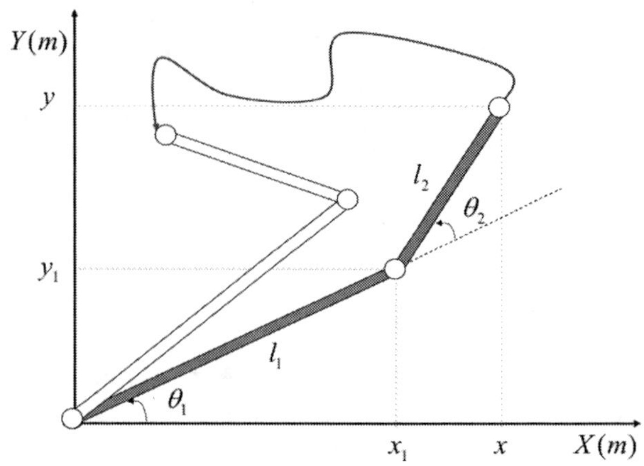

Figure 2. Planar scheme of a SCARA robot [47, 48].

For this type of robot, an analytical closed form solution already exists. The SCARA robot is here used only as a test. Classical numerical methods, considered in this paper, consist on linearizing the inverse kinematic equations, setting up an initial position and then using Newton's method, a generalization of Newton's method or another iterative methods [49-51]. Numerical methods can be applied to analyze any manipulator topology, but they can be affected by uncertainty in obtaining all possible solutions [52].
A set of charts, Fig. 3, based on numerical simulations shows when classical algorithms do not converge to a correct solution. The simulations are structured on an initial position to set up the numerical algorithm, then it is imposed to the robot to achieve a collection of positions uniformly distributed on the work space, starting from the same point. The non converging points are plotted in a diagram, as shown in Fig. 3.

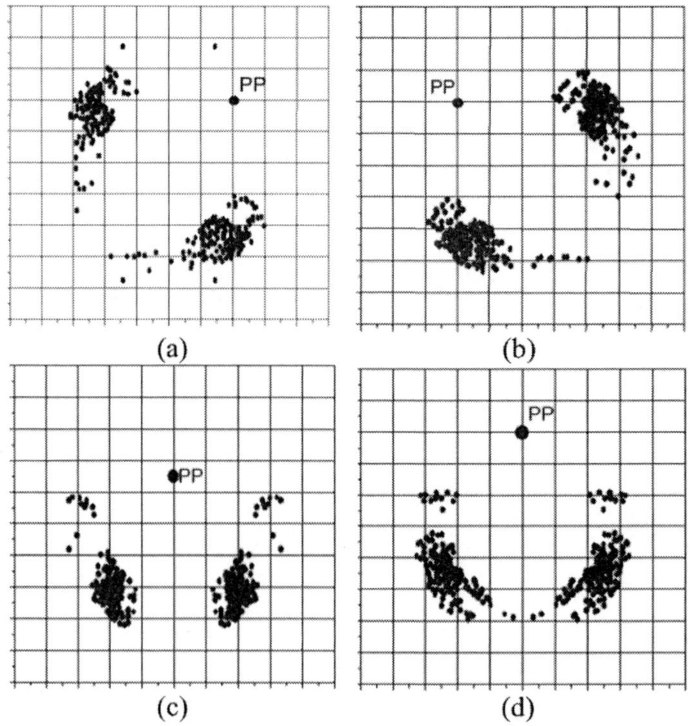

Figure 3. Numerical simulations for the Newton-Raphson algorithm.

In Fig. 3, PP is the starting position while the black points are non converging points for the Newton-Raphson algorithm, as follows:

- Non converging points (NCPs) are gathered in non-converging regions (NCRs);
- The feature of NCRs seems simple at a first approach, nevertheless micro-characteristics are difficult to model;
- Rotating the starting point around the center of the work space, the NCRs undergo the same rotation (Fig. 3a and Fig. 3b);
- Translating the starting point along a radius of the work space, the non converging regions change their features in a complex way, but approximately the radial coordinate of the non converging region barycenter remains constant (Fig. 3c and Fig. 3d).

The first and the second observations have general validity while the third and the fourth conclusions are based on the SCARA robots.

Due to the uncertainty in the aforementioned observations, a soft-computing approach is convenient in the definition and treatment of non converging regions [53-55]. The use of soft-computing techniques in the modeling of mechanical phenomena is well known [56-58]. In this study, a fuzzy approach is proposed to define and evaluate non converging regions.

3 A fuzzy logic to solve Inverse Kinematics problem

Fuzzy systems are the most popular constituent of the soft computing area and they are used in modeling uncertainty and vagueness [59-61]. Many studies have demonstrated the power of this approach contributing to transform a pioneering approach into a scientific method [62, 63].

Under specific hypotheses, a fuzzy system S given by n rules of the form is defined as (compact form):

$$\{R_i\colon If\ and_{j=1}^{m}\left(x \in A_{ij}\right)\ then\ y \in B_i\}_{i=1,2,\dots,n} \tag{2}$$

Where:

- $A_{ij} \in F(R)$ and $B_j \in F(R)$ are fuzzy sets representing linguistic variables (*i.e.* dark or light, long or short, black or light)
- A_{ij} is characterized by a membership function $\mu_{A_{ij}}$ with convenient features where:

$$\mu_{A_i}(x) = T_{j=1}^{m}\mu_{A_{ij}}(x_j) \tag{3}$$

for some *t-norm T*.

A further simplification is introduced in (2) in a specific case: if x_j is the assertion that $X_j = \bar{X}_{ij}$ and A_{ij} is the set of true assertion that $X_j = \bar{X}_{ij}$, (2) is written as (4) or (5):

$$\{R_i\colon If\ and_{j=1}^{m}\left(X_j = \bar{X}_{ij}\right)\ then\ y \in B_i\}_{i=1,2,\dots,n} \tag{4}$$

$$\{R_i\colon If\ and_{j=1}^{m}\left(X_j = \bar{X}_{ij}\right)\ then\ Y = \bar{Y}_i\}_{i=1,2,\dots,n} \tag{5}$$

Starting from these assumptions, the new algorithm based on Fuzzy logic is now presented. The first step is the definition of the starting point to every iteration of the algorithm. At the first iteration, this point is chosen in accordance with a rule independent from the proposed algorithm. At the following steps, the starting point of the *t-th* iteration is defined by the arrival point of the *(t-1)th* iteration. The algorithm terminates, when the arrival point of the *t-th* iteration is close to the final point P_f.

A set of starting points P_i is taken as example for the fuzzy algorithm. The first fuzzy system selects a correct example:

$$\{R_i\colon If\ and_{j=1}^{2}\left(x_j = x_{ij}\right)\ then\ P = P_i\}_{i=1,2,\dots,n} \tag{6}$$

where: x_j are the coordinates of the fuzzy starting point P and x_{ij} are the coordinates of each example of starting point P_i. There is a non-converging region A_i associated to every starting point P_i. More

specifically $A_i = < \hat{A}_1 \ \hat{A}_2 \ \ldots \ \hat{A}_{ALG} >$ is a vector of non-converging regions, one for every type of considered classical numerical algorithm.
The second fuzzy system is used to define the non-converging region A.

$$\{R_i: If\ P = P_i\ then\ A = A_i\}_{i=1,2,\ldots,n} \quad (7)$$

If the desired final point P_f for the numerical method is out of the non-converging region $\hat{A}_i$, associated with the numerical algorithm a_i, it is possible to use the algorithm a_i:

$$\{R_i: If\ P_f \notin \hat{A}_i\ then\ use\ a_i\}_{i=1,2,\ldots,n} \quad (8)$$

This algorithm is robust with a fail probability close to 0.02%. Nevertheless, to avoid these issues, a further fuzzy system is introduced. In case the original algorithm has not reached the solution in $\widetilde{T}$ iterations, the rules (8) are denied:

$$\{R_i: If\ P_f \notin \hat{A}_i\ then\ use\ -a_i\}_{i=1,2,\ldots,n} \quad (9)$$

4 Conclusions

This paper aims at presenting a new general fuzzy iterative algorithm based on classical algorithms to solve the inverse kinematic problem on a serial manipulator. The effectiveness of the algorithm was compared with the classical methods on a SCARA robot.

References

[1] T. Hondo and I. Mizuuchi, *Proceedings - IEEE International Conference on Robotics and Automation*, 2015, 756-761

[2] S. R. Munasinghe, M. Nakamura, S. Aoki, S. Goto, and N. Kyura, *Proceedings of the IEEE International Conference on Systems, Man and Cybernetics*, 1999, II-854 - II-859

[3] L. A. Zadeh, *IEEE Transactions on Fuzzy Systems,* **4**, 103-111, 1996

[4] J. J. Hopfield, *Proceedings of the National Academy of Sciences of the United States of America,* **79**, 2554-2558, 1982

[5] M. E. Tipping, *Journal of Machine Learning Research,* **1**, 211-244, 2001

[6] A. Borboni, R. Bussola, R. Faglia, P. L. Magnani, and A. Menegolo, *Journal of Mechanical Design, Transactions of the ASME,* **130**, 0823011-0823016, 2008

[7] A. Borboni, F. Aggogeri, and R. Faglia, *International Journal of Advanced Robotic Systems,* **10**, 1-10, 2013

[8] C. Amici, A. Borboni, P. L. Magnani, and D. Pomi, *Proceedings of EUCOMES 2008 - The 2nd European Conference on Mechanism Science,* **2008**, 487-493, 2009

[9] C. Amici, A. Borboni, P. L. Magnani, and D. Pomi, *Proceedings of EUCOMES 2008 - The 2nd European Conference on Mechanism Science,* **2008**, 479-485, 2009

[10] C. Amici, A. Borboni, R. Faglia, D. Fausti, and P. L. Magnani, *IEEE/RSJ International Conference on Intelligent Robots and Systems IROS,* **2008**, 735-740, 2008

[11] C. Amici, A. Borboni, and R. Faglia, *Advances in Mechanical Engineering,* **2**, 1-9, 2010

[12] M. Tiboni, A. Borboni, M. Mor, and D. Pomi, *Proceedings of the Institution of Mechanical Engineers. Part I: Journal of Systems and Control Engineering,* **225**, 443-451, 2011

[13] A. Borboni, R. Faglia, and M. Palpacelli, *MESA 2014 - 10th IEEE/ASME International Conference on Mechatronic and Embedded Systems and Applications,* **1**, 1-7, 2014

[14] A. Borboni and R. Faglia, *Procedia Engineering*, 2014, 1378-1381

[15] A. Borboni, F. Aggogeri, N. Pellegrini, and R. Faglia, *Advanced Materials Research,* **590**, 405-410, 2012

[16] A. Borboni and M. Lancini, *Journal of Vibration and Acoustics, Transactions of the ASME,* **137**, 1-9, 2015

[17] A. Borboni, R. Faglia, and M. Mor, *ASME 2014 12th Biennial Conference on Engineering Systems Design and Analysis ESDA 2014,* **1**, 1-10, 2014

[18] A. Borboni and R. Faglia, *Journal of Applied Mechanics Transactions ASME,* **80**, 1-7, 2013

[19] A. Borboni, D. De Santis, and R. Faglia, *ASME 2010 10th Biennial Conference on Engineering Systems Design and Analysis ESDA2010,* **2**, 99-106, 2010

[20] A. Borboni and D. De Santis, *Meccanica,* **49**, 1327-1336, 2014

[21] A. Borboni, F. Aggogeri, N. Pellegrini, and R. Faglia, *Advanced Materials Research,* **590**, 399-404, 2012

[22] F. Aggogeri, A. Borboni, R. Faglia, A. Merlo, and S. De Cristofaro, *Applied Mechanics and Materials,* **336-338**, 1170-1173, 2013

[23] N. L. Vatin, J. Havula, L. Martikainen, A. S. Sinelnikov, A. V. Orlova, and S. V. Salamakhin, *Advanced Materials Research* **945-949**, 1211-1215, 2014

[24] B. Stamatovic, R. Upadhyay, and N. Vatin, *Procedia Engineering*, 2015, 660-667

[25] B. Šćepanović, M. Knežević, D. Lučić, Informes de la Construccion, **66 (1)**, 1-13 (2014)

[26] V.V Okrepilov, Ivanova, G. Standarty i Kachestvo, **12**, 62-68. (2004)

[27] V.V Okrepilov, Standarty i Kachestvo, **10**, 52-55 (2005)

[28] F. Aggogeri, A. Borboni, A. Merlo, and N. Pellegrini, *Advanced Materials Research,* **590**, 252-257, 2012

[29] F. Aggogeri, A. Borboni, and R. Faglia, *Applied Mechanics and Materials,* **373-375**, 130-133, 2013

[30] D. Niculescu and B. Nath, *Conference Record / IEEE Global Telecommunications Conference*, 2001, 2926-2931

[31] D. Niculescu and B. Nath, *Proceedings - IEEE INFOCOM*, 2003, 1734-1743

[32] G. Alici and B. Shirinzadeh, *Mechanism and Machine Theory,* **40**, 879-906, 2005

[33] Y. S. Kwoh, J. Hou, E. A. Jonckheere, and S. Hayati, *IEEE Transactions on Biomedical Engineering,* **35**, 153-160, 1988

[34] J. P. Merlet, *Journal of Mechanical Design, Transactions of the ASME,* **128**, 199-206, 2006

[35] W. K. Veitschegger and C. H. Wu, *IEEE Journal on Robotics and Automation,* **2**, 171-179, 1986

[36] J. Angeles and C. S. Lopez-Cajun, *International Journal of Robotics Research,* **11**, 560-571, 1992

[37] M. L. Husty, M. Pfurner, and H. P. Schröcker, *Mechanism and Machine Theory,* **42**, 66-81, 2007

[38] M. Gautier and W. Khalil, *IEEE Transactions on Robotics and Automation,* **6**, 368-373, 1990

[39] R. He, Y. Zhao, S. Yang, and S. Yang, *IEEE Transactions on Robotics,* **26**, 411-423, 2010

[40] A. Mohan and S. K. Saha, *Multibody System Dynamics,* **17**, 291-319, 2007

[41] J. Angeles, *International Journal of Robotics Research,* **4**, 21-37, 1985

[42] L. Sciavicco and B. Siciliano, *IEEE Journal on Robotics and Automation,* **4**, 403-410, 1988

[43] G. Taveggia, J. H. Villafañe, F. Vavassori, C. Lecchi, A. Borboni, and S. Negrini, *Journal of Manipulative and Physiological Therapeutics,* **37**, 242-252, 2014

[44] M. Follmer, P. Hehenberger, and K. Zeman, *Lecture Notes in Computer Science (including subseries Lecture Notes in Artificial Intelligence and Lecture Notes in Bioinformatics)* **6928 LNCS**, 105-112, 2012

[45] N. Hammouda, G. Habchi, C. Barthod, J. Lottin, and O. Duverger, *10th France-Japan Congress, 8th Europe-Asia Congress on Mecatronics, MECATRONICS 2014*, 2014, 41-46

[46] C. P. Kiran, S. Clement, and V. P. Agrawal, *International Journal of Reliability and Safety,* **7**, 128-155, 2013

[47] S. S. Ge, T. H. Lee, and G. Zhu, *IEEE/ASME Transactions on Mechatronics,* **3**, 138-144, 1998

[48] A. Omodei, G. Legnani, and R. Adamini, *Journal of Robotic Systems,* **17**, 291-307, 2000

[49] S. Abbasbandy, *Applied Mathematics and Computation,* **145**, 887-893, 2003

[50] M. J. Lindstrom and D. M. Bates, *Journal of the American Statistical Association,* **83**, 1014-1022, 1988

[51] T. J. Ypma, *SIAM Review,* **37**, 531-551, 1995

[52] C. J. Freitas, *Applied Mathematical Modelling,* **26**, 237-248, 2002

[53] R. M. Aron, D. Pérez-García, and J. B. Seoane-Sepúlveda, *Studia Mathematica,* **175**, 83-90, 2006

[54] G. Contopoulos, C. Efthymiopoulos, and A. Giorgilli, *Journal of Physics A: Mathematical and General,* **36**, 8639-8660, 2003

[55] T. E. Croley Ii and B. Hunt, *Journal of Hydrology,* **49**, 121-138, 1981

[56] M. Cacciola, F. La Foresta, F. C. Morabito, and M. Versaci, *NDT and E International,* **40**, 357-362, 2007

[57] J. Ghaboussi, *Structural Engineering and Mechanics,* **11**, 485-502, 2001

[58] O. Kaynak and I. Rudas, *Computing and Control Engineering Journal,* **6**, 68-72, 1995

[59] J. S. R. Jang, *IEEE Transactions on Systems, Man and Cybernetics,* **23**, 665-685, 1993

[60] C. C. Lee, *IEEE Transactions on Systems, Man and Cybernetics,* **20**, 404-418, 1990

[61] T. Takagi and M. Sugeno, *IEEE Transactions on Systems, Man and Cybernetics,* **SMC-15**, 116-132, 1985

[62] Q. Liang and J. M. Mendel, *IEEE Transactions on Fuzzy Systems,* **8**, 535-550, 2000

[63] H. O. Wang, K. Tanaka, and M. F. Griffin, *IEEE Transactions on Fuzzy Systems,* **4**, 14-23, 1996

MATEC Web of Conferences **53**, 01061 (2016)
DOI: 10.1051/matecconf/20165301061

Review on Shape Memory Materials

Maurizio Mor[1,a]

[1]*Polibrixia, R&D Department, 25123 Brescia, Italy*

Abstract. In this work, the author tries to trace a route of shape memory materials from their definition, in terms of experimental observation and physical model to the application, especially in mechanical actuators and in device adopting this type of actuators.

1 Introduction

An interesting introduction to shape memory effect and to different types of shape memory actuators can be find in [1].

According to the type of stimulus, SMAs fall in two categories, one is thermo-responsive, and the other is magneto-responsive (i.e., ferromagnetic SMA) [2].

As such, the former can only be activated thermally by heating, while the latter can also be activated by applying a static magnetic field (magneto-responsive) or an alternating magnetic field (inductive heating) to trigger the SME. In both types, the reversible martensitic transformation is the underlying driving mechanism.

Thermo-responsive SMA has matured much more from the real engineering application point of view but many commercial applications are far from being realized. Therefore, author will only focused on thermo-responsive SMAs in this paper. Thermo-responsive SMA has matured much more from the real engineering application point of view and many commercial applications have, so far, been realized. Therefore, author will only focus on thermo-responsive SMAs in this paper.

There are three major types of SMA systems at present, namely Cu-based (mainly CuAlNi and CuZnAl), NiTi-based, and Fe-based (e.g., FeMnSi, FeNiC and FeNiCoTi). The first two are more suitable for engineering applications, while the last one is traditionally seldom used. SMAs based on NiTi are also more reliable and can be highly biocompatible. NiTi-based SMAs are more reliable and can be highly biocompatible.

Although the term "shape memory polymer" can be controversial, it is common in literature and it is associated to polymeric materials that, under a thermal trigger, are able to recover a shape. There is a variety of these examples; they are usually characterized by low exerted mechanical work, low speed and low cost with respect to shape memory metals. An interesting work [3] show also the possibility of combination of different materials (polymeric and metallic) to realize a multifunctional smart material system.

2 State of the art on materials

[a] Corresponding author : maurizio.mor@polibrixia.it

An interesting field of research is on the constitutive model of materials.

Zhu Y., Zhang Y., and Zhao D. [4] shows a micromechanical constitutive model for phase transformation of NiTi polycrystalline SMA. By the assumption of laminated microstructure and perfect interfacial relationship for NiTi single crystal transformation, microscopic strain for each phase can be transformed to overall strain of respective volume element, and expression of phase transformation driving force is derived, then control equation of phase transformation is constructed. Based on the single crystal model, the NiTi polycrystalline constitutive model is constructed by Taylor assumption.

Zhou B. and Yoon S.H. [5] developed a macroscopic constitutive model of shape memory alloy considering cyclic effects. The progressive decrease of phase transformation critical stresses with the increased cycle number in such materials is assumed to be a result from the progressive increase of phase transformation critical temperatures upon the cyclic effects. A cyclic evolution equation is supposed to describe the influences of cycle effects on the material properties of the SMA under cyclic loading. A phase transformation equation expressing the phase transformation behaviors of the SMA under cyclic loading is established based on the differential relationship between martensitic volume fraction and the free energy increment of phase transformation. A mechanical constitutive equation predicting the mechanical characteristics of the SMA under cyclic loading is developed on the basis of thermodynamics and continuum mechanics.

Song Z., Dai H.H., and Sun Q.P. [6] worked on the propagation stresses in phase transitions of an SMA wire with new analytical formulas based on an internal-variable model. Based on the constitutive model in the literature with the specific Helmholtz free energy and rate of mechanical dissipation, the three-dimensional model is formulated. Identifying the characteristic axial strain as the small parameter, they arrive at the asymptotic one-dimensional equation which involves the stress, the axial strain and the phase state variable. By considering the evolution law of the phase state variable, the stress-strain relations corresponding to austenite, martensite and phase transition (phase mixture) regions are obtained. Although they take the dissipation into consideration, they find that each of the pure loading and unloading processes is equivalent to the Ericksen's bar problem with three branches in the stress-strain curve. As a result, they successfully deduce the analytical formulas for the nucleation stresses and propagation stresses.

Fonseca M.A., et al. [7], in their work, described a new class of shape-memory materials (SMMs) for applications in active control and morphing structures. These SMM can be metallic alloys (SMAs), piezoelectric, and polymers such as polyurethanes (SMPUs). The latter detain higher recovery rates but better processability; however, the reaction time is longer when compared with the SMA. The addition of carbon nanotubes (CNTs) to SMPU seems to improve its overall properties with a great deal of potential in what concerns improved shape memory.

Revathi A., et al. [8] described the effect of strain on the thermomechanical behavior of epoxy based shape memory polymers. They studied the effect of deformation strain on the thermomechanical (TM) behavior of a commercially available bifunctional epoxy resin modified with 5 % carboxyl-terminated butadiene acrylonitrile (CTBN). The results were compared with those for the unmodified epoxy SMP. The cyclic thermomechanical tests performed here prove that the CTBN-epoxy SMP has a better endurance than the unmodified epoxy SMP, as well as enhanced shape-recovery characteristics.

Aggogeri F., et al [9] used Phase Change Materials in Machine Tools (MTs) thermostabilization in an inverse way to maintain the shape over a temperature exchange. By considering the main issues related to the thermal stability, authors presented the application of novel multifunctional material to Machine Tools structures. A set of advanced materials are considered: aluminum foams, corrugate-core sandwich panels and polymeric concrete beds. The adopted solutions have been infiltrated by phase change materials (PCMs) in order to maintain the thermal stability of MTs when the environmental temperature is perturbed.

Wang H.L., et al. [10] studied the stochastic bifurcation and first-passage failure of shape memory alloy beam subjected to stochastic excitation. Based on the isostrain loading and unloading experimental data of SMA (shape memory alloy), the van der Pol cycle model was used to characterize the hysteretic nonlinearity of material. According to the theory of elastic and Galerkin's method, the vibration model was built for the beam excited by the Gaussian noise. The Hamilton function was taken to describe the one dimensional diffusion process by using stochastic average method for quasi-non-integrable-Hamiltonian systems. Using the largest Lyapunov exponent and the singular boundary theory simultaneously, both the local and global stability of the system were studied. The stochastic Hopf bifurcation was analyzed by observing the changes of stationary and joint probability densities. Finally, the Backward Kolmogorov equation for the reliability function and the first-passage time was established and the numerical results were given under designated boundary and initial conditions.

Interesting stochastic models were build up for different materials [11, 12] incorporating the stochastic characteristic in the parameters of the material, that are considered as imperfectly known. This error in the knowledge of such parameters is then transferred to the prediction of the stochastic behavior of the material.

These innovative materials [13] are complex to modelize and often FEM approach are used [14] non-linear consitutive laws must be considered. A main lack of these materials for mechanical power generation ist he low enrgy efficiency associated with the thermomechanical conversion, thus proper energy saving tecniques must be implemented to obtain further interesting results [15, 16]. Furthermore the finishing of the material can play a role in thermic exchange thus proper evaluations of surface treatments must be taken in consderation [17].

3 State of the art on systems

Shape memory materials are usually used to realize movements [18], where these movements can be, under some considerations, designed with the well-known approaches just used for different industrial devices [9, 19-23]. Due to their compactness, shape memory systems used in micro-devices, where the connection between elements can be properly realized with flexure hinges [24] or, more generally, with flexible elements [25-29].

Degeratu et al. [30] proposed a thermal study of a shape memory alloy (SMA) spring actuator designed to insure the motion of a barrier structure.

Borboni et al. [31-33] designed a modular SMA actuator. The actuator is able to perform linear movements, while the modularity allows force and/or stroke improvements.

Clausi et al. [34] realized a robust actuation of silicon MEMS using SMA wires integrated at wafer-level by nickel electroplating. The integration method provides both high bond strength and electrical connections in one processing step, and it allows mass production of microactuators having high work density. SEM observation showed an intimate interconnection between the SMA wires and the silicon substrate. The variation of the actuators' performance across the wafer was evaluated on three 4.5 mm × 1.8 mm footprint devices, proving repeatable results. The actuators showed a mean hot state deflection of 536 μm and a mean stroke of 354 μm at low power consumption (less than 70 mW). One actuator was tested for m150 × 10^3 cycles, and it demonstrated a highly reliable long-term performance, showing neither material degradation, nor failure of the nickel anchors.

Borboni et al. [19] proposed a fiber shaped micro-actuator for robotic gripping. This prototype is based on a shape memory alloy (SMA) actuator, on a dedicated PI controller and on software to command a desired motion profile for the actuator. The proposed micropositioning system is characterized by a 4 mm stroke, a 1 μm resolution and a 70 g nominal force and can be commanded directly by a personal computer and without a human retroaction. The closed loop positioning resolution (1 μm) is obtained in spite of an inaccurate system behavior during its movement.

SMA actuators are used in different applications to realize complex machines.

Furst et al. [35] realized a bat-inspired flapping-flight platform using shape memory alloy muscles and joints.

Amici et al. [25, 36] designed a parallel compliant meso-manipulator for finger rehabilitation treatments. Tiboni et al proposed a pneumatic mini-valve [37]. Kheirikhah et al. [38] proposed a new biomimetic earthworm robot for endoscopy actuated by SMA wires. Taveggia et al studied a rehabilitation device realized with SMA components [39, 40].

4 Conclusions

The research on shape memory materials and their applications is gaining new important results. The most new results on constitutive models are especially devoted to implement non-linear complex behaviors and stochastic characteristics. New materials show the possibility of combination of different materials (polymeric and metallic) to realize a multifunctional smart material system. Different applications are present in the most recent literature especially in robotics, welfare applications and bio-mimetics.

References

[1] Z. Wang, X. Zu, and Y. Fu, International Journal of Smart and Nano Materials, **2**, 101-119, // 2011
[2] N. Pellegrini, **651**, 42-48, 2013
[3] P. Ghosh, A. Rao, and A. R. Srinivasa, 2012
[4] Y. Zhu, Y. Zhang, and D. Zhao, Jinshu Xuebao/Acta Metallurgica Sinica, **49**, 123-128, // 2013
[5] B. Zhou and S. H. Yoon, Science China: Physics, Mechanics and Astronomy, 1-10, // 2013
[6] Z. Song, H. H. Dai, and Q. P. Sun, International Journal of Plasticity, **42**, 101-119, // 2013
[7] M. A. Fonseca, B. Abreu, F. A. M. M. Gonçalves, A. G. M. Ferreira, R. A. S. Moreira, and M. S. A. Oliveira, Composite Structures, **99**, 105-111, // 2013
[8] A. Revathi, S. Rao, K. V. Rao, M. M. Singh, M. S. Murugan, S. Srihari, et al., Journal of Polymer Research, **20**, 1-10, // 2013
[9] F. Aggogeri, A. Borboni, A. Merlo, and N. Pellegrini, Advanced Materials Research, **590**, 252-257, 2012
[10] H. L. Wang, S. L. Yao, G. Ge, and J. Xu, Zhendong yu Chongji/Journal of Vibration and Shock, **31**, 24-28, // 2012
[11] A. Borboni, D. De Santis, and R. Faglia, Proceedings of 8th Biennial ASME Conference on Engineering Systems Design and Analysis, 2006
[12] A. Borboni and R. Faglia, Journal of Applied Mechanics, **80**, 021003-021003, 2013
[13] B. Stamatovic, R. Upadhyay, and N. Vatin, Procedia Engineering, 2015, 660-667
[14] V.V Okrepilov, V.N. Krutikov, G.I. El'kin, Measurement Techniques, **57 (2)**,109-116 (2014)
[15] V.L. Kvint, V.V Okrepilov, Herald of the Russian Academy of Sciences, **84 (3)**, 188-200 (2014)
[16] V.V Okrepilov, V.L. Makarov, A.R. Bakhtizin, S.N Kuzmina, Economy of Region, **2**, 301-313 (2015)
[17] V.V Okrepilov, M.V. Leonidovich, Asian Social Science, **11 (7)**, 312-325 (2015)
[18] H. Yaguchi, K. Kamata, and K. Sasaki, 2012, 630-634
[19] F. Aggogeri, A. Borboni, R. Faglia, A. Merlo, and S. De Cristofaro, Applied Mechanics and Materials, **336-338**, 1170-1173, 2013
[20] A. Borboni, F. Aggogeri, N. Pellegrini, and R. Faglia, Advanced Materials Research, **590**, 399-404, 2012
[21] A. Borboni, R. Bussola, R. Faglia, P. L. Magnani, and A. Menegolo, Journal of Mechanical Design, Transactions of the ASME, **130**, 0823011-0823016, 2008
[22] F. Aggogeri, A. Borboni, and R. Faglia, Applied Mechanics and Materials, **373-375**, 130-133, 2013
[23] A. Borboni and M. Lancini, Journal of Vibration and Acoustics, Transactions of the ASME, **137**, 1-9, 2015

[24] N. Lobontiu, M. Cullin, M. Ali, and J. Hoffman, Mechanics Based Design of Structures and Machines, **41**, 202-221, // 2013

[25] C. Amici, A. Borboni, and R. Faglia, Advances in Mechanical Engineering, **2010**, // 2010

[26] A. Borboni and D. De Santis, Meccanica, **49**, 1327-1336, 2014

[27] A. Borboni, D. De Santis, and R. Faglia, ASME 2010 10th Biennial Conference on Engineering Systems Design and Analysis ESDA2010, **2**, 99-106, 2010

[28] C. Amici, A. Borboni, P. L. Magnani, and D. Pomi, Proceedings of EUCOMES 2008 - The 2nd European Conference on Mechanism Science, **2008**, 487-493, 2009

[29] C. Amici, A. Borboni, P. L. Magnani, and D. Pomi, Proceedings of EUCOMES 2008 - The 2nd European Conference on Mechanism Science, **2008**, 479-485, 2009

[30] S. Degeratu, P. Rotaru, S. Rizescu, and N. G. Bîzdoacă, Journal of Thermal Analysis and Calorimetry, **111**, 1255-1262, // 2013

[31] A. Borboni, F. Aggogeri, N. Pellegrini, and R. Faglia, Advanced Materials Research, **590**, 405-410, 2012

[32] A. Borboni, R. Faglia, and M. Palpacelli, MESA 2014 - 10th IEEE/ASME International Conference on Mechatronic and Embedded Systems and Applications, **1**, 1-7, 2014

[33] A. Borboni and R. Faglia, Procedia Engineering, 2014, 1378-1381

[34] D. Clausi, H. Gradin, S. Braun, J. Peirs, G. Stemme, D. Reynaerts, et al., Sensors and Actuators, A: Physical, **189**, 108-116, // 2013

[35] S. J. Furst, G. Bunget, and S. Seelecke, Smart Materials and Structures, **22**, // 2013

[36] C. Amici, A. Borboni, R. Faglia, D. Fausti, and P. L. Magnani, IEEE/RSJ International Conference on Intelligent Robots and Systems, 2008, 735-740

[37] M. Tiboni, A. Borboni, M. Mor, and D. Pomi, Proceedings of the Institution of Mechanical Engineers. Part I: Journal of Systems and Control Engineering, **225**, 443-451, 2011

[38] M. M. Kheirikhah, A. Khodayari, and S. M. Nikpey, 2012, 148-151

[39] G. Taveggia, J. H. Villafañe, F. Vavassori, C. Lecchi, A. Borboni, and S. Negrini, Journal of Manipulative and Physiological Therapeutics, **37**, 242-252, 2014

[40] A. Borboni, R. Faglia, and M. Mor, ASME 2014 12th Biennial Conference on Engineering Systems Design and Analysis ESDA 2014, **1**, 1-10, 2014

MATEC Web of Conferences **53**, 01062 (2016)
DOI: 10.1051/matecconf/20165301062

Review on Upper Limb Continuous Passive Motion Devices

Francesco Ragazzo[1,a]

[1]*Polibrixia, R&D Department, 25123 Brescia, Italy*

Abstract. The paper is devoted to a survey on the state of the art of elements and parts for the upper limb rehabilitation. As a matter of fact, the use of technological, and specifically of robotic, devices is entering in the habits of clinical approaches, due to their ability to work efficiently and to be able to obtain, at least, the same rehabilitation results of manual therapy. At the same time, the therapists can change his/her role in rehabilitation activity from a physical contribution to an intellectual/motivational one.

1 Introduction

As shown in literature, mobilization is effective in rehabilitation to improve the motor function and to reduce pain [1, 2]. These results can be obtained for different pathologies: stroke [3], cerebral palsy [4], osteoarthritis [5, 6], etc. Under these hypotheses, Salter, who proposed the continuous passive motion (CPM), derived this concept on the basis of a series of experimental investigations and well thought-out rationale. Salter and Field [7] showed in 1960 that immobilization of a rabbit knee joint under continuous compression, provided by either a compression device or forced position, resulted in pressure necrosis of the cartilage. In 1965, Salter et al. reported deleterious effects of immobilization on the articular cartilage of rabbit knee joints and the resultant lesion that they termed "obliterative degeneration of articular cartilage". Salter believed that "The relative place of rest and of motion is considerably less controversial on the basis of experimental investigation than on the basis of clinical empiricism". He reasoned that because immobilization is obviously unhealthy for joints, and if intermittent movement is healthier for both normal and injured joints, then perhaps continuous motion would be even better. Because of the fatigability of skeletal muscle, and because a patient could not be expected to move his or her own joint constantly, he concluded that for motion to be continuous it would also have to be passive.

2 State of the art

CPM devices are mechatronics systems [8], or, in the most complex solutions, a robotic device [9-14] able to implement automatically a CPM rehabilitation. A large number of devices and methods for CPM rehabilitation is well known in literature. The aim of this paper is to show the most interesting devices currently used for the rehabilitation of wrist and hand. Some of these devices are designed to perform both a coordinate movement and different exercises. Many of these do not realize a complete

[a] Corresponding author : f.ragazzo86@gmail.com

movement, for example, only flexor–extensor or abdo–adduction. Two list of CPM devices are shown (Table 1-2), respectively, for the wrist and for the hand rehabilitation.

Table 1. List of devices for CPM wrist rehabilitation. E.D. Electrical driver.

Device name	Ref.	Year	Exercises	ROM	Other
Hand Mentor - Kinetic muscles, inc.	[15]	2008	Flex and ext wrist and fingers.	Flexor limit 90°, extensor limit 90°.	Control security system, rotation reverse. P. D.
Patent US 5,458,560	[16]	1995	Flex and ext wrist	-	E.D.
Patent US 7,618,381	[17]	2009	Flex-ext abdo-addu of wrist	-	E.D.

Table 2. List of CPM devices for hand rehabilitation. E.D. Electrical driver, V.F. visual feedback. CPM continuous passive motion, CAM continuous active motion.

Device name	Ref.	Year	Exercises	ROM	Other
Gloreha	[18]	2011	Flex and ext finger.	Compete of flex or ext exercise	Moves fingers smoothly, both individually and simultaneously. V.F.
Tyromotion Amadeo	[19]	2011	Flex and ext fingers (Active and passive exercises).	Compete of flex or ext exercise	V.F., individual and contemporary exercises, forces calibration.
Artromot-F	[20]	2008	Flex and ext finger.	Compete of flex or ext exercise	LCD display for monitoring, microcomputer for control system.
Patent	[21]	1987	Flex and ext finger	Compete of flex or ext exercise	Security system control. E.D.
Patent	[22]	1990	Flex and ext finger	Compete of flex or ext exercise	Spiral motion, Security control system E.D.
Patent	[23]	2010	Flex ext finger CPM CAM	90° Flex-Ext	Security control system, E.D

Different actuator systems are shown in literature [24-26] and are classified with different criteria; one of the most common classifications is based on the used technology: piezoelectric devices [27, 28], electric motors [29, 30], shape memory actuators [31-37], etc. The most commonly used devices are electric ones that represent also the most common choice in rehabilitation devices.

3 Electrical driver's rehabilitation

When a designer starts a project, the first specification is on electrical drivers. There is two possible solution, on the patient body or separate from body. For the first one, the electrical drivers have to be light and small, because the size can impede the movement. For the second solution, the driver's size is not important but transmission is essential. On the market there are two principal families of electrical drivers: rotary motors and linear motors.

Rotary motors. The use brushless motors is more frequent in biomechanics [29] because they are efficient with intermittent movement. Usually the electrical devices are used with a speed reducer [30]. Nowadays these devices are cheap. Another kind of rotary motors are DC drivers. These drivers are usually employed for their easy control. Asynchronous motors are used for repetitive movements because changing the speed is difficult and expensive. The final results are shown in Table 3.

Table 3. Difference between rotary motors.

Type	Max torque	Movement type	Driver electronic	Cost	Maintenance
Brushless	Average-high torque	Not periodic movements	Necessary and complex electrical circuit	High	Sometimes
DC	Low torque	Not periodic movements	Not necessary and easy	Low	Frequent
Asynchronous	High torque	Periodic movements	Necessary for speed regulation	Low	Rarely

Linear motors. The relative position of the armature and stator of a linear motor is controlled using the output of two optional transducers in phase quadrature from an optical grid disposed along the length of the stator. This device is very expensive and little used.

4 Mechanical transmission

The movement's generation is based on electrical drivers with high-speed. Transmission is used for reducing and changing movement of end-effectors. In sequence, after the motor there is a speed-reducer. The most used are divided into main classes: ordinary reducer, epicyclical reducer, Harmonic-Drive reducer and RV-A reducer. The first is less used for low efficiency. The most used system is epicyclic reducer because it is cheap and it has a good efficiency. The Harmonic-Drive has a high ratio but it does not allow the retrograde movement. The final results are in Table 4.

Table 4. Difference between reducers.

Type	Ratio	Efficiency	Retrograde motion	Cost
Ordinary reduc.	1/10 max	Very low	Yes	Low
Epicyclic reduc.	1/10 ÷ 1/200	Sufficient	Yes	Intermediate
Harmonic-Drive	1/50 ÷ 1/400	High	No	High
RV-A reduc.	1/50 ÷ 1/400	Very high	No	Very High

After the reducer usually there is the transmission's movement. For this problem there are a lot of solution: belt, machanisms system, cogwheel, flexible shafts [38] and chain. Interesting solutions are characterized by a fluid media as power transmission system [39-41]. Flexible shafts can divide the part where there are the electrical motor and the limb of patient.

5 Future developments

The main problem is that the science is not secure of methods and exercises for a correct rehabilitation. This uncertainty does not allow a singular way for the designing rehabilitation devices. The first work is defining correct exercises and methods, for specifics pathologies with the collaboration of medical staff. Once they will be defined, research could start the mechatronic design. Nowadays we are still looking for these suggestions from medical staff to implement a proper solution for rehabilitation designing a complete device, both for the wrist and for the hand. Two important characteristics are:

- the coordinating exercise, that is when the patient use more than one degree of freedom in an exercise, i.e. to move the wrist in flexo-extension and abdo-adduction together.
- the visual feedback of exercise for the patient is very important in rehabilitation, because, if the patient watches the virtual movement on the monitor, it is possible to activate intact neuronal systems that provide direct stimulation to motor areas affected by brain lesions.

An interesting demanded function for these systems is the possibility of realizing a specific motion profile [42-46], whereas the common state of the art is based on the possibility of assigning an actuation motion, thus a proper system must realize a kinematic inversion [47-49] to transform a finger motion in an actuator motion and, then, properly optimize [50-53] the motion profile of the actuator to realize the desired finger motion, with an adequate control system [52-54].

6 Conclusions

In conclusion of this paper, on the market there is not a complete device for the wrist and hand rehabilitation with coordinated movements, that could be used for a large range of pathologies from neuronal to orthopedic disease. The rehabilitative research is very important, because only they are able to suggest the exact way for realizing instruments and methods for rehabilitation.

References

[1] J. H. Villafane, G. B. Silva, M. D. Bishop, and J. Fernandez-Carnero, *Archives of physical medicine and rehabilitation,* **93**, 396-403, Mar 2012

[2] J. H. Villafane, G. B. Silva, S. A. Diaz-Parreno, and J. Fernandez-Carnero, *Journal of manipulative and physiological therapeutics,* **34**, 547-56, Oct 2011

[3] A. Sundseth, B. Thommessen, and O. M. Ronning, *Journal of stroke and cerebrovascular diseases : the official journal of National Stroke Association,* May 13 2013

[4] D. P. Grogan, M. S. Lundy, and J. A. Ogden, *Journal of pediatric orthopedics,* **7**, 338-40, May-Jun 1987

[5] J. H. Villafane, M. D. Bishop, C. Fernandez-de-Las-Penas, and D. Langford, *Journal of physiotherapy,* **59**, 25-30, Mar 2013

[6] J. H. Villafane, J. A. Cleland, and C. Fernandez-de-Las-Penas, *The Journal of orthopaedic and sports physical therapy,* **43**, 204-13, Apr 2013

[7] R. B. Salter and P. Field, *The Journal of Bone & Joint Surgery,* **42**, 31-90, 1960

[8] F. Aggogeri, A. Borboni, and R. Faglia, *Applied Mechanics and Materials,* **373-375**, 130-133, 2013

[9] C. Amici, A. Borboni, and R. Faglia, *Advances in Mechanical Engineering,* **2010**, 706023, 2010 2010

[10] C. Amici, A. Borboni, R. Faglia, D. Fausti, and P. L. Magnani, *Proceedings of the IEEE/RSJ International Conference on Intelligent Robots and Systems*, 2008, 735-740

[11] G. Taveggia, J. H. Villafañe, F. Vavassori, C. Lecchi, A. Borboni, and S. Negrini, *Journal of Manipulative and Physiological Therapeutics,* **37**, 242-252, 2014

[12] A. Borboni, R. Faglia, and M. Mor, *ASME 2014 12th Biennial Conference on Engineering Systems Design and Analysis ESDA 2014,* **1**, 1-10, 2014

[13] C. Amici, A. Borboni, P. L. Magnani, and D. Pomi, *Proceedings of EUCOMES 2008 - The 2nd European Conference on Mechanism Science,* **2008**, 487-493, 2009

[14] C. Amici, A. Borboni, P. L. Magnani, and D. Pomi, *Proceedings of EUCOMES 2008 - The 2nd European Conference on Mechanism Science,* **2008**, 479-485, 2009

[15] L. Rosenstein, A. L. Ridgel, A. Thota, B. Samame, and J. L. Alberts, *The American journal of occupational therapy : official publication of the American Occupational Therapy Association,* **62**, 28-35, Jan-Feb 2008

[16] R. T. Kaiser, R. Zambelli, G. Telepko, B. C. Kalustyan, and V. Ricci, "Continuous passive motion device for a wrist," U.S. Patent 5,458,560, Ottobre 1995, 1995

[17] H. I. Krebs, N. Hogan, D. Williams, and J. Celestino, "Wrist and upper extremity motion," U.S. Patent 7,618,381, Novembre 2009, 2009

[18] K. I. Ites, E. J. Anderson, M. L. Cahill, J. A. Kearney, E. C. Post, and L. S. Gilchrist, *Journal of geriatric physical therapy,* **34**, 109-16, Jul-Sep 2011

[19] J. Stein, L. Bishop, G. Gillen, and R. Helbok, *American journal of physical medicine & rehabilitation / Association of Academic Physiatrists,* **90**, 887-94, Nov 2011

[20] M. Syrko and J. Jablecki, *Chirurgia narzadow ruchu i ortopedia polska,* **73**, 257-8, Jul-Aug 2008

[21] J. B. Yates and L. M. Lubbers, "Hand exerciser," U.S. Patent 4,644,938, Febbraio 1987, 1987

[22] D. Shamir and M. W. Groves, "Continuous passive motion device for imparting a spiral motion to the digits of the hand," U.S. Patent 4,962,756, Ottobre 1990, 1990

[23] E. J. W. Park, B. J. Birch, C. E. Haslam, N. Dechev, and I. Heerah, "Continuous passive and active motion device and method for hand rehabilitation," U.S. Patent 12,539,565, Agosto 2009, 2009

[24] A. Borboni, D. De Santis, and R. Faglia, *10th Biennial Conference on Engineering Systems Design and Analysis ESDA2010*, 2010, 99-106

[25] F. Aggogeri, A. Borboni, R. Faglia, A. Merlo, and S. De Cristofaro, *Applied Mechanics and Materials,* **336-338**, 1170-1173, 2013 2013

[26] A. Borboni and D. De Santis, *Meccanica,* **49**, 1327-1336, 2014

[27] A. Borboni, D. De Santis, and R. Faglia, *Proceedings of 8th Biennial ASME Conference on Engineering Systems Design and Analysis*, 2006

[28] A. Borboni and R. Faglia, *Journal of Applied Mechanics,* **80**, 021003-021003, 2013

[29] H. R. Johnson, "Permanent Magnet Motor," U.S. Patent 4,151,431, Aprile 1979, 1979

[30] R. J. Tolmie Jr, "Integrated driver-encoder assembly for brushless motor," Patent US 4,988,905, Gennaio 1991, 1991

[31] A. Borboni, Aggogeri, F.; Faglia, R., *International Journal of Advanced Robotic Systems,* **10**, 1-10, 2013

[32] K. K. Ahn and N. B. Kha, *Mechatronics,* **18**, 141-152, Apr 2008

[33] A. Borboni, F. Aggogeri, N. Pellegrini, and R. Faglia, *Advanced Materials Research,* **590**, 405-410, 2012

[34] A. Bhattacharyya, W. L. Ables, G. Kumara, and M. A. Qidwai, *Smart Structures and Materials 2004 Active Materials Behavior and Mechanics*, 2004, 118-127

[35] A. Borboni, R. Faglia, and M. Palpacelli, *MESA 2014 - 10th IEEE/ASME International Conference on Mechatronic and Embedded Systems and Applications,* **1**, 1-7, 2014

[36] A. Borboni and R. Faglia, *Procedia Engineering*, 2014, 1378-1381

[37] M. Tiboni, A. Borboni, M. Mor, and D. Pomi, *Proceedings of the Institution of Mechanical Engineers. Part I: Journal of Systems and Control Engineering,* **225**, 443-451, 2011

[38] B. Kovačič, R. Kamnik, Z. Kapović, Geodetski List, **63** (4), 315-327 (2009)

[39] B. Kovačič, R. Kamnik, A. Štrukelj, Promet - Traffic - Traffico, **22** (1), 15-22 (2010)

[40] V.V Okrepilov, Standarty i Kachestvo, **3**, 94-96. (2003)

[41] V.V Okrepilov, Ivanova, G. Standarty i Kachestvo, **12**, 62-68. (2004)

[42] A. Borboni, F. Aggogeri, N. Pellegrini, and R. Faglia, *Advanced Materials Research,* **590**, 399-404, 2012

[43] H. Li, M. D. Le, Z. M. Gong, and W. Lin, *Ieee-Asme Transactions on Mechatronics,* **14**, 264-269, Apr 2009

[44] K.-H. Rew and K.-S. Kim, *Ieee Transactions on Industrial Electronics,* **57**, 2499-2506, Jul 2010

[45] A. Borboni and M. Lancini, *Journal of Vibration and Acoustics, Transactions of the ASME,* **137**, 1-9, 2015

[46] A. Borboni, R. Bussola, R. Faglia, P. L. Magnani, and A. Menegolo, *Journal of Mechanical Design, Transactions of the ASME,* **130**, 0823011-0823016, 2008

[47] A. Borboni, *IEEE International Conference on Fuzzy Systems*, 2001, 336-339

[48] D. N. Nenchev, *Proceedings of the IEEE International Conference on Robotics and Automation*, 1994, 2698-2703

[49] A. Ramdane-Cherif, B. Daachi, A. Benallegue, N. Levy, and I. Ieee, *Proceedings of the Ieee/Rsj International Conference on Intelligent Robots and Systems*, 2002, 1904-1909

[50] M. Antonini, A. Borboni, R. Bussola, and R. Faglia, *Proceedings of 8th Biennial ASME Conference on Engineering Systems Design and Analysis,* 95122, 2010 2006

[51] M. Antonini, A. Borboni, R. Bussola, and R. Faglia, *Proceedings of 8th Biennial ASME Conference on Engineering Systems Design and Analysis,* 95123, 2010 2006

[54] Y. Shi, J. Huang, and B. Yu, *Ieee Transactions on Industrial Electronics,* **60**, 5864-5874, Dec 2013

[55] F. Aggogeri, A. Borboni, A. Merlo, and N. Pellegrini, *Advanced Materials Research,* **590**, 252-257, 2012

[56] F. Xu, L. Shi, and Y. Li, *Ieee Transactions on Power Electronics,* **28**, 5665-5672, Dec 2013

MATEC Web of Conferences **53**, 01063 (2016)
DOI: 10.1051/matecconf/20165301063

Discretization Analysis in FEM Models

Giulia Martello[1,a]

[1]*Università degli Studi di Brescia, Mechanical and Industrial Engineering Department, Brescia, Italy*

Abstract. Discretization allows to switch from an infinite number of elements in a finite number and thanks to that the infinite degrees of freedom of the system become a finite number. The discretization of the domain consists of a subdivision into subdomains, which form a lattice or mesh. This work deals on different types of mesh and propose a choice for a specific problem.

1 Introduction

The discretisation [1-4] of the solution domain produces a numerical description of the computational domain [5-9]. The space is divided into a finite number of discrete regions or cells [10, 11], which form a mesh [12, 13].
This allows to switch from an infinite number of elements in a finite number and thanks to that the infinite degrees of freedom of the system become a finite number [14].
The discretization with Finite Element Method (FEM) [15-19] is based on the principle which is called with the latin expression "*divide et impera*" (thus "divide and rule"). This means that a complex problem is decomposed into many simpler problems, whose solutions lead to the solution of the general problem [20-23]. In this case the model of difficult solution is decomposed into many elements that can be solved more easily.
Alongside spatial discretization, also temporal discretization is, for transient simulations, fundamental as well, thus the time interval is also split into a finite number of time-steps [24-33]. This leads to less computational efforts and an analysis of smaller deformation of the general problem.

[a] Corresponding author : giulia.mrt@gmail.com

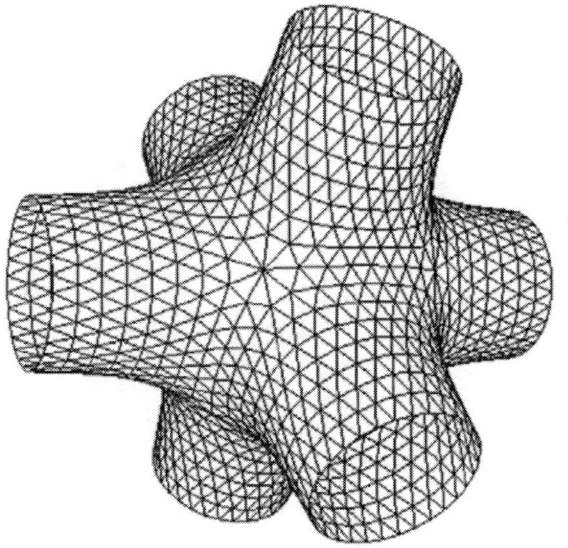

Figure 1. Mesh of a complex geometry.

The mesh generation is a fundamental step in the simulation because it is an integral part of the numerical solution and must satisfy certain criteria to ensure a valid, and hence accurate, solution [34-39]. The mesh size [40-42] (defined for each direction and identified by the characteristic length *h*) has to be optimized [43-48], indeed it should be fine enough to make the problem easily resolvable and at the same time it has not to be too fine because the higher number of cells, the higher computational costs. Therefore the mesh density [49] depends also on the machine which is used for computations. After an accurate observation of the model shape and of gradients involved, a proper mesh can be obtained by refining it in the regions with physical needs, for example close to geometrical details that imply high gradients of the variable evaluated.
In the beam considered the variable which has a fundamental role is the stress component S12, parallel to the *y* axis.

2 Types of meshing

Typically, cells used in FEM are simple geometric figures (in the one-dimensional problem segments, triangles or quadrilaterals in 2D problems, tetrahedrons or hexagonal in 3D problems). The topology is defined by special points called *nodes*.
The main types of cells that one can choose in a three-dimensional model follow.
• HEX: this type of elements are polyhedrons with 6 faces and quadrilateral faces, therefore these elements have 8 knots. An example of a mesh constructed by using these elements is shown in Figure 2.

Figure 2. Mesh with cells type HEX.

• HEX-dominated: HEX elements dominate, nevertheless it is also allowed to use some triangular base prisms (wedge) in the transition regions. An example in Figure 2.4.

Figure 3. Mesh with HEX-dominated elements.

• TET: composition of elements only tetrahedral, thus with 4 knots, as shown in figure 4

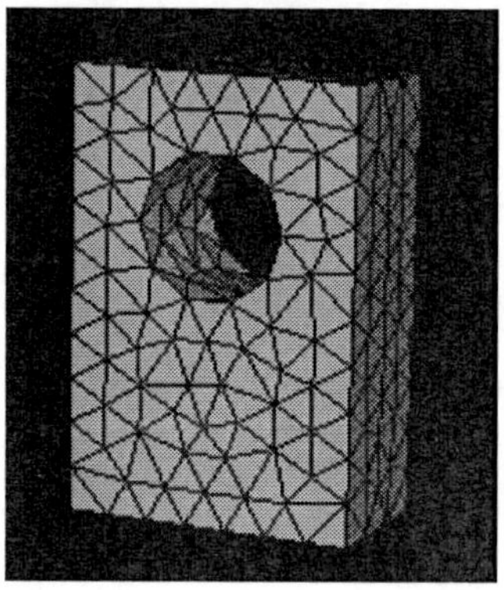

Figure 4. Mesh with tetrahedral elements.

• WEDGE: composition of wedge-shaped elements with 6 knots. More precisely, it refers to prisms with a triangular base, as shown in figure 5.

Figure 5. Element with WEDGE shape.

3 Meshing effects and selection

Given these four possible forms of item, one must choose the most appropriate geometry treated. It can be noted that HEX-dominated and HEX types differ only if the geometry is not completely regular; for example a triangular base prisms may appear if one uses this type of elements for a square plate with smoothed corner or with a hole in the middle. Since the geometry considered is extremely regular, this option is not taken into account. Also the tetragonal form of elements can be discarded, since these elements can be used to simplify a complex geometry but if the starting shape is already regular the use of these elements can lead to a more complicated problem. The more plausible possibilities are therefore HEX and WEDGE.

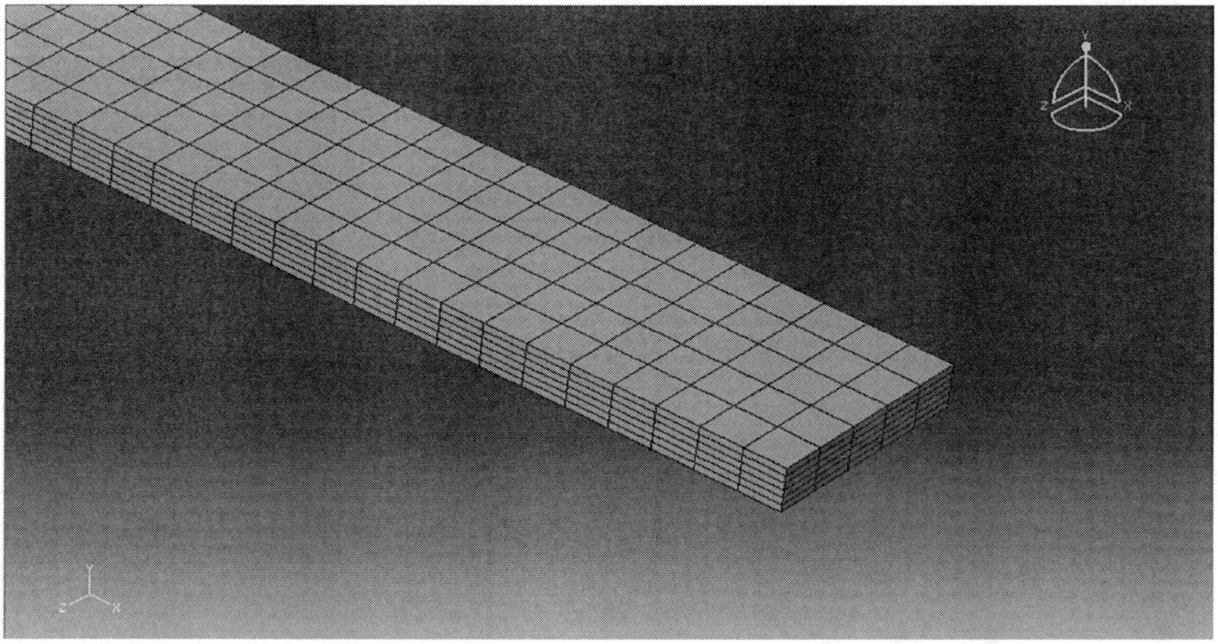

Figure 6. Discretization with HEX cells option.

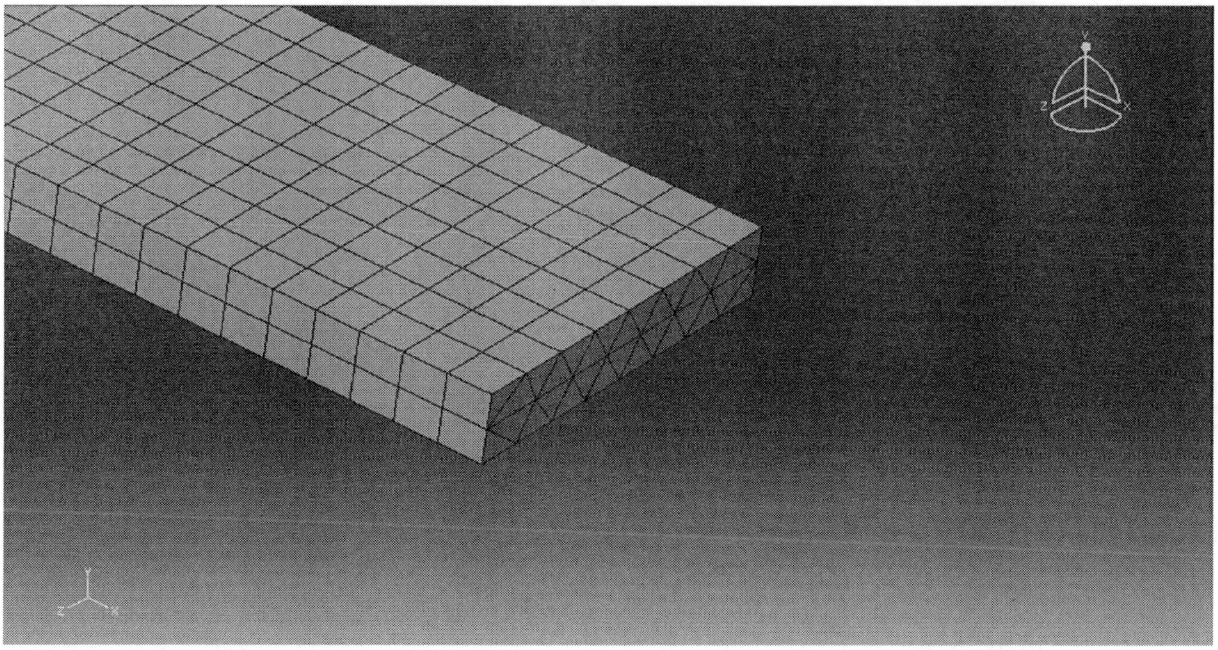

Figure 7. Discretization with WEDGE cells option.

Concerning elements with triangular base, the concept explained regarding case tetrahedral is still true, even if to a lesser degree. Indeed, also WEDGE elements introduce some irregularity in a regular model.

HEX mesh, or in cubes or generic parallelepipeds, is the simplest and most intuitive mesh which perfectly reflects the geometry of the model. Nevertheless before approving it permanently also boundary conditions and loads must be taken into account, since they could make irregular stresses even if within a regular geometry.

As seen, the beam is analyzed by using a fixed joint at one end and the other end a load in axis with the symmetry axis of the beam is applied. The model then is symmetric also taking into account constraints and loads. It can therefore be assumed that the discretization with HEX cells does not limit the analysis on the beam.

Moreover, results obtained with the two types of meshes are comparable, since the sample tests carried out reveal an error of about 1%.

Based on these considerations, it was decided to use a rectangular mesh (HEX type).

4 Conclusions

This work describes an introduction to the discretization problem and to different types of elements. Furthermore, after a discussion on different type of meshes, we propose a solution for a specific discretization problem.

References

[1] T. Nguyen-Van and N. Hori, *Journal of Computational and Nonlinear Dynamics,* **11**, 2016

[2] A. Pascau and M. Arici, *Computers and Fluids,* **125**, 101-115, 2016

[3] A. Borboni and D. De Santis, *Meccanica,* **49**, 1327-1336, 2014

[4] A. Borboni, D. De Santis, and R. Faglia, *ASME 2010 10th Biennial Conference on Engineering Systems Design and Analysis ESDA2010,* **2**, 99-106, 2010

[5] A. Caprara, M. Locatelli, and M. Monaci, *Computational Optimization and Applications,* 1-21, 2016

[6] J. Gai, Z. H. Xia, and Q. D. Cai, *Chinese Physics B,* **24**, 2015

[7] C. Amici, A. Borboni, and R. Faglia, *Advances in Mechanical Engineering,* **2**, 1-9, 2010

[8] C. Amici, A. Borboni, P. L. Magnani, and D. Pomi, *Proceedings of EUCOMES 2008 - The 2nd European Conference on Mechanism Science,* **2008**, 487-493, 2009

[9] C. Amici, A. Borboni, P. L. Magnani, and D. Pomi, *Proceedings of EUCOMES 2008 - The 2nd European Conference on Mechanism Science,* **2008**, 479-485, 2009

[10] N. Nguyen-Minh, T. Nguyen-Thoi, T. Bui-Xuan, and T. Vo-Duy, *Applied Mathematics and Computation,* **266**, 212-234, 2015

[11] M. H. Nguyen-Thoi, L. Le-Anh, V. Ho-Huu, H. Dang-Trung, and T. Nguyen-Thoi, *Journal of Contemporary Physics,* **50**, 341-358, 2015

[12] M. Brdar and H. Zarin, *Journal of Computational and Applied Mathematics,* **292**, 307-319, 2016

[13] Y. Yuan, R. Wang, J. Huang, Y. Jia, and H. Bao, *Computers and Graphics (Pergamon),* **54**, 135-144, 2016

[14] A. Borboni, F. Aggogeri, N. Pellegrini, and R. Faglia, *Advanced Materials Research,* **590**, 399-404, 2012

[15] A. Bautista, J. Montesinos, and P. Pintado, *Mechanism and Machine Theory,* **97**, 100-111, 2016

[16] G. Buffa, J. Hua, R. Shivpuri, and L. Fratini, *Materials Science and Engineering A,* **419**, 381-388, 2006

[17] C. Farhat, *Computers and Structures,* **28**, 579-602, 1988

[18] A. Borboni, R. Faglia, and M. Mor, *ASME 2014 12th Biennial Conference on Engineering Systems Design and Analysis ESDA 2014,* **1**, 1-10, 2014

[19] N. L. Vatin, J. Havula, L. Martikainen, A. S. Sinelnikov, A. V. Orlova, and S. V. Salamakhin, *Advanced Materials Research* **945-949**, 1211-1215, 2014

[20] D. C. Rizos and Z. Wang, *Engineering Analysis with Boundary Elements,* **26**, 877-888, 2002

[21] G. R. Liu, T. Nguyen-Thoi, H. Nguyen-Xuan, and K. Y. Lam, *Computers and Structures,* **87**, 14-26, 2009

[22] A. Borboni, F. Aggogeri, and R. Faglia, *International Journal of Advanced Robotic Systems,* **10**, 1-10, 2013

[23] A. Borboni, R. Faglia, and M. Palpacelli, *MESA 2014 - 10th IEEE/ASME International Conference on Mechatronic and Embedded Systems and Applications,* **1**, 1-7, 2014

[24] X. Guan, B. Kang, N. Shu, Q. Yan, and Z. Li, *Applied Computational Electromagnetics Society Journal,* **30**, 1223-1231, 2015

[25] W. Huang, S. Koric, X. Yu, K. J. Hsia, and X. Li, *Nano Letters,* **14**, 6293-6297, 2014

[26] H. K. Kiran Kumar and D. Ajad, *International Journal of Applied Engineering Research,* **10**, 44383-44386, 2015

[27] A. Borboni, F. Aggogeri, N. Pellegrini, and R. Faglia, *Advanced Materials Research,* **590**, 405-410, 2012

[28] M. Tiboni, A. Borboni, M. Mor, and D. Pomi, *Proceedings of the Institution of Mechanical Engineers. Part I: Journal of Systems and Control Engineering,* **225**, 443-451, 2011

[29] F. Aggogeri, A. Borboni, A. Merlo, and N. Pellegrini, *Advanced Materials Research,* **590**, 252-257, 2012

[30] A. Borboni and M. Lancini, *Journal of Vibration and Acoustics, Transactions of the ASME,* **137**, 1-9, 2015

[31] L. Muravyeva and N. Vatin, *Applied Mechanics and Materials* **635-637**, 451-456, 2014

[32] S. Ognjenovic, R. Donceva, and N. Vatin, *Procedia Engineering*, 2015, 556-563

[33] N. Vatin, K. I. Strelets, and N. Kharkov, *Applied Mechanics and Materials* **635-637**, 17-21, 2014

[34] D. L. Michels, J. P. T. Mueller, and G. A. Sobottka, *Computers and Graphics (Pergamon),* **53**, 136-146, 2015

[35] J. Yan and D. Jiao, *IEEE Antennas and Propagation Society, AP-S International Symposium (Digest)*, 2015, 342-343

[36] J. Yan and D. Jiao, *IEEE Antennas and Propagation Society, AP-S International Symposium (Digest)*, 2015, 1830-1831

[37] A. Borboni, R. Bussola, R. Faglia, P. L. Magnani, and A. Menegolo, *Journal of Mechanical Design, Transactions of the ASME,* **130**, 0823011-0823016, 2008

[38] C. Amici, A. Borboni, R. Faglia, D. Fausti, and P. L. Magnani, *IEEE/RSJ International Conference on Intelligent Robots and Systems IROS,* **2008**, 735-740, 2008

[39] F. Aggogeri, A. Borboni, R. Faglia, A. Merlo, and S. De Cristofaro, *Applied Mechanics and Materials,* **336-338**, 1170-1173, 2013

[40] E. Ágoston-Szabó, K. Schöll, A. Kiss, and M. Dinka, *Hydrobiologia,* 1-16, 2015

[41] P. K. Pandey and B. D. Pandey, *Boletim da Sociedade Paranaense de Matematica,* **34**, 9-27, 2016

[42] C. Wendeler and A. Volkwein, *Natural Hazards and Earth System Sciences,* **15**, 2597-2604, 2015

[43] Y. Sakamoto, Y. Bonkobara, T. Kondou, K. Abe, and T. Taniguchi, *Nihon Kikai Gakkai Ronbunshu, C Hen/Transactions of the Japan Society of Mechanical Engineers, Part C,* **77**, 4367-4381, 2011

[44] S. Wei, Z. Lei, and Q. Du, *Jixie Qiangdu/Journal of Mechanical Strength,* **32**, 859-864, 2010

[45] A. Borboni and R. Faglia, *Procedia Engineering*, 2014, 1378-1381

[46] F. Aggogeri, A. Borboni, and R. Faglia, *Applied Mechanics and Materials,* **373-375**, 130-133, 2013

[47] A. Borboni and R. Faglia, *Journal of Applied Mechanics Transactions ASME,* **80**, 1-7, 2013
[48] M. Adjemout, N. Brunetiere, and J. Bouyer, *Proceedings of the Institution of Mechanical Engineers, Part J: Journal of Engineering Tribology,* **229**, 1132-1144, 2015

MATEC Web of Conferences **53**, 01064 (2016)
DOI: 10.1051/matecconf/20165301064

The Use of Design Models of Wind-Electric Set with a Horizontal Axis of Rotation of the Wind Wheel for Dynamic Calculations at Urban Development

Igor Konstantinov[1], Alexey Savchenko[1], Artem Frolov[1,a], Luka Akimov[1] and Mihail Holmin[1]

[1]*Peter the Great Saint-Petersburg Polytechnic University, 29 Polytechnicheskaya st., St.Petersburg, 195251, Russian Federation*

Abstract. The issues of modern urban development raise a significant question about an environmental cleanliness of progressing cities. Energy sources which are running on fuel cause tremendous harm to the atmosphere. Therefore, special attention is paid to the rational use of natural renewable resources such as wind and solar energy. Wind-electric sets, or wind turbines, are able to work autonomously, which is also important for the development of modern "smart" cities. Currently, the most commonly used design of wind turbines is the system which has the form of a tower of circular cross section (also called pipe), which carries at the upper end a nacelle with wind wheel. When such a system is being designed in urban conditions the wind pulsation and seismic calculations are added to the standard calculations. These added calculations are dynamic loads. It is known that in the process of solution of dynamic tasks design models of various levels of approximation can be used. It occurs due to stages of the design and other factors. The question of errors, which are associated with the use of a dissected, or partitioned, design scheme, raises.

1 Introduction

In today's urban construction it is very important to pay attention to renewable energy sources [1-4] because of their ability to harmless and autonomously supply power to rapidly developing progressing cities [5-7]. Multifactorial assessment of wind energy potential of urban development area allows to use the wind energy wisely [8]. One of the most commonly used renewable energy sources is wind turbine which has the form of a tower of circular cross section (also called pipe), which carries at the upper end a nacelle with wind wheel. The wind wheel has two or three blades and it is rotated about a horizontal axis or an axis deviated from the horizontal by a small angle (in one type of the wind-electric set called “Raduga” this angle is 5°). The lower end of the tower rests on massive reinforced

[a] Corresponding author: artemfrolov007@mail.ru

concrete foundation or reinforced concrete grillage of pile foundation, which is relying on the subgrade.

When such a system is being designed in urban conditions some calculations of dynamic loads, such as wind pulsation and seismic, are added to the standard calculations. Therefore, it is necessary to solve a number of dynamic problems including the definition of natural frequencies (NF) and forms of natural oscillations (FNO) of tower at its bending fluctuations [9-10]. It is known that in the process of solution of dynamic tasks computational schemes of various levels of approximation can be used. It occurs due to stages of the design and other factors. Thus, at the stage of conceptual design fairly simple schemes can be applied, and then these schemes can be refined. Factor that influenced on the design schemes of wind turbines now is really established practice in our country of designing the system "wind electric set-foundation-base", when the part of the system located above the foundation (the actual wind turbine) is designed in design department of engineering type (eg, "Raduga" in Dubna) and the foundation with the base are designed in building type design organization (for example, Lengidroproekt in St. Petersburg).

Analysis of calculation of projected wind-electric set shows that the design scheme of the system "turbine-foundation-base" at dynamic calculations is divided into the base of the tower at the bottom and the part above the foundation is treated separately from the foundation as a tower console clamped on her soles. Efforts by acting loads, which are acting on the part above the foundation and which are defined in the seals, are transmitted as a load in the form of forces and moments (Fig. 1) for calculating the foundation and the base.

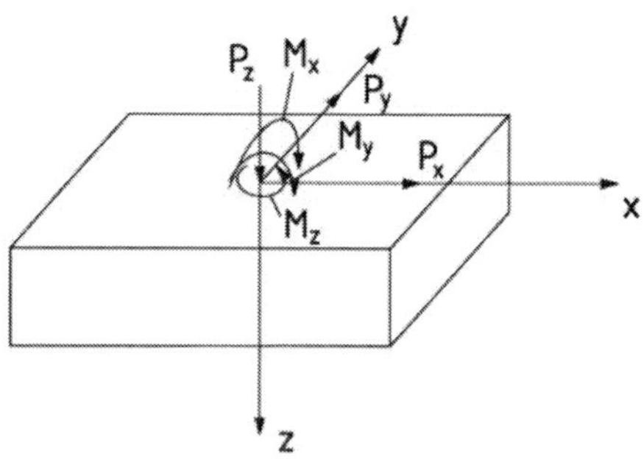

Figure 1. Load in the form of forces and moments in calculation of the foundation and base.

The question of errors, which are associated with the use of a dismembered design scheme, raises. There are no known specific studies on this issue. In [11] it is suggested to estimate the additional inertial loads approximately, these loads are caused because of the fluctuation of the mass of the elements above the foundation of the system and because of deformation of the base by the mass of the foundation. To estimate this additional inertial loads the calculation scheme in the form of a rigid system "wind turbine-foundation (or grillage pile foundation)" on a linearly deformable soil foundation under the action of harmonic loads (resulting from the calculation of the elements above the foundation of the wind turbine) on the gasket face of the foundation are used.

2 Description of research

To simplify the research in [11] it is suggested to consider a number of simple tasks to said rigid of the system in the form of steady vertical harmonic oscillations of the load $P_{z\,dyn} = P_z \sin\overline{\omega}t$, rotational oscillations of the load $M_{z\,dyn} = M_z \cos\overline{\omega}t$, and horizontal rotational fluctuations in planes XOZ and YOZ. For example, in the plane XOZ harmonic steady-state oscillations are considered of the loads $P_{x\,dyn} = P_x \sin\overline{\omega}t$ and $M_{y\,dyn} = M_y \cos\overline{\omega}t$. The amplitudes of the loads and frequency $\overline{\omega} = k\omega$, where ω is the angular frequency of rotation of the propeller, k is a certain coefficient, are set.

In this paper we present the results of these calculations which were performed on the Department of structural mechanics and theory of elasticity SPbGPU on the instructions of Lengidroproekt for wind turbines such as "Raduga 1000". The calculations are performed in accordance with the recommendations of [11] for disturbing frequencies $\overline{\omega} = \omega$ and $\overline{\omega} = 3\omega$, where $\omega = 4.4$ s^{-1} is the angular frequency of rotation of the propeller at the maximum operating speed (42 rev/min). Metal tower of wind turbine has a height of 36 m, the mass of 6116 (kgf·s^2·m^{-1}), mass moment of inertia of the tower around the axis of the tower, passing through its center of gravity which is perpendicular to the plane of oscillation $\theta_{\phi T} = 0.660 \cdot 10^6$ (kgf·s^2·m). The mass of the nacelle with the wind wheel and its moment of inertia about the axis, which is passing through the center of gravity and perpendicular to the oscillation plane, are respectively: $m_n = 5066$ (kgf·s^2·m^{-1}); $\theta_{\phi n} = 0.108 \cdot 10^6$ (ktf·s^2·m).

Massive foundation is a square in plan concrete slab with height $h_f = 3$ m and side length $b_f = 11$ m. Weight of foundation $m_f = 92500$ (kgf·s^2·m^{-1}), mass moment of inertia around the foundation around the axis of the plate, passing through its center of gravity which is perpendicular to the plane of oscillation, $\theta_{\phi f} = 1.002 \cdot 10^6$ (kgf·s^2·m^{-1}).

The base is loam with deformation modulus $E = 1.5 \cdot 10^6$ (kgf·m^{-2}). Stiffness coefficients of the base are calculated by the (Eq. 1-3) of [12]:

$$\boldsymbol{K_z} = \boldsymbol{C_z} \cdot \boldsymbol{A} = 2.317 \cdot 10^6 \cdot 121 = 0.280 \cdot 10^9 \text{ (kgf·m}^{-1}\text{)}, \quad (1)$$

$$\boldsymbol{K_x} = \boldsymbol{C_x} \cdot \boldsymbol{A} = 1.622 \cdot 10^6 \cdot 121 = 0.196 \cdot 10^9 \text{(kgf·m}^{-1}\text{)}, \quad (2)$$

$$\boldsymbol{K_\phi} = \boldsymbol{C_\phi} \cdot \boldsymbol{A} = 4.645 \cdot 10^6 \cdot 1220 = 5.655 \cdot 10^9 \text{ (kgf·m}^{-1}\text{)}. \quad (3)$$

The angular frequency of vertical oscillations of the considered rigid system was found to be 52s^{-1}. When $P_Z = 3900$ kg, the additional vertical inertial force in view of the weight of the foundation under steady harmonic oscillation with frequency $\omega = 4.4$ s^{-1} was found to be 30 kg. With steady-state horizontal rotational harmonic oscillations with a frequency $\overline{\omega} = 3\omega = 13.2$ s^{-1} the additional moment of inertia on the sole of foundation was about 47% of the set value of the amplitude $M_{yb} = 159300$ (ktf · m).

An even greater effect of an additional inertial moment on the sole of the pile foundation grillage of the considered type of wind turbine occurs when the first natural frequency of the rigid system is close to the disturbing frequency ($\omega_t = 11.83$ s^{-1} at $\overline{\omega} = 3\omega = 13.2$ s^{-1}).

In this case, the values of P_{xbadd} и M_{ybadd} at the sole grillage are close to a predetermined amplitude of loads P_{xb} and M_{yb} (calculation is given in the report of the department in 1993 to Lengidroproekt).

Thus, if in the designing of wind turbines dissected scheme is used and loads on the foundation are set without inertial forces arising from the mass fluctuations of the foundation due to the deformation of the base, it is necessary to calculate the additional inertial forces using the settlement scheme in the form of a rigid system "wind turbine-foundation (or grillage pile foundation)" on a linearly deformable natural (or pile) basis.

It is also interesting to estimate the effect of partitioning the design scheme in determining the NF and FNO flexural fluctuations of the tower of the wind turbine.

Let's carry out a study, which would help us to evaluate the impact of partitioning the design scheme on the first NF of bending fluctuations of the tower of the wind turbine. We would use for this purpose the simplest calculation schemes when the distribution mass of the tower is replaced by discrete masses (Fig. 2a). In determining the discrete masses based on trapezoidal rule for numerical integration replacing the mass of the nacelle with a wind wheel with a point mass we would obtain (with equal portions of the partition of the tower r) the following results: $m_1 = m_r + 0.5h \cdot m_{T1}$, $m_2 = 0.5h \cdot (m_{T1} + m_{T2})$ etc. With only one portion ($r = 1$), we would obtain the calculation scheme shown in Fig. 2b, wherein $m_1 = m_r + 0.5\, m_T$ and $m_2 = 0.5\, m_T$, where m_T is a full mass of the tower. Introducing the reinforced concrete foundation (or grillage pile) as a hard array, based on the natural (or pile) linearly deformable base in the latter case we would obtain a system with three degrees of freedom.

As it was mentioned above, to determine the NF and FNO of bending fluctuations of the tower using partitioned design scheme oscillation of mass of the foundation is not considered, and the tower in section (2) shall be considered as clamped in a rigid or pliable base (Fig. 2c).

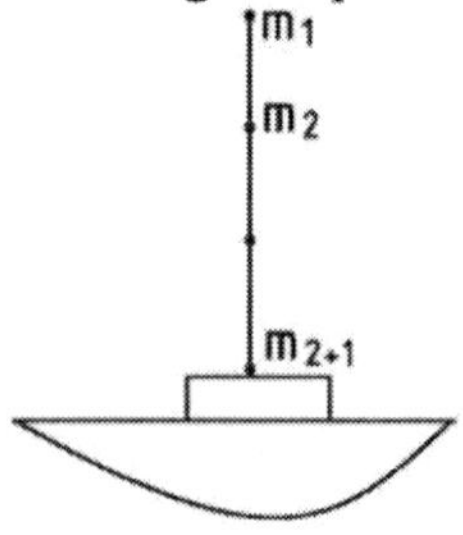

Figure 2a. The simplest calculation scheme when the destribution mass of the tower is replaced by discrete masses.

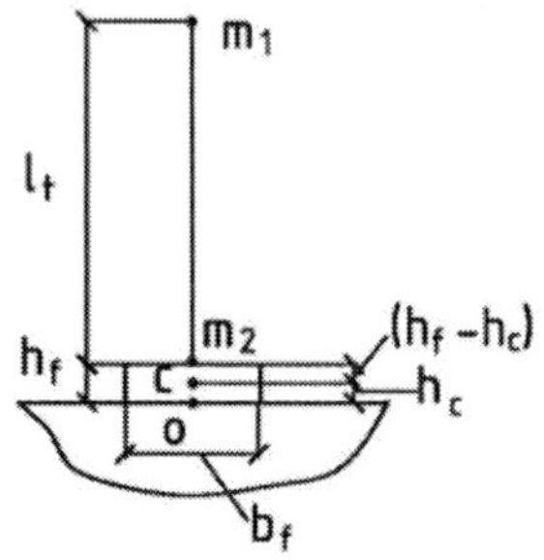

Figure 2b. The calculation scheme with only one portion (r = 1).

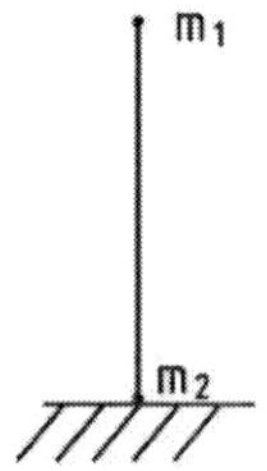

Figure 2c. The calculation scheme with a rigid or pliable base.

With rigid fixing we would have a system with one degree of freedom with mass $m_1 = m_r + 0.5\,m_T$ (scheme 1, a in Tab. 1). More accurate results in determining the frequency ω_1 in the case of rigid fixing would be obtained when the equivalent mass, replacing the distribution mass of the tower will be equal $33/140\ m_T$ [13] (scheme 1, b, Tab. 1).

Let's consider two cases with supple termination in section (2) for comparison. The first one: the oscillation mass $m_2 = 0.5\,m_T$ is taken into account (as well as the mass of the foundation), then we would have a system with one degree of freedom (Fig. 1, c) and as we done it before we would consider the reduced mass of the tower in two senses: $0.5\,m_T$ (scheme 2, a, Tab. 1) and $(33/140)m_T$ (scheme 2, b, Tab. 1). In the second case, with supple termination in section (2) fluctuation of mass $m_2 = m_T$ would be taken into account (scheme 2, c, Tab. 1) then we would have a system with two degrees of freedom.

To assess the impact of fluctuations of the mass of the foundation on the first natural frequency, let's consider the calculation scheme in view of the foundation of a rigid array. To account for the distributed mass of the tower we should still take the simplest case, when $r = 1$. Then for bending fluctuations of the tower and the horizontal rotational fluctuations of rigid foundation in the plane XOZ we would have a system with three degrees of freedom: horizontal movement of the masses m_1 and m_c and the angle of rotation of the foundation around the center of gravity (C) of the total mass

$m_c = (m_f + m_2)$, where m_f is the mass of the foundation, $m_2 = 0.5\,m_T$ (Fig. 2,b). Such a calculation scheme with $m_1 = m_n + 0.5\,m_T$, $m_2 = 0.5\,m_T$ and $\theta_{\varphi c} = m_1\,[l_T + (h_\varphi - h_c)]^2 + 0.5\,m_T\,(h_\varphi - h_c)^2 + [\theta_{\varphi f} + m_f\,(h_2 - 0.5h_\varphi)^2]$ we would identify as scheme 3.

Table 1. The values of the first natural frequency f1 of bending fluctuations of the tower.

Number of design scheme	Value of f_l [Hz] at		
	$E = \infty$	$E = 1.5 \cdot 10^6$ [kgf·m^{-2}]	$E = 0.6 \cdot 10^6$ [kgf·m^{-2}]
1,a	1.10		
1,b	1.20		
2,a		1.05	0.97
2,b		1.15	1.06
2,c		1.05	0.97
3		1.04	0.94

The Tab. 1 shows the values of the first natural frequency f_l (in hertz) of bending fluctuations of the tower for the considered design schemes of one variation of the wind turbine. The calculations were performed for the following values of the masses of elements of the design scheme: the mass of the nacelle with the wind wheel $m_r = 5066$ (kgf·s^2·m^{-2}), the mass of the tower (only pipe) $m_T = 4534$ (kgf·s^2·m^{-1}), mass of square in plan concrete slab with a side $b_f = 11$ m, and a thickness $h_f = 3$ m, $m_f = 92500$ (kgf·s^2·m^{-1}). The basis for the foundation is loam with the deformation modulus $E = 1.5 \cdot 10^6$ (kgf·m^{-2}). The base was considered as Winkler and the characteristics of its stiffness (Cz, Cϕ, Cx, kϕ, kx) under the sole of foundation were calculated by the formulas (4) and (5), (9), (10) in [12].

Comparing the values of f_1 (Hz) for the calculation schemes 1, a; 2, a; 2, c; 3, it is seen that for the base with module $E = 1.5 \cdot 10^6$ (kgf·m^{-2}) and a reinforced concrete foundation of these sizes an account of deformation of the base and of the mass of the foundation has a very little effect on the lowest natural frequency of bending flrctuations of the tower. Therefore, in this case it is permissible to use the calculation scheme only of the part above the foundation in calculating ω_1 and f_l ($f_1 = \omega_1/2\pi$) and consider the cross-section (2) as rigid or resiliently deformable.

The calculations has been done for fl (Hz), when the soil deformation modulus was decreased by 3 times, to investigate the effect of increasing the compliance of the base. $E = 0.5 \cdot 10^6$ (kgf·m^{-2}) (minimum value is shown in Tab. 3 of Annex I of [14]). Assume that the size of the foundation and the coefficient b_0 in the equation (4) in [12] remain unchanged. Then, in accordance with a decrease of E by three times all values of C_z, C_φ, C_x, k_φ, k_x decrease by three times too, and the coefficients of compliance $\delta_{xc} = 1 / k_x$ and $\delta_{\varphi c} = 1 / k_\varphi$ increase in three times.

Comparing the results of calculations for schemes 1, a; 2, a; 2, c; 3 (Tab. 1) it is seen that the transition from the rigid base ($E = \infty$) to very compliant (minimum value of E in Tab. 3 of Annex I of [14]) $E = 0.5 \cdot 10^6$ (kgf·m-2) changes the value of f_l by about 13%. Accounting of mass of foundation with more pliable base decreases the value of the first frequency much greater than the more rigid base (scheme 2, a; 2, c; and 3).

Some calculations related with wind turbines would be held in future in a view of:

- geometric nonlinearity [15],
- thin-walled cross-sections [16],
- sustainability [17],
- elastic plasticity [18].

The held calculations are relevant for the diversity of renewable energy sources which are actively developing in our time [19-23].

3 Conclusions

It is proved that the inclusion of deformation of the base and the mass of the foundation has a very little effect on the lowest natural frequency of bending fluctuations of the tower. The transition from the rigid base to very pliable base changes the value of f1 by about 13%. The studies show that the use of dissected calculation scheme for determining the load on the foundation and the base, and the natural frequencies and shapes of bending fluctuations of the tower is allowed, especially at the initial stage of designing wind turbines. In this case unaccounted inertial forces associated with the fluctuation of the system due to the deformation of the base can be approximately obtained by the rigid system "Wind Turbine + foundation (or grillage pile foundation)" on a natural (or pile) linearly deformable foundation.

The first natural determined frequency at a rigid termination on the sole of the tower should be lowered in view of studies, shown in Tab. 1.

References

1. N. Vatin, N. Lavrov, A. Shipilov, Procedia Engineering, **117 (1)**, 374-380 (2015)
2. O. V. Shepovalova En. Proced., **74**, 1551-1560 (2015)
3. K.N. Shxinek, A.I. Alximenko, A.A. Alximenko, Altern. energ. i ekol., **11**, 29-35 (2014)
4. T.V. Miroliubova, Vest. Perm. un. (Ekon.), **4(7)**, 78-86 (2010)
5. Q.S. Li, F.B. Chen, Y.G. Li, Y.Y. Lee, J. of Wind Eng. and Ind. Aerod., **116**, 70-82 (2013)
6. Tao Ma, Hongxing Yang, Lin Lu, Jinqing Peng, Appl. En., **160**, 728-736 (2015)
7. Ayşenur Erdil, Hikmet Erbiyik, Proced. - Soc. and Beh. Sc., **207**, 669-679 (2015)
8. E. V. Solomin, Vest. Iuj.-Ural. gos. un. (Ekon. i men.), **20 (196)**, 32-36 (2010)
9. Fahad jamil, Saad bin zia, Parvez ali abbasi, Moiz ansari, *To Study of Wind Resistant Stability of Tubular Wind Turbine Tower* (NED university of engineering and technology, Karachi, 2012)
10. Ching-Wen Chien, Jing-Jong Jang, Proc. of the Sev. As.-Pac. Conf. on Wind Eng., **8 (12)**, 1-8 (2009)
11. I.A. Konstantinov, Yu.A. Ippa, N.V. Miklashevich, *Raschetnye sxemy oistemy "VEU-fundament-osnovanie" dlya opredeleniya raschetnyx nagruzok na fundament i osnovanie i amplitud ix kolebanij* (Predso-93, 1994).
12. SP 26.13330.2012. Updated edition of SNiP 2.02.05-87 Foundation of machine with dynamic loads
13. S.P. Timoshenko. *Kolebaniya v inzhenernom dele* (Stroyizdat, Moscow, 2006)
14. SP 22.13330.2011. Updated edition of SNiP 2.02.01-83* Foundations of buildings and constructions.
15. A.V. Yavarov, Zhil. stroy., **5**, 51-54 (2013)
16. Arash Sahraei, Liping Wu, Magdi Mohareb, Thin-Wall. Struct., **89**, 212-226 (2015)
17. D.A. Kushova, Appl. Mech. and Mat., **617**, 181-186 (2014)
18. B.E. Melnikov, A.S. Semenov, ZAMM, **76 (2)**, 615-616 (1996)
19. J. Ćetković, S. Rutešić, M. Zarković, M. Knežević, N. Vatin, Procedia Engineering, **117 (1)**, 780-790 (2015)
20. Cheng-Dar Yue, Chung-Sheng Chen, Yu-Chen Lee, Ren. En., **86**, 930-942 (2016)
21. Dany Perwita Sari, Wida Banar Kusumaningrum, En. Proced., **47**, 29-36 (2014)
22. Z. Bonić, G.T. Ćurčć, M. Trivunić, N. Davidović, N. Vatin, (2015) Procedia Engineering, **117 (1)**, 424-435 (2015)
23. Z.Popović, L. Lazarevic, N. Vatin, Procedia Engineering, **117 (1)**, 846-853 (2015)

MATEC Web of Conferences **53**, 01065 (2016)
DOI: 10.1051/matecconf/20165301065

Experimental studies of Steel Corrugated Constructions

Yuriy Lazarev[1], Alisher Akhatuly[1,a], Raushan Alzhanova[1] and Alla Panfilova[1]

[1] *Peter the Great Saint-Petersburg Polytechnic University, 29 Polytechnicheskaya st., St.Petersburg, 195251, Russian Federation*

Abstract. The purpose of this particular article is to assess existing calculations of steel corrugated constructions. Steel Corrugated Construction is a perspective type of constructions, which is exhibiting numerous advantages in comparison with one that currently applied in automobile and railroad networks (reinforced concrete water-throughput pipes, reinforced concrete frame bridges). The evaluation of experimental data on models of constructions of this particular type has been carried out in order to improve calculations of Steel Corrugated Constructions.

1 Introduction

A considerable share in road construction is occupied by small artificial constructions – water-throughput pipes and small bridges. In the last decade, there was an extensive circulation of steel corrugated pipes (SCP), construction of which has significantly reduced labor input and material capacity in contrast with reinforced concrete pipes, which in its turn lead to cost reductions in construction.

Nowadays there is a large number of the programs allowing solving numerally equations of mathematical physics with a method of finite elements. There are special programs solving a narrow class of tasks: calculation of construction designs, calculation of problems of mechanics of soil, hydrodynamics. In addition, there are universal programs allowing modeling any systems and the phenomena, but demanding high qualification of the researcher. Whether universal or special, there are programs not only commercial but also free of charge, which are just as efficient and functional [1, 2, 5].

It is possible to claim that currently modeling virtually of any system is available to an engineer, including filling constructions from corrugated steel considering all aspects of work. However, for a number of reasons practical application of programs that are solving the equations of mathematical physics for calculation of corrugated steel constructions is rather complicated.

Formulas for calculation are borrowed from VSN 176-78, nevertheless it is necessary to expand a scope of formulas from the appendix 11 for calculation of a design for limit balance. The coefficient of increasing of bearing capacity K has to be modified by means of a technique stated in chapter III of the book "Steel Corrugated Pipes under Embankments" (authors of Bells N.M., Yankovsky O. A., Shcherbina K.B., Chernyakhovsk S.E Transport) and section VI.3 of the book "Water-Throughput Pipes under Embankments", considering all contemporary nomenclature of corrugations and of steel.

[a] Corresponding author: akhatuly_alisher@mail.ru

Calculated bearing capacity of a pipe out of q_l soil, has to be specified, bearing in mind actual durability of applied steel [9, 18, 22].

Linear calculation of corrugated constructions via method of finite elements is acceptable to be executed in any modern calculated complex. Therefore, it is unnecessary to be limited to purely linear calculation. High flexibility of a corrugated construction permits it to work beyond elasticity limit, which is potential to be carried out in programs of linear calculation due to the elementary iterative procedure. The latter formed the basis of obtaining a formula that allows determining of bearing capacity recommended by VSN (The instruction on design and construction of steel corrugated water-throughput pipes). According to standards, it is necessary to consider two ways of modeling the soil surrounding the construction: linearly - deformable environment and elastic connections. For the first approach, it is necessary to make recommendations for assigning the sizes of soil mass model, the module of elasticity and Poisson's coefficient. For the second approach, it is necessary to provide a technique of determination of rigidity of connections and an order of their deactivation while carrying out iterative procedure [13, 17, 9].

Nonlinear calculation via method of finite elements is based on application of reliable models of functioning of the pipe material and the soil surrounding it considering all stages of work. It is essential to specify in standards and accurately to describe the recommended models of soil and material of corrugation, and of an area of their application. Parameters of any suggested model have to be attached to real parameters of the applied soil and steel, which is checked and provided in the course of construction [29, 32].

Other problem capable to nullify all efforts of scientists and designers on creation of calculated and theoretical basis is inadequate quality of work constructing a corrugated pipe.

Small artificial constructions (water-throughput pipes, cattle crossing, etc.) make the main part of a number of bridge constructions built on automobile and railroad networks. There are various materials applied for their production. For many decades before and nowadays, the core material for water-throughput pipes is reinforced concrete. However, there are several reasons that lead to defects and damages of forced and corrosive character bringing those to pre-emergency and emergency states. Among them:

1) Reinforced concrete water-throughput pipes in use are often exposed to joint influence of loading (power influences), and hostile environment (environmental influences).

2) Insufficient attention is paid to a problem of effective operation of reinforced concrete constructions on automobile roads [22, 25].

As a result, integrity of an overlying soil embankment is broken, the roadbed collapses and regular traffic on the road is interrupted. One of disadvantages of reinforced concrete constructions is complexity of their strengthening and repair, especially if the purpose of repair is to increase loading capacity of a construction [15, 18].

Lately, a special attention is given to a problem of design and calculation of water-throughput pipes considering their factual conditions of exploitation. Especially interesting are findings of Ivanov A.V., Ovchinnikova I.G., Belyaev V. S., Yakovlev L.S., Kalinovsky M. I., Migunov V. N., Petrov E.N., etc. [3, 5, 12].

Due to a considerable reduction of durability of reinforced concrete water-throughput pipes in aggressive service conditions, a question of how to increase durability of such constructions arises. [24,27]

The most effective way to increase durability of water-throughput pipes, according to a number of researchers, is an employment of corrugated steel while producing water-throughput pipes. [6,8]

The main advantages of SCC:

• Small weight of constructions elements;

• Simplicity of assembly; smaller, in comparison with reinforced concrete designs;

• Construction time;

• Attractive appearance;

• The cost of building constructions from corrugated steel is substantially lower than the cost of the small and average bridge constructions with a similar scope [1, 2, 19].

2 Description of research

During an assessment of this experiment, the following objectives are achieved (Figure 1, 2):
1. Comparison of experimental and calculated data of SCC
2. Determination of accuracy and specifications of SCC calculation in intended model of the Method of Finite Elements.

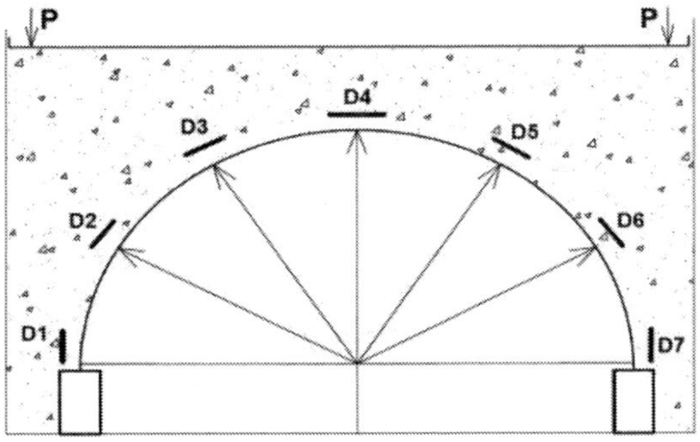

Figure 1. Tests of a semi-circular arch steel corrugated construction.

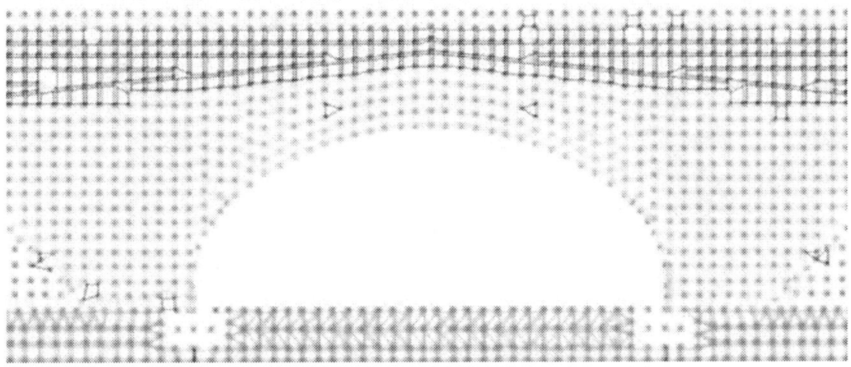

Figure 2. SCC design model.

Object of the current research is an arch SCC of a circular outline with a corner of circular sector φ=90° (a semi-circular arch). SCC built with corrugated sheets by a profile of 381x140 mm, 7 mm thick. Tests were carried out in the camera of volume loading, the sizes 8.0×4.5×3.5 of m having one mobile wall. Movement of a mobile wall was carried out by means of the battery of the hydro jacks having total effort~2700 t.f. and the course of rods to 1.0 m of SCC I was in the camera of volume loading in horizontal position. Fixing of heels of arch SCC–hinged. The distance between walls of the camera and SCC was filled with coarse-grained sand with layer-by-layer consolidation.

In total 28 sensors, including 7 sensors of soil pressure, 14 sensors of relative deformations (gage), 7 sensors of deflections were installed. All sensors of deformations are focused along the corrugation. In Fig. 1 are shown the installed sensors of relative deformations and the sensor of relative movements in the SCC model.

The layer-by-layer filling with sandy soil in space between walls of the load camera and SCC was carried out, and the necessary height of a filling of 2.6 m was reached by laying of eight layers of soil on 0.25-0.35 m.

Tests were completed at the appendix of the mobile wall of loading distributed on all area which size is determined by indications. Indications specified were averaged and transferred to the height of a column of conditional soil with the specific weight of 20000 N/m^3 [1, 2, 3].

Table 1. Values of intensity of equivalent loading of SCC

q [kN/m]	Height of a conditional filling [m]
467	30

Table 2. Value of the characteristic of materials

Name	Soil	Steel
Module of elasticity [MPa]	22.2	$2*10^6$
Poisson's coefficient	0.3	0.3
Density [Kg/m^3]	1700	7800
Calculated resistance on strength [MPa]	-	300
Coupling coefficient	0.005	-
Corner of internal friction	37	-

Results of test and comparison of experimental data with calculated (Figure 3-5, Table 3):

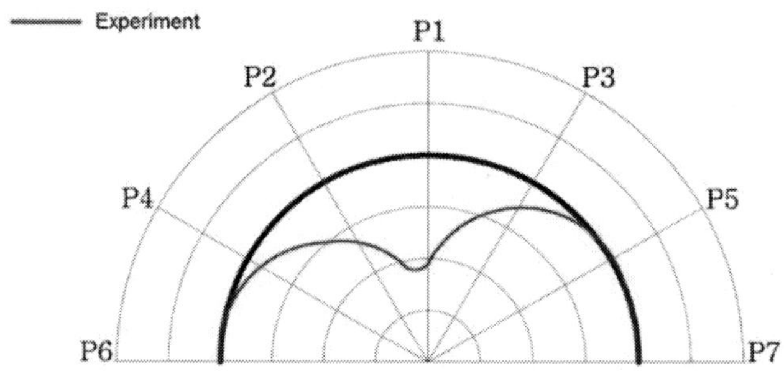

Figure 3. An experimental form of deformation of SCC (1 radial division – 50 mm)

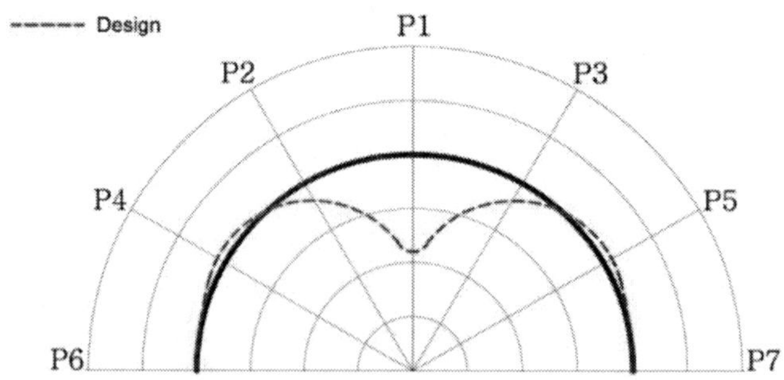

Figure 4. Design form of deformations of SCC

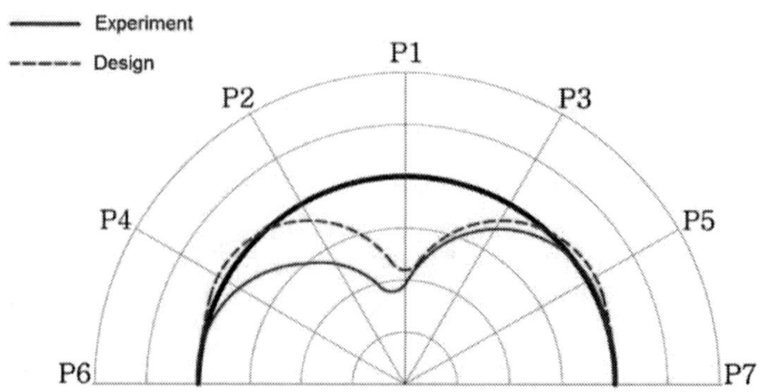

Figure 5. Experimental and design form of deformation of SCC

(1 radial division of-50 mm)

Table 3. Comparison of results of the experiment

Method of determination of tension	Location of sensors	Values of normal tension [MPa]							Average %
		t11-12	t7-8	t3-4	t1-2	t5-6	t9-10	t13-14	
Experiment	Corrugation top	+169	+147	-400	-609	+298	+209	+158	
Method of finite elements		-103	+67	-314	-300	-314	+67	-103	
Accuracy	%	39	54	22	51	-5	68	35	38
Experiment	Corrugation hollow	-512	-538	-58	+806	-262	-542	-557	
Method of finite elements		-185	-318	+146	+268	+146	-318	-185	
Accuracy	%	64	41	15	67	44	41	67	48

3 Conclusion

Results of test and comparison of experimental data with design are shown in Fig. 3-5.

Therefore, evaluating factual and calculated results it is possible to draw the following conclusion:

1. Accuracy of calculations for a method of finite elements corresponds 40-50%, from experimental data.

2. In calculated model, there is no input of a possibility of asymmetrical behavior, and actual data demonstrates asymmetrical behavior of corrugated constructions as visible in fig. 3.

3. Considering it, it is crucial to recommend insuring uniformity of soil of the filling, identical and symmetric level of its merging over a structure while building actual SCC filled with soil, in order to exclude or to reduce probability of occurrence of an asymmetrical nature of deformations in constructions.

In any way, the constructions completed from steel corrugated sheets retain high reliability and possess operational qualities while loaded surpassing calculated ones by 2.5 times.

References

1. A.V. Ivanov, I.G. Ovchinnikov, Regional'naya arkhitektura i stroitel'stvo , **1(2)**, 43-52 (2007)
2. I.I. Ovchinnikov, M.I. Kalinovskiy, Dorogi i mosty, Sbornik statey FGUP Rosdor **22(2)**, 186-200 (2009)
3. M.I. Kalinovskiy, I.I. Ovchinnikov, Stroitel'nyye materialy, **10**, 15-17 (2010)
4. I.I. Ovchinnikov, V.N. Migunov, I.G. Ovchinnikov, Zhilishchnoye stroitel'stvo, **10**, 6-8 (2012)
5. V.S. Belyayev, L. S. Yakovlev, I.G. Ovchinnikov, I. A. Osokin, Naukovedeniye, **6** (2013)
6. Osokin, I.A., A.S. Permikin, NTU (2012)
7. A.C. Mak, R.W.I. Brachman, I.D. Moore, Washington D.C, **9**, 14.
8. D. Beben, The Baltic journal of road and bridge engineering, **4 (1)**, 13-21 (2009)
9. N. Vatin, J. Havula, L. Martikainen, A. S. Sinelnikov, A. Orlova, S. Salamakhin Advanced Materials Research, 945–949, 1211–1215 (2014)
10. I.G. Ovchinnikov, *Naukovedeniye*, **6** (2013)
11. V. Lalin, V. Rybakov, A. Sergey, Applied Mechanics and Materials, 578–579, 858–863 (2014)

12. Y. Lazarev, G. Sobko, M. Chakir, Applied Mechanics and Materials, 584-586, 1681-1686 (2014)
13. N.I. Vatin, O. Gamayunova, Applied Mechanics and Materials, 670-671, 1612-1615 (2014)
14. V. Rybakov, A. Panteleev, G. Sharbabchev, E. Epshtein, Applied Mechanics and Materials, 584–586, 1876–1880 (2014)
15. N. Vatin, J. Havula, L. Martikainen, A. Sinelnikov, L. L. Shurovkina, Applied Mechanics and Materials, 725–726, 722–727 (2015)
16. D. Trubina, D. Abdulaev, E. Pichugin, V. Rybakov, Applied Mechanics and Materials, 633–634, 1133–1139 (2014)
17. M. Garifullin, D. Trubina, N. Vatin, Applied Mechanics and Materials, 725–726, 697–702 (2015)
18. M. Seslija, N. Radovic, N. Vatin, Applied Mechanics and Materials, 725-726, 596-604 (2015)
19. Y.G. Lazarev, SPbGASU, 93 (2013)
20. Y.G. Lazarev, Y.B. Sinitsyna, Tekhniko-tekhnologicheskiye problemy servisa, **2(24),** 92-93, (2013)
21. N. Vatin, A. Sinelnikov, M. Garifullin, D. Trubina, Applied Mechanics and Materials 633–634, 1037–1041 (2014)
22. V.N. Bukhartsev, M.R. Petrichenko, Power Technology and Engineering, vol.**44**, 5, 374-377 (2012)
23. M.R. Petrichenko, N.S. Kharkov The Russian Journal of Applied Physics, **54 (7),** 1063-1065 (2009)
24. V.N. Bukhartsev, M.R. Petrichenko, Power Technology and Engineering, **46**,185-189 (2012)
25. V.N. Bukhartsev, M.R. Petrichenko, Power Technology and Engineering, **46(1)**, 1-3 (2012)
26. M.R. Petrichenko, Heat transfer–Soviet Research, **23(5),** 703-714 (1991)
27. N. Vatin, J. Havula, L. Martikainen, A. Sinelnikov, L.L. Shurovkina, Applied Mechanics and Materials, 725–726, 722–727 (2015)
28. F. Peterson, R. Fruchter, M. Fisher, T. Hartmann, Automation on construction, **20 (2),** 115-125 (2011)
29. M. V. Ananina, N. A. Beresneva, L. L. Shurovkina, Construction of Unique Buildings and Structures, **7 (22),** 54–70 (2014)
30. N. Vatin, A. Sinelnikov, M. Garifullin, D. Trubina, Applied Mechanics and Materials, 633–634, 1037–1041 (2014)

MATEC Web of Conferences **53**, 01066 (2016)
DOI: 10.1051/matecconf/20165301066

Research of the Fire Resistance of Translucent and Composite Facade System

Oleg Nedryshkin[1,a], Artem Krivtcov[1] and Valerii Pershakov[2]

[1]*St. Petersburg State Polytechnical University, Polytechnicheskaya, 29, St. Petersburg, Russia, 195251, Russia*
[2]*National Aviation University, Kosmonavta Komarova, 1, Kiev, 03680, Ukraine*

Abstract. The paper aims at researching fire resistance of a prototype facade system "Technocom" (type Alucobond A2). Experimental and theoretical research of fire hazard facade system is carried out. The objectives of the study are to determine compliance with the applicable front of special technical requirements. The status of problem reducing fire hazard facade system is reviewed. The method developed by compensatory measures is applied.

1 Introduction

Nowadays we are witnessing a fast growth in the construction of buildings including facade systems. For the last 10 years translucent and composite facade system used in Russia over than 4 million m2[1, 2].

The technical regulations about the requirements of fire safety, the limits of fire resistance of construction are defined under the conditions of standard tests. The dynamics of burning, however, is not always possible to standardize due to the variety of architectural, space-planning and construction solutions. Requirements vary depending on classes of buildings fire functional hazard [3-12].

Research papers [1, 2, 13-15] is consider the natural ventilation of buildings and features some translucent ventilated facades. The problem is suggested for discussion of fireproof glass structures in [3-13], focused on different methods of mathematical modeling [10, 12, 16-18].

The current regulatory requirements don't apply to test method of exterior curtain walls of building's facade system. The current Russian standards concerning the methods of fire tests don't address specific construction solutions for exterior curtain walls with translucent elements and associated possible scenarios of fire, as well as the orientation of the thermal effect on such a structure [3, 4].

Objectives:

- to study fire resistance of translucent combined facade systems "Tehnokom" type Alucobond A2 projected on an object-"41-storey office building" in Moscow;
- to define possible fire hazard reduction facade system "Tehnokom".

A glass unit block was made of translucent single-chamber glass unit 38 mm thick. Execution scheme: external glass triplex 6.6.2 Stopsol Supersilver Clear (SSC) (produced by firm "Glaverbel" Belgium), spacer 16 mm, inner glass-transparent glass "Pilkington" 10M1 tempered. Frame built from blocks of aluminum profiles HUEK (manufactured by "Eduard Huek & co. KG GmbH", Germany).

The translucent filling decorative element is executed according to the following scheme:

sandwich panel type Alucobond A2, 4 mm thickness;

[a] Corresponding author : nedryshkin@gmail.com

sheet zinced steel thickness 0.55 mm;
mineral plates of Rockwool, 150 mm thick, density 45 kg/m3.
Rubber is laid along the length of the butt joints between blocks sample.

2 Methods

This system configuration made it possible to formulate a program to determine the feasibility of the application of the investigated translucent and combined facade systems in a building and to establish their conformity with the requirements of the specifications [18]. The work included of the following main steps:

1. The limits of fire resistance and fracture mechanism of the construction are explored under the conditions of standard tests;
2. The application of compensatory measures are established:

- the device for sprinkling (water) of a glass unit of a translucent construction from the inside on the floor of the fire;
- the aluminum structure, facings for fire protection of fasteners of a facade and decrease in temperature of a flame of the external fire (fig. 1 shows);
- the device in internal cavities of decorative elements of fire retardant belts and fireproof data processing of cavities for the purpose of prevention of distribution of a flame;

3. Fire tests are conducted for fragments of facade system with compensating activities and determining whether application of the investigated system.
4. Identify the possibility of spreading fire in the floors of the building.

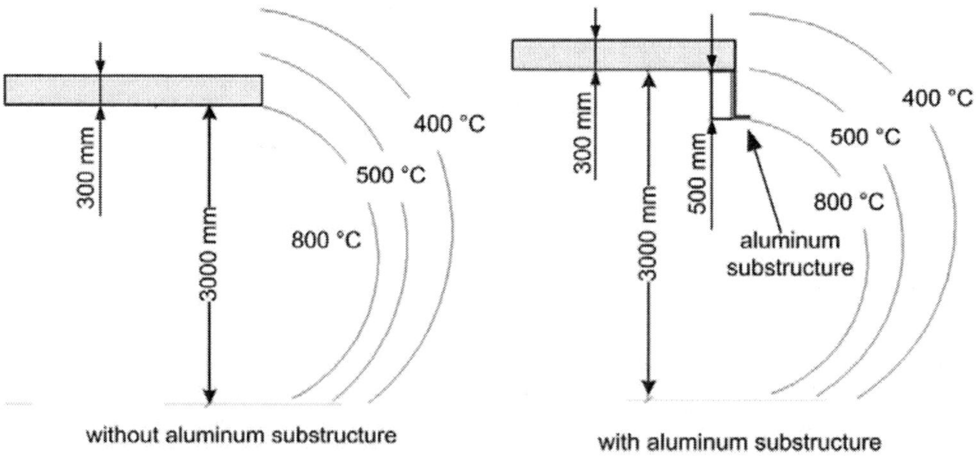

Figure 1. Aluminum substructure for decrease heat temperatures influence on the facade systems.

Fire tests give experimental data about of fire resistance limit of translucent constructions.

The purpose of the fire test is to establish the time from the beginning of fire exposure before the limit state (loss of integrity).

It was assumed that the integrity of the facade glass unit on the floor where there is a fire was trouble and facade of the upstream floors subjected to fire exposure. Fire resistance of facade system in mode external fire was investigated. The aim is to investigate ways of distribution of fire from floor to floor.

Fragment of facade set up upside towards the fire effects. After the beginning of the tests the temperature regime calculated by equation 1 [3-5]:

$$T = 660(1 - 0.687 \times e^{-0.32t} - 0.313 \times e^{-3.8t}) + 20 \qquad (1)$$

where: T is the temperature in the furnace, the appropriate time t,° c;
t-time, calculated from the start of the test, min.

3 Results

From fire testing are identify what own fire resistance limit of the translucent facade construction on the grounds of loss of integrity during the external fire without the use of compensating measures be E9. The limit state (loss of bearing capacity) of elements of the mullion-transom constructions doesn't occur.

On the basis of the external temperature ratio and standard fire in case of fire inside the building limit of fire resistance of structures on the basis of loss of integrity will be no more E4 ... E6 [3]. The results give right to believe that the fire is able to spread on the facade of the building with a speed of not less than 1 floor for 13-15 minutes.

The purpose of the device for sprinkling of the glass from the inside during a fire is: reduce heat flow affecting the construction, increase the limit of facade system resistance by reducing room temperature. After the destruction of the glass unit on the floor of fire continued sprinkling water reduces the temperature of the flame, affecting the higher floor.

To test for fire resistance was applied the kerosene. 5 burners installed in the furnace, plug in a phased manner to increase the temperature. The calculation showed that convective flow through opening, with account for sprinkling water equal to 328.4° C (fig. 2 shows). The data obtained are not contrary to the calculations made by other methods [16, 17]. Flame temperature of external fire affecting the facade block floors above the fire floor is accepted maximum and equal to 328.4 °C.

The amount of heat released during the combustion of kerosene (KO-30):

$$Q = q \times m = 46 \times 10^6 \times 0.042 = 1.9 \text{ MJ/s} \tag{2}$$

where: q - low heating value, MJ/kg;

m – kerosene consumption, kg/s.

The amount of heat from the flame per unit time taking into account the heat losses attributable to cladding structures (wall, floor, ceiling).

$$Q_\Sigma = Q_c + Q_p + Q_{per} \tag{3}$$

$$Q_c = a_c(t_m - t_c)F_c \tag{4}$$

$$Q_p = a_p(t_m - t_p)F_p \tag{5}$$

$$Q_{per} = a_{per}(t_m - t_{per})F_{per} \tag{6}$$

where: Q – quantity of heat, J;

$a_{c,p,per}$ – average heat transfer coefficients, W/m^2 °C;

t_m – average temperature volume of the gas environment in the room, °C;

$t_{c,p,per}$ – average temperature of the surfaces respectively of the walls, floor and ceilings, °C.

$$a_c = 0.2\lambda \div H \times (GrPr)^{\frac{1}{3}} \{Pr^{\frac{2}{3}} [2.14(1+N) + Pr^{\frac{2}{3}}]\} \times (1+N) \tag{7}$$

where: Gr – Grashof number;

Pr – Prandtl number;

N – analogue of the number Kirpicheva characterizing radiation-conductive heat transfer in the boundary layer;

λ – heat-conduction coefficient, W/mK.

Under splintering on the floor of water flow of 0.9…1.1 l/s (90% of surface area):

$$a_{per} = 1.4\frac{\lambda}{r} Pr^{\frac{1}{3}} Re^{\frac{1}{2}} + 13.3(\frac{\sigma F \tau}{B}) \tag{8}$$

where: $\sigma = 5.7 \times 10^{-8}$ – blackbody coefficient, W/(m^{2}K^4);

B – analogue of the Bouguer number, which characterizes the effective optical density in the boundary layer;

r – distance from frontal point on the surface of the slab, m;

Re – Reynolds number (Re = 2350).

$$a_p = 0.7 \times a_c \tag{9}$$

The surface temperature of the plane wall:

$$t_{c2} = t_{cl} - \frac{\dfrac{Q_{light}}{\lambda F \tau}}{\sigma} \tag{10}$$

where: λ – coefficient of thermal conductivity, is made under the assumption that a plane wall consists of a mixture of water and air, W/mK;

τ – time to take in the calculation, c;

σ – planar wall thickness, m;

t_{c1} – the high temperature inside the room in case of fire, °C;

Q_{light} – amount of heat coming through the light, J;

F – conventional flat wall area, increased by the total surface area of the drops of water supplied to irrigation, m^2.

$$F = S_{light} + \sum S^{drops} \tag{11}$$

$$N_k = V_{water} / V_{drops} \tag{12}$$

$$V_{drops} = \frac{4}{3}\pi R^3 \tag{13}$$

$$\sum S^{drops} = \frac{4}{3}\pi R^2 N_k \tag{14}$$

$$Q_{light} = Q - Q\Sigma \tag{15}$$

where: S_{light} – area of light, m^2;

$\sum S^{water}$ – total surface area of water drops applied to the irrigation of the window opening, m^2;

V_{water} – volume of water supplied for irrigation per unit of time, c;

V_{drops} – volume of one drop of water, m^3;

Q_{light} – amount of heat transported through the light per unit time, J/s×m^2 [3].

Thus:

- flame temperature near glass unit floor from upper the floor of the fire after the destruction of the glass unit on the floor of the fire will not exceeds 328.4 °C.
- for carrying out tests temperature on all surface of the tested fragment (in the fire camera of the test furnace) is accepted by the highest (equal to the temperature at the bottom) – 330 ±5 °C.

The fire resistance is equal to 65 E, under splintering on the floor of water flow of 0.9…1.1 l/s. This is according to specifications.

The calculation showed that the application of the regular glass unit with sprinkling device allows to reduce the temperature of the flame of the fire affecting to the upper glass unit floor to a value lower than the loss of bearing capacity aluminum mullion-transom constructions.

Failsafety sprinkler should be not less than 0.99 [3]. To compensate for the risk of release of sprinklers inoperable and reducing the temperature of the flame over the opening on the floor of the fire are encouraged to install in ceiling fire barrier of each floor. In this case, the calculated temperature distribution near the opening with destroyed glass unit if there is fire countermeasure shut-off failure of sprinkling systems is presented in Figure 2. As a fire-retardant coating was selected composition of the intumescent type. The thickness of a layer of a fireproof covering has to provide filling coked cellular material the gap which is formed as a result of deformations and transfer material of a decorative element to group not extending a flame on a surface (RP 1) [3].

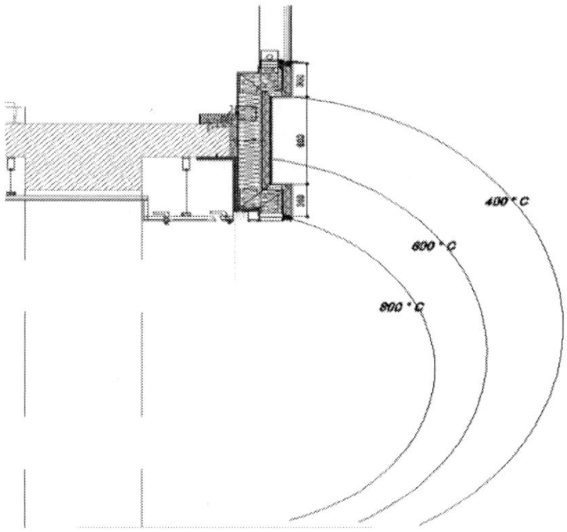

Figure 2. Temperature distribution of convective flow near the doorway with destroyed glass unit.

4 Discussion

Presented at the fire tests of samples of translucent facade construction fall short of specifications for the following reasons:

1) own limit of fire resistance of construction under external fire tests – E 9, when a fire indoors E 4 – E 6, i.e. from one floor to another building fire may spread over 13 ... 15 min;
2) own limit of fire resistance of metal fasteners facade constructions is R 14, which could lead to the collapse of mullion-transom constructions;
3) decoration of the facade can serve as a source of flame propagation for hidden cavities, which makes it difficult to extinguish the fire and contributes to its spread along the facade.

To ensure compliance with the requirements of the specifications developed the following compensating event:

1) sprinkling not less than 90% of the surface covering of the doorway on each floor with water by an automatic sprinkler system with 0.9…1.1 l/s;
2) device in the ceiling part of the fire the cutoff of a width not less than 500 mm, the limit of fire resistance must not be less than the value REI 60;
3) provision of fire resistance limit node of an adjunction to overlappings facade system building at least REI 60;
4) providing the limit of fire resistance of the fastening elements of façade system not less than R 60;
5) device inside fire isolate leaf of steel with a thickness of 0.55 mm and steal covered with a coat of paint intumescent flame retardant composition 1 mm thick.

5 Conclusions

The research of translucent combined facade systems "Tehnokom" with the compensatory activities has shown that the implementation of these activities ensures compliance with regulatory requirements of facade systems.

References

1. Yu. Nikitin, V. Murgul, N. Vatin, V, Pukhkal, AMM, **680**, 481-485 (2014)
2. D. Nemova, V. Murgul, A. Golik, E. Chizhov, V. Pukhkal, N. Vatin, JAES, **12(1)**, 37-44 (2014)

3. O. Lamkin, M. Gravit, O. Nedryshkin, Journal "Construction of Unique Buildings and Structures", **11(38)**, 42-58 (2015)
4. M. Gravit, V. Gumenyuk, O. Nedryshkin, Procedia Engineering, **117**, 114-118 (2015)
5. A. Krivtcov, V. Kazakova, I. Mingalimov, P. Bogdanov, I. Nitsa, Journal "Construction of Unique Buildings and Structures", **6(33)**, 34-46 (2015)
6. M. Kaziev, E. Zubkova, V. Bezborodov, Tekhnologii tekhnosfernoj bezopasnosti, **6 (58)** (2014)
7. A. Gusev, V. Zigern-Korn, I. Molchadskiy, Tekhnologii stroitelstva, **4 (59)**, 20-24 (2008)
8. I. Khasanov, I. Molchadskiy, K. Goltsov, A. Fire safety, **5**, 36-47 (2006)
9. M. Gravit, Fire and Explosion Safety, **11**, 42-45 (2014)
10. M. Heinisuo, M. Laasonen, J. Outinen, J. Hietaniemi, Application of Structural Fire Construction, 405-410 (2011)
11. M. Lazarevska, M. Cvetkovska, M. Knezevic, A. Gavriloska, M. Milanovic, V. Murgul, N. Vatin, Applied Mechanics and Materials, **627**, 276-282 (2014)
12. I. Romanenkov, V. Zigern-Korn, *Fire resistance of building structures of effective materials,* (Moscow, (1984)
13. J. Hana, L. Lin, J. Penga, Y. Hongxing, Energy and Buildings, 204–209 (2013)
14. L. Gaillard, S. Giroux-Julien, C. Menezo, H. Pabiou, Journal of Fundamentals of Renewable Energy and Applications, 5 (2012)
15. N. Vatin, A. Gorshkov, D. Nemova, A. Staritcyna, D. Tarasova, AMR, **941 – 944**, 905-920 (2014)
16. S. Puzach, A. Smagin, O. Lebedchenko, Ye. Abakumov. *New ideas about the calculation of the required time to evacuate people and the effectiveness of the use of portable filter self-rescuers for the evacuation fires. Monograph*, (Moscow, (2007)
17. Yu. Koshmarov, M. Bashkirtsev. *Thermodynamics and heat transfer in the fire case,* (Moscow, (1987)
18. O. Halyavin, M. Gravit, A. Pryanikova, Materials Science Forum with international participation, **34**, 22-25 (2015)

MATEC Web of Conferences **53**, 02001 (2016)
DOI: 10.1051/matecconf/20165302001

Modern Beer Factory Interwoven in Urban Texture

Violeta Gjorgjevska[1, a] and Aleksandar Radevski[1]

[1]*Ss Cyril and Methodius University, Faculty of Architecture, Skopje, Macedonia*

Abstract. The main idea of this research is to present a concept of integration of a beer factory in a new urban reality. Nowadays, modern development of industrial architecture enables beer factories to leave their typical locations in industrial zones and draw nearer to their customers. To design a modern industrial complex incorporating the ideas of the contemporary way of designing an industrial facility with a clear and indicative architectural expression of its hybrid character requires great skilfulness. Using a linear grid, the concept of this project will be developed in a composition consisting of several strips. This kind of structure creates a compositional equivalence from strip to strip, integrating the natural and the urban, the public and the private.

1 introduction

Striped forms have been part of modern urbanism since the 1920s when a specific type of linear city was formulated by the Soviet planner N.A. Miliutin for the new city of Magnitogorsk. More recently, striped urbanism was revived by OMA [1] in their competition proposal for Parc La Villette in 1982. Although limited as a one-dimensional abstraction, striped urbanism remains a viable alternative to block and street urbanism [2-6].

The aim of this paper is to present the idea how a beer factory can be integrated in strips and thus create a new urban texture [7-12].

2 Previous experiences - urban strips

The proposed project for "Parc La Villette " by Rem Koolhaas does not actually refer to any particular park, but rather it presents a method that - combining programmatic instability with architectural specificity - will eventually generate a park. [1]

In La Villette, the strategy of strip operates the whole process to design a social condenser by means of tactical design acts. The strips manage the construction process of the park by means of operational tactics. They are divided and directed in such a manner as to prepare a generic ground on which programmatic elements are distributed, added, and connected. The mutual relationship between strategy and tactics is essential for the design process of La Villette.

The idea is comprised of 5 steps (Figure 1):

[a] Corresponding author : violeta.gorgevska@yahoo.com

1. The major programmatic components are distributed in horizontal bands across the site, creating a continuous atmosphere in its length and perpendicular, rapid change in experience.
2. Some facilities - kiosks, playgrounds, barbecue spots are distributed mathematically according to different point grids.
3. The addition of a "round forest" as architectural elements.
4. Connections
5. Superimpositions

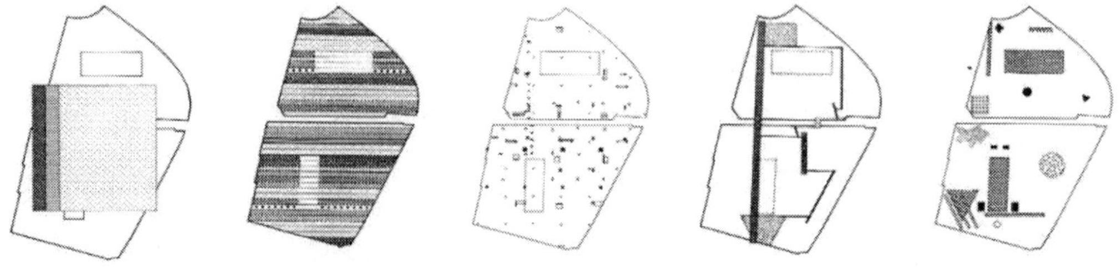

Figure 1. Layers of Parc de la Villette project

3 Research results - architectural concept

The initial steps in the development of the concept of urban brewery were the "urban strips". Based on the analysis of all parameters of the urban location and its surroundings, several guiding principles were defined, which articulate the whole area and define the urban settlement [2]:

1. Set the linear grid - forming an urban park in strips
2. Penetration of prisms in the strips - beer factory
3. Differentiation of pedestrian traffic from vehicle traffic, completely separating them by levels.

The primary point in the articulation of the site is its urban settlement as an urban - ecological park. The concept starts with setting a linear grid. It appears that one side of the site significantly dominates over others in length, and hence the linear grid was placed over that side. The grid used a module of 6m, which is suitable for the constructive solution of the object. Nevertheless, the 6 meter module is one of the main design modules that are used in architectural design. From situational view, the object position occupies two thirds of the parcel and it does not interfere in the space intended for future housing. The other half consists of recreational sports fields, a kids' park, access for visitors, hiking trails ... This part in essence represents a kind of buffer zone between the beer factory and the environment. The whole setting gives an impression that the greenery is interwoven with the building (Figure 2a, b).

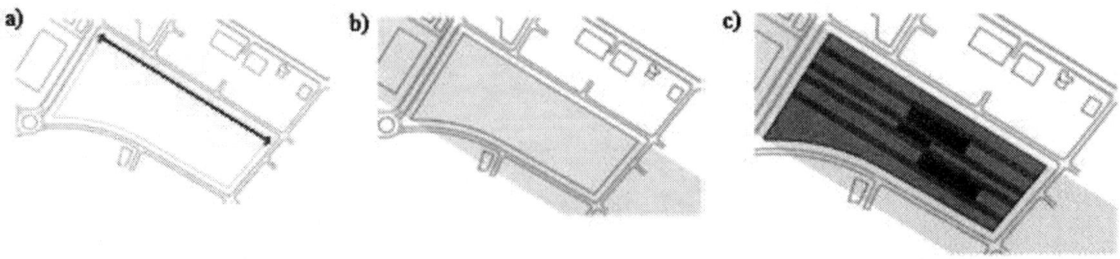

Figure 2.a,b - Linear - forming an urban park in strips; c- Penetration of prisms in strips - beer factory

Based on this scheme 7.4ha were divided into 37 parallel landscape strips, with equal thickness, which cover the entire site. Each strip contains various program activities such as sports and games, and at the same time also different types and positioning of trees, shrubs, fruit orchards, as a kind of open botanical gardens that are not based on scientific classifications, but intended for spontaneous exploration for pure enjoyment. This is a new type of park - ecological park without any stereotypes, perfect for socializing of the visitors. In a park like this, every visit is a unique experience. For example, moving through the strips gives you a sequence of landscape experience, while if you cut through the tape, a completely different sequence of perception will be revealed. This type of park can be called a proto- or supra- urban landscape, as its experience evokes qualities that we would like to encounter in all settlements [13-16].

The second step involves penetration of the volumes in the form of individual prisms (Fig.3). They are in a linear disposition. Initially, the whole program was contained in a single long linear volume in accordance with the morphology of the site. This volume was further segmented into three separate volumes, translated into each other to evoke a sense of movement. The tendency of this concept is to interpolate greenery and architecture, to weave one into the other. To create a composition where the prisms will penetrate from the inside, but will not interfere with the environment [17, 18].

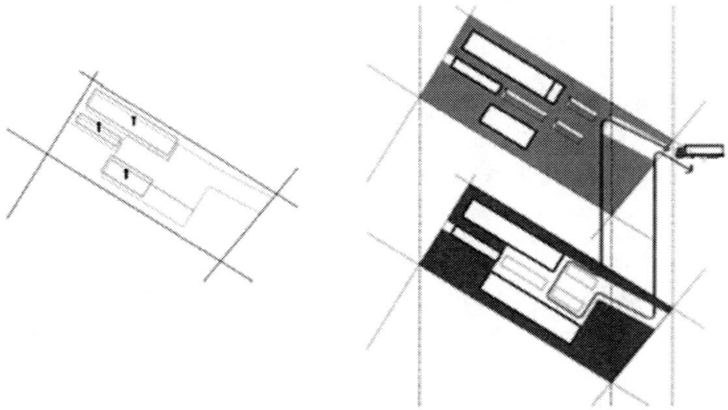

Figure 3. Differentiation of pedestrian traffic from vehicle traffic

The third step includes the genesis of the traffic solution proposed by the project. Namely, the traffic solution is of utmost importance for the functioning of any industrial complex. The aim of this project is to separate the vehicle from the pedestrian traffic. This will be achieved by leaving linear paths on the ground, connected with the prisms on both sides, while the vehicle traffic will be moved one level below ground level. There is a ramp to access the lower plateau, which connects the three linear prisms that penetrate the surface. This is the way that leads to the storage areas, warehouses and the manufacturing plant [2].

4 Genesis of ground solution

The project for urban beer factory is not a project that proposes a definite industrial architecture. The idea behind this project is to apply a method that combines industrial architecture and environment, in order to attain an urban, ecological park (Figure 4).

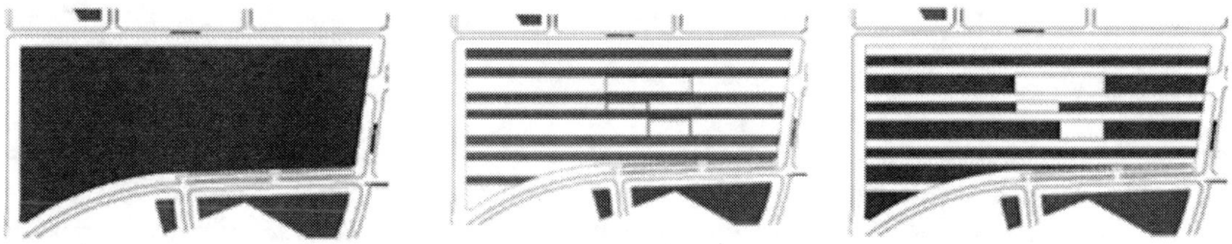

Figure 4. Layers of urban solution

By setting the linear grid, one type of texture is formed from the free, inbuilt surface on the site. A number of bands are reserved for the visitors, to be used as pedestrian bands, while the remaining are intended for greenery. Thus, the concept for an urban beer factory is finalized. By combining of all functions a single harmonic composition is formed. A scenario for the concept's operation is provided below [4].

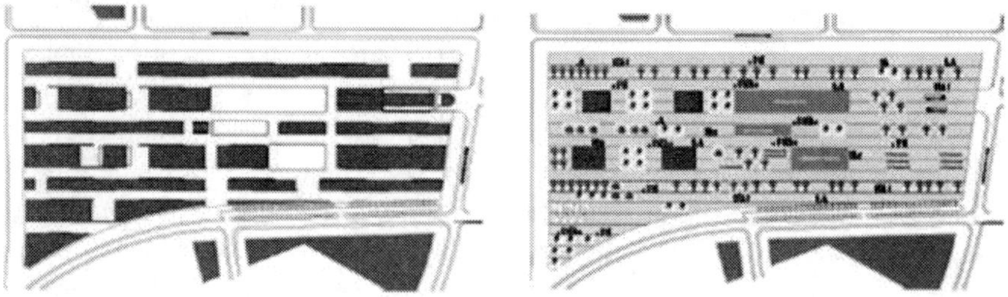

Figure 5. Scenario for functioning of the locations

5 Designing process - product

The project for urban beer factory is graphically presented (Fig.6, 7, 8). Everything is designed in strips. That is the aim of the concept. The whole building, which is composed of three prisms that are visible on the ground level, is formed from the same material, concrete and steel panels [2].

Figure 6. Facade of beer factory

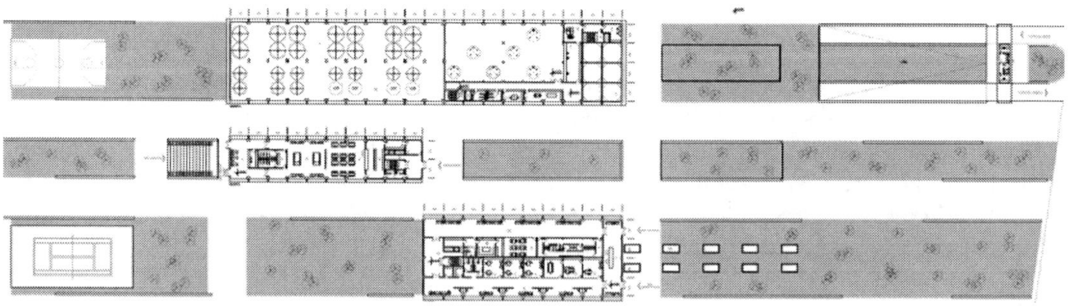

Figure 7. Plan of beer factory

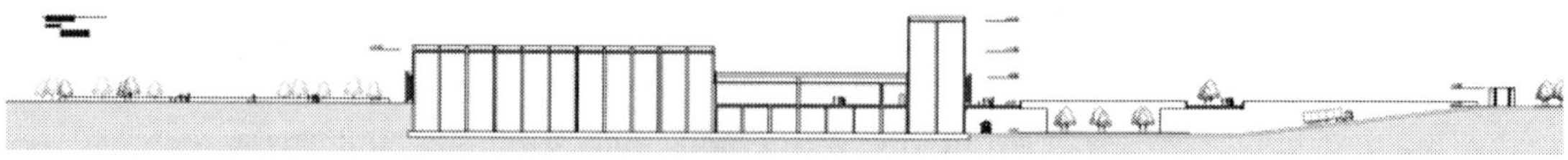

Figure 8. Section of beer factory

6 Summary

In the course of its entire development, industrial architecture has undergone countless changes. The contemporary ways of developing industrial facilities enable their relocation from the typical industrial zones into urban areas where they are closer to their customers. Beer factories are examples that demonstrate the changes in the perceptions of the overall picture of the industrial facility, which used to represent a symbol of production, closed within its designed limits. The changing lifestyle, human needs and creativity of architects lifted the brewery to a more advanced level not only in view of the manufacturing plant itself, but also in view of the rest areas, entertainment, recreation, and spaces that can provide for tourism development.

The aim of this research is to highlight, examine and adjust these production facilities, beer factories, as a part of the new urban reality. To design a modern industrial complex incorporating the ideas of the contemporary way of designing an industrial facility with a clear and indicative architectural expression of its hybrid character requires great skilfulness. Every city needs this type of architecture. The result will be a modern industrial facility, which would be considered as a novelty in architecture in general.

In this sense, this study shows one of the many possible variants for intertwining of such innovative ideas that would lead to a new ideological basis from which would stem modern improved industrial facilities and ultimately, improved architecture in general.

References

1. Information from: www.oma.eu [7.8.2015].
2. V. Gjorgjevska, Master Thesis, (Architectural Faculty - Skopje, 2014)
3. A. Christopher, *The Timeless Way Of Building* (Oxford University Press, 1979)
4. Ch. Datz, Ch. Kullman, *Winery Design* (Neues, Roma, 2006)
5. A. Jüren, K. Hausmann, F. Jüttner, *A Design Manual Industrial Buildings* (Birkhäuser; Berlin, 2004)
6. H.R. Franzke, *Industriebau [Industrial building]* (Birkhauser; Berlin, 2001)
7. M. Webb, *Adventurous Wine Architecture* (Images Publishing, Sidney, 2005
8. R. Krier, *Architectural Composition* (Rizzoli, New York, 1988)
9. G. Radovic, V. Murgul, N.I. Vatin, Applied Mechanics and Materials, **584-586**, 564-569 (2014)
10. V. Murgul, Journal of Applied Engineering Science, **12 (2)**, 121-128 (2014)
11. J. Ćetković, M. Knežević, M. Žarković, V. Murgul, N. Vatin, Applied Mechanics and Materials, **638-640**, 2465-2470 (2014)
12. Y. Nikitin, V. Goryunov, V. Murgul, N. Vatin, Applied Mechanics and Materials, **680**, 504-509 (2014)
13. R. Alihodzic, V. Murgul, N. Vatin, Applied Mechanics and Materials, **680**, 494-498 (2014)
14. G. Radović, V. Murgul, M. Cvetkovska, E. Aronova, N. Vatin, Journal of Applied Engineering Science, **12 (4)**, 277 – 284 (2014)
15. G. Radovic, V. Murgul, N. Vatin, Applied Mechanics and Materials, **641-642**, 634-638 (2014)
16. M. Penića, S. Golovina, V. Murgul, Procedia Engineering, **117**, 883-890 (2015)
17. V. Murgul, N. Vatin, I. Zayats, Procedia Engineering, **117**, 824-829 (2015)
18. V. Goryunov, I. Zayats, T. Konjkova, V. Murgul, Procedia Engineering, **117**, 825-831 (2015)

MATEC Web of Conferences **53**, 02002 (2016)
DOI: 10.1051/matecconf/20165302002

Orientational Reaction as a Way of Adjusting to Space

Ema Alihodzic Jasarovic[1, a] and Rifat Alihodzic[1]

[1]*Faculty of Architecture, University of Montenegro, 81 000, Podgorica*

Abstract. This paper defines basic terms of movement and system of reference points in a city. We can conclude that primary difference in the perception of space and spatial orientation depends on the fact whether the subject lives in the city or he is a foreigner to that space. In this respect, we can conclude that when it comes to space orientation, foreigners mostly rely on visually and dimensionally strong and visible reference points, which can simultaneously be historical or morphological in terms of their character. On the one hand, the subject who is familiar with the area typically moves within a system of individual reference points. Individual system of reference points does not necessarily require any of the physical features that recommend an object as a reference point. On the other hand, there were numerous theories among which Lynch, after conducting a series of experiments related to space perception, pointed out that orientation depends solely on one's ability. Based on the analysis of the system of reference points, position of focus points and their relations, we can conclude that Podgorica, as most towns in Montenegro has a problem in terms of visual readability.

1 Introduction

Readability of a city image represents an ability through which a city could be seen as a whole which is made of several independent units and parts, indicating a physical clarity of the image [1]. Also, the entire landscape could be visually perceived as a complex composition of recognized symbols, which are used to identify certain parts of the city. The value of a city is measured by the amount of pleasure it can give to the persons who use it. It represents a need for a space that is suitable for people, in terms of its organisation that fits perceptive features of modern man. The role of the city image as a guide through space is very significant, because one gets to know and masters the space through movement. In accordance with this, the function of a visually well-organised surroundings is to facilitate movement. Also, the user of the space feels the constant need to be positioned and oriented, to create an image about their surroundings. The vividness does not necessary imply that the image of a city is apparent at first sight, or even noticeable and generally known or simple. A well-organised image of an urban surroundings is emphasised as the main precondition for using a city. Accordingly, discomfort resulting from a poor space organisation is not that uncommon. Kevin Lynch wrote about space recognition, noting that a city image has its value, not only by acting directly as a map used for orientation, but it also has a more general sense by serving as a general coordination system where individuals act or arrange their experiences and knowledge [1]. As a process of self-confirmation and self-determination in a space, orientation is just a segment in exploring possibilities of designing a space suitable for people.

[a] Corresponding author : aema@t-com.me

2 Movement and importance of a goal

The topic of this paper is orientation of the subjects during their movement, their visual impressions when switching places, where dynamic characteristics are ascribed to the subject, in relation to the observed, immovable – static surroundings. Movement represents a change of position in a certain space during time, where perception of the movement cannot be perceived outside this context and that is when we are referring to a space spatiotemporality. By moving through space, one accomplishes a visual communication between the space user and the space itself, which mechanically presents a process that resulted from actions of visual forces. This paper deals exclusively with pedestrian movement, because every other type of movement would change the character and concept of the paper. Subject is the user of the space, he is its part, he goes through it, experiences and conquers it. The man, as a homo viator, is motivated with permanently movable and restless thinking [2]. At the beginning, it is necessary to make a difference between the subject who is a foreigner in a space, who does not have a defined starting point and who is not familiar with image of the city, and the one who lives in the city, and has a memory – mental image of the surroundings. That is why they have different perception of the city's image, distances, directions and proportions, and they use different ways for space orientation. Hodological space where the observer moves is usually, accidentally or potentially called a path, and it is always determined by forces that are created between the known and the unknown. Cornelia Faragode defined the path as non-place [2]. Christian Norberg-Shulz linked the path with one of the basic characteristics of human existence [3]. We always link direction with the path. Gordon Cullen described the space division of the beginning and the end through drama of space relations of what is "here" and what is "there" [4]. "Here" presents a static position of the observer, place – position of the current existence, where the being of the subject is defined as distance between presence and absence. In that case, the goal presents a focal point, to which we strive, a place of gathering, some type of a centre. Goal as a place of occupation, a focal point that we strive towards, affects the subject, motivates the movement and makes it focused. Therefore, the existence of a goal enables progress in movement and stimulates the state of expectation.

3 Orientational reaction, disorientation

In the process of movement, it is often not enough to define the starting point and the goal. Several different paths often lead to the same destination. Labyrinth, the old game of searching for the right path to the goal represents an example of path where only starting and finishing points are known. Such space without landmarks, leaves us without any support from the space and sensory signals that the surrounding sends. In the previous part, it was defined that the existence of a goal stimulates movement. On the other side, inexistence of forces of attraction and repulsion causes a panic because of the effect of objects' forces. Experiments were conducted in ideally artificial situation of laboratory labyrinth. Due to the lack of strong visual reference point, the interviewees reached for simple reference point like a gnarly board that they would recognize. In one such theory characterised by spatial divisions, differentiations, limitations, distances and demarcations, desert space succumbs to constant changes and unrecognisable paths, which leads to a limited and undirected movement. Accordingly, it could be concluded that orientation implies knowing our position in space and constant improvement of relations between us and the space. The lack of perceptual powers causes disorientation, which leads to fear and anxiety. Rudolf Arnheim stated that, in order to be successfully oriented, one must establish a number of markings in the exact timeframe and try to get a general insight that will show context and relations, alternatives, distances [5]. Disorientation is caused by a disturbed relation of perceptual powers owned by every object in space, to a greater or lesser degree. It could be caused by a collision of visual forces without any order, which makes it impossible to determine a place and spatial function of objects in the field of vision. This is most frequent in spaces that do not have a dominant space reference point whose visual forces are competitive to all of the structures built in space.

4 Space orientation reference systems

View of a city is not constant, but fragmented, in order to accomplish clarity with which parts of the city can be recognised and organised in a whole. When a path has qualities such as direction or a defined and clear goal, we can talk about its proportion, and the space user could have a sense of how far has he gone and how much longer will it take to cross the path. Sense of proportion usually appears as a result of numerous famous markings along a path. In this case, we are referring to a subject that knows the locality and has a mental image of the city. When heading towards the goal, one conquers the path through stages, and by determining space in accordance with their personal sensibility. This way, subject form sections in their image which make short-term goals, and they always know where one phase of the path ends and another begins. Kevin Lynch introduced a term district [1], and Christian Norberg – Shulz mapped such points as places [3]. This way the path becomes a system of familiar point, where the trajectory becomes divided into a part before and after a point, i.e. what is behind and in front of us. This way, user of the space forms a system of individual reference points in relation to which they are oriented and which determine their homological space. Shulz believed that the entire person's space is centred subjectively and that every person's personal life has its centre [3]. As analogy to this, we can conclude that a foreigner cannot have a developed perception of the distance and sense of proportion in an unknown space, unless they have previous experiences and memories of that space. Unlike the subject that is familiar with the locality, a foreigner does not have a dimensional or temporal relation to the city. A foreigner does not know or feels the proportion. Those who visit the city for the first time form an impression based on the image that is experienced as a certain given thing. This impression is mostly unbiased, unburdened by those emotional elements that local residents have. As such, it can be very realistic in evaluation of the quality of spatial relations. A foreigner needs a readable image in order to get around in space more efficiently. That image is formed as an abstract and simplified system of visual references. It demands a strong visual reference point – a visual and spatial identification, in order for their movement to be clearly directed towards the goal [6].

Exploration and determination of the reference point system of a city analytically digs deeply into the zone of urban and spatial relations, architectural compositions and its elements from the aspect of perception, as well as personal abilities of the subject – participant in the space [7]. When writing about the ways of orientation, Arnheim pointed out that the subject is very frequently oriented in relation to the street [5]. He believes that buildings are parts of sequences and they barely express their three-dimensional quality in that structure. Figural character formed by walls of the buildings is the dominant characteristic of spatial actions. This is even more visible when driving a car, where the increased speed accentuates penetration into empty space and focuses attention on what is happening within that space. It should be pointed out that the character of the street is greatly affected by its width and height of the objects that form it. Only in the case of well-positioned height and width proportions, street could be established as a visual object with its personal visual centre, which actively resists forces that come from the buildings that form it. In many streets, orientation points cannot be recognised, and the streets do not have a figural character that was previously discussed, because the centres of visual balance were disturbed due to reckless setting of objects and their mutual relations. It points to a fact that a street, in terms of design, should provide expressive properties that state direction, in order to enable easier orientation and movements through space [8, 9].

Sometimes the image of surroundings is not organised by a system of dominant directions, but through one or more foci, which are used to measure distances and determine directions. These reference points also include morphological reference points such as mountains, rivers, hills... Urban planning of Podgorica is determined in relation to two rivers, which divide it into three clearly defined zones: Stara Varoš, Nova Varoš and Novi grad. The river represents an important spatial reference point within this spatial organisation.

One of the most common ways of orientation is finding previously imposed, adopted and already known reference points, which are usually historical reference points. The first contact with an unfamiliar city includes spatial orientation and positioning. We do this by locating things in relation to already familiar city points. Since we are strangers, we do not have earlier spatial experiences related to that place and so we reach for information related to those areas or buildings in the city which represent its landmarks and which we use to recognise that place. They become our zero –starting point, from which we perceive the city image. Landmarks such as specific buildings, especially those important for the city, towers or sculptures, squares and bridges, they all help to determine the reference points and emphasize the hierarchy of places. The Fig. 1 shows Acropolis (a strong morphological and visual reference point) and Eiffel tower (historical and visual reference point).

Figure 1. a) Acropolis; b) Eiffel tower.

For example, when we arrive in Paris for the first time, the first thing we look for is the Eiffel tower, a historical, architectural, monumental symbol which we will use to position ourselves in space, and it will become our main orientation and reference point as well as a starting point for any further movement, orientation and reconnaissance of space. This is how we come to a situation where we determine a place by using a building, by forming limited areas which make that place distinctive. From the analytical point of view of perception, Rudolf Arnheim considers these places as a basis for dynamic action [5]. He points out that in a dynamic sense, a tower which stands in an empty field or a ceramic vase on a table represent centres of visual energy which transmit vectors in all directions. However, it is very rare that a building is alone in an empty space. Usually, the building is in touch with the surroundings and neighbouring buildings which also have vector forces that affect each other, influence the observer's eye and disrupt the centricity of the composition. Therefore, every spatial composition represents a configuration of forces of perception, and the reference point's task is to overcome competitive vector forces and become dominant. This visually dominant position in space represents a feature of visually distinctive buildings which are easily recognised in space [10].

In the previous section we have repeatedly emphasized the importance of space that is suitable for people, it terms of its physical and visual identification, which enables easy movement and orientation, and makes the city image easy to perceive. A survey was performed as a part of this paper and it included citizens of Montenegro, coming from different urban structures and city's levels of development. When asked what is the reference point of a city, most of the participants of different age and education structure connected that characteristic place with a place where they meet, which they use for orientation and to locate other places of lesser significance. In most case, this was not a high and dimensionally strong benchmark, but instead their answers often included fountains, bridges, monuments, rivers, parks, squares, stations, benches, stairs, shops, advertisements. They often pointed out the place of residence, a house, as a focus point and explained that this represented a place which they left and returned to on a daily basis. In the given moment, they located their position in relation to the strongest reference point they listed. When doing this, interviewees demonstrated good knowledge of dimensions, distances, scale and often added time dimension to describe the distance from the benchmark, thus indicating experiential category. We can conclude that every interviewee had an individual system of reference points in which he moved and functioned. In the previous part of the term paper it was pointed out that when it comes to a person who knows the city, the orientation system was usually reduced to an individual system of reference points. In terms of perception, these points can be completely unexpected things which do not have an important role in the overall townscape [7, 8].

5 Physical characteristics of reference points

By analysing physical characteristics of buildings that may become potential visual reference points in an area, we can conclude that one of the most significant features is the dimension of the building. When it comes to architectural composition and its elements in terms of perception, the spiritual component of human aspiration towards God is most easily accepted as something that bears an individual shaping of distinct dominance in pure nature. Such forms are easily perceived by a human eye, and they are accepted as accents, landmarks and signposts. Organisation with the help of these reference points proved to be very useful in practice. It also creates an environment which is easy to perceive [11-13]. With their elevation above the uniform structures, the silhouette of high structures can be used to highlight key locations. This type of orientation bears a risk of losing the reference point in case the position or distance is changed, because we perceive things from different points. We often notice a reference point from a large distance and start moving towards it, or we orient in relation to that point. The problem occurs when we go inside the city and lose the sight of the reference point because of the visual influences from the surrounding buildings, which "block" the reference point. In this way, it loses its spatial role in the city's system of reference points [14]. All this indicates another important category which implies the importance of positioning of a reference point within an urban area as well as a constant road tracing and opening of vistas towards the focus point. The Fig. 2 shows Obelisk, (Washington) - a strong and dimensional reference point and Egyptian pyramids – purity of shapes.

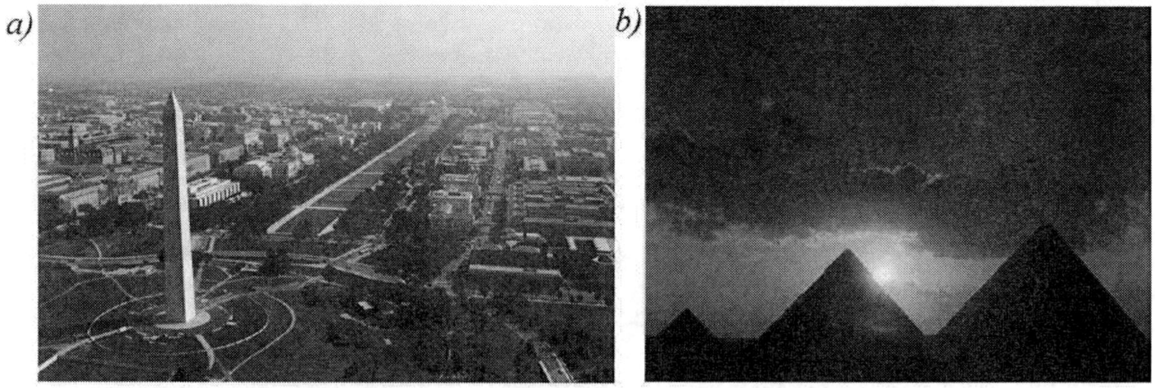

Figure 2. a) Obelisk, Washington; b) Egyptian pyramids.

In addition to the height characteristic, shape component represents a prominent feature of a reference point. In this case we are talking about individuality of an object which stands out by being different, and aims to make the object easier to remember and find. This is where we leave the primary area of readiness to receive stimuli, and move to a higher level of remembering and association processes. When it comes to a visual experience of the architectural form in terms of perception, we can conclude that visually strong, dynamic and aggressive objects are key for perception of a certain urban image. Lynch notes that the object is also more remarkable if it has a clarity of general form, as does a column or a sphere [1].

Another important category in terms of readability of reference points is the quantity – frequency of reference points in a certain area. Sometimes it is enough to have one reference point which has a good position in terms of dimension, shape and space in order to make a space an easily perceivable surrounding, which will be easy to orient and move. On the other hand, to many reference points can easily create a visual imbalance, due to a disturbed relations between forces possessed by every object in space. In the spatial concept of megacities, the position of reference points becomes compacted and frequent, thus losing this effect (Fig. 3). These cities do not have suitability of space, they disturb every sense of size and scale, and the subject sets for himself a new, suitable system of reference points for orientation [14, 15].

Figure 3. Saturation with visual reference points New York [10].

The principle of perception of an object stands out in further analysis of features that suggest an object as a reference point of a city. Here, an important role is played by a principle of figure and background, where dynamic features of the background do not make it passive. It can accentuate the object or completely diminish its visual effect and power [16]. Therefore, the position of the reference point in relation to the background represents is a very important principle. Its role is to favour the figure by keeping it the foreground. The Fig. 4 shows Istanbul skyscrapers which disturb the vista over the most famous city reference point and simplicity principle [6].

a)

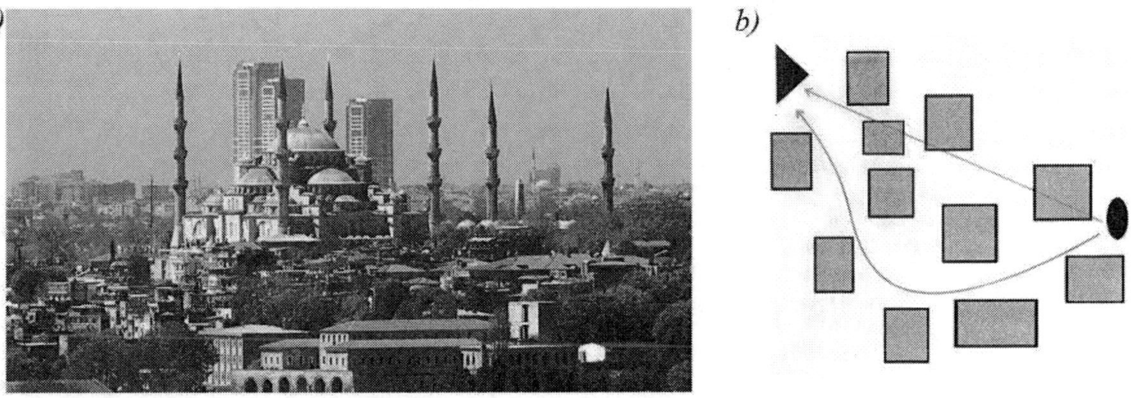

Figure 4. a) Istanbul; a) Simplicity principle.

Another gestalt principle of perception includes the principle of simplicity, which explains that a person functions in a three-dimensional space in every given moment. When walking towards an element which has emerged as a focus, our spatial relation remains a straight line despite the fact that there are numerous buildings between us and the reference point [17, 18]. This difference between the mental and real image of the city can often cause confusion or disorientation.

6 Summary

The dynamics of composition, vibrancy and coherence of the urbane image is emphasized as a central and main condition for the use of a city, because the user of a space seeks for a visual drama and not only a gallery of moving images. It is also important to emphasize that we orient by remembering a series of distinct details. There are undeniable benefits from spending time in an enriched environment in terms of harmonious development of person's sphere of psychological knowledge. In addition to its function in movement and orientation, the importance of the system of reference points in the image of a city is that it can significantly contribute to the visual identity of a space. Visual identity of the urban image is desirable and it can contribute to achieving a balanced relation between its global recognisability and local authenticity [9].

References

1. K. Lynch, *Image of the City*, (DTB, Belgrade, (1974)
2. K. Farago, K. *Dynamics of space, movement places,* (DTB, Belgrade, 2007)
3. K. N. Šulc, *Existence, space and architecture*, (DTB, Belgrade, 2002)
4. G. Cullen, *City landscape*, (DTB, Belgrade, 1990)
5. R. Arnhajm, Dynamics architetectural form, (DTB, Belgrade, 1995)
6. Information from: http://gorkorg.blogspot.co.uk/ (26.12.2015)
7. R. Alihodzic, *Defining the primary aspects of psychological experience of architectural space and form,* (Ulcinj, 2007)
8. R. Alihodzic, V. Murgul, N. Vatin, Applied Mechanics and Materials, **680**, 494-498 (2014)
9. G. Radovic, V. Murgul, N.I. Vatin, Applied Mechanics and Materials, **584-586**, 564-569 (2014)
10. S. Popović, N. Djurovic, V. Murgul, Procedia Engineering, **117**, 832-840 (2015)
11. G. Radovic, V. Murgul, N. Vatin, Applied Mechanics and Materials, **641-642**, 634-638 (2014)
12. I. Yamshanov, V. Goryunov, V. Murgul, Procedia Engineering, **117**, 675-684 (2015)
13. I. Zayats, V. Murgul, Procedia Engineering, **117**, 706-711 (2015)
14. Y. Nikitin, V. Goryunov, V. Murgul, N. Vatin, Applied Mechanics and Materials, **680**, 504-509 (2014)
15. V. Goryunov, I. Zayats, T. Konjkova, V. Murgul, Procedia Engineering, **117**, 825-831 (2015)
16. V. Murgul, N. Vatin, I. Zayats, Procedia Engineering, **117**, 824-829 (2015)
17. I. Yamshanov, V. Goryunov, V, Murgul, Procedia Engineering, **117**, 663-674 (2015)
18. Noth, M., Borning, A., Waddell, P. Computers, Environment and Urban Systems, **27(2)**, 181-203, (2003)

MATEC Web of Conferences **53**, 02003 (2016)
DOI: 10.1051/matecconf/20165302003

Decentralization as a Cause of Spatial Segregation

Ema Alihodzic Jasarovic[1,a], Dragan Komatina[1] and Sanja Paunovic Zaric[1]

[1]*Faculty of Architecture, University of Montenegro, 81 000, Podgorica,*

Abstract. City represents an incomplete dynamic process prone to the expansion with a causal link between urban expansion and socio-spatial segregation. The socio-spatial distribution in the city is mostly related to the increased social polarization and inequality. There is a clear connection between divided society and divided city: if society is divided, urban space must be divided. It is the question of the relations between the social inequalities on one hand, and spatial segregation on the other. In the last 10 years, Podgorica is the city that shows alarming statistic values when it comes to demographic trends and the influx of the residents from the northern municipalities, which necessarily causes the city sprawl. Past experiences show that city is unevenly expanding, creating new functions and zones expressed by socio-spatial differences. The beginning of this process lies in modernist conception of the city, by which city was mostly developed, while the current functional organization is based on the same concept. With the first urban plans, which carried similarproblems mentioned in previous section, Podgorica was divided into three clearly differentiated zones: Stara Varoš, Nova Varoš and Novi grad, which became a platform for hierarchical divisions within the space, reflecting them in the society.

1 Introduction

This paper analyzes the issue of socio-spatial transformation of the modern city that has emerged as a result of very complex processes, starting with the industrialization and development of the tertiary activities, which have provoked uncontrolled urbanization process. Dynamic demographic growth, which is reflected in the immigration population in the cities, a new lifestyle and increase in life standards, introduction of new traffic networks, the causes which influence spatial expansion of the city and increase the distance and the proximity of the city [1, 2, 3]. These and similar processes cause structuring of the city into smaller parts, with distinct centers and sub-centers, deployed in accordance with the spatial organization of the city. In modern urban concepts, the prevailing network system of the centers is polycentrism, which is the opposite approach to former monocentric organization of the city. Polycentric systems emerge as a modern city needs to create a number of urban centers, whereby such spatial organization is "functional or sociological imperative that creates a balance in life of the city" [1].

The initiated process of globalization as a form of postcolonial imperialism is increasing social inequality. The question of socio-economic and spatial separation in cities is one of the most popular research topics in the last ten years. The key question relates to whether the evolution of changes through the expansion of the city and the creation of a polycentric system necessarily leads to spatial

---

[a] Corresponding author : aema@t-com.me

segregation? Numerous authors have attempted to explain the relationship between the occurrence of a new social order and spatial divisions, particularly in global cities. The topic of research is directed towards questioning whether due to all these processes, our cities changed the physiognomy of the class, and what is the structure of the modern city [4]. By observing the historical genesis of the development of the city, there is no city in the history of architecture which has not been divided into rich and poor. Having this fact in mind, spatial separation is a logical outcome of social polarization and inequality. Often the differences are occupied to a larger extent, which leads to the marginalization of area and creation of spatial stigma and is labeling a group of people belonging to it completely [5, 6]. This paper also questions whether there are positive aspects of this phenomenon, considering that spatial segregation can contribute significantly to the local or regional recognition. Taking into account that the complexity of the city lies in the spatial relationship of opposed social groups, thus inequality becomes an indelible part of the urban area, and very often has a dominant characteristic [7, 8].

Nowadays the cities play a strategic role in the new urban hierarchy, and an increasing amount of literature is dedicated to the understanding of the links between the contemporary globalization and socio-spatial transformation which is particularly evident in cities over the past 30 years. Actuality, but also the significance of the subject of polycentric city and decentralization, and the city's role in promoting its development and competitiveness of its parts, currently deserves special attention and a meaningful urban concept [9-11].

2 Expansion of the city and the emergence of decentralization. Criticism of the functional city

Up until the seventeenth century the cities were static, enclosed by walls, with no possibility of expansion within that area. The walls were a kind of urban development regulator with an important role in the design of their building. As population increased, many cities felt the need to expand, whereby walls obstructed the further urban development and expansion of the city. The decline of the city walls, in symbolic and functional sense, marks the beginning of an uncontrolled expansion of the city towards an open free space, causing the displacement of the center, and the emergence of decentralization [12]. The most significant from that period is a radical reconstruction of Paris, then Vienna and Barcelona, which solved the problem of urban expansion by demolishing the walls and introducing the new traffic system [13, 14].

The nineteenth century will be remembered as the period in which the turbulent economic and social situations happened, closely associated with urbanization, which marked a new era. The industrial revolution created the tremendous changes in demographics of population, as in the very concept of the functional organization of the new city. Most noticeable consequence of this whole process is the transformation of the city in terms of migration of population from rural to urban areas. Industrial urbanization denied the traditional values of the city, moving geographical expansion and leading to the creation of a new, industrial city. Henri Lefebvre indicates that the "critical stage" in the foundation of industrial urbanization has crucially changed the nature of the city and transformed it into a new, far more complex entity: urban [2].

On the fourth CIAM congress (Congress International d'Architecture Moderne) in 1933, the Athens Charter was adopted when the area of interest of CIAM was spreading to urban planning and principles of "functional city", carrying out the analysis by using the example of 33 cities. The result of this research suggested that cities need to be organized according to the principle of strict functional zoning, while the fragmentation process is reflected in the five key functions in urban areas, namely: residence, work, recreation, transportation and central functions. The fragmented areas of the city, with a certain role and function, are mutually classified by the hierarchy [15].

The modernist concept of the city eliminates the former principles of urban city space. The concept of a functional city is often criticized because of the large number of failures that have been proven in practice. New functionalist vision of the city based on the accurate and defined redistribution of the functions, leads to the disappearance of certain urban areas, and this is usually reflected through the loss of the main square and streets. By analyzing the problems of the city in the twentieth century, on a concrete example of Frankfurt, Shadrach Woods said that it is not possible to create city by an urban zoning; with the mere separation of functions, their interdependence is ignored [3]. The spacial continuum of the city is disturbed by the new distances and proximities, new proportions which cause the new means of transport, as a result of zoning. Therefore, the loss of continuity of urban space could mean its symbolic destruction. During the last congress of CIAM in Dubrovnik in 1956, Alison and Peter Smithson raised objections to the dogmatism of the Athens Charter and highlighted the problem of the alienation of the population in the new modernist settlements [16, 17, 18].

The modernist urban planning is based on the idea that the city should be transformed into a single homogeneous entity divided into functional zones. However, this concept has turned entirely into something else. Instead of overcoming the differences, the forms of functional fragmentation of space became the perfect tool for creating the inequality. Demolition of the housing block "Pruitt- Igo" (Pruitt Igoe Housing architect Minoru Yamasaki), is a typical illustration of the idea in which the most important postulates of modern planning were applied and what Charles Jencks views as a moment in which the modern dies [4]. Figure 1.

Figure 1. Iconic demolition of the Pruitt-Igoe housing project in St. Louis [22].

Although this residential complex was designed in accordance with the most progressive ideals of CIAM, the explanations for the failure of Pruitt Igoa are very complex. The project has been known for a notable segregation policy, which included other problems in crime, poverty, problems of anonymity, public-private relationships, and many others. After years of devastation, in the end it resulted in demolition. Pruitt-Igoe is one word for a social experiment [19].

Today - the dynamic city is subjected to a constant change. Due to the continued growth and fluidity of cities, there is a constant adjustment of functions of the city, and the creation of new urban zones. In this constant variability, the question is how to maintain the continuity in the functional space and socio-economic transformations. Constant Nieuwenhuys believes the expansion of the city cannot be stopped as it would be an attempt to stop the overall historical development, and that the reasons which led to the expansion of the city still continue to exist, regardless of the redistribution of the functions [6]. Ebenezer Howard said when the city that once reached the optimal size no longer needs to be enlarged, but it needs to fit into a complex system of balance and functional completeness. He was against the chaotic growth of the city, and believed that the city should be planned and considered as a whole [1].

In the mid XX century there have been multi-million cities as a historical novelty. New technological developments have caused different concepts of building cities, questioning the experiences so far. In writing about the disappearance of urban space by building cities of the XX century, many theorists are trying to oppose their views to this uncontrolled development of the city. The reasons for the problems in the city expansion are to be found in the functional structures that are able to withstand the degree of mobility that requires the ten-million city. The goal is always the same, to find a successful model that will lead to the creation of a new urban spatial order, which will maintain an open organization and spontaneous mobility of modern society [20].

3 Division of space - Polycentric system of organization of the city

As mentioned previously, with the expansion of the city, the first divisions of space were formed. The division of the city had different phases throughout the history. The first significant division of space was on village and town, city or suburb, which dates back to the ancient city. Looking at the genesis of the city, we can come to the conclusion that the city was, even in Mesopotamia, divided into quarters. Furthermore, the Roman city was divided into quarters, with their smaller centers, while medieval towns were divided into residential districts or parishes [21]. Even in the thirties, the theories suggested that the city should be separated into groups of distinct area and shape, which are part of the hierarchical whole. Michel Foucault notes that in the Middle Ages, there was a hierarchical system of places that were sacred, profane and protected, and in contrast, open and undefended, city areas [8]. This mixture of places in the hierarchy is characteristic of the medieval urban space in particular.

Today, with the growth and expansion of the city, the development of suburbs is gaining more significance and the city ceases to be an accurately limited spacial entity. Between 1940 and 1980 the city was divided to the center and the periphery, where, the rich and middle classes lived in the center, and the poor on the outskirts and dangerous periphery. The city center has always been the focal point in the life of the city and a place of exchange of social activities. Formerly it was symbolized by the agora, churches, and cathedrals, and was most vividly described in the medieval towns, where this space represented a type of a set for various civic and religious ceremonies .

The process of densification and expansion (implosion and explosion) of cities created new spatial formations and changed the previous concept of modern urban city. With urban growth and population, spatial differences were decreasing, and these parts of the city were getting closer. These and similar processes condition the city to structure into smaller parts, with special centers and sub-centers, deployed in accordance with the spatial organization of the city. „ The new concepts stimulate the diffused polynuclear development of the cities, so that the new centers become functionally structured, in order to generate a certain hierarchy order and some sort of continuity with the central urban area “, writes Mirko Maretić [1].

The concept of polycentric organization of space occurs even in medieval towns. Each square represents the city center of a specific function, carrying the visual accent that is represented by an architectural object. Today the theory of multiple centers is a tendency, with distributed activities or functions, whereat the centers are placed in the system of hierarchy, according to their roles and size. The functionalized city zones are made in such a manner, which creates a single organism, and within that organism they have their spatial autonomy and recognition. In order the city to function on the basis of a polycentric organization, the communication between its constituent parts must not be compromised. This is implying that within this structure, there is a dominant center, and that all others are in such a connection that does not create a competitive image, and ruins the basic functions of the city. However, spacial situations are not always predictable, they occur as a synergy of different circumstances. With such conscious or unconscious (random) formation of secondary centers, decentralization leads to relocation of certain content, or creates new ones, which often entails specific target groups and space users, which in most cases may lead to social segregation.

4 Space as a reflection of the society. Socio – spacial segregation

Based on the assumption that the city is a mirror of society and social activities, and that all phenomena in the city are dependent from each other, Ranko Radović stated that the social connections and the lifestyle are the main generators of urban form, he says: “City expresses life “[9]. The space is a reflection of society and social activities, which implies that the hierarchy in space is closely connected with the hierarchy of the society, which is a basis for the spacial segregation. The spatial practices, representations of space and spatial representation, appear as interdependent, simultaneous variables / dimensions that create the totality of social space [10]. The question is whether the expansion of the city, that knowingly or unknowingly, causes decentralization and the creation of a polycentric system necessarily leads to spatial segregation? Frederico Roman Ramos and Ciro Biderman in Urban sprawl and spatial segragation in Sao Paulo, with the use of numerous measurable results, showed that the urban sprawl causes segregation, where the indexes indicate that the emerging parts enhance the segregation [11].

Even Socrates in his "Republic" wrote on the emergence of spatial segregation in the ancient Greece, where the polis, was divided into two cities poor and rich. In contrast, in the ancient Rome there was a spatial integration between rich and poor as a result of special anti-separational spatial planning policy. Sreten Vujović on the development of modern cities and the segregation claims it is the inevitable result of this process, which moves mainly in two directions: spontaneous and organized [12]. It can be said that the segregation inside Montenegrin towns that previously existed based on ethnic and religious grounds, now gives way to a segregation which mainly stems from the socio-economic inequalities. In Podgorica urbanism, we have witnessed an emphasized form of that phenomenon in the last 20 years which is represented with the housing blocks in which the inhabitants live in poverty, and new housing blocks where visibly affluent class population lives. The choice of location in the city imposes the investment cost, which quite naturally defines the clienteles and customer financial solvency, consciously producing a concept of socio - spatial inequality and segregation. One of the greatest examples of socio - spatial segregation in the world is Sao Paulo; the walls were erected between rich and poor blocks, so today it is the city walls, that Teresa Caldeira writes about [13]. (Figure 2). The established enclaves were created that represent a new alternative for urban livingof the middle and high classes and as such are considered as status symbols, deepening the social distance and encouraging social differences and inequalities. They are physically and functionally introverted and turned inward, defining the outer world as negative.

Figure 2. Sao Paulo [23].

The modern city is represented as an incomplete process which brings class, ethnic and racial segregation by growing, and thus forming socially fragmented cities. The city is socially heterogeneous from the village, and therefore all forms of social differentiation are highlighted in the urban than in rural areas. In microcontext and emerging urban planning of Podgorica, the process of social stratification in commodity and monetary economy, it is mostly reflected in urban sphere through the phenomenon of segregation. Consequently today we have parts of the city such as Konik where Roma and refugee populations live, then Stari Aerodrom with the sole function of "dormitory" without any additional content, labeled for a group of residents who immigrated from the north; and the elite parts the city with ambassies, residential complexes of politicians and wealthy citizens as Malo Brdo or the latest residential and business complex Atlas Capital center. (Figure 3). Segregation is defined as space and content in the form of the reserved part of the city. Setha Low writes about "divided city", where this term, in the anthropological sense, refers to the hidden racial and class barriers written in metaphors such as "residential neighborhood" (uptown), "the business part of the city" (downtown), "tony parts" and "ghetto" [15-18].

This image cannot but remind on the Berlin Wall as the most famous example of the socio-economic divisions between eastern and western parts of the city. Spacio-social sense divisions are particularly marked in the cities of the United States, between Afro-American and Caucasian race.

a) b)

Figure 3. a) Konik, Podgorica [24]; b) Atlas Capital Centar, Podgorica.

The dynamics of widening the differences and division of space with respect to them can create the ethnic city, which is the area that can be economically, socially and linguistically completely independent. Segregation along ethnic lines is also visibly expressed in our cities, particularly during the period after the '90s war. The examples are Gypsy mahalas, then individual locations of Albanians from Kosovo, but also Serbs and Montenegrins in Kosovo, Bosnians in Slovenia etc.

5 Summary

The city is a complex, constant and incomplete process. Dynamic changes in the demographic and spatial growth of the modern city, affect its functional organization. Consequently, cities with the specific expansion both vertically and horizontally, change the urban concept over the time. In this regard, the paper will highlight the problem of spatial segregation and alienation of the population which is the idea of a functional city, as it continues to exist through the concept of polycentric cities. This was a clear message that the rationality of the organization of the city did not offer good results. This principle of urbanism is characterized as a new form of organizing the social differences and creation of segregation, contrary to the idea of urbanism that turns the city into a single homogeneous entity eliminating differences. Along with the aforesaid, the neoliberal globalization process, emphasizing the hierarchical divisions, deepens and inaugurates a new concept of the divided city. Due to the extreme inequalities, the town itself produces a new urbanism that is reflected in the significant spatial divisions and forms of behavior in cities. Socioeconomic polarization and inequality pollute the space giving birth to a new idea of a city. The city becomes a complex process and structure that is imprisoned in the model of duality between conflicting social spaces. All this implies an unbreakable bond between the divided society and the divided city.

References

1. M. Maretić, *Gradski centri*, (Zagreb (1996)
2. N. Lefebvre, *From the City to Urban Society*, (Orion art, Belgrade, (2003)
3. S. Woods, *Frankfurt: The Problems of a City in the Twentieth Century*, (Studio Vista, London, (1964)
4. C. Jencks, *Nova paradigma u arhitekturi*, (Orion art, Beograd (2007).
5. http://www.thepolisblog.org/2012/01/management-as-design-in-urban-housing.html
6. C. Nieuwenhuys, G. Debord, *Architecture and arts*, (Skira editore, Milan, (2004)
7. M. Perovic, *Antologija teorije arhitekture XX vijeka*, (Građevinska knjiga, Belgrade (2009)
8. M. Foucault, *Of Other Spaces, Utopias and Heterotopias,* (Diacritics, Belgrade, 2010)
9. G. Radović, *Forma grada. Osnove, teorija i praksa*, (Građevinska knjiga, Beograd (2009)

10. H. Lefebvre, *The production of space*, (Blackwell Publishers, Oxford (1991)
11. G. Radovic, V. Murgul, N.I. Vatin, Applied Mechanics and Materials, **584-586**, 564-569 (2014)
12. S. Popović, N. Djurovic, V. Murgul, Procedia Engineering, **117**, 832-840 (2015)
13. V. Murgul, Journal of Applied Engineering Science, **12 (2)**, 121-128 (2014)
14. G. Radovic, V. Murgul, N. Vatin, E. Aronova, Applied Mechanics and Materials, **627**, 357-364 (2014)
15. Y. Nikitin, V. Goryunov, V. Murgul, N. Vatin, Applied Mechanics and Materials, **680**, 504-509 (2014)
16. R. Alihodzic, V. Murgul, N. Vatin, Applied Mechanics and Materials, **680**, 494-498 (2014)
17. G. Radovic, V. Murgul, N. Vatin, Applied Mechanics and Materials, **641-642**, 634-638 (2014)
18. V. Murgul, N. Vatin, I. Zayats, Procedia Engineering, **117**, 824-829 (2015)
19. V. Murgul, Procedia Engineering, **117**, 808-818 (2015)
20. V. Goryunov, I. Zayats, T. Konjkova, V. Murgul, Procedia Engineering, **117**, 825-831 (2015)
21. I. Zayats, V. Murgul, Procedia Engineering, **117**, 706-711 (2015)
22. http://lsecities.net/media/objects/articles/worlds-set-apart/en-gb/ (12.01.2016)
23. http://www.scielo.br/img/revistas/ccrh/v23n59/05f01.jpg (12.01.2016)
24. http://ngm.nationalgeographic.com/ngm/photo-contest/2011/entries/122822/view/ (12.01.2016)

MATEC Web of Conferences **53**, 02004 (2016)
DOI: 10.1051/matecconf/20165302004

The Liberty Style - Italian Art Nouveau Architecture

Vasilii Goriunov[1,a]

[1]*St. Petersburg State University of Architecture and Civil Engineering, 2-Krasnoarmejskaja, 4, Saint-Petersburg, 190005, Russia*

Abstract. This article refers the architecture of Italy of the end of the 19- the beginning of the 20 centuries. It shoes the origin of the term "Italian Liberty architecture", its main centers, its peculiarities and the buildings of its leading representatives. The assessment of importance of such studies provide the right understanding of the processes in European architecture of this time.

1 Introduction

All historians of architecture know very well that the term Art Nouveau has a lot of synonyms, accepted by researches of different countries. Most of these terms do not reflect the essence of the things and they are used conditionally. The exception is the term " Liberty Stile" although its origin also seems accidental like the origin of the majority of its synonyms. This term came from the name of the founder of British trading company which specialized on artworks from Japan and China.

This was the reason for the fact that many researchers are not quite accurately called this company one of the conductors of Japanese influence on European art. This was true as long as the company was engaged in the export of the works of art only. Soon, however the company began to cooperate with young British artists of applied art who created company s own style, which brought it European s fame. The basic principle of Liberty Style was the Celtic ornament-oriental influence was not very noticeable.

2 The Italian Liberty Style architecture

The greatest success of company Liberty was connected with Milan in the North of Italy where representative office of this firm was located. After the International Exhibition in Turin in 1902 " The Liberty Style Architecture" became the symbol of what we would call the Italian Art Nouveau. At the same time it was forgotten very quickly that initially "Liberty" was the family name and soon " Liberty stile" became for Italian architects a style of Liberty. The fact is that in Italy "Liberty Style" acquired a new sense was not accidentally [1-3]. The main reason was in the emotional atmosphere which originated in Italy at the end of the 19-beginning of the 20 centures. Almost during the whole 19 century a fierce struggle for unification of Italy was going on. This process was resulted with unification of the country in 1871.At the same time within the centralized state some centrifugal

[a] Corresponding author : vsgorunov@rambler.ru

tendencies were still strong enough. Cultural differences between the various parts of the country and contradictions between the North and the South even now are very serious (2)

In this regard the Italian architecture is marked by the processes typical for other European countries. The " Thirst for style" seen in the architecture of the most European countries was realized in the attempts to create a new formal language of architecture for general use. This " thist" was minimal in Italy because from one side we find great historical tradition, and from another there was strong aspiration for creative freedom. Thus the unique situation emerged when the eclectics was the most suitable way for the development of architecture in that country. New forms of Art Nouveau were admitted first of all as the enrichment of decorative means and often were combined with traditional elements of architectural forms. Even the concept of " Liberty Stile" contained a certain contradiction. The point is that the term " Style" always means " regulation", " order", strictly speaking-something contrary to freedom. In Italian architecture aspiration to freedom dominated over aspiration to order. Thus the " thirst" for style which inspired the representatives of anti-eclectic movement of the end of the 19 century was very weak in the Italian applied art and architecture where prevailed individualism combined with personal tastes and preferences. This circumstancies together with very deep artistic roots of Italian culture resulted mostly eclectic character of " Liberty Style" which quickly approached to what Zeldmaier called " aesthetic Anarchism" [4, 5].

The Art Nouveau architects failed to create international style because their historical tastes , theoretic basis, even purposes were very diverse. Art Nouveau never was a single style but it was a collection of local styles which contrasted to each other being at the same time in the stage of interaction.(3). In one of our works we justify the position that of all stylistic trends in the Art Nouveau Architecture the closest to the ideal of the Great style was Finnish architecture (4). "Liberty Style" was the most remote variant [6-8].

The transition from classical eclecticism to the stage of Art Nouveau in Italy was relatively late and so decisive as in many other countries. The eclectic architecture in Italy never seemed decadent and was received favorably both by critics and by public. "Liberty style" in Italy included both traditional eclectic motives and those invented in the end of the 19 century. Thus eclectic architecture in Italy reached its last limit, using new elements [9]. This was possible because the eclectic and Art Nouveau architecture in Italy were more united than divided. The proximity to eclecticism and vague stylistic characteristics of " Liberty Style" in Italy led to the fact that Italian architecture of this period was out of sight of researchers of European architectural history. Meanwhile its uniqueness, originality, high aesthetic quality cause the respect and deep interest." Liberty Style" in Italy was widespread. We can find the examples of its structures in all regions of Italy. Nevertheless there were several centers where the " Liberty style architecture was concentrated. First of all is Turin, where The International Exhibition of 1902 was located [10]. Its main architect was Raimondo D Aronco- the builder of the pavilions (Figure 1) .

Figure 1. The main pavilion of the International Exhibition. 1902. D. Aronco

Known as a restorer of Saint-Sophia Cathedral in Istambul, D Aronco created one of the most widespread variants of " Liberty Style" which sometimes is called " Floreal". Its soarces can be found in the style of the Viennese Art Nouveau school in the interpretation of Josef Olbricht and Otto Wagner [11-13]. However this Style has not become the object of direct imitation but remained a personal "handwriting" of its creators, which inspired sometimes other architects for experiments in the field of artistic décor, The results of these experiments characterize the " Liberty "architecture in Turin.

Another important center of "Liberty" was Milan. It was here that the Italian office of British company " Liberty" was located. It was here that the most architects of Italian Art Nouveau worked- Pirovano, Bossi Kompanini, Arta and certainly Giuseppe Somaruga famous beyond the borders of Italy. Milan represented the stylistic diversity of " Liberty " architecture [14-17].

" Palazzo Castiglione" by Sommaruga built in 1901 together with the main pavilions of the Turin Exhibition represent Italian architecture of Art Nouveau period in Italy and abroad. (Figure 2)

Figure 2. "Palazzo astiglione" by Sommaruga.

For understanding of " Liberty" architecture the great interest represent The Raman quarter named after its creator architect Gino Cappede (Figure 3)

Figure 3. The Raman quarter Cappede, 1920

The architecture of this quarter as well as all buildings of Cappede and his brother Adolfo is very eclectic basing on classical tradition. At the same time it is sufficiently original including some details typical for Art Nouveau. Italian Art historians call this buildings the example of " Liberty Style". Built in 1920-ies The Coppede quarter formed the bridge to the Art Deco architecture which inherited much from " histgoricism" of the past period [18, 19].

To understand better the development of Italian architecture one has to turn to the Sicilian architecture. This large island was always the territory of collision of different peoples and civilizations. So the architecture of Sicily was a mixture of various traditions, forming the " delightful caos of eclectics". It seems that the influence of the traditional architecture of Sicily made eclecticism acceptable and attractive in the eyes of Italian architects of Liberty period. Palermo became a centre of opposition to Futurism and avant-guard [20-24].

One of the most famous and influentional figures was Sicilian architect Ernesto Basile who worked in Palermo. Commissioned by the family Florio and constructed by Basile "Grand Hotel" in Palermo was considered an example of comfort and at the same time it became an example of creative freedom within the forms and motives of eclecticism. (Figure 4)

Figure 4. "Grand Hotel" in Palermo, 1898. Basile

Italian art critics considered Basile the founder of" Liberty style" of late 19 century. What was called in Italy " Liberty " architecture was closely connected with eclecticism and this allows us to suppose that the architecture of Art Nouveau period with all its anti-eclectic trends was the concluding stage of the great process that started at the time of Renaissance in Europe and was imbuied with " historicism". At the same time it may be called the result and the synthesis of 2 periods-of the Antiquity and of the Middle ages. The great change in the development of European architecture was caused the appearance of avant-guard with its tecnicism and anti-histiricism. Probably here we may find the answers on the main problem: what was the real place of Art Nouveau Architecture. Whether it continued the previous period or perhaps it is the beginning of new historic stage-the first stage of the contemporary architecture. Perhaps it presents independent part of European architecture differing from eclectic and from avant-guard [25-27].

In1912 in Milan a new group was founded-"New Tendecies"-an union of the Italian Art Nouveau artists. Among the members were two architects-Gino Coppede(the triumphant eclectic) and Antonio Sant-Elia(the young futurist). Two years later the" Manifesto of Futurist architecture" will be published, starting the new epoch.

References

1. V. Lintner, *Italy. The history of the country* (SPB, 2008)
2. C. Kin, *Italy in the end of 19 century* (M,1978)
3. V. Gorunov, M. Tubli, *The architecture of the period of Art Nouveau* (SPB, 1992)
4. V. Gorunov, *The Nothern Art Nouveau and the Problems of Style* (SPB, 2014)
5. L. Caramel, Longati, *Antonio Sant-Elia* (Como 1962)
6. N. Lipovac, A.M. Jandriček, Prostor, **19 (2)**, 443-455 (2011)
7. N. Lipovac, Prostor, **19 (1)**, 2-17 (2011)
8. V. Murgul, Procedia Engineering, **117**, 808-818 (2015)
9. N. Lipovac, Prostor, **18 (2)**, 268-287 (2010)
10. N. Lipovac, Prostor, **5 (1)**, 1-33 (199
11. S. Kosanović, A. Fikfak, Energy and Buildings, (2015)
12. I. Čerpes, N. Pandol, A. Fikfak, European Countryside, **6 (3)**, 225-242 (2014)
13. C. Zappulla, C. Suau, A. Fikfak, Journal of Architecture and Urbanism, **38 (4)**, 247-264 (2014)
14. A. Fikfak, Journal of Architecture and Urbanism, **37 (4)**, 257-267 (2013)
15. S. Popović, N. Djurovic, V. Murgul, Procedia Engineering, **117**, 832-840 (2015)
16. I. Yamshanov, V. Goryunov, V. Murgul, Procedia Engineering, **117**, 675-684 (2015)
17. A. Olaj, P. Gabrijelčič, A. Fikfak, Geodetski Vestnik, **56 (1)**, 151-168 (2012)
18. I. Zayats, V. Murgul, Procedia Engineering, **117**, 706-711 (2015)
19. V. Goryunov, I. Zayats, T. Konjkova, V. Murgul, Procedia Engineering, **117**, 825-831 (2015)
20. V. Murgul, N. Vatin, I. Zayats, Procedia Engineering, **117**, 824-829 (2015)
21. V. Murgul, V. Pukhkal, Procedia Engineering, **117**, 891-899 (2015)
22. R. Alihodzic, V. Murgul, N. Vatin, Applied Mechanics and Materials, **680**, 494-498 (2014)
23. I. Yamshanov, V. Goryunov, V, Murgul, Procedia Engineering, **117**, 663-674 (2015)
24. Y. Nikitin, V. Goryunov, V. Murgul, N. Vatin, Applied Mechanics and Materials, **680**, 504-509 (2014)
25. G. Radovic, V. Murgul, N.I. Vatin, Applied Mechanics and Materials, **584-586**, 564-569 (2014)
26. A. Fikfak, G. Mrak, Glasnik SED, **50 (1-2)**, 48-56 (2010)
27. A. Fikfak, M.Z. Senegačnik, Glasnik SED, **48 (1-2)**, 2-67 (2008)

MATEC Web of Conferences **53**, 02005 (2016)
DOI: 10.1051/matecconf/20165302005

Reconstruction of the Historical Heritage (Based on the Example of Azerbaijan Hamam)

Grigoriy Garmanov[1,a] , Mikhail Volkov[1] and Tatiana Spiridonova[1]

[1] *St. Petersburg State Polytechnical University, Politekhnicheskaya, 29, Saint-Petersburg, 195251, Russia*

Abstract. The article discusses the history and features of recovery of Azerbaijan hamams. Restoration of the traditional architecture with the reference to the history is considered to be a significant factor for the national culture of Azerbaijan. The history of the hamams are reviewed in this article. The impact and importance of hamams in the development of settlements has been analyzed in this document. Besides aforesaid topics, the issue deals with the reconstruction and rehabilitation of hamams.

1 Introduction

Restoration of the traditional architecture with the reference to the history is considered to be a significant factor for the national culture conservation against the background of the modern massive development of the city of Baku, the capital of Azerbaijan [1, 2]. Among the historical buildings as well as prominent constructions preserved within the territory of the Absheron peninsular hamams appear to be of interest. Hamams are traditional public baths closely related to the lifestyle of the Azerbaijan population [3].

Hamams are quite often located near hieratic constructions due to requirements and peculiarities of the Muslim religion having such rituals which can be associated with washing and a number of other procedures [4].

2 Hamams arrangement in a settlement planning structure

As cultural and social buildings hamams were widely spread on the Absheron peninsular both in populated inhabitant places and garden areas where the basic population of Baku moved to adjacent settlements from April up to November (Figure 1).

Hamams arrangement in a settlement planning structure is caused by people's attraction to hieratic centres. These interrelated complementary objects form volumetric and spatial image of the Absheron settlements. Many hamams were built on the Absheron peninsular back in the medieval ages but a few of them have been conserved up to nowadays. There is a hamam of the year of 1388 in the settlement of Nardaran among them [2].

Archaeological and Ethnographic Museum Complex Gala contains many interesting examples hamams (Fig. 2-3). In the village of Gala and the adjacent territory, there are 215 archaeological sites. There remained a historical context, referring to the III millennium century BC In the village of Gala

[a] Corresponding author : gri.garmanov@gmail.com

are 5 mosques, 3 hamam (bath), houses, buildings, agricultural lands, tombs, mausoleums, burial mounds, remains of the fortress, have not lost their historical significance to the III millennium XX century.

Figure 1. Hamam. Boradigah, Azerbaijan.

In 2008, on the initiative and under the direct supervision of the Heydar Aliyev Foundation in the territory of Archeology and Ethnography Reserve Gala was created "first Historical-Ethnographic Museum Complex under the open sky." The complex has collected and restored monuments of architecture found on the Absheron Peninsula [5-7]. (Figure 2, 3)

Figure 2. "Gum hamam" ("sand bath"), Azerbaijan. Figure from [7]

Figure 3. a) "Shore hamam" ("salt bath") b) "Bej hamam", Azerbaijan.. Reconstruction made in 1881. Figure from [7]

There was basic construction of hamams at the turn of XIX – XX centuries when the settlements became economically developed and population increased. It is an important period in the history of the Absheron settlements development where inhabitants were engaged into oil boom and took advantage of economic challenges for construction of hamams and other social and public constructions. It was that very time when hamams stated to play a significant role in settlements planning and development [8-10].

As to city planning hamams have their own peculiarities defined by their specific location. Land plots, where hamams are located, are quite varied. Volumetric and spatial system of hamams was elicited depending on their location [3].

The general trend in hamams arrangement lies in striving for making this large-scale construction significant and highlighting the fact it is a social and public building. It's quite important in the context of typical traditional solid fences and walls of private housing. Hamams show specifics of public constructions in mass development meeting the requirements for construction site management with due account for water supply sources (water wells, water reservoirs, open pools) and heat supply sources as well (storage pits for keeping oil and etc.).

Certain approaches of hamams arrangement on the construction sites in relation to street network have been found due to investigation of the existing hamams. This has afforded ground for their initial classification [11, 12]. The most popular approaches of hamams arrangement can be the following ones:

1) hamams take active planning nodes and become dominant in construction development;
2) hamams are located in the through streets of a settlement;
3) hamams in a row development;
4) hamams in the outskirts.

All this contributed to bringing hamams to the positions of the significant civil facilities as well as hieratic objects [4-6].

3 Types of Hamam volumetric layouts

The volumetric layouts of Absheron hamams have been developing under the traditional functional pattern: entering room, dressing room, intermediate passing chamber for keeping heating regime between dressing and bathing rooms, 'hazna' (storage for hot and cold water).

There are 4 types of the Absheron hamams.

Considerable areas to ensure capacity for a large number of visitors were offered in the first type. This can be achieved by increasing a span between longitudinal walls resulting in big costs and technical complications. Four columns were set up in the central part in dressing and bathing rooms that allowed for 9.36 m to 12.80 m in length. The length of hamams normally reached 30 m.

The second type of the Absheron hamams was more popular than the previous one. Keeping basic layouts points and technical approaches of the first type hamams the second type has its own architecture and composition features as well as volumetric spatial ones.

The hamams of this type with extended length had fixed rectangular shape and were clearly split into supplementary and main ones according to their layouts. The dominating elements of the hamams were the following: dressing and bathing rooms were the halls with joint space, having no intermediate supports, with non-deep enclaves along the axis. The rooms in the layout formed a shape of a cross; the corners were actual supports bearing under-dome space.

The third type of the hamams had layouts with displaced longitudinal axis relatively to dressing and bathing rooms. But its technological working cycle associated with water supply an

The constructions which have complex and combined layouts with independent volumetric and spatial rooms can be related to the fourth type.

The dressing room was represented in a whole volume under a relatively large dome, and bating room was split into parts on the internal network of the aisles owing to central supports gaining the dome of a small size. Such volumetric and spatial solution made it possible to use varied methods which combined basic composition elements. In this context we may compare the type under consideration with the previous one where an axial line of the hamam layout was displaced and volumes became more dynamic. According to this classification all the types given herein can vary significantly in the context of both layouts and volumetric and spatial solutions, as well as of construction approaches and decoration elements [13-15].

The Absheron hamams have been developing throughout the last century contributing to certain architectural and planning approaches which were specified in the classification given. This classification makes it possible to elicit all the general relations in the hamam construction evolution giving a wide range of the various types of buildings. Regardless the fact there is quite a definite typology there are no similar hamams. Each hamam has its own features. Conserving as well as research and contemporary use Absheron hamams is an integral part of the architectural heritage and civil culture. (Figure 4)

Figure 4. The medieval hamam. Engraving

4 Temperature and humidity regimes of Hamams

The most important element of hamams requiring specific concern when reconstructing and repairing objects is a conventional and historically developed heating and ventilation supply system. Historical heating and natural ventilation methods ensured a unique microclimatic regime in rooms.

The creators of hamams 'tuned' internal microclimate in the rooms as if it was a musical instrument'. An attempt to connect hamams to contemporary engineering networks inevitably leads to a temperature and humidity regime violation.

When making restoration and reconstruction works for this type of construction architectural layouts, construction features and unique temperature and humidity regime in the rooms must be considered. Since hamams have quite a complicated engineering system, which can be restored or re-built on the basis of the customs and experience, a reasonable engineering estimation appears to be a must and an evident issue.

The traditional heating and ventilation systems meant for hamams is a basic construction component. And along with that construction materials underwent a special attention. The builders of those times selected specific waterproof limestone that played a great role when exploiting hamams over several centuries. At the same time construction materials had to meet all the hygienic and esthetical requirements.

5 Summary

The territory of Azerbaijan is rich in archaeological sites, proving belonging to ancient settlements. From the standpoint of historical monuments Absheron Peninsula is of particular value.

The hamams within the Absheron territory are presented by a varied range of architectural layouts and construction approaches. This makes the grounds for an assumption that they can be distributed to other areas with due account for research and restoration and maintenance experience.

The unique heating and water supply system assigned for hamams called for quite a complicated individual «setting up» based on the experience and long life traditions. There are many cases of upgrading hamams and using modern methods to ensure water supply, heating and ventilation, as well as new construction materials which have finally resulted in deteriorating of historically developed functioning system applied in these objects. Restoration of hamams is a complex research and practical issue nowadays.

Regretfully these interesting civil and social facilities, which were an important part of local settlement development, have lost its significance in our days. In many cases these land plots are still idle and empty giving an example of descendants' neglect regarding their national architectural and cultural heritage.

There are about 50% of hamams remained on the Absheron peninsular. And they make it possible to evaluate their significance as civil architexctural monuments having a large scope of shapes and volumes.

It's also important to make them an integral part of city planning and state their contemporary significance in the settlement planning structures.

References

1. Dadashev, S., Usejnov, *Arhitektura Azerbajdzhana 3 -19 vekov*, (Stroyizdat, Moscow (1983)
2. Bretanickij, L.S. Salamzade. G. *Istorija arhitektury Azerbajdzhana*, (Stroyizdat, Moscow (1963)
3. Fatullaev-Figarov, S.S., Zodchestvo Apsherona. *Issledovanija zamkov, mechetej, hamamov, ovdanov, karavan-saraev, djevjelikov, mavzoleev i vill Apsherona,* (Stroyizdat, Moscow (2003)
4. Fatullaev-Figarov, S.S. *Gradostroitel'stvo i arhitektura Azerbajdzhana 19 – nachala 20 veka,* (Stroyizdat, Moscow (1986)
5. Alihodzic, R., Applied Mechanics and Materials, **680**, 494-498 (2014)
6. Huseynov, F. E. Oglu, Procedia Engineering, **21**, 534-542 (2011)
7. http://www.trend.az/life/history/1932698.html date accessed 18.10.2014
8. Innovative Technologies in Development of Construction Industry 10.4028/www.scientific.net/AMM.725-726
9. Azerbaijan Hamam, the History and Restoration Specifics 10.4028/www.scientific.net/AMM.725-726.1079
10. N. Lipovac, A.M. Jandriček, Prostor, **19 (2)**, 443-455 (2011)
11. N. Lipovac, Prostor, **19 (1)**, 2-17 (2011)
12. N. Lipovac, Prostor, **18 (2)**, 268-287 (2010)
13. Lipovac, Prostor, **5 (1)**, 1-33 (199
14. Čerpes, N. Pandol, A. Fikfak, European Countryside, **6 (3)**, 225-242 (2014)
15. C. Zappulla, C. Suau, A. Fikfak, Journal of Architecture and Urbanism, **38 (4)**, 247-264 (2014)

MATEC Web of Conferences **53**, 02006 (2016)
DOI: 10.1051/matecconf/20165302006

Neogothic Orthodox Churches and Chapels Built Near St. Petersburg and in the Russian Province (the Second Half of XVIII – the Early XIX Centuries)

Igor Yamshanov[1, a]

[1]*St. Petersburg State Polytechnical University, Politekhnicheskaya ul., 29, 195251, Saint Petersburg, Russia*

Abstract. The article is devoted to the development of religious buildings of Neogothic style and their stylistic features in the second half of the XVIII century. During the preromantic period the hobby for a Neogothic style in Russia has been connected with idealized idea of customers about the Middle Ages. The gothic style attracted them with exotic shapes. The desire of customers to create illusion of travel to the past over the different countries, typical for a sentimentalism era. was one of the reasons for the emergence of neogothic constructions in the Russian Empire.

1 Introduction

In the first period of Neogothic style in the Russian Empire on the territory of its capital - St. Petersburg was not built a single church in the Neogothic forms, but those can be seen in the vicinity of the city. It should also be noted that the religious buildings of St. Petersburg in its appearance virtually no features characteristic of Old Russian architecture, differing from that of Moscow.

2 The Neogothic orthodox churches and chapels built near St. Petersburg in the 2nd half of the XVIII century

The first of them is the church of St John the Baptist (Figure 1) on Kamenniy Island - Orthodox church built near the city border, (now it is the centre of St. Petersburg) in the style of the early Russian Gothic Revival in the 1776-1778 by architect Y.M. Felten. It has a cross-shaped plan and its architectural image is similar to the Catholic churches of the Middle Ages, without any details typical to the baroque or Russian Style in its appearance. The original church was built for the hospital of the Baltic Fleet. But then it was transferred to the Order of Malta, after Emperor Paul I became its head. In 1799 Knights of Malta Cemetery was arranged near the church. In this church Cavaliers took the oath. The red brick walls of the church are bare without plaster coating. Window openings have lancet shape and they are decorated with carved white stone details. Dome of the church, built on a high drum with lancet windows, has a conical shape. Above the entrance is single-tiered bell tower, terminated by four-sided tent, supported by four square pillars. The church interior is enclosed with serried lancet Gothic walls. Initially tents, crowning the church and the bell tower, had a concave

[a] Corresponding author : dartigor@mail.ru

shape, which gave them an even greater resemblance to the churches of the Middle Ages [217 S. 323]. Near the church of St. John the Baptist the building of the Ushakovsky bridge watch was built. Now it is the chapel of the icon "Vsetsaritsa". The building of red brick is a round in plan is situated on the banks of the Big Nevka. Its walls are cut with Gothic lancet windows, stylistically similar to the Church of St. John the Baptist. The upper part of the building walls is crowned with a decorative frieze and above it - the architrave decorated with the chain-like sequence of interconnected undulating arcs. Building dome has the shape of a nearly flat roof, the center of which is decorated with brick drum, topped by a cupola of onion shape with a cross. This building is an example of centric composition which is typical characteristic of classicism.

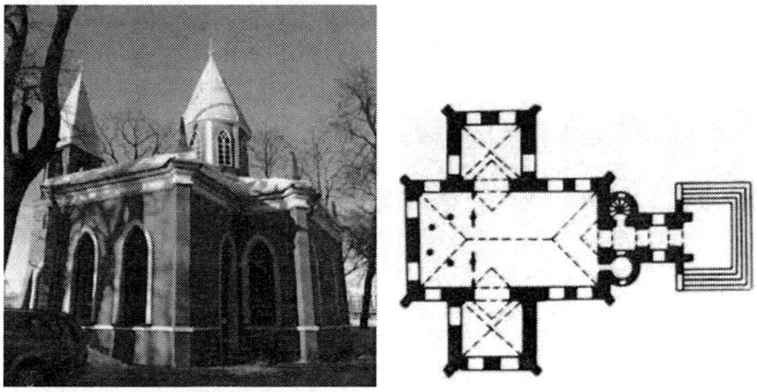

Figure 1. The Church of the Nativity of John the Baptist on Kamenniy Island. St. Petersburg. 1776-78. Y.M. Felten. View from the south-east. Photo by the author, 2011. Plan.

A genuine masterpiece of religious architecture, erected near St. Petersburg, is the Cesma Church (1777-1780) designed by Y.M. Felten. This church got its name in the honor of the victory over Turkey. In the plan it has the shape of quatrefoil as an example of the centric composition, with a strong Baroque influence. The volume of the temple is completed with five domes, each of the domes is topped with a slender spire. Four domes, each of which is surrounded by four pinnacles, placed on the cardinal directions and have throughout openings, and therefore not connected with the interior space of the church. At the same time the central dome has a drum with four big windows, which gives an extra light to the interior. High arched windows, a large number of pinnacles, the vertical division of the facades - all this gives the Chesma church special vertical aspiration and delicacy, close to authentic Gothic buildings. Its interior space filled with light is also close to the medieval concept of the Divine light pouring through the windows, which in the XII century the abbot Suger wrote about. This special role of light, which had a primary importance to the medieval cathedral, is partly typical for Chesma Church. Near the church had previously been two gates and a wooden pavilion, also built in the Neogothic forms, but not extant. The pavilion burned down in 1908, and the gates were dismantled during the restructuring of the Chesma palace in the XIX century. Chesma ensemble can be called the largest and most expressive buildings that were built near St. Petersburg in the early period of Neogothic style in Russia.

3 Neogothic orthodox churches of the preromantic period in the Russian province

Further we will pass to the churches built in the Russian province. They show, in general, the same tendency, as temples of Moscow – a mixture both with Russian Old Russian style, and with classicism, and in some cases with baroque. Very often churches were a part of country manors or estates. As a rule, it is explained by following to the general trend: the exotic imitating of antiquity constructions going from the Imperial Court, and supported by the nobility. In the province copies of the Chesma church built by Felten near St. Petersburg were created. Some customers, in particular Batashev, belonged to Masonic lodges that was also affected by a choice of stylistics for the churches built on the territories of their estates. It is also necessary to note that both in Moscow, and in country towns of Russia construction of cult buildings in forms typical to second half of XVIII century proceeds and in the XIX century. Thus, early Russian Neogothic style on time of distribution coincided with application of forms of already mature Neogothic style, and almost for 70 years as forms of the all-European Neogothic style in the capital of the empire start being applied about 1800.

One of early churches in the Neogothic style constructed in the Russian province, having in the space-planning solution the features of baroque is the church of the Sign (in Veshalovka of the Lipetsk region). It was built in Tatishchev's manor in 1768-1784, allegedly, on Bazhenov's project. It is one-storey church, which main volume is square in the plan. From the East it is adjoined by a rectangular apse, and from the West the rectangular ante church connected by a refectory to a belltower. A four-level belltower with the quadrangle basis and three cylindrical circles of almost identical diameter is topped with a high stone spire. The ante church and an apse are finished by eight pitched roofs. The main volume of church is closed by the close vault which is cut through by lukarn windows and topped with a high spire on the drum surrounded with four pinnacles. Corners of the main volume are strengthened by square towers with sharp stone spires. This church, in my opinion, is advanced version of Archangel Michael Church with richer decoration. The planning structure and the volume decision of church in Veshalovka are typical for the Baroque style.

We will in more detail consider manor copies of Chesma church. The earliest of them is the St. Nicholas Church constructed in the village Posadnikovo of the Novo-Rzhev district in the Pskov province. It was built in the estate of the favourite of Catherine II A. D. Lanskoy in 1782-1784. Some excerpts from earlier documents are given in the artcle about this church published in the Zodchy magazine in 1892. In particular from the church inventory of 1831 where it is told: "The church for the sake of St. Nikolay, stone, Gothic art, semi-circular, is constructed in 1784 for Alexander Dmitriyevich Lanskoy. It is covered with white iron, painted outside and inside by lime, figures above, five-domed. There are 24 quadrangular columns with wooden spires on them, between domes on encircling wall, covered by white iron, on corners finished with white iron apples; crosses with stars above domes, under the crosses are copper domes gilded in pure gold".

The church in Posadnikovo was under construction right after Chesma therefore there is a high probability that works were performed under supervision of Felten, who at the same time was working on reorganization of A. D. Lanskoy house in St. Petersburg. Also a few early mentionings about this monument remained, one of them is in the church inventory of 1831, others are brought together in the case of an estimated demolishion of St. Nicholas Church, stored in archive of the Imperial archaeological commission.

The second one for time of construction is the church of the Transfiguration in the village Krasnoe, constructed in Poltoratsky's manor in 1790. There is an opinion that for construction of this church Poltoratsky ordered someone from architects to copy Felten's drawings to construct the building. As a result a number of changes, in particular, which concerned architectural details in finishings of a portal, sculptures of angels and a form of lancet arches was brought in the project. Arches of church in Krasnoe have more pointed form. In Poltoratsky's archive the drawing different in details from Chesma church, but thus more corresponding to church in Krasnoe remained.

In the same years churches with planning decision, typical for classicism are built. Bright confirmation to that is Church of the Transfiguration in the village the Velikaya Topal (Figure 3) (Bryansk region) built in 1780 in the estate of the count P. A. Rumyantsev-Zadunaysky to replace more ancient wooden temple.

Figure 2. The Church of the Transfiguration. Velikaya Topal. 1780 North façade.

The main volume of the church is a rotunda with two tiers of windows, covered by a cupola, complete with a small octagonal drum and smaller dome. Rotunda is surrounded by four square towers, the northern pair of towers, topped with tents surrounded by gable crowns, significantly higher than the south pair completed with simple tetrahedral tents slightly concave in shape. The appearance of the church is clearly visible mixture of Neogothic and classic features. In general facade detailing is typical for classical buildings, there are also Neogothic pointed arches of door and window openings, as well as the spires and gable crown. Church of the Assumption, built in the S.S. Apraksin's village Peremilovo, perhaps designed by Francesco Kamporezi in 1792. If church in the village of Velikaya Topal (Figure 3) mentioned above, shows the use of the classic composition, but with some original interpretation, in the case of the Ascension church in the village Peremilovo in the plan decision and facades there is absolute adherence to the classical canons, disturbed only by using gothic decorations. The volume of the temple is a massive quadrangle, surmounted by a hemispherical dome on a low drum, with the completion of a small drum with a decorative cupola. Similar composition with Neogothic details and tripartite division of the facade demonstrates the baptismal church of the Kazan Church in Puchkovo village, built in 1802 by Khitrovo landowners, as well as the church of St. Nicholas in Rzhavki (in Moscow) started in 1805 and completed in the 1818-1820, when Andrei Nikolaevich Dolgorukov owned the estate. The church in Rzhavki was supplemented by a free-standing three-tiered bell tower with rectangular extensions. All of these three monuments demonstrate the classic earthiness and a complete lack of vertical aspiration typical for the Gothic style.

Then let us take a closer look at a few of the provincial Neogothic churches with a marked influence of the Baroque style, and above all, Novo-Nikolsky cathedral in Mozhaisk (Fig. 3), built on the site of the Nicholas Gate in the Kremlin of Mozhaisk in 1779-1814. It is quite exceptional work of early Russian Gothic Revival with strong baroque features. The cathedral was completed by the dome rotunda (lost during the Second World War), surrounded by four towers with a small bulbous dome. These four towers remind belvedere of Tsaritsyno Palace and the tower of the church in the village Bykovo. The four-tiered high belltower, adjacent to the cathedral from the west topped by a sharp spire, surrounded by pinnacles, is full of classical and baroque elements. In silhouette, it is close to the early works of M.F. Kazakov.

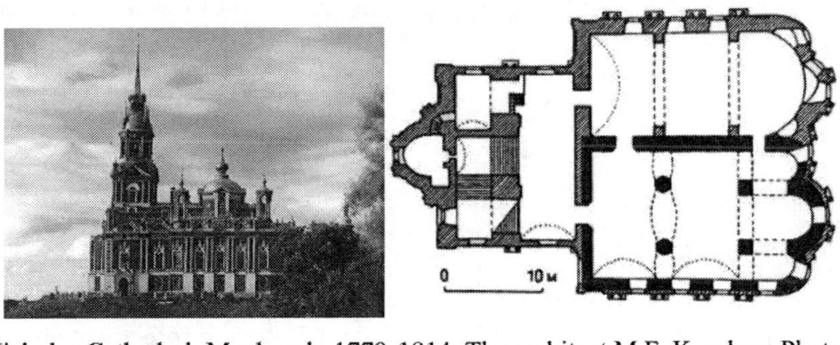

Figure 3. New Nicholas Cathedral. Mozhaysk. 1779-1814. The architect M.F. Kazakov. Photo XX century. Plan.

Church of Our Lady of Vladimir in Bykovo (Figure 4) was designed by Vasily Bazhenov in 1789. The church is different quite significantly from the original project (Figure 6), in which there was no grand staircase leading to the main entrance. The church building consists of two churches - the top in honor of the Vladimir Icon of the Mother of God and the bottom – of Christ. The bulk of the temple oval in plan, is delivered on high basement; from the west a rectangular refectory with two belltowers at the corners was attached to it. The bulk of the light rotunda is completed with high spire at the top. It is surrounded by a crown composed of the same spires. On each side of the oval in the main volume of the church are tall lancet windows, between which are placed decorative porches with columns flanking the side entrances. This temple is rare for the period of the early Russian Neogothic example of composition with two-towered western facade, typical for the Western European, mainly French, Middle Ages. Moreover, the temple was built of stone, which was also relatively rare.

Figure 4. The Church of Our Lady of Vladimir. Village Bykovo. Project V.I. Bazhenov 1782. Bell tower of the XIX century.

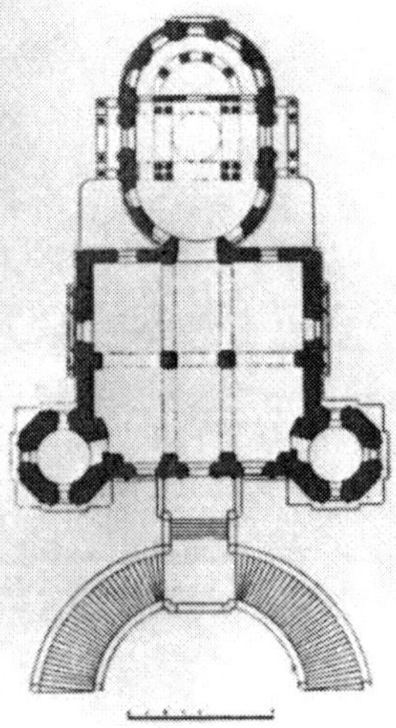

Figure 5. Our Lady of Vladimir Church. Village Bykovo. Architect V. I. Bazhenov. 1789. View from the northeast. Plan. V. I. Bazhenov's project of 1782.

Holy Trinity Church in Batashev's manor, Gus-Zhelezny (Figure 10) of 1802-1868. This two-storied church, also as well as church in Bykovo, is built of a white stone. On the planning structure it is crosswise, however this cross form is created by three apses surrounding its east part. From the West the church is added with a refectory and a belltower with a rectangular ante church. Construction works on this temple were actively conducted till 1825, by this time a refectory with a belltower were brought to eaves level, and the main building of church – to the level of the main dome. Further works were resumed in 1847, and came to the end only by 1868. This church as some researchers consider, is the largest rural temple in the territory of Russia. In the shape the church combines features of baroque, classicism and an early Russian Neogothic style. Who from architects possesses the project of church still isn't established: on the one version the author of the project is I. S. Gagin, on another – V. I. Bazhenov, the third names M. F. Kazakov as the author. In my opinion, the project was executed by I. S. Gagin. My point of view is based on that Bazhenov and Kazakov's constructions doesn't possess such laconicism and at the same time monumentalism and ponderousness. Despite considerable height this church makes impression of massiveness and monumentalism that is partly connected with laconicism in use of the front detailing.

Figure 6. a) Holy Trinity Church. Batashev's manor. City of Gus-Zhelezny (1802-1868), b) Church of Saints Peter and Paul. Mosviyka. Estate Volkhovsky. Ukraine. 1808. View from the southeast

Influence of early Russian Neogothic style can be observed as well in Ukraine that is confirmed by existence of Church of Saints Peter and Paul in Mosviyka (Figure 7), (The Cherkassk Region, earlier Piryatinsky district of the Poltava province, Ukraine) built in 1808 as a part of complex of constructions of Volkhovsky's estate. It shows a mixture of Gothic, classical and baroque elements. This centric building with the square two-storey belltower attached from the West and three semicircular apses located on cardinal directions. Lancet arches in its shape are combined with domes, characteristic for a baroque era, and also with constructive system of classicism. The central part of church is covered by a hemispherical dome on pendentives, western – the semi-circular vault. Most likely, that initially the temple was Catholic, and from the middle of the XIX century became the Orthodox Christian. This church has very considerable sizes: its length is 27 m, width is 19,4 m.

In some cases it is possible to observe examples of construction of stand alone belltowers with application of Neogothic shapes. Its construction is the same as earlier built orthodox churches. The phenomenon of this which began in the XVIII century will proceed during the second stage of development of Neogothic style in Russia in the XIX century. One of early examples of such constructions is the belltower in Pyotr Dmitriyevich Eropkin's (Fig. 8) estate (Grabtsevo, Kaluga region) 1791. This belltower is separate, with a rich set of various Gothic architectural details. In many aspects reminds a belltower of the Zachatyevsky monastery, however documents on Kazakov's authorship concerning this complex didn't remain.

One more rather expressive sample of a belltower, constructed near ancient cathedral is the belltower of Assumption Cathedral in Vladimir (Fig. 9) built on the project of the architect A. N. Vershinsky in 1810. 56 m in height it has four levels, the first three of which are quadrangle, and cylindrical at the top. Lower, more massive, quadrangle level on corners is finished by four round turrets with gear parapets. Corners of the second level are decorated with pinnacles. The belltower is topped with a gilded dome with a high graceful spire. In its shape Neogothic, classical and Old Russian motives are clearly visible.

In the Russian province in the 2nd half of the XVIII century as well as in Moscow, there was a tendency on unusual Gothic constructions. Rebuilding of more ancient temples in Neogothic style was sometimes carried out. This phenomenon is demonstrated by the church of the Assumption of the Blessed Virgin in the village Bernovo (Fig. 8) built in 1687-1699 in Baroque style, which later received the three level Neogothic belltower.

Figure 7. A belltower in P. D. Eropkin's estate. Grabtsevo. 1791. Architect M. F. Kazakov. Photo of 1905. General view.

Figure 8. a) Assumption Cathedral belltower. City of Vladimir. Architect A. N. Vershinsky (1810), b) Church of the Assumption of the Blessed Virgin. Village Bernovo. 1687-1699

Special type of constructions in style of an early Neogothic are monastic walls with towers. In their appearance we see mixture of the gothic and classic shapes. One of the most expressive monuments of this kind are the towers of the walls of Spaso-Yakovlensky Dmitriyev (Figure 9) monastery in the Rostov region constructed in the 2nd half of the XVIII century. Towers of the monastery have two levels, and second level is octahedral, completed with different tents or domes with spires. The towers with domes located in the central parts of walls are decorated with pinnacles, which are on the first level quadrangle, and on the second level - octagonal. The first level of towers have features of classicism: semi-circular arches, simple round windows, and one of towers has even a Doric portico.

Figure 9. Towers of a wall of Spaso-Yakovlensky Dmitriyev of the monastery. Rostov region. 2nd half XVIII century.

Walls of the Staro-Golutvin's monastery (Fig. 10). Kolomna. 90th of the XVIII century. Neogothic towers of a wall made of a red brick have a Gothic and classical details of a white stone. Their first level has a cylindrical shape. The towers are different: some of them low, almost without windows, topped with domes, some – high with the stone conic truncated tents decorated with a white stone details. Over these tents is the top cylindrical level which is richly decorated, topped with conic spires. However as all Neogothic towers built by M. F. Kazakov it have rather rich filigree detailing like towers of the Petrovsky palace.

Figure 10. Walls of the Staro-Golutvin's monastery. Kolomna. 1790th. Architect M. F. Kazakov.

4 Conclusion

In the Russian Empire within the Age of Enlightenment worldview to the middle of the XVIII century have developed such common cultural situation which has allowed a Gothic subject, inherently alien to the Russian art, to be shown brightly also in architectural works. It has been directly connected with the Western European influence which has amplified due to establishment of more close cross-cultural ties. Gothic images, getting on the Russian soil, as a rule, are enriched and supplemented with various elements, intertwine with national Russian, classical, baroque and mannerist motives.

References

1. *Architect Yury Felten. To the 250 anniversary since birth.* Exhibition catalog: Art (1982) 55 p.
2. M. G. Baranova, *"Gothic style" in the Russian architecture of the second half of XVIII (on the example of monuments of St. Petersburg and its suburbs)*: RGPU of Herzen (2005) 242 p.
4. N. A. Evsina, *Russian architecture during Catherine II's era* (1994) Moskva: Science, 223 p.
5. S. I. Razgonov, *Vasily Ivanovich Bazhenov* (1985) Moskva: Art, 168 p.
6. I. Yamshanov, V. Goryunov, V. Murgul, Procedia Engineering, **117**, 675–684 (2015)
7. I. Yamshanov, V. Murgul, N. Vatin, Applied Mechanics & Materials, **740**, 1192 (2014)
8. I. Yamshanov, V. Goryunov, V. Murgul, Procedia Engineering, **117**, 663–674 (2015)
9. I. Yamshanov, V. Goryunov, V. Murgul, N. Vatin. Advanced Materials Research, **1065-1069**, 2669-2673 (2015)

MATEC Web of Conferences 53, 02007 (2016)
DOI: 10.1051/matecconf/20165302007

Religious Buildings of the Period of an Early Russian Neogothic Style (the Second Half of XVIII – the early XIX centuries)

Igor Yamshanov[1, a] and Olena Oliynyk[2]

[1]St.Petersburg State Polytechnical University, Politekhnicheskaya ul., 29, 195251, SaintPetersburg, Russia
[2]National Aviation University, Kosmonavta Komarova, 1, Kiev, 03680, Ukraine

Abstract. The article is devoted to the development of religious buildings of Neogothic style and their stylistic features in the second half of the XVIII century. Each new building of this period has an unusual combination of new compositional techniques that illustrates the continuity of architects creativity who wanted to create something new and different, but at the same time causing the association with the Middle Ages.

1 Introduction

For cult buildings of early Neogothic style in the second half XVIII – the early XIX century was unusual interpretation of Gothic shapes. According to the volume and planning decisions they had almost nothing in common with Gothic cathedrals of the Middle Ages. During this period construction of orthodox churches with the use of Neogothic forms obviously prevailed.

2 The first orthodox churches of Neogothic style in the Russian Empire

The first orthodox churches of Neogothic style in the Russian Empire appeared in Moscow region. It is possible that the earliest of them is St. Nicholas church on Starki's (Figure 1-2) country churchyard in Cherkizovo built possibly on Bazhenov's project in 1759-1763. However, at this point it is not determined exactly when the temple got Neogothic features. Whether there was it at construction in the late fifties, or it got them as a result of rebuilding after construction of a belltower (Figure 1) in the 1780th. The decorative lancet arches decorating a drum of the dome, blind lancet arcade at the top of the quadrangle and pillars, trailing angles of the same quadrangular, the crown of pinnacles around the dome with a sharp spire belong to Neogothic elements of church.

With regard to the structural system of religious buildings in the style of early Russian Gothic Revival features are the following: building walls, made mostly of brick (stone was rarely used); vaults are often either closed, or the type of groined; dome usually coated with metal sheets are bulbous, hemispherical or baroque form.

The decision of the facade, without exception, of the churches were built in the 2nd half of the XVIII century, maintains strict symmetry, indicating the significant influence of the classicism.

[a] Corresponding author : dartigor@mail.ru

Facade compositions with completely different themselves, bearing different unique features. There are churches with one and five domes, with a bell tower, included in the volume of a temple or stand-alone, without the bell tower at all or with two ones. The same diversity is observed in the planning solution. There are oval, rectangular, cross-shaped plan, in the form of a quatrefoil plan, simple and complex.

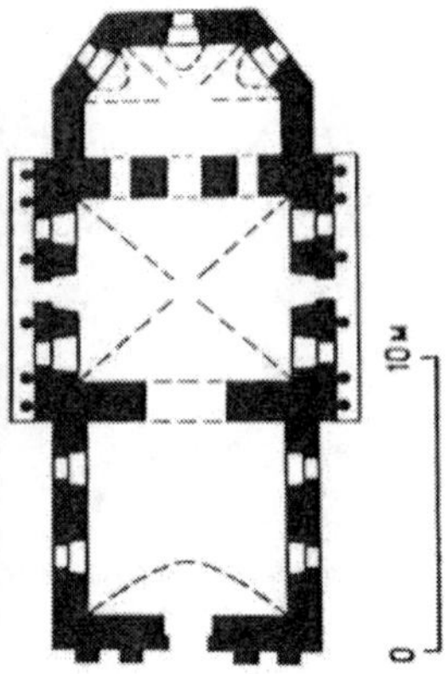

Figure 1, 2. The Church of St. Nicholas, 1759-1763. Cherkizovo-Starki. The architect Vasily Bazhenov (?). View from the southwest. Plan.

In this historical period, we almost do not see a direct continuity in development of the style in the form of planning and facade solutions of the projects of earlier buildings, with the exception of the Church of the Archangel Michael in Podzhigorodovo (Figure 3, 4) (1763) in Moscow, and in the church Signs Veshalovka (Figure 5) (1768-1784) in the Lipetsk region, where a commonality in space-planning, and the facade decision is clearly visible. This is an example of how a simple project has evolved and been enriched in the plastic, and the three-dimensional solutions. The main volume of the church in Veshalovka was supplemented by four corner towers with high spires, and the porch of the apse and gable not blocked and eight-pitched roofs, and are accordingly not one, but three gable, decorative pointed arches. Windows-lucarnes, cutting through the arch, are more ornate frames, and the whole facade decoration is much more finely crafted and is more expressive.

Figure 3. Church of the Archangel Michael. Podzhigorodovo. 1763 View from the southeast.

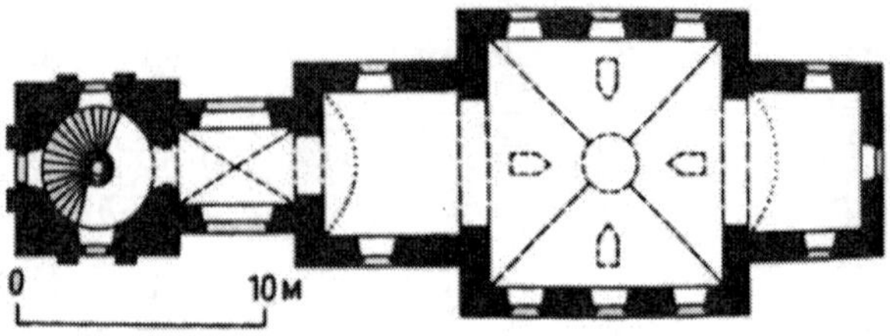

Figure 4. The Church of the Archangel Michael. Podzhigorodovo. 1763 Plan.

Figure 5. Church of the Sign. Veshalovka. 1768-1784. The architect Vasily Bazhenov (?). North façade.

The facade of the church has a vertical three-part division, and it is stored in the structure and elements of its form. So facade itself consists of a central part, complete lancet kokoshnik and two towers topped with spiky octagonal spires. Towers also have tripartite division, their corners are decorated with pilasters, and its central part is further decorated with graceful turret with a spire and pinnacles. The vertical aspiration is vividly expressed in every part. With great finesse made the central part of the side facades. They have one portal flanked by a pair of lancet windows, above which is located a pair of circular openings, and the central part of the portal is decorated with vimpergs made of red brick, which is located on the lancet window. This window has a frame in the form of rings of the "white stone" with adjoining archivolt. This archivolt with a belt rings form a sort kokoshnik completing the central part of the facade. The crowning element of the central part of the facade, the final total vertical operation, is a window-lucarnes lancet form, located above kokoshnik. The total composite structure of the facade is characterized by a dynamism in the construction of forms.

At the beginning of the XIX century, a type of centric church building, typical to the early Russian Neogothic style, receives a distribution. This type of temple has square plan, topped by a hemispherical dome on a low drum. However, in this case the continuity possible to speak of only in part, because of a completely different interpretation of the façade decoration. Four churches of this type have been identified. The first of them - is the Church of the Ascension in the village Peremilovo (Figure 6, 7.), it was built in 1792. The second is a baptismal church of the Kazan Church Puchkovo village (Figure 8.), built in 1802. The third is the church of St. Nicholas in Rzhavki (Figure 9), (1805, 1818-1820), supplemented by a free-standing bell tower (Figure 9). The last of them is the church of St. Nicholas in the Simonov monastery, built in 1834, has a lateral extension which housed communal cells.

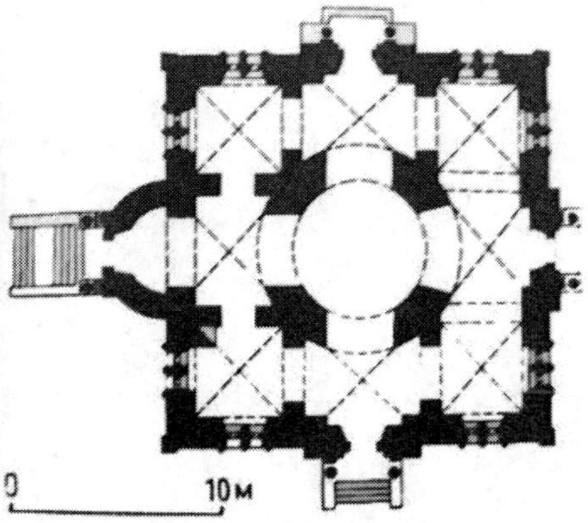

Figure 6, 7. The Church of the Ascension. Village Peremilovo of S.S. Apraksin. 1792. Architect F. Kamporezi. The western facade. Plan.

Figure 8. Baptismal Church of the Kazan church. Village Puchkovo. 1802.

Figure 9. Church of St. Nicholas. Rzhavki. 1805 1818-1820.

There are examples of almost identical buildings, and only decorative details vary. Thus was copied famous Chesma Church (Figure 10-12) in St. Petersburg, designed by Y.M. Felten. And it was copied twice. The first copy of the church, built in the 1781-1784, was in the village Posadnikovo (13,14) Pskov province Manor A.D. Lanskoy (to this day has not been preserved), while it was supplemented by a free-standing bell tower (Figure 13, 14), that was absent in the original; the second is located in the village Krasnoe (15-18) Tver region - it is the Church of the Transfiguration, the work of Felten, built in 1790 in the estate Poltoratsky. All this indicates the absence of a single historical prototype, which could serve as a basis for the design of the Orthodox churches in the early Gothic Revival style. Also, there is almost no direct continuity in the evolution of forms of Gothic Revival architecture of religious buildings in the period under review. Each new building has an unusual combination of new compositional techniques that illustrates the continuity of creativity of architects who wanted to create something new and different, but at the same time causing the association with the Middle Ages. In some cases, creating the impression of antiquity, in the other - the impression of solemnity.

Figure 10, 11, 12. Cesma Church. St. Petersburg. 1777-80. The architect Y.M. Felten. View from the southwest. Interior. Photo by author 2010. Plan.

Figure 13, 14. The St. Nicholas Church. Posadnikovo. Manor A.D. Lanskoy 1782-84. General view. Plan.

Figure 15, 16. Savior Transfiguration Church. Krasnoe Village. View from the west. The project of the western facade, the drawing late XVIII century. (Poltoratsky archive)

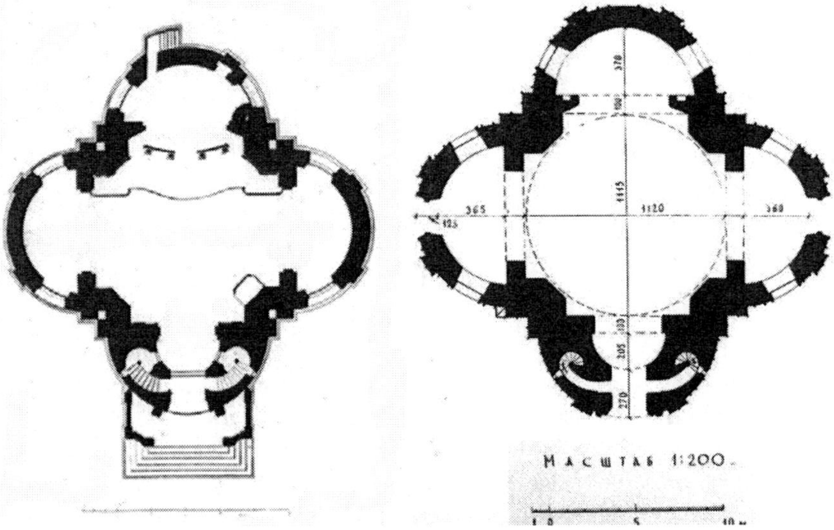

Figure 17, 18. Savior Transfiguration Church. Krasnoe Village. The project plan, the drawing late XVIII century - left. (Poltoratsky archive) modern plan – right.

The construction of churches of other Christian confessions is practically not conducted, with rare exceptions, an illustration of this is a Lutheran church of St. Peter in the village Malye Kolpany (Figure 19, 20) at Gatchina, completed in 1800. It was built by unknown architect since 12 July 1789 but work was suspended. Its construction was completed in the 1799-1800 under the leadership of Andreian Dmitrievich Zakharov. The consecration of the church took place on February 2 1802. Kirch was entirely built of stone. It belongs to the nasal type, with rectangular apse and a square plan bell tower, which once topped with high iron spire, the top of which is decorated with a gold-plated ball and cock, made a drawing by Zakharov in 1800. Of all the religious buildings built in the first period of the development of the Neogothic style in Russia it is closest to its historical prototype - European medieval parish church.

Figure 19, 20. Lutheran Church of St. Peter. village Malye Kolpany at Gatchina. 1800. The architect A.D. Zakharov. Photo of the early XX century. West façade.

3 Orthodox church of the early Russian Neogothicperiod, erected in Moscow and Moscow region

A typical example of the application of Neogothic forms in Russian architecture is the church of St. Nicholas in the churchyard Starkey (Figure 1, 2) in the village of Cherkizovo, built in the forms close to the works of V.I. Bazhenov in 1759-1763. There is also speculation that the project could have belonged to someone of his students. This church was built at the expense of Prince Cherkassky P.B. It is a small cube-shaped temple, covered with closed vault topped with a hollow decorative drum that completed high spire. The walls of the bulk of the temple are decorated with Gothic arches and niches, and the north and south facades are also false arcades. From the west the temple has the refectory of smaller height, the entrance to which is decorated with a portico of Tuscan order. Inside the church is preserved wooden iconostasis, also designed in the Gothic Revival style. The free-standing bell tower was build in 1780 (Figure 1), a constructive basis of it consists of four pylons, it completed with high spire. St. Nicholas Church was not rebuilt in the following years, thereby its original layout structure was preserved. An example of an ascetic approach to the decoration of facades is the church of St. Archangel Michael (Figure 3,4) in Podzhigorodovo. V.G. Lisowski dates the construction of the church by 1763. The construction of the church of the Archangel Michael, was started by order of the owners of the village, the brothers A.M. and P.M. Jurjev. I personally, as well as M.G. Baranova, support the opinion of V. Lisowski, since his research is supported by the inscription on the wall of the church wich dates it by the year 1763. The temple project was attributed to the famous Russian architect Vasily Bazhenov, but there is also the likelihood that the project belonged to someone from the architect's students, but no documentary evidence for this was revealed. In my opinion, this project could not be executed by Bazhenov, because all decorative details are too ascetic, and there are no elements of Russian Style that does not correspond to the master's creative manner. Built in the style of early Russian Neogothic two-storey brick church was erected on the white-stone plinth. The building is square in plan with an apse and square porch attached to the volume of the church. The apse is completed with pediment. The building is crowned with a dome, lucarnes and light drum, where instead of the traditional cupolas raised gothic stone steeple with a cross. A high bell tower crowned by a similar spire, which is a cylindrical two-tiered tower with a circular balcony, is standing on the first tier bushel, resembling a tower of a medieval castle. Decorative finishing of the temple is ascetic, we do not see here any lush decorations. The lower tier of the church closed with barrel vault, and the upper two tiers with windows and a lucarnes and light drum closed with closed vault. The architectural trend of the period was the reason for the restructuring in the Gothic Revival style a number of the older churches. Thus was rebuilt the church of St. Nicholas (Figure 21), built in 1697-1699 years in the traditional Russian style, and in 1802 received a new Gothic decoration in the form of lancet arches on the facade. At the same time the upper openings quadrangle tier bell tower and square openings of the drum cupolas crowning the main volume of the church, also received lancet arches. (This church was closed in 1932 and later demolished, it was located in an alley Mokrinskiy).

Figure 21. Church of St. Nicholas. Moscow. 1697-99, rebuilt in 1802.
Figure 22. Christmas chapel of Old Believers on Rogozhskoe cemetery. Moscow. 1804. Architect I. Zhukov.

Noteworthy is the fact that in Moscow the churches in early Russian Neogothicstyle continue to be built in the first and the 10-s years of the XIX century, with an influence of the Baroque, in particular the general formation principle of three-dimensional structure of religious buildings. An example of this is the Old Believer Christmas chapel on Rogozhskoe cemetery (Figure 22) (1804, architect I. Zhukov), as well as the Cathedral of the Immaculate Conception of St. Anna Conception monastery (Figure 23), built in 1804-1807. This was the main monastery church, which replaced an older building. Author of the project is believed to be M.F. Kazakov, but there is no documentary evidence of this. In its appearance the temple combines elements of Neogothicand Baroque (it was completely destroyed in the 1930s.). It was quite large and majestic cathedral, square in plan with a semicircular apse. From the west the main volume of its two tiers of windows was adjoined with the refectory and the octagonal three-storey bell tower. The cathedral was completed by massive dome on a square drum with truncated corners. Drum was topped with a tiny cupola and surrounded by four towers with spires. The appearance of the cathedral was heavy and impressive. A more slender proportions had only its bell tower.

Figure 23. Cathedral of the Immaculate Conception of St. Anna Conception monastery. Moscow. 1804-1807. The architect M.F. Kazakov. Photo of the early XX century.

The Baroque style influenced to the front compositions of some buildings in the style of early Russian neo-gothic, for example the Church of the Saviour in Ivanteevka (Figure 24-27), presumably erected by Alexander Bakarev, in E.O. Batashev manor in 1808. It also shows the mixed gothic and baroque elements, which is especially noticeable in the shape of windows, a very different shape - lancet, curly, three-blade, circular. This church cruciform in plan, was completed with low muffled drum with a dome, cut fanciful lucarnes, and surmounted by a stone spire. Similarly, square bell tower is complete with a bunk and a spire, connected with the main volume of the church through the refectory.

Figure 24, 25. The Church of the Saviour. Ivanteyevka. E.O. Batashev manor. 1808 Architect A.N. Bakarev (?). View from the southwest (left). View from the southeast.

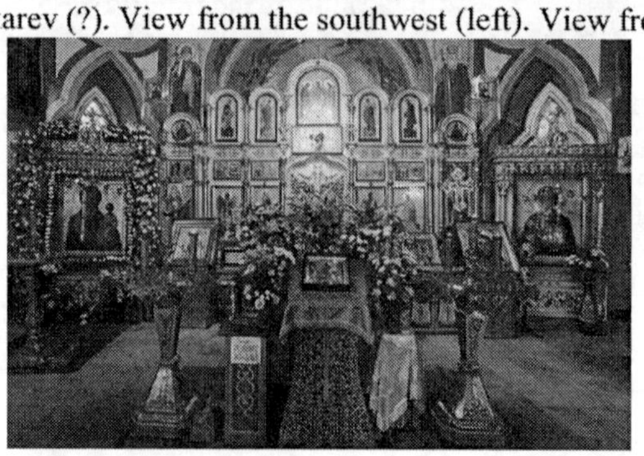

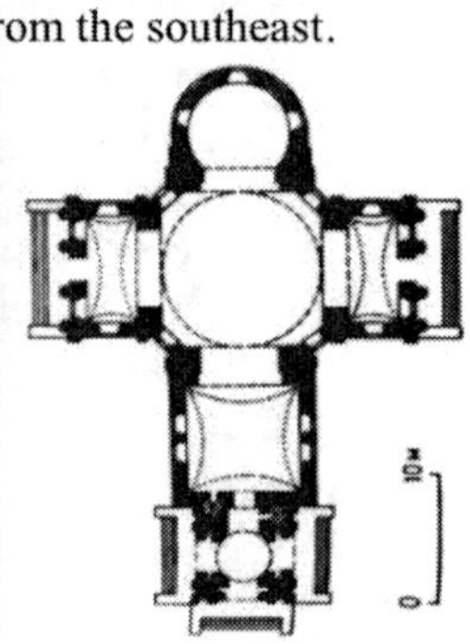

Figure 26, 27. The Church of the Saviour. Ivanteyevka. E.O. Batashev manor. 1808 Architect A.N. Bakarev (?). Interior and plan.

Another notable example of the connection of baroque planning structure with gothic decorative design is St. Nicholas Church in the village of Tsarevo (1812-1815) (28,29), built in three years, the author of the project is considered to be a pupil of V.I. Bazhenov and M.F. Kazakov – I.V. Egotov. This church was built by the order of the owner of the village Nikolai Durasov. Planning structure of the temple is fairly complex and very unusual: the main church hall is round, from east to adjoin the semicircular apse and the west - two oval aisle and rectangular porch. The main dome rests on eight pillars supporting the central hall. Church completed with spire surrounded by four small spires, mounted on a small octagonal drum resting on the central dome, erected on a high major drum with lancet windows. On the west side of the main dome is located a belfry, octagonal-shaped cupola, completed with five spires. The same turret, but not carrying the belfry functions, located on the east side. Unusual is also a classical frieze encircling the perimeter of the church, and depicts a scene from the Bible, made by sculptor T. Zamaraev. This frieze is supplemented by Gothic frieze below, giving the composition greater integrity.

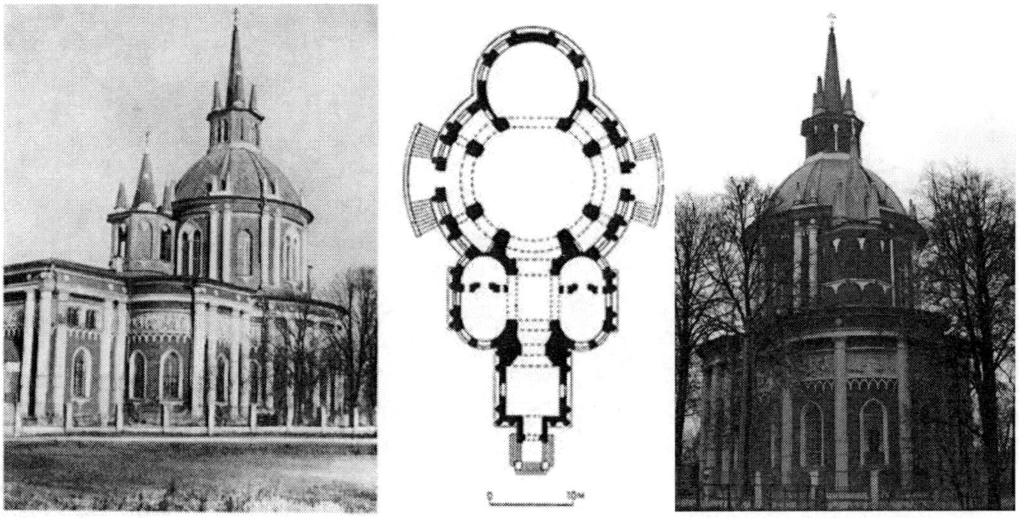

Figure 28.29. St. Nicholas Church. The village of Tsarevo. 1812-1815 years. Architect I.V. Egotov (?). View from the southwest. Plan. View from the east.

Next, you need to mention also the Orthodox chapels, showing the use of Gothic shapes. One of them was erected in 1804 on the project of Fedor Kirillovich Sokolov. It Nicholas Chapel at the Transfiguration Cemetery (Figure 30, 31) in Moscow. Previously it was thought that it was designed by Vasily Bazhenov in 1772. This chapel belongs to a complex of buildings of the Old Believers' almshouse built after the plague in 1771 on the expense of Ilya Alekseevich Kavylin. Buildings in the complex included a chapel, the church of St. Nicholas (Figure 32), the bell tower (Figure 33) and the area of the cemetery that was surrounded by a wall with Neogothic towers. This chapel is differs special richness of decorative finishes and is an example of centric composition. Square in plan, it completed with a hemispherical dome, which is crowned with five elegant white stone spires, the central of which is significantly superior to the other in height. Angles of chapel detailed with filigree reinforced cylindrical columns, and topped with tall spires. Its appearance combines Gothic, Old Russian and baroque motifs.

Figure 30, 31. St. Nicholas Chapel at the Transfiguration Cemetery 1804 Architect F.K. Sokolov.

Figure 32. Church of St. Nicholas in the Transfiguration Cemetery 1784-1790. Architect F.K. Sokolov.
Figure 33. The bell tower of the church of St. Nicholas. 1784-1790. Architect F.K. Sokolov.

Separately, it must be said about the rebuilding of the Kremlin at the end of the XVIII - beginning of XIX century, and in particular on the porch annex to the refectory of the monastery Chudov (Figure 34) in the late 1770s under the direction of M.F. Kazakov. Porch is showing a synthesis of old Russian, classical and gothic motifs, completed with two tetrahedral dome tents and adorned Ivanov area of the Kremlin. later the Moscow Kremlin will receive a lot of Gothic elements.

Figure 34. Porch of refectory of the Chudov monastery. 1770s. The architect M.F. Kazakov.

In 1800-s architect I.V. Egotov creates Resurrection Gate project in Moscow (Figure 35). The gates of these as well as the porch of the refectory of Chudov monastery show a combination of Gothic, classical and Russian styles. Extremely interesting was also unfulfilled bell tower restoration project of the Ivan the Great (Figure 36) made by A.N. Bakarev. This project involves a completely new construction on the site of the Belfry and the Filaret outbuildings, which were destroyed by the retreating French army in October 1812. According to the architect a new bell tower was supposed to be a kind of monument to the victory of Russian arms. At the same time the appearance of the bell tower was supposed to be a fundamentally eclectic: classic Doric colonnade with triangular gables are combined with Gothic pointed arches and rose windows, and the whole composition was completed with Russian bulbous dome. Façade of belfry was supposed to be decorated with the numerous sculptures and bas-reliefs. This project once again confirms the common perception of the European and Russian Middle Ages. In Moscow it continued to 1820s.

Figure 35. Resurrection Gate project in Moscow. Architect I.V. Egotov. 1800s.
Figure 36. The project of bell tower Ivan the Great restoration. Architect A.N. Bakarev. 1812.

4 Conclusion

Concluding the observation of the early Russian Neogothicreligious buildings we need to note the main features: in the stylistic aspect, it is the impact of Russian Style, classicism and baroque architectural appearance on church buildings; typology can be distinguished as Orthodox, Lutheran and Catholic churches, as well as the Orthodox chapels, free-standing bell towers and monastery walls.

References

1. *Architect Yury Felten. To the 250 anniversary since birth*. Exhibition catalog: Art (1982) 55 p.
2. M. G. Baranova, *"Gothic style" in the Russian architecture of the second half of XVIII (on the example of monuments of St. Petersburg and its suburbs)*: RGPU of Herzen (2005) 242 p.
4. N. A. Evsina, *Russian architecture during Catherine II's era* (1994) Moskva: Science, 223 p.
5. S. I. Razgonov, *Vasily Ivanovich Bazhenov* (1985) Moskva: Art, 168 p.
6. I. Yamshanov, V. Goryunov, V. Murgul, Procedia Engineering, **117**, 675–684 (2015)
7. I. Yamshanov, V. Murgul, N. Vatin, Applied Mechanics & Materials, **740**, 1192 (2014)
8. I. Yamshanov, V. Goryunov, V. Murgul, Procedia Engineering, **117**, 663–674 (2015)
9. I. Yamshanov, V. Goryunov, V. Murgul, N. Vatin. Advanced Materials Research, **1065-1069**, 2669-2673 (2015)

AUTHOR INDEX

AUTHOR INDEX